Signals and Communication Technology

This series is devoted to fundamentals and applications of modern methods of signal processing and cutting-edge communication technologies. The main topics are information and signal theory, acoustical signal processing, image processing and multimedia systems, mobile and wireless communications, and computer and communication networks. Volumes in the series address researchers in academia and industrial R&D departments. The series is application-oriented. The level of presentation of each individual volume, however, depends on the subject and can range from practical to scientific.

Indexing: All books in "Signals and Communication Technology" are indexed by Scopus and zbMATH

For general information about this book series, comments or suggestions, please contact Mary James at mary.james@springer.com or Ramesh Nath Premnath at ramesh.premnath@springer.com.

Mohammed El Ghzaoui · Sudipta Das ·
Varakumari Samudrala ·
Nageswara Rao Medikondu
Editors

Next Generation Wireless Communication

Advances in Optical, mm-Wave, and THz Technologies

Editors
Mohammed El Ghzaoui
Faculty of Sciences Dhar El Mahraz
Sidi Mohamed Ben Abdellah University
Fez, Morocco

Sudipta Das
Department of Electronics
and Communication Engineering
IMPS College of Engineering
and Technology
Malda, West Bengal, India

Varakumari Samudrala
Department of Electronics
and Communication Engineering
NRI Institute of Technology
Vijayawada, Andhra Pradesh, India

Nageswara Rao Medikondu
School of Mechanical and Civil Sciences
KLEF Deemed to be University
Guntur, Andhra Pradesh, India

ISSN 1860-4862 ISSN 1860-4870 (electronic)
Signals and Communication Technology
ISBN 978-3-031-56143-6 ISBN 978-3-031-56144-3 (eBook)
https://doi.org/10.1007/978-3-031-56144-3

This Springer imprint is published by the registered company Springer Nature Switzerland AG
The registered company address is: Gewerbestrasse 11, 6330 Cham, Switzerland

Preface

The primary focus of this book proposal is to highlight advancements in the design of sophisticated components and systems catering to optical, mm-wave, and THz technologies. The exploration of these domains commenced over a century ago, even before these terms were coined. During that era, scientist began delving into the unexplored region of the electromagnetic spectrum situated between infrared and microwave frequencies. Today, these fields of research have gained substantial popularity, witnessing the development of numerous wireless components and experiencing continued expansion. This comprehensive book encompasses the entire scope of the emerging and interdisciplinary field of the electromagnetic spectrum in a concise format. This book could eventually work as a textbook for engineering and communication technology students or science master's programs and for researchers as well. The content includes a thorough exploration of the physical phenomena and cutting-edge wireless components and technologies.

This book explores recent and upcoming technological breakthroughs within the optical, millimeter-wave (mm-wave), and terahertz (THz) frequency ranges. Encompassing a substantial portion of the electromagnetic spectrum, the scope extends up to the conclusion of the near-IR spectrum (i.e., 450 THz). Notably, this frequency span captures a crucial transition zone in technology, marking the shift from electronics to photonics. The focus of the book is on recent advancements and various research challenges pertaining to materials, antennas, detectors, passive circuits, as well as advanced signal processing algorithms tailored for optical, mm-wave, and THz frequency bands. Catering to a broad readership, the book addresses individuals ranging from those with a foundational understanding of basic science to technological experts and research scholars. This comprehensive resource serves as a valuable guide to the diverse aspects of technological evolution in the specified frequency spectrums, making it relevant for a wide audience, including both novices and seasoned professionals.

This book compiles scientific and technological innovations from the academic, industry, and research sectors. Catering to a diverse readership, including undergraduate and master's degree students, research scholars, microwave engineers, biomedical engineers, and professionals in electronics and electrical engineering, the book is poised to serve as an indispensable reference highlighting advanced ideas and concepts in mm-wave, THz, and optical communication technology. It is envisioned as an ideal choice for those seeking to explore fundamental and pivotal advancements in the domain of advanced communication technology, particularly relevant to microwave, electrical, and communication engineers engaged in shaping the next generation of technology.

Fez, Morocco — Mohammed El Ghzaoui
Malda, West Bengal, India — Sudipta Das
Vijayawada, Andhra Pradesh, India — Varakumari Samudrala
Vijayawada, Andhra Pradesh, India — Nageswara Rao Medikondu

Contents

Part I
Millimeter Wave (mm-Wave) Technology and Its Applications

Chapter 1
Compact MIMO Antenna Design with Enhanced Isolation for mm-Wave Applications

Navneet Kaur, Aarti Bansal, Surbhi Sharma, and Jaswinder Kaur

1.1 Introduction

The millimeter (mm)-wave-based technology for wireless communication has rapidly gained significant importance owing to its high data rate, high channel capacity to accommodate video streaming, 5G cellular, and mobile communication applications [1, 2]. Millimeter-wave (mm-wave) technology is authorized to be used free of license and highly resistant to atmospheric attenuation with compatibility toward recent 5G technology [3]. Some of the challenges in its implementation are atmospheric attenuation, reflection, and diffraction of electromagnetic waves (EM) leading to multipath fading and low transmission quality in the dense medium [4, 5]. Also, wearable antenna design demands a flexible substrate having the advantage of bending easily on a curved surface [6]. Recently designed conformal antennas have employed [7, 8] conductive textiles, liquid material, and polydimethylsiloxane (PDMS) as their substrates. These materials offer the advantage of being portable supporting the mobility of a person. However, these materials still pose challenges such as integration into application boards or soldering different components. Further, the proposed antenna's performance is severely deteriorated when bending over the surface due to significant multipath reflections and scattering [9, 10]. This further leads to a reduced data rate. Therefore, to overcome these challenges there is a need to increase frequency spectrum and power resources which are restricted. Hence, mm-wave band MIMO antenna has been proposed to increase diversity and spatial multiplexing. However, the MIMO antenna must employ a suitable mutual coupling reduction technique to increase isolation.

N. Kaur · S. Sharma · J. Kaur
Thapar Institute of Engineering and Technology, Patiala, India

A. Bansal (✉)
Chitkara University, Institute of Engineering and Technology, Chitkara University, Patiala, Punjab, India
e-mail: aarti.bansal@chitkara.edu.in

M. El Ghzaoui et al. (eds.), *Next Generation Wireless Communication*, Signals and Communication Technology, https://doi.org/10.1007/978-3-031-56144-3_1

Several isolation techniques were introduced to minimize the coupling in MIMO antennas [11–17]. But, still, these are not suitable to be employed in compact wearable antennas. Moreover, optimal placement of antenna elements is required to achieve miniaturization [18]. MIMO antenna design discussed in [19] to achieve resonance in the ISM band. The designed antenna covers a bandwidth of 20% and utilizes a ground plane at its back serving as its radiating element. Further, a wearable antenna designed in [20] uses jeans as its substrate covering a 2.7–12.33 GHz frequency range. To add further, the MIMO antenna designed in [21] exhibits linear polarization in the range from 2.4 to 2.49 GHz. Circularly polarized (CP) MIMO antenna designs enhance communication performance and reduce interference due to multipath propagation while the wearer device is on the move [22]. To summarize, there is a need to design antenna structures with novel isolation techniques and better diversity performance to attain extreme data rates and radiation characteristics. To further enhance performance, MIMO technology is employed, achieving up to a thousand-fold increase in data rates through spatial diversity and multiplexing techniques. The biggest task in the MIMO antenna is to minimize coupling to obtain the acceptable value of isolation within antenna elements.

Here, a compact flexible wearable MIMO antenna with dual-element exhibiting good isolation is designed for 5G mm-wave operating band applications. The suggested design for the proposed antenna utilizes a circular-shaped patch antenna with sectored slots etched around its periphery exciting the specific resonating mode. The slot size and its position are chosen for optimizing its resonance frequency. Also, multiple slots with similar dimensions lead to attaining wide bandwidth resonance. Further, we have optimized distance between antenna elements to minimize the coupling between its individual elements. Also, elements are strategically placed to significantly enhance isolation between them, resulting in optimal diversity. Furthermore, an in-depth evaluation of the suggested MIMO antenna has been done, encompassing analysis of S-parameters and MIMO performance parameters like ECC, diversity gain, etc. From the results, the designed MIMO antenna is justified to be considered as a compelling contender for seamless integration into 5G mm-wave applications such as radar, military, and radio astronomy. The chapter is organized into different sections as detailed here: MIMO antenna design in Sect. 1.2, Sect. 1.3 presents the various significant result parameters such as reflection coefficient, transmission coefficient, ECC, and DG for the proposed antenna design. At last, Sect. 1.4 presents the conclusion of the presented work.

1.2 MIMO Antenna Design

A dual-element MIMO antenna for wideband millimeter-wave operation is constructed with two circular radiating patches positioned on a top plane of the middle substrate layer, as depicted in Fig. 1.1. Individual unit of antenna elements incorporates four sectored slots and excited with a 50-Ω microstrip-fed. Further, dimensions of suggested antenna have been determined through fundamental equations. The final design of antenna has been fabricated on a RT/duroid 5880 a semi-flexible characterized by a ε_r of 2.2, tan δ of 0.0004, and a thickness (h) of 0.508 mm. In Fig. 1.1a, the perspective plane of designed MIMO antenna is presented. On the bottom side, a partial ground plane is integrated for wider bandwidth as portrayed in Fig. 1.1b. The size of the suggested MIMO antenna is 22.5 × 36.0 × 0.508 mm^3, and the optimized value of different dimensions parameters has been tabulated in Table 1.1.

Evolution of Suggested MIMO Antenna

The MIMO antenna design evolution is depicted in Fig. 1.2. The simulation of proposed design has been carried out using CST microwave EM software which involves different design steps as Design-1 (Fig. 1.2a), Design-2 (Fig. 1.2b), and Design-3 (Fig. 1.2c).

This design comprises dual circular-shaped radiating patches fed by simple microstrip feed lines while maintaining the same ground plane as substrate which is shown in Fig. 1.2a. After that design-1 ground plane has been modified to partial

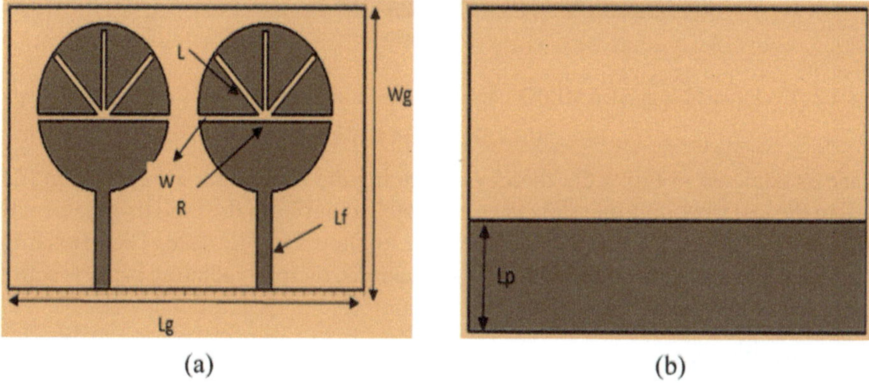

Fig. 1.1 MIMO antenna design: **a** perspective plane, **b** back plane

Table 1.1 Optimized parametric value of suggested MIMO antenna

Parameters	Value (mm)
L	7
W	0.5
R	6.73
Lf	7.7
Lg	22.5
Wg	36
Lp	7.5

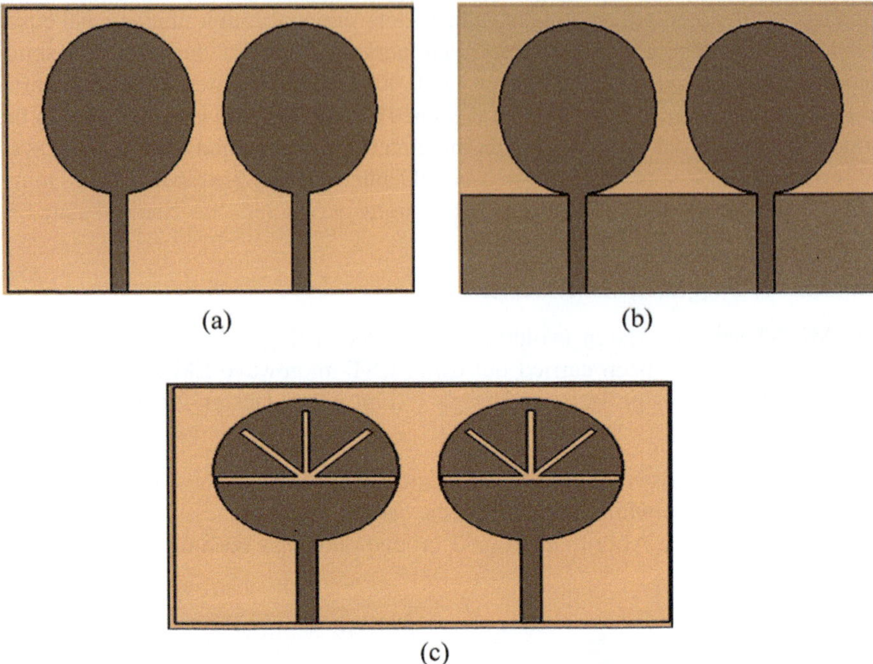

Fig. 1.2 Evolution of suggested MIMO antenna **a** Design step-1, **b** Design step-2, **c** Design step-3

plane as depicted in Fig. 1.2b, thereby enhancing the bandwidth as reported in [23, 24]. In the last step, multiple slots have been etched out from the radiating element of antenna for desired band of 19.7–29.02 GHz as shown in Fig. 1.2c. The resonating frequency (fr) has been determined using equations, thereby radiating patch is acting as a circular waveguide [25–27].

Parametric Studies

The geometry of suggested MIMO antenna has been optimized with parametric sweep of software. Therefore, parametric studies of some important parameters have been discussed in this section.

- **Length of Partial Ground Plane (Lp)**

In the designed part of antenna, partial ground plane plays a crucial role to attain the wider bandwidth of 2 GHz. Thus, a parametric sweep in the CST window has been performed with step size of 0.5 mm from 6.5 to 7.5 mm. As depicted in Fig. 1.3, optimized results have been obtained at Lp equal to 7.5 mm.

- **Length of Slots (*L*)**

Slots have been etched out from the proposed MIMO antenna. These slots are helping in alteration of surface current, and designed antenna is resonating in the desired band.

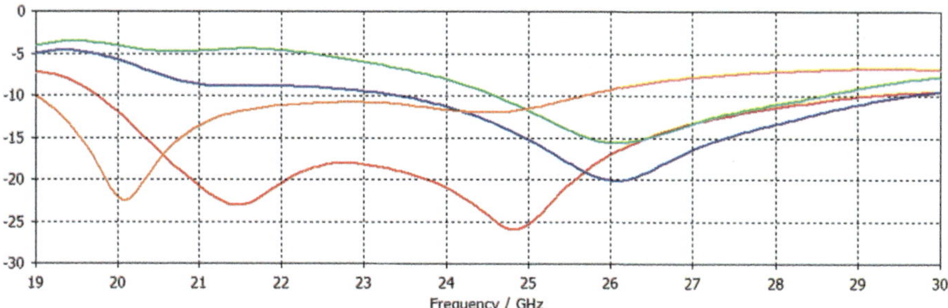

Fig. 1.3 Parametric of partial ground plane

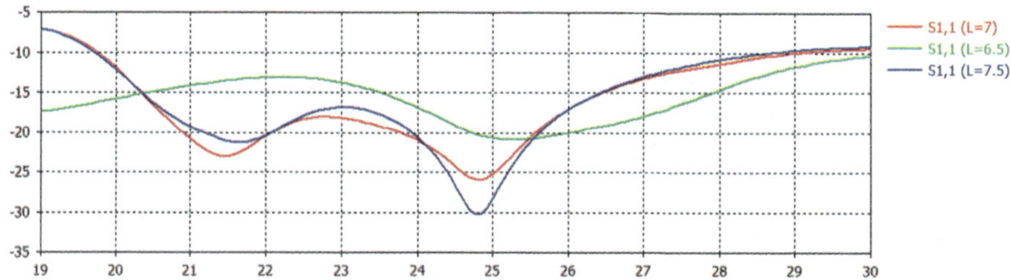

Fig. 1.4 Parametric on length of slots

Thus, a parametric analysis is depicted in Fig. 1.4 with a step size of 0.5 mm. The desired results of proposed antenna are achieving at L equal to 7 mm.

1.3 Results and Discussions

The suggested two element MIMO antenna is simulated, and the performance has been analyzed for reflection coefficient, transmission coefficient, ECC, and diversity gain. It is observed from the reflection coefficient results in Fig. 1.5 that the designed antenna resonates in the millimeter range and exhibits bandwidth of around 10 GHz from 19.7 to 29 GHz. Further, the transmission coefficient is displayed in Fig. 1.6 to observe the isolation between adjacent elements of proposed MIMO antenna design. It is revealing that the S_{12} parameter lies below $-$ 10 dB in the complete resonance band between $-$ 19 and $-$ 29 GHz frequency range. This validated that the proposed elements are isolated with each other.

Transmission Coefficient (S_{12})

Envelope Correlation Coefficient (ECC)

The relationship between the individual elements of antenna has been described with envelope correlation coefficient (ECC). It is computed by either S-parameters given

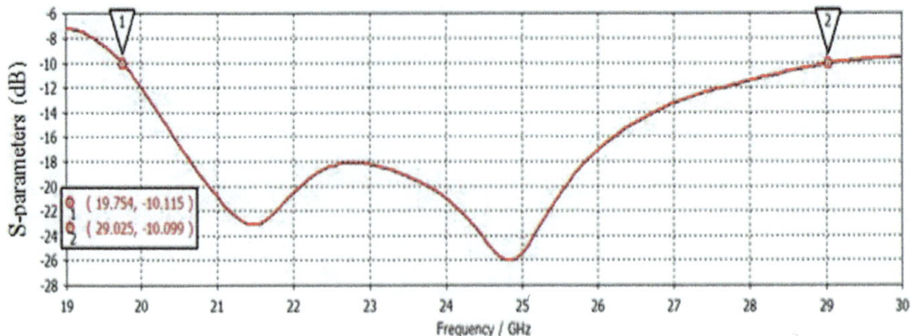

Fig. 1.5 Reflection coefficient of suggested MIMO antenna

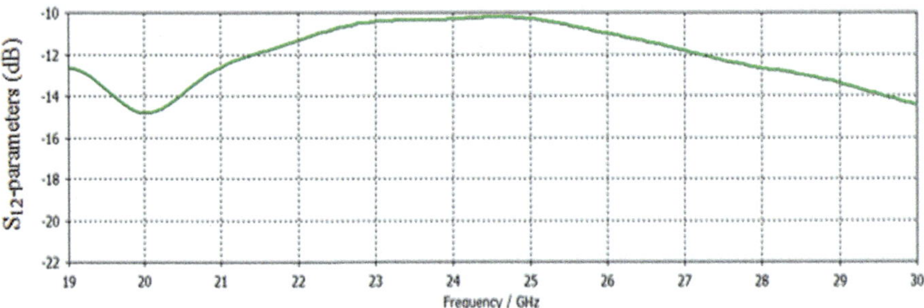

Fig. 1.6 Transmission coefficient of suggested MIMO antenna

in (1.1) or field pattern equation in [2] (Fig. 1.7).

$$\mathrm{ECC_S}(\rho_s) = \frac{|S_{11}^{*}S_{12} + S_{21}^{*}S_{22}|^2}{\left(1 - |S_{11}|^2 - |S_{21}|^2\right)\left(1 - |S_{22}|^2 - |S_{12}|^2\right)} \tag{1.1}$$

$$\mathrm{ECC_F}(\rho_F) = \frac{|\iint [F_i(\theta,\phi)\cdot F_j(\theta,\phi)]\mathrm{d}\Omega|^2}{\iint |F_i(\theta,\phi)|^2\mathrm{d}\Omega \iint |F_j(\theta,\phi)|^2\mathrm{d}\Omega} \tag{1.2}$$

Here, ECCs stands for the envelope correlation coefficient obtained through S-parameters, while $\mathrm{ECC_F}$ represents the envelope correlation coefficient determined using far-field methods. Furthermore, the symbol ρ denotes the correlation coefficient specifically associated with MIMO antennas. In case of an ideal MIMO system, this is recommended to keep the ECC value less than 0.5 for optimal performance.

Diversity Gain

Diversity gain (DG) refers to the enhancement in signal-to-noise ratio for the multiple antenna system with respect to the individual antenna. General equations for the calculation of diversity gain are given below [20]:

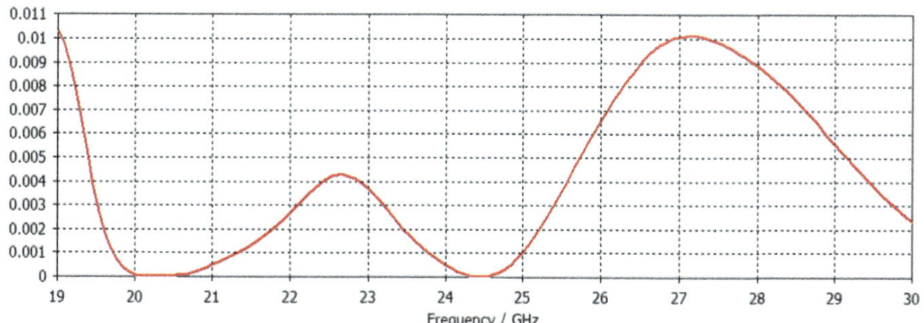

Fig. 1.7 ECC characteristics of MIMO antenna

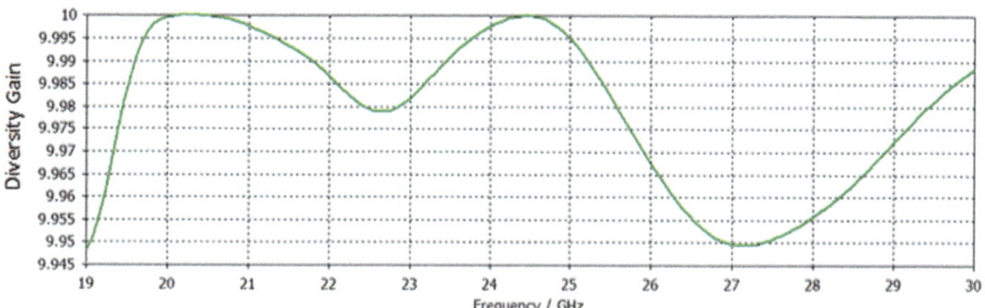

Fig. 1.8 Diversity gain characteristic of MIMO antenna design

$$DG_S = 10\sqrt{1 - (ECC_S)^2} \quad (1.3)$$

$$DG_F = 10\sqrt{1 - (ECC_F)^2} \quad (1.4)$$

Here, DG_s and DG_F are referred for diversity gain using reflection coefficient (*S*-parameters) and far-field methods, respectively. For an ideal MIMO antenna, the value of diversity gain should be equal to 10 (Fig. 1.8).

1.4 Conclusion

MIMO antennas are designed to overcome the challenges of deterioration in data rate, and multipath reflections in conformal antenna design. However, these antennas demand reduced mutual coupling resulting in enhanced isolation amid its individual antenna components. This work discusses and explores various techniques for addressing the issue of mutual coupling. The proposed antenna has been meticulously engineered to deliver exceptional features, including compact size, high isolation, and better diversity performance achieved through precise alignment of radiating elements of the MIMO antenna. This antenna system has demonstrated impressive performance metrics, featuring minimal coupling below − 10 dB, lower ECC (< 0.005), and an ideal diversity gain of 10 dBi. Furthermore, its compact and cost-effective design characteristics make it particularly well-suited for mm-wave 5G applications, facilitating seamless integration into various applications of mm-wave. Additionally, the recommended MIMO design being portable and high performing is a suitable choice for consideration in 5G mm-wave broad spectrum-oriented applications.

References

1. Rappaport, T.S., Sun, S., Mayzus, R., Zhao, H., Azar, Y., Wang, K., Wong, G.N., Schulz, J.K., Samimi, M., Gutierrez, F.: MillimeterWave mobile communications for 5G cellular: it will work! IEEE Access **1**, 335–349 (2013)
2. Hasan, M., Faruque, M.R.I., Islam, M.T.: Dual band metamaterial antenna for LTE/Bluetooth/ WiMAX system. Sci. Rep. **8**, 1240 (2018)
3. Ur-Rehman, M., Malik, N.A., Yang, X., Abbasi, Q.H., Zhang, Z., Zhao, N.: A low profile antenna for millimeter-wave body-centric applications. IEEE Trans. Antennas Propag. **65**(12), 6329–6337 (2017). https://doi.org/10.1109/TAP.2017.2700897
4. Abbas, M.A., Allam, A., Gaafar, A., Elhennawy, H.M., Sree, M.F.: Compact UWB MIMO antenna for 5G millimeter-wave applications. Sensors **23**(5), 2702 (2023)
5. Nadeem, I., Choi, D.-Y.: Study on mutual coupling reduction technique for MIMO antennas. IEEE Access **7**, 563–586 (2018)
6. Pi, Z., Khan, F.: An introduction to millimeter-wave mobile broadband systems. IEEE Commun. Mag. **49**, 101–107 (2011)
7. Katalinic, A., Nagy, R., Zentner, R.: Benefits of MIMO systems in practice: increased capacity, reliability and spectrum efficiency. In: Proceedings of the Proceedings ELMAR 2006, Zadar, Croatia, 7–10 June 2006, pp. 263–266 (2006). ess2020, 8, 100337–100345 [CrossRef]
8. Kumar, S., Dixit, A.S., Malekar, R.R., Raut, H.D., Shevada, L.K.: Fifth generation antennas: a comprehensive review of design and performance enhancement techniques. IEEE Access **8**, 163568–163593 (2020)
9. Molisch, A., Win, M.: MIMO systems with antenna selection. IEEE Microw. Mag. **5**, 46–56 (2004)
10. Paulraj, A., Gore, D., Nabar, R., Bolcskei, H.: An overview of MIMO communications—a key to gigabit wireless. Proc. IEEE **92**, 198–218 (2004)
11. Roshna, T.K., Deepak, U., Sajitha, V.R., Vasudevan, K., Mohanan, P.: A compact UWB MIMO antenna with reflector to enhance isolation. IEEE Trans. Antennas Propag. **63**(4), 1873–1877 (2015). https://doi.org/10.1109/TAP.2015.2398455
12. Deng, J.Y., Li, J., Zhao, L., Guo, L.: A dual-band inverted-F MIMO antenna with enhanced isolation for WLAN applications. IEEE Antennas Wirel. Propag. Lett. **16**, 2270–2273 (2017). https://doi.org/10.1109/LAWP.2017.2713986
13. Wang, L., et al.: Compact UWB MIMO antenna with high isolation using fence-type decoupling structure. IEEE Antennas Wirel. Propag. Lett. **18**(8), 1641–1645 (2019). https://doi.org/10.1109/LAWP.2019.2925857
14. Xu Mao, C., Zhou, Y., Wu, Y., Soewardiman, H., Werner, D.H., Jur, J.S.: Low-profile strip-loaded textile antenna with enhanced bandwidth and isolation for full-duplex wearable applications. IEEE Trans. Antennas Propag. **68**(9), 6527–6537 (2020). https://doi.org/10.1109/TAP.2020.2989862
15. Mirmozafari, M., Zhang, G., Fulton, C., Doviak, R.J.: Dualpolarization antennas with high isolation and polarization purity: a review and comparison of cross-coupling mechanisms.

IEEE Antennas Propag. Mag. **61**(1), 50–63 (2019). https://doi.org/10.1109/MAP.2018.2883032. Tiwari, R.N., Singh, P., Kanaujia, B.K., Kumar, P.: Compact circularly polarized MIMO printed antenna with novel ground structure for wideband applications. Int. J. RF Microw. Comput.-Aided Eng. **31**(8) (2021). https://doi.org/10.1002/mmce.22737

16. Ren, Z., Zhao, A., Wu, S.: MIMO antenna with compact decoupled antenna pairs for 5G mobile terminals. IEEE Antennas Wirel. Propag. Lett. **18**(7), 1367–1371 (2019). https://doi.org/10.1109/LAWP.2019.2916738
17. Iqbal, A., Smida, A., Alazemi, A.J., Waly, M.I., Mallat, N.K., Kim, S.: Wideband circularly polarized MIMO antenna for high data wearable biotelemetric devices. IEEE Access **8**, 17935–17944 (2020). https://doi.org/10.1109/ACCESS.2020.2967397
18. Li, H., Sun, S., Wang, B., Wu, F.: Design of compact singlelayer textile MIMO antenna for wearable applications. IEEE Trans. Antennas Propag. **66**(6), 3136–3141 (2018). https://doi.org/10.1109/TAP.2018.2811844
19. Biswas, A.K., Chakraborty, U.: Compact wearable MIMO antenna with improved port isolation for ultra-wideband applications. IET Microw. Antennas Propag. **13**(4), 498–504 (2019). https://doi.org/10.1049/iet-map.2018.5599
20. Jilani, S.F., Munoz, M.O., Abbasi, Q.H., Alomainy, A.: Millimeterwave liquid crystal polymer based conformal antenna array for 5G applications. IEEE Antennas Wirel. Propag. Lett. **18**(1), 84–88 (2019). https://doi.org/10.1109/LAWP.2018.2881303
21. Wen, D., Hao, Y., Munoz, M.O., Wang, H., Zhou, H.: A compact and low-profile MIMO antenna using a miniature circular highimpedance surface for wearable applications. IEEE Trans. Antennas Propag. **66**(1), 96–104 (2018). https://doi.org/10.1109/TAP.2017.2773465
22. Qu, L., Piao, H., Qu, Y., Kim, H.-H., Kim, H.: Circularly polarised MIMO ground radiation antennas for wearable devices. Electron. Lett. **54**(4), 189–190 (2018). https://doi.org/10.1049/el.2017.4348
23. Ullah, U., Al-Hasan, M., Koziel, S., Mabrouk, I.B.: A series inclined slot-fed circularly polarized antenna for 5G 28 GHz applications. IEEE Antennas Wirel. Propag. Lett. **20**(3), 351–355 (2021). https://doi.org/10.1109/LAWP.2021.3049901
24. Xia, Z., et al.: A wideband circularly polarized implantable patch antenna for ISM band biomedical applications. IEEE Trans. Antennas Propag. **68**(3), 2399–2404 (2020). https://doi.org/10.1109/TAP.2019.2944538
25. Sharma, M.: Design and analysis of MIMO antenna with high isolation and dual notched band characteristics for wireless applications. Wirel. Pers. Commun. **112**, 1587–1599 (2020). https://doi.org/10.1007/s11277-020-07117-4
26. Garg, R., Bhartia, P., Bahl, I., Ittipiboon, A.: Microstrip Antenna Design Handbook. Artech House, London, UK (2001)
27. Tayyab, U., Kumar, A., Petry, H.P., Asghar, M.E., Hein, M.A.: Dual-band nested circularly polarized antenna array for 5G automotive satellite communications. Appl. Sci. **13**(21), 11915 (2023)

Chapter 2
Security Threats and Privacy Challenges in Millimeter-Wave Communications

A. Amsaveni and M. Bharathi

2.1 Introduction

This chapter aims to provide an extensive overview of the security threats and privacy challenges faced in the realm of Millimeter-wave communications, addressing both current concerns and potential future issues. It explores various vulnerabilities, and potential exploits, and, importantly, offers a range of mitigation strategies to safeguard networks and protect individual privacy in the rapidly advancing landscape of wireless communication technologies.

2.1.1 Evolution of mmWave Communications

Millimeter-wave (mmWave) communications represent a revolutionary leap in wireless technology. Traditionally confined to lower frequency bands, the integration of mmWave has unleashed a new era of high-speed data transmission, enabling applications in 5G networks, point-to-point communication, and various other wireless technologies. These extremely high frequencies, ranging from 30 to 300 GHz, have unlocked the potential for faster data rates and low-latency connectivity, contributing significantly to digital transformation across various industries. Initial research in this field aimed to explore the feasibility of mmWave frequencies for wireless communications, emphasizing their ability to carry large amounts of data due to their high bandwidth.

A. Amsaveni (✉) · M. Bharathi
Department of Electronics and Communication Engineering, Kumaraguru College of Technology, Coimbatore, Tamilnadu 641049, India
e-mail: amsaveni.a.ece@kct.ac.in

M. Bharathi
e-mail: bharathi.m.ece@kct.ac.in

M. El Ghzaoui et al. (eds.), *Next Generation Wireless Communication*, Signals and Communication Technology, https://doi.org/10.1007/978-3-031-56144-3_2

Advancements in antenna design, semiconductor technologies, and signal processing have played a crucial role in shaping the evolution of mmWave communications. Breakthroughs in phased array antennas and beamforming techniques have enabled the efficient focusing and directing of signals, compensating for the propagation limitations associated with mmWave frequencies [1].

The advent of 5G technology marked a pivotal point in integrating mmWave frequencies into commercial networks. 5G networks incorporated mmWave bands, leveraging their potential for ultra-fast, high-bandwidth applications. Initial commercial deployments, primarily in dense urban areas, aimed to exploit the high data rates offered by mmWave frequencies [2].

2.1.2 Scope and Significance of Security Threats

The implementation of mmWave communications has expanded the scope of connectivity and introduced a new frontier of security threats. With its unique characteristics, such as shorter transmission ranges and higher susceptibility to signal attenuation, mmWave technology faces a spectrum of security challenges. Threats such as eavesdropping, interception, man-in-the-middle attacks, and vulnerability at the physical layer pose significant risks to the confidentiality, integrity, and availability of data transmitted over mmWave networks. Understanding and mitigating these threats are imperative to ensure the reliability and security of these networks in our increasingly interconnected world [3].

2.1.3 Privacy Challenges in mmWave Networks

In tandem with security threats, mmWave technology confronts substantial privacy challenges. The high data transfer speeds and the ability to support numerous connected devices intensify concerns regarding user privacy. Issues like location tracking, identity theft, and unauthorized surveillance raise significant concerns about individual privacy rights. The seamless and pervasive nature of mmWave networks heightens the risk of personal information exposure, demanding effective measures to protect users' sensitive data while maintaining a balance between innovation and privacy.

As the utilization of mmWave communications becomes more prevalent, an in-depth understanding of the security threats and privacy challenges is essential. This chapter aims to explore these critical issues, providing insights into the vulnerabilities, potential exploits, and strategies to safeguard the integrity and privacy of mmWave networks in this rapidly evolving technological landscape [4].

2.2 Security Threats in mmWave Communications

2.2.1 *Eavesdropping and Interception*

Eavesdropping and interception represent critical threats in millimeter-wave (mmWave) communications due to the vulnerability of these high-frequency signals to interception, potentially compromising the confidentiality of transmitted data [5]. MmWave signals are more prone to line-of-sight transmission, making interception relatively easier. Malicious actors can exploit this vulnerability to passively intercept data transmitted between devices, compromising the confidentiality of information. The directional nature of mmWave transmission, often used in point-to-point and point-to-multipoint communications, can be advantageous for high-speed data transfer. Still, it also increases the susceptibility of intercepted signals [6].

Encryption plays a crucial role in mitigating eavesdropping threats. Implementing robust encryption algorithms, such as Advanced Encryption Standard (AES) or Elliptic Curve Cryptography (ECC), ensures that data is protected during transmission, preventing unauthorized access and interception of sensitive information [7].

Additionally, advancements in beamforming technology have introduced adaptive beamforming techniques, enabling better signal control and directionality. Secure beamforming strategies play a role in reducing the vulnerability to eavesdropping by limiting the accessibility of signals to intended recipients and reducing the exposure to potential eavesdroppers [8].

However, ensuring the security of mmWave communications against eavesdropping (as in Fig. 2.1) requires a holistic approach, incorporating encryption, secure beamforming, and continual advancements in cryptographic protocols to counter potential threats to the integrity and confidentiality of data transmitted over mmWave networks.

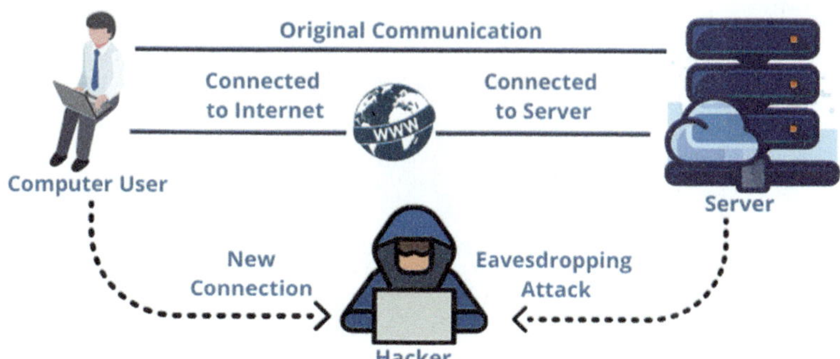

Fig. 2.1 Illustration of eavesdropping attack

2.2.2 *Man-in-the-Middle Attacks*

Man-in-the-middle (MitM) attacks pose a serious threat in millimeter-wave (mmWave) communications, exploiting the characteristics of these high-frequency transmissions. As shown in Fig. 2.2, MitM attacks occur when an attacker intercepts communication between two parties and may either passively eavesdrop on the transmission or actively modify the data being exchanged [9]. The directional nature of mmWave transmission, typically employed in point-to-point and point-to-multipoint communication, presents a particular vulnerability to such attacks due to the focused transmission patterns.

Secure beamforming, a fundamental technology in mmWave networks, is integral in reducing the susceptibility to MitM attacks. Properly implemented beamforming enables precise control over the direction of the transmitted signal, limiting its accessibility to intended recipients and reducing the chance of interception by unauthorized entities [10].

The integration of robust authentication protocols is critical in preventing MitM attacks. Utilizing strong authentication mechanisms, such as mutual authentication and digital signatures, ensures the integrity and authenticity of communication channels. Implementing secure key exchange protocols, such as the Diffie–Hellman key exchange or Public Key Infrastructure (PKI), fortifies the communication channel against potential manipulation or eavesdropping by unauthorized entities [11].

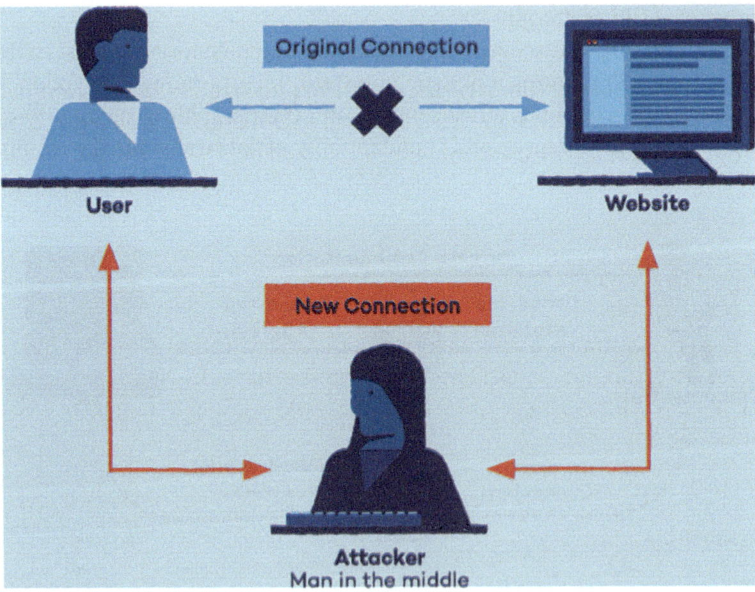

Fig. 2.2 Illustration of meet-in-the middle attack

Moreover, continuous monitoring using intrusion detection systems in mmWave networks is vital. These systems can identify anomalous behaviors and potential threats, providing a layer of defense against MitM attacks by detecting any deviations in the normal transmission patterns [12].

While these measures contribute to mitigating the risks associated with MitM attacks in mmWave communications, a comprehensive security approach is necessary, combining secure beamforming, robust authentication, secure key exchange, and active monitoring to protect against these sophisticated attack vectors.

2.2.3 Denial-of-Service (DoS) Attacks

Denial-of-Service (DoS) attacks pose a significant threat to millimeter-wave (mmWave) communications, potentially disrupting the availability and reliability of these networks. These attacks aim to overwhelm a network's resources, rendering it inaccessible to legitimate users as depicted in Fig. 2.3 [13].

The vulnerability of mmWave signals to atmospheric conditions and physical obstacles can be exploited in DoS attacks. Adversaries can manipulate these vulnerabilities to create signal obstructions or introduce interference, leading to disruptions in the transmission of mmWave signals [14].

Implementing traffic filtering mechanisms and rate limiting is crucial in mitigating DoS attacks. These mechanisms help in identifying abnormal traffic patterns and restricting excessive incoming traffic that may overwhelm the network infrastructure [15].

Advanced anomaly detection systems, particularly Intrusion Detection and Prevention Systems (IDPSs), play a pivotal role in identifying and mitigating DoS attacks. These systems monitor network behavior and patterns, identify unusual

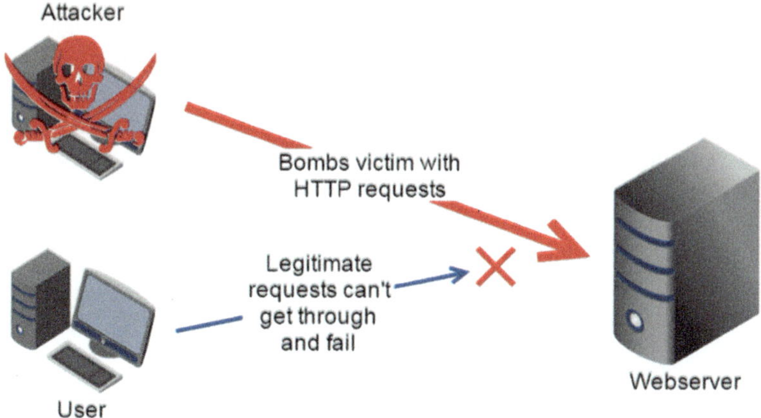

Fig. 2.3 Illustration of Denial-of-Service attack

traffic and behavior that could indicate a potential DoS attack, and then take preventive measures to mitigate the impact of such attacks.

Moreover, dynamic spectrum access techniques can be employed to mitigate the impact of DoS attacks. Dynamic frequency allocation and spectrum management allow for efficient utilization of available frequency bands, enabling the network to switch to less congested or interference-free frequency channels when an attack is detected [16].

The combined application of these strategies—traffic filtering, anomaly detection systems, and dynamic spectrum access—can effectively mitigate the impact of DoS attacks, ensuring the availability and reliability of mmWave networks.

2.2.4 Physical Layer Attacks

Physical layer attacks in millimeter-wave (mmWave) communications exploit vulnerabilities in the underlying transmission medium, potentially compromising the integrity and reliability of the communication. The susceptibility of mmWave signals to environmental conditions, such as rain, humidity, and atmospheric absorption, can be manipulated to launch physical layer attacks. Adversaries might utilize signal obstructions or interference to disrupt the transmission of mmWave signals, causing signal degradation, packet loss, or complete communication failure [17].

Radio Frequency (RF) interference is a significant threat to the physical layer. Adversaries might deploy unauthorized transmitters or intentionally generate interference that disrupts the mmWave signal, causing communication degradation or Denial of Service [18].

To mitigate these threats, implementing secure physical layer transmission techniques is essential. Beamforming technology, when used securely, can direct signals precisely, mitigating interference and ensuring more reliable communication paths.

Employing advanced modulation and coding schemes can also enhance resilience against physical layer attacks. Using error-correcting codes and robust modulation schemes increases the system's ability to tolerate interference and maintain communication integrity even in the presence of noise or interference.

Furthermore, adaptive and cognitive radio techniques offer a way to dynamically adjust to changing environmental conditions and counteract physical layer attacks by allowing devices to dynamically change their transmission parameters to optimize communication in the presence of interference [19].

By integrating secure transmission techniques, advanced modulation schemes, and adaptive radio techniques, mmWave systems can improve their resilience against physical layer attacks, ensuring reliable and robust communication.

2.2.5 *Authentication and Authorization Vulnerabilities*

Authentication and authorization vulnerabilities in millimeter-wave (mmWave) communications can potentially compromise the security and integrity of these networks. Weaknesses in these mechanisms can lead to unauthorized access, data breaches, and various security threats.

Weak Authentication Protocols: Flaws in authentication protocols can open gateways for attackers to gain unauthorized access to the network as shown in Fig. 2.4. Inadequate authentication mechanisms or weak password policies can allow adversaries to infiltrate the system [20].

Lack of Mutual Authentication: One-way authentication in mmWave communications might leave the network susceptible to impersonation attacks. Lack of mutual authentication mechanisms could enable attackers to impersonate legitimate users or devices, gaining access to the network and compromising its security [21].

Inadequate Authorization Controls: Authorization mechanisms might not be sufficiently robust, allowing unauthorized users or devices to access sensitive data or network resources. Lack of stringent access control can lead to unauthorized usage or manipulation of network resources.

Implementing robust authentication and authorization measures is crucial to reinforce mmWave networks against these vulnerabilities. Utilizing strong encryption, two-way authentication protocols, and stringent access control policies can help mitigate the risks associated with these vulnerabilities, safeguarding the integrity and confidentiality of communication in mmWave systems.

These vulnerabilities necessitate a comprehensive approach to strengthen authentication and authorization protocols, ensuring the resilience of mmWave networks against potential security breaches.

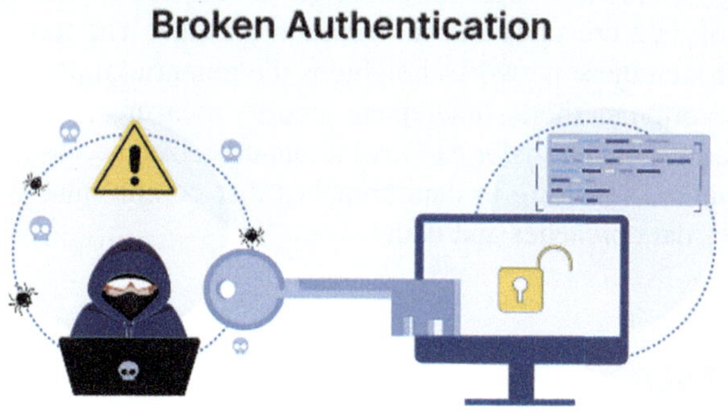

Fig. 2.4 Broken authentication

2.3 Privacy Concerns in mmWave Networks

2.3.1 Location Tracking and Profiling

One of the foremost privacy concerns in mmWave networks revolves around location tracking and user profiling. The directional and focused nature of mmWave signals, integral to achieving high data rates, also brings about the potential for precise location tracking. The use of phased array antennas and beamforming techniques for signal transmission and reception enables more accurate localization of devices within the network. This precision could lead to detailed and specific tracking of device locations.

While precise location tracking can offer various benefits, it also raises significant privacy concerns. The ability to accurately track and trace the movements of devices within the coverage area poses potential threats to individual privacy. Adversaries could potentially exploit this characteristic to track user movements, creating detailed profiles of an individual's activities, routines, and habits. This information could be misused for targeted advertising, invasion of personal space, or even physical security threats.

The detailed information gathered through location tracking might facilitate the creation of user profiles and behavioral patterns. Profiling users based on their movement, habits, and frequent locations could result in the potential misuse of this data for targeted advertising, surveillance, or even invasive tracking without user consent. The sample of this is shown in the Fig. 2.5.

2.3.2 Data Breaches and Information Leakage

The high-speed, high-bandwidth capabilities of mmWave networks also expose users to increased risks of data breaches and information leakage. The transfer of large volumes of data through these networks heightens the potential impact of a breach. Weaknesses in encryption methods, inadequate security measures, or vulnerabilities in the protocols used for data transfer can lead to unauthorized access and exposure of sensitive information, resulting in data breaches that compromise user privacy. Figure 2.6 compares data breaches and data leaks.

2.3.3 Identity Theft and Impersonation

MmWave networks, if compromised, can become a breeding ground for identity theft and impersonation. Attackers, by exploiting vulnerabilities in the network, could gain access to user identities or impersonate legitimate users. Such breaches not only compromise the individual's personal information but could also lead to

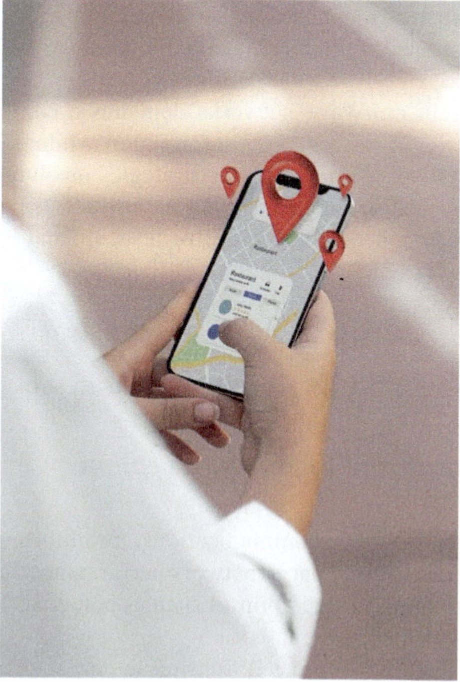

Fig. 2.5 Location tracking

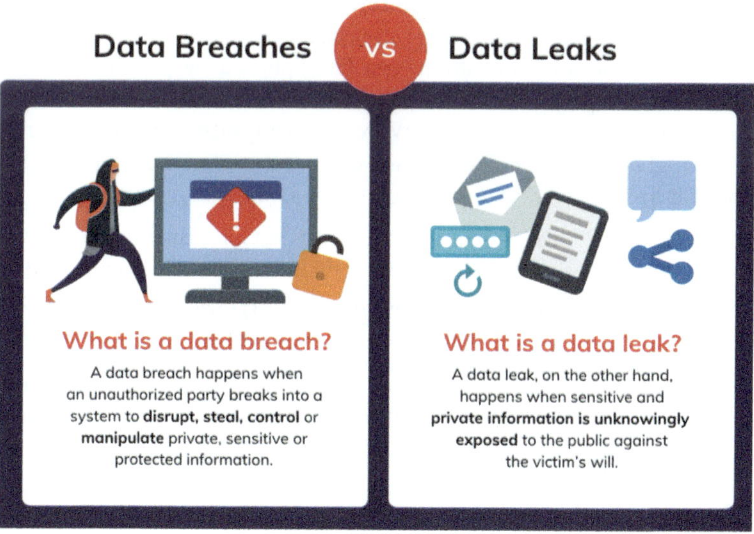

Fig. 2.6 Data breaches versus data leaks

further exploitation, such as financial fraud, manipulation of personal data, or unauthorized access to restricted resources, potentially causing severe consequences for the affected individuals.

Protecting privacy in mmWave networks demands a comprehensive approach that prioritizes the safeguarding of user information while balancing the technological benefits of high-speed data transfer. Strategies involving robust encryption, strict access controls, and enhanced user-aware privacy measures are pivotal in mitigating these privacy concerns and fostering a secure and trustworthy mmWave communication environment.

2.4 Vulnerabilities and Exploitable Weaknesses

2.4.1 Beamforming and Signal Interception

In millimeter-wave (mmWave) communications, beamforming technology is a double-edged sword, offering enhanced network performance through focused signal transmission and introducing vulnerabilities such as potential signal interception by unauthorized entities within the signal's path.

Beamforming facilitates the directed transmission of wireless signals, enabling highly focused transmissions from a transmitter to a specific receiver or area as shown in Fig. 2.7. This technology optimizes signal strength and improves overall network performance by concentrating the signal toward the intended recipient.

Focused Signal Transmission: The concentrated nature of mmWave signals, although efficient for communication, can be exploited by potential attackers for signal interception within the focused path of transmission.

Line-of-Sight Interception: As mmWave signals typically require clear line-of-sight for optimal transmission, interception can occur within the direct path between the transmitter and receiver, potentially compromising data integrity.

2.4.2 Spectrum Sharing and Frequency Hijacking

In millimeter-wave (mmWave) communications, the concept of spectrum sharing and the vulnerability of frequency hijacking pose critical security concerns due to the nature of shared frequency bands and potential unauthorized access to these spectrums.

mmWave technology often operates within shared frequency bands. This sharing of frequency resources among multiple users or systems increases the potential risk of interference or unauthorized access. Shared frequency bands can lead to interference issues, where multiple systems attempt to use the same frequencies concurrently, resulting in signal degradation or disruption.

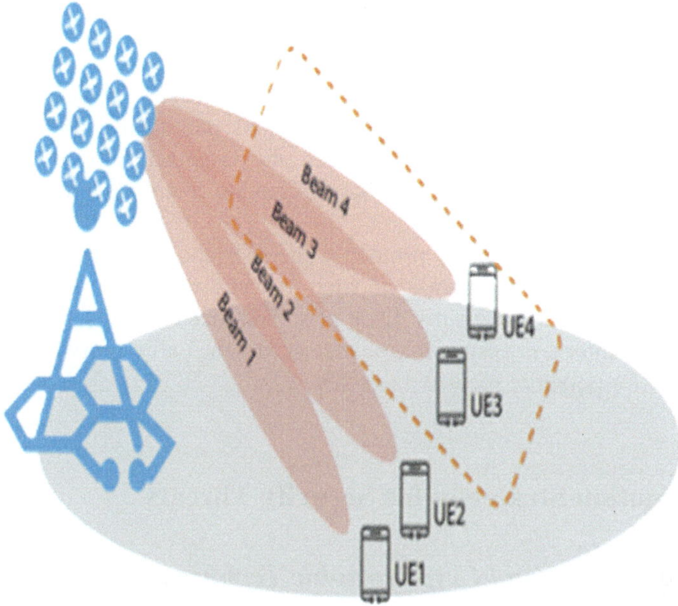

Fig. 2.7 Beamforming technique

Frequency hijacking refers to the unauthorized occupation or control of a specific frequency band by an entity not entitled to use it. Unauthorized entities hijacking frequency bands can lead to various security breaches, including the interception of sensitive data or disruption of legitimate communication.

The susceptibility of shared frequency bands to frequency hijacking necessitates robust access control measures, spectrum sensing techniques, and dynamic spectrum access protocols to mitigate the risks of interference and unauthorized access.

2.4.3 Inadequate Authentication Mechanisms

Inadequate authentication mechanisms present critical vulnerabilities in mmWave networks. Weak authentication methods or poorly implemented access control measures can lead to unauthorized access by malicious entities. Insufficient authentication allows attackers to penetrate the network, posing risks of data breaches, unauthorized information access, and other malicious activities that compromise the network's security and users' privacy.

2.4.4 Lack of Standardized Security Protocols

The absence of uniform security protocols across mmWave devices and systems results in a diverse range of security implementations, making it challenging to maintain consistency and interoperability. Devices and systems developed with differing security standards face interoperability issues, creating potential security gaps, which malicious actors could exploit.

Without standardized security protocols, inconsistencies in implementation and practices might lead to vulnerabilities, such as weak encryption methods or ineffective authentication mechanisms. Lack of standardized security protocols makes it difficult to monitor and ensure compliance across the entire mmWave ecosystem, leaving room for potential security breaches.

2.5 Mitigation Strategies for Security Threats

2.5.1 Encryption and Cryptographic Techniques

Encryption is a fundamental cornerstone in securing millimeter-wave (mmWave) communications against various threats. Robust encryption and cryptographic techniques play a pivotal role in safeguarding the confidentiality and integrity of data transmitted over these networks. The following are the various mitigation techniques for security threats:

1. Encryption algorithms: Implementing strong encryption algorithms, such as Advanced Encryption Standard (AES), Rivest–Shamir–Adleman (RSA), or Elliptic Curve Cryptography (ECC), ensures that data remains secure during transmission. These algorithms encode the information, making it unreadable to unauthorized entities [22].
2. End-to-End Encryption: Employing end-to-end encryption mechanisms ensures that data is secured from the point of origin to the destination, preventing unauthorized access and interception during transmission [23].
3. Public Key Infrastructure (PKI): Implementing a PKI allows for secure key exchange, verifying the authenticity of communicating parties, and ensuring the confidentiality and integrity of transmitted data. It employs digital certificates and public–private key pairs to validate the identity of entities involved in communication [24].
4. Cryptographic Hash Functions: Utilizing cryptographic hash functions aids in ensuring the integrity of transmitted data. Hash functions generate fixed-size unique checksums, verifying that the data remains unchanged during transmission and that it has not been tampered with [25]. Figure 2.8 lists the popular cryptographic techniques used for encryption.

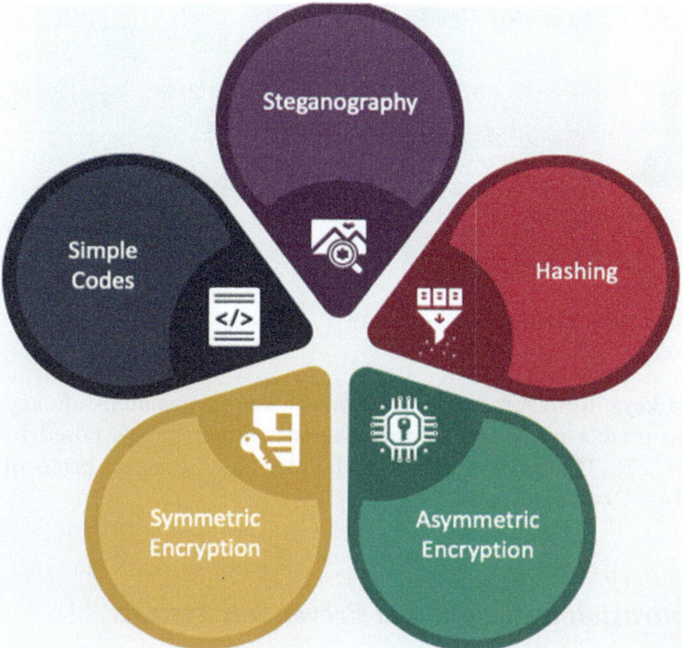

Fig. 2.8 Popular cryptographic techniques

Integrating strong encryption and cryptographic techniques in mmWave communications is essential to reinforce the security and privacy of the transmitted data. These strategies ensure that sensitive information remains confidential and untampered throughout the communication process, mitigating potential threats and vulnerabilities.

2.5.2 Secure Key Exchange Protocols

Secure key exchange protocols play a pivotal role in ensuring the confidentiality and integrity of data transmitted over mmWave networks. Implementing robust key exchange protocols such as Diffie–Hellman (DH), Elliptic Curve Diffie–Hellman (ECDH), or Transport Layer Security (TLS) enhances the security of the key exchange process [26]. This prevents potential attackers from intercepting or manipulating exchanged keys, maintaining the confidentiality and integrity of communication.

Incorporating forward secrecy mechanisms ensures that compromising one session's key does not compromise previous or subsequent keys, thereby enhancing overall security. Periodic key updates and re-negotiation of keys between communicating entities strengthen security by minimizing the window of vulnerability for

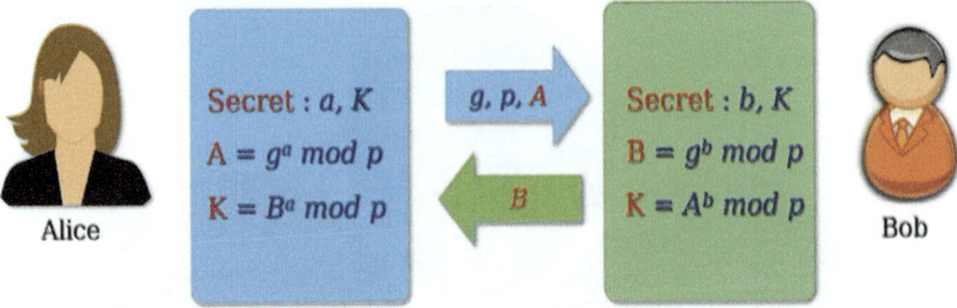

Fig. 2.9 Diffie–Hellman key exchange protocol

intercepted keys. Research and development focus on quantum-safe key exchange methods to protect mmWave networks against future threats posed by quantum computing [27]. The process in Diffie-Hellman key exchange protocol is shown in Fig. 2.9.

2.5.3 Intrusion Detection and Prevention Systems

Intrusion Detection and Prevention Systems (IDPSs) play a crucial role in protecting millimeter-wave (mmWave) communications against various security threats by actively monitoring network activities, detecting anomalies, and preventing potential intrusions. The following are the various IDPS Systems:

1. Behavioral Anomaly Detection: IDPS monitors the behavior of the network and devices to identify abnormal patterns. This involves analyzing deviations from standard behavior and raising alerts for potentially malicious activities, such as sudden increases in traffic, unusual access patterns, or unrecognized devices trying to connect to the network [28].
2. Signature-Based Detection: IDPS employs signature-based detection by comparing network activities with a database of known attack patterns or signatures. If a match is found, the system can take predefined actions to prevent or mitigate the attack [29].
3. Real-Time Monitoring and Response: IDPS systems continuously monitor the network in real-time, allowing for immediate response to identified threats. The system can automatically block or isolate suspicious devices or traffic, preventing potential attacks from compromising the network's security [30].
4. Machine Learning and AI: Implementing machine learning and artificial intelligence within IDPS can improve the system's capability to adapt and recognize new threats. By learning from patterns and data, AI-driven IDPS can detect and respond to emerging threats effectively [31].

By leveraging sophisticated monitoring, analysis, and response capabilities, IDPS acts as a proactive defense mechanism, providing continuous surveillance and instant

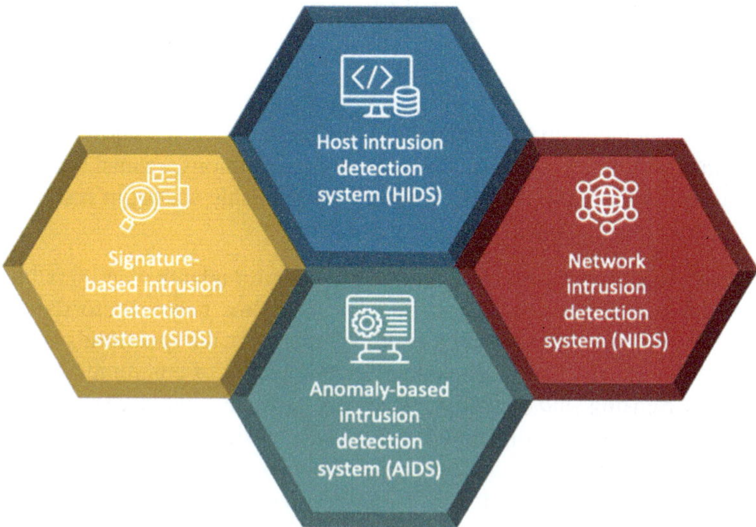

Fig. 2.10 Intrusion detection and prevention systems

responses to potential threats, thereby keeping the security posture of mmWave networks. Figure 2.10 lists the various types of Intrusion detection systems.

2.5.4 Enhanced Authentication Mechanisms

Enhanced authentication mechanisms are critical to protect mmWave networks. Multi-factor authentication, biometric authentication, and strong user verification processes add layers of security.

1. Multi-factor Authentication (MFA): Implementing MFA, combining multiple authentication factors such as passwords, biometrics, or tokens, significantly strengthens the authentication process and mitigates the risks associated with single-factor authentication [32].
2. Biometric Authentication: The integration of biometric authentication, utilizing unique biological traits such as fingerprints or facial recognition, provides a highly secure and reliable authentication method.
3. Continuous Monitoring and Anomaly Detection: Employing continuous monitoring and anomaly detection systems aids in identifying irregularities in authentication attempts, preventing potential breaches.

Implementing strict access controls and robust authentication measures ensures that only authorized users can access the network, reducing the risk of unauthorized access and potential security breaches.

2.5.5 Spectrum Management and Dynamic Frequency Allocation

Spectrum management and dynamic frequency allocation techniques are pivotal in protecting millimeter-wave (mmWave) communications against various security threats and optimizing the utilization of the available spectrum resources. The mitigation strategies for spectrum management are,

1. Dynamic Spectrum Access (DSA): DSA allows for adaptive and flexible usage of available frequency bands. It enables mmWave networks to dynamically select the most suitable frequency bands by analyzing the spectrum's availability, minimizing interference, and maximizing the quality of communication [33].
2. Spectrum Sensing and Cognitive Radio: Cognitive radio technology enables intelligent spectrum utilization by detecting available frequency bands and adapting to changes in the spectrum environment. Spectrum sensing techniques allow devices to detect and utilize underutilized or less congested frequency bands, enhancing the overall efficiency of the network [34].
3. Interference Mitigation: By employing techniques to mitigate interference in mmWave communications, such as beamforming, power control, and interference alignment, these systems can effectively reduce the impact of external interference, ensuring reliable and secure communication channels.
4. Resource Allocation Optimization: Dynamic frequency allocation optimizes the utilization of available resources by allocating frequency bands based on real-time network conditions and user demands. This ensures that mmWave devices can utilize the spectrum efficiently without causing or succumbing to interference [35].

By implementing dynamic spectrum access and efficient frequency allocation techniques, mmWave networks can adapt to changing environmental conditions, avoid interference, and optimize spectrum utilization, thereby enhancing reliability, security, and the overall performance of the network.

Adopting a combination of these mitigation strategies (as shown in Table 2.1) is imperative to safeguard mmWave networks from a spectrum of security threats. Implementing robust encryption, secure key exchange, advanced authentication mechanisms, and dynamic spectrum management will enrich the integrity and confidentiality of data transmitted over mmWave networks, ensuring a secure and reliable communication environment.

Table 2.1 Various mitigation strategies for security threats in millimeter-wave communication

Security threat	Mitigation strategies
Eavesdropping and interception	1. **Encryption**: Implement end-to-end encryption using strong cryptographic algorithms
	2. **Secure Key Exchange**: Use secure key exchange protocols like Diffie–Hellman or Elliptic Curve Diffie–Hellman
	3. **Intrusion Detection**: Employ intrusion detection systems to monitor and detect unauthorized access
Man-in-the-middle attacks	1. **Multi-factor Authentication**: Implement multi-factor authentication mechanisms for user verification
	2. **Secure Communication Channels**: Use protocols like TLS to ensure secure communication between entities
	3. **Continuous Monitoring**: Employ continuous monitoring to detect and prevent man-in-the-middle attacks in real-time
Denial-of-Service (DoS) attacks	1. **Traffic Filtering**: Implement traffic filtering mechanisms to identify and block malicious traffic
	2. **Rate Limiting**: Apply rate limiting to mitigate the impact of excessive requests and potential DoS attacks
	3. **Redundancy**: Establish redundant systems to ensure service availability in the event of a DoS attack
Physical layer attacks	1. **Secure Transmission**: Implement physical layer security techniques to protect against unauthorized access
	2. **Beamforming Security**: Ensure secure beamforming techniques to prevent unauthorized signal interception
	3. **Tamper Detection**: Deploy tamper detection mechanisms for physical mmWave equipment to detect and respond to physical attacks
Authentication and authorization	1. **Biometric Authentication**: Utilize biometric methods like fingerprints or facial recognition for secure authentication
Vulnerabilities	2. **Regular Security Audits**: Conduct periodic security audits and vulnerability assessments
	3. **Standardized Protocols**: Adhere to standardized security protocols to ensure consistency and robust security measures

2.6 Future Perspectives and Emerging Trends

The continuous development and integration of advanced security solutions, AI-driven threat detection, evolving privacy standards, and anticipation of challenges in next-generation mmWave networks are key elements in shaping the future of secure and privacy-focused mmWave communications.

2.6.1 Advancements in mmWave Security Solutions

As millimeter-wave (mmWave) communications evolve, continuous advancements in security solutions are critical to address emerging threats and vulnerabilities. Innovative approaches and technologies are being developed to protect the security of mmWave networks.

Research in quantum-resistant cryptographic algorithms aims to future-proof security in anticipation of quantum computing threats (NIST Post-Quantum Cryptography Project). Development of multifactor authentication and biometric-based authentication methods for more robust identity verification in mmWave networks [36]. Implementation of threat intelligence systems provides real-time data on potential threats, facilitating proactive responses to emerging security risks [37]. Implementation of adaptive security solutions that dynamically adjust in response to evolving threats ensures ongoing protection against new attack vectors.

2.6.2 Integration of AI and Machine Learning for Threat Detection

Leveraging AI to analyze user and network behaviors enables the identification of anomalies and potential security breaches in mmWave networks [38]. Machine learning algorithms predict potential threats based on historical data, enhancing proactive threat prevention.

2.6.3 Evolving Privacy Standards and Policies

Anticipated global privacy standards and policies aim to harmonize data protection regulations, improving consistency and protection across regions (European Data Protection Board). Emphasis on integrating privacy into the design and default settings of mmWave systems aligns with principles of data protection and user privacy (GDPR guidelines).

2.6.4 Anticipated Challenges in Next-Generation mmWave Networks

Securely integrating a massive number of IoT devices in mmWave networks ensures security and scalability while managing diverse device types and functionalities. It is important to ensure high reliability in adverse environmental conditions and maintain network resilience across diverse terrains and weather conditions.

2.7 Conclusion

Millimeter-wave (mmWave) communications present a new era of connectivity, accompanied by distinct security threats and privacy challenges. Eavesdropping, man-in-the-middle attacks, and the potential for DoS attacks pose significant risks to data integrity and user privacy. Concerns such as location tracking, data breaches, and vulnerabilities in authentication mechanisms highlight the importance of addressing privacy challenges within mmWave networks.

Eavesdropping and interception, man-in-the-middle attacks, Denial-of-Service (DoS) attacks, physical layer attacks, authentication and authorization vulnerabilities, and general system vulnerabilities collectively constitute a multifaceted threat landscape. These challenges highlight the critical need for comprehensive security measures to protect the integrity, confidentiality, and availability of mmWave communication.

The scope of privacy challenges encompasses location tracking, data breaches, and identity theft, posing risks to individuals' personal information and potentially compromising user trust. As mmWave networks become integral to various sectors, addressing these privacy concerns becomes paramount for fostering a secure and trustworthy communication ecosystem.

To navigate these challenges, adopting advanced encryption protocols, robust authentication mechanisms, and regular security audits is imperative. Integrating privacy by design, implementing user-controlled privacy measures, and adhering to evolving global privacy standards contribute to a resilient and privacy-centric mmWave communication framework.

The future of mmWave communication holds immense promise for transforming connectivity across various sectors. Advancements in technology, integration of AI, expanded applications in smart cities and IoT, and reinforced security and privacy measures are poised to revolutionize the communication landscape. Embracing global privacy standards, complying with regulations, and addressing scalability and network resilience challenges will be pivotal in fostering a secure, efficient, and privacy-conscious mmWave ecosystem.

References

1. Rappaport, T.S., Xing, Y., Kanhere, O., Ju, S., Madanayake, A., Mandal, S., Alkhateeb, A., Trichopoulos, G.C.: Wireless communications and applications above 100 GHz: opportunities and challenges for 6G and beyond. IEEE Access **7**, 78729–78757 (2019)
2. Niu, Y., Li, Y., Jin, D., Su, L., Vasilakos, A.V.: A survey of millimeter wave communications (mmWave) for 5G: opportunities and challenges. Wirel. Netw. **21**, 2657–2676 (2015)
3. Dejen, A., Ridwan, M., Anguera, J., Jayasinghe, J. (2022). Millimeter wave cellular communication performances and challenges: a survey. EAI Endors. Trans. Mob. Commun. Appl. **7**(21)

4. Faisal, G.M., Alshadoodee, H.A.A., Abbas, H.H., Gheni, H.M., Al-Barazanchi, I.: Integrating security and privacy in mmWave communications. Bull. Electr. Eng. Inform. **11**(5), 2856–2865 (2022)
5. Mestoui, J., et al.: A survey of NOMA for 5G: implementation schemes and energy efficiency. Lect. Notes Electr. Eng. **745**, 949–959 (2022)
6. Rappaport, T.S., Sun, S., Mayzus, R., Zhao, H., Azar, Y., Wang, K., Wong, G.N., Schulz, J.K., Samimi, M., Gutierrez, F.: Millimeter wave mobile communications for 5G cellular: it will work! IEEE Access **1**, 335–349 (2013)
7. MacCartney Jr., G.R., Sun, S., Rappaport, T.S., Xing, Y., Yan, H., Koka, J., Wang, R., Yu, D.: Millimeter wave wireless communications: new results for rural connectivity. In: Proceedings of the 5th Workshop on All Things Cellular: Operations, Applications and Challenges, Oct 2016, pp. 31–36 (2016)
8. Kadambar, S., Goyal, A., Chavva, A.K.R.: Millimeter wave multi-beam combining algorithm for efficient 5G cell search. In: 2020 IEEE 17th Annual Consumer Communications & Networking Conference (CCNC), Jan 2020, pp. 1–6. IEEE (2020)
9. He, S., Zhang, Y., Wang, J., Zhang, J., Ren, J., Zhang, Y., Zhuang, W., Shen, X.: A survey of millimeter-wave communication: physical-layer technology specifications and enabling transmission technologies. Proc. IEEE **109**(10), 1666–1705 (2021)
10. Alkhateeb, A., El Ayach, O., Leus, G., Heath, R.W.: Channel estimation and hybrid precoding for millimeter wave cellular systems. IEEE J. Sel. Topics Signal Process. **8**(5), 831–846 (2014)
11. Nisar, K., Jimson, E.R., Hijazi, M.H.A., Welch, I., Hassan, R., Aman, A.H.M., Sodhro, A.H., Pirbhulal, S., Khan, S.: A survey on the architecture, application, and security of software defined networking: challenges and open issues. Internet Things **12**, 100289 (2020)
12. Ferdiana, R.: A systematic literature review of intrusion detection system for network security: research trends, datasets and methods. In: 2020 4th International Conference on Informatics and Computational Sciences (ICICoS), Nov 2020, pp. 1–6. IEEE (2020)
13. Zhao, J., Jing, X., Yan, Z., Pedrycz, W.: Network traffic classification for data fusion: a survey. Inf. Fusion **72**, 22–47 (2021)
14. Rappaport, T.S., Xing, Y., MacCartney, G.R., Molisch, A.F., Mellios, E., Zhang, J.: Overview of millimeter wave communications for fifth-generation (5G) wireless networks—with a focus on propagation models. IEEE Trans. Antennas Propag.Propag. **65**(12), 6213–6230 (2017)
15. Denko, M.K.: Detection and prevention of denial of service (DOS) attacks in mobile ad hoc networks using reputation-based incentive scheme. J. Syst. Cybern. Inform. **3**(4), 1–9 (2005)
16. Chkirbene, Z., Hamdi, N.: A survey on spectrum management in cognitive radio networks. Int. J. Wirel. Mob. Comput.Wirel. Mob. Comput. **8**(2), 153–165 (2015)
17. Sun, S., Rappaport, T.S., Shafi, M., Tang, P., Zhang, J., Smith, P.J.: Propagation models and performance evaluation for 5G millimeter-wave bands. IEEE Trans. Veh. Technol. **67**(9), 8422–8439 (2018)
18. Banday, Y., Mohammad Rather, G., Begh, G.R.: Effect of atmospheric absorption on millimetre wave frequencies for 5G cellular networks. IET Commun.Commun. **13**(3), 265–270 (2019)
19. Paul, A.K., Sato, T.: Localization in wireless sensor networks: a survey on algorithms, measurement techniques, applications and challenges. J. Sens. Actuator Netw.Netw. **6**(4), 24 (2017)
20. Wang, D., Zhang, Z., Wang, P., Yan, J., Huang, X.: Targeted online password guessing: an underestimated threat. In: Proceedings of the 2016 ACM SIGSAC Conference on Computer and Communications Security, Oct 2016, pp. 1242–1254 (2016)
21. Liu, Y., Zhao, B., Zhao, P., Fan, P., Liu, H.: A survey: typical security issues of software-defined networking. China Commun. **16**(7), 13–31 (2019)
22. Menezes, A.J., Van Oorschot, P.C., Vanstone, S.A.: Handbook of Applied Cryptography. CRC Press (2018)
23. Saani, J.I.: Management Information Systems. Published by Intellectual Capital Enterprise Limited ICE Kemp House, pp. 152–160 (2019)
24. Vacca, J.R.: Public Key Infrastructure: Building Trusted Applications and Web Services (2007)

25. Schneier, B.: Applied Cryptography: Protocols, Algorithms, and Source Code in C. Wiley (2007)
26. Boneh, D., Shoup, V.: A graduate course in applied cryptography. Draft 0.5 (2020)
27. Yao, J., Zimmer, V., Yao, J., Zimmer, V.: Cryptography. Building Secure Firmware: Armoring the Foundation of the Platform, pp. 767–823 (2020)
28. Sekar, R., Bendre, M., Dhurjati, D., Bollineni, P.: A fast automaton-based method for detecting anomalous program behaviors. In: Proceedings 2001 IEEE Symposium on Security and Privacy. S&P 2001, May 2001, pp. 144–155. IEEE (2000)
29. Liu, X., Zhu, P., Zhang, Y., Chen, K.: A collaborative intrusion detection mechanism against false data injection attack in advanced metering infrastructure. IEEE Trans. Smart Grid **6**(5), 2435–2443 (2015)
30. Garcia-Teodoro, P., Diaz-Verdejo, J., Maciá-Fernández, G., Vázquez, E.: Anomaly-based network intrusion detection: techniques, systems and challenges. Comput. Secur. **28**(1–2), 18–28 (2009)
31. Elrawy, M.F., Awad, A.I., Hamed, H.F.: Intrusion detection systems for IoT-based smart environments: a survey. J. Cloud Comput. **7**(1), 1–20 (2018)
32. Zhong, S., Zhong, H., Huang, X., Yang, P., Shi, J., Xie, L., Wang, K.: Security and Privacy for Next-Generation Wireless Networks. Springer (2019)
33. Attiah, M.L., Isa, A.A.M., Zakaria, Z., Abdulhameed, M.K., Mohsen, M.K., Dinar, A.M.: Independence and fairness analysis of 5G mmWave operators utilizing spectrum sharing approach. Mob. Inf. Syst. **2019** (2019)
34. Haykin, S.: Cognitive radio: brain-empowered wireless communications. IEEE J. Sel. Areas Commun.Commun. **23**(2), 201–220 (2005)
35. Le, N.T., Tran, L.N., Vu, Q.D., Jayalath, D.: Energy-efficient resource allocation for OFDMA heterogeneous networks. IEEE Trans. Commun.Commun. **67**(10), 7043–7057 (2019)
36. Chow, M.C., Ma, M.: A lightweight traceable D2D authentication and key agreement scheme in 5G cellular networks. Comput. Electr. Eng.. Electr. Eng. **95**, 107375 (2021)
37. Lindström, O.: Next Generation Security Operations Center (2018)
38. Liu, Q., Li, P., Zhao, W., Cai, W., Yu, S., Leung, V.C.: A survey on security threats and defensive techniques of machine learning: a data driven view. IEEE Access **6**, 12103–12117 (2018)

Chapter 3
The Performance Analysis on Channel Estimation in Millimeter-Wave Communication and Their Challenges

K. Shoukath Ali, Arfat Ahmad Khan, T. Perarasi, and M. Leeban Moses

3.1 Introduction

Thanks to rapid technological progress, a lot of customers are connecting to the internet daily. This surge has led to congestion within available lower frequency bands. As a solution, researchers are investigating mmWave frequency ranges to meet the needs of current and future network generations. However, the mmWave frequency band is plagued by substantial path losses, which can be mitigated by implementing large antenna arrays, especially for high-data transmission purposes [1–5]. Additionally, in MIMO systems, individual dedicated RF chains are assigned to each antenna, intensifying power consumption and hardware complexity. This poses a significant challenge for mmWave Massive MIMO systems. Innovative beamforming techniques, facilitated by advanced lens antenna arrays, offer a potential avenue for cost reduction [6–12]. Simultaneously, the Intelligent Reflecting Surface (IRS) is a good prospective for future networks owing to its ability to enhance capacity coverage and energy efficiency [13–16]. Combining IRS with the mmWave band presents a propitious strategy for future wireless communication systems. Essentially, IRS comprises a cost-effective passive reflective component featuring reconfigurable angle and peak signal reflections. Tailoring array elements enables the reflection of electromagnetic waves in specific directions, yielding gain between receivers and transmitters [17]. Moreover, the accurate design of beamforming hinges on Channel

K. Shoukath Ali
Presidency University, Itgalpura, Rajanukunte, Yelahanka, Bengaluru, Karnataka 560064, India

A. Ahmad Khan
Department of Computer Science, College of Computing, Khon Kaen University, Khon Kaen 40002, Thailand
e-mail: arfatkhan@kku.ac.th

T. Perarasi (✉) · M. Leeban Moses
Bannari Amman Institute of Technology, Erode, Tamil Nadu 638401, India
e-mail: perasweety@gmail.com

M. El Ghzaoui et al. (eds.), *Next Generation Wireless Communication*, Signals and Communication Technology, https://doi.org/10.1007/978-3-031-56144-3_3

State Information (CSI), an essential requirement for IRS realization. Nonetheless, amassing CSI poses substantial challenges due to the intricate processing and sensing of pilot signals. With numerous, sizable elements, each IRS entails higher pilot overhead for channel estimation in cascaded channels. Existing methodologies fail to deliver a solution for achieving minimal pilot overhead and enhanced accuracy in IRS-aided systems [18]. In a related study [19], authors explore an optimal IRS pattern incorporating antenna gain, showcasing reduced estimation errors compared to on/off-based methods [20]. Yet, this approach encounters elevated pilot overhead, constraining IRS performance [21]. To address this challenge, a Compressive Sensing (CS) scheme, specifically Orthogonal Matching Pursuit (OMP), is employed to reduce pilot overhead while capitalizing on channel spatial characteristics [22]. However, due to the cascading nature and sparsity of the channel, OMP's channel estimation accuracy remains unsatisfactory [23]. A potential strategy for pilot overhead reduction involves the utilization of Double-Structured Orthogonal Matching Pursuit (DS-OMP). This configuration, reliant on the traditional OMP for channel estimation, is discussed in [24]. By jointly estimating complete rows and partial columns for different users within DS-OMP, the pilot overhead is effectively diminished [25, 26].

In the OMP and DS-OMP frameworks, the escalating signal dimension leads to heightened computational complexity, increased pilot overhead, and elevated channel estimation errors. To counter these challenges, in [27–29], the authors proposed an AMP technique, an iterative CS technique. While AMP reduces computational complexity in high-dimensional signals, it falls short in achieving satisfactory channel estimation accuracy. Particularly in mmWave MIMO systems, the fixed shrinkage parameters of AMP indirectly compromise system performance. The shrinkage parameter predicament finds resolution in the Learned AMP (LAMP) network, where distinct shrinkage parameters are assigned to each layer. Recent work [30, 31] introduces a modified LAMP algorithm, employing a deep neural network (DNN) to handle the parameters of shrinkage (linear and nonlinear) in terms of optimization. Although the LAMP network incorporates a soft threshold shrinkage function, ensuring adequate accuracy for channel estimation remains a challenge. In [32], the authors explore an AMP-based network with deep residual learning with the aim of computing channel. This novel approach integrates two key components: the LAMP network and the deep residual learning network. Initial channel estimation matrices are derived from the LAMP network, with the residual learning network contributing noise reduction and enhanced refinement. The authors identify two pivotal concerns within the LAMP network: the utilization of fixed shrinkage parameters in each level and the significant usage of memory. To address these concerns, [33] introduces a novel solution termed as HNR-LAMP. LAMP encounters performance challenges when handling non as well as i.i.d. Gaussian matrices, interpreting the learned parameters which remains unclear. Consequently, the LAMP algorithm fails to achieve satisfactory accuracy in IRS-assisted mmMIMO. The efficacy of this approach is underscored through the introduction of VAMP, improved AMP algorithms that successfully address LAMP network limitations. These advancements enhance the results of AMP across a wider spectrum of large random matrices.

The simulation outcomes highlight the success of AMP-based channel estimation within IRS-assisted communication systems, motivating the adoption of AMP-based algorithms for accurate results [34, 35]. VAMP extends the scope of AMP's effectiveness from i.i.d. Gaussian matrices to encompass rotationally invariant matrices. We utilized the VAMP technique with Expectation Maximization (EM) extensions for the channel estimation algorithm, bypassing the need to specify exhaustive priors on channel distribution and noise variance. The EM-VAMP algorithm delivers precise estimates with reasonable complexity in large-scale estimation scenarios, outperforming particularly well with a GM distribution as the prior for the proposed sparse channel at higher signal-to-noise ratios (SNRs) [36]. Estimation technique and equalization are used also in to overcome the challenges mentioned above [37–39]. By taking into the account of VAMP outstanding results, the authors utilized this technique in [40–42]. This paper showcases the extension of VAMP-SBL-EM. Numerical findings attest to the remarkable resilience of the proposed VAMP-SBL-EM against measurement matrix ill-conditioning, surpassing the performance of OMP, AMP, VAMP, and EM-VAMP.

3.2 System Model

Figure 3.1 depicts an IRS-assisted Uplink mmWave multi-user MIMO systems, in which K mobiles and a Base Station (BS) with M Uniform Planner Array (UPA). The IRS N-Reflecting elements are deployed between the K single antenna user and BS in order to have proper communication among them, and BS controls these elements wirelessly [43]. We consider Frequency Division Duplex (FDD) scheme in the IRS-assisted system. The reciprocity of channels does not exist in FDD scheme.

$\mathbf{G} \in \mathbb{C}^{M \times N}$ is the link established among the IRS and BS. In the uplink process, user sends information to the BS. At the receiver side, the received pilot signal is represented as:

$$\mathbf{y}_k = s_k \mathbf{w}_k \mathbf{G} \mathbf{\Phi} \mathbf{f}_k + n_k. \tag{3.1}$$

$\mathbf{y}_k$ denotes the gotten symbols at the BS, $\mathbf{\Phi} \in \mathbb{C}^{N \times N}$ refers to the IRS reflections, $\mathbf{w}_k \in \mathbb{C}^{1 \times M}$ is a pre-coding vector located at the user, $s_k \in \mathbb{C}$ is the symbols from mobiles, and $n_k \in \mathbb{C}$ denotes the additive noise. Let the reflecting matrix $\mathbf{\Phi} = \text{diag}(\phi_1, \phi_2, \ldots, \phi_N)$, and ϕ_n specifies the sending back data of the nth IRS element. We can use Saleh-Valenzuela channel model for $\mathbf{G}$ and $\mathbf{f}_k$, and $\mathbf{G}$ is expressed in Eq. (3.2).

$$\mathbf{G} = \sqrt{\frac{MN}{L_G}} \sum_{l_1}^{L_G} \alpha_{l_1}^{G} \mathbf{b}(\vartheta_{l_1}^{G_t}, \psi_{l_1}^{G_t}) \mathbf{a}(\vartheta_{l_1}^{G_r}, \psi_{l_1}^{G_r})^T. \tag{3.2}$$

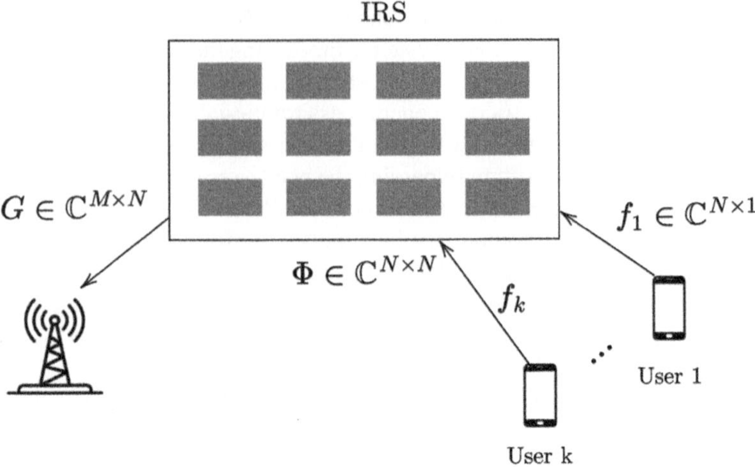

Fig. 3.1 IRS-assisted multi-user system

The total paths among the BS and IRS are represented by L_G. The value of $\mathbf{G}$ does not change for all user, even when the IRS and when the IRS is installed in the cell.

Likewise, $\mathbf{f}_k$ can be expressed as:

$$\mathbf{f}_k = \sqrt{\frac{N}{L_k}} \sum_{l_2=1}^{L_k} \alpha_{l_2}^k \mathbf{a}(\vartheta_{l_2}^k, \psi_{l_2}^k). \tag{3.3}$$

The total paths among the IRS and kth mobile are denoted by L_k, $\alpha_{l_2}^k$, $\vartheta_{l_2}^k$, and $\psi_{l_2}^k$ which refer the gain and angles at the IRS for the l_2th path, respectively. $\mathbf{b}(\vartheta, \psi) \in \mathbb{C}^{M\times 1}$ and $\mathbf{a}(\vartheta, \psi) \in \mathbb{C}^{N\times 1}$ refer to the vector of steering linked with BS as well as IRS. Under the scenario of $N1 \times N2$ ($N = N1 \times N2$) UPA, $\mathbf{a}(\vartheta, \psi)$ is expressed as:

$$\mathbf{a}(\vartheta, \psi) = \frac{1}{\sqrt{N}} \left[\mathrm{e}^{-j2\pi d \cos(\psi) n1/\lambda} \right] \otimes \left[\mathrm{e}^{-j2\pi d \sin(\psi) n2/\lambda} \right], \tag{3.4}$$

where $\mathbf{n}_1 = [0, 1, \ldots, N_1 - 1]^T$ and $\mathbf{n}_2 = [0, 1, \ldots, N_2 - 1]^T$, and λ expressed the wavelength with the antenna spacing d and keeping its distance as $d = \lambda/2$.

By denoting $\phi = [\phi_1, \phi_2, \ldots, \phi_N]$, Eq. (3.1) is expressed as:

$$\mathbf{y}_k = s_k \mathbf{w}_k \mathbf{G} \mathrm{diag}(\mathbf{f}_k) \phi + n_k, \tag{3.5}$$

$$\mathbf{y}_k = s_k \mathbf{w}_k \mathbf{H}_k \phi + n_k, \tag{3.6}$$

where $\mathbf{H}_k = \mathbf{G}\text{diag}(\mathbf{f}_k)$. As IRS has no ability for independent signal processing, the estimation of cascaded channel $\mathbf{H}_k$ is done with two separate channels of $\mathbf{G}$ and $\mathbf{f}_k$ [44, 45].

We decompose $\mathbf{H}_k \in \mathbb{C}^{M \times N}$ by utilizing virtual angular domain representation:

$$\mathbf{H}_k = \mathbf{H}_M \tilde{\mathbf{H}}_k \mathbf{U}_N^T, \tag{3.7}$$

where $\tilde{\mathbf{H}}_k$ represents the cascaded channel with the dimension of $M \times N$. The dictionary unitary matrices are represented by $\mathbf{U}_M$ having dimension $M \times M$ and $\mathbf{U}_N$ having dimension $N \times N$ at the BS and the IRS, respectively [40]. As scatters are limited around the BS and the IRS, $\tilde{\mathbf{H}}_k$ being an angular element has few elements that are not null and they exhibit the sparsity.

3.2.1 Problem Formulation

The uplink cascaded channel $\tilde{\mathbf{H}}_k$ is calculated with the help of known pilot signals that are sent from users. According to Eq. (3.5), the signals are denoted by:

$$\mathbf{y}_{k,q}^p = p_{k,q} \mathbf{w}_k \mathbf{H}_k \phi_q + n_{k,q}, \tag{3.8}$$

where $p_{k,q}$ denotes the pilot signals sent from users. After the transmission of pilots at Q time slots, the $Q \times 1$ dimensional expected vectors of pilots $\mathbf{y}_k^p = \left[\mathbf{y}_{k,1}^p, \mathbf{y}_{k,2}^p, \ldots, \mathbf{y}_{k,Q}^p\right]^T$ by assuming $p_{k,q} = 1$ can be written as:

$$\mathbf{y}_k^p = \mathbf{w}_k \mathbf{H}_k \mathbf{\Theta} + n_k, \tag{3.9}$$

where $\mathbf{\Theta} = \left[\phi_1^T, \phi_1^T, \ldots, \phi_Q^T\right]^T$ and $\mathbf{n}_k = \left[n_{k,1}, n_{k,2}, \ldots, n_{k,Q}\right]^T$.

As $\text{vec}(\mathbf{ABC}) = (\mathbf{C}^\mathbf{T} \otimes \mathbf{A})\text{vec}(\mathbf{B})$, Eq. (3.9) can be written as

$$\mathbf{y}_k^p = (\mathbf{\Theta}^\mathbf{T} \otimes \mathbf{w}_k)\text{vec}(\mathbf{H}_k) + \mathbf{n}_k. \tag{3.10}$$

The $\text{vec}(\mathbf{H}_k)$ can be further written as $(\mathbf{U}_M^* \otimes \mathbf{U}_N)\mathbf{h_k}$, and Eq. (3.10) can be re-written as:

$$\mathbf{y}_k^p = (\mathbf{\Theta}^\mathbf{T} \otimes \mathbf{w}_k)(\mathbf{U}_M^* \otimes \mathbf{U}_N)\mathbf{h}_k + \mathbf{n}_k. \tag{3.11}$$

By denoting $\mathbf{\Psi}_k = (\mathbf{\Theta}^\mathbf{T} \otimes \mathbf{w}_k)(\mathbf{U}_M^* \otimes \mathbf{U}_N)$ and $\mathbf{h}_k \in \mathbb{C}^{MN \times 1}$, we can write:

$$\mathbf{y}_k^p = \mathbf{\Psi}_k \mathbf{h}_k + \mathbf{n}_k, \tag{3.12}$$

where $\boldsymbol{\Theta}^{\mathrm{T}} \in \mathbb{C}^{Q\times N}$, $\mathbf{y}_k^p = [y_{k,1}, y_{k,2}, \ldots, y_{k,Q}] \in \mathbb{C}^{Q\times 1}$, $\boldsymbol{\Psi}_k = [\Psi_{k,1}, \Psi_{k,2}, \ldots, \Psi_{k,Q}] \in \mathbb{C}^{Q\times MN}$ and $\mathbf{n}_k = [n_{k,1}, n_{k,2}, \ldots, n_{k,Q}] \in \mathbb{C}^{Q\times 1}$.

Literally, the received pilot vector can be re-written as

$$\mathbf{y}^p = \boldsymbol{\Psi}\mathbf{h} + \mathbf{n}. \tag{3.13}$$

It is observed that $\mathbf{w}_k$ and $\boldsymbol{\Theta}$ are pre-designed and stay unchanged during the estimation of channel. In contrast to the traditional MIMO channel, the cascaded channel comes out to be N times greater, which enhances the overhead of pilots. The traditional LS technique must hold $Q \geq$ NM for Q time slots. The reduction of pilot overhead can be done by using iterative-based CS schemes, such as AMP and VAMP algorithm, and it is explained in the next section.

3.3 Iterative Methods

Generally, the compressed sensing methods are used to reconstruct high-dimensional signals by exploiting the signal properties. The accurate reconstruction is possible using iterative methods.

3.3.1 Approximate Message Passing (AMP) Algorithm

The Iterative AMP technique helps in the retrieving of sparse symbols from many directions with the reduced computational complexity. The AMP approach solves under-determined problems, that is, it can recover sparse signals from few samples. The AMP algorithm is discussed for the $\ell 1$ problem, and it is explained in Algorithm 1.

Algorithm 1 AMP

Input Parameters: $\boldsymbol{\Psi}_k$, $\mathbf{y}_k$, Number of iteration T for each user.

Initialization: $\mathbf{z}_k - 1 = \mathbf{0}$, $u_0 = 0$, $\hat{\hat{\mathbf{h}}}_0 = 0$.

For $t = 0, 1, \ldots, T-1$ *do*.

1. $\mathbf{z}_{k,t} = \mathbf{y}_k - \boldsymbol{\Psi}_k \hat{\hat{\mathbf{h}}}_{k,t} + u_{k,t}\mathbf{z}_{k,t-1}$
2. $\sigma_{k,t}^2 = \frac{1}{M}\left\|\mathbf{z}_{k,t}\right\|_2^2$
3. $\mathbf{r}_{k,t} = \hat{\hat{\mathbf{h}}}_{k,t} + \boldsymbol{\Psi}_k^T \mathbf{z}_{k,t}$
4. $\hat{\hat{\mathbf{h}}}_{k,t+1} = \eta_{st}(\mathbf{r}_{\mathbf{k},\mathbf{t}}; \lambda_{k,t}, \sigma_{k,t}^2)$
5. $u_{k,t+1} = \frac{1}{Q}\sum_{i=1}^{MN} \frac{\partial \eta_{st}(r_{k,t,i}; \lambda_{k,t}, \sigma_{k,t}^2)}{\partial r_{k,t,i}}$

End for

Output: $\hat{\hat{\mathbf{h}}}_{k,t} = \hat{\hat{\mathbf{h}}}_{K,T}$.

In [30], the authors recently opted for AMP in terms of solving $\ell 1$ problem. Comparing AMP-$\ell 1$ to ISTA, $\mathbf{z}_{k,t}$ denotes the residue component of AMP which adds the Onsager correction term $u_{k,t}\mathbf{z}_{k,t-1}$ to the optimization problem. The Step 4 is the quite important, which denotes $\hat{\hat{\mathbf{h}}}_{k,t+1} = \boldsymbol{\eta}_{k,st}(\mathbf{r}_{k,t}; \lambda_{k,t}, \sigma^2_{k,t})$. In this step, $\hat{\hat{\mathbf{h}}}_{k,t+1}$ is calculated with the help of soft threshold shrinkage $\boldsymbol{\eta}_{k,st}$.

Under the case of ith, the $r_{k,t,i} = |r_{k,t,i}|e^{jw_{k,t,i}}$ $(i = 1, 2, \ldots, MN)$ of r_t can be written as:

$$\begin{aligned}\left[\boldsymbol{\eta}_{k,st}(\mathbf{r}_{k,t}; \lambda_{k,t}, \sigma^2_{k,t})\right]_i &= \boldsymbol{\eta}_{k,st}\left(|\mathbf{r}_{k,t,i}|e^{jw_{k,t,i}}; \lambda_{k,t}, \sigma^2_{k,t}\right) \\ &= \max\left(|\mathbf{r}_{k,t,i}| - \lambda_{k,t}\sigma_{k,t}, 0\right)e^{jw_{k,t,i}},\end{aligned}$$

where $\mathbf{r}_{k,t,i}$ is a complex-value element having phase $w_{k,t,i}$, $\lambda_{k,t}$ denotes the AMP shrinkage parameter, and $\sigma^2_{k,t}$ denotes the noise variance and updated in Step 2. By using (3.13), the peak of complex-valued inputs can be reduced to zero with the help of $\boldsymbol{\eta}_{k,st}$. Moreover, the $u_{k,t+1}$ is calculated by taking derivate, as shown in the algorithm. Finally, $\hat{\hat{\mathbf{h}}}_{k,t+1}$ is estimated using soft threshold shrinkage function $\boldsymbol{\eta}_{k,st}$ and $\hat{\hat{\mathbf{h}}}_k$, where $\hat{\hat{\mathbf{h}}}_{k,t+1}$ denotes the sparse channel vector at $(t + 1)$th iteration. The Iterative AMP algorithm helps in the recovery of sparse signals with reduced computational complexity particularly with high-dimensional sparse signals [18]. The AMP approach solves under-determined problems, that is, to recover sparse signals from few samples. However, to resolve the issues of AMP algorithm, the Learned AMP (LAMP) is presented as given in the next section.

3.3.2 Learned Approximate Message Passing (AMP) Algorithm

LAMP algorithm provides a solution to the first problem of AMP algorithm, which optimizes the shrinkage parameter λ_t during every iteration where $\lambda_{k,t}$ is given in Eq. (3.14).

$$\lambda_{k,t} = \frac{1}{N_T}\|\mathbf{z}_{k,t}\|_2^2\alpha. \tag{3.14}$$

In LAMP network, the noise variance is assumed similar to AMP as given in Eq. (3.15).

$$\sigma^2_{k,t} = \frac{1}{N}\|\mathbf{z}_{k,t}\|_2^2. \tag{3.15}$$

The layered architecture of general AMP algorithm known as Learned AMP (LAMP) is given in Fig. 3.2. The estimation accuracy is improved due to the layered

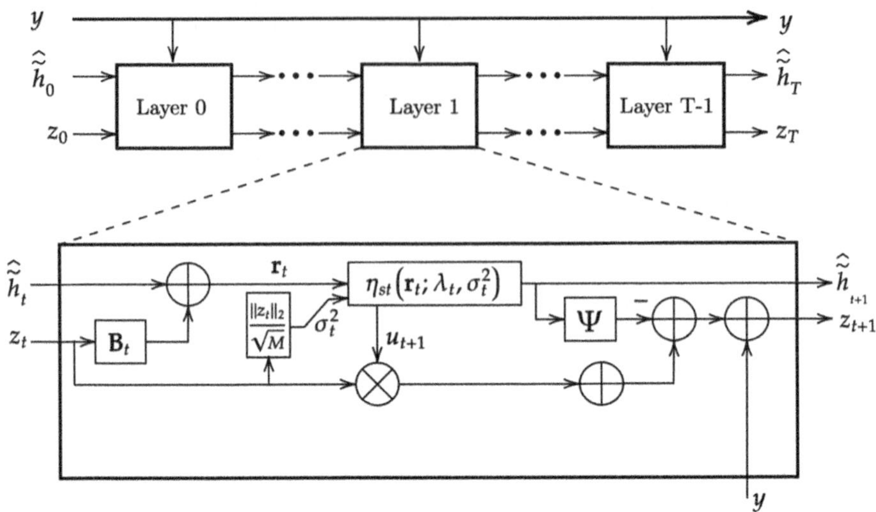

Fig. 3.2 Learned approximate message passing (LAMP) algorithm

iteration method. In the figure, $\mathbf{y}_k, \hat{\tilde{\mathbf{h}}}_{k,t}, \mathbf{z}_{k,t}$ defines the input of the tth layer and $\hat{\tilde{\mathbf{h}}}_{k,t+1}$, $\mathbf{z}_{k,t+1}$ defines the output of the $(t-1)$th. Instead of $\boldsymbol{\Psi}_k^T$ dictionary matrix, there are varieties of linear coefficients like $\mathbf{B}_{k,t}$ which is used during tth layer of LAMP network to find the sparse vector $\hat{\tilde{\mathbf{h}}}_{k,t}$. The size of the matrix for linear coefficient $\mathbf{B}_{k,t}$ is $MN \times Q$. The tth level in LAMP is written in Eqs. (3.16)–(3.18).

$$\hat{\tilde{\mathbf{h}}}_{k,t+1} = \eta_{k,st}(\mathbf{r}_{k,t}; \lambda_{k,t}, \sigma_{k,t}^2), \tag{3.16}$$

$$\mathbf{z}_{k,t+1} = \mathbf{y}_k - \boldsymbol{\Psi}_k \hat{\tilde{\mathbf{h}}}_{k,t} + u_{k,t+1}\mathbf{z}_{k,t}, \tag{3.17}$$

$$\mathbf{r}_{k,t} = \hat{\tilde{\mathbf{h}}}_{k,t} + \mathbf{B}_{k,t}\mathbf{z}_{k,t}. \tag{3.18}$$

3.3.3 Learning the LAMP Parameters

The complete training procedure in the offline estimation is divided among T sub-procedures. During the offline estimation, the supervised learning method is considered for the training purpose. For this method, the dataset of training is denoted as $\{\mathbf{y}_k^d, \hat{\tilde{\mathbf{h}}}_k^d\}_{d=1}^D$, where D denotes the total sets in terms of training, $\mathbf{y}_k^d$ denotes the provided signal in LAMP networks, and $\hat{\tilde{\mathbf{h}}}_k^d$ is the corresponding label. To handle the overfitting problem, layer-by-layer technique with Adam optimizer is utilized along with the backpropagation for optimizing the final value. In LAMP-$\ell 1$ network, two cases of LAMP network are considered. Case (i) is the "tied" case which provides the exact transformation in every level, and Case (ii) is the "untied" case that does not perform exact transformation. LAMP-$\ell 1$ parameters are then $\mathbf{\Omega}_k = \{\mathbf{B}_k, \{\boldsymbol{\theta}_{k,i}\}_{i=1}^{t}\}$ in the tied case, and $\mathbf{\Omega}_k = \{\mathbf{B}_{k,t}, \boldsymbol{\theta}_{k,t}\}_{t=0}^{T-1}$ under untied case. Layer-by-Layer Tied LAMP-l1 and Layer-by-Layer Untied LAMP-l1 process is discussed further. For the "tied" case of LAMP-$\ell 1$, $\mathbf{\Omega}_{k,T-1}^{\mathbf{tied}} = \{\mathbf{B}_k, \{\boldsymbol{\theta}_{k,i}\}_{t=0}^{T-1}\}$ is learned to minimize the Minimum Square Error (MSE) during training data with loss function as

$$L_{k,t}(\mathbf{\Omega}_{k,t}) = \frac{1}{J}\sum_{j=1}^{J}\left\|\hat{\tilde{\mathbf{h}}}_{k,t+1}^{j}(\mathbf{y}_k^j, \mathbf{\Omega}_{k,t}) - \tilde{\mathbf{h}}_k^j\right\|_2^2. \tag{3.19}$$

As a first attempt, we set $\mathbf{B}_k = \mathbf{\Psi}_k^T,\ \boldsymbol{\theta}_0 = \boldsymbol{\theta}^0$. Due to overfitting, failure in analysis happens. Thus, we can tackle this issue by using the combination of "layer-wise" and "global" optimization avoid this problem. To be more precise, the main aim is to learn $\mathbf{\Omega}_{k,0}^{\mathbf{tied}}$, then $\mathbf{\Omega}_{k,1}^{\mathbf{tied}}$, until $\mathbf{\Omega}_{k,T-1}^{\mathbf{tied}}$. The complete details of LAMP network are explained in Algorithms 2 and 3.

Algorithm 2 Layer-by-Layer Tied LAMP-l1 Training Method
Set: $\mathbf{B} = \mathbf{\Psi}_k^{\mathbf{T}},\ \boldsymbol{\theta}_{k,0} = \boldsymbol{\theta}_k^0$
Learn $\mathbf{\Omega}_{k,0}^{\mathbf{tied}} = \{\mathbf{B}_k, \boldsymbol{\theta}_{k,0}\}$
When $t = 1, \ldots, T-1$ *do*
Set: $\boldsymbol{\theta}_{k,t} = \boldsymbol{\theta}_{k,t-1}$
Learn $\boldsymbol{\theta}_{k,t}$ with fixed $\mathbf{\Omega}_{k,t-1}$
Re-Learn $\mathbf{\Omega}_{k,t}^{\mathbf{tied}} = \{\mathbf{B}_k, \{\boldsymbol{\theta}_{k,i}\}_{i=1}^{t}, \boldsymbol{\theta}_{k,0}\}$
End for
Output: $\mathbf{\Omega}_{k,T-1}^{\mathbf{tied}}$

Algorithm 3 Layer-by-Layer Untied LAMP-l1 Training Method
Compute $\{\mathbf{\Omega}_{k,t}^{\mathbf{tied}}\}_{t=1}^{T-1}$ using Algorithm 2
Initialization: $\mathbf{B}_{k,0} = \mathbf{\Psi}_k^{\mathbf{T}},\ \boldsymbol{\theta}_{k,0} = \boldsymbol{\theta}_k^0$
Learn $\mathbf{\Omega}_{k,0}^{\mathbf{tied}} = \{\mathbf{B}_k, \boldsymbol{\theta}_{k,0}\}$
When $t = 1, \ldots, T-1$ *do*
Set: $\mathbf{B}_{k,t} = \mathbf{B}_{k,t-1},\ \boldsymbol{\theta}_{k,t} = \boldsymbol{\theta}_{k,t-1}$
Calculate $\{\mathbf{B}_{k,t}, \boldsymbol{\theta}_{k,t}\}$ with constant $\mathbf{\Omega}_{k,t-1}^{\text{untied}}$
Set $\mathbf{\Omega}_{k,t}^{\text{untied}} = \{\mathbf{B}_{k,i}, \{\boldsymbol{\theta}_{k,i}\}_{i=1}^{t}\}$
If $\mathbf{\Omega}_{k,t}^{\text{tied}}$ performs better than $\mathbf{\Omega}_{k,t}^{\text{untied}}$
End if
Re-learn $\mathbf{\Omega}_{k,t}^{\text{untied}}$
End for
Return $\mathbf{\Omega}_{k,T-1}^{\text{untied}}$

The performance of LAMP network during non-i.i.d.-Gaussian matrices $\mathbf{\Psi}_k$ is degraded. To handle this issue, we move to the next algorithm.

3.3.4 Vector Approximate Message Passing (VAMP) Algorithm

VAMP extends the knowledge of AMP to vector-valued signals, where VAMP uses a more sophisticated message passing scheme to propagate information between the different components of the vector signal, taking into account the correlation between them. For signal recovery scenarios, VAMP has been shown better performance compared to AMP, especially when the signal has a high degree of correlation between its components. However, the VAMP algorithm has higher computational complexity than AMP, and its performance can depend on the specific covariance structure of the signal being recovered. The VAMP algorithm is discussed in Algorithm 4.

Algorithm 4 Vector AMP

Need $\tilde{\boldsymbol{\eta}}(.; \tilde{\sigma}, \tilde{\boldsymbol{\theta}})$ from (3.20), $\boldsymbol{\eta}(.; \sigma, \boldsymbol{\theta})$, T, $\{\boldsymbol{\theta}_t\}_{t=1}^{T}$ and $\tilde{\boldsymbol{\theta}}$. for each user or all user.
1: set $\tilde{r}_1$ and $\tilde{\sigma}_1 > 0$.
2: **During** $t = 1, 2, \ldots, T$ **do**
3: //LMMSE Working
4: $\tilde{\boldsymbol{h}}_t = \tilde{\boldsymbol{\eta}}(\tilde{\boldsymbol{r}}_t; \tilde{\sigma}_t, \tilde{\boldsymbol{\theta}})$
5: $\tilde{z}_t = \langle \tilde{\boldsymbol{\eta}}'(\tilde{\boldsymbol{r}}_t; \tilde{\sigma}_t, \tilde{\boldsymbol{\theta}}) \rangle$
6: $\boldsymbol{r}_t = (\tilde{\boldsymbol{h}}_t - \tilde{z}_t \tilde{\boldsymbol{r}}_t)/(1 - \tilde{z}_t)$
7: $\sigma_t^2 = \tilde{\sigma}_t^2 \tilde{z}_t/(1 - \tilde{z}_t)$
8: // Shrinkage stage:
9: $\hat{\boldsymbol{h}}_t = \boldsymbol{\eta}(\boldsymbol{r}_t; \sigma_t, \boldsymbol{\theta}_t)$
10: $z_t = \langle \boldsymbol{\eta}'(\boldsymbol{r}_t; \sigma_t, \boldsymbol{\theta}_t) \rangle$
11: $\tilde{\boldsymbol{r}}_{t+1} = (\hat{\boldsymbol{h}}_t - z_t \boldsymbol{r}_t)/(1 - z_t)$
12: $\tilde{\sigma}_{t+1}^2 = \sigma_t^2 z_t/(1 - z_t)$
End for
Return $\hat{\boldsymbol{h}}_T$

The VAMP Algorithm 4 is consisting of two stages such as LMMSE and shrinkage stage.

In the first stage, LMMSE technique is used and it is written as

$$\tilde{\boldsymbol{\eta}}(\tilde{\boldsymbol{r}}_t; \tilde{\sigma}_t, \tilde{\boldsymbol{\theta}}) \triangleq V\left(\text{Diag}(s)^2 + \frac{\sigma_w^2}{\tilde{\sigma}_t^2}\boldsymbol{I}_R\right)^{-1}\left(\text{Diag}(s)^2\boldsymbol{U}^T\boldsymbol{y} + \frac{\sigma_w^2}{\tilde{\sigma}_t^2}\boldsymbol{V}^T\tilde{r}_t\right). \tag{3.20}$$

It is linked with $\mathbf{y}$ and can be expressed as

$$\tilde{\boldsymbol{\theta}} \triangleq \{\boldsymbol{U}, \boldsymbol{s}, \boldsymbol{V}, \sigma_w\}. \tag{3.21}$$

The shrinkage function is used in Eq. (3.22) and as same we used in AMP algorithm.

$$\hat{\mathbf{h}}_{t+1} = \boldsymbol{\eta}(\hat{\mathbf{h}}_t + \boldsymbol{\Psi} z_t; \sigma_t, \boldsymbol{\theta}_t). \tag{3.22}$$

In terms of average, the elements of diagonal in $\tilde{\boldsymbol{\eta}}(.; \tilde{\sigma}_t, \tilde{\boldsymbol{\theta}})$ and and $\boldsymbol{\eta}(.; \sigma_t, \boldsymbol{\theta}_t)$ are computed in Lines 5 and 10 of Algorithm 1, respectively.

$$\langle \boldsymbol{\eta}'(\boldsymbol{r}_t; \sigma_t, \boldsymbol{\theta}_t) \rangle \triangleq \frac{1}{MN} \sum_{j=1}^{MN} \frac{\partial [\boldsymbol{\eta}(\boldsymbol{r}; \sigma, \boldsymbol{\theta}]_j}{\partial_{r_j}}. \tag{3.23}$$

From (3.20), the Jacobian of $\tilde{\boldsymbol{\eta}}(.; \tilde{\sigma}_t, \tilde{\boldsymbol{\theta}})$ is expressed in Eq. (3.24) and take average for diagonal matrices and it is shown in Eq. (3.25).

$$\frac{\sigma_w^2}{\tilde{\sigma}_t^2} \mathbf{V} \left(\mathrm{Diag}(s)^2 + \frac{\sigma_w^2}{\tilde{\sigma}_t^2} \mathbf{I}_R \right)^{-1} \mathbf{V}^T, \tag{3.24}$$

$$\langle \tilde{\boldsymbol{\eta}}'(\tilde{\boldsymbol{r}}_t; \tilde{\sigma}_t, \tilde{\boldsymbol{\theta}}) \rangle = \frac{1}{N} \sum_{i=1}^{R} \frac{1}{s_i^2 \tilde{\sigma}_t^2 / \sigma_w^2 + 1}. \tag{3.25}$$

The first-stage LMMSE estimator is defined in Eq. (3.20) $\tilde{\boldsymbol{\eta}}(.; \tilde{\sigma}_t, \tilde{\boldsymbol{\theta}})$, where $\boldsymbol{h}^0$ is computed using MMSE estimator under the likelihood function

$$p(\boldsymbol{y} | \boldsymbol{h}^0) = N(\boldsymbol{y}; \boldsymbol{\Psi} \boldsymbol{h}^0, \sigma_w^2 \boldsymbol{I}), \tag{3.26}$$

where $w \sim N(0, \sigma_w^2 \boldsymbol{I})$ and the pseudo-prior $\boldsymbol{h}^0$ are assumed from Eq. (3.12).

$$\boldsymbol{h}^0 \sim N(\tilde{r}_t, \tilde{\sigma}_t^2 \boldsymbol{I}). \tag{3.27}$$

Equation (3.27) is called "pseudo" prior due to being built within the VAMP during convergence. The MMSE estimator of $\boldsymbol{h}$ is expressed as conditional mean $E\{\boldsymbol{h} | \boldsymbol{y}\}$. Combined to Eqs. (3.26)–(3.27), Eq. (3.28) can be written as

$$\left(\boldsymbol{\Psi}^T \boldsymbol{\Psi} + \frac{\sigma_w^2}{\tilde{\sigma}_t^2} \mathbf{I}_N \right)^{-1} \left(\boldsymbol{\Psi}^T \mathbf{y} + \frac{\sigma_w^2}{\tilde{\sigma}_t^2} \tilde{r}_t \right). \tag{3.28}$$

Substituting $\boldsymbol{\Psi}$ in (3.28) with SVD from (3.29), we get the expression in (3.20).

$$\boldsymbol{\Psi} = \mathbf{USV}^T. \tag{3.29}$$

In the second stage, shrinkage function $\boldsymbol{\eta}(.; \sigma_t, \boldsymbol{\theta}_t)$ is used to estimate $\hat{\boldsymbol{h}}_t = \boldsymbol{\eta}(\boldsymbol{r}_t; \sigma_t, \boldsymbol{\theta}_t)$, which removes the error in vector.

$$\boldsymbol{r}_t = \boldsymbol{h}^0 + N(0, \sigma_t^2). \tag{3.30}$$

The denoised under the case of $\boldsymbol{\eta}$:

$$\boldsymbol{\eta}(\boldsymbol{r}_t;\sigma_t,\boldsymbol{\theta}_t)=E\{\boldsymbol{h}^0|r_t\}. \tag{3.31}$$

3.3.5 EM-VAMP Algorithm

In this section, EM-VAMP is derived mathematically to analyze the performance of proposed work.

From the received output measurements vector, recover the unknown channel vector using EM-VAMP algorithm and it is explained below, where $\mathbf{h}$ follows a $p(\boldsymbol{h}|\theta_{h1})$ prior density.

In contrast to other AMP-based methods such as GAMP, the broader applicability of VAMP's state evolution encompasses a wider range of matrices $\boldsymbol{\Psi}$, while no restrictions are imposed on $\mathbf{U}_\mathrm{A}$ and $\mathbf{S}_\mathrm{A}$. This fundamental difference empowers the VAMP algorithm to surmount AMP's primary drawback and maintain numerical stability even in the presence of ill-conditioned matrices $\boldsymbol{\Psi}$. Thus, VAMP emerges as an ideal choice for integration within our proposed IRS-aided mmWave channel system.

The ensuing discussion elucidates the details of the tth iteration within the EM-VAMP technique, as expounded in Algorithm 5.

Algorithm 5 EM-VAMP Algorithm

Require: measurement matrix $\boldsymbol{\Psi}\in\mathbb{C}^{Q\times MN}$, Observed vector $\mathbf{y}\in\mathbb{C}^{Q\times1}$, the number of iterations T, the prior $p_h(\boldsymbol{h}|\theta_{h1})$ and $p_{y/h}(\mathbf{y}|\mathbf{h},\theta_{h2})$;

Define: $\mathbf{g}_{h1}(\mathbf{r}_1,\gamma_1)$, $\mathbf{g}_{h2}(\mathbf{r}_2,\gamma_2)$
1: Initialize $\mathbf{r}_1^0,\gamma_1^0,\theta_{h1}^0,\theta_{h2}^0$
2: **for** $t=0,1,\ldots,T$ **do**
3: //Input Denoising
4: $\hat{\mathbf{h}}_1^t=g_{h1}(\mathbf{r}_1^t,\gamma_1^t,\theta_{h1})$
5: $1/\eta_{h1}^t=\langle g'_{h1}(\mathbf{r}_1^t,\gamma_1^t,\theta_{h1})\rangle/\gamma_1^t$
6: $\gamma_2^t=\eta_{h1}^t-\gamma_1^t$
7: $\mathbf{r}_2^t=\left(\eta_{h1}^t\mathbf{h}_1^t-\gamma_1^t\mathbf{r}_1^t\right)/\gamma_2^t$
8: $\hat{\theta}_{h1}^{t+1}=\arg\max_{\theta_{h1}}\mathrm{E}\left[\ln p_h(\mathbf{h}|\theta_{h1})|r_{1k},\gamma_{1k},\hat{\theta}_{h1}^{t+1}\right]$
9: //LMMSE Denoising
10: $\hat{\mathbf{h}}_2^t=g_{h2}(\mathbf{r}_2^t,\gamma_2^t,\theta_{h2})$
11: $1/\eta_{h2}^t=\langle g'_{h2}(\mathbf{r}_2^t,\gamma_2^t,\theta_{h2})\rangle/\gamma_2^t$
12: $\gamma_1^{t+1}=\eta_{h2}^t-\gamma_2^t$
13: $\mathbf{r}_1^{t+1}=\left(\eta_{h2}^t\mathbf{h}_2^t-\gamma_2^t\mathbf{r}_2^t\right)/\left(\eta_{h2}^t-\gamma_2^t\right)$
14: $\hat{\theta}_{h2}^{t+1}=\arg\max_{\theta_{h2}}\mathrm{E}\left[\ln p_{y/h}(\mathbf{y}|\mathbf{h}|\theta_{h2})|r_{2k},\gamma_{2k},\hat{\theta}_{h2}^t\right]$
15: Return $\hat{\mathbf{h}}_1$

The final outcome of $\boldsymbol{r}$ in AMP technique is denoted as:

$$g_{h1}(\mathbf{r}_1, \gamma_1) \triangleq \frac{\int \mathbf{h} p_h(\mathbf{h}|\theta_{h1}) N(\mathbf{h}; \mathbf{r}_1, \mathbf{I}/\gamma_1) d\mathbf{h}}{\int p_h(\mathbf{h}|\theta_{h1}) N(\mathbf{h}; \mathbf{r}_1, \mathbf{I}/\gamma_1) d\mathbf{h}}, \tag{3.32}$$

$$\langle g'_{h1}(\boldsymbol{r}_1^t, \gamma_1^t)\rangle = \frac{1}{N} \mathrm{tr}\left\{\frac{\partial g_i(\boldsymbol{r}, \gamma)}{\partial \boldsymbol{r}}\right\}, \quad \text{for } i = 1, 2. \tag{3.33}$$

The LMMSE under of scenario of $\mathbf{h}$ can be written as:

$$\mathbf{g}_{h2}(\mathbf{r}_{2k}, \mathbf{p}_{2k}, \gamma_{2k}, \tau_{2k}, \theta_{h2}) = \mathbf{V}\mathbf{D}_k(\theta_{h2}\tau_{2k}\mathbf{S}^T\mathbf{U}^T\mathbf{p}_{2k} + \gamma_{2k}\mathbf{V}^T\mathbf{r}_{2k}), \tag{3.34}$$

$$[\boldsymbol{D}_k]_{nn} \triangleq (\theta_{h2}\tau_{2k}s_n^2 + \gamma_{2k})^{-1}, \tag{3.35}$$

$$\mathbf{g}_{h2}(\mathbf{r}_{2k}, \mathbf{p}_{2k}, \gamma_{2k}, \tau_{2k}, \theta_{h2}) = \mathbf{r}_{2k} + \mathbf{V}\mathbf{S}^T\left(\theta_{h2}\frac{\gamma_{2k}}{\tau_{2k}}\mathbf{I} + \mathbf{S}\mathbf{S}^T\right)^{-1}(\mathbf{U}^T\mathbf{p}_{2k} - \mathbf{S}\mathbf{V}^T\mathbf{r}_{2k}), \tag{3.36}$$

$$\partial \mathbf{g}_{h2}/\partial \mathbf{r}_{2k} = \gamma_{2k}\mathbf{V}\mathbf{D}_k\mathbf{V}^T, \tag{3.37}$$

$$\alpha_{2k} = \left\langle \mathbf{g}'_{h2}(\mathbf{r}_{2k}, \mathbf{p}_{2k}, \gamma_{2k}, \tau_{2k}, \theta_{h2})\right\rangle = \frac{1}{MN}\sum_{n=1}^{MN}\frac{\gamma_{2k}}{(\theta_{h2}\tau_{2k}s_n^2 + \gamma_{2k})}. \tag{3.38}$$

3.4 Proposed VAMP-SBL-EM Algorithm

Here, we formulate the suggested VAMP-SBL technique by integrating the VAMP approach for sparse signal estimation $\mathbf{h}$ with the simultaneous utilization of the EM method as well as auto-tuning of variance within the SBL structure to learn the unknown parameters (γ_1, θ_{h1}) and (γ_2, θ_{h2}). VAMP-SBL-EM Algorithm 5 is discussed below.

Require: measurement matrix $\boldsymbol{\Psi} \in \mathbb{C}^{Q\times MN}$, Observed vector $\mathbf{y} \in \mathbb{C}^{Q\times 1}$, the number of iterations T, the prior $p_h(\mathbf{h}|\theta_{h1})$ and $p_{y/h}(\mathbf{y}|\mathbf{h}, \theta_{h2})$;

Define: $\mathbf{g}_{h1}(\mathbf{r}_1, \gamma_1)$, $\mathbf{g}_{h2}(\mathbf{r}_2, \gamma_2)$
1: Initialize $\mathbf{r}_1^0, \gamma_1^0, \theta_{h1}^0, \theta_{h2}^0$
2: for $t = 0, 1, \ldots, T_o$ **do**
3: Initialize $\theta_{h1}^1 \leftarrow \theta_{h1}^{t-1}, \theta_{h2}^1 \leftarrow \theta_{h2}^{t-1}, r_1^1 \leftarrow r_1^{t-1}, \gamma_1^1 \leftarrow \gamma_1^{t-1}$
4: for $\tau = 0, 1, \ldots, T_E$ **do**
5: $\hat{\mathbf{h}}_1^\tau = g_{h1}(\mathbf{r}_1^\tau, \gamma_1^\tau, \theta_{h1})$
6: $1/\eta_{h1}^\tau = \langle g'_{h1}(\mathbf{r}_1^\tau, \gamma_1^\tau, \theta_{h1})\rangle/\gamma_1^\tau$

7: $1/\gamma_1^\tau = \frac{1}{N}\left\|h_1^\tau - r_1^\tau\right\|^2 + 1/\eta_{h1}^\tau$
8: End for
9: $\eta_{h1}^t \leftarrow \eta_{h1}^\tau, 1/\gamma_1^t \leftarrow 1/\gamma_1^\tau, r_1^t \leftarrow r_1^\tau, h_1^t \leftarrow h_1^\tau$
10: $\gamma_2^t = \eta_{h1}^t - \gamma_1^t$
11: $\mathbf{r}_2^t = \left(\eta_{h1}^t \mathbf{h}_1^t - \gamma_1^t r_1^t\right)/\gamma_2^t$
12: $\hat{\mathbf{h}}_2^t = g_{h2}(\mathbf{r}_2^t, \gamma_2^t, \theta_{h2})$
13: $1/\eta_{h2}^t = \langle g_{h2}'(\mathbf{r}_2^t, \gamma_2^t, \theta_{h2})\rangle/\gamma_2^t$
14: $1/\gamma_2^\tau = \frac{1}{N}\left\|h_2^\tau - r_2^\tau\right\|^2 + 1/\eta_{h2}^\tau$
14: $\gamma_1^{t+1} = \eta_{h2}^t - \gamma_2^t$
15: $\mathbf{r}_1^{t+1} = \left(\eta_{h2}^t \mathbf{h}_2^t - \gamma_2^t \mathbf{r}_2^t\right)/\left(\eta_{h2}^t - \gamma_2^t\right)$
17: $\hat{\theta}_{h1}^{t+1} = \arg\max_{\theta_{h1}} \mathrm{E}\left[\ln p_h(\mathbf{h}|\theta_{h1})|r_{1k}, \gamma_{1k}, \hat{\theta}_{h1}^{t+1}\right]$
18: $1/\hat{\theta}_{h2}^{t+1} = \frac{1}{M}\left[\left\|y - \Psi h_2^t\right\|^2 + \sum_{k=1}^{R} \frac{s_k^2}{\theta_2^k s_k^2 + \gamma_2^t}\right]$
19: if $\left\|h_1^t - h_1^t\right\|^2/\left\|h_1^t\right\|^2 < \varepsilon$, break
20: end for
21: Return $\hat{\mathbf{h}}_1$

3.5 Simulation

The results are obtained by simulating the environment using Python within the TensorFlow framework. We use $M = 64$ ($M1 = 8$, $M2 = 8$), $N = 256$ ($N1 = 16$, $N2 = 16$), and $K = 16$ for the simulations. For the analysis work, the links among IRS and BS are $L_G = 5$ and $L_k = 8$ which refer to the total links owing RIS and mobiles. The channel $\mathbf{f}_k$ from the kth user and IRS is generated with using the Saleh-Valenzuela channel model in (3.3). We set the frequency be 28 GHz.

Where we can write NMSE as

$$\mathrm{NMSE} = \frac{\mathrm{E}\left\{\sum_{k=1}^{K} \left\|\hat{\mathbf{h}} - \mathbf{h}\right\|_2^2\right\}}{\mathrm{E}\left\{\sum_{k=1}^{K} \|\mathbf{h}\|_2^2\right\}}$$

The performance evaluation depicted in Fig. 3.3 shows the NMSE against SNR outcomes under the case of VAMP-SBL-EM technique and weighs up with alternative approaches such as OMP, AMP, and VAMP-EM. Notably, the results reveal that VAMP-EM and VAMP-SBL-EM algorithms exhibit superior performance at elevated SNRs, while the OMP algorithm excels in lower SNR scenarios. However, the efficacy of OMP is comparatively compromised, primarily stemming from its sensitivity to step size and decision threshold selection. Notably, in i.i.d. Gaussian matrix $\boldsymbol{\Psi}$, both VAMP-EM algorithms underperform when contrasted with the VAMP-SBL-EM algorithm. These findings reinforce the preference for the proposed algorithm in addressing the IRS-aided mmWave MIMO channel estimation challenge, showcasing its advantages over existing methods. The comparison of NMSE performance with

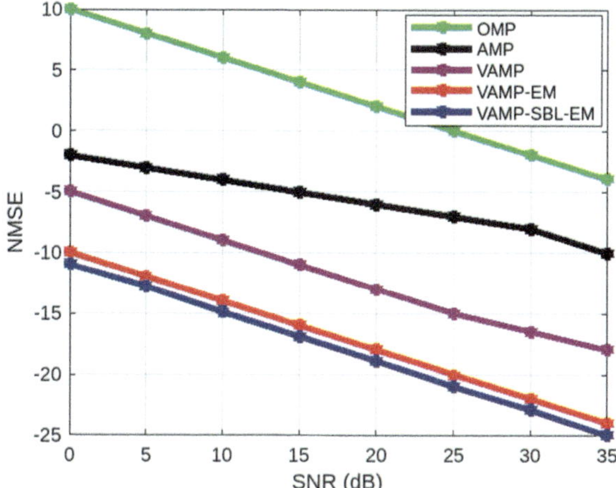

Fig. 3.3 NMSE versus SNR

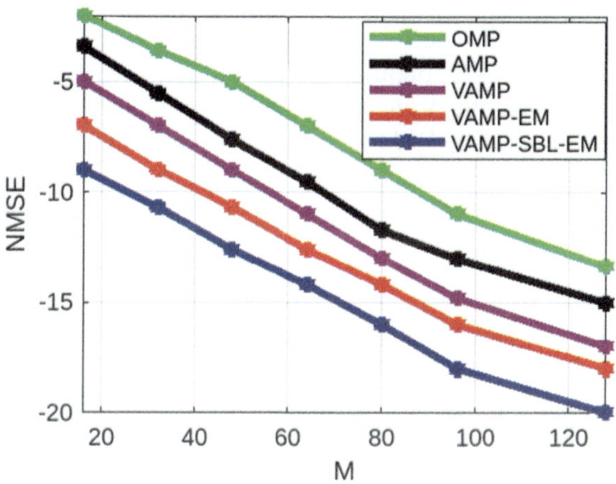

Fig. 3.4 NMSE versus M

varying numbers of BS antennas, denoted as M, is graphically represented in Fig. 3.4 for both the proposed and VAMP-SBL-EM algorithms. Notably, as the count of BS antennas, M, experiences augmentation, a noticeable trend emerges and VAMP-SBL-EM algorithms consistently exhibit reduced NMSE values. This behavior can be attributed to the inherent characteristics of sparser environments, wherein the fixed scattering paths lead to a consistent number of substantial nonzero elements. Consequently, an increase in M yields enhanced estimation outcomes, particularly within contexts of sparser conditions. Furthermore, it is worth highlighting that the proposed VAMP-SBL-EM algorithms consistently yield lower NMSE values in comparison to existing algorithms. This further underscores the efficacy of our proposed algorithms, particularly in scenarios involving the presence of larger numbers of BS antennas within the sparser environment.

Fig. 3.5 NMSE versus N

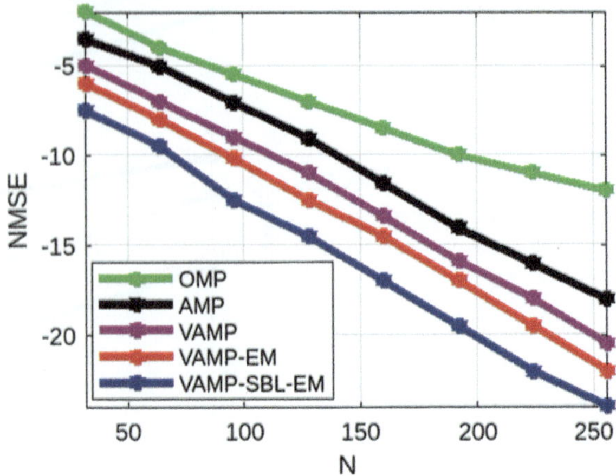

Figure 3.5 illustrates the variation of NMSE in relation to the number of IRS elements, denoted as N, for the proposed and VAMP-SBL-EM algorithm, across different SNR conditions. Notably, a discernible trend emerges as the quantity of IRS elements, N, is augmented, wherein VAMP-SBL-EM algorithm consistently exhibits reduced NMSE values. This phenomenon can be attributed to the persistent nature of the fixed scattering paths, thereby leading to a noticeable enhancement in the accuracy of estimation outcomes within the framework of the proposed algorithms.

3.6 Conclusion

This paper addresses the channel calculation challenge in an IRS-assisted MIMO multi-user mmWave communication system. We introduce the VAMP-SBL-EM algorithms for efficient channel vector estimation within this context. Leveraging the computational efficiency of the VAMP-SBL-EM algorithm, we achieve accurate channel estimation without necessitating key parameters such as noise variance and prior distribution parameters. Through extensive simulations, we compare the performance of our proposed VAMP-SBL-EM algorithms against existing methods like VAMP-EM, VAMP, AMP, and OMP. Our results demonstrate that, particularly at higher SNRs, the VAMP-SBL-EM algorithm offers superior estimation accuracy with swift convergence, reinforcing the merits of our approach and the validity of our analysis.

References

1. Han, S., I, C.-L., Xu, Z., Rowell, C.: Large-scale antenna systems with hybrid analog and digital beamforming for millimeter wave 5G. IEEE Commun. Mag. **53**(1), 186–194 (2015)
2. Raghavan, V., Cezanne, J., Subramanian, S., Sampath, A., Koymen, O.: Beamforming tradeoffs for initial UE discovery in millimeter-wave MIMO systems. IEEE J. Sel. Topics Signal Process. **10**(3), 543–559 (2016)
3. Gao, X., Dai, L., Han, S., I, C.-L., Wang, X.: Reliable beamspace channel estimation for millimeter-wave massive MIMO systems with lens antenna array. IEEE Trans. Wirel. Commun. **16**(9), 6010–6021 (2017)
4. He, H., Wen, C.-K., Jin, S., Li, G.Y.: Deep learning-based channel estimation for beamspace mmWave massive MIMO systems. IEEE Wirel. Commun. Lett. **7**(5), 852–855 (2018)
5. Yang, J., Wen, C.-K., Jin, S., Gao, F.: Beamspace channel estimation in mmWave systems via cosparse image reconstruction technique. IEEE Trans. Commun. **66**(10), 4767–4782 (2018)
6. Ahmad Khan, A., Almuzaini, K.K., Daniel, V., Ojo, S., Minchula, V.K., Roy, V.: MaReSPS for energy efficient spectral precoding technique in large scale MIMO-OFDM. Phys. Commun. (2023) (SCI E/ISI) (IF = 2.37)
7. Uthansakul, P., Anchuen, P., Uthansakul, M., Ahmad Khan, A.: Estimating and synthesizing QoE based on QoS measurement for improving multimedia services on cellular networks using ANN method. IEEE Trans. Netw. Serv. Manag. **17**(1), 389–402 (2020) (SCI-E/ISI) (IF = 4.682)
8. Uthansakul, P., Ahmad Khan, A.: On the energy efficiency of millimeter wave massive MIMO based on hybrid architecture. Energies **12**(11), 2227 (2019) (SCI-E/ISI) (IF = 3.04)
9. Uthansakul, P., Ahmad Khan, A.: Enhancing the energy efficiency of mmWave massive MIMO by modifying the RF circuit configuration. Energies **12**(22), 4356 (2019) (SCI-E/ISI) (IF = 3.04)
10. Uthansakul, P., Ahmad Khan, A., Duangmanee, P., Uthansakul, M.: Energy efficient design of massive MIMO based on closely spaced antennas: mutual coupling effect. Energies **11**(8), 2029 (2018) (SCI-E/ISI) (IF = 3.04)
11. Ahmad Khan, A., Uthansakul, P., Duangmanee, P., Uthansakul, M.: Energy efficient design of massive MIMO by considering the effects of nonlinear amplifiers. Energies **11**(5), 1045 (2018) (SCI-E/ISI) (IF = 3.04)
12. Ahmad Khan, A., Uthansakul, P., Uthansakul, M.: Energy efficiency design of massive MIMO by incorporating with mutual coupling. Int. J. Commun. Antenna Propag. (IRECAP) **7**(3) (2017) (Scopus)
13. Wu, Q., et al.: Intelligent reflecting surface-aided wireless communications: a tutorial. IEEE Trans. Commun. **69**(5), 3313–3351 (2021)
14. Huang, C., et al.: Reconfigurable intelligent surface assisted multiuser MISO systems exploiting deep reinforcement learning. IEEE J. Sel. Areas Commun. **38**(8), 1839–1850 (2020)
15. Jensen, T.L., Carvalho, E.D.: An optimal channel estimation scheme for intelligent reflecting surfaces based on a minimum variance unbiased estimator. In: Proceedings of the IEEE International Conference on Acoustics, Speech and Signal Processing (ICASSP), May 2020, pp. 5000–5004
16. Huang, C., Zappone, A., Alexandropoulos, G.C., Debbah, M., Yuen, C.: Reconfigurable intelligent surfaces for energy efficiency in wireless communication. IEEE Trans. Wireless Commun. **18**(8), 4157–4170 (2019)
17. Mishra, D., Johansson, H.: Channel estimation and low-complexity beamforming design for passive intelligent surface assisted MISO wireless energy transfer. In: Proceedings of the IEEE International Conference on Acoustics, Speech and Signal Processing (ICASSP), May 2019, pp. 4659–4663
18. Wang, P., et al.: Compressed channel estimation for intelligent reflecting surface-assisted millimeter wave systems. IEEE Signal Process. Lett. **27**, 905–909 (2020)
19. Liu, S., et al.: Deep denoising neural network assisted compressive channel estimation for mmWave intelligent reflecting surfaces. IEEE Trans. Veh. Technol. **69**(8), 9223–9228 (2020)

20. Shoukath Ali, K., Sampath, P.: Sparse Bayesian learning Kalman filter-based channel estimation for hybrid millimeter wave MIMO systems: a frequency domain approach. IETE J. Res. **69**(7), 4243–4253 (2023). https://doi.org/10.1080/03772063.2021.1951367
21. Shoukath Ali, K., Perarasi, T., Phihlip, S.P., Leeban Moses, M., Poongodi, C.:GM-LAMP with residual learning network for millimetre wave MIMO architectures. In: 2022 Smart Technologies, Communication and Robotics (STCR), Sathyamangalam, India, pp. 1–5 (2022). https://doi.org/10.1109/STCR55312.2022.10009163
22. Shoukath Ali, K., Perarasi, T., Poongodi, C., Deepa, D., Sampath, P.:Approximate message passing for mmWave massive MIMO architecture using optimal hybrid precoder/combiner. In: 2021 Smart Technologies, Communication and Robotics (STCR), Sathyamangalam, India, 2021, pp. 1–4. https://doi.org/10.1109/STCR51658.2021.9588908
23. Shoukath Ali, K., Sampath, P.: Time domain channel estimation for time and frequency selective millimeter wave MIMO hybrid architectures: sparse Bayesian learning-based Kalman filter. Wirel. Pers. Commun. **117**, 2453–2473 (2021). https://doi.org/10.1007/s11277-020-07986-9
24. Ali, K.S., Sampath, P., Poongodi, C.: Symbol error rate performance of hybrid DF/AF relaying protocol using particle swarm optimization based power allocation. In: 2019 International Conference on Advances in Computing and Communication Engineering (ICACCE), Sathyamangalam, India, pp. 1–5 (2019). https://doi.org/10.1109/ICACCE46606.2019.9079970
25. Tropp, J.A., Gilbert, A.C.: Signal recovery from random measurements via orthogonal matching pursuit. IEEE Trans. Inf. Theory **53**(12), 4655–4666 (2007)
26. Dai, L., Wei, X.: Distributed machine learning based downlink channel estimation for RIS assisted wireless communications. IEEE Trans. Commun. **70**(7), 4900–4909 (2022). https://doi.org/10.1109/TCOMM.2022.3175175
27. Wei, X., Shen, D., Dai, L.: Channel estimation for RIS assisted wireless communications—Part I: fundamentals, solutions, and future opportunities. IEEE Commun. Lett. **25**(5), 1398–1402 (2021). https://doi.org/10.1109/LCOMM.2021.3052822
28. Wei, X., Shen, D., Dai, L.: Channel estimation for RIS assisted wireless communications—Part II: an improved solution based on double-structured sparsity. IEEE Commun. Lett. **25**(5), 1403–1407 (2021). https://doi.org/10.1109/LCOMM.2021.3052787
29. Lecun, Y., Bengio, Y., Hinton, G.: Deep learning. Nature **521**(7553), 436–444 (2015)
30. Borgerding, M., Schniter, P., Rangan, S.: AMP-inspired deep networks for sparse linear inverse problems. IEEE Trans. Signal Process. **65**(16), 4293–4308 (2017)
31. Borgerding, M., Schniter, P.:Onsager-corrected deep learning for sparse linear inverse problems. In: 2016 IEEE Global Conference on Signal and Information Processing (GlobalSIP), pp. 227–231 (2016). https://doi.org/10.1109/GlobalSIP.2016.7905837
32. Wei, X., Hu, C., Dai, L.: Deep learning for beamspace channel estimation in millimeter-wave massive MIMO systems. IEEE Trans. Commun. **69**(1), 182–193 (2021). https://doi.org/10.1109/TCOMM.2020.3027027
33. Wei, Y., Zhao, M.-M., Zhao, M., Lei, M., Yu, Q.: An AMP-based network with deep residual learning for mmWave beamspace channel estimation. IEEE Wirel. Commun. Lett. **8**(4), 1289–1292 (2019). https://doi.org/10.1109/LWC.2019.2916786
34. Rangan, S., Schniter, P., Fletcher, A.K.: Vector approximate message passing. IEEE Trans. Inf. Theory **65**(10), 6664–6684 (2019). https://doi.org/10.1109/TIT.2019.2916359
35. Schniter, P., Rangan, S., Fletcher, A.K.: Vector approximate message passing for the generalized linear model. In: 2016 50th Asilomar Conference on Signals, Systems and Computers, Pacific Grove, CA, USA, pp. 1525–1529 (2016). https://doi.org/10.1109/ACSSC.2016.7869633
36. Ruan, C., Zhang, Z., Jiang, H., Dang, J., Wu, L., Zhang, H.: Approximate message passing for channel estimation in reconfigurable intelligent surface aided MIMO multiuser systems. IEEE Trans. Commun. **70**(8), 5469–5481 (2022). https://doi.org/10.1109/TCOMM.2022.3182369
37. Mestoui, J., et al.: BER performance improvement in CE-OFDM-CPM system using equalization techniques over frequency-selective channel. Procedia Comput. Sci. **151**, 1016–1021 (2019)

38. Belkadid, J., et al.: PAPR reduction in CE-OFDM system for numerical transmission via PLC channel. Int. J. Commun. Antenna Propag. **3**(5), 267–272 (2013)
39. Elaage, S., et al.: MB-OOK transceiver design for terahertz wireless communication systems. Int. J. Syst. Control Commun. **12**(4), 309–326 (2021)
40. Wu, Q., Zhang, R.: Intelligent reflecting surface enhanced wireless network: joint active and passive beamforming design. In: Proceedings of the IEEE Global Communication Conference (GLOBECOM), Abu Dhabi, United Arab Emirates, Dec 2018, pp. 1–6
41. Ruan, C., Zhang, Z., Jiang, H., Dang, J., Wu, L., Zhang, H.: Vector approximate message passing with sparse Bayesian learning for Gaussian mixture prior. China Commun. **20**(5), 57–69 (2023). https://doi.org/10.23919/JCC.2023.00.005
42. Ali, K.S., Khan, A.A., Perarasi, T., Ur Rehman, A., Ouahada, K.: Learned-SBL-GAMP based hybrid precoders/combiners in millimeter wave massive MIMO systems. PLoS ONE **18**(9), e0289868 (2023). https://doi.org/10.1371/journal.pone.0289868
43. Mishra, D., Johansson, H.: Channel estimation and low complexity beamforming design for passive intelligent surface assisted MISO wireless energy transfer. In: Proceedings of the IEEE International Conference on Acoustics, Speech and Signal Processing (ICASSP), Brighton, UK, May 2019, pp. 4659–4663
44. Nadeem, Q.-U.-A., Alwazani, H., Kammoun, A., Chaaban, A., Debbah, M., Alouini, M.-S.: Intelligent reflecting surface-assisted multiuser MISO communication: channel estimation and beamforming design. IEEE Open J. Commun. Soc. **1**, 661–680 (2020)
45. Zymnis, A., Boyd, S., Candès, E.J.: Compressed sensing with quantized measurements. IEEE Signal Process. Lett. **17**(2), 149–152 (2010)

Chapter 4
Innovative mm-Wave Compact Dual-Port MIMO Antenna with Inherent Wideband Isolation at 28 GHz for 5G Wireless Networks

Asma Khabba, Lahcen Sellak, Jamal Amadid, Zakaria El Ouadi, Saida Ibnyaich, and Abdelouhab Zeroual

4.1 Introduction

As cellular networks continue to rapidly evolve and expand, the volume of data traffic over the internet has surged to unprecedented levels. Consequently, in accordance with the directives of the Federal Communications Commission (FCC), there is a growing imperative to alleviate congestion through the assignment of dedicated frequency segments in the millimeter wave (mmW) range beyond 24 GHz. These mm-wave frequencies are being harnessed to facilitate the deployment of fifth-generation (5G) systems. The FCC has designated various mm-wave frequency bands for this purpose, with prominent examples including the 26.5–29.5 GHz and 27.5–28.35 GHz bands, with a central frequency of approximately 28 GHz for both [1, 2]. Operating within the mm-Wave band offers a pivotal advantage in its diminutive wavelength, facilitating the seamless amalgamation of numerous antenna elements within a constrained spatial domain. Nevertheless, it is imperative to recognize that the spectrum of higher frequency range displays a heightened susceptibility to atmospheric variations, leading to substantial signal attenuation in the mmW frequency ranges. This stands in contrast to lower-frequency bands, where attenuation can be mitigated through strategies such as small cell deployments and the use of high-gain antennas [3–8].

A. Khabba (✉) · J. Amadid · Z. El Ouadi · S. Ibnyaich · A. Zeroual
I2SP Research Team, Department of Physics, Faculty of Sciences Semlalia, Cadi Ayyad University, Marrakesh, Morocco
e-mail: asma.khabba@edu.uca.ac.ma

L. Sellak
LISAD Research Team, Industrial Engineering Department, National School of Applied Sciences, Ibn Zohr University, Agadir, Morocco

M. El Ghzaoui et al. (eds.), *Next Generation Wireless Communication*, Signals and Communication Technology, https://doi.org/10.1007/978-3-031-56144-3_4

When it comes to the core objective of enhancing data rates and channel capacity, Multi-Antenna Systems (MAS), recognized by MIMO antennas, stand out as a prime strategy with immense potential to boost the quality of data transmission in the realm of mm-wave frequencies [9–12]. Consequently, the scientific community has witnessed a proliferation of studies in the current literature, each proposing distinct MIMO antenna configurations for deployment in fifth-generation wireless networks operating within the mm-wave spectrum. For example, in a recent study [13], researchers proposed a MIMO antenna featuring a planar configuration with dual elements, designed for operation at 28 GHz. This antenna showcased a reasonably respectable gain attaining 7.88 dB. Nevertheless, it demonstrated a comparatively limited bandwidth and encompassed a substantial physical area. In another noteworthy endeavor [14], researchers introduced an Electromagnetic Band Gap (EBG)-based MIMO antenna featuring two conventional patch elements, designed primarily for wearable technologies operating at 24 GHz. The incorporation of the EBG structure was aimed at enhancing gain. Nonetheless, the physical size of the antenna system proved to be quite substantial, measuring 21.6×21.6 mm^2, leaving room for further miniaturization. Additionally, the bandwidth achieved was limited to 0.8 GHz. In the study outlined in [15], a double-band MIMO planar antenna comprising two elements was presented for applications at both 27 and 39 GHz. While this antenna exhibited wideband characteristics, its physical size remained relatively large. Additionally, the assessment of MIMO performance focused solely on the ECC, well known by Envelope Correlation coefficient. Similarly, in a separate investigation detailed in [16], researchers introduced a modified patch MIMO antenna equipped with dual ports, designed to operate in a dual-band configuration. The antenna demonstrated reasonable bandwidth capabilities at 28 and 38 GHz, offering 2.55 and 2.1 GHz, respectively. However, it's important to note that the highest achievable gain with this antenna was limited to 1.83 dB. In another research effort discussed in [17], the authors leveraged a substrate-integrated waveguide (SIW) structure to develop a 2-port MIMO antenna tailored for fifth-generation systems operating within the 28 GHz mmW spectrum. This MIMO system boasted a commendable gain of up to 6.9 dB and a worst-case isolation of 17 dB. Nevertheless, the achieved bandwidth was limited to a mere 0.4 GHz. In a separate study documented [18], researchers have unveiled a MIMO antenna solution comprising 4 T shaped Coplanar Waveguide (CPW) elements, specifically designed to cater to the demands of ultra-wide band (UWB) purposes. However, it is important to note that this specific configuration covered a significant surface area, measuring 50.8×12 mm^2.

As we critically assess the existing body of literature, we discern several limitations inherent in the reported antenna designs. These limitations predominantly manifest as large physical dimensions, restricted bandwidth, modest gain, and intricate structural complexities, collectively diminishing their suitability for fifth-generation systems. Consequently, an imperative remains to conceptualize a novel MIMO antenna design that seamlessly integrates the most coveted attributes, including compact dimensions, extensive bandwidth, uncomplicated layout, favorable radiation characteristics, and exemplary diversity performance. This constitutes the central objective of our endeavor. This manuscript introduces a novel MIMO antenna design featuring dual elements and a common ground plane structure, designed to cater to 5G applications within the innovative 28 GHz frequency band. Each discrete antenna component consists of a radiating element with a trapezoidal design meticulously placed on the RTRogers Duroid 5880 substrate (with tan$\delta = 0.0009$ and $\varepsilon_r = 2.2$). A slotted ground plane is incorporated to facilitate miniaturization and enhance the antenna's bandwidth capabilities. Distinguished by its simplicity, compact form factor, and diminutive size, this design exhibits impressive performance attributes. It boasts a wide bandwidth, favorable radiation characteristics, and robust MIMO capabilities, rendering it a compelling candidate for integration into fifth-generation networks and systems.

The subsequent sections of this research work respect the following structure: Sect. 4.2 offers a comprehensive examination of the antenna's design and the accomplishments of its individual units. Section 4.3 delves into the implications of assembling the antenna in a MIMO configuration, detailing the overall performance and outcomes. In Sect. 4.4, we evaluate the proposed design by conducting a comparative analysis with previously published works. Finally, Sect. 4.5 offers a concise summary and conclusion of this study.

4.2 Structure and Performance Analysis of the Submitted Individual Antenna

4.2.1 Submitted Antenna Layout

The proffered antenna's geometric configuration is presented in Fig. 4.1. In this design, a radiation element in the shape of a trapezoid serves as the core of the antenna, with a microstrip 50 Ω feed-line connected to the patch for excitation purposes. This innovative design is implemented on a 0.8 mm-thick laminate substrate, specifically the RT Rogers Duroid 5880 material, which is characterized by dielectric constant of 2.2 and a minimal loss tangent of 0.0009. To support the entire structure, a metal ground plane is used, featuring a trapezoidal slot that has been wisely etched into it. Notably, the antenna boasts compact dimensions, measuring only $6 \times 5 \times 0.8$ mm^3, with the precise values of other critical parameters illustrated in Fig. 4.1. It is of significance to note that the overall antenna design was carefully and precisely crafted, with each parameter refined to achieve optimal performance. Advanced electromagnetic simulation software, such as computer simulation technology (CST), played a pivotal role in fine-tuning the antenna's geometry and ensuring its effectiveness.

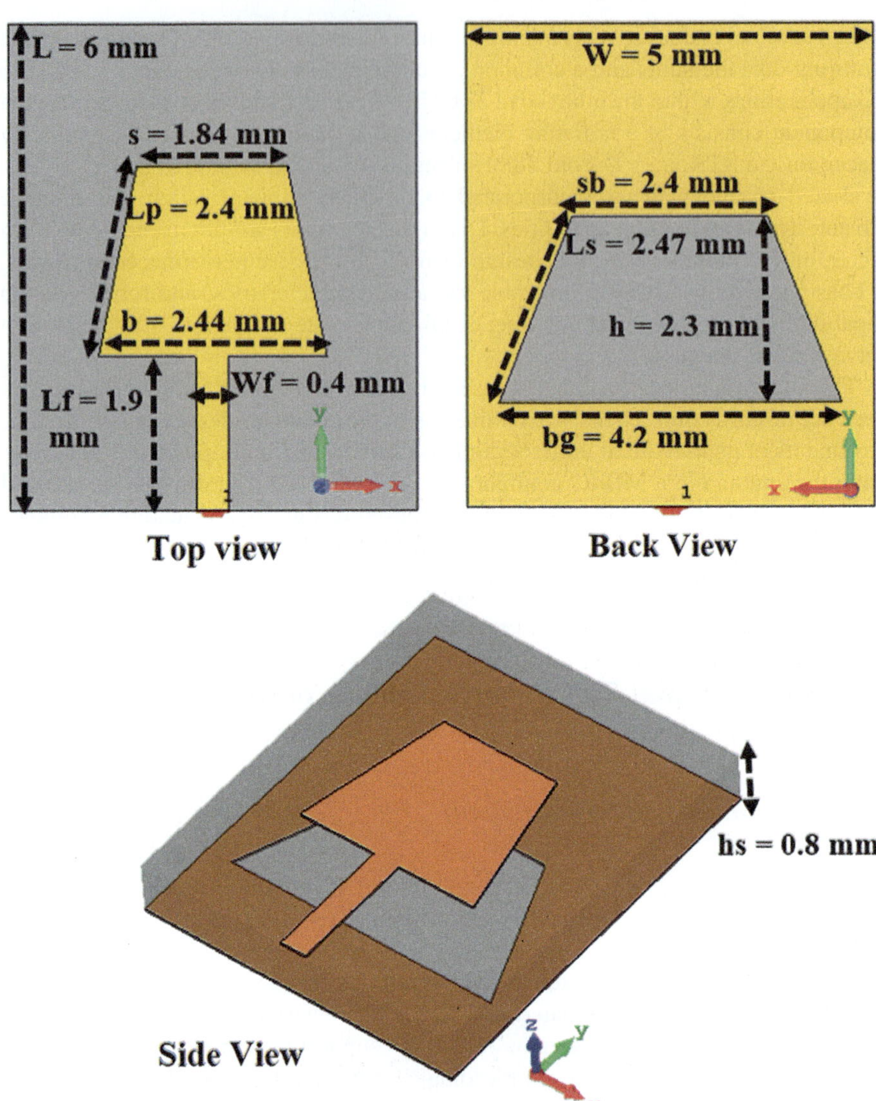

Fig. 4.1 Recommended antenna display

4.2.2 *Antenna Evolutionary Phases*

Figure 4.2 depicts the progressive development of the antenna's design alongside its associated frequency response. With respect to Fig. 4.2a, a standard antenna comprised of a rectangular patch measuring 2.2 mm in length L_p and 2.4 mm in width W_p, reinforced by a consistent ground plane, was meticulously constructed within

CST simulator, adhering to the foundational principles (4.1)–(4.4) [19]. As shown in Fig. 4.2b, the subsequent stage entailed the elimination of triangular segments from the lateral aspects of the radiation patch, thus metamorphosing the rectangular-shaped antenna into a trapezoidal configuration. The last part of the process, as exposed in Fig. 4.2c, involved the precise creation of a trapezoidally shaped slot within the ground plane underneath the radiant patch, with adjustments made to ensure the desired resonance characteristics.

In order to gain a comprehensive insight into the influence of each alteration on the antenna performance, Fig. 4.2d provides the antenna frequency response at various developmental phases. In the outset, the baseline antenna design demonstrated resonance at 34 GHz, showcasing an overall bandwidth spanning 2.8 GHz, while maintaining the scattering parameter $|S_{11}| < -10$ dB. This broadband feature was achieved due to the relatively small patch dimensions while comparing to the ground plane. Nevertheless, even with its wide bandwidth, the baseline antenna did not align with the intended 28 GHz frequency band, thus prompting the need for

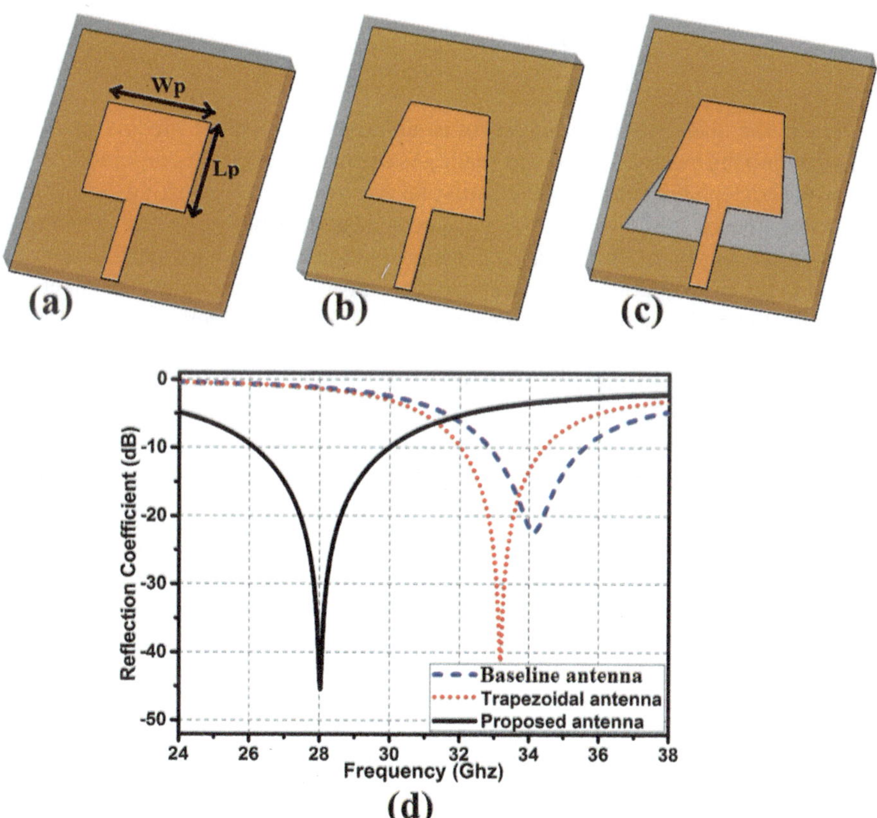

Fig. 4.2 Developmental phases leading to the eventual recommended design: **a** baseline antenna, **b** antenna with trapezoidal patch, **c** suggested antenna, and **d** corresponding reflection coefficient

additional geometric improvements. Thus, in the following phases, the objective is to transition the antenna initial resonant frequency, which was initially set at 34 GHz, toward the lower end of the frequency spectrum. As a consequence, the resonant frequency stabilized at 33.2 GHz during the second phase, bringing about improved matched impedance and a full operational band of 2.2 GHz, consequently bringing it into better alignment with the target 28 GHz frequency band [20].

$$W_{\mathrm{p}} = \frac{c}{2 f_{\mathrm{r}} \sqrt{\frac{\varepsilon_{\mathrm{r}}+1}{2}}} \tag{4.1}$$

$$\varepsilon_{\mathrm{eff}} = \frac{\varepsilon_{\mathrm{r}}+1}{2} + \frac{\varepsilon_{\mathrm{r}}-1}{2}\left[1 + 12\left(\frac{h_{\mathrm{s}}}{W_{\mathrm{p}}}\right)\right]^{-1/2} \tag{4.2}$$

$$\Delta \mathrm{L} = 0.421 h_{\mathrm{s}} \frac{(\varepsilon_{\mathrm{eff}} + 0.3)\left(\frac{W_{\mathrm{p}}}{h_{\mathrm{s}}} + 0.264\right)}{(\varepsilon_{\mathrm{eff}} - 0.258)\left(\frac{W_{\mathrm{p}}}{h_{\mathrm{s}}} + 0.8\right)} \tag{4.3}$$

$$L_p = \frac{c}{2 f_r \sqrt{\varepsilon_{eff}}} - 2\Delta \mathrm{L} \tag{4.4}$$

Hence, the final step in this endeavor is undertaken to achieve precise resonance tuning within the intended frequency range. As demonstrated, the incorporation of the slot with a trapezoidal contour played a pivotal role in effectively shifting the resonant frequency toward the lower end of the spectrum until it was precisely centered at the target frequency, namely 28 GHz. This adaptation led to a broadened bandwidth extending by 4 GHz, alongside an impressive improvement in impedance matching, marked by the $|S11|$ reaching a low point of − 45 dB. Hence, the devised configuration effectively attained a broad frequency response encompassing the 28 GHz band, all the while maintaining its original compact dimensions.

To gain deeper insights into the antenna behavior, Fig. 4.3 provides an illustration of the distribution of electrical current along the trapezoidal formed antenna, both before and after inserting the trapezoidal slot. Upon a comprehensive analysis of the surface current distribution depicted in Fig. 4.3a, a conspicuous phenomenon comes to the fore: the inclusion of lateral cuts emerges as a pivotal factor that amplifies the propagation of electric current along the periphery of the radiating patch.

This extension, in turn, significantly augments the electrical length, yielding a reduction in the resonant frequency. Additionally, it is worth noting that prior to the etching of the slot with trapezoidal form, the coupling effect predominantly directs the concentration of electric current toward the ground immediately under the patch. The discernible impact of the engraved slot on the current dispersion is conspicuously evident in Fig. 4.3b. As illustrated, following the insertion of the slot, the current shows a propensity to escalate along the slot perimeter, as these boundaries offer a favorable conduit. Hence, the incised slot introduces supplementary boundaries for current dispersion, thus amplifying the antenna electrical size and correspondingly diminishing the antenna resonating frequency.

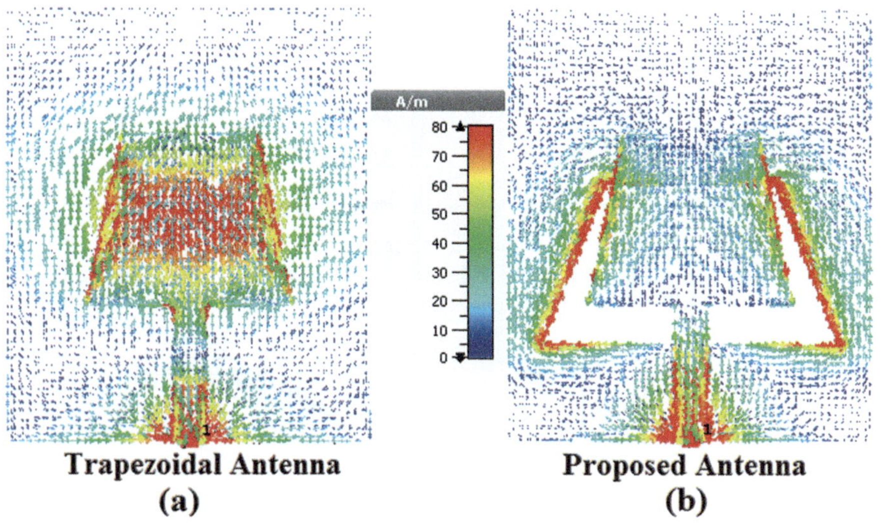

Fig. 4.3 Electrical current dispensation at the following: **a** 33.2 GHz for trapezoidal-shaped antenna and **b** 28 GHz for the proposed antenna

4.2.3 Exploration of Trapezoidal Slot Parameters Through Comprehensive Analysis

To thoroughly investigate the drilled slot's influence on the behavior of antenna frequency response, we turn to Fig. 4.4, which presents a parametric study involving different slot parameters, which are the height (h), long base (bg), and short base (sg). As visually represented in Fig. 4.4a, adjusting the extended base, denoted by "bg", within the interval of 3.8–4.4 mm brings about a noticeable modulation in the resonance frequency. In particular, the resonance frequency is undergoing a change, departing the 28.5 GHz and approaching a lower frequency. The sought-after resonance condition is attained at a "bg" value of 4.2 mm.

Figure 4.4b sheds light on the sensitivity of the frequency response to changes in the height (h) of the trapezoidal slot. It becomes evident that variations in "h", ranging from 2.1 to 2.5 mm, have a slight impact on the resonant frequency. In contrast, there is a notable effect on impedance matching, with the best matching occurring at an "h" value of 2.3 mm. Finally, Fig. 4.4c explores the effectiveness of the short base (sg) on the resonating frequency. The short base variation from 2 to 2.8 mm is shown to exert a significant impact on the resonance frequency. A noticeable phenomenon emerges, wherein the resonance frequency experiences a decrement, transitioning out of 28.9 down to 27.5 GHz, while the optimal result is attained when "sg" is configured to 2.4 mm. Hence, the parametrical analysis conducted emphasizes the significant influence wielded by the dimensions of the carved slot in molding the antenna frequency characteristics. Indeed, meticulous

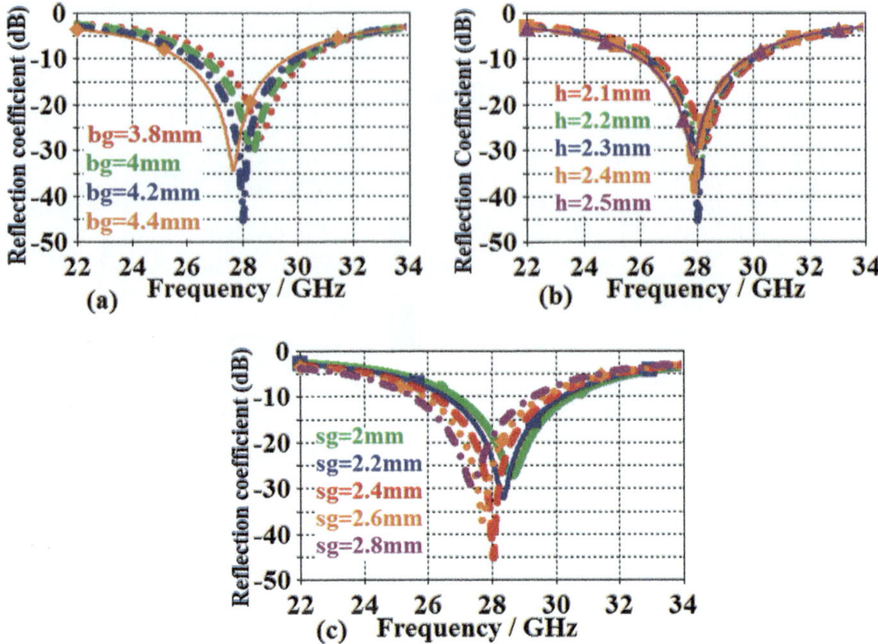

Fig. 4.4 Examination of different factors associated with the drilled slot: **a** long base "bg", **b** height "*h*", and **c** short base "sg"

tuning of these slot dimensions is instrumental in attaining the best possible antenna performance.

4.2.4 Findings and Analysis

In Fig. 4.5a, the reflection coefficient ($|S_{11}|$) of the proposed antenna is displayed. This data was obtained through simulations using two separate software tools, namely CST Simulator and HFSS. This twofold simulation strategy operates as a validation process. As observed, the resultant curves exhibit remarkable agreement with only a negligible discrepancy, which can be attributed to the inherent variations in the computational methods employed by both simulators. The antenna demonstrates robust performance characteristics, resonating precisely at the designated 28 GHz frequency, which is a prominent band for its intended application. Moreover, it boasts an extensive bandwidth spanning from 26 to 30 GHz, encompassing a substantial 4 GHz range. Notably, this bandwidth encompasses numerous auspicious bands for 5G networks, such as 26.5–29.5, 27.5–29.5, and 27.5–28.35 GHz. Consequently, the suggested antenna exhibits the versatility to operate effectively across diverse geographical regions.

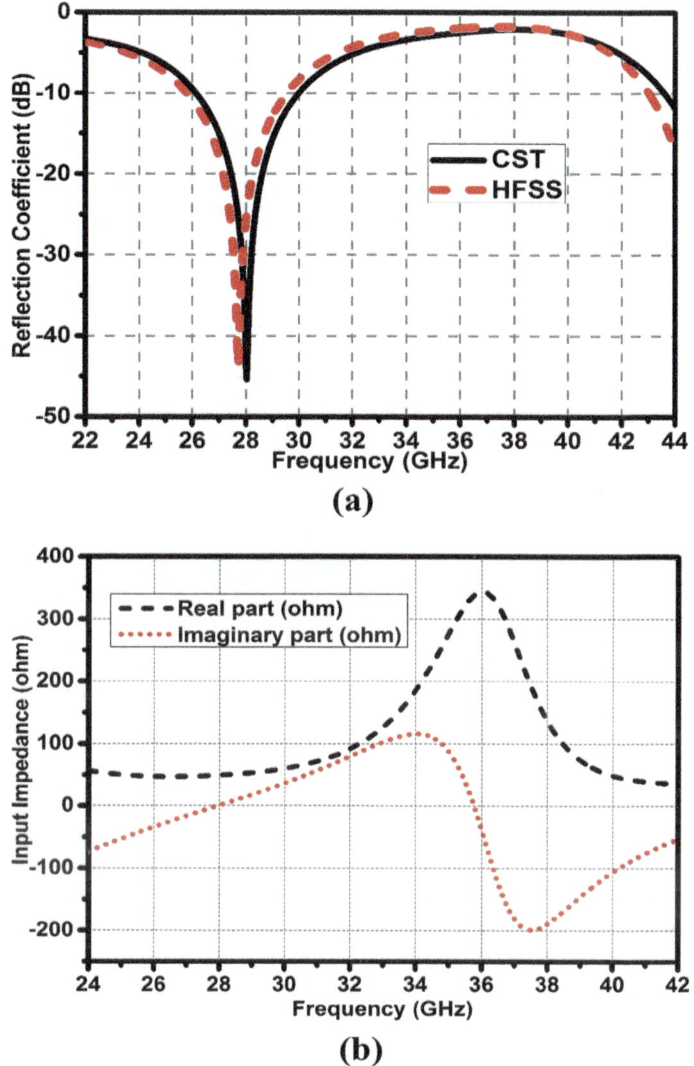

Fig. 4.5 Antenna electrical impedance traits: **a** return loss and **b** input impedance

Moving on to Fig. 4.5b, the attention is turned to another critical impedance characteristic for the proposed antenna, namely the input impedance. This characteristic is represented by a number in the complex plane, where the imaginary and real components correspond to the antenna reactance and resistance, respectively. At a particular frequency, to attain a properly matched antenna, the input impedance should ideally approach a value in proximity to 50 Ω. This entails the real component being in the vicinity of 50 Ω, while the imaginary component hovers around zero. Figure 4.5b reveals that the real and imaginary parts of the input impedance measure 49.4 and –

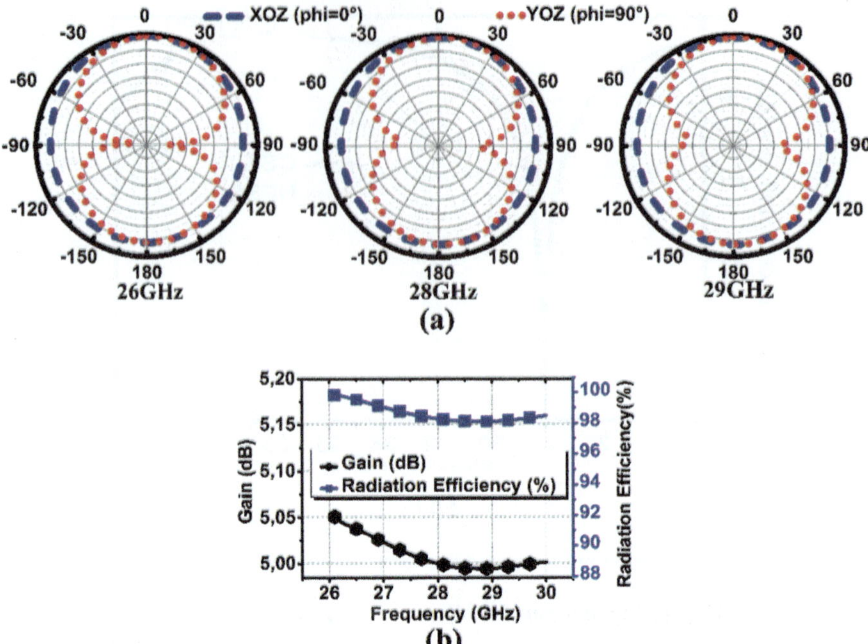

Fig. 4.6 Suggested antenna radiation behavior: **a** two-dimension radiation pattern and **b** radiation efficiency along with gain versus frequency

0.34 Ω, respectively, at the resonance frequency of 28 GHz, underscoring the antenna favorable matching characteristics. The radiative behavior of the proffered antenna is exposed in Fig. 4.6. In this illustration, precisely in Fig. 4.6a, polar representations of the antenna pattern are showcased for YoZ (*E*) and XoZ (*H*) planes, covering distinct frequencies, specifically 26, 28, and 29 GHz.

As the results demonstrate, the proposed antenna consistently showcases advantageous radiative qualities in the *E*-plane (YOZ) as well as the *H*-plane (XOZ) across all three frequencies under scrutiny. Notably, the *H* plane maintains an omnidirectional pattern, while the *E* plane showcases a bidirectional pattern. This consistent behavior underscores the antenna stability and uniformity in radiation performance. Figure 4.6b offers further insights into the antenna gain and radiation efficiency as a function of frequency. Throughout the entire operational range, the antenna consistently demonstrates exceptional radiative qualities, maintaining a constant gain of 5 dB and a radiation efficiency spanning from 98 to 99%.

4.3 Antenna Configuration in MIMO Composition

4.3.1 Layout and Scattering Parameters

Recognizing the essential role played by MIMO systems in the deployment of 5th-gen mobile technology and their pivotal role in achieving swift data transmission, this section is dedicated to the intelligent refinement of MIMO systems, building upon the analysis of individual antenna units conducted earlier. In Fig. 4.7, a visual depiction of the MIMO antenna configuration is fined, comprising two individual antenna units, arranged orthogonally on the identical RTRogers 5880 substrate. Together, they occupy a compact total area of 6×11 mm^2. The orthogonal placement of these antenna units is a strategic choice designed to impart polarization diversity, thereby enhancing isolation. Furthermore, this configuration upholds the ground plane with shared geometry, a prerequisite for real deployment and seamless incorporation with wireless devices components. In the figure provided, it is evident that the inclusion of the MIMO composition does not adversely affect the individual antenna behavior. The dual elements continue to provide robust performance characteristics with respect to mutual coupling and reflection coefficient. A wideband operation spanning 4 GHz within the range 26–30 GHz is maintained, with transmission coefficient remaining below − 26 dB across the entire spectrum.

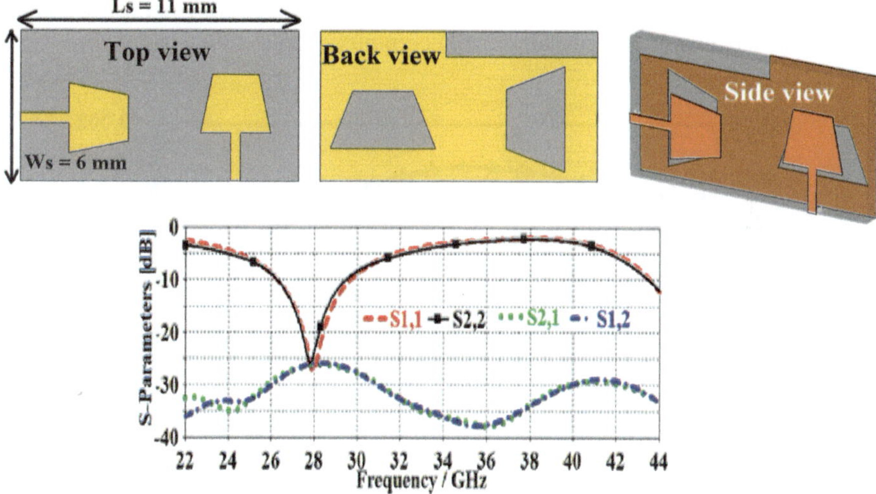

Fig. 4.7 Antenna configuration in MIMO arrangement and resulting S-parameters

4.3.2 Characteristics of Diversity

For a high comprehensive exploration of the proposed MIMO system, we have delved into its performance in terms of various metrics, including ECC (envelope correlation coefficient) and DG (diversity gain). These metrics provide crucial insights into the system's capabilities and effectiveness. Figure 4.8 presents a detailed overview of these parameters, shedding light on their significance. ECC, which stands for envelope correlation coefficient, plays a pivotal role in assessing the correlation between the MIMO elements. For a MIMO system to exhibit strong performance and high uncorrelation, ECC should not exceed 0.5. This important metric is calculated based on the S-parameters, as defined in Eq. (4.5) [1, 3]. As revealed, the diversity antenna excels in the realm of ECC, exhibiting an impressively low value of < 0.001.

This achievement aligns well with industrial requirements and underscores the antenna's ability to maintain minimal correlation between its elements, a key attribute for robust MIMO performance. Diversity gain (DG) is another critical metric under scrutiny, as it quantifies the enhancement in SNR ratio achievable in a diversity plan scenario [7]. DG can be easily achieved through ECC, which is expressed by Eq. (4.6) [9]. Based on this equation, a smaller ECC corresponds to a greater DG. In the quest for a MIMO with excellent properties, DG ought to target an achievement of around 10 dB. As Fig. 4.8 illustrates, DG maintains a consistent performance above 9.99 dB throughout the entire frequency spectrum, reaching the highest amount of 10 dB when resonating at 28 GHz. This exceptional performance validates the antenna's high MIMO diversity capabilities. These outcomes convincingly establish the antenna's capacity to facilitate reliable communication.

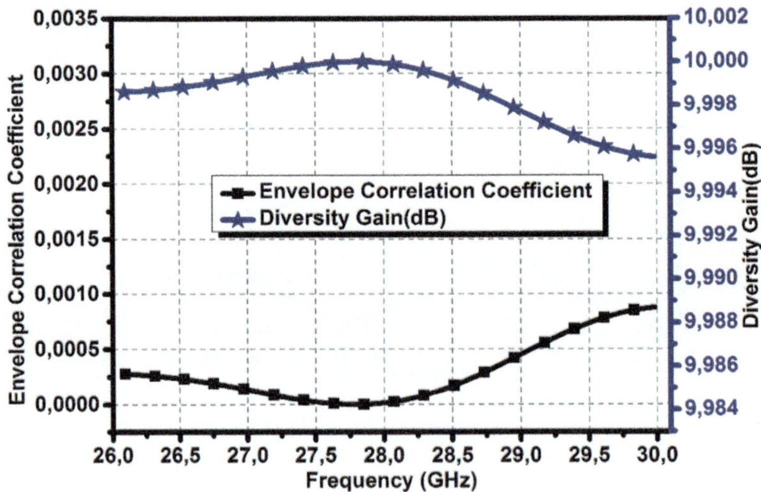

Fig. 4.8 Diversity antenna performance along the full band of operation

$$\mathrm{ECC} = \frac{\left|S_{11}^{*}S_{12} + S_{21}^{*}S_{22}\right|^{2}}{\left(1 - \left(|S_{11}|^{2} + |S_{21}|^{2}\right)\right)\left(1 - \left(|S_{22}|^{2} + |S_{12}|^{2}\right)\right)} \tag{4.5}$$

$$\mathrm{DG} = 10\sqrt{1 - ECC^{2}} \tag{4.6}$$

4.3.3 Evaluation Through Relevant Earlier Research

An exhaustive evaluation of the proffered MIMO antenna is undertaken in this sub-section with comparison to numerous designs that have been previously documented, evaluating them against a range of standards and criteria. The outcomes of this comparative analysis are succinctly summarized in Table 4.1. Undoubtedly, the realm of published research boasts a vast array of reports presenting diverse designs of antenna with MIMO composition, tailored for mmW 5G networks. However, a closer examination of the comparative table below reveals the undeniable superiority of the proposed design in several crucial aspects. In reality, our proposed MIMO design distinguishes itself with its compact footprint, elevated radiation efficiency, expansive bandwidth, and robust isolation capabilities. Notably, it maintains competitive gain and satisfactory diversity performance, a noteworthy achievement considering that many configurations lack evaluation in terms of channel capacity loss (CCL). Furthermore, the practicality of some previously reported works, such as [15, 16], is constrained due to their omission of a common ground structure. This constraint reinforces the accomplishments of our proposed design, highlighting its immense potential as an ideal candidate for 5G applications, thanks to its superior performance metrics, especially when juxtaposed with larger-sized counterparts.

4.4 Conclusion

The fundamental aim of this research project was to examine the viability of developing a concise and uncomplicated MIMO antenna with modest size. The MIMO antenna surpassed expectations, showcasing commendable radiation properties and boasting an extensive operational range centered around 28 GHz. Notably, the MIMO elements sustain a remarkable level of isolation, further enhancing the system's diversity performance. Comparative analysis against existing works in the field substantiates the suitability of our proposed multiple antenna design for seamless integration into MIMO systems tailored for 5G mm-wave applications.

Table 4.1 Performance evaluation

Refs.	Babu et al. [8]	Iqbal et al. [14]	Ali et al. [15]	Hasan et al. [16]	Usman et al. [17]	This research
Size (mm^3)	35 × 11 × 1.6	19.04 × 15.06 × 0.254	26 × 11 mm^2	14 × 26 × 0.38	33 × 27.5 × 0.76	**11 × 6 × 0.8**
Elements	3	2	2	2	1 (2 ports)	**2**
Frequency range (GHz)	26–40	24 GHz band	25–29	26.65–29.2	28 GHz band	**26–30**
Gain (dB)	> 5	> 4	> 3.5	> 0.5	6.9 (at 28 GHz)	**5 >**
Efficiency (%)	> 70	> 70	> 97	> 76	87 (at 28 GHz)	**> 98**
Isolation (dB)	> 20	> 30	> 20	> 20	> 17	**> 26**
ECC	< 0.01	< 0.24	< 0.001	< 0.01	< 0.2	**< 0.001**
DG (dB)	> 9.9	> 9.6	> 9.99	> 9.99	> 9.7	**> 9.99**

Bold values represents our proposed Antenna results

References

1. Khabba, A., Wakrim, L., Ibnyaich, S., Hassani, M.M.: Beam-steerable ultra-wide-band miniaturized elliptical phased array antenna using inverted-L-shaped modified inset feed and defected ground structure for 5G smartphones millimeter-wave applications. Wireless Pers. Commun. **125**, 3801–3833 (2022)
2. Errahili, S., Khabba, A., Ibnyaich, S., Hassani, M.M.: New frequency reconfigurable patch antenna for wireless communication. In: Proceedings of the 3rd International Conference on Networking, Information Systems and Security, pp. 1–4. ACM, Morocco (2020)
3. Khabba, A., Amadid, J., Ibnyaich, S., Zeroual, A.: Pretty-small four-port dual-wideband 28/38 GHz MIMO antenna with robust isolation and high diversity performance for millimeter-wave 5G wireless systems. Analog Integr. Circ. Sig. Process **112**, 83–102 (2022)
4. Khabba, A., Errahili, S., Ibnyaich, S., Hassani, M.M.: A novel design of miniaturized leaf shaped antenna for 5G mobile phones. In: Proceedings of the 3rd International Conference on Networking, Information Systems and Security. pp. 1–4. ACM, Morocco (2020)
5. Khabba, A., Amadid, J., Ouadi, Z.E., Wakrim, L., Ibnyaich, S., Al-Gburi, A.J., Zeroual, A.: Micro-sized graphene-based UWB annular ring patch antenna for short-range high-speed terahertz wireless systems. In: Patel, S.K., Taya, S.A., Das, S., Vasu Babu, K. (eds) Recent Advances in Graphene Nanophotonics: Advanced Structured Materials, vol. 190, pp. 227–247. Springer, Cham (2023)
6. Lchhab, T., El Ghzaoui, M.: A circularly polarized wideband high gain antenna for THz wireless applications. Opt. Quant. Electron. **54**(12), 787 (2022)
7. Khabba, A., Amadid, J., Mohapatra, S., El Ouadi, Z., Ahmad, S., Ibnyaich, S., Zeroual, A.: UWB dual-port self-decoupled O-shaped monopole MIMO antenna with small-size easily extendable design and high diversity performance for millimeter-wave 5G applications. Appl. Phys. A (2022). https://doi.org/10.1007/s00339-022-05881-7
8. Babu, K.V., Das, S., Ali, S.S., Madhav, B.T.P., Patel, S.K.: Broadband sub-6 GHz flower-shaped MIMO antenna with high isolation using theory of characteristic mode analysis (TCMA) for 5G NR bands and WLAN applications. Int. J. Commun. Syst. **36**(6), e5442 (2023)

9. Elabd, R.H., Al-Gburi, A.J.: SAR assessment of miniaturized wideband MIMO antenna structure for millimeter wave 5G smartphones. Microelectron. Eng. **282**, 112098 (2023)
10. Khabba, A., Amadid, J., El Ouadi, Z., Sellak, L., Wakrim, L., Ibnyaich, S., Zeroual, A.:. A novel high gain tri-band three-dimensional phased array for 24/26/28 GHz smartphone applications. In: 2023 IEEE 3rd International Maghreb Meeting of the Conference on Sciences and Techniques of Automatic Control and Computer Engineering (MI-STA), pp. 628–633, IEEE, Libya (2023)
11. Zhang, Y.-M., Yao, M., Zhang, S.: Wide-band decoupled millimeter-wave antenna array for massive MIMO Systems. IEEE Antennas Wirel. Propag. Lett. **22**(11), 2680–2684 (2023)
12. Sellak, L., Khabba, A., Chabaa, S., Ibnyaich, S., Sarosh, A., Zeroual, A., Baddou, A.: ANFIS-based 4 × 4 dual band Circular Mimo Antenna Design with pretty-small size and large bandwidth for 5 g millimeter-wave applications at 28/38 GHz. J. Infrared Millimeter Terahertz Waves **44**, 551–601 (2023)
13. Marzouk, H.M., Ahmed, M.I., Shaalan, A.-E.H.: Novel dual-band 28/38 GHz MIMO antennas for 5G Mobile Applications. Prog. Electromagn. Res. C **93**, 103–117 (2019)
14. Iqbal, A., Basir, A., Smida, A., Mallat, N.K., Elfergani, I., Rodriguez, J., Kim, S.: Electromagnetic bandgap backed millimeter-wave MIMO antenna for wearable applications. IEEE Access **7**, 111135–111144 (2019)
15. Ali, W., Das, S., Medkour, H., Lakrit, S.: Planar dual-band 27/39 GHz millimeter-wave MIMO antenna for 5G applications. Microsyst. Technol. **27**, 283–292 (2020)
16. Hasan, M.N., Bashir, S., Chu, S.: Dual band omnidirectional millimeter wave antenna for 5G communications. J. Electromagn. Waves Appl. **33**, 1581–1590 (2019)
17. Usman, M., Kobal, E., Nasir, J., Zhu, Y., Yu, C., Zhu, A.: Compact SIW fed dual-port single element annular slot MIMO antenna for 5G mmWave applications. IEEE Access **9**, 91995–92002 (2021)
18. Jilani, S.F., Alomainy, A.: Millimetre-wave t-shaped MIMO antenna with defected ground structures for 5G Cellular Networks. IET Microwaves Antennas Propag. **12**, 672–677 (2018)
19. Khabba, A., Mohapatra, S., Wakrim, L., Ez-zaki, F., Ibnyaich, S., Zeroual, A.: Multiband antenna design with high gain and robust spherical coverage using a new 3D phased array structure for 5G mobile phone MM-wave applications. Analog Integr. Circ. Sig. Process **110**, 331–348 (2021)
20. El Ghzaoui, M., Das, S., Lenka, T.R., Biswas, A.: Terahertz Wireless Communication Components and System Technologies (2022), pp. 1–310

Chapter 5
Design of Ka-Band Power Amplifier and Low-Noise Amplifier for 5G Communication Systems

Deergha Agarwal, Zuber M. Patel, and Kirti Inamdar

5.1 Introduction

Fifth-generation (5G) wireless technology encompasses a variety of technical specifications and standards that define its capabilities and performance characteristics. Here are some key technical specifications of 5G:

- Frequency Bands: 5G revolves around frequency bands, including: sub-6 GHz and mm-Wave frequencies. Former provides broader coverage and good penetration through obstacles. This includes frequencies below 6 GHz (3.5 and 2.4 GHz bands), whereas latter are higher frequency bands above 24 GHz, such as 26 GHz and 32 GHz. mm-Wave offers extremely high data rates but has limited coverage and is sensitive to blockage by buildings and foliage.
- Data Rates: The design of 5G aims to offer significantly higher data rates compared to 4G LTE, i.e., download speeds up to 20 Gbps and upload speeds up to 10 Gbps (peak).
- Latency: 5G networks aim for extremely low latency, reaching as low as 1 ms (ms) or even less. Such minimal latency is essential for applications such as autonomous vehicles and real-time remote control.

A basic yet complete front-end architecture is shown in Fig. 5.1, where the highlighted portion forms the major design concern at front. All the devices exhibit diverse requirement, sizes, and power characteristics as per design methodology. The main focus always lies on designing the amplifier section for both the channels, i.e., transmitter and receiver, as they are the last and first block respectively for the signal and plays a very important role.

D. Agarwal (✉) · Z. M. Patel · K. Inamdar
Sardar Vallabbhai National Institute of Technology, Surat, India
e-mail: d20ec006@eced.svnit.ac.in

M. El Ghzaoui et al. (eds.), *Next Generation Wireless Communication*, Signals and Communication Technology, https://doi.org/10.1007/978-3-031-56144-3_5

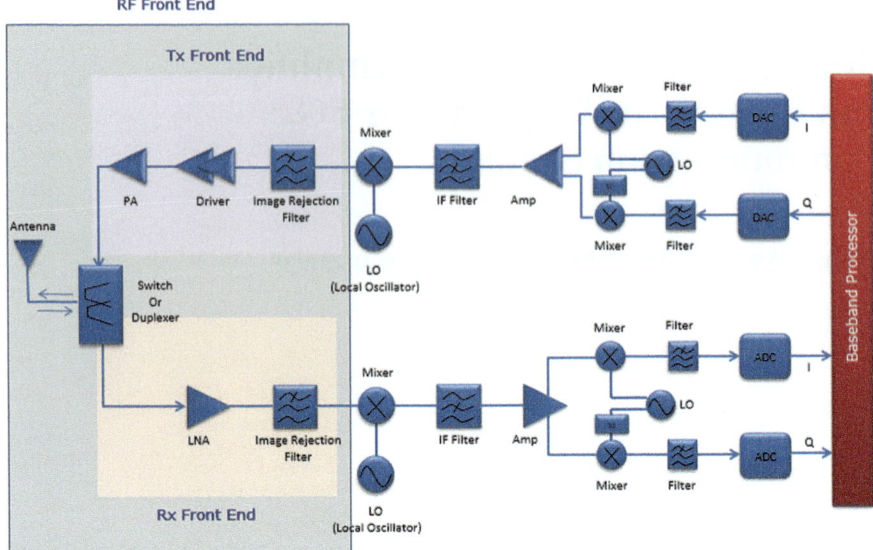

Fig. 5.1 Complete front-end architecture for 4G/5G with amplifier sections for transmitter and receiver

5.1.1 Power Amplifier

Amplifiers are electronic devices designed to increase the strength or amplitude of a signal, typically an electrical voltage or current. Amplifiers are grouped based on their input characteristics, distinguishing between current, voltage, and PAs. Specifically, a power amplifier is crafted to amplify input power to such extent that various loads like speakers, headphones, or RF transmitters can be driven. To ensure effective amplification, the input to the PA is typically pre-amplified until it reaches a specific threshold. The PA plays a vital role in the transmitter section of any front-end architecture.

The classification of power amplifiers is further determined by the manner in which input power is supplied, resulting in the categorization of audio PAs, DC power amplifiers, and RF power amplifiers. In the context of 5G technology, the RF power amplifier is particularly well-suited as it takes in RF signals as input and delivers amplified RF signals at the output to energize the antenna.

The RF power amplifier stages or modules include the following components:

- Input Matching Network: In order to match the input line having 50 Ω with the input impedance of device, specialized network is to be designed.
- Output Matching Network: This is employed to align the amplifier device's output impedance with the output 50 Ω impedance line.

- Amplifiers (single or multiple stages): Depending on the desired gain in the circuit, one or more amplifier devices are utilized. In the case of multiple devices, they are connected either in parallel or in cascade, considering the P1dB requirement.
- Biasing Network: This network is used to supply bias/voltage to the device.
- Accessories Network: These comprise various methods and facilities aimed at enhancing the stability and linearity characteristics of the amplifier.

In addition to these modules, PA classes are defined as the biasing point choice for gate and drain of device used for amplification. There are many classes available on the basis of linear or switched mode operation and classified as shown in Table 5.1.

For effective PA design, certain parameters are to be kept in mind to be measured as discussed below:

- Frequency of operation—mm wave provides low gain and output power.
- Bandwidth—large for mm-wave amplifiers.
- Linearity—determined by modulation scheme.
- P_{1dB} compression point—measure of linearity.
- Peak power output—given by standards.
- Efficiency—most power-hungry block.
- Power gain—determined by the number of stages.
- High efficiency at back off—determined by high PAPR of OFDM.
- Peak-to-average ratio (PAR).
- Adjacent channel power.
- Error vector magnitude—determines how much error free transmission in there within constellation.

Any design can be made effective by properly choosing class of operation, following the module design procedure and analyzing all the parameters.

Table 5.1 Classes of power amplifier

Class of operation	Peak swing at drain	Maximum theoretical efficiency (%)	Linearity	Normalized output power handling capacity
A	$2V_{dd}$	50	Maximum	0.125
B	$2V_{dd}$	78.5	Moderate	0.125
AB	$2V_{dd}$	50–78.5	Moderate	1.4π
C	$2V_{dd}$	79–99	Reduces as $\theta \sim 0$	0.125–0.3
D	V_{dd}	100	Nonlinear	0.32
E	$3.6V_{dd}$	100	Nonlinear	0.098
F	$2V_{dd}$	100	Nonlinear	0.16

5.1.2 Low-Noise Amplifier

To achieve reduced noise levels, a high degree of amplification in the initial stage of receiver is essential. For this purpose, JFETs and HEMTs are frequently employed. These devices effectively minimize shot noise, despite of operating in high-current regime, which are generally not energy-efficient. Also designing input and output matching for such narrow band circuits is crucial. A common-source amplifier is a single-stage FET amplifier configuration acting as s a voltage or transconductance amplifier, identified by signal entering through gate and exiting through drain, leaving the source terminal sole. This common-source FET circuit can be conveniently adapted to HEMTs.

This configuration as a transconductance amplifier interprets the input voltage to modulate the current supplied to the load. Current flowing through the FET amplifies the voltage across output, and this in turn deviates the FET output resistance form ideal characteristics, lacking the necessary infinite resistance for an optimal transconductance amplifier or zero resistance for a proficient voltage amplifier. Another notable limitation is its constrained high-frequency response. For practical applications, the output is often directed through either a voltage follower known as common-drain or a current follower known as common-gate, to achieve more favorable output and appropriate frequency characteristics. The amalgamation of these two topologies is termed a cascade amplifier, which will be explored further.

5.1.3 GaN HEMT

GaN technology possesses crucial characteristics such as a high-saturated drift velocity, leading to increased saturation currents and elevated power densities in GaN, making it a good contestant for high-power devices. Consequently, GaN finds predominant use in high-electron-mobility transistor (HEMT) devices, featuring a junction between two materials with distinct bandgaps. GaN can be developed on various substrates, including Si, SiC, and Sapphire. The fundamental layering of a GaN HEMT device is illustrated in Fig. 5.2.

Devices based on typical AlGaN developed on GaN exhibit superior current handling capabilities. Due to SiC's higher thermal conductivity compared to GaN, it is frequently the preferred substrate for manufacturing GaN devices. GaN on SiC provides high reliability and performance, particularly at large power levels, proving to be effective in design space and thermal dissipation. These attributes collectively make GaN-based HEMT an ideal choice for designing high-power devices operating at significantly higher frequencies.

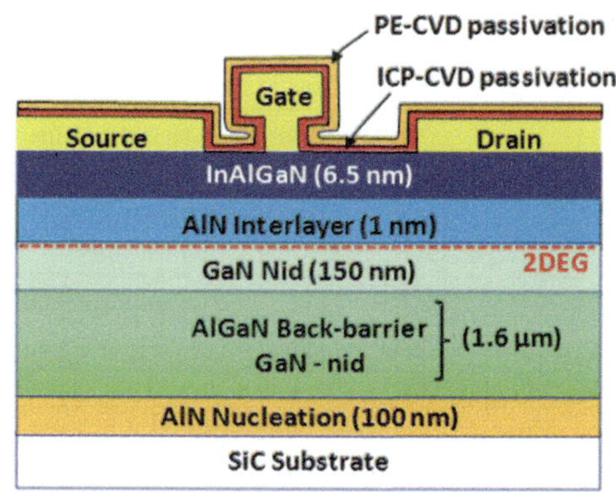

Fig. 5.2 Layer arrangement in AlGaN/GaN HEMT

5.1.4 Objective

The basic objective behind designing these designs is to provide a simple yet effective solution for linear amplifier designing using nonlinear devices. The application specific analog circuit design always demands accuracy as well as optimum performance. The designed power and low-noise amplifiers are based on 150 nm GaN HEMT technology in order to achieve the following parameters:

- Output power for PA > 30 dBm.
- PAE of PA > 70%.
- Gain for LNA > 10 dB.
- NF for LNA ~ 2.5–3.5 dB.
- Lesser chip size.
- Reduced complexity of circuit.

All these specifications can be met by proper design methodology as:

- Characterization of device.
- Load/source pull simulations for device in order to get input and output device impedance.
- Designing of effective input and output matching networks. In this L-type matching is used in order to reduce chip size and circuit complexity.
- Designing of biasing network as per characterization values for both gate and drain voltage in order to pass only DC to terminals and bypass any RF to the voltage source.
- Complete amplifier design and result analysis as per requirement.
- Layout level designing for fabrication purpose.

All these points are kept intact for designing for 28 GHz with minimum 400 MHz bandwidth as per channel requirement.

Table 5.2 Summary of work surveyed for 5G RF front end

References	Year	Design	Performance	Technology	Freq. (GHz)
[1]	2019	LNA	18.7 dB gain 0.85 dB NF	130 nm RFSOI CMOS	
[2]	2019	PA	27 dBm P_{out} 33% PAE	CMOS	Sub-6
[3]	2019	PA	Topology studies	CMOS	Sub-6
[4]	2018	Review for various PA topologies and their performance			

5.2 Literature Survey

The very first step in designing any analog circuit is to study the background for the work already done, various topologies available, devices effective for design, and other connecting blocks available. For the design of amplifiers to be used in 5G front end, blocks of importance are:

- Study of basic front-end requirements.
- Study of device to be used.
- Study of power amplifier.
- Study of low-noise amplifier.
- Study of biasing and matching networks.

All these points are surveyed from available literature and presented in the following section.

5.2.1 5G RF Front End

A front-end architecture comprises either transmitter section (power amplifier + other blocks) or receiver section (low-noise amplifier + other blocks) or an integration of both. These other blocks are basically switches, filters, etc. Some of the studies are summarized in Table 5.2.

5.2.2 Power Amplifier

For transmitter section, power amplifiers are very important. Many factors impact their design such as device, matching network, topology, and target specifications. For the chosen work, device selection was GaN, so many of these have been studied and summarized in Table 5.3. All the works are for Ka band and have demonstrated various topologies such as multi-stage, current combiner, and Doherty architecture for achieving desired results of output power, power-added efficiency, and gain.

Table 5.3 Summary of work surveyed for power amplifier

References	Year	Tech	P_{out} (dBm)	Gain (dB)	PAE (%)	BW (GHz)
[5]	2018	GaN	A review work on PA designing for cm and mm-wave frequencies			
[6]	2005	GaN	33	8	8	28
[7]	2013	GaN	21.4	11	18	28
[8]	2015	GaN	45.9	18	35	28
[9]	2015	GaN	37	22	29.6	28
[10]	2016	GaAs	37	20	26	31–41
[11]	2017	GaN	43.4	10	19.8	25–32
[12]	2018	GaN	40	22	25	28
[13]	2018	GaN	35	15	30	26
[14]	2019	GaN	40	21	25	31
[15]	2020	GaN	36	–	28	25–29
[16]	2021	GaN	30	20	34.1	26–30
[17]	2021	GaN	38.5	22	30–38	20.5–25.5
[18]	2018	GaN	41	22	34	28

5.2.3 *Low-Noise Amplifier*

For receiver section, low-noise amplifier is important in order to differentiate between noise and received signal. Different topologies and devices are being worked upon for Ka-band LNA design and summarized in Table 5.4. Basic common-source, cascade, multi-stage, single-stage designs are presented with effective gain and noise figure values.

Table 5.4 Summary of work surveyed for low-noise amplifier

References	Year	Tech	Gain (dB)	NF (dB)	BW (GHz)
[19]	2014	GaAs	33	1.8	26–36
[20]	2021	GaN	26	2	33–38
[21]	2018	CMOS	12.8	1.4	14–31
[22]	2016	SiGe HBT	32	4	27.5
[23]	2019	GaN	14–17	0.8–1.1	23–30
[24]	2019	GaN	33	2.3	35–36
[25]	2021	GaN	20	2.5	26.5–29.5

5.2.4 *Biasing Network*

In [26, 27], a novel approach to designing a concurrent multi-band biasing network for a multi-band RF power amplifier is introduced. The biasing network (BN) consists of a transmission line with high-impedance quarter-wave open stubs (QWOS) strategically positioned along the transmission line and butterfly stub, respectively.

5.3 Design of Power Amplifier

The primary objective behind amplifier design is achieving highest possible output power and power efficiency. All these should be done keeping the stability criteria intact, i.e., unconditionally stable > 1 in the whole operational range. To ensure optimal power transfer through device, source impedance and output impedance are in terms with internal impedance conjugatively.

It is important to select a power amplifier that is well-matched to the specific requirements of the application, considering factors such as power output, efficiency, and the nature of the input signal. The choice of amplifier class and design depends on the particular needs of the system in which it will be employed, additionally ensuring that the input and output impedances of the amplifier match the source and load impedances, respectively. This helps in maximum power transfer and minimizes signal reflections.

5.3.1 *Design Consideration*

Power gain is defined as product of current gain and voltage gain as shown.

$$\text{Power gain} = A_v A_i, \tag{5.1}$$

where

$A_v = \frac{V_{\text{out}}}{V_{\text{in}}}$ and $A_i = \frac{I_{\text{out}}}{I_{\text{in}}}$.

Any amplifier design should be able to achieve following set of parameters as per specifications:

- Gain and gain flatness (in dB).
- Operating frequency and bandwidth (in Hz).
- Output power (in dBm).
- Power supply requirement (in V and A).
- Power-added efficiency.
- Harmonics (in V and A).

5.3.2 Proposed Power Amplifier Using UMS GH15-10

With the help of discussed design equations and proposed biasing point, power amplifier is designed using UMS GH15-10 GaN HEMT in ADS tool of Keysight. Various variations and modifications are being done to achieve desired output.

Single-Stage Power Amplifier for Class-B Biasing

Initially, a single-stage power amplifier is designed on UMS GH15-10 with biasing network and biasing points as shown in Fig. 5.3. All the dimensions are optimized for 26 GHz band operation with biasing network, matching network, and stability network designed using GH15 components.

The topology is operational for 25–29 GHz of band with less than 10 dB input return loss and 9 dB small signal gain for the same range as in Fig. 5.4. Thus, the designed amplifier is having around 4 GHz of bandwidth with gain varying between 7 and 8.5 dB in that range. The single-stage topology is able to produce power-added efficiency and drain efficiency of 25% and 35%, respectively, as in Fig. 5.5. The concept that the DC-RF conversion or drain efficiency should follow PAE is seen to valid in obtained results. Also, for a single-stage amplifier, these values are considered good.

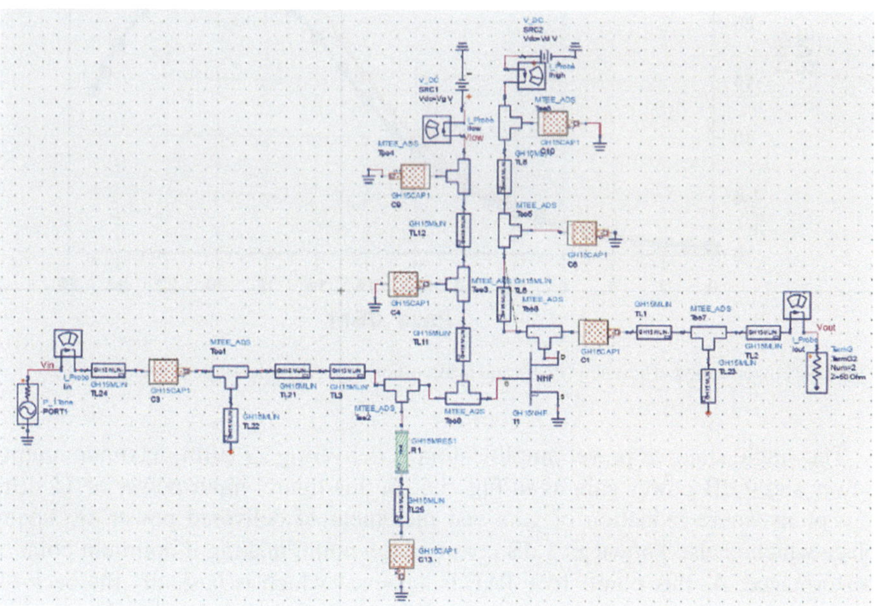

Fig. 5.3 Single-stage power amplifier designed with biasing network at drain voltage 20 V and gate voltage − 1.2 V

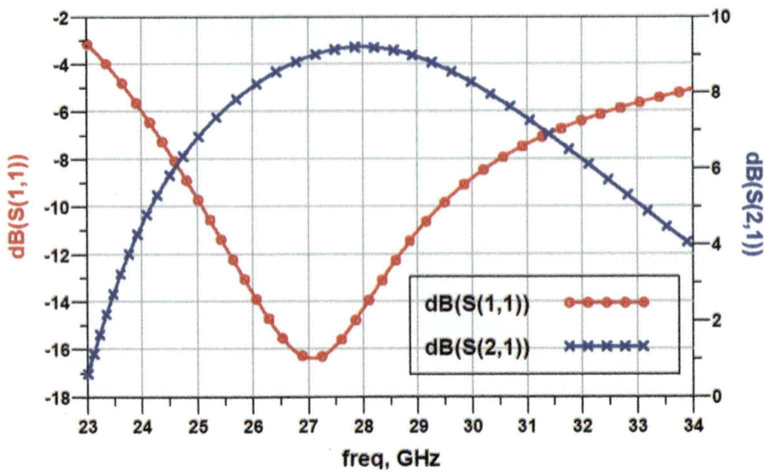

Fig. 5.4 *S*-parameters of single-stage power amplifier with 16.586 dB isolation and 10.7 dB gain

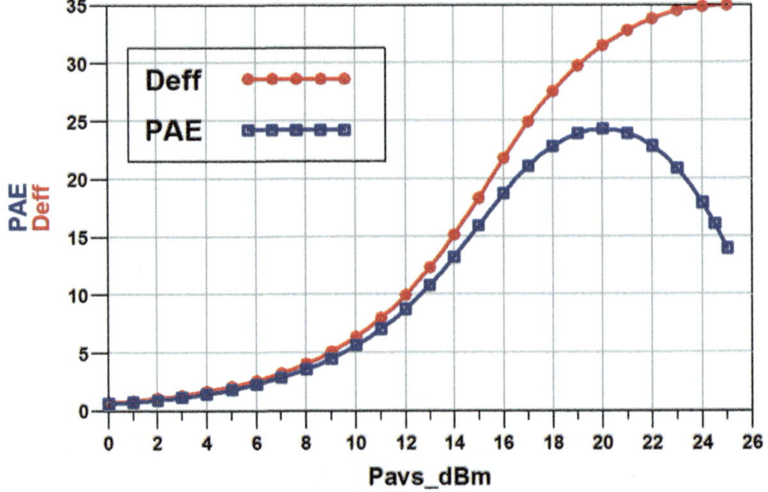

Fig. 5.5 Power-added efficiency (PAE) and drain efficiency of single-stage power amplifier

The single stage of power amplification is delivering 27 dBm maximum output power and 9 dB power gain as in Fig. 5.6. In this figure, input power of 15 dBm is a point where reduction of gain and increment of delivered power are taking place which is also known as 1 dB compression point, making it optimum point of performance. At this point, 16% PAE is achieved which is basically the back-off power. Another important concept in amplifier performance is its linearity which is very much clear from output power vs. input power curve as the former follows latter linearly (it is a straight line with slope almost 1).

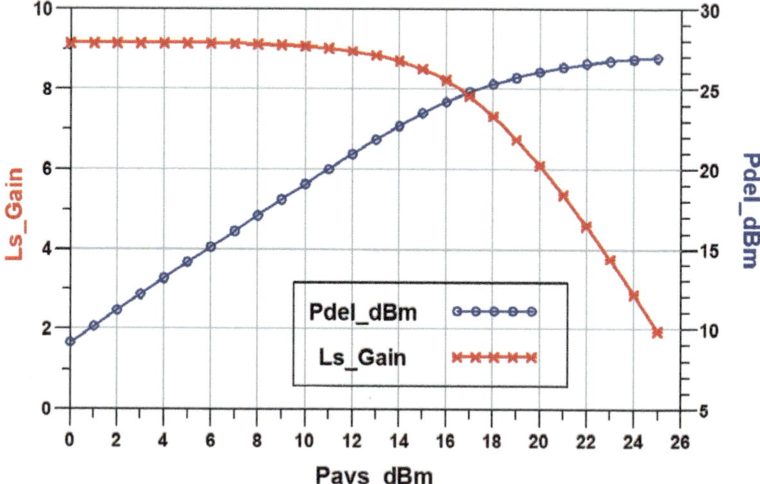

Fig. 5.6 Output power (Pdel) and power gain for single-stage power amplifier

Layout and EM Simulation of Single-Stage PA

The single-stage power amplifier designed is converted to layout for post-layout simulations and EM simulation so as to check the reliability of designed schematic according to foundry rules for verification. Initially, the layout is being generated by ADS itself as in Fig. 5.7. The *S*-parameter results are available in Fig. 5.8, where S_{11} and S_{21} are present but not good due to larger dimension of transmission lines creating coupling and long electrical length.

As all the TLs are at their original dimensions making the die size large enough for mm-wave, step by step all TLs are mitered so as to reduce chip consumption and improve performance as in Fig. 5.9. All the active components such as transistor, capacitors, and resistors are removed for EM simulation and further added in schematic which is simulated with EM model of layout. The schematic can be seen in Fig. 5.10, where all the capacitors, resistor, device, and simulation parameters are applied to obtain *S*-parameters and power amplifier parameters.

This bending of transmission lines is done keeping in mind the coupling distance between two bends and best possible configuration which provides better *S*-parameters as seen in Fig. 5.11 with input return loss of 13 dB and gain of 5.1 dB at 31 GHz. This frequency shift and lesser gain are to be improved by more precise TL placement and bending.

Dual-Stage PA

To enhance the gain and other parameters, dual-stage amplifier is designed as shown in Fig. 5.12.

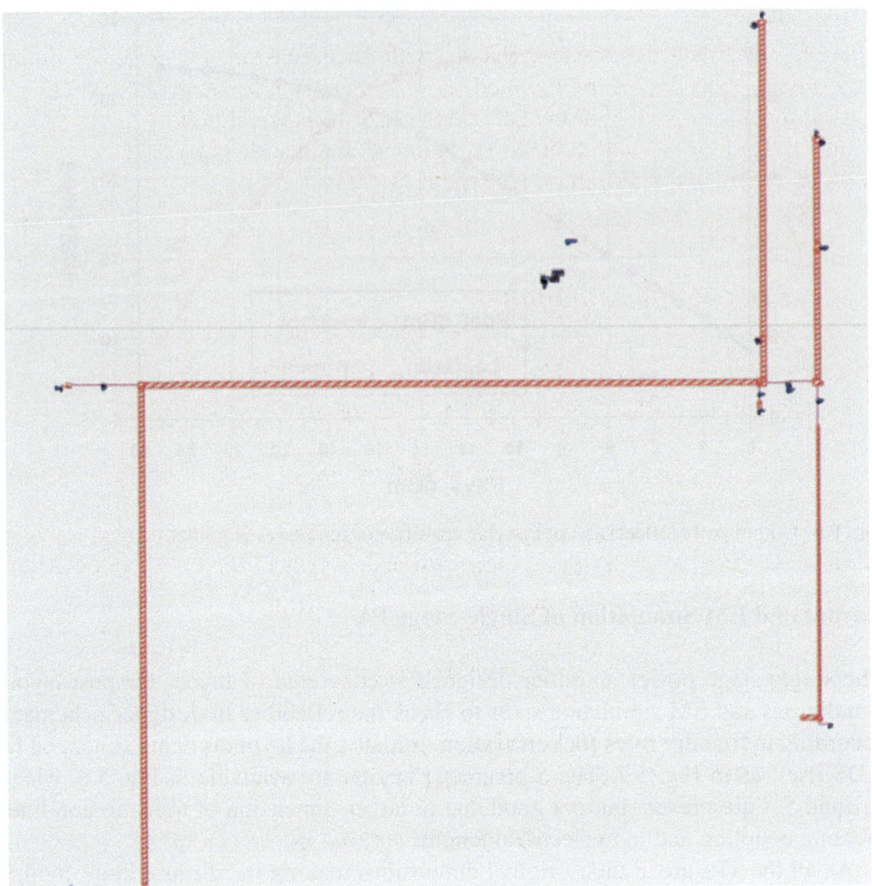

Fig. 5.7 Initial layout as being generated by ADS without the actives

In the design, basic biasing is first used comprising lumped C and strip lines. It is done in order to establish a premise for dual-stage amplification. For designing two stages, first and second stages are connected using interstage matching so as to completely transfer power from first stage to second. First stage is known as driver stage and second one as power stage. First stage known as driver stage is used to provide amplified output to gate of second stage (power stage) in addition to bias, so as to improve gain and efficiency. This improvement is affected by the device size. In the power stage, device size is increased by approximately 5 times. This gate periphery enhancement helps in amplifying the RF signal due to large current flow but may affect gain requirements. As the number of gate fingers increases, more parasitic inductance will be generated which in turn increases phase shift between fingers. For this reason, recommended number of maximum gate fingers is 8 with unit length not exceeding 100 μm.

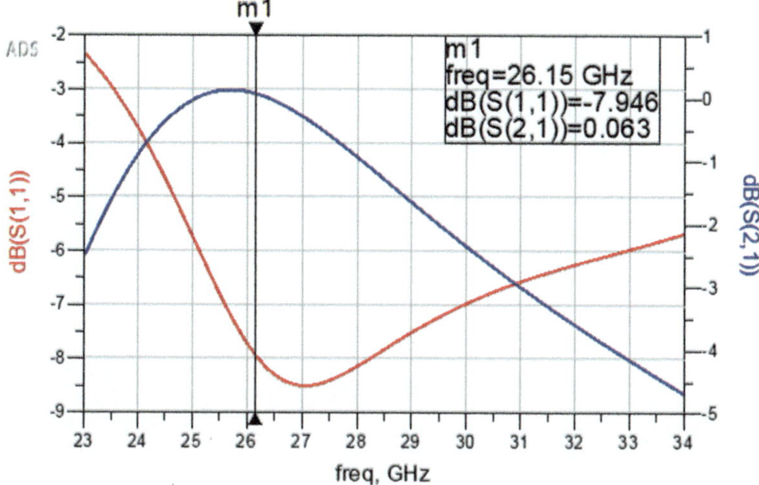

Fig. 5.8 *S*-parameter EM simulation result for initial layout where return loss and gain are available but low due to large TL

S-parameter performance can be seen in Fig. 5.13 having higher input return loss in the range of 25.5–29.25 GHz with value more than 16 dB and small signal gain of 24 dB. For amplifier operation, power-added efficiency (PAE) and drain efficiency are achieved 77.78% and 95%, respectively, in Fig. 5.14. Figure 5.15 displays the delivered power and power gain for dual-stage amplifier. Increment of power stage is helping in delivering around 31 dBm output power with a power gain of 24 dB. As it is clear from the gain trend, that linearity is compromised a bit which can be improved further. It results in shifting of 1 dB compression point to 11 dBm input power where the power gain is 17 dB and output power is 27 dBm.

5.4 Low-Noise Amplifier

An amplifier is recognized for boosting both the signal and the inherent noise at its input, along with introducing its own noise during the process. Low-noise amplifiers (LNAs) are specifically designed to minimize this additional noise. Achieving this involves selecting components with low-noise characteristics, optimizing operating points, and employing appropriate circuit topologies. The pursuit of noise reduction needs to be balanced with other design goals, such as achieving power gain and ensuring proper impedance matching.

A typical LNA is expected to provide a power gain of 100 (equivalent to 20 decibels or 20 dB) while causing a reduction in the signal-to-noise ratio by no more than a factor of two (resulting in a 3 dB noise figure or NF). Even though LNAs

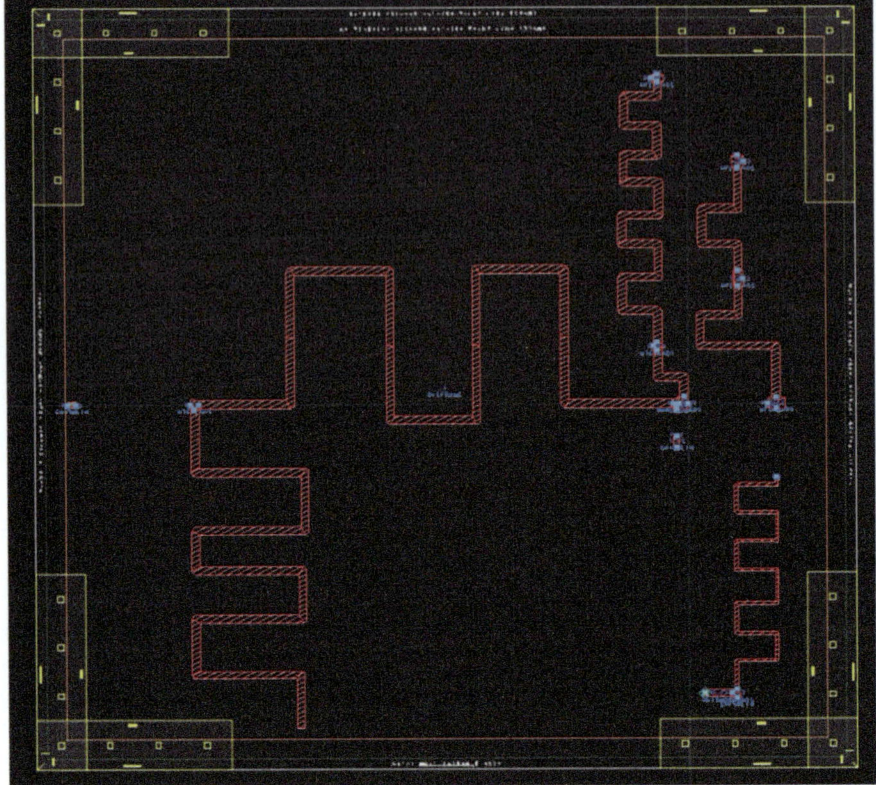

Fig. 5.9 Layout of single-stage power amplifier with all TL meandered to reduce die area and improve performance

primarily amplify weak signals just above the noise floor, they must also consider the presence of stronger signals that can result in intermodulation distortion.

5.4.1 Design Consideration

Low-noise amplifiers serve as the foundational components in communication systems and devices. The key specifications or characteristics of low-noise amplifiers (LNAs) include:

- Gain.
- Noise figure.
- Linearity.
- Maximum RF input.

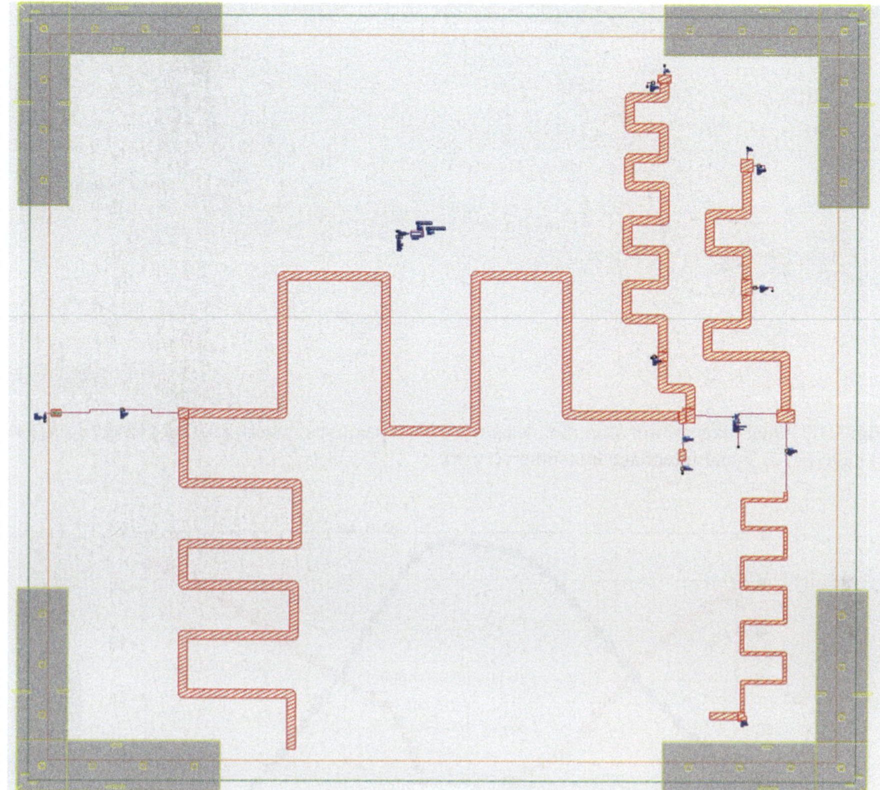

Fig. 5.10 Schematic of updated layout with passives attached

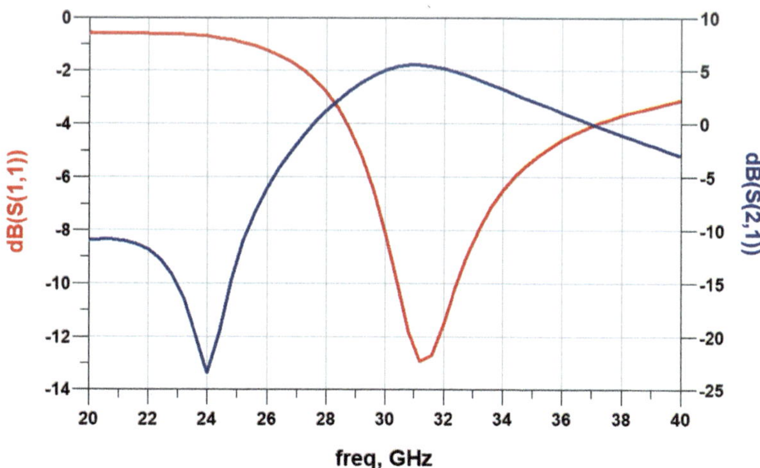

Fig. 5.11 *S*-parameter results for updated layout EM simulation with input return loss as 13 dB and gain of 5.1 dB

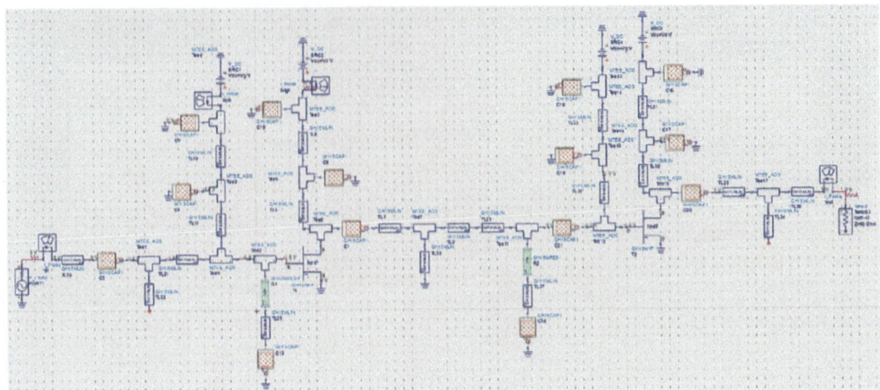

Fig. 5.12 Dual-stage power amplifier designed with capacitive network at drain voltage 20 V, gate voltage − 1.2 V and interstage matching network

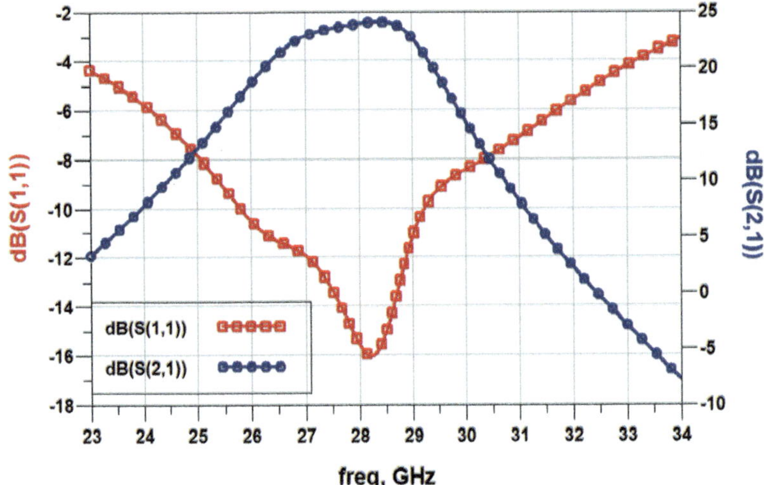

Fig. 5.13 *S*-parameters for dual-stage power amplifier yielding 16 dB input return loss and 24 dB gain for 26 GHz band

A high-quality LNA is defined by a low-noise figure (e.g., 1 dB), ample gain for signal amplification (e.g., 10 dB), and an intermodulation and compression point (IP3 and P1dB) capable of handling the required tasks. Additional specifications encompass the LNA's operating bandwidth, gain uniformity, stability, as well as its input and output voltage standing wave ratio (VSWR). Considering all these attributes and the desired performance criteria, multiple LNAs are designed and subsequently compared in the following sections, utilizing the matching networks developed earlier.

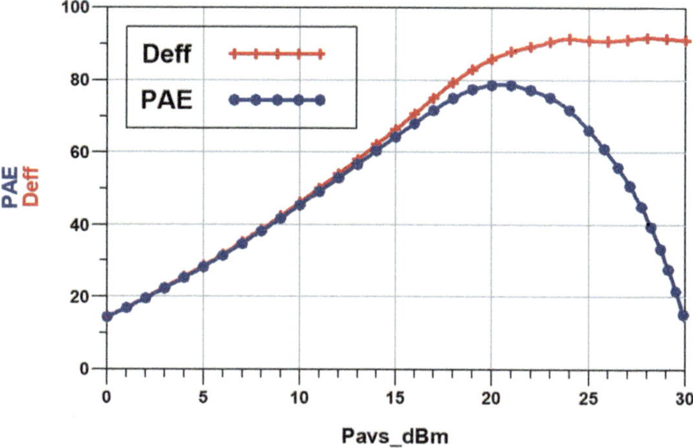

Fig. 5.14 Power-added efficiency (PAE) and drain efficiency of dual-stage power amplifier

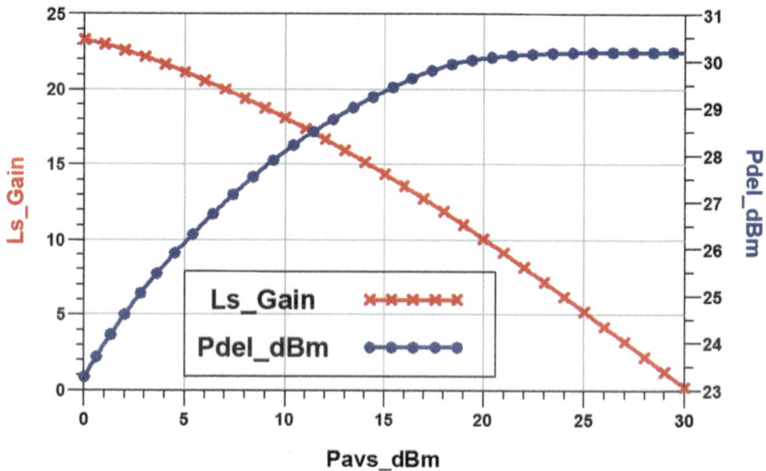

5.15 Output power (Pdel) and power gain for dual-stage power amplifier

5.4.2 Proposed Low-Noise Amplifier Design Using UMS GH15-10

Linear Noise FET (LNF) Characterization

The foundry has provided a separate device for low-noise applications. It is a linear noise model of "hot-FET" available with the PDK. The device is ideal for low-noise applications due to linearity of noise when operated for 4, 6, 8 fingers with each finger

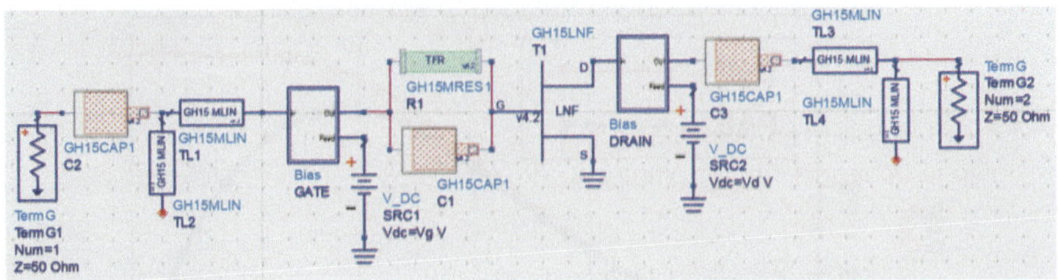

Fig. 5.16 Single-stage LNA using the previous biasing network and operating points defined from characterization of device

size of 20–50 μm. The device with mentioned size of 6 × 50 μm is characterized with variable drain and gate voltages applied for optimum operating point choice.

The various values of V_{ds} and V_{gs} are applied, in order to obtain approximately 150 mA quiescent current and maintaining the drain voltage to 20 V (for easy integration with the power amplifier), and gate voltage is selected to be 1.2 V as below 0 V no current situation occurs. So, for multi-stage LNA also voltages in the range of 0–1.2 V are considered.

Single-Stage LNA

After deciding the operating points of voltages and current, biasing network similar to power amplifier is designed. Also, with the help of load pull analysis, input and output matching network and interstage matching networks are designed. For LNA designing, all the above networks with optimum size device are combined to form an amplifier as in Fig. 5.16.

Single device is not generating sufficient return loss all over the frequency range for input as well as output as in Fig. 5.17. Also, the gain of LNA is very low with 5 dB value in Fig. 5.18. The noise figure margin is of 5 dB as in Fig. 5.19.

Three-Stage LNA

As a single device is not able to perform appropriately, a greater number of devices are added so that gain and noise figure are cascaded for enhancement. The schematic is in Fig. 5.20 where three devices are used in cascading for common-source configuration in source degeneration mode. This combination adds up for gain and combines for NF on the basis of individual stage values. The device size is constant with 6 × 50 μm, and for source degeneration, inductor is used at the source end. Interstage matching is added in order to properly transmit the signal from one stage to other.

The *S*-parameter for return losses' performance of the three-stage LNA can be seen in Fig. 5.21. Input return loss is − 20 dB with operating range according to

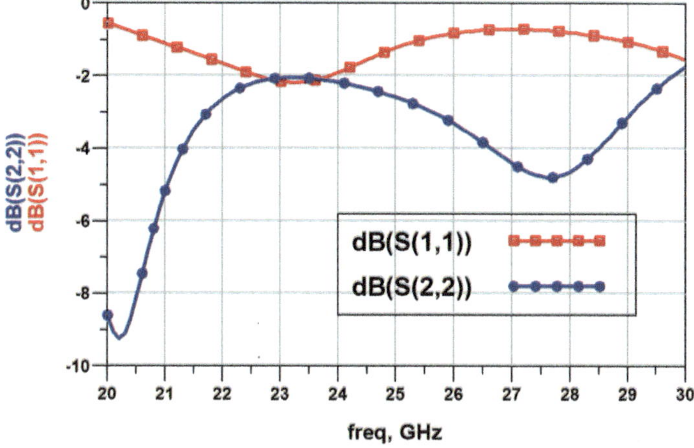

Fig. 5.17 Input and output return losses for single-stage LNA where the values are very low

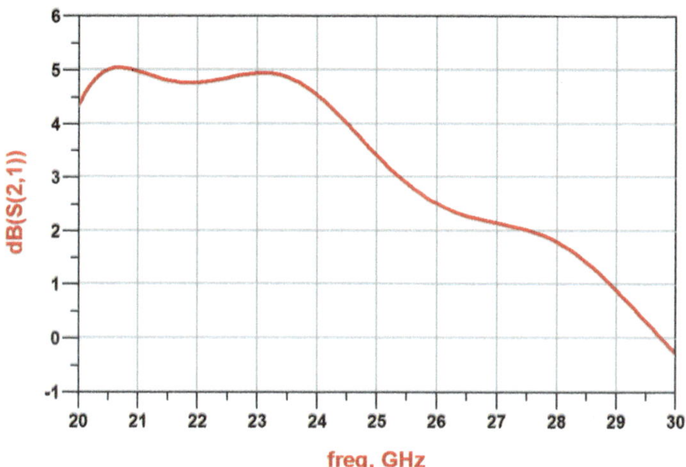

Fig. 5.18 Small signal gain for LNA with maximum value of 5 dB for single-stage LNA

10 dB criteria lying from 28.1 to 28.7 GHz. Also, the output return loss is in good compromise with the input one indicating about proper matching of the devices.

As the return losses are not lying in the 26 GHz band of 5G communication, this improvement in frequency shifting is required. Considering one of the main factors for LNA performance, i.e., gain, Fig. 5.22 shows a maximum value of 10 dB and a decreasing slope in the obtained frequency band. The second important factor of LNA is in Fig. 5.23 with a minimum of 5.4 and rising in 28.1–28.7 GHz range.

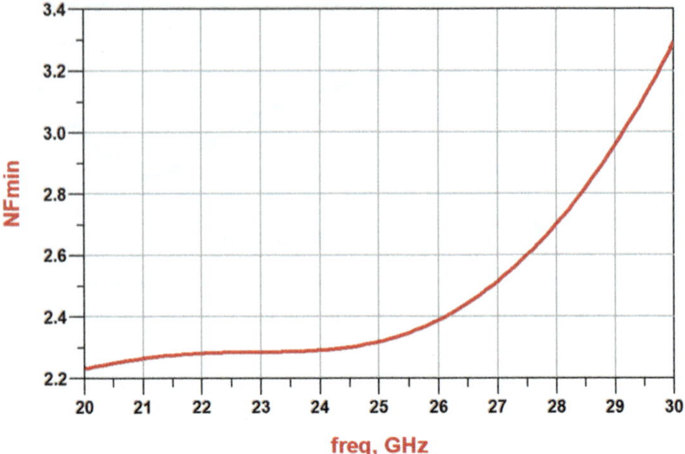

Fig. 5.19 Noise figure NF for single-stage LNA with a value of 2.2–3.3 dB

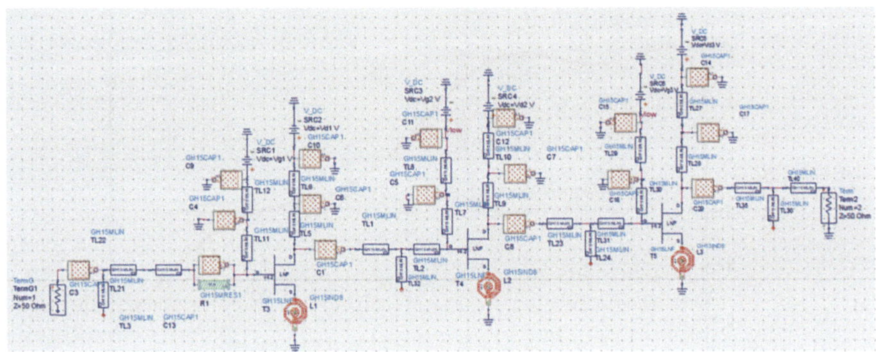

Fig. 5.20 Three stages cascaded with input, output, and interstage matching networks and a biasing network

5.5 Summary

In order to keep up with the growing demand for increased bandwidth and efficiency, the design of high-power amplifiers (HPAs) and low-noise amplifiers (LNAs) are very important. These designs are demanding in terms of power, cost, efficient, and effective topologies. This makes analog RF circuit design a challenging task, especially for high frequencies, i.e., 26 GHz band.

In this research, the UMS GH15-10 GaN High-Electron-Mobility Transistor (HEMT) device is chosen due to its exceptional performance in millimeter-wave applications. The device is characterized for Class-B operation using specific biasing values. Moreover, load pull analysis, stability analysis, and harmonic analysis are conducted to ensure the proper operation of the device and its suitability for circuit

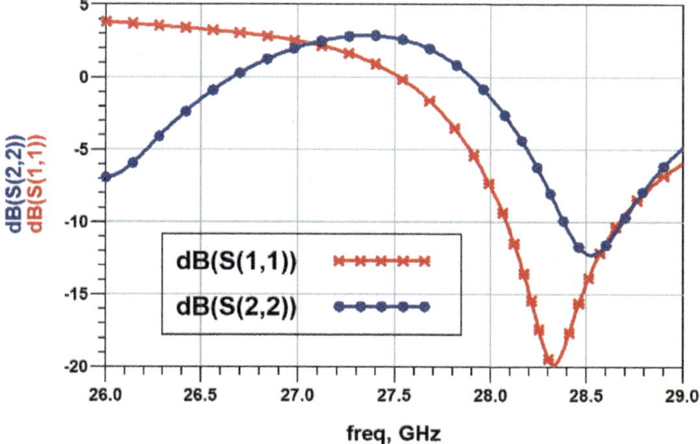

Fig. 5.21 Input and output return losses for three-stage LNA

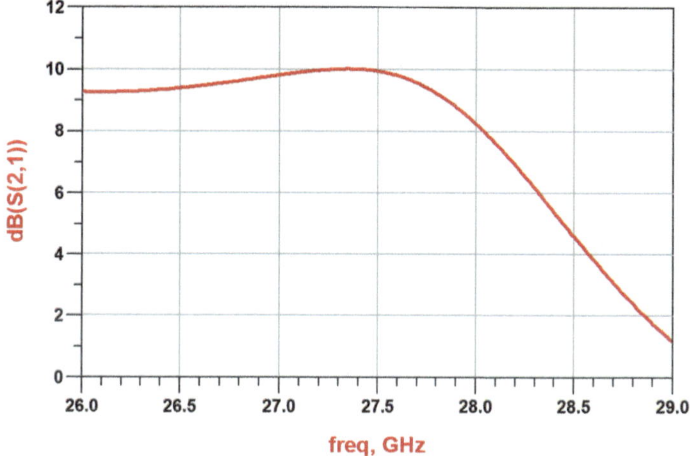

Fig. 5.22 Gain of 9–10 dB for three-stage LNA

use. Based on the results of these studies, a biasing network is designed using transmission lines and stubs. The network's stability and impedance, both with and without the transistor, are thoroughly examined, and the most suitable design is chosen for further development.

Utilizing the device characterization and the designed biasing network, single-stage and two-stage power amplifiers are created, adhering to MMIC design principles. The designed PAs are able to provide 25% and 80% PAE for single-stage and dual-stage PAs, respectively. Both the designs are more than 25 dBm output power amplifiers.

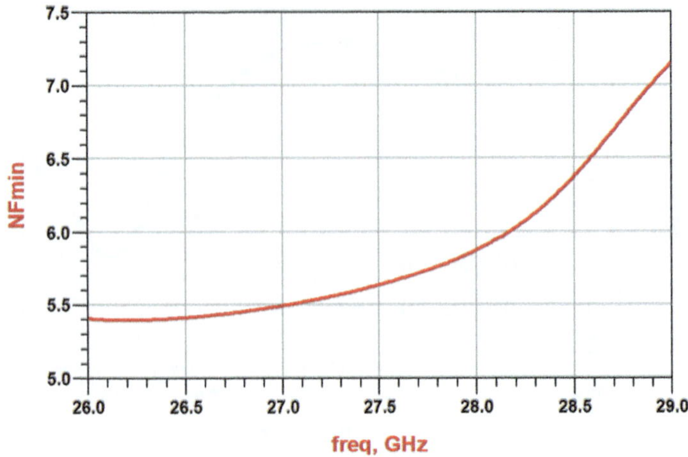

Fig. 5.23 Linear NF for the operational range of LNA with values from 5.4 to 7 dB

Furthermore, a linear noise device available in the Process Design Kit (PDK) is characterized to enable the design of a single-stage and three-stage LNA. The single-stage LNA is providing 6 dB gain and 2.2–3.3 dB noise figure, whereas for the three-stage LNA design, gain achieved is 10 dB gain and 5 dB noise figure in desired frequency range.

To enhance overall performance, further work can be carried out, including the investigation of Doherty power amplifiers, the conversion of schematics into efficient layouts, and the creation of an integrated MMIC for both transmitter and receiver applications.

References

1. Dai, R., Ren, J., He, J., Xiao, J., Kong, W.: RFFE integration design for 5G in 0.13μm RFSOI technology. In: 2019 China Semiconductor Technology International Conference (CSTIC) (2019)
2. Balteanu, F.: RF front end module architectures for 5G. In: 2019 IEEE BiCMOS and Compound Semiconductor Integrated Circuits and Technology Symposium (BCICTS) (2019)
3. Balteanu, F., Drogi, S., Modi, H., Choi, Y., Lee, J., Khesbak, S., Agarval, B.: New architecture elements for 5G RF front end modules. In: 2019 IEEE Asia-Pacific Microwave Conference (APMC) (2019)
4. Lozhkin, A.N., Maniwa, T., Shimizu, M.: RF front-end architecture for 5G. In: 2018 IEEE 29th Annual International Symposium on Personal, Indoor and Mobile Radio Communications (PIMRC) (2018)
5. Lie, D.Y.C., Mayeda, J.C., Li, Y., Lopez, J.: A review of 5G power amplifier design at cm-wave and mm-wave frequencies. Wirel. Commun. Mob. Comput. **2018** (2018)
6. van Heijningen, M., van Vliet, F.E., Quay, R., van Raay, F., Kiefer, R., Muller, S., Krausse, D., Seelmann-Eggebert, M., Mikulla, M., Schlechtweg, M.: Ka-band AlGaN/GaN HEMT high

power and driver amplifier MMICs. In: European Gallium Arsenide and Other Semiconductor Application Symposium, GAAS 2005 (2005)
7. Yue, C., Guo, D., Luo, X., Zhou, L., Du, S.: Design of a power amplifier based on GaN HEMTs at Ka-band. In: 2013 IEEE International Conference on Microwave Technology & Computational Electromagnetics (2013)
8. Din, S., Wojtowicz, M., Siddiqui, M.: High power and high efficiency Ka band power amplifier. In: 2015 IEEE MTT-S International Microwave Symposium (2015)
9. Noh, Y., Choi, Y.-H., Yom, I.: Ka-band GaN power amplifier MMIC chipset for satellite and 5G cellular communications. In: 2015 IEEE 4th Asia-Pacific Conference on Antennas and Propagation (APCAP) (2015)
10. Hosseinzadeh, N., Medi, A.: Wideband 5 W Ka-band GaAs power amplifier. IEEE Microwave Wirel. Compon. Lett.Wirel. Compon. Lett. **26**, 622–624 (2016)
11. Yamaguchi, Y., Kamioka, J., Hangai, M., Shinjo, S., Yamanaka, K.: A CW 20W Ka-band GaN high power MMIC amplifier with a gate pitch designed by using one-finger large signal models. In: 2017 IEEE Compound Semiconductor Integrated Circuit Symposium (CSICS) (2017)
12. Potier, C., Piotrowicz, S., Patard, O., Gamarra, P., Altuntas, P., Chartier, E., Dua, C., Jacquet, J.C., Lacam, C., Michel, N., Oualli, M., Delage, S.L., Chang, C., Gruenenpuett, J.: First results on Ka band MMIC power amplifiers based on InAlGaN/GaN HEMT technology. In: 2018 International Workshop on Integrated Nonlinear Microwave and Millimetre-wave Circuits (INMMIC) (2018)
13. Samis, S., Friesicke, C., Feuerschütz, P., Lozar, R., Maier, T., Brückner, P., Quay, R., Jacob, A.F.: A 5 W AlGaN/GaN power amplifier MMIC for 25–27 GHz downlink applications. In: 2018 11th German Microwave Conference (GeMiC) (2018)
14. Potier, C., Piotrowicz, S., Chang, C., Patard, O., Trinh-Xuan, L., Gruenenpuett, J., Gamarra, P., Altuntas, P., Chartier, E., Jacquet, J.-C., Lacam, C., Michel, N., Dua, C., Oualli, M., Delage, S.L.: 10W Ka band MMIC power amplifiers based on InAlGaN/GaN HEMT technology. In: 2019 49th European Microwave Conference (EuMC) (2019)
15. Samis, S., Friesicke, C., Maier, T., Quay, R., Jacob, A.F.: Broadband high-efficiency power amplifiers in 150 nm AlGaN/GaN technology at Ka-band. In: 2020 IEEE Asia-Pacific Microwave Conference (APMC) (2020)
16. Peng, L., Lin, S., Chen, J., Li, J., Zhang, G.: Ka-band 1 W GaN MMIC power amplifier design. In: 2021 IEEE International Conference on Signal Processing, Communications and Computing (ICSPCC) (2021)
17. Bhavsar, M.L., Srivastava, P., Singh, D.K., Parikh, K.S.: K-band 8-watt power amplifier MMICs using 150nm GaN process for satellite transponder. In: 2021 IEEE MTT-S International Microwave and RF Conference (IMARC) (2021)
18. Di Giacomo-Brunel, V., Byk, E., Chang, C., Grünenpütt, J., Lambert, B., Mouginot, G., Sommer, D., Jung, H., Camiade, M., Fellon, P., et al.: Industrial 0.15-μm AlGaN/GaN on SiC technology for applications up to Ka band. In: 2018 13th European Microwave Integrated Circuits Conference (EuMIC) (2018)
19. Cuadrado-Calle, D., George, D., Fuller, G.: A GaAs Ka-band (26–36 GHz) LNA for radio astronomy. In: 2014 IEEE International Microwave and RF Conference (IMaRC) (2014)
20. Florian, C., Traverso, P.A., Santarelli, A.: A Ka-Band MMIC LNA in GaN-on-Si 100-nm technology for high dynamic range radar receivers. IEEE Microwave Wirel. Compon. Lett.Wirel. Compon. Lett. **31**, 161–164 (2021)
21. Li, C., El-Aassar, O., Kumar, A., Boenke, M., Rebeiz, G.M.: LNA design with CMOS SOI Process-l.4dB NF K/Ka band LNA. In: 2018 IEEE/MTT-S International Microwave Symposium—IMS (2018)
22. Candra, P., Xia, T.: SiGe HBT X-band and Ka-band switchable dual-band low noise amplifier. In: 2016 IEEE International Symposium on Circuits and Systems (ISCAS) (2016)
23. Tong, X., Zhang, S., Zheng, P., Xu, J., Wang, R.: Ka band LNA and PA based on 100 nm GaN/Si HEMT process. In: 2019 Compound Semiconductor Week (CSW) (2019)
24. Pace, L., Ciccognani, W., Colangeli, S., Longhi, P.E., Limiti, E., Leblanc, R.: A Ka-band low-noise amplifier for space applications in a 100 nm GaN on Si technology. In: 2019 15th Conference on Ph.D. Research in Microelectronics and Electronics (PRIME) (2019)

25. Seo, J.-S., Hwang, J.-H., Kim, K.-J., Ahn, G.-H.: High linearity Ka-band GaN Hemt low noise amplifier. In: 2021 International Conference on Information and Communication Technology Convergence (ICTC) (2021)
26. Sadeque, M.G., Yusoff, Z., Roslee, M., Hashim, S.J., Mohd Marzuki, A.S.: Analysis and design of the biasing network for 1 GHz bandwidth RF power amplifier. IJEECS **24**, 308–316 (2021)
27. Xuan, A. N., Negra, R.: Design of concurrent multiband biasing networks for multiband RF power amplifiers. In: 2012 42nd European Microwave Conference (2012)

Chapter 6
Advanced MIMO Antenna Design with Defected Ground Structure for 5G NR (N75 and N77) Applications

K. Vasu Babu, Gorre Naga Jyothi Sree, Tanvir Islam, Sudipta Das, and Bhuma Anuradha

6.1 Introduction

Mobile and wireless communication devices have experienced a significant growth taken analog to digital communications. The modern design of antenna system facing the challenges are interoperability, space constraints, multiple frequency bands, hearing aid compatibility, and specific absorption rate. Multi-band system for MIMO [1], half-circular MIMO system with compact size [2], a six-port wideband MIMO system for LTE regions [3], fractal loaded deep learning assisted MIMO system [4], a co-axial feeding fractal-based MIMO system [5], for the array of MIMO systems reduced the mutual coupling analysis [6], for the region of THz easily designed 4-port system with wide bandwidth [7], for multi-band operations designed MIMO system

K. V. Babu (✉)
Department of Electronics and Communication Engineering, BVRIT HYDERABAD College of Engineering for Women, Bachupally, Telangana, India
e-mail: vasubabuece@gmail.com

G. N. J. Sree
Department of Electronics and Communication Engineering, Vasireddy Venkatadri Institute of Technology, Guntur, Andhra Pradesh, India

T. Islam
Department of Electrical and Computer Engineering, University of Houston, Houston, TX 77204, USA

S. Das
Department of Electronics and Communication Engineering, IMPS College of Engineering and Technology, Malda, West Bengal, India

B. Anuradha
Department of Electronics and Communication Engineering, S V University, Tirupati, Andhra Pradesh, India

M. El Ghzaoui et al. (eds.), *Next Generation Wireless Communication*, Signals and Communication Technology, https://doi.org/10.1007/978-3-031-56144-3_6

using DGS [8], for mm-wave future 5G system with 4-elements [9], a I-shaped flexible antenna designed for 5G systems [10]. Various MIMO antenna design with optimization parameters and PIFA designs are explained in [11–22]. In [23] explained the fractal-based MIMO for the 5G regions [24], a CPW type-fed MIMO system using parasitic arrays [25], a broadband graphene material absorber designed [26], a semi-circular type MIMO designed [27], using the approach of deep learning assisted MIMO system [29], a spiritual shape compact MIMO system [30], fractal type MIMO structure for suppression characteristics [31], a two-port flower structure with applications of Wi-Max [32], nonagonal patch MIMO design using four elements [33], fractal-based MIMO system experimental verification and designed for sub 6-GHz [34], a crescent shape MIMO structure designed for 5G systems [35], dual-band MIMO radiator with dual-band operation for 5G and radar systems [36], a graphene based micro-sized MIMO system designed [37], a penta-band compact monopole antenna designed for communication systems [38], interference inherent MIMO design with DGS [39], miniaturized multi-frequency band-printed system for WLAN region [40], using the technique of DGS designed UWB system [41], a half-ring structure MIMO system [42], a system of dual-band with dual-element designed with loaded type loops [43]. In the same manner, it designed various MIMO configurations using defected ground structure, EBG structure, spacing among the elements, and various other explanations reported [44–53].

6.2 Geometry of the MIMO System

The designed system with compact size 40 mm $\times$ 40 mm is considered shown in Fig. 6.1. The front side of the design shown in Fig. 6.1a using FR-4 substrate and patch and ground considered as perfect electric conductor material. The designed patch consists of 4 rectangular slots are considered and arranged one parallel to other on the substrate edges based on the optimization selection procedure with proper element spacing based on the mathematical back ground. The back side of plane shown Fig. 6.1b which consists of four rectangular patches at front side arranged the four rectangular ground planes at the bottom side in the form of defected ground structure for producing the better isolation and improve antenna performance at resonant band frequency. To arrange rectangular slots on the substrate material considered the element spacing is a crucial criterion among patches arrangement. Table 6.1 shows designed structure dimensions are measured in terms of mm.

Figure 6.2a shows the current designed system arrangement S-parameters comparison which identified at the required frequency of 3.3 GHz obtained S_{11} is – 34.37 dB and its corresponding S_{21} is – 32.42 dB. The obtained S-parameters are used in the applications wavelength of n 77 (90–70 mm) and n 78 (90–70 mm) bands. Figure 6.2b shows the proposed system VSWR and at the resonate wavelength its standing wave ratio is ≤ 2. Figure 6.3 shows the radiation patterns of 4-port rectangular slots inserted MIMO system E-filed and H-filed representations in xz plane and yz planes. These radiation patterns are obtained by considering in first case phi $= 0°$ and phi $= 90°$.

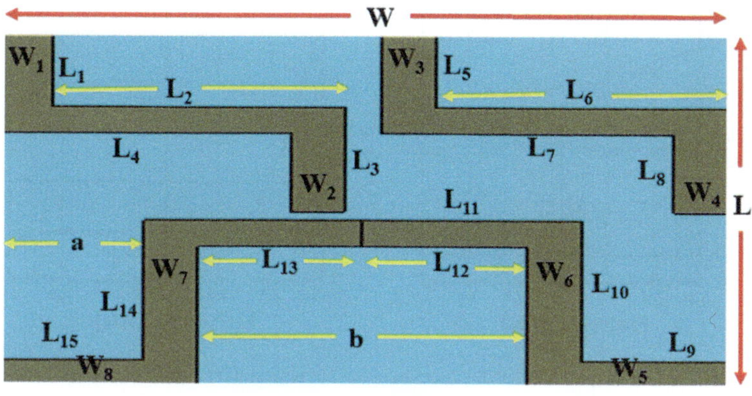

(a) Front-side

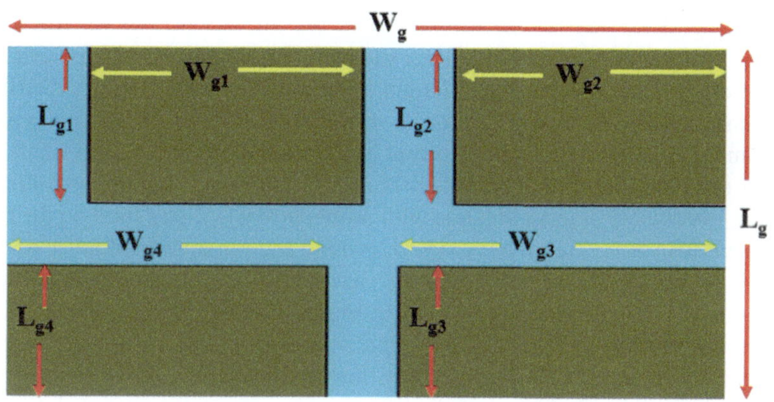

(b) Back-view

Fig. 6.1 Four-port rectangular slots MIMO design

Table 6.1 Dimensions of 4-element MIMO antenna design

Symbol	W	L	W_1	W_2	W_3	W_4	W_5	W_6	W_7	W_8	W_{g1}	W_{g4}	L_{g3}
Value	40	40	3	3	3	3	2	3	3	2	20	25	15
Symbol	L_1	L_2	L_3	L_4	L_5	L_6	L_7	L_8	L_9	L_{10}	W_{g2}	L_{g1}	L_{g4}
Value	8	16	8	18	8	16	18	8	8	10	20	15	15
Symbol	L_{11}	L_{12}	L_{13}	L_{14}	L_{15}	a	b	L_g	W_g	ε_r	W_{g3}	L_{g2}	h
Value	12	8	8	10	8	7	15	40	40	4.3	25	15	1.6

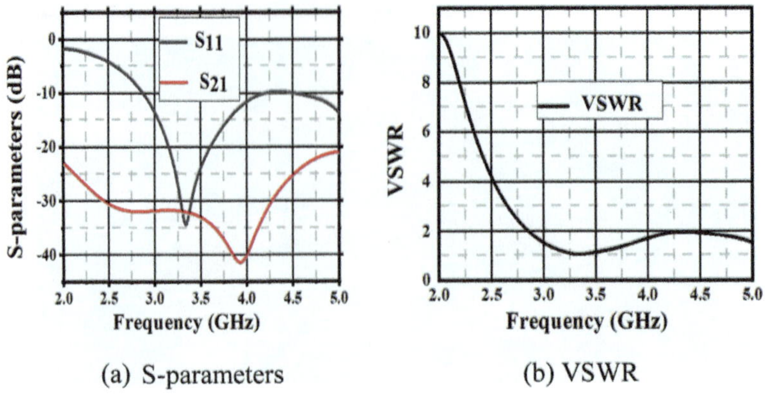

Fig. 6.2 Designed structure antenna parameters

The output of radiation patterns is look like an omnidirectional pattern and nearly figure of eight at the resonate wavelength at 90 mm. Figure 6.4 shows distribution related surface current wavelength at 90 mm. Figure 6.4a identifies that port_1 at front plane with more appreciated current is distributed on the 'L' strip shape structure and remaining all other ports less current is observed. Figure 6.4b identifies that port_1 is excited at ground plane with more current is distributed on the rectangular shape strip structure on back side of ground and remaining all other ports less current is observed. Figure 6.4c identifies that port_2 is excited at front plane with more current is distributed on the 'L' strip shape structure and remaining all other ports less current is observed. Figure 6.4d identifies that port_1 is excited at ground plane with more current is distributed on the rectangular shape strip structure on back side of ground and remaining all other ports less current is observed. Similarly, the same observation is identified that for the remaining ports 3 and 4 on front and back side of the planes more and a smaller number of currents are flows through the designed structure.

6.3 MIMO Analysis

Figure 6.5 represents analysis of current designed structure MIMO system. Figure 6.5a shows combination of real, imaginary part of designed structure at the resonant wavelength at 90 mm are 80 ohms and − 20 ohms which produce the better values. Figure 6.5b shows ECC of design versus frequency. The ECC among the two arbitrary radiating elements of *I* and *j* which can evaluated using *S*-parameters and complex radiation patterns with Eqs. (6.1)–(6.2) as [45]. The diversity gain also derived from the analysis of ECC which can be evaluated using Eq. (6.3) as [44].

E-filed
H-filed
Co-Polarization
Cross-Polarization

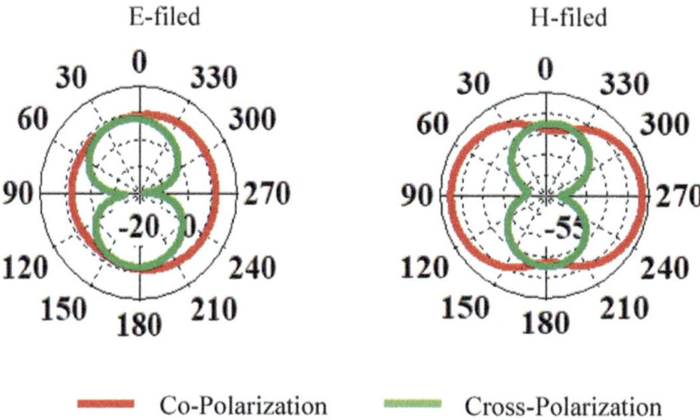

Fig. 6.3 Radiation patterns at wavelength of 90 mm

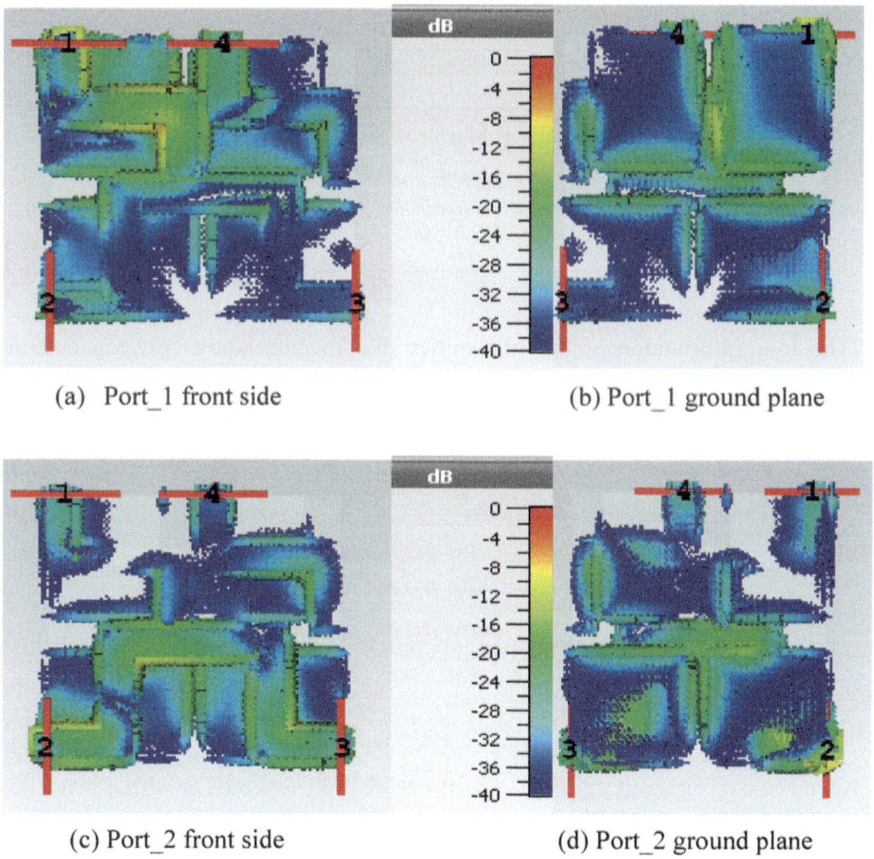

dB
(a) Port_1 front side
(b) Port_1 ground plane
(c) Port_2 front side
(d) Port_2 ground plane

Fig. 6.4 Surface current distributions of 4-element rectangular slots design

$$\mathrm{ECC}_{i,j} = \frac{\left|\iint_{4\pi}\left[\vec{E}_i(\theta,\phi)\cdot\vec{E}_j(\theta,\phi)\mathrm{d}\Omega\right]\right|^2}{\iint_{4\pi}\left|\vec{E}_i(\theta,\phi)\right|^2 d\Omega \iint_{4\pi}\left|\vec{E}_j(\theta,\phi)\right|^2 \mathrm{d}\Omega} \tag{6.1}$$

$$\mathrm{ECC}_{i,j} = \frac{\left|S_{ii}^* S_{ij} + S_{ji}^* S_{jj}\right|^2}{\left(1 - |S_{ii}|^2 - \left|S_{ji}\right|^2\right)\left(1 - \left|S_{jj}\right|^2 - \left|S_{ij}\right|^2\right)} \tag{6.2}$$

$$\mathrm{DG}_{i,j} = 10\sqrt{1 - \left|\mathrm{ECC}_{i,j}\right|^2} \tag{6.3}$$

For the performance of MIMO system, TARC is an important factor which considered into account of all S-parameters which can be evaluated by using Eqs. (6.4–6.5) as [49].

$$\mathrm{TARC} = \frac{\sqrt{\sum_{i=1}^{N}\left|S_{lm} + S_{lm}e^{j\theta_{m-1}}\right|}}{\sqrt{N}} \tag{6.4}$$

$$\mathrm{TARC} = \frac{\sqrt{\begin{array}{l}\left|S_{11} + S_{12}e^{j\theta_1} + S_{12}e^{j\theta_2} + S_{14}e^{j\theta_3}\right|^2 \\ +\left|S_{21} + S_{22}e^{j\theta_1} + S_{23}e^{j\theta_2} + S_{24}e^{j\theta_3}\right|^2 \\ +\left|S_{31} + S_{32}e^{j\theta_1} + S_{33}e^{j\theta_2} + S_{34}e^{j\theta_3}\right|^2 \\ +\left|S_{41} + S_{42}e^{j\theta_1} + S_{43}e^{j\theta_2} + S_{44}e^{j\theta_3}\right|^2\end{array}}}{\sqrt{4}} \tag{6.5}$$

The loss information at the transmitter due to correlation between antenna elements represents channel capacity loss which can be derived by using Eqs. (6.6–6.9) as [51].

$$C_{\mathrm{loss}} = -\log_2 \det\left(\Psi^R\right) \tag{6.6}$$

$$\Psi^R = \begin{bmatrix} \rho_{11} & \rho_{12} & \rho_{13} & \rho_{14} \\ \rho_{21} & \rho_{22} & \rho_{23} & \rho_{24} \\ \rho_{31} & \rho_{32} & \rho_{33} & \rho_{34} \\ \rho_{41} & \rho_{42} & \rho_{43} & \rho_{44} \end{bmatrix} \tag{6.7}$$

$$\rho_{ii} = 1 - \left(\sum_{j=1}^{4}\left|S_{ij}\right|^2\right) \tag{6.8}$$

$$\rho_{i,j} = -\left(S_{ii}^* S_{ij} + S_{ji}^* S_{jj}\right) \tag{6.9}$$

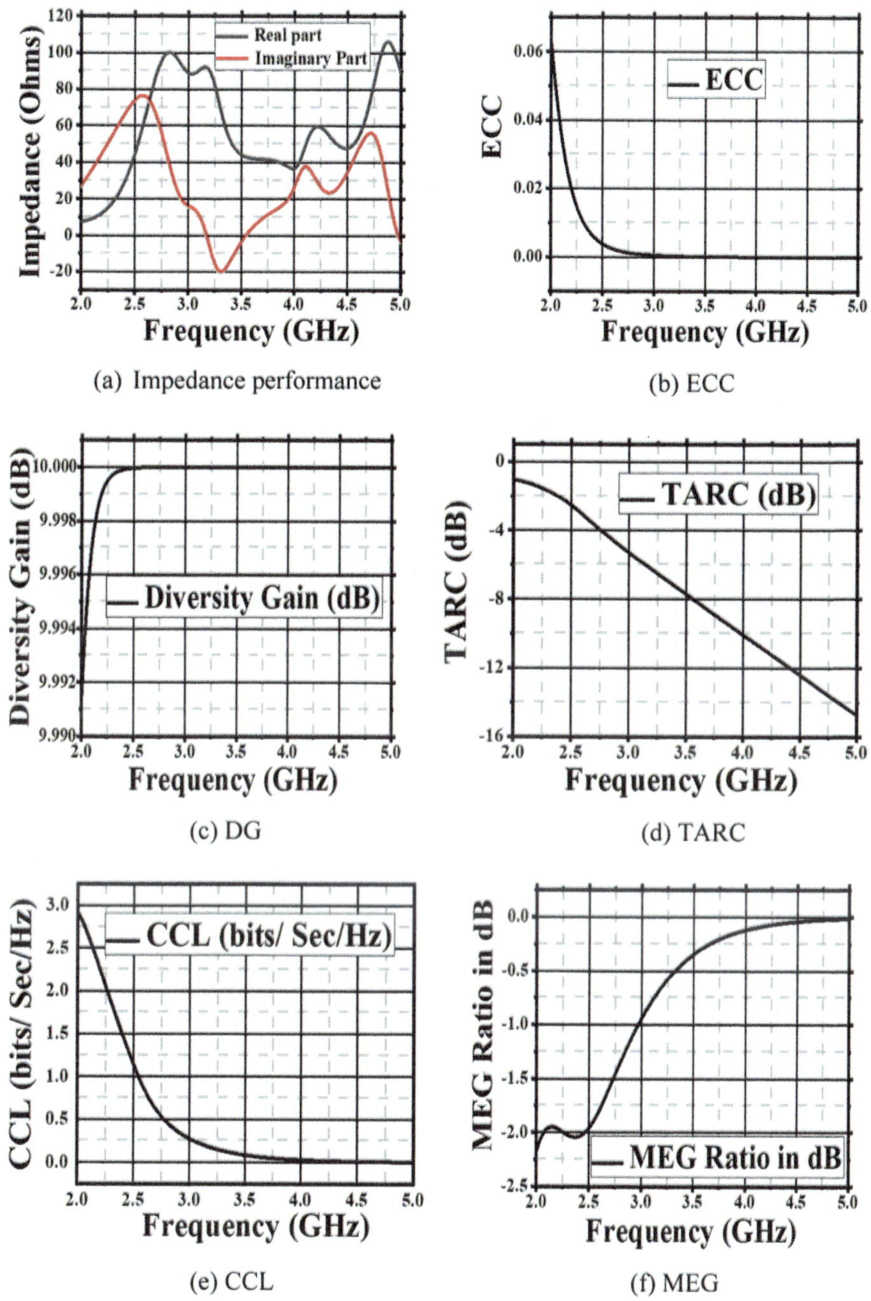

Fig. 6.5 MIMO representation of current structure

Mean effective gain of the system for the 4 antenna elements can be evaluated using the expression (6.10) as [52].

$$\mathrm{MEG}_i = 0.5\eta_{i,\mathrm{rad}} = 0.5\left[1 - \sum_{j=1}^{M} \left|S_{ij}\right|^2\right] \tag{6.10}$$

6.4 Conclusion

The 4-element rectangular slots of the current design have been investigated by placing the all four rectangular slots in the form of orthogonal fashion for the better mutual coupling as well as for better MIMO analysis system. The back plane of the designed system has been investigated the monopole defected ground plane of the system under the patch elements. The acceptable limit of its S-parameters S_{11} is − 34.37 dB and S_{21} is − 32.42 dB has been identified at resonant band frequency. For designed structure, it have been noticed that wavelength at 90 mm the obtained MIMO parameters.

References

1. Babu, K.V., Anuradha, B.: Design of multi-band minkowski MIMO antenna to reduce the mutual coupling. J. King Saud Univ. Eng. Sci. **32**(1), 51–57 (2020)
2. Babu, K.V., et al.: Compact dual-band design and analysis of half-circular U-shape MIMO radiator for wireless applications. Microsyst. Technol. **29**(4), 501–514 (2023)
3. Yadav, V., et al.: Dual and wideband 6-port MIMO antenna for WiFi, LTE and carrier aggregation systems applications. AEU-Int. J. Electron. Commun. **162**, 154576 (2023)
4. Babu, K. V., et al.: Deep learning assisted fractal slotted substrate MIMO antenna with characteristic mode analysis (CMA) for Sub-6 GHz n78 5 G NR applications: design, optimization and experimental validation. Phys. Script. **98**(11), 115526 (2023)
5. Babu, K.V., et al.: Design and analysis of fractal-based THz antenna with co-axial feeding technique for wireless applications. In: Recent Advances in Graphene Nanophotonics, pp. 351–358. Springer Nature Switzerland, Cham (2023)
6. Chae, S.H., Oh, S., Park, S.-O.: Analysis of mutual coupling, correlations, and TARC in WiBro MIMO array antenna. IEEE Antennas Wirel. Propag. Lett. **6**, 122–125 (2007)
7. Raj, U., et al.: Easily extendable four port MIMO antenna with improved isolation and wide bandwidth for THz applications. Optik **247**, 167910 (2021)
8. Babu, K.V., Anuradha, B.: Design and analysis of multi-band circle shape MIMO antenna using defected ground structure to reduce mutual coupling. Int. J. Ultra Wideband Commun. Syst. **4**(1), 32–40 (2019)
9. Khan, M.A., et al.: mmWave four-element MIMO antenna for future 5G systems. Appl. Sci. **12**(9), 4280 (2022)
10. Kulkarni, N., Linus, R.M., Bahadure, N.: A small wideband inverted l-shaped flexible antenna for sub-6 GHz 5G applications. AEU Int. J. Electron. Commun. **159**, 154479 (2023)

11. Babu, K.V., Anuradha, B., Naga Jyothisree, G.: Improved return loss and reduction of mutual coupling of microstrip MIMO antenna for C-band applications. Int. J. Electr. Electron. Eng. Telecommun. **6**(2), 43–49 (2017)
12. Yuan, X.-T., et al.: A wideband PIFA-pair-based MIMO antenna for 5G smartphones. IEEE Antennas Wirel. Propag. Lett. **20**(3), 371–375 (2021)
13. Ishfaq, M.K., et al.: Multiband split-ring resonator based planar inverted-F antenna for 5G applications. Int. J. Antennas Propag. **2017**, 1–7 (2017)
14. Babu, K.V., Sree, G.N.J.: Design and circuit analysis approach of graphene-based compact metamaterial-absorber for terahertz range applications. Opt. Quantum Electron. **55**(9), 769 (2023)
15. Deng, J., et al.: A dual-band inverted-F MIMO antenna with enhanced isolation for WLAN applications. IEEE Antennas Wirel. Propag. Lett. **16**, 2270–2273 (2017)
16. Babu, K.V., et al.: Design of monopole ground graphene disc-inserted THz antenna for future wireless systems. In: Recent Advances in Graphene Nanophotonics, pp. 305–312. Springer Nature Switzerland, Cham (2023)
17. Kumar, N., Khanna, R.: A two element MIMO antenna for sub-6 GHz and mm wave 5G systems using characteristics mode analysis. Microw. Opt. Technol. Lett. **62**(2), 587–595 (2021)
18. Yacoub, A., Khalifa, M., Aloi, D.N.: Wide bandwidth low profile PIFA antenna for vehicular sub-6 GHz 5G and V2X wireless systems. Prog. Electromag. Res. C **109**, 257–273 (2021)
19. Babu, K.V., Anuradha, B.: Design of nine-shaped MIMO antenna using parasitic elements to reduce mutual coupling. In: Optical and Wireless Technologies: Proceedings of OWT 2019. Springer Singapore, Singapore (2020)
20. Gundumalla, A., Agrawal, S., Parihar, M.S., Miniaturized active stepped impedance planar inverted-F antenna using common ground. AEU Int. J. Electron. Commun. **83**, 233–239 (2018)
21. Chattha, H.T.: 4-Port 2-element MIMO antenna for 5G portable applications. IEEE Access **7**, 96516–96520 (2019). https://doi.org/10.1109/ACCESS.2019.2925351
22. Zhang, X., et al.: Ultra-wideband 8-port MIMO antenna array for 5G metal-frame smartphones. IEEE Access **7**, 72273–72282 (2019)
23. Ali, W.A.E., Ibrahim, A.A.: A compact double-sided MIMO antenna with an improved isolation for UWB applications. AEU Int. J. Electron. Commun. **82**, 7–13 (2017)
24. Sree, G.N.J., Nelaturi, S.:Opportunistic control of fractal-based MIMO antenna for sub-6-GHz 5G applications. Int. J. Commun. Syst. **34**(17), e4991 (2021)
25. Babu, K.V., et al.: Design and analysis of a CPW-fed fractal MIMO THz antenna using an array of parasitic elements. In: Terahertz Devices, Circuits and Systems: Materials, Methods and Applications, pp. 53–60. Springer Nature Singapore, Singapore (2022)
26. Babu, K.V., et al.: Design of graphene-based broadband metamaterial absorber with circuit analysis approach for Terahertz region applications. Opt. Quant. Electron. **55**(13), 1188 (2023)
27. Sree, G.N.J., Nelaturi, S.: Semi-circular MIMO patch antenna using the neutralization line technique for UWB applications. In: Microelectronics, Electromagnetics and Telecommunications, pp. 175–182. Springer, Singapore (2020)
28. Sree, G.N.J., et al.: Design and optimization of a deep learning algorithm assisted stub-loaded dual band four-port MIMO antenna for sub-6GHz 5G and X band satellite communication applications. AEU Int. J. Electron. Commun. 155074 (2023)
29. Babu, K.V., et al.: Design and implementation of MIMO graphene patch antenna to improve isolation for THz applications. Microsyst. Technol. 1–11 (2023)
30. Sree, J., Naga, G., Nelaturi, S.: Spiritual leaf shape compact MIMO patch antenna for 5G lower sub-6 GHz applications. Frequenz **77**(3–4), 185–201 (2023)
31. Babu, K.V., et al.: Design and analysis of fractal type MIMO radiator for the applications of sub 6-GHz 5G systems. In: Microelectronics, Circuits and Systems: Select Proceedings of Micro2021, pp. 243–250. Springer Nature Singapore, Singapore (2023)
32. Sree, G.N.J., Suman, N.: Design of two-port flower shaped MIMO antenna suppression characteristics with Wi-MAX applications. J. Instrum. **15**(04), P04018 (2020)
33. Addepalli, T., et al.: Design and analysis of nonagonal patch unite with rectangular shaped 4-element UWB-MIMO antenna for portable wireless device applications. Analog Integ. Circ. Sig. Process. **114**(3), 459–473 (2023)

34. Sree, G.N.J., Nelaturi, S.:Design and experimental verification of fractal based MIMO antenna for lower sub 6-GHz 5G applications. AEU Int. J. Electron. Commun. **137**, 153797 (2021)
35. Sree, G.N.J., Nelaturi, S.: Opportunistic control of crescent shape MIMO design for lower sub 6 GHz 5G applications. Microw. Opt. Technol. Lett. **64**(5), 896–904 (2022)
36. Babu, K.V.: Kokkirigadda, S., Das, S.: Design and simulation of dual-band MIMO antenna for radar and sub-6-GHz 5G applications. In: Futuristic Communication and Network Technologies: Select Proceedings of VICFCNT 2020. Springer Singapore, Singapore (2022)
37. Vasu Babu, K., et al.: A micro-scaled graphene-based tree-shaped wideband printed MIMO antenna for terahertz applications. J. Comput. Electron. **21**(1), 289–303 (2022)
38. Das, S., et al.: A compact penta-band printed monopole antenna for multiple wireless communication systems. In: Futuristic Communication and Network Technologies: Select Proceedings of VICFCNT 2020. Springer Singapore, Singapore (2022)
39. Babu, K.V., Anuradha, B.: Design of MIMO antenna to interference inherent for ultra wide band systems using defected ground structure. Microw. Opt. Technol. Lett. **61**(12), 2698–2708 (2019)
40. Das, S., et al.: Analysis of a miniaturized modified multifrequency printed antenna with broadband characteristics for WLAN application. In: Advances in Power Systems and Energy Management: Select Proceedings of ETAEERE 2020. Springer Singapore, Singapore (2021)
41. Babu, K.V., Anuradha, B.: "Design of UWB MIMO antenna to reduce the mutual coupling using defected ground structure. Wirel. Pers. Commun. **118**(4), 3469–3484 (2021)
42. Babu, K.V., Anuradha, B.: Design of half-ring MIMO antenna to reduce the mutual coupling. In: Proceedings of International Conference on Artificial Intelligence, Smart Grid and Smart City Applications: AISGSC 2019. Springer International Publishing (2020)
43. Sharawi, M.S., Numan, A.B., Aloi, D.N.: Isolation improvement in a dual-band dual-element MIMO antenna system using capacitively loaded loops. Prog. Electromag. Res. **134**, 247–266 (2013)
44. Babu, K.V., Anuradha, B.: A dual-band Minkowski-shaped MIMO antenna to reduce the mutual coupling. In: Optical and Wireless Technologies: Proceedings of OWT 2018. Springer Singapore, Singapore (2020)
45. Li, Y., Yang, G.: Dual-mode and triple-band 10-antenna handset array and its multiple-input multiple-output performance evaluation in 5G. Int. J. RF Microw. Comput. Aided Eng. **29**(2), e21538 (2019)
46. Sun, L., et al.: Compact 5G MIMO mobile phone antennas with tightly arranged orthogonal-mode pairs. IEEE Trans. Antennas Propag. **66**(11), 6364–6369 (2018)
47. Babu, K.V., Anuradha, B.: Design of annular ring MIMO antenna using defected ground structure. In: 2019 TEQIP III Sponsored International Conference on Microwave Integrated Circuits, Photonics and Wireless Networks (IMICPW). IEEE (2019)
48. Babu, K.V., Anuradha, B.: Design & isolation reduction of circle inserted MIMO antenna. In: 2019 IEEE International Conference on System, Computation, Automation and Networking (ICSCAN). IEEE (2019)
49. Babu, K.V., Anuradha, B.: Design of dual-band MIMO antenna for LTE 2500, Wimax, and C-band applications to reduce the mutual coupling. In: Microelectronics, Electromagnetics and Telecommunications: Proceedings of the Fourth ICMEET 2018. Springer Singapore, Singapore (2019)
50. Zhao, A., Ren, Z.: Size reduction of self-isolated MIMO antenna system for 5G mobile phone applications. IEEE Antennas Wirel. Propag. Lett. **18**(1), 152–156 (2018)
51. Chattha, H.T., et al.: Polarization and pattern diversity-based dual-feed planar inverted-F antenna. IEEE Trans. Antennas Propag. **60**(3), 1532–1539 (2011)
52. Babu, K.V., Anuradha, B.: Design of rectangular MIMO antenna for bluetooth and WLAN applications to reduce the mutual coupling. In: Engineering Vibration, Communication and Information Processing: ICoEVCI 2018, India. Springer Singapore, Singapore (2019)
53. Babu, K.V., Anuradha, B.: Design of dual-band MIMO antenna for LTE 2500, Wimax, and C-band applications to reduce the mutual. In: Microelectronics, Electromagnetics and Telecommunications: Proceedings of the Fourth ICMEET 2018, vol. 521, p. 107 (2018)

Chapter 7
Beamforming Array Failure Correction for mm-Wave Synthetic Aperture Radar Applications

Hina Munsif, Raja Aasim Bin Saleem, Arslan Ali Shah, Shahid Khattak, Ali Imran Najam, Benjamin D. Braaten, and Irfanullah

7.1 Introduction

Millimeter-wave (mm-wave) beamforming arrays are highly advanced antennas due to their small size and wider bandwidth. Designed carefully to control electromagnetic waves with great precision, operating within the frequency spectrum spanning between 30 and 300 GHz. These arrays find versatile applications in many fields such as unmanned aerial vehicles (UAVs) communication [1], 5G technology [2],

H. Munsif · R. A. B. Saleem · A. A. Shah · S. Khattak · Irfanullah (✉)
Department of Electrical and Computer Engineering, COMSATS University Islamabad, Abbottabad Campus, Islamabad, Pakistan
e-mail: eengr@cuiatd.edu.pk

H. Munsif
e-mail: hinaa3387@gmail.com

R. A. B. Saleem
e-mail: rajaasim05@gmail.com

A. A. Shah
e-mail: shaharslan884@gmail.com

S. Khattak
e-mail: skhattak@cuiatd.edu.pk

A. I. Najam
RF & Microwave Department, National Electronics Complex of Pakistan (NECOP), Islamabad, Pakistan
e-mail: alimranajam@gmail.com

B. D. Braaten
Department of Electrical and Computer Engineering, North Dakota State University, Fargo, ND 58102, USA
e-mail: benjamin.braaten@ndsu.edu

M. El Ghzaoui et al. (eds.), *Next Generation Wireless Communication*, Signals and Communication Technology, https://doi.org/10.1007/978-3-031-56144-3_7

MIMO antenna arrays [3], and synthetic aperture radar (SAR) [4]. The recent rise in mm-wave technology has gained significant importance in the field of SAR applications. The primary focus of mm-wave SAR system is to overcome the inherent challenges associated with achieving high-resolution imaging and increased sensitivity. It is important to note that in radar systems the distance between the transmitting antenna array and the receiving locations is affected by the incident angle. When an antenna array with a wider half-power beamwidth (HPBW) is used as a transmitter, a greater quantity of radiated power is required to cover the outer edges in the field of view (FoV). As a result, the received power at nearer points exceeds the actual requirements, reducing the entire system's power efficiency dramatically. To address the significant propagation loss and issues involved with generating significant radiated power inside the mm-wave bands, customized antennas with a specifically designed cosecant square pattern were devised. This pattern shape is suitable for "one-way" communication applications where distance between transmit array and target is R, but the received power is inversely related to R^2.Please check and confirm if the author names and initials are correct both in the author group and in the references section.ConfirmedKindly update the author name "Irfanullah" with initial(s).The author has a single name 'Irfanullah'Please check and confirm the inserted city name is correct.Confirmed

7.1.1 Significance of Cosecant 4th Power Pattern

In one-way communication applications like broadcasting systems, as shown in Fig. 7.1a, the received power is inversely proportional to the square of the distance ($\mathbf{R}^2$) between the radar and the target, as demonstrated in [4]. In this "one-way" radar system, the array is used only for transmitting the signals. Therefore, to send a strong signal from the radar to the receiving side, the cosecant power pattern is used, which directs the signal in a specific direction to have constant received power in FoV and gradually reduces its strength as receiving point move away from that direction to limit the interference. Using the cosecant power pattern for transmission, the received power at two points (**a1**, **b1**) within the FoV will be same, if both receiving antennas have the same gain. In Fig. 7.1a, the distances from transmitting array to receiving point **a1** and the receiving point **b1** are $\mathbf{R}_{\mathbf{a}1}$ and $\mathbf{R}_{\mathbf{b}1}$, respectively. Even though $\mathbf{R}_{\mathbf{a}1}$ is slightly shorter than $\mathbf{R}_1$, if both points have receiving antennas with the same gain, then received power at both points will be same. When the cosecant power pattern is used in a one-way communication system, the change in the distances (**R**) from the transmitter to receiving points within the FoV have a negligible effect on received power. This happens because in one-way communication, the receiving array directly receives transmitted signal from the transmitter without any scattering.

On the other hand, in "two-way" communications like typical SAR, both transmitting and receiving arrays are placed on the same side of UAV, as shown in Fig. 7.1b. In this case, the signals travel from the transmitting antenna to target and then reflect back to the receiving antenna. As a result, the received power inversely relates with

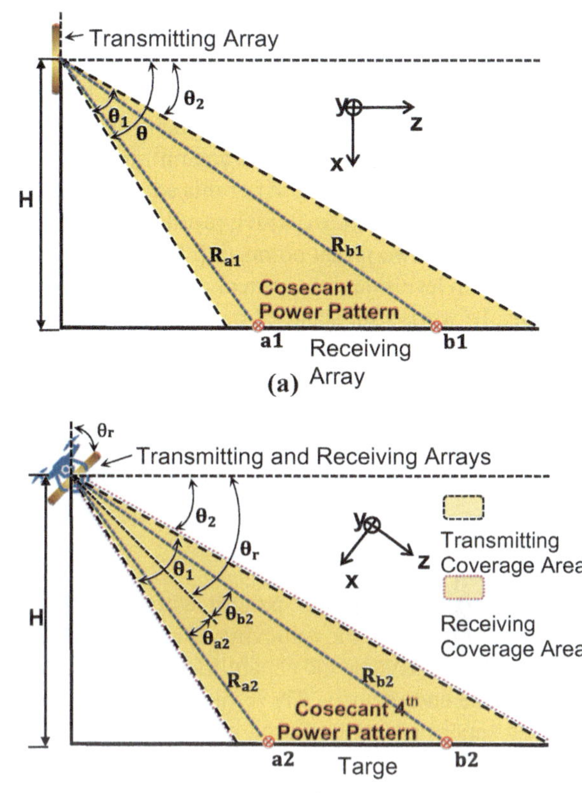

Fig. 7.1 **a** $\mathbf{R^2}$ (One-way) communication with cosecant power pattern, **b** $\mathbf{R^4}$ (two-way) communication with cosecant 4th power pattern (**H** is height of array position, $\boldsymbol{\theta}$ is any angle between the normal direction of array to cosecant power area, $\boldsymbol{\theta}_1$ is end angle and $\boldsymbol{\theta}_2$ is starting angle of cosecant/cosecant 4th power area, $\boldsymbol{\theta}_r$ is rotating angle of radar in cosecant 4th power area, and $\boldsymbol{\theta}_{a2}$ and $\boldsymbol{\theta}_{b2}$ are received power angles from target point **a2** and **b2**)

$\mathbf{R^4}$, where the power changes more rapidly even with a small change in the distance from any point within FoV to receiving array. In such communications, the power received due to reflection of the signals between (**a2**, **b2**) target points cannot be identical, because the transmitting array uses a cosecant power pattern, which is associated with $\mathbf{R^2}$ communication. This can lead to poor image resolution and decrease sensitivity. Therefore, to meet the practical demands of mm-wave SAR systems, it is advantageous to have a more concentrated and narrower beam than the cosecant power pattern, capable of providing uniform received power distribution within the FoV. To address this issue, the cosecant power pattern is modified to a precisely designed cosecant 4th power pattern in [4]. Cosecant 4th power pattern provides, (1) high distance sensitivity: enhance the sensitivity with a change in distance to measure highly accurate distance which is very important in mm-wave SAR for generating high resoluntion images and detecting targets, (2) highly focused narrow beam: cosecant 4th power pattern has narrower HPBW than the cosecant power pattern, to enhance mm-wave SAR ability to detect and generate high-resolution images, and (3) beamforming: used to better control over the direction and strength of the transmitted signal, because it helps with creating images and keeping track of target.

In SAR applications, the areas covered by the transmitting and receiving arrays do not overlap due to vertical placement of the radar [4]. To make the receiving coverage area overlap with the transmitting coverage area, the radar is tilted at an angle $\boldsymbol{\theta}_{\mathbf{r}}$, shown in Fig. 7.1b. After this adjustment, the transmitting array sends out a cosecant 4th power pattern, which results in a uniform received power from scattered signals through two target points (**a2**, **b2**) within FoV, if **a2** and **b2** have same radar cross-section levels, and the gain of receiving array antenna at two angles $\boldsymbol{\theta}_{\mathbf{a}2}$ and $\boldsymbol{\theta}_{\mathbf{b}2}$ is same. To work with scanned pattern of transmitting array, receiving array needs a wider HPBW in *xyz*-plane, to capture more scattered signals from target. Since, the received power from the FoV is identical, therefore beamforming only used in transmitting array.

7.1.2 Element Failure

The cosecant fourth power pattern with low-profile microstrip patch array offers promising prospects for achieving enhanced radar resolution and sensitivity. This advancement enhances the capabilities of mm-wave systems in critical applications like defense and security. These radar arrays consist of multiple antenna elements, provide superior signal control, and offer significant advantages. mm-wave SARs operate optimally when all elements within the arrays are working in an active state.

The radar systems utilize semiconductor technology such as CMOS/Bi-CMOS in electronic components like transceivers, low-noise amplifiers/power amplifiers (LNAs/PAs) and TRMs, which offers compelling advantages like minimized power consumption [5]. Furthermore, these transistors facilitate continuous transitions between high-power pulse transmission and the reception of delicate echo signals, ensuring operational flexibility [6]. mm-Wave SAR technology, known for its adaptability in high resolution and sensitivity for generating narrow beams and maintaining low side lobe level (SLL) through TRMs by efficiently managing the weights comprising amplitudes and phases. However, the research work in [7–11] suggest that the adoption of Bi-CMOS/CMOS-based radar design introduces potential challenges to compromise reliability. These challenges involve mechanisms that can compromise solid-state devices, including semiconductors, RF components, and integrated circuits. Notably, CMOS transistor switching within TRMs can cause potential failures within the radar architecture, such as overvoltage, overcurrent, hot carrier injection, electrostatic discharge (ESD), gate oxide breakdown, thermal stress, noise, and cross-talk. These complications may lead to premature or permanent failure of antenna array elements, thereby affecting the complete radiation pattern. This can subsequently result in the rise of SLLs, wider beam width, and a decrease in gain, while depending upon the number and arrangement of the defected elements. This affects the primary goal of mm-wave SARs to provide high-resolution and increased sensitivity.

These radar systems are mostly employed in demanding environments like battlefields, aerospace, and complicated communication applications, where the feasibility of real-time repair of damaged elements within the antenna array becomes challenging. Hence, hardware replacement is not a feasible solution for restoring array availability in such applications. Additionally, the occurrence of one or more defected/failed element in an antenna array make it a non-symmetric array, thus limiting the feasibility of implementing traditional analytical techniques.

7.2 Literature Review

Existing literature broadly uses two strategies to address antenna array element failures, first is the hardware solution and the second one is as software-based solution. In hardware-based solutions, discussed in [12, 13], the redundant/idle/backup TRMs are employed. The hardware connections pertaining to the switching of input and output signals, module detection, and power distribution systems are removed from faulty TRM and linked to backup TRMs. But, normally the hardware solutions are not preferred for pattern correction in an antenna array with failed elements, due to higher costs of the components and the complexity in real-time implementation. Therefore, software-based real-time solutions are identified. Researchers have investigated the self-restoration techniques, mainly based on optimization to restore the antenna array's desired radiation pattern with failed/inactive elements present. Consequently, a methodology was devised to restore the characteristics of the desired radiation pattern. This was achieved through the re-synthesis or recalculation of the weights (i.e., excitation coefficients) of the active array elements by using optimization algorithms, like particle-swarm optimization (PSO), genetic-algorithm (GA), bacteria-foraging (BFO) algorithm and iterative-Fourier technique (IFT). These algorithms offer software-based solutions for the recovery of desired parameters in the presence of faulty antenna elements.

7.2.1 Software-Based Antenna Array Element Failure Correction Techniques

Within the scope of discussed software-based approaches, most of the algorithms involve recalculation of the excitations (i.e., amplitude and phase) of active antenna elements in a defected array to reconstruct the desired radiation pattern. In [14], the computer coding is implemented to compute the fresh excitation angles of active/operational phase shifters. This aims to reinstate a wider null of 10° with a depth of − 66 dB in a specified direction for a 28-element linear isotropic array. The feasibility of null recovery with random element failures is demonstrated, but at the cost of increased SLL. Khan and Rahim [15] introduces a novel technique named the matrix pencil technique (MPT) for the recovery of SLL and Nulls with edge elements failure in linear array antenna configuration. Similarly, Klhan et al. [16] proposed a methodology for the recovery of patterns in the symmetrical element failures scenario within a linear array. In [17], an investigation is conducted on a linear array consist of 50 and 80 isotropic elements. The objective is to resynthesize SLL (− 45 dB) with 8 randomly failed antenna elements. Yadav et al. [18] explores inverse fast Fourier transform (IFFT) approach on a 1×34 isotropic array to recover wider multiple nulls with a desired sidelobe level. Extending this concept, Munsif et al. [19] implemented IFFT on a 1×8 linear array with 25% defected elements, to recover various SLL values (− 20, − 30, and − 40 dB) in both broadside ($\theta = 0°$) and beam steering ($\theta = 20°$ and $40°$) patterns. In the study of element failure recovery, Lozano et al. [20] carefully examines SLL and nulls recovery in the presence of 4 edge elements failure within a 1×40 array, utilizing the simulated annealing (SA) technique. Mandal et al. [21] applied the differential evolution (DE) algorithm to optimize the spacing between antenna elements in 1×30 linear half-wave dipole array, with the purpose of pattern recovery. Guney et al. [22] investigated the recovery of SLL of − 35 dB, addressing edge element failures within a 1×32 linear array, implementing an alternative search algorithm. Notably, Grewal et al. [23] used the firefly algorithm (FA) to redistribute excitation amplitudes within linear arrays. The results demonstrate a recovery of − 35 dB SLL with symmetrical failure of 10 edge elements within electronically scanned arrays. Hamici [24] introduces the recursive intelligent optimizer (RIO), a novel signal processing algorithm capable of operating in stochastic and deterministic modes. This complicated approach involves tasks like beam steering, rapid beamforming, fault tolerance for array elements, and interference rejection. These tasks are made possible through digitally controlled phase shifters. GA in [25], recovered the optimal pattern with a SLL of − 20 dB for linear isotropic arrays consisting of 18 and 44 elements. This recovery involves the resynthesize of excitations for the functional elements in the presence of edge element failure. Yeo and Lu [26] reported − 35 dB SLL recovery, while [27] showed − 25 dB SLL recovery in edge element failures of linear arrays. The comparison of two significant optimization techniques, PSO and BFO, is discussed in [28]. The re-optimization of excitation amplitudes within arrays having failed elements, resulting a successful restoration of SLLs and nulls of linear array. Acharya et al. [29] emphasized the restoration of SLL and nulls position in the desired direction through an amplitude correction technique. This research is executed on a 1×32 elements linear array operating at a frequency of 2.33 GHz. Showing the flexibility of the PSO algorithm, Acharya et al. [30–32] presented SLL recovery with symmetric and asymmetric nulls restoration in random element failure 1×20 linear array.

Similarly, in [33] PSO is implemented for the recovery of SLL, SLL with symmetric nulls, and SLL with asymmetric nulls of 1×8 and 1×32 linear arrays, with a primary focus on edge element failure. In [34, 35], main emphasis is on recovery of SLL and gain in both broadside and scanned patterns. This recovery is achieved through the perturbation of phases at random positions in linear arrays. However, this approach involved measuring the output power of the array, but this required calibration hardware setup. Jiang et al. [36] redistributed amplitudes across functional TRMs, recovering asymmetrical nulls at − 50° and 45° in linear arrays through the non-uniform norm constrained least mean square (NU-NCLMS) technique and the genetic algorithm, while compromising the SLL. Khan et al. [37] emphasized on restoration of single and double symmetrical nulls. This correction is investigated by the cultural algorithm in conjunction with differential evolution for linear arrays. An enhanced version of bat algorithm is introduced in [38], which is designed for random element failure recovery, specifically targeting SLL recovery of − 20 dB within linear array. Greda et al. [39] used particle-swarm optimization algorithm and Engroff et al. [40] used neural network technique to explore the restoration of a scanned pattern with SLL of − 20 dB and a cosecant squared shape pattern.

A significant portion of the existing academic literature is primarily concerned with addressing element failures within linear arrays, where various optimization techniques are implemented to reconfigure complex weights among functional elements or TRMs. The ultimate objective of these research works is to recover array pattern parameters such as gain, SLLs, and nulls in linear arrays, while the application of mm-wave SAR using planar array, which remains relatively less explored in terms of recovery of SLL, HPBW and gain in the presence of element. A novel approach is introduced in [41], where pattern is recovered by implementing the sparse-recovery methodology on 17 × 17 planar array configuration. This approach succeeded in achieving SLL values of − 20 and − 40 under conditions of random edge element failures. However, the study did not report gain/HPBW recovery, and scanned pattern recovery. Furthermore, Keizer [42] conducted a study exploring the SLL (− 45 dB) and gain of a rectangular planar array consisting of 5800 elements. The investigation applied the inverse Fourier transform technique in the presence of randomly defected 290 elements. However, the positions of the failed elements were not explicitly mentioned, and the study did not include scenarios involving scanning with element failures. Edge elements failure recovery through SA technique is reported in [43] for linear and planar arrays. The study achieved two outcomes: first, a redistribution of amplitude-only leading to an SLL improvement of − 34.5 dB; second, a phase-only redistribution resulting in an SLL improvement of − 28 dB. However, the study did not explore the recovery of gain or the restoration of scanned patterns. In [44], efforts were directed toward improving Signal-to-Noise Ratio (SNR) in scenarios involving random element failure within planar arrays. To achieve this, GA, PSO, and SA algorithms were investigated. Notably, the study did not investigate SLL and gain recovery. Yang and Stark [45] proposed a method involving vector-space projections (VSP) for pattern recovery. The study emphasized that complete recovery is not feasible in small linear arrays (fewer than 50 elements). However, in a linear array comprising 50 elements, a 3 dB SLL improvement (SLL = − 14 dB) was achieved, while in a 100-element array, a 7 dB SLL improvement was attained along with an SLL of − 21 dB. However, in planar array SLL = − 22 is recoverable, but the study did not lay emphasis on gain or HPBW recovery within broadside or scanned patterns in planar arrays. In the context of Active Phased Antenna Arrays (APAAs), Chen and Tsai [46] proposed an innovative approach known as the Cumulative Sum Scheme (CUSUM). This technique is designed for monitoring phase shifters and attenuators located within TRMs. Its main function is to detect and correct failures, particularly in scenarios where a single bit is malfunctioning in attenuator. However, the study did not specifically examine complete TRM failures. It is noteworthy that a comprehensive comparison of the literature's findings related to element failure in planar arrays is detailed in Table 7.1.

In the literature, it was investigated that element failure in linear arrays are extensively explored as compared to the planar arrays, that are less studied. To the best of our knowledge, there is no literature available regarding the recovery of failed elements in 4th power pattern. Consequently, our proposed work, adopted this absence of information as a starting point and introduced the concept of element failure within [4]. The desired pattern can be recovered by redistribution of complex weights (amplitudes & phases) on working elements using non-evolutionary or evolutionary optimization algorithms. Evolutionary algorithms are preferred for element failure correction in antenna arrays due to their ability to efficiently explore large search spaces and perform global optimization. These properties make evolutionary algorithms more valuable than non-evolutionary algorithms. In evolutionary algorithms, GA provides optimal solutions than others, therefore in this research; the proposed GA is used for antenna elements failure problem on 6×12 isotropic planar array in MATLAB, to investigate SLL, gain, and HPBW recovery in cosecant fourth power pattern of base paper. Whereas 8.33–25% elements failure recovery is reported with edge element failure. The proposed GA is designed to optimize the complex weights to get desired SLL, gain and HPBW of the desired cosecant fourth power pattern for mm-wave SAR applications. These complex weights are then verified by 6×12 microstrip patch series-fed planar array in CST studio 2019, to include the real environment effects like mutual coupling, efficiency of array, and element pattern of patch antenna.

7.3 Microstrip Patch Series-Fed Antenna

Following the guidelines from [4], a 6×1 microstrip patch series-fed antenna is designed to operate at 24.3 GHz, by using the dimensions given in Fig. 7.2a. Bandwidth of single element is 1.44 GHz (23.49–24.93) GHz, presented in Fig. 7.2b, while the radiation pattern is in Fig. 7.2c. The chosen substrate for this antenna had a thickness of 0.508 mm and was made of Taconic TLY-5 material ($\varepsilon_r = 2.2$ and $\tan\delta = 0.0009$).

Additionally, this antenna is extended to planar array, composed of 6×12 microstrip patch series-fed antennas shown in Fig. 7.2d, with $\lambda_o/2$ inter-element spacing (where λ_o is free-space wavelength at 24.3 GHz). The performance parameters of 6×1 microstrip patch series-fed antenna and 6×12 microstrip patch series-fed array are reported in Table 7.2. These single antenna and array are designed/simulated in CST studio 2019 under the following setup given in Table 7.3.

Table 7.1 Critical review of existing failure correction techniques for planar antenna arrays

Research work	Techniques used	Composite factor	Array type	SLL recovery	Gain investigation	HPBW investigation	Scanning
[41]	Sparse recovery technique	Amplitude-only	Planar	✓	×	×	×
[42]	Iterative Fourier method	Amplitude-only	Planar	✓	✓	✓	×
[43]	Simulated annealing technique	Amplitude-only	Planar	✓	×	×	×
[44]	GA/PSO/SA/PS	–	Planar	×	×	×	×
[45]	Vector-space projections (VSP)	Complex weight	Linear/planar	✓	×	×	×
[46]	Cumulative sum scheme	Complex weight	Linear/planar	✓	×	×	✓
Proposed work	Genetic algorithm (GA)	Complex weight	Planar	✓	✓	✓	✓

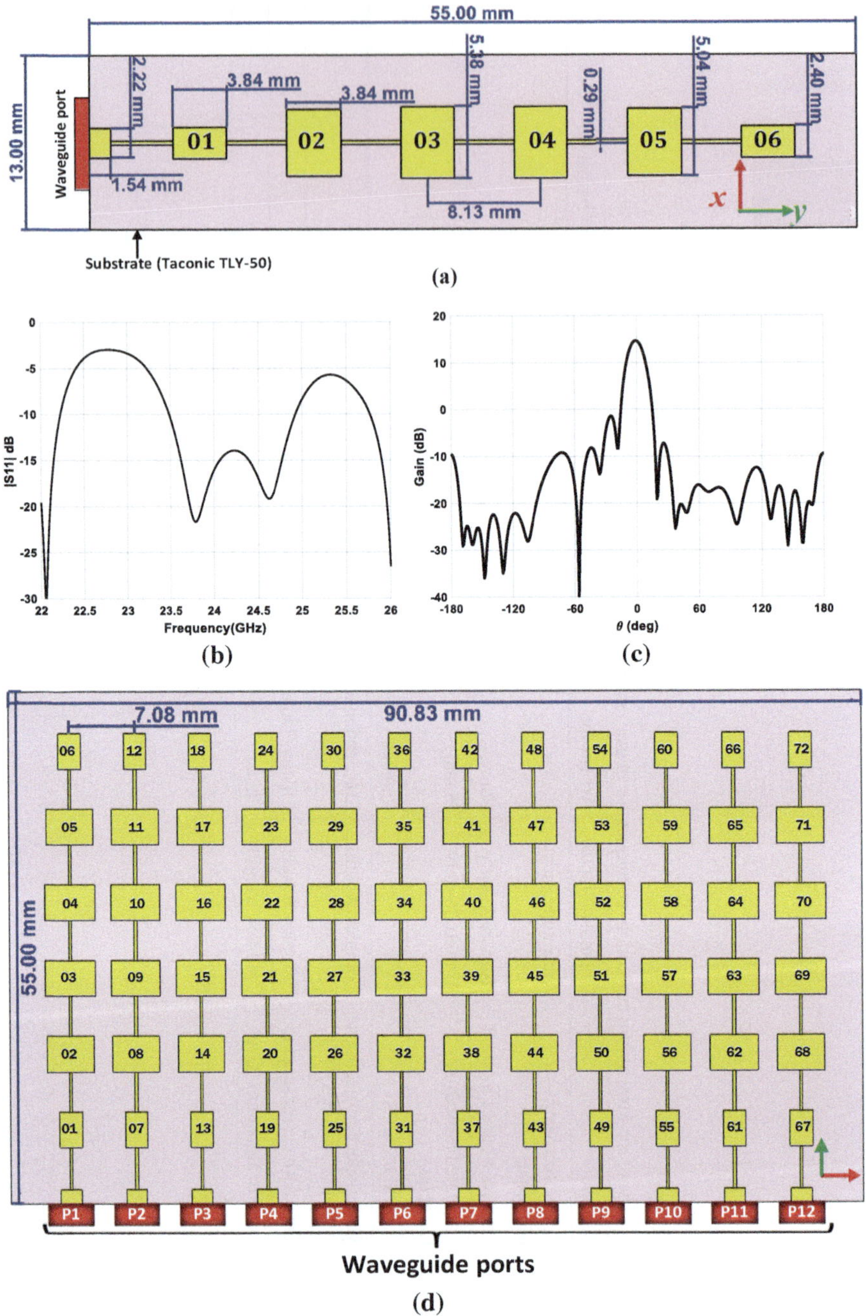

Fig. 7.2 Microstrip patch series-fed antenna array design in CST. **a** 6 × 1 microstrip patch series-fed antenna and **b** 6 × 12 microstrip patch series-fed array

Table 7.2 Characteristics of 24.3 GHz proposed series-fed microstrip patch antenna (MPA)

Performance parameters	Value (single antenna)	Value (6 × 12 array)
Bandwidth	1.44 GHz	1.44 GHz
Gain	14.7 dBi	22.9 dBi
HPBW	14.3°	9.8°
SLL	– 16.1 dB	– 20.3 dB
Directivity	15.3 dBi	23.3 dBi

Table 7.3 CST setup parameters

Setup parameter	Value
Solver	Time domain
Accuracy	– 40 dB
Number of pulses	20
Mesh type	Hexahedral
Cell/wavelength	15
Cell/maximum model box edge	20
Fraction of maximum cell near to model	20

7.4 Proposed Genetic Algorithm for Element Failure Correction

The genetic algorithm (GA) is a computational optimization method inspired by the process of natural evolution. It is an abstraction of real biological evolution. GA utilizes the basic genetic principles of selection, crossover, and mutation to drive the population toward increasingly better solutions over successive generations. By exploring and exploiting the search space effectively, it aims to find optimal or near-optimal solutions to complex optimization problems. The algorithm can be described as follows:

Initialization. A population of potential solutions (often called individuals or chromosomes) is randomly generated. Everyone represents a possible solution to the problem at hand, encoded in a specific format, such as a binary string, a complex-valued vector, or a permutation.

Evaluation. Everyone in the population is evaluated and assigned a fitness value, which quantifies its quality or suitability as a solution to the problem. The objective function measures how well an individual solves the optimization problem, and it is always problem specific.

Selection. Individuals are selected from the current population based on their fitness scores, using a selection method such as tournament selection, roulette wheel selection, or rank-based selection. The idea is to favor individuals with higher fitness, increasing their chances of being chosen as parents for the next generation.

Reproduction. It is basically to retain the best chromosomes (say 30 percent) for the next generation. The other chromosomes are replaced by new chromosomes formed through processes of crossover and mutation. This reproduction process is termed "rank-based selection" and allows only the elite and best fitted chromosomes to proceed to the next iteration. Key processes involved in reproduction are:

Crossover. Selected individuals (parents) are combined through crossover (also known as recombination) to create new offspring. Crossover involves exchanging genetic material (represented by the encoded problem solutions) between parents to produce new solutions. The specific crossover method depends on the encoding scheme and can be performed in various ways, such as single-point crossover, two-point crossover, or uniform crossover.

Mutation. After crossover, the offspring may undergo random changes or mutations in their genetic material. Mutation introduces diversity into the population, allowing exploration of new regions in the search space that might contain better solutions. Mutation typically involves random alterations of individual genes or problem-specific modifications.

Replacement. The new offspring, along with some individuals from the previous generation, form the next population. The replacement strategy determines which individuals are selected to survive and become part of the next generation. Elitism is often employed to preserve the best individuals, ensuring that the overall quality of the population does not deteriorate.

Termination. The algorithm continues iterating through selection, crossover, mutation, and replacement until a termination condition is met.

This condition can be a predefined number of generations, a satisfactory fitness level, or a computational time limit. The whole procedure is then repeated for a predefined number of generations (iterations) to produce a final solution. The complete procedure is illustrated by the flow diagram in Fig. 7.3.

The proposed genetic algorithm (GA) is implemented on 6×12 isotropic elements antenna array in MATLAB to obtain optimal or suboptimal weight vector (W). The algorithm is implemented, and parameters are modified for a specific antenna array geometry, i.e., 6×12 planar array. This antenna array optimization is a convex optimization problem where we must find the optimal or suboptimal solution corresponding to the minimum cost function. An initial population of chromosomes is generated at random, and these are decoded to obtain the corresponding problem parameters and are given to the system model. The algorithm runs, performing reproduction operations and all the chromosomes within a population are evaluated based on an objective (cost) function. The cost value corresponding to each chromosome tells the convergence of algorithm to find an optimal solution. Therefore, chromosomes after evaluation are sorted based on cost (fitness) value to select the new population and best chromosomes in each generation. Resultantly, selecting the best fitted chromosome for a particular optimization problem corresponding to the smallest fitness value.

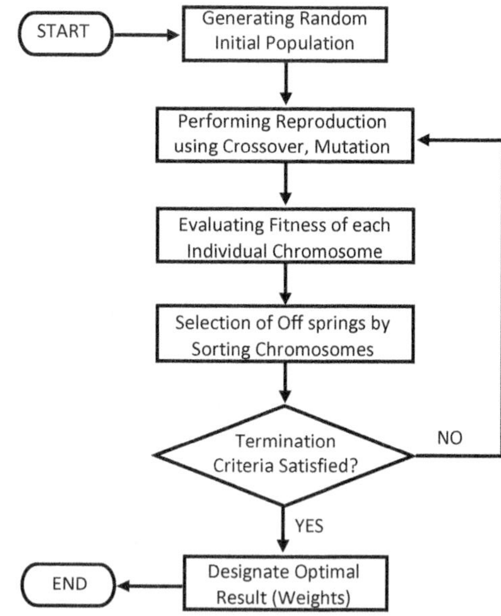

Fig. 7.3 Flow diagram explaining the working principle of a genetic algorithm

7.5 Results and Discussion

Evolutionary algorithms are preferred for element failure correction in antenna arrays due to their ability to efficiently explore large search spaces [48] and perform global optimization [49]. These properties make evolutionary algorithms more valuable than non-evolutionary algorithms. In evolutionary algorithms, GA provides optimal solutions than others, therefore in this research, the proposed GA shown in Fig. 7.3 is used for antenna element failure problem in planar array, to investigate SLL, gain and HPBW recovery in cosecant 4th power pattern. Whereas 8.33–25% edge elements failure recovery is reported. The proposed GA is designed to get desired SLL for original/corrected patterns on main axis $\varphi = 0°$ plane. In corrected patterns of planar array, SLL is successfully achieved in all cases of edge element failure with a negligible gain loss, as the number of failed elements increases the gain lose also increases.

A 6 × 12 isotropic elements planar array with uniform inter-element spacing of $\lambda_0/2$ at operating frequency of 24.3 GHz, is considered for the recovery of SLL = − 20 dB, gain and HPBW in the presence of failed edge elements. The proposed GA given in Fig. 7.3 is implemented to optimize the complex weights (amplitudes & phases) for original pattern (without element failure) with a SLL of − 20 dB and a HPBW of 9°. The failure correction analysis for a 6 × 12 planar array is discussed in three different failure scenarios: (A) 8.33% element failure correction in 4th power pattern scanned at − 20°, (B) 16.67% element failure correction in 4th power pattern scanned at − 20°, and (C) 25% element failure correction in 4th power pattern scanned at − 20°. For each of these cases, the obtained optimum weights from the

06	12	18	24	30	36	42	48	54	60	66	72
05	11	17	23	29	35	41	47	53	59	65	71
04	10	16	22	28	34	40	46	52	58	64	70
03	09	15	21	27	33	39	45	51	57	63	69
02	08	14	20	26	32	38	44	50	56	62	68
01	07	13	19	25	31	37	43	49	55	61	67

Fig. 7.4 Positioning and orientation of antenna elements in a 24.3 GHz, 6 × 12 planar array

genetic algorithm are used in a 6 × 12 microstrip patch series-fed antenna array, designed in CST studio 2019 to provide more practical environment in validating the MATLAB results. Because CST offers element pattern and mutual coupling effects, that come across real world applications. For more detail about other techniques and components one can refer to [50].

In this section, "Original" represents the initial pattern with all array elements are active, "Defected" means the pattern with some specific inactive elements with their weights set to zero and they are not participating in array pattern, while the "Corrected" means the pattern for which the proposed GA has redistributed the complex weights on correctly working (active) elements of defected array, to reconstruct the original (optimal) beam. Thus, these weights are used to make a corrected pattern. Figure 7.4 shows the orientation and positioning of antenna elements in a 6 × 12 planar array.

A. **Random 8.33% Edge Elements Failure Correction with Cosecant 4th Power Pattern**

The original pattern of 6 × 12 planar array achieved SLL = – 20.30 dB, with a gain of 22.6 dB and HPBW = 9.8°, as mentioned in Table 7.3, and is used as a reference pattern for this case. The resulting pattern is referred to as "defected pattern" when 8.33% elements (i.e., 7, 8, 9, 10, 11, 12) from the edge of the array are failed (set their amplitudes to zero). These defective elements affect the overall beam shape and has increased SLL values ranging from – 20.3 dB to nearly – 14.8 dB, while the gain has reduced to 22.5 dB which is slightly lower than the gain of original pattern. The proposed GA is applied on the 8.33% defective elements of 6 × 12 planar array to correct these defects by adjusting the amplitudes and phase of remaining active elements. Thus, resulting in an improved sidelobe level in the corrected pattern, i.e., – 20.9 dB, which is a notable improvement with respect to the sidelobe level of defected beam. However, in comparison to the original pattern's gain of 22.6 dB, the corrected pattern's gain is 22.3 dB. Additionally, the corrected pattern's HPBW of 12.0° is slightly higher than the original pattern due to the inverse relation between gain and HPBW. The SLL recovery is our key concern, but it is to be considered that the resolution of main beam might not get much affected. The original, defected, and corrected patterns are shown in Fig. 7.5, where MATLAB provides the ideal scenario and CST validations to test in a more practical environment.

The defected pattern for 8.33% (19th–24th) element failure exhibits an increased SLL value of – 9.8 dB with a reduced gain of 22.4 dB. The gain reduction and increase

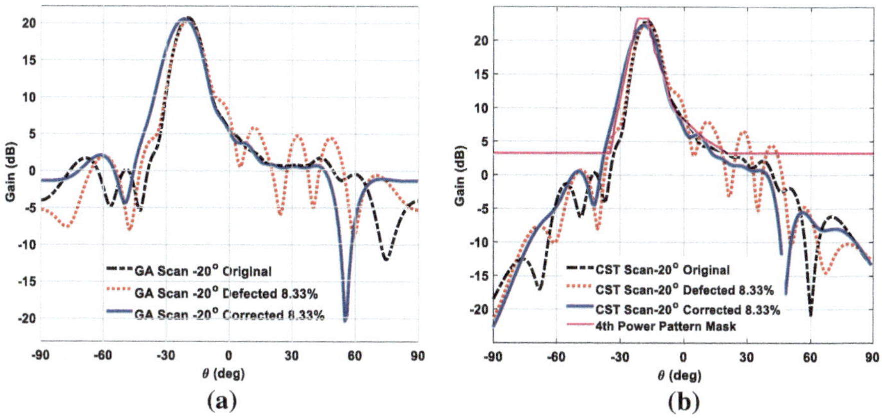

Fig. 7.5 A comparison of 4th power radiation patterns with 8.33% (7th–12th element) failure correction in a 24.3 GHz 6 × 12 planar array. **a** MATLAB and **b** CST

in SLL in this failure case is slightly worse than previous failure case, because these elements are approaching near to the center of the array and have more contribution in generating the radiation pattern. Now, the weights are to be recalculated by applying the proposed algorithm, to construct a new radiation pattern comprising of active elements only. The corrected pattern then successfully recovered the desired SLL of − 20.0 dB and providing a maximum gain of 22.1 dB which is nearly 0.5 dB less than the original pattern's gain. This reduction in gain has increased the HPBW in corrected pattern to 11.4°, which is still manageable while resolving the target beam. Resultantly, the GA has provided the optimal weights to recover the 8.33% element failure with a desired SLL of − 20.0 dB. Figure 7.6 shows the radiation patterns for SLL recovery, both in MATLAB and CST.

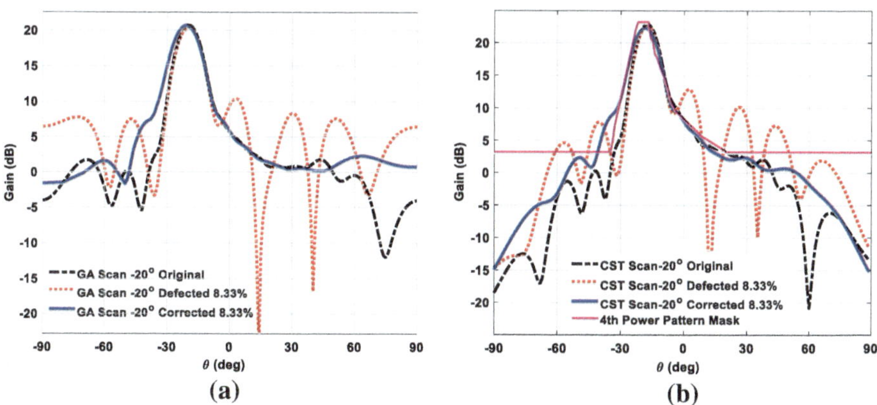

Fig. 7.6 A comparison of 4th power radiation patterns with 8.33% (19th–24th element) failure correction in a 24.3 GHz 6 × 12 planar array. **a** MATLAB and **b** CST

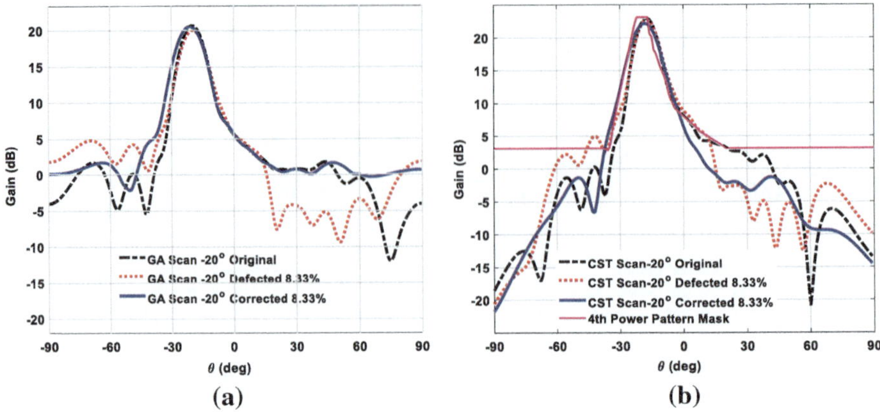

Fig. 7.7 A comparison of 4th power radiation patterns with 8.33% (55th–60th element) failure correction in a 24.3 GHz 6 × 12 planar array. **a** MATLAB and **b** CST

Considering another failure case for 8.33% defected elements from far end of the array, i.e., from 55 to 60th elements, the defected radiation pattern has reported the increased SLL of − 17.8 dB with a gain loss of 0.5 dB. Also increasing the HPBW to 10.1° in comparison to 9.8° in the original pattern with zero failure. To recover the desired pattern with this failure, GA calculates the optimal weights by utilizing only active elements such that it might reconstruct the original radiation pattern. Resultantly, the recovered radiation pattern has reduced the SLL to − 23.5 dB, with a gain of 22.1 dB and the HPBW of 12.1°. In this failure case, the algorithm has reduced SLL to nearly − 3 dB than the original pattern, making the main beam look finer while compensating at resolving the target. Figure 7.7 shows the comparison of radiation patterns for SLL recovery in MATLAB and CST.

In the above three scenarios of 8.33% failure of array elements at multiple positions, the proposed algorithm has recovered a SLL of − 20 dB while managing the HPBW as low as possible. A specific 4th power pattern mask has helped us to minimize the HPBW, which resultantly has improved the resolution of this antenna array in target detection, as the finer the beam, give more resolution. In the presence of multiple antenna elements' failure, the recovery of the SLL and a controlled HPBW makes it notable. It is noted that, as we move toward the center of the array, i.e., the failure of 19th–24th elements, the defected pattern has higher SLL and more gain reduction than the failure of 7th–12th elements. Therefore, recovery of SLL in 19th–24th elements failure cost more gain loss and wider HPBW as compared to SLL recovery in 7th–12th elements' failure. The performance parameters for the above 8.33% edge elements failure are reported in Table 7.4, while their corresponding optimized complex weights are given in Table A1. It is important to note that recovering the SLL using the proposed GA is possible for the specific cases of 8.33% elements failure from the edge in the 6 × 12 planar array. The cost of this SLL recovery is reduction in gain and slightly wider HPBW, because lowering the SLL will also reduce the gain [47], which results wider HPBW.

Table 7.4 Performance parameters comparison in 24.3 GHz 6 × 12 planar array 4th power pattern for 8.33% failure correction of edge element

Failed elements	Achieved SLL (dB)			Gain (dB)			HPBW (degree)		
	Original	Defected	Corrected	Original	Defected	Corrected	Original	Defected	Corrected
8.33% (7th–12th)	− 20.3	− 14.8	− 20.9	22.6	22.5	22.3	9.8	10.3	12.0
8.33% (19th–24th)	− 20.3	− 9.8	− 20.0	22.6	22.4	22.1	9.8	9.7	11.4
8.33% (55th–60th)	− 20.3	− 17.8	− 23.5	22.6	22.5	22.1	9.8	10.1	12.1

B. **Random 16.67% Edge Element Failure Correction with Cosecant 4th Power Pattern**

This case has presented 16.67% elements (7th–12th and 61st–66th) failure correction in 6 × 12 planar array with 4th power pattern. In an array of 72 elements, 12 elements have failed (i.e., amplitude set to zero) to recover SLL, gain and HPBW. The original/reference pattern is same as used in Case A, generated by implementing the proposed GA given in Fig. 7.3, on a 6 × 12 isotropic elements array. The presence of defective elements resulted in a defected pattern affecting the radiation pattern plot and has reported an increased SLL of − 14.5 dB, with main lobe magnitude (gain) reduced to 22.3 dB, lower than the original pattern's gain. The corrected pattern obtained from the optimized weights with 12 defective elements, has improved sidelobe level to − 19.87 dB, an improvement of − 5.3 dB from defected pattern. However, the reported gain is 22.0 dB and the half-power beam width of 12.5° which is nearly 2.5° higher than the original pattern, due to the trade-off between SLL and HPBW, as we always must compensate on one parameter to optimize the other, but the further increase might result in poor resolution for the target. The comparison of radiation patterns is shown in Fig. 7.8 for MATLAB and CST separately. The performance parameters for the above failure cases are reported in Table 7.5, while their corresponding optimized complex weights are provided in Table A2.

C. **Random 25% Edge Element Failure Correction with Cosecant 4th Power Pattern**

In another failure case, the 25% antenna elements (1st–6th, 13th–18th, and 67th–72nd) have their amplitudes set to zero. In the presence of these 18 defective elements, the radiation pattern deteriorated more than the Case A and B, because of more percentage of elements failure. Thus, providing an increased SLL of − 10.8 dB and HPBW of 12.6°, along with a reduced gain of 21.8 dB. Just like the previous

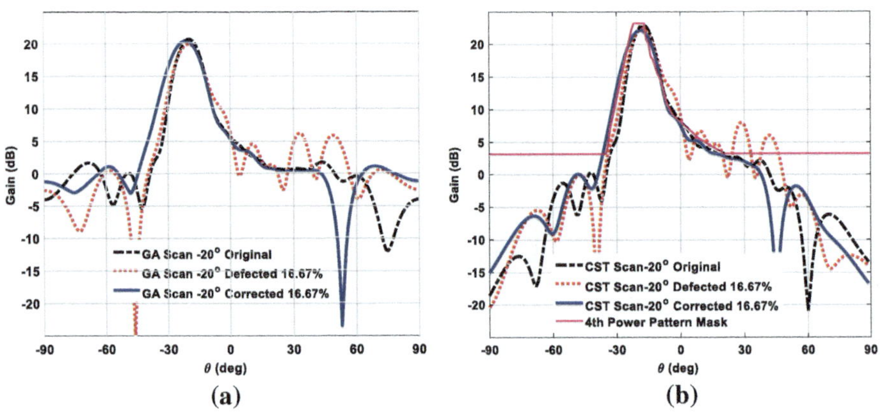

Fig. 7.8 A comparison of 4th power radiation patterns with 16.67% (7th–12th and 61st–66th element) failure correction in a 24.3 GHz 6 × 12 planar array. **a** MATLAB and **b** CST

Table 7.5 Performance parameters comparison in 24.3 GHz 6 × 12 planar array 4th power pattern for 16.67 and 25% failure correction of edge element

Failed elements	Achieved SLL (dB)			Gain (dB)			HPBW (degree)		
	Original	Defected	Corrected	Original	Defected	Corrected	Original	Defected	Corrected
16.67% (7th–12th) (61st–66th)	− 20.3	− 14.5	− 19.8	22.6	22.3	22.0	9.8	10.8	12.5
25% (1st–6th) (13th–18th) (67th–72nd)	− 20.3	− 10.8	− 19.8	22.6	21.8	21.3	9.8	12.6	15.0

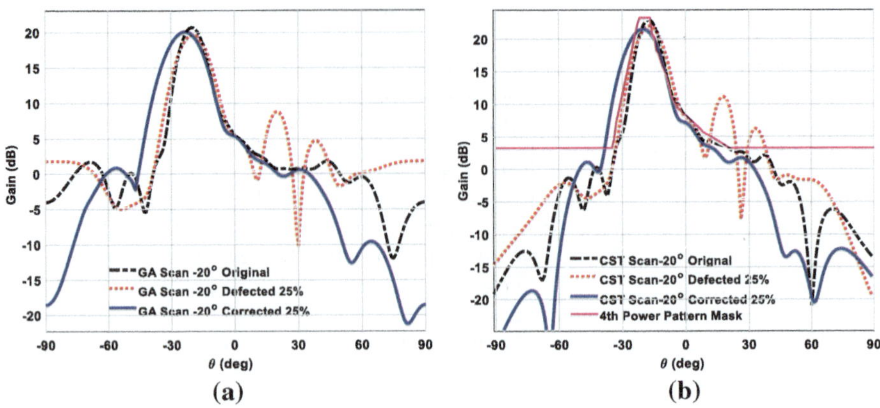

Fig. 7.9 A comparison of 4th power radiation patterns with 25% (1st–6th, 13th–18th and 61st–66th element) failure correction in a 24.3 GHz 6 × 12 planar array. **a** MATLAB and **b** CST

discussed cases, the gain, HPBW, and SLL has shown similar kind of behavior, while recovering the desired (original) radiation pattern. The corrected pattern constructed using the optimized weights from our proposed algorithm, has successfully reported a recovery for the desired SLL of − 20 dB, by achieving − 19.8 dB but at the cost of an increased HPBW of around 15° and an added loss of 1.3 dB in gain of 21.3 dB. A comparison of radiation patterns for desired, defected, and corrected patterns in MATLAB and CST are shown in Fig. 7.9. The performance parameters for the above failure cases are reported in Table 7.5, while their corresponding optimized complex weights are provided in Table A2.

The reported results present a clear picture for the SLL recovery in the presence of multiple failure of edge elements (up to 25%), using the proposed genetic algorithm. But this recovery costs some increased HPBW and a reduction in gain of the main lobe. It is noted that as the number of failed elements increases, the SLL recovery is costing in wider HPBW and reduced gain. This can be acceptable to an extent such that it might not affect the overall resolution for the target. Therefore, the literature and technical experts suggest going for hardware replacement with 30 percent and more elements failed, rather than trying to recover through the proposed evolutionary optimization algorithm. Concluding that the proposed GA has made it possible to recover the antenna array radiation pattern in the presence of 25% failed elements. Moreover, the use of 4th power pattern mask to limit the resolution of target has provided an added benefit because under normal scenario this recovery undergoes massive increase in the HPBW and therefore affecting the overall resolution for the target in practical applications.

7.6 Summary

The discussed literature presents two main approaches for antenna array failure correction: hardware solutions involving redundant modules and software solutions, based on optimization techniques where antenna elements' excitations are recalculated to reconstruct the original/desired radiation patterns. These methods aim to recover SLL, nulls, and gain in the presence of element failures. Literature mainly addresses linear array failure recovery [19–45], research on planar arrays remains limited [46–50] Few studies have explored element failure effects and recovery strategies for SLL, main lobe magnitude, and half power beamwidth (HPBW) in planar arrays. Evolutionary algorithms, particularly GA, are favored for their efficiency in exploring large search spaces. In this research work, the proposed GA is used to recover SLL, radiation pattern gain, and HPBW by reoptimizing complex weights on correctly working antenna elements in MATLAB for a 6×12 planar array with equidistant isotropic elements operating at 24.3 GHz. The study addresses element failure recovery for an electronically scanned beam at $-20°$ and successfully achieved a desired SLL of -20 dB while managing a minimum HPBW using 4th power pattern mask, but with a slight gain loss because of an increased number of failed elements. Since, MATLAB provides an ideal scenario for antenna array element failure problem. It is mandatory to validate the obtained weights in a more practical environment, where mutual coupling, miss-match impedance losses and dielectric losses affect the radiation pattern. Therefore, the generated MATLAB-based results are validated with CST Studio 2019. This implementation of genetic algorithm for antenna array failure correction provides an efficient recovery mechanism in mm-wave SAR applications, offering reduced interference, high resolution, and improved signal-to-clutter ratios.References [51] are cited in the text but not provided in the reference list. Please provide the respective references in the list or delete these citations.[51] is deleted. Please note that "*" has been removed from "Table A1" and "Table A2" naming. Kindly check and confirm.ok confirmed

Acknowledgements This work is funded by the HEC-NRPU, Pakistan via Project No. 20-14696/NRPU/R&D/HEC/2021.

Appendix

See Tables A1 and A2.

Table A7.1 Case A Computed weights using proposed GA algorithm for SLL = − 20 dB of 24.3 GHz 6 × 12 planar antenna array having 8.33% random edge elements failure correction with cosecant 4th power pattern

Element No	For original pattern		8.33% (7th–12th)		8.33% (19th–24th)		8.33% (55th–60th)	
	Amplitudes	Phases (°)	Amplitudes	Phases (°)	Amplitudes	Phases (°)	Amplitudes	Phases (°)
1	0.1818	296.9618	0.0021	181.6295	0.2177	257.9210	0.0066	8.3252
2	0.3817	20.1553	0	0	0.4346	328.9885	0.0975	4.8359
3	0.5194	88.7912	0.2273	292.0345	0.3398	37.3499	0.2800	75.2744
4	0.8143	145.9367	0.6440	7.7536	0	0	0.5847	142.2225
5	0.9909	203.2894	0.9487	68.0515	0.8251	165.6037	0.8424	211.7807
6	0.9530	266.9531	1.0000	130.7914	0.8346	223.5703	0.9653	272.6143
7	1.0000	321.4742	0.6659	193.1880	1.0000	284.9518	1.0000	334.3518
8	0.8184	8.0628	0.6116	263.0052	0.9019	351.0298	0.7419	32.2007
9	0.1925	69.6048	0.5797	348.9254	0.6175	55.3497	0.5051	99.4237
10	0.4206	175.5872	0.3750	28.6415	0.3874	127.6722	0	0
11	0.2826	184.1635	0.0181	88.4051	0.1272	211.1781	0.5999	236.8422
12	0.4928	278.0006	0.3111	172.9527	0.0035	216.8123	0.2359	295.8568

Table A7.2 Case B and C: Computed weights using proposed GA algorithm for SLL = − 20 dB of 24.3 GHz 6 × 12 planar antenna array having 16.66% and 25% random edge elements failure correction with cosecant 4th power pattern

Element No	16.67% (7th–12th and 61st–66th)		25% (1st–6th and 13th–18th and 67th–72nd)	
	Amplitudes	Phases (°)	Amplitudes	Phases (°)
1	0.2964	173.3447	0	0
2	0	0	0.0506	121.9503
3	0.3750	310.1445	0	0
4	0.5980	355.7821	0.5233	223.3433
5	0.5698	80.1164	1.0000	265.0749
6	0.6783	145.4831	0.8984	341.3929
7	1.0000	211.6624	0.8344	50.6219
8	0.9130	270.2405	0.6612	127.3070
9	0.6667	330.8529	0.5176	194.0341
10	0.2312	50.9791	0.3393	277.3938
11	0	0	0.1272	4.4584
12	0.0023	132.1925	0	0

References

1. Xiao, Z., Zhu, L., Liu, Y., Yi, P., Zhang, R., Xia, X.G., Schober, R.: A survey on millimeter-wave beamforming enabled UAV communications and networking. IEEE Commun. Surv. Tutorials **24**(1), 557–610 (2021)
2. Mestoui, J., et al.: A survey of NOMA for 5G: implementation schemes and energy efficiency. Lecture Notes Electr. Eng. **745**, 949–959 (2022)
3. Taghikhani, P., Buisman, K., Fager, C.: Hybrid beamforming transmitter modeling for millimeter-wave MIMO applications. IEEE Trans. Microw. Theory Tech. **68**(11), 4740–4752 (2020)
4. Yu, Y., Jiang, Z.H., Zhang, H., Zhang, Z., Hong, W.: A low-profile beamforming patch array with a cosecant fourth power pattern for millimeter-wave synthetic aperture radar applications. IEEE Trans. Antennas Propag. **68**(9), 6486–6496 (2020)
5. Skolnik, M.I.: Introduction to Radar Systems. New York (1980)
6. Öztürk, E., Genschow, D., Yodprasit, U., Yilmaz, B., Kissinger, D., Debski, W., Winkler, W.: A 120 Ghz Sige Bicmos monostatic transceiver for radar applications. In: 2018 13th European Microwave Integrated Circuits Conference (EuMIC), pp. 41–44. IEEE (2018)
7. Issakov, V., Ciocoveanu, R., Weigel, R., Geiselbrechtinger, A., Rimmelspacher, J.: Highly integrated low-power 60 GHz multichannel transceiver for radar applications in 28 Nm CMOS. In: 2019 IEEE MTT-S International Microwave Symposium (IMS), pp. 650–653. IEEE (2019)
8. NEviANi, A.: Design of Fully Integrated High-Resolution Radars in CMOS and BiCMOS Technologies
9. Van Thillo, W., Giannini, V., Guermandi, D., Brebels, S., Bourdoux, A.: Impact of ADC Clipping and Quantization on Phase-Modulated 79 GHz CMOS Radar. In: 2014 11th European Radar Conference, pp. 285–288. IEEE (2014).
10. Paulino, H., Goes, J., Garção, A.S.: Low Power UWB CMOS Radar Sensors. Springer Science & Business Media.
11. Giannini, V., Nuzzo, P., Chironi, V., Baschirotto, A., Van der Plas, G., Raninckx, J.: An 820μw 9b 40MS/S noise-tolerant dynamic-SAR ADC in 90 nm digital CMOS. In: 2008 IEEE International Solid-State Circuits Conference-Digest of Technical Papers, pp. 238–610. IEEE (2008)
12. Agrawal, A.K., Holzman, E.L.: Active phased array design for high reliability. IEEE Trans. Aerosp. Electron. Syst. **35**(4), 1204–1211 (1999)
13. Zhu, S., Han, C., Meng, Y., Xu, J., An, T.: Embryonics based phased array antenna structure with self-repair ability. IEEE Access **8**, 209660–209673 (2020)
14. Liu, S.C.: Phase-only sidelobe sector nulling for a tapered array with failed elements. In: Proceedings of IEEE Antennas and Propagation Society International Symposium and URSI National Radio Science Meeting, vol. 1, 136–139 (1994)
15. Khan, S.U., Rahim, M.K.A.: Correction of failure in antenna array using matrix pencil technique. Chin. Phys. B **26**(6), 1–8 (2017)
16. Klhan, S.U., Qureshi, I., Shoaib, B., Naveed, A.: Recovery of failed element signal with a digitally beamforming using linear symmetrical array antenna. J. Inf. Sci. Eng. **32**(3), 611–624 (2016)
17. Keizer, W.P.: Low-sidelobe pattern synthesis using iterative Fourier techniques coded in MATLAB [EM programmer's notebook]. IEEE Antennas Propag. Mag. **51**(2), 137–150 (2009)
18. Yadav, K., Rajak, A.K., Singh, H.: Array failure correction with placement of wide nulls in the radiation pattern of a linear array antenna using iterative fast fourier transform. In: 2015 IEEE International Conference on Computational Intelligence & Communication Technology, pp. 471–474 (2015)
19. Munsif, H., Braaten, B.D., Irfanullah.: Element failure correction techniques for phased array antennas in future terahertz communication systems. In: Advances in Terahertz Technology and Its Applications, pp. 19–46 (2021)

20. Lozano, M.V., Rodriguez, J.A., Ares, F.: Recalculating linear array antennas to compensate for failed elements while maintaining fixed nulls. J. Electromagn. Waves Appl. **13**(3), 397–412 (1999)
21. Mandal, S.K., Patra, S., Salam, S., Mandal, K., Mahanti, G.K., Pathak, N.N.: Failure correction of linear antenna arrays with optimized element position using differential evolution. In: 2016 IEEE Annual India Conference (INDICON), pp. 1–5 (2016)
22. Guney, K., Durmus, A., Basbug, S.: Antenna array synthesis and failure correction using differential search algorithm. Int. J. Antennas Propag. (2014) (Art. no. 276754)
23. Grewal, N.S., Rattan, M., Patterh, M.: A linear antenna array failure correction using firefly algorithm. Prog. Electromagn. Res. M **27**, 241–254 (2012)
24. Hamici, Z.: Fast beamforming with fault tolerance in massive phased arrays using intelligent learning control. IEEE Trans. Antennas Propag. **67**(7), 4517–4527 (2019)
25. Han, J.H., Lim, S.H., Myung, N.H.: Array antenna TRM failure compensation using adaptively weighted beam pattern mask based on genetic algorithm. IEEE Antennas Wirel. Propag. Lett. **11**, 18–21 (2011)
26. Yeo, B.K., Lu, Y.: Array failure correction with a genetic algorithm. IEEE Trans. Antennas Propag. **47**(5), 823–828 (1999)
27. Rodriguez, J.A., Ares, F., Moreno, E., Franceschetti, G.: Genetic algorithm procedure for linear array failure correction. Electron. Lett. **36**(3), 196–198 (2000)
28. Acharya, O.P., Patnaik, A.: Antenna array failure correction [antenna applications corner]. Antennas Propag. Mag. **59**(6), 106–115 (2017)
29. Acharya, O.P., Patnaik, A., Sinha, S.N.: Limits of compensation in a failed antenna array. Int. J. RF Microw. Comput. Aided Eng. **24**(6) 635–645 (2014)
30. Acharya, O.P., Patnaik, A., Sinha, S.N.: A Healing System for Failed Antenna Array using PSO (2010)
31. Acharya, O.P., Patnaik, A., Sinha, S.N.: Null steering in failed antenna array. In: IEEE 2011 National Conference on Communications (NCC), pp. 1–4 (2011)
32. Acharya, O.P., Patnaik, A., Sinha, S.N.: Null steering in failed antenna arrays. In: Applied Computational Intelligence and Soft Computing, pp. 1–9 (2011)
33. Munsif, H., Braaten, B.D., Irfanullah.: Sidelobe level and nulls control of defected linear antenna arrays using an optimization algorithm. In: 19th International Bhurban Conference on Applied Sciences and Technology (IBCAST), pp. 984–989 (2022)
34. Djigan, V., Kurganov, V.: Antenna array calibration algorithm based on phase perturbation. In: IEEE East-West Design & Test Symposium (EWDTS), pp. 1–5 (2019)
35. Djigan, V.I., Kurganov, V.V.: Antenna array calibration algorithm without access to channel signals. Radioelectron. Commun. Syst. **63**(1), 1–14 (2020)
36. Jiang, X., Xu, P., Jiang, T.: Comparison between NU-CNLMS and GA for antenna array failure correction. In: 2018 IEEE International Symposium on Antennas and Propagation & USNC/URSI National Radio Science Meeting, pp. 2205–2206 (2018)
37. Khan, S.U., Qureshi, I.M., Zaman, F., Shoaib, B., Naveed, A., Basit, A.: Correction of faulty sensors in phased array radars using symmetrical sensor failure technique and cultural algorithm with differential evolution. Sci. World J. (2014) (Art. no. 852539)
38. Singh Grewal, N., Rattan, M., Singh Patterh, M.: A linear antenna array failure correction using improved Bat algorithm. Int. J. RF Microw. Comput. Aided Eng. **27**(7), e 21119 (2017)
39. Greda, L.A., Winterstein, A., Lemes, D.L., Heckler, M.V.: Beamsteering and beamshaping using a linear antenna array based on particle swarm optimization. IEEE Access **7**, 141562–141573 (2019)
40. Engroff, A.M., Greda, L.A., Magalhães, M.P., Winterstein, A., Pereira, L.S., Girardi, A.G., Heckler, M.V.: Comparison of beamforming algorithms for retro-directive arrays with faulty elements. In: 2016 10th European Conference on Antennas and Propagation (EuCAP). IEEE, 1–5 (2016)
41. Migliore, M.D., Pinchera, D., Lucido, M., Schettino, F., Panariello, G.: A sparse recovery approach for pattern correction of active arrays in presence of Element failures. Antennas Wirel. Propag. Lett. **14**, 1027–1030 (2014)

42. Keizer, W.P.: Element failure correction for a large monopulse phased array antenna with active amplitude weighting. IEEE Trans. Antennas Propag. **55**(8), 2211–2218 (2007)
43. Rodriguez, J.A., Ares, F.: Optimization of the performance of arrays with failed elements using the simulated annealing technique. J. Electromagn. Waves Appl. **12**(12), 1625–1638 (1998)
44. Boopalan, N., Ramasamy, A.K., Nagi, F., Alkahtani, A.A.: Planar array failed element (s) radiation pattern correction: a comparison. Appl. Sci. **11**(19), 9234 (2021)
45. Yang, Y., Stark, H.: Design of self-healing arrays using vector-space projections. IEEE Trans. Antennas Propag. **49**(4), 526–534 (2001)
46. Chen, Y.S., Tsai, I.L.: Detection and correction of element failures using a cumulative sum scheme for active phased arrays. IEEE Access **6**, 8797–8809 (2018)
47. Stutzman, W.L., Thiele, G.A.: Antenna Theory and Design. Wiley (2012)
48. Vincent, S., John Francis, S.A., Kumar, O.P., Raimond, K.: Recovery of a failed antenna element using genetic algorithm and particle swarm optimization for MELISSA. In: Advances in Signal Processing and Intelligent Recognition Systems: 4th International Symposium SIRS 2018, Bangalore, India, September 19–22, 2018, Revised Selected Papers, vol. 4, pp. 217–228. Springer Singapore (2019)
49. Huang, H., Hoorfar, A., Lakhani, S.: A comparative study of evolutionary programming, genetic algorithms and particle swarm optimization in antenna design. In: 2007 IEEE Antennas and Propagation Society International Symposium, pp. 1609–1612 IEEE (2007)
50. El Ghzaoui, M., Das, S., Lenka, T.R., Biswas, A. Terahertz wireless communication components and system technologies. In: Terahertz Wireless Communication Components and System Technologies, pp. 1–310 (2022)

Chapter 8
A Novel, Compact, Broadband Band-Stop Filter for Rejecting 5G Millimeter-Wave Communications

Fatima Kiouach, Bilal Aghoutane, Mohammed El Ghzaoui, and Rachid El Alami

8.1 Introduction

The rising need for RF and microwave filters is a notable trend, driven by their growing significance in various telecommunications applications. They hold a central position in contemporary wireless communication and radar systems, underscoring their fundamental importance in the current technological landscape. Notably, RF and microwave filters are not limited to high-tech communication; their utilization extends to public address systems and speaker systems, where their application ensures the delivery of high-quality audio [1]. Besides, millimeter-wave communications is a promising technology that entails the utilization of radio frequencies within the millimeter-wave spectrum, typically ranging from 30 to 300 GHz. This innovative approach holds immense potential for various applications, spanning wireless communication networks, high-speed data transmission, next-generation mobile networks (5G and beyond), and emerging technologies like the Internet of Things (IoT) and smart cities [2–4].

The unique characteristics of millimeter-wave frequencies, including wider bandwidths and higher data rates, contribute to the efficiency and capacity of communication systems. The shorter wavelength allows for the integration of numerous small-sized antennas, enabling the development of compact and high-performance communication devices. Moreover, millimeter-wave technology facilitates the deployment of dense networks with increased data-carrying capacity, making it a cornerstone for addressing the burgeoning demand for high-speed and low-latency communication services. In the context of 5G networks, millimeter-wave communication plays

F. Kiouach (✉) · M. El Ghzaoui · R. El Alami
Faculty of Sciences Dhar El Mahraz, Sidi Mohamed Ben Abdellah University, Fez, Morocco
e-mail: fatima.kiouach@usmba.ac.ma

B. Aghoutane
Faculty of Sciences, Ibn Tofail University, Kenitra, Morocco

M. El Ghzaoui et al. (eds.), *Next Generation Wireless Communication*, Signals and Communication Technology, https://doi.org/10.1007/978-3-031-56144-3_8

a pivotal role in unlocking the full potential of ultra-fast, low-latency connectivity. The technology facilitates the implementation of massive MIMO (multiple input, multiple output) [5, 6] systems and beamforming techniques, enhancing the overall network performance and enabling the deployment of advanced applications such as augmented reality (AR), virtual reality (VR), and high-definition video streaming.

Microwave and millimeter-wave circuits have been subjected to numerous theoretical propositions aimed at meeting the escalating demands for filtering specific frequency bands and suppressing harmonics [7]. Categorically, there are four types of filters: band-stop filters (BSF), band-pass filters (BPF), high-pass filters (HPF), and low-pass filters (LPF) [8]. Each of these filter types holds paramount importance in the telecommunications sector. Band-stop filters, in particular, emerge as vital component in wireless communication systems. In practical applications, there is a persistent demand for compact devices, necessitating the miniaturization of components.

Several suggestions have been put forth for the realization of band-stop filters at this point [9]. As telecommunications technology continues its advance, filters are expected to evolve, adapting to the ever-changing landscape of modern communication systems.

Reference [10] introduces an ultra-wideband band-stop filter featuring three transmission zeros and one reflection zero, with a size of 13 mm × 24.30 mm. The suggested filter demonstrates an impressive ultra-wide bandwidth spanning from 3.7 to 9.8 GHz and attains a rejection depth of 20. In Ref. [11], BSF designed for wireless applications is introduced. The proposed BSF exhibits advantageous characteristics, including a small size and a wide stopband. With a stopband of 5.5 GHz, and a center frequency of 6.2 GHz, the band-stop filter achieves an S11 (return loss) exceeding 0.4 dB and an S21 (insertion loss) of less than − 30 dB. In Ref. [12], a broadband BSF is introduced, featuring seven transmission zeros and an extended upper passband. This BSF exhibits a 3-dB frequency range spanning from 6.5 to 23 GHz, with a substantial 92.62% stop band fractional bandwidth at 20 dB. The proposed BSF demonstrates efficacy in segregating the microwave frequencies of 4G and the millimeter wave frequencies of 5G, catering to the evolving landscape of mobile communication applications. Furthermore, it possesses the capability to effectively isolate signals from sub-6 GHz and 28 GHz Next Generation Node B (gNBs), facilitating the deployment of both base stations concurrently at the same location if necessary. In Ref. [13], a wideband BSF is introduced, featuring an expansive stop band spanning from 3.52 to 12.38 GHz. The filter is compact, measuring 7.4 mm × 15 mm, and is designed for applications in the C and X bands. With a center frequency of 8 GHz and a fractional bandwidth of 110%, the proposed filter incorporates two resonator types: a Quad mode resonator and an L-shaped resonator. In Ref. [14], an adjustable band-stop filter operating in the 22–43 GHz range is introduced. The chapter provides a comprehensive explanation of the underlying theory and design approach. The filter utilizes silicon-micromachined evanescent-mode cavity resonators, which are adjusted by low-power electrostatic actuators. In Ref. [15], a cost-effective, continuously tunable quasi-absorptive band-stop filter operating in the K/Ka-band is introduced. The filter is fabricated using PCB technology, and its measured 3-dB bandwidth exhibits variability, ranging from 1.47% at 22 GHz to 4.65% at 42 GHz.

This chapter provides an overview of filters, with a specific emphasis on band-stop filters, and details the design of an innovative band-stop filter customized for the rejection of 5G millimeter-wave communications. The filter exhibits a rejected bandwidth of 6.3 GHz, a center frequency at 27.2 GHz, and a fractional bandwidth (FBW) of 23%, showcasing a substantial rejection of 41 dB. To assess the filter's effectiveness, simulations, and evaluations were conducted utilizing the HFSS (high-frequency structure simulator), and its equivalent circuit was modeled using the ANSYS circuit editor. The acquired results highlight the filter's appropriateness for applications operating at frequencies both below 24 GHz and above 30 GHz, to function efficiently.

8.2 Fundamentals of Filters

Filters encompass a vast domain. As per Matthaei [16], they, along with related techniques, serve diverse purposes such as signal filtering (filtering), their primary task, impedance adaptation (adaptation), combining or separating signals with varying frequencies (multiplexing), and creating delays or phase shifts, among other applications.

The filters can be grouped into four main categories based on how they handle different frequencies of a signal. These broad categories are low-pass filters, high-pass filters, band-pass filters, and band-stop filters.

8.2.1 Filter Categories

Low-Pass Filters (LPF). Low-pass filters are specifically crafted to enable the transmission of frequencies below a specified threshold. Their essential role lies in suppressing undesirable high frequencies, effectively isolating the desired signal (Fig. 8.1).

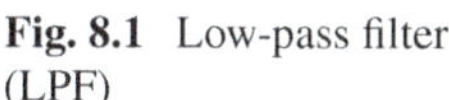

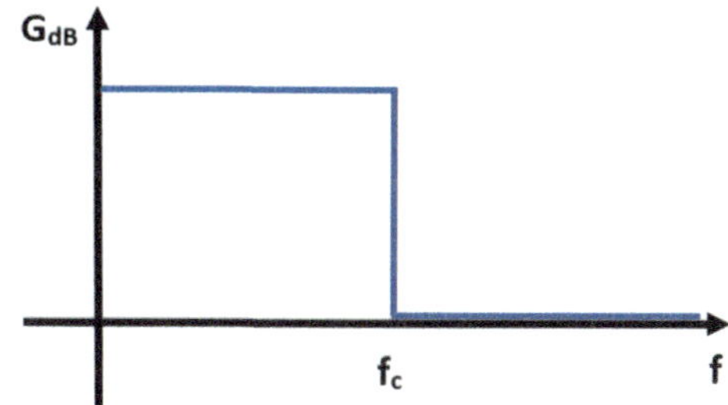

Fig. 8.1 Low-pass filter (LPF)

High-Pass Filters (HPF). Unlike low-pass filters, high-pass filters allow the passage of frequencies higher than a specified value, eliminating low-frequency components. These filters are essential for the transmission of high-frequency signals without distortion from low-frequency interference (Fig. 8.2).

Band-Pass Filters (BPF). The band-pass filters provide a selective window, allowing only the passage of a specific range of frequencies [17]. This category of filters is commonly used to isolate specific communication channels in environments where multiple signals coexist (Fig. 8.3).

Band-Stop Filters (BSF). The band-stop filters are configured to block a specific range of frequencies, allowing only the passage of signals located outside of this range (Fig. 8.4).

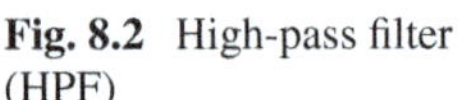

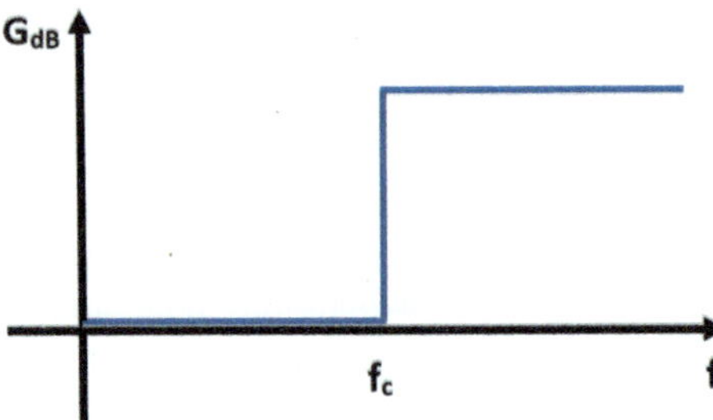

Fig. 8.2 High-pass filter (HPF)

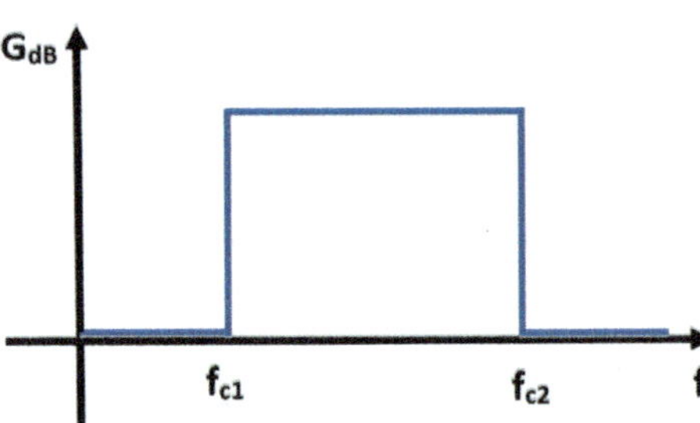

Fig. 8.3 Band-pass filter (BPF)

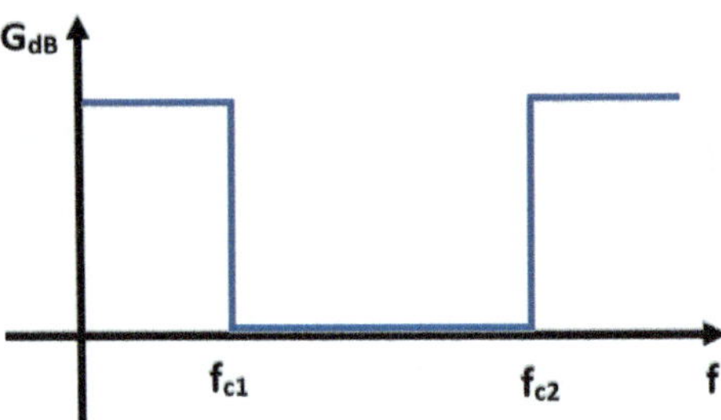

Fig. 8.4 Band-stop filter (BSF)

8.2.2 Filter Classes

The response of an ideal filter would be zero attenuation in the passband and infinite attenuation in the stopband. Moreover, the transition from the passband to the stopband would occur instantaneously, meaning that the slope of the filter at the cutoff frequency would be infinite.

Obviously, this response cannot be achieved with a physically realizable circuit. Therefore, it is necessary to use functions that will give us a response that approximates more or less the ideal characteristics.

This is how several functions, or classes of filters, have been studied. Among the most well-known are Butterworth filters and Chebyshev filters. However, no function allows for the maximization of all the characteristics of an ideal filter simultaneously. Consequently, some functions focus on a constant response in the passband (Butterworth), others opt for a high cutoff slope (Chebyshev and Cauer), others for a constant group delay in the passband (Bessel), etc.

Let's now examine in a bit more detail the three main classes of filters. Firstly, the Butterworth filters provide an approximately constant response within the bandwidth. However, the transition between the passband and the stopband occurs very slowly.

On the other hand, Chebyshev filters exhibit an undulating response within the passband. Moreover, the response of these filters is more sensitive to errors in the components than that of Butterworth filters. However, the transition from the passband to the stopband is faster than that of Butterworth filters.

In terms of Cauer filters, these are actually Chebyshev filters to which transmission zeros have been added to increase the steepness of the slope at the cutoff frequency. Consequently, these filters are even more sensitive to errors in the components.

8.2.3 Importance of Filters in Communication Systems

The filters play a fundamental role at the heart of communication systems, acting as silent guardians of signal quality. Their importance lies in their ability to discriminate and regulate the different frequencies present in an electromagnetic spectrum. In a world where bandwidth is a precious resource, filters become the architects of the spectrum, defining the paths to take and eliminating undesirable noise. Filters are essential to ensure the clarity of signals, minimize interference, and maximize the efficiency of the available spectrum. Whether in mobile networks, satellite links, or short-range wireless communications, filters are key elements ensuring reliable and high-quality data transmission. Their silent but crucial presence in communication systems reflects the importance of understanding and optimizing these devices to meet the growing challenges of modern networks.

8.2.4 Filtenna

The progression of communication technology has seen an intriguing fusion between filters and antennas, resulting in the emergence of a groundbreaking concept referred to as Filtenna. This integration evolved organically from the essential role filters play in molding signal characteristics and the pivotal function that antennas serve in propagating these signals.

As explored in earlier sections, filters play a crucial role in signal processing by enabling the precise manipulation and refinement of signals according to their frequency parameters. Acknowledging the interdependence of signal filtering and propagation, researchers and engineers embarked on a quest to seamlessly integrate these two functions.

The culmination of this exploration gives rise to the Filtenna, an innovative amalgamation seamlessly blending the meticulous frequency control of filters with the signal propagation prowess of antennas.

Through the integration of these traditionally separate components, Filtennas present a synergistic solution that holds significant potential for shaping the future of communication systems.

8.3 Fundamentals of Filters

8.3.1 Types of Band-Stop Filters

Notch Filters. Notch filters represent a specific category within band-stop filters, characterized by their capacity for highly pronounced attenuation within a narrow frequency range. Engineered to target and eliminate a singular frequency, these filters find widespread application in radio receivers, where their precision is instrumental in suppressing undesired interference from a specific radio station. Leveraging a narrow bandwidth, notch filters excel at isolating and eliminating the unwanted signal, ensuring the passage of the remaining signal components without alteration.

Wideband Stop Filters. Wideband notch filters, alternatively known as broadband stop filters, exhibit the capability to attenuate a more extensive range of frequencies. Specifically engineered to eliminate noise and interference spanning a broad spectrum, these filters find frequent application in audio and telecommunications systems. Their usage is particularly advantageous in scenarios where noise and interference can manifest across a diverse range of frequencies, allowing for effective signal enhancement and improved overall system performance.

8.3.2 Characteristics of Band-Stop Filters

Center Frequency (f_o). The center frequency or resonance frequency is a crucial parameter in the design of band-stop filters. It represents the frequency at which the frequency response of the filter reaches its midpoint.

Bandwidth (BW). The bandwidth represents the frequency range over which the filter exerts its cutoff effect. It is defined as the difference between the frequencies at the points with -3 dB of maximum attenuation.

Quality factor (Q). The quality factor, denoted as Q, is a crucial indicator for assessing the selectivity of the band-stop filter. It is defined as the ratio of the center frequency (f_0) to the bandwidth (BW).

8.3.3 Applications and Uses

Band-stop filters play a pivotal role in numerous electronic systems, finding application across a diverse range of fields. Apart from their integral role in audio, radio frequency systems, and telecommunications, as well as in signal-processing applications, band-stop filters also serve crucial functions in various other domains, including:

Wireless Communication. In the realm of wireless communication systems, band-stop filters emerge as crucial tools to thwart interference issues arising from multiple frequency bands. Consider the scenario of a mobile phone: Within this context, a band-stop filter proves indispensable. Its primary function is to safeguard against the potential disruption caused by the transmission signal of the mobile phone, ensuring that it doesn't impede the reception of signals on closely situated frequency bands. This strategic implementation of band-stop filters plays a pivotal role in maintaining the integrity and efficiency of wireless communication, particularly in densely populated frequency spectrums.

Medical Device. Band-stop filters play a vital role in medical equipment, particularly in enhancing the accuracy of diagnostic tools. In medical applications such as ECG (electrocardiogram) machines, band-stop filters are strategically employed to eliminate undesirable noise and interference. This precise filtration ensures that the signals captured by medical instruments remain clear and uncontaminated, contributing to the reliability of diagnostic readings in health-care settings.

Radar Systems. In radar systems, the integration of band-stop filters is instrumental in mitigating undesired interference originating from other radar systems or sources of electromagnetic radiation. The deployment of band-stop filters in this context is crucial as such interference has the potential to induce erroneous readings and compromise the overall accuracy of the radar system. The strategic use of these filters ensures that radar systems operate with heightened precision.

Power Electronics. Within the domain of power electronics, band-stop filters serve as essential components for minimizing the EMI (electromagnetic interference) produced through the switching of power devices. This interference, if unaddressed, has the potential to create disruptions in various electronic systems. By integrating band-stop filters, the adverse effects of EMI are effectively mitigated, ensuring the smooth operation and reliability of interconnected electronic devices.

8.4 Design and Analysis of a Band-Stop Filter

The filter design is applied to a Rogers TMM 10i (tm) substrate with a relative permittivity of 9.8 and a dielectric loss tangent of 0.002 using the High Frequency Structure Simulator (HFSS). The filter design is positioned on the upper surface of the substrate, with a partial ground plane taking the form of a rectangular shape on the lower side. As shown in Fig. 8.5, the filter's overall dimensions are 14.8 × 3.8 × 0.2 mm^3. Figure 8.6 depicts the proposed filter module in the HFSS simulator.

8.4.1 *Parametric Study*

The parameters presented in this research underwent a thorough assessment using the HFSS simulator to ensure their accuracy. In this section, we thoroughly analyzed

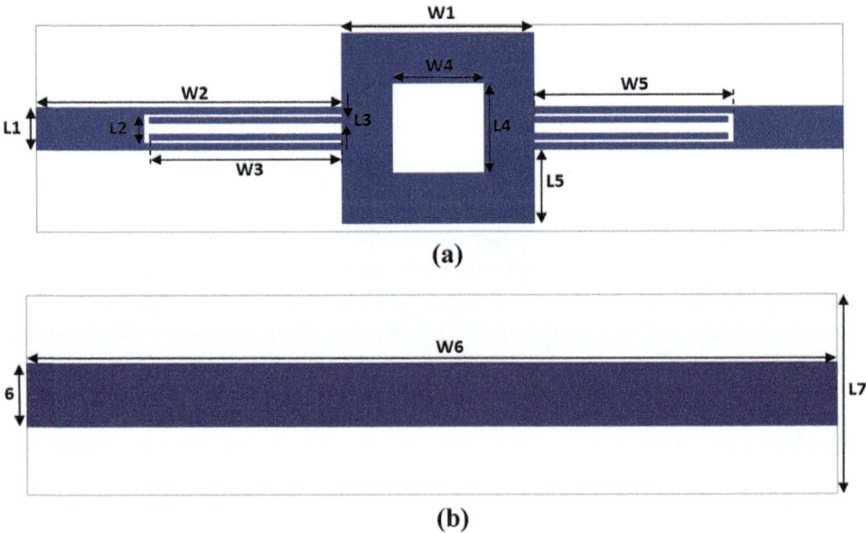

Fig. 8.5 Suggested band-stop filter (BSF) design **a** top view, and **b** back view

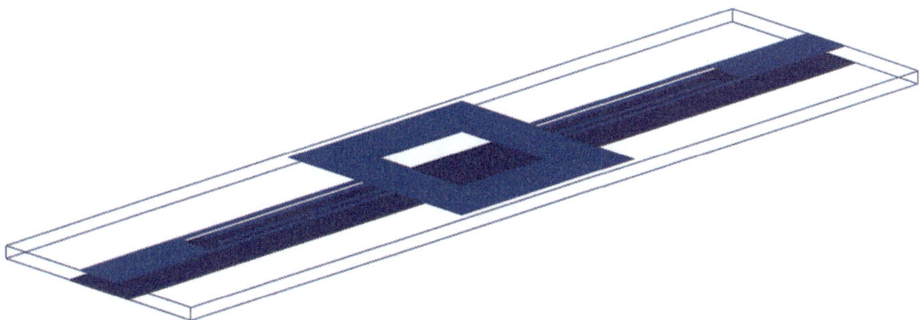

Fig. 8.6 Suggested band-stop filter within the HFSS simulator

a subset of these parameters to elucidate how they influence the overall performance of the filter.

This parametric study initially concentrates on W1, exploring variations from 3.3 to 3.5 mm. All other parameters remain constant throughout this investigation. Figure 8.7 depicts the corresponding S11 (return loss) and S21 (insertion loss) against frequency for the variation in the value of the parameter W1. Figure 8.7 clearly demonstrates that as W1 increases, the rejected band's central frequency increases, aligning with a rise in the insertion loss. Meanwhile, the return loss remains relatively stable despite variations in the W1 value.

The second examination involved modifying the W3 value from 3.34 to 3.4 mm, while keeping all other parameters constant. Figure 8.8 clearly illustrates that an increase in W3 leads to a corresponding rise in the central frequency of the rejected band.

The influence of parameter W4 is illustrated in Fig. 8.9. Upon analyzing the graph, it becomes apparent that altering the value of W4 noticeably influences the insertion

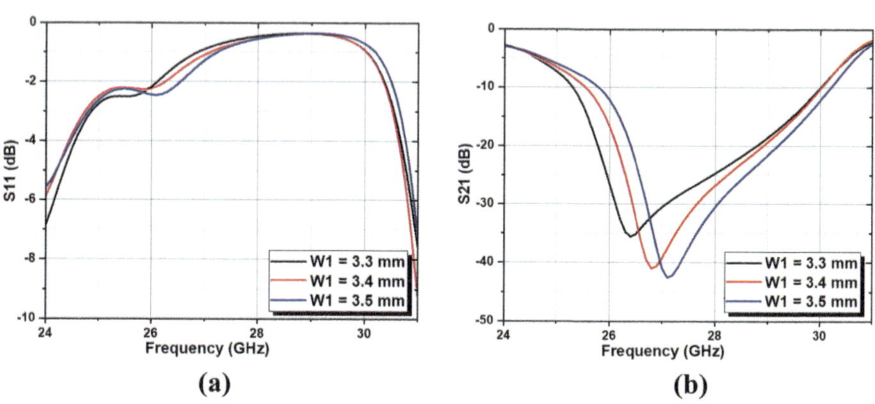

Fig. 8.7 Simulated *S*-parameters of the proposed filter corresponding to parameter W1: **a** S11, and **b** S21

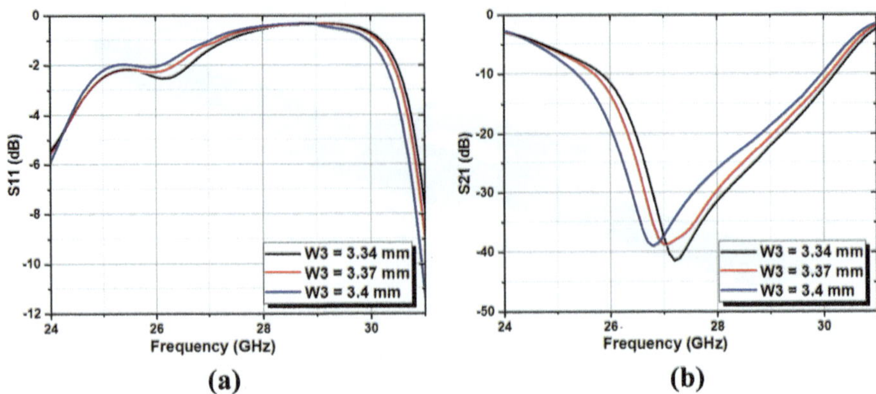

Fig. 8.8 Simulated *S*-parameters of the proposed filter corresponding to parameter W3: **a** S11, and **b** S21

loss values. Precisely, a rise in the magnitude of W4 leads to a decline in the S21 value, while the central frequency's position remains constant with variations in the W4 parameter.

The influence of parameter L3 on the filter's performance is depicted in Fig. 8.10. As illustrated in the figure, L3 has a minimal impact on the S21 value without affecting the center frequency or the S11. When the L3 value increases from 0.1 to 0.11 mm, a slight decrease in the S21 value is observed. Subsequently, with an increase in L3 to 0.12 mm, there is a modest rise in the S21 value.

To assess the influence of parameter L4 on the filter's performance, we varied its value from 1.6 to 1.66 mm. Figure 8.11 clearly indicates that L4 has an effect on the S21 value without altering the center frequency. As L4 increases, the S21 value also increases, while the S11 remains stable despite changes in the L4 value.

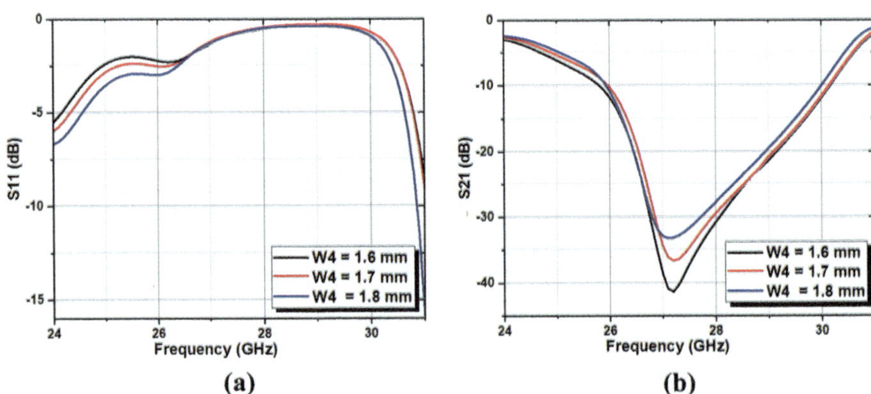

Fig. 8.9 Simulated *S*-parameters of the proposed filter corresponding to parameter W4: **a** S11, and **b** S21

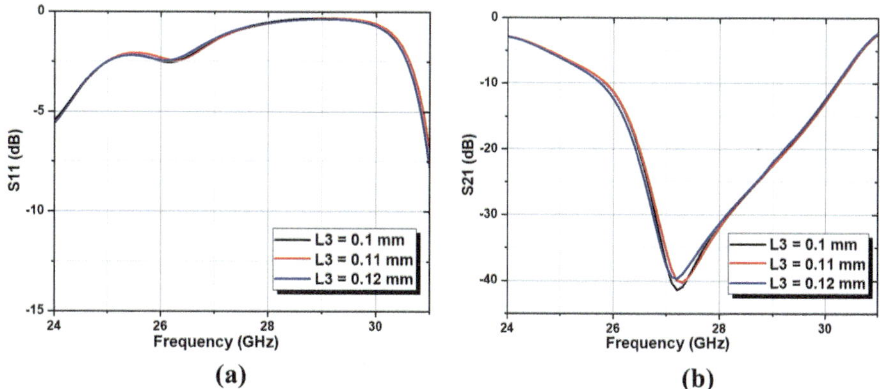

Fig. 8.10 Simulated *S*-parameters of the proposed filter corresponding to parameter L3: **a** S11, and **b** S21

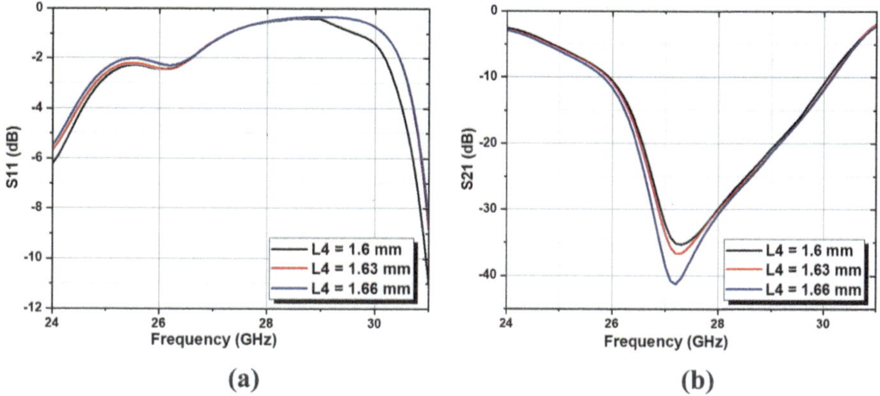

Fig. 8.11 Simulated *S*-parameters of the proposed filter corresponding to parameter L4: **a** S11 and **b** S21

The influence of the L5 parameter on the performance of the band-stop filter is illustrated in Fig. 8.12. Illustrated in Fig. 8.12b, variations in the L5 parameter exert an influence on both the S_{21} value and the center frequency. In particular, with an increase in the L5 value, there is an elevation in the S_{21} value, coupled with a minor reduction in the center frequency. On the flip side, with an augmentation of the L5 value to 1.37 mm, a marginal upswing is observed in the S_{21} value, coupled with a slight elevation in the center frequency. Figure 8.12a demonstrates that changes in the L5 parameter do not impact the S_{11}.

The final parameter under investigation is L6, representing the length of the partial ground. As depicted in Fig. 8.13, an increase in the value of L6 results in an elevated S21 value without affecting the center frequency. Meanwhile, the S11 remains stable across variations in the L6 parameter.

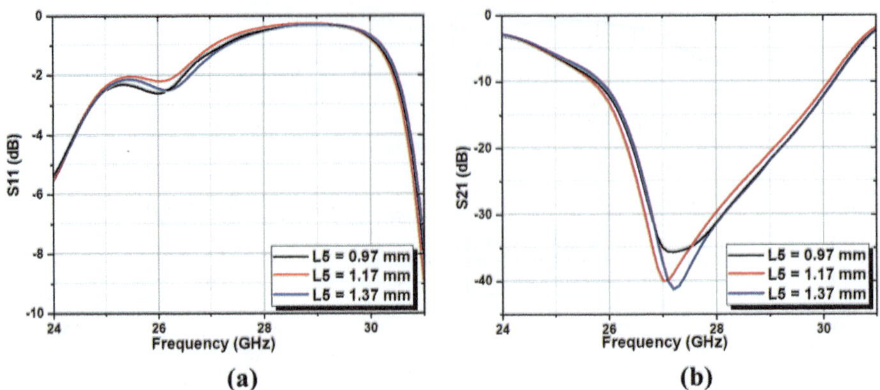

Fig. 8.12 Simulated *S*-parameters of the proposed filter corresponding to parameter L5: **a** S11, and **b** S21

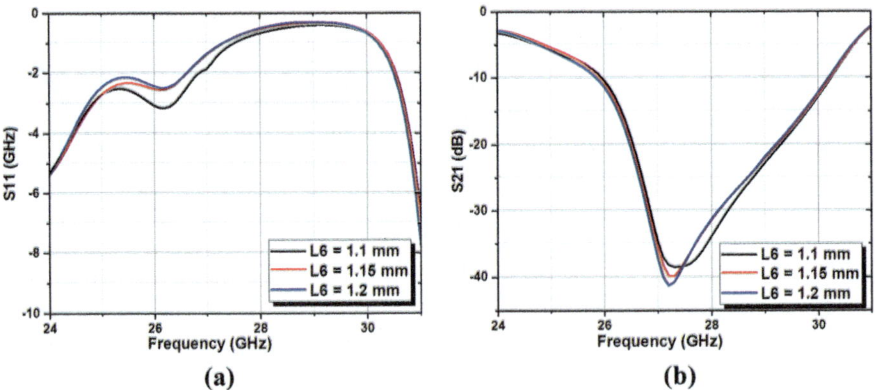

Fig. 8.13 Simulated *S*-parameters of the proposed filter corresponding to parameter L6: **a** S11 and **b** S21

Table 8.1 depicts the appropriate parameter values for the suggested filter. Figure 8.14 presents the conclusive *S*-parameters of the suggested Band-Stop Filter. The filter displays a center frequency at 27.2 GHz, showcasing a rejected bandwidth spanning 6.3 GHz (24.5–30.8 GHz). Notably, it achieves a substantial rejection of 41 dB. The stopband fractional bandwidth (FBW) of this filter is 23%, calculated using equation [18]:

$$\text{FBW} = \frac{f_{c2} - f_{c1}}{f_0}$$

In Fig. 8.15, we simulate and plot the group delay to control the signal transit time through our filter in relation to frequency. The results from the simulation indicate a group delay response of under 5.3 ns throughout the rejected frequency band.

Table 8.1 The proposed stop band filter's dimensions

Filter parameters	Value (mm)
W1	3.5
W2	5.64
W3	3.34
W4	1.6
W5	3.35
W6	14.8
L1	0.76
L2	0.56
L3	0.1
L4	1.66
L5	1.37
L6	1.2
L7	3.8

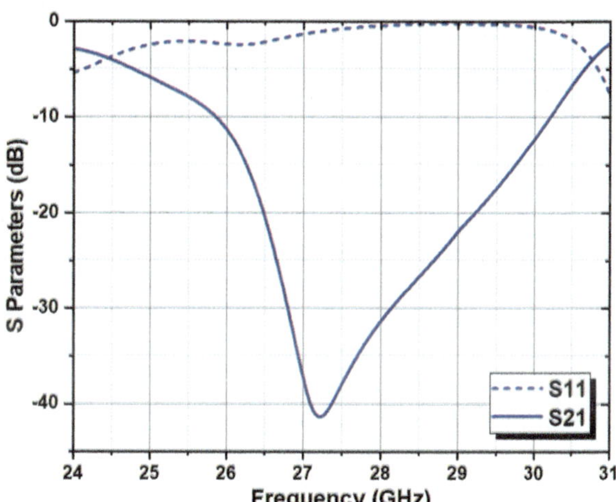

Fig. 8.14 The *S*-parameters for the band-stop filter

Figure 8.16 illustrates the surface current distribution of the suggested band-stop filter across frequencies of 26.2, 27.2, and 28.2 GHz. The diagram showcases the transmission behavior of signals through the filter at different frequency points.

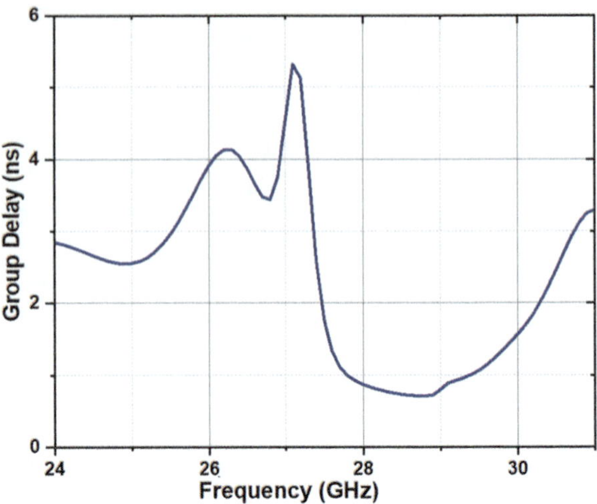

Fig. 8.15 Simulated group delay for suggested band-stop filter

8.4.2 Band-Stop Filter Equivalent Circuit

To model the equivalent circuit of the proposed band-stop filter, we employed the Filter Solution tool within Ansys Electronics Desktop 2022 R1 for simplified calculation of the values of the RLC components in the circuit. Initially, 'Band stop' was selected in the Band Classes, followed by choosing 'Butterworth approximation' in the Shape Types, and subsequently opting for 'Lumped Synthesis' in the Implementations. For General Requirements, values for Pass Band Attenuation, Stop Band Attenuation, and impedance were inputted. Additionally, for Band Stop Requirements, we specified the Center Frequency, Bandwidth, and the filter's order. Upon completion, the equivalent circuit was exported to ANSYS circuit editor.

Figure 8.17 illustrates the equivalent circuit, with its elements carefully tuned for optimal alignment between the simulated and equivalent circuit responses. Figure 8.18 presents the simulation outcomes of the suggested equivalent circuit model, subject to a comprehensive comparison with the initially simulated results. The remarkable proximity between the responses serves as validation for the effectiveness of the proposed equivalent circuit model.

The effectiveness of the band-stop filter presented in this study is assessed through a comparative analysis with findings from other research endeavors, as outlined in Table 8.2. The comparison encompasses: center frequency, rejected band, stop-band FBW (fractional bandwidth), IL (insertion loss), RL (return loss), and the size of the filter. As indicated in Table 8.2, it is clear that the filter demonstrates a broad stop-band, distinguished by a substantial degree of attenuation.

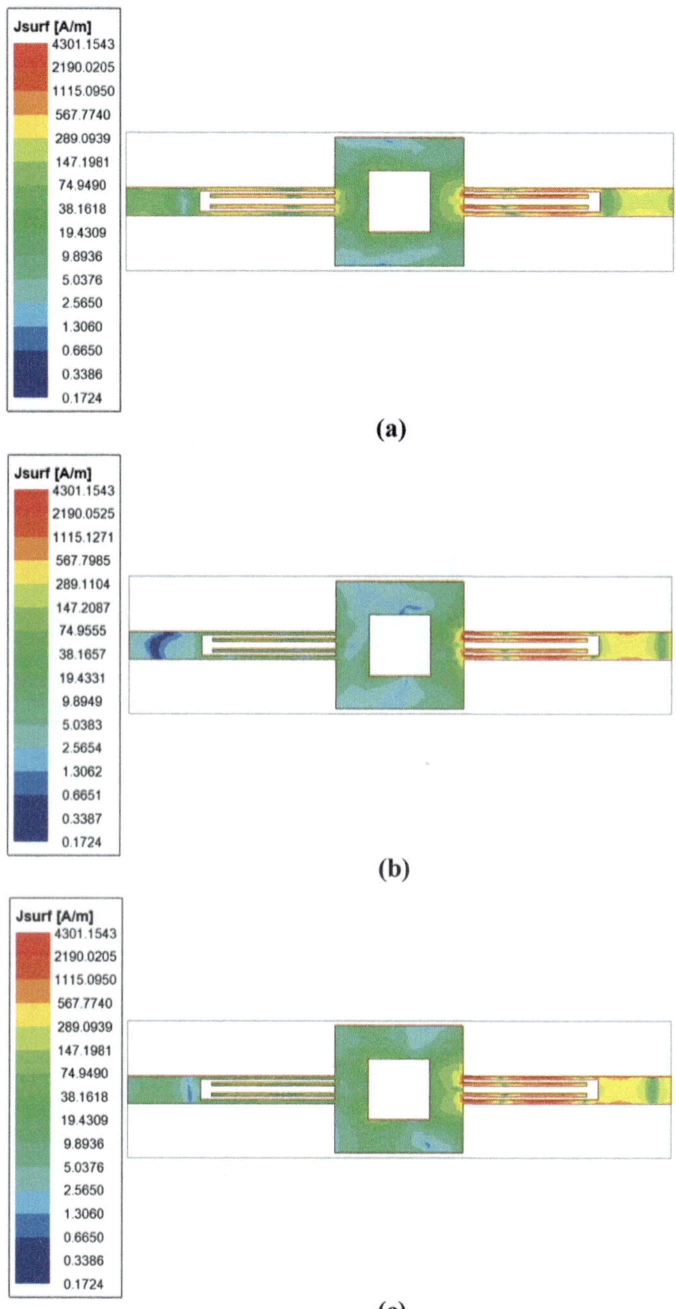

Fig. 8.16 Surface current distribution of the band-stop filter at **a** 26.2 GHz, **b** 27.2 GHz and **c** 28.2 GHz

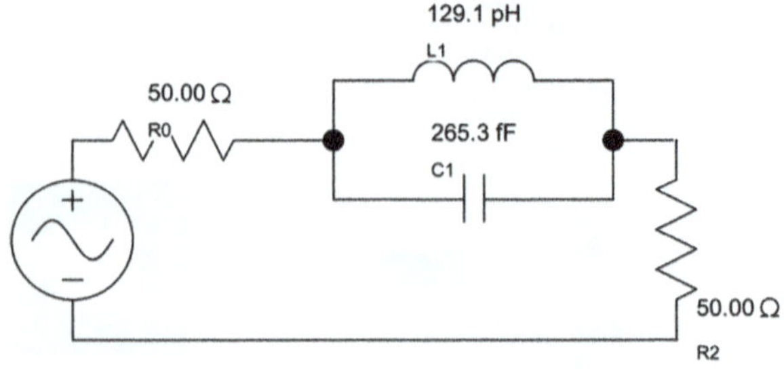

Fig. 8.17 Equivalent circuit of proposed band-stop filter

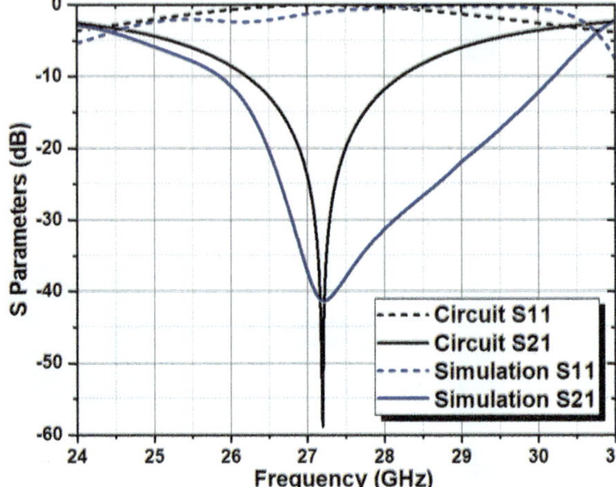

Fig. 8.18 Comparison of the *S*-parameters (S11 and S21) of proposed band-stop structure and its equivalent circuit

Table 8.2 Comparisons the proposed work with other studies

Reference/year of publication	Center frequency (GHz)	Rejected band (GHz)	FBW (%)	IL (dB)	RL (dB)	Size (mm^2)
5 (2021)	6.75	3.7–9.8	90.37	–	–	13 × 24.3
6 (2020)	6.2	3.5–9	88.7	30	0.4	8 × 8
7 (2021)	12.3	6.5–23	92.62	22	1.7	13.35 × 12.27
8 (2020)	8	3.52–12.38	110	23	1.14	7.4 × 15
9 (2016)	31	22–43	67.74	60	2	–
10 (2018)	30	22.45–42.39	66.47	50	0.9	–
This work (2023)	27.2	24.5–30.8	23	41	1.1	11.22 × 13

8.5 Conclusion

This chapter begins by providing general information about filters, with a specific focus on band-stop filters. It then introduces an innovative and compact microstrip band-stop filter designed with a Rogers TMM 10i (tm) substrate. The equivalent circuit of the filter was modeled using the ANSYS circuit editor and the design underwent simulation using the high-frequency structure simulator (HFSS). Notable features of the suggested filter include its small size and a wide stop-band. The band-stop filter effectively rejects signals in the 24–30 GHz band, commonly utilized by 5G communication. This rejection enables applications using frequencies below 24 GHz and above 30 GHz to function efficiently.

References

1. Falcone, F., Lopetegi, T., Baena, J.D., Marqués, R., Martin, F., Sorolla, M.: Effective negative ε stopband microstrip lines based on complementary split ring resonators. IEEE Microwave Wirel. Compon. Lett.Wirel. Compon. Lett. **14**, 280–282 (2004)
2. Elaage, S., El Ghzaoui, M., Mrani, N., Das, S.: Optimum GMSK based transceiver model for cellular IoT networks. Simul. Model. Pract. Theory **125**, 102756A (2023)
3. Ghazaoui, Y., El Ghzaoui, M., Bri, S., Kumari, S.V., Das, S.: A flower-shaped quad-port dual-band MIMO antenna of 29/39 GHz millimeter-wave for 5G applications. J. Nano- Electron. Phys. **14**(3), 03023 (2022)
4. Aghoutane, B., Meskini, N., El Ghzaoui, M., Faylali, H.E.: Millimeter-wave microstrip antenna array design for future 5G cellular applications. In: 2018 International Conference on Electronics, Control, Optimization and Computer Science, ICECOCS 2018, 8610507 (2018)
5. Babu, K.V., Das, S., Ali, S.S., Madhav, B.T.P., Patel, S.K.: Broadband sub-6 GHz flower-shaped MIMO antenna with high isolation using theory of characteristic mode analysis (TCMA) for 5G NR bands and WLAN applications. Int. J. Commun. Syst. **36**(6), e5442 (2023)
6. El Ghzaoui, M., Das, S., Lenka, T.R., Biswas, A.: Terahertz wireless communication components and system technologies. Terahertz Wireless Communication Components and System Technologies, pp. 1–310 (2022)
7. Devi, N.M., Maity, S.: Metamaterial-based miniaturized CPW band stop filter design on silicon substrate for microwave applications. In: 2014 International Conference on Control, Instrumentation, Communication and Computational Technologies (ICCICCT), July 2014, pp. 171–174. IEEE (2014). https://doi.org/10.1109/ICCICCT.2014.6992950
8. Todoran, G., Holonec, R., Dragomir, N.D., Tarnovan, I.G.: A modeling method for analog high-pass and band-stop type filter by using stripline operators. Acta Electrotechnica (2004)
9. Adoum, B.A., Wen, W.P.: Miniaturized matched band-stop filter based dual mode resonator. In: 2011 National Postgraduate Conference, Sept 2011, pp. 1–3. IEEE (2011). https://doi.org/10.1109/NatPC.2011.6136440
10. Rajput, A., Chauhan, M., Mukherjee, B.: An ultra-wideband bandstop filter with circularly etched stub resonator. Microw. Opt. Technol. Lett.. Opt. Technol. Lett. **63**(12), 2958–2963 (2021). https://doi.org/10.1002/mop.33001
11. Ibrahim, A.A., El Shafey, O.K., Abdalla, M.A.: Compact and wideband microwave bandstop filter for wireless applications. Analog Integr. Circ. Sig. ProcessIntegr. Circ. Sig. Process **104**, 243–250 (2020). https://doi.org/10.1007/s10470-020-01675-0
12. Dardeer, O.M., Elsadek, H.A., Elhennawy, H.M., Abdallah, E.A.: Ultra-wideband bandstop filter with multiple transmission zeros using in-line coupled lines for 4G/5G mobile applications. AEU-Int. J. Electron. Commun. **131**, 153635 (2021). https://doi.org/10.1016/j.aeue.2021.153635
13. Gupta, A., Chauhan, M., Rajput, A., Mukherjee, B.: Wideband bandstop filter using L-shaped and Quad mode resonator for C and X band application. Electromagnetics (2020). https://doi.org/10.1080/02726343.2020.1726003
14. Hickle, M.D., Sinanis, M.D., Peroulis, D.: Design and implementation of an intrinsically-switched 22–43 GHz tunable bandstop filter. In: 2016 IEEE 17th Annual Wireless and

Microwave Technology Conference (WAMICON), April 2016, pp. 1–3. IEEE (2016). https://doi.org/10.1109/WAMICON.2016.7483828
15. Adhikari, P., Yang, W., Wu, Y.C., Peroulis, D.: A PCB technology-based 22–42-GHz quasi-absorptive bandstop filter. IEEE Microwave Wirel. Compon. Lett.Wirel. Compon. Lett. **28**(11), 975–977 (2018). https://doi.org/10.1109/LMWC.2018.2867108
16. Matthaei, G.L., Young, L., Jones, E.M.: Microwave Filters, Impedance-Matching Networks, and Coupling Structures. McGraw-Hill (1964)
17. Kiouach, F., Aghoutane, B., El Ghzaoui, M.: Novel microstrip bandpass filter for 5G mm-wave wireless communications. e-Prime—Adv. Electr. Eng. Electron. Energy **6**, 100357 (2023)
18. Chen, W.C.W., Zhao, Y., ZhouxiaoJun, Z.: Compact and wide upper-stopband triple-mode broadband microstrip BPF. TELKOMNIKA (Telecommun. Comput. Electron. Control) **10**(2), 353–358 (2012). https://doi.org/10.12928/telkomnika.v10i2.805

Chapter 9
A Compact Two-Port Semi-flexible Dual-Band Circularly Polarized MIMO Antenna Structure for Millimeter-Wave 26/31 GHz 5G Applications

Saïd Douhi, Adil Eddiai, Tanvir Islam, Sudipta Das, Omar Cherkaoui, and M'hammed Mazroui

9.1 Introduction

Researchers globally are growing increasingly interested in wearable wireless technologies, driven by their diverse applications in health care, fitness, navigation, entertainment, sports, military, and energy harvesting [1, 2]. The widespread utilization of these technologies has resulted in a significant surge in device adoption, escalating from 20 million in 2015 to 187.2 million in 2020. The widespread application of such technologies has contributed to a remarkable rise in the adoption of devices, increasing from 20 million in 2015 to 187.2 million in 2020. Within the research community, wearable communication devices and antennas designed for on-body use are preferred options due to their flexibility, lightweight construction, user-friendly design, durability, and customizable features [3]. The key characteristic of portable antennas is their flexibility, allowing for easy placement on

S. Douhi (✉) · A. Eddiai · M. Mazroui
Laboratory of Physics of Condensed Matter (LPMC), Faculty of Sciences Ben M'Sik, Hassan II University of Casablanca, Casablanca, Morocco
e-mail: said.douhi.15@gmail.com

S. Douhi · O. Cherkaoui
REMTEX Laboratory, Higher School of Textile and Clothing Industries (ESITH), Casablanca, Morocco

T. Islam
Department of Electrical and Computer Engineering, University of Houston, Houston, TX 77204, USA

S. Das
Department of Electronics and Communication Engineering, IMPS College of Engineering and Technology, Malda, West Bengal 732103, India

M. El Ghzaoui et al. (eds.), *Next Generation Wireless Communication*, Signals and Communication Technology, https://doi.org/10.1007/978-3-031-56144-3_9

curved body surfaces [4]. Various materials and fabrication methods, such as inkjet-printed antennas [5], textiles [6, 7], polydimethylsiloxane (PDMS) substrate [8], Kapton polyimide (PI) substrate [9], conductive fabrics [10], and mesh-like structures [11], have been employed to create flexible antennas. The selection of these materials is based on their distinctive characteristics to meet the requirements of diverse applications, ensuring seamless adaptation for on-body antennas. However, antenna component integration and soldering techniques pose problems due to their complexity, despite the improved mobility they offer. In addition, curvature effects on the antenna considerably impair the robustness of the communication connection, leading to reflections and scattering of multipath radio waves as they propagate over curved body surfaces. Consequently, this degradation in the reliability of the communication connection considerably reduces the data transmission rate.

In recent years, the demand has risen for communication technologies operating in the mm-wave band that provide high capacity, high speed, and low latency. This demand has propelled advancements for the next generations of these technologies. The potential to achieve data speeds in the gigabytes per second range is substantial. Moreover, the scarcity of spectrum available in the current commercial and defense application bands has prompted the shift toward the millimeter-wave band [12, 13]. The impressive data rates achievable at millimeter-wave frequencies can be credited to the abbreviated wavelengths, expansive impedance bandwidth, and minimal signal interference. Nevertheless, it is crucial to acknowledge that at elevated frequencies, especially in the millimeter-wave band and beyond, there is a notable increase in path loss attributed to its diminished resilience to atmospheric variations. However, the millimeter-wave band presents a distinct advantage with smaller circuits, particularly for antennas. This size flexibility enables the design of high-gain antennas with directional beams, addressing challenges related to path loss and obstructions effectively. In addition to various other applications, the mm-wave band is being explored for portable applications [14].

This research investigates a semi-flexible MIMO antenna designed for dual-band and dual-circular polarization (CP) in the 26/31 GHz 5G mm-wave band applications. The suggested antenna structure consists of two identical decagon-shaped radiating patch elements, each designed with a pair of E-shaped slots on both the left and right sides, complemented by three strategically positioned rectangular slots in the central region. Utilizing Rogers RO4003C as the dielectric substrate, which possesses a thickness of 0.508 mm, the slender profile enables a slight flexing of the antenna, making it well-suited for wearable applications. The investigation employs CST Microwave Studio (2019) and validation through HFSS (2020 R2). The MIMO antenna operates in dual bands, exhibiting impedance bandwidths of 320 MHz (26.65–26.97 GHz) and 350 MHz (30.86–31.21 GHz). It achieves peak gains of 8.74 and 6.62 dBi, covering the 5G bands within those frequency ranges. The antenna provides isolation levels exceeding 20 and 30 dB, along with an ECC below 0.0001 and a DG exceeding 10 for the 26 and 31 GHz bands. Consequently, the presented MIMO demonstrates its competence as an mm-wave antenna suitable for advanced 26/31 GHz applications in the fifth-generation (5G) context.

9.2 MIMO Antenna Design and Discussion

The geometric representation of the suggested miniaturized dual-band mm-wave MIMO antenna is illustrated in Fig. 9.1. The top layer showcases a radiating patch with a decagonal shape, which is fed through a 50 Ω microstrip line. E-shaped slots are incorporated into the left and right sides of the radiating element, complemented by three strategically positioned rectangular slots within the central region of the patch. The slot design provides three key features: wide impedance matching, CP properties, and miniaturization. A complete ground plane is located on the rear face of the dielectric substrate. The envisioned antenna is developed on a semi-flexible thin substrate (Rogers RO4003C) with a dielectric constant (ε_r) of 3.55, a loss tangent ($\tan \delta$) of 0.00027, and a thickness (h) of 0.508 mm. The antenna elements are connected to a 50-Ω feed transmission line, with the width determined using fundamental antenna design equations. The specified dimensions of the suggested MIMO antenna configuration are $22.5 \times 30.0 \times 0.508$ mm^3, with the optimized parameters detailed in Table 9.1.

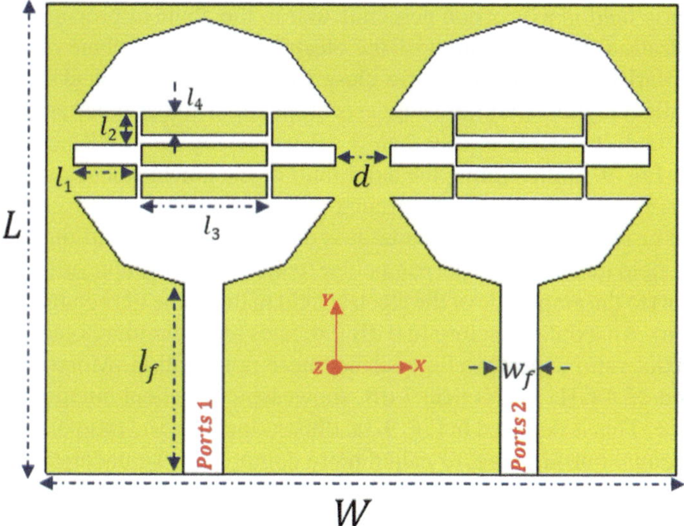

Fig. 9.1 Schematic of the proposed MIMO antenna geometry

Table 9.1 Antenna design parameters (dimensions in mm)

Parameter	Value (mm)	Parameter	Value (mm)	Parameter	Value (mm)
L	22.5	w_f	1.80	l_2	1.5
W	30	D	2.63	l_3	6
l_f	9.04	l_1	3	l_4	1

9.3 Simulated Results and Discussion

9.3.1 Impedance Characteristics

In Fig. 9.2a, the antenna exhibits resonance at two distinct frequencies within the millimeter-wave band, specifically at 26.81 and 31 GHz, highlighting favorable impedance characteristics. Results obtained from two distinct electromagnetic software packages consistently converge across a broad frequency spectrum. Any minor discrepancies observed are likely attributed to variations in the numerical methods employed by the respective software tools. Assuming complete similarity between the two antenna elements, it is presumed that $S11$ equals $S22$, and $S12$ equals $S21$. Subsequently, Fig. 9.2b illustrates the simulated parameters $|S21|$, which are intended for assessing the mutual coupling effects between adjacent antenna elements. Notably, the isolation effect is observed to be below − 20 dB in both operating bands, indicating effective separation between the antenna elements. The axial ratio of an antenna is a measure of the circular polarization of an electromagnetic wave transmitted or received by the antenna. It is defined as the ratio of the strength of the electric field in a direction perpendicular to the plane of propagation (orthogonal polarization) to the strength of the electric field in the plane of propagation (parallel polarization). An axial ratio close to 0 dB indicates ideal circular polarization, while a higher axial ratio suggests elliptical or linear polarization. Besides, the AR around 26.4 GHz is less than 3 dB, achieving good CP performance, as depicted in Fig. 9.3a illustrating the axial ratio of the antenna. The axial ratio of an antenna serves as a measure of the circular polarization of an electromagnetic wave transmitted or received by the antenna. It is defined as the ratio of the amplitude of the electric field in a direction perpendicular to the plane of propagation (orthogonal polarization) to the amplitude of the electric field in the plane of propagation (parallel polarization). An axial ratio close to 0 dB indicates ideal circular polarization, while a higher axial ratio suggests elliptical or linear polarization. Moreover, the axial ratio around 26.4 GHz is less than 3 dB, showcasing excellent circular polarization performance. This is depicted in Fig. 9.3a, illustrating the axial ratio of the dual-band MIMO antenna. Furthermore, to gain a more comprehensive understanding of how the MIMO antenna system operates, the distribution of surface currents at 26.81 and 31 GHz is obtained and depicted in Fig. 9.4. The representation indicates that the surface current concentrates in and around the first antenna without any spillover to the second antenna, given that the port linked to antenna 1 is stimulated, and the other ports are terminated with loads having a matching impedance of 50 Ω. This observation offers valuable insights into the enhanced isolation between neighboring antenna components.

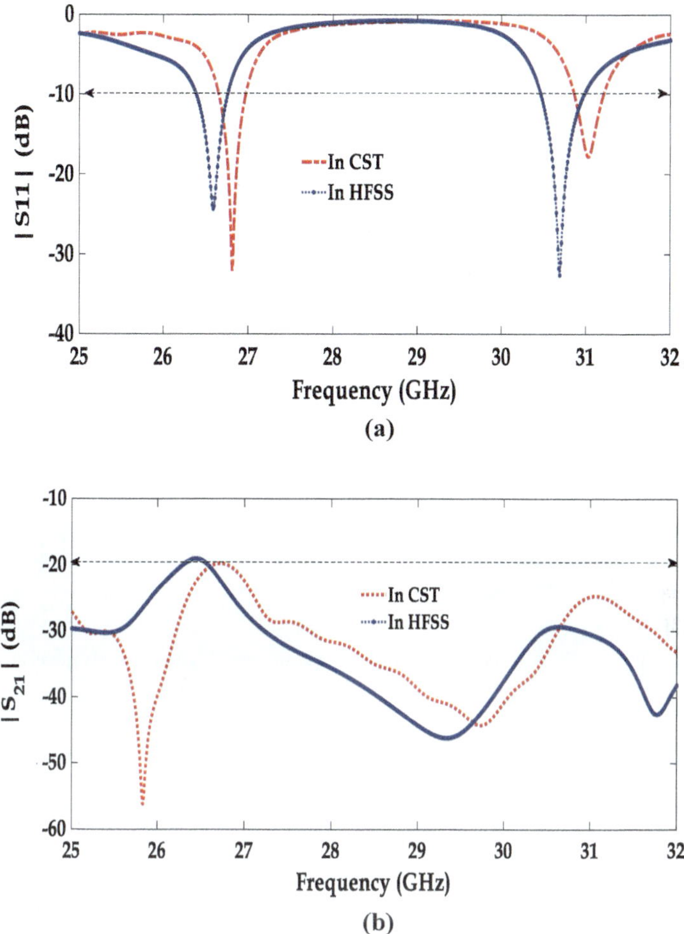

Fig. 9.2 Simulated results of the suggested antenna in CST and HFSS, **a** reflection coefficient (S_{11}) and **b** isolation (S_{12}/S_{21})

9.3.2 *Bending Analysis*

One of the pivotal attributes inherent to flexible antennas lies in their flexibility and conformal nature. To gauge these characteristics, the performance is scrutinized under bending conditions along the x- and y-directions, delineated by radii R_x and R_y, as depicted in Fig. 9.5. In Fig. 9.6a, the simulated bandwidth of − 10 dB |$S11$|, which encompasses the resonant peaks at the specified frequencies, undergoes a nuanced frequency adjustment with an increase in the degree of bending along the x-direction. Notably, this shift transpires without discernible detuning in performance. Additionally, a comparable frequency shift toward lower values, coupled with negligible detuning evident in the overall outcomes, is observed in cases of substantial

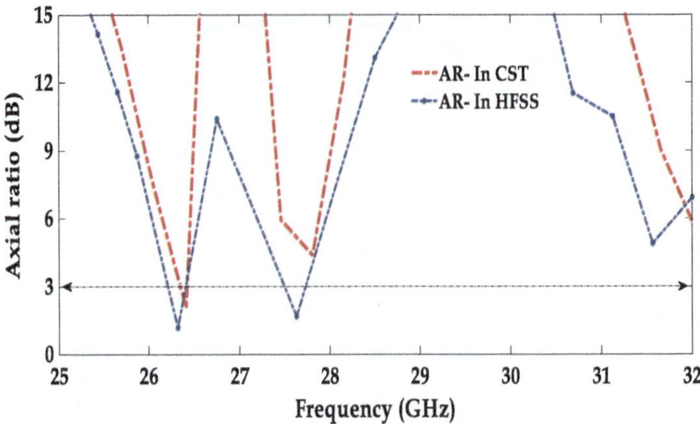

Fig. 9.3 Simulated axial ratio (AR) for proposed mm-wave MIMO antenna

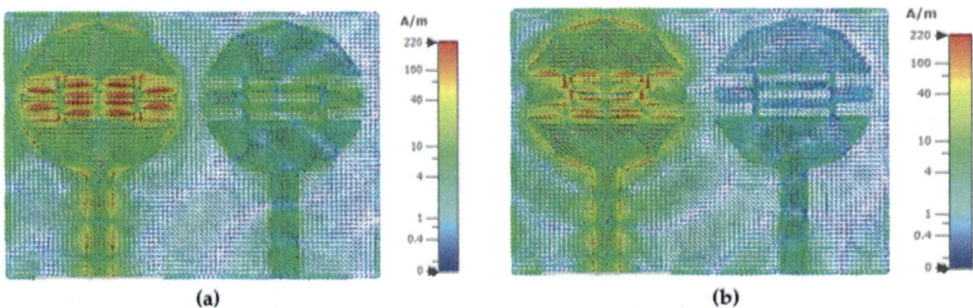

Fig. 9.4 Simulated surface current when the first antenna excited at **a** 26.81 GHz, and **b** 31 GHz

bending along the *y*-direction, as depicted in Fig. 9.6b. This anticipated shift in frequency, concerning the flat configuration, is attributed to the elongation of the current trajectory on the radiating element induced by bending in both directions.

9.3.3 Radiation Characteristics

Figure 9.7 illustrates the performance of the MIMO antenna in dual bands, showcasing both its gain and radiation efficiency. Within the initial frequency band (26.81 GHz), the antenna attains a maximum gain of 8.76 dBi, while in the second band (31 GHz), it achieves a peak gain of 6.39 dBi. Concurrently, the simulated radiation efficiency is 76% at 26.81 GHz, and in the second band, it reaches a peak of 77%. These notable efficiency values stem from the utilization of high-performance Rogers RO4003C dielectric material, known for its low loss tangent of 0.0027. The three-dimensional radiation patterns for the proposed antenna's gain are computed

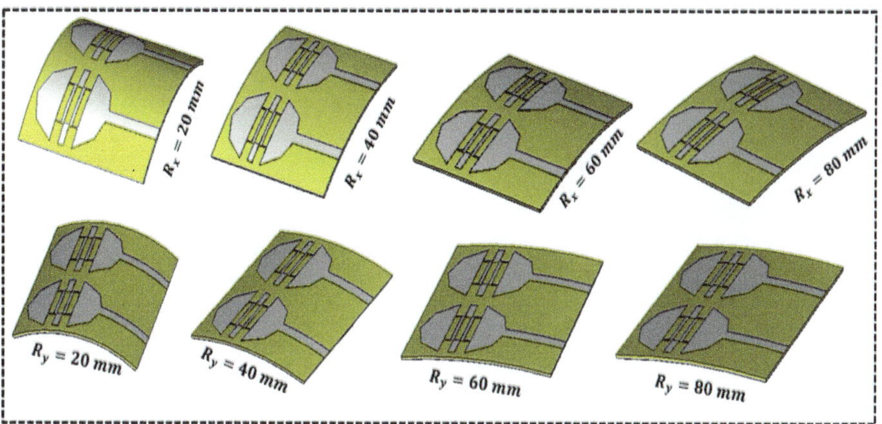

Fig. 9.5 Proposed antenna in a bent configuration, with radii R_x and R_y applied along the x- and y-directions, respectively

using CST. As illustrated in Fig. 9.8, the antenna demonstrates peak gains of 8.2 dBi at 26.81 GHz and 6.18 dBi at 31 GHz, respectively. Subsequently, Fig. 9.9 illustrates the simulated radiation pattern of the antenna, showcasing the E-plane ($\phi = 0°$) and H-plane ($\phi = 90°$), observed at frequencies of 26.81 GHz and 31 GHz. The radiation patterns showcase a distinct dumbbell shape in the E-plane while exhibiting omnidirectional characteristics in the H-plane.

9.3.4 Diversity Performance

Beyond conventional antenna performance parameters, which encompass bandwidth, resonance properties, surface current, circular polarization characteristics, gain, radiation patterns, and efficiency, an array of parameters must be considered to comprehensively evaluate the diverse performance characteristics of a MIMO antenna configuration. Parameters such as ECC and DG are widely utilized to assess the diversity features of a MIMO antenna system. The ECC serves as a performance metric, indicating the degree of isolation or correlation between communication channels. Figure 9.10a illustrates the simulated diversity performance of the MIMO antenna system using CST. The ECC consistently remains below 10^{-4} across the two operational bands of the suggested MIMO antenna, well below the accepted threshold of 0.5. Additionally, the simulated diversity gain exhibits an approximate variation of 10 dB across the MIMO two operational bands, as depicted in Fig. 9.10b.

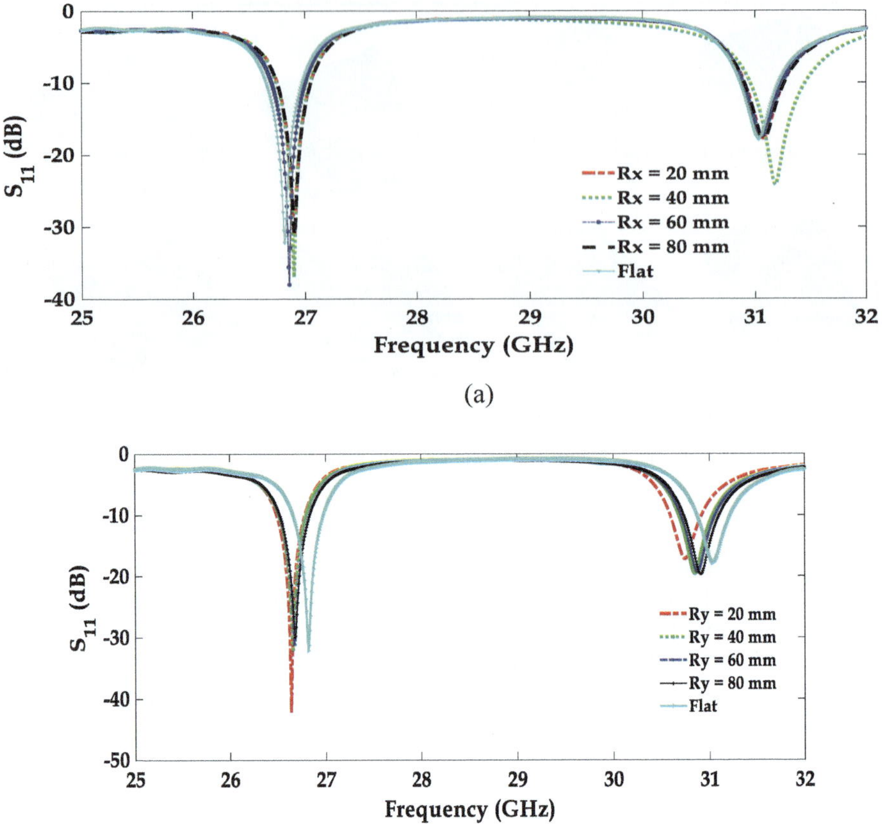

S_{11} (dB)

Rx = 20 mm
Rx = 40 mm
Rx = 60 mm
Rx = 80 mm
Flat

Frequency (GHz)

(a)

S_{11} (dB)

Ry = 20 mm
Ry = 40 mm
Ry = 60 mm
Ry = 80 mm
Flat

Frequency (GHz)

(b)

Fig. 9.6 S_{11} curves of the suggested semi-flexible antenna in various bending radii. **a** Bending along the x-axis and **b** bending along the y-axis

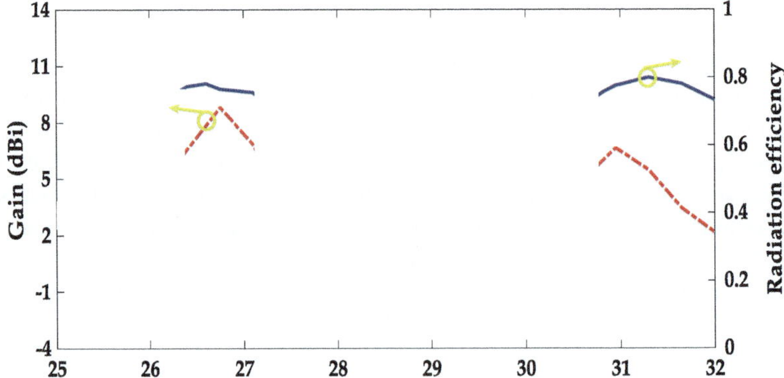

Fig. 9.7 Simulated gain and radiation efficiency for the proposed millimeter-wave MIMO antenna configuration

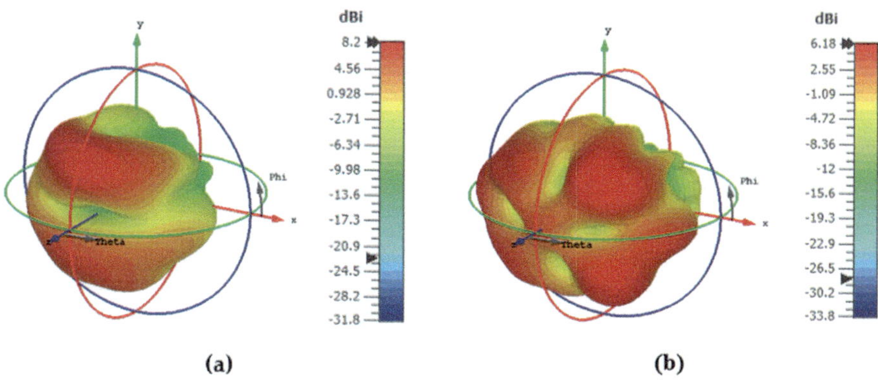

Fig. 9.8 Simulated 3D gain radiation patterns **a** 26.81 GHz and **b** 31 GHz

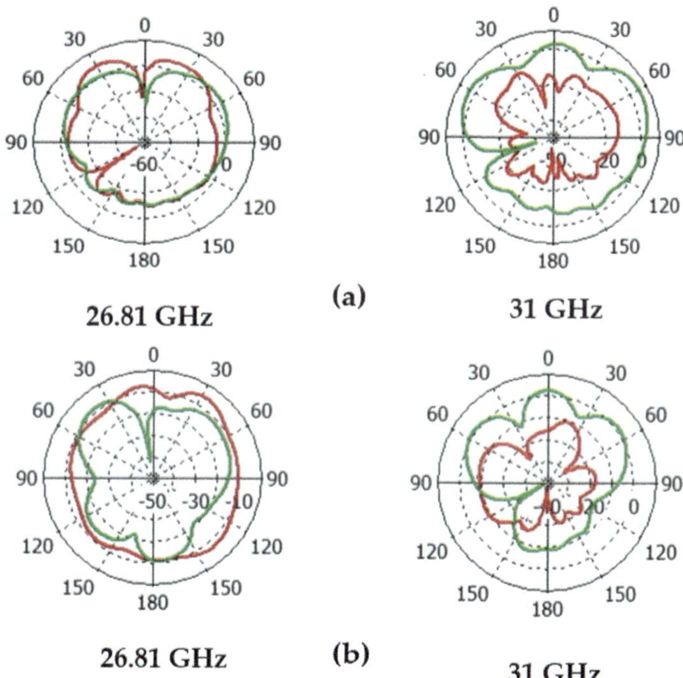

Fig. 9.9 Simulated far-field radiation pattern of the antenna **a** E-plane and **b** H-plane

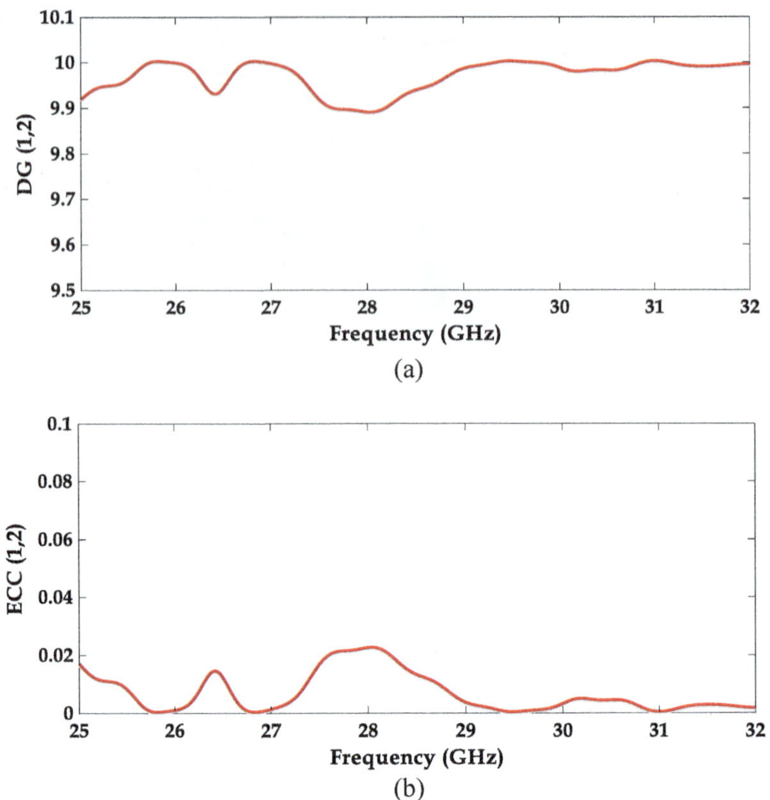

Fig. 9.10 MIMO performance analysis of the proposed millimeter-wave antenna across frequencies, **a** ECC and **b** DG

9.4 Conclusion

In this study, we present a circularly polarized, dual-port, semi-flexible MIMO antenna configuration specifically designed for millimeter-wave spectrum applications, optimized for operation at 26 and 31 GHz. Notably, the axial ratio (AR) around 26.4 GHz remains below 3 dB, indicating excellent circular polarization (CP) performance. By incorporating suggested geometrical modifications, our proposed MIMO antenna system facilitates dual-band operation, showcasing remarkable attributes such as high gain, efficiency, and CP properties, along with outstanding MIMO performance characteristics. The achieved impedance bandwidths of 320 MHz (26.65–26.97 GHz) and 350 MHz (30.86–31.21 GHz) contribute to the adaptability of the proposed antenna. With a commendable high gain of approximately 8.74 dB at 26 GHz and 6.62 dB at 31 GHz, the antenna exhibits favorable radiation characteristics. Further assessments of MIMO performance, including ECC and diversity gain (DG), reveal practically acceptable values, ensuring exceptional MIMO performance. The strategic use of Rogers 5880 dielectric substrate, featuring a thickness of 0.508 mm, allows the antenna to conform comfortably to curved surfaces when worn. Its thin profile permits slight bending, rendering it well-suited for applications in the field of wearables. In conclusion, the proposed semi-flexible CP two-port MIMO antenna system emerges as a promising candidate for the next-generation 26 GHz (26.65–26.97 GHz) and 31 GHz (30.86–31.21 GHz) 5G bands. It offers a comprehensive combination of dual-band operation, high gain, and reliable MIMO performance, making it a potential solution for advanced 5G applications.

Acknowledgements We want to thank the Moroccan Ministry of Higher Education, Scientific Research and Innovation, and the OCP Foundation, which funded this work through the APRD research program.

References

1. Douhi, S.: Investigation of SAR reduction and gain enhancement using an all-textile antenna with metamaterial structure for wireless body area network applications. Mater. Today (n.d.)
2. Douhi, S., Eddiai, A., Das, S., Madhav, B.T.P., Meddad, M., Cherkaoui, O., Mazroui, M.: Design of a compact super wideband all-textile antenna for radio frequency energy harvesting and wearable devices. Opt. Quantum Electron. **55**, 1189 (2023). https://doi.org/10.1007/s11082-023-05498-x
3. Ullah, U., Koziel, S., Pietrenko-Dabrowska, A.: Design and characterization of a planar structure wideband millimeter-wave antenna with wide beamwidth for wearable off-body communication applications. IEEE Antennas Wirel. Propag. Lett. **21**, 2070–2074 (2022). https://ieeexplore.ieee.org/abstract/document/9829278/. Accessed 25 Dec 2023
4. Yang, H., Liu, X., Fan, Y., Xiong, L.: Dual-band textile antenna with dual circular polarizations using polarization rotation AMC for off-body communications. IEEE Trans. Antennas Propag. **70**, 4189–4199 (2022). https://ieeexplore.ieee.org/abstract/document/9670701/. Accessed 25 Dec 2023
5. Yang, W., Cheng, X., Guo, Z., Sun, Q., Wang, J., Wang, C.: Design, fabrication and applications of flexible RFID antennas based on printed electronic materials and technologies. J. Mater. Chem. C **11**, 406–425 (2023)
6. Douhi, S., Islam, T., Eddiai, A., Das, S., Cherkaoui, O.: Design of a flexible rectangular antenna array with high gain for RF energy harvesting and wearable devices. J. Nano-Electron. Phys. **15**, 03010-1–03010-6 (2023). https://doi.org/10.21272/jnep.15(3).03010
7. Douhi, S., Prasad, G.R.K., Eddiai, A., Cherkaoui, O., Mazroui, M., Das, S.: A miniaturized wearable textile UWB monopole antenna for RF energy harvesting. J. Nano-Electron. Phys. **15**, 01028-1–01028-6 (2023). https://doi.org/10.21272/jnep.15(1).01028
8. Janapala, D.K., Nesasudha, M., Mary Neebha, T., Kumar, R.: Design and development of flexible PDMS antenna for UWB-WBAN applications. Wirel. Pers. Commun. 1–17 (2022)
9. Wang, Z., Qin, L., Chen, Q., Yang, W., Qu, H.: Flexible UWB antenna fabricated on polyimide substrate by surface modification and in situ self-metallization technique. Microelectron. Eng. **206**, 12–16 (2019)
10. Zhang, K., Soh, P.J., Yan, S.: Design of a compact dual-band textile antenna based on metasurface. IEEE Trans. Biomed. Circuits Syst. **16**, 211–221 (2022)
11. Kumar, P., Pathan, S., Vincent, S., Kumar, O.P., Yashwanth, N., Kumar, P., Shetty, P.R., Ali, T.: A compact quad-port UWB MIMO antenna with improved isolation using a novel mesh-like

decoupling structure and unique DGS. IEEE Trans. Circuits Syst. II Express Briefs **70**, 949–953 (2022)
12. Tan, Q., Fan, K., Yu, W., Liu, L., Luo, G.Q.: A parallel folded dipole antenna with an enhanced bandwidth for 5G millimeter-wave applications. IEEE Trans. Antennas Propag. (2023)
13. Wagih, M., Hilton, G.S., Weddell, A.S., Beeby, S.: Broadband millimeter-wave textile-based flexible rectenna for wearable energy harvesting. IEEE Trans. Microw. Theory Tech.Microw. Theory Tech. **68**, 4960–4972 (2020)
14. Tiwari, R.N., Kaim, V., Singh, P., Khan, T., Kanaujia, B.K.: Semi-flexible diversified circularly polarized millimeter-wave MIMO antenna for wearable biotechnologies. IEEE Trans. Antennas Propag. (2023)

Chapter 10
Extension of Indoor MmW Link Radio Coverage in Non-line-of-Sight Conditions

Mbissane Dieng, Gheorghe Zaharia, Ghaïs El Zein, Raphaël Gillard, and Renaud Loison

10.1 Introduction

The high-data-rate applications have experienced significant growth thanks to the integration of millimetre-band wireless communications. mmWave frequencies, ranging from 30 to 300 GHz, ensure such data rates and increase network capacity. However, due to their short wavelengths, these millimetre waves have high propagation losses, high attenuation by penetration or obstruction and are very attenuated by blocking, which can be caused by the human body.

To provide solutions to these problems, recent research is focused on the integration of massive MIMO and beamforming techniques. In the same time, the environment could be equipped with passive reflectors and/or arrays of active and reconfigurable surfaces. These objects will be placed at intermediate points capable of improving radio coverage and beamforming [1, 2]. In [3], the propagation channel was investigated in order to increase the received power at 28 GHz in NLOS configuration. Reflectors with different shapes were used: flat squares, cylindrical and spherical. When using flat reflectors, it was observed that their orientation is more important than their size. In [4], a cylindrical reflector was used at 60 GHz in an *L*-shaped corridor for a *Tx*–*Rx* separation of 124 m. In [5], the authors designed and experimentally validated a passive reflector operating in order to increase the received power at 60 GHz in a *T*-shaped corridor. In [6], the effect of passive reflectors was studied to improve NLOS coverage of a mmWave system based on IEEE 802.11ad. The results show that metal reflectors can reduce the path loss (PL) by more than 10 dB in some indoor environments. Moreover, one of the first studies of the human blocking [7] was conducted by measurements at 60 GHz for wireless LANs within an office-like environment. Losses of around 20 dB were noted with omnidirectional antennas and 30 dB when using directional antennas [8, 9]. In this experimental

M. Dieng (✉) · G. Zaharia · G. E. Zein · R. Gillard · R. Loison
UMR 6164, Univ. Rennes, INSA Rennes, CNRS, IETR, 35000 Rennes, France
e-mail: mbissane.dieng@insa-rennes.fr

M. El Ghzaoui et al. (eds.), *Next Generation Wireless Communication*, Signals and Communication Technology, https://doi.org/10.1007/978-3-031-56144-3_10

study, we consider the optimal position and orientation of the *Rx* directive antenna according to the type of reflector and the dimensions of different corridors.

Initially, the objective of this work is to increase the power received at 60 GHz in NLOS scenarios, first using a passive reflector in an *L*-shaped corridor, then using an array of 80 metallic grooves in a *T*-shaped corridor. Next, the study focuses on the blocking losses introduced by the human body at 60 GHz.

In order to achieve these objectives, measurements and simulations are carried out in these environments. The results obtained, in terms of path loss or channel impulse response, lead to a comparative analysis with and without the reflectors or with and without a human blocker.

The chapter is organized as follows: Sect. 10.2 describes a study of passive reflectors in different indoor environments at 60 GHz, then highlights the improvement in radio coverage and the influence of the reflector arrangement. Section 10.3 focuses on the study of the impact of blocking the radio link by the human body and the possibility of implementing beamforming as a solution against blocking. Finally, Sect. 10.4 summarizes several conclusions.

10.2 Propagation Using Passive Reflectors at 60 GHz

10.2.1 Use of a Passive Reflector in an L*-Shaped Corridor*

In this part, we evaluate by measurements the impact of a metal reflector panel on radio coverage in an indoor NLOS scenario at the central frequency $f_c = 60$ GHz.

Measuring System

The used measurement system (Fig. 10.1) is based on a vector network analyzer (VNA). $N = 401$ frequency points were considered over a bandwidth of $B = 2$ GHz, around the intermediate frequency IF = 3.5 GHz. Therefore, the frequency step is $\Delta f = 5$ MHz. The transmitted power is $P_t = 0$ dBm.

More details on the measurement system and calculation of the frequency response of the indoor radio channel are given in [10]. Table 10.1 gives the principal parameters of the used system.

For these measurements, two types of antennas were used (Fig. 10.2).

An omnidirectional antenna, with a gain of 2 dBi in azimuth and an opening at − 3 dB of 30° in elevation, was used on the *Tx* side. The *Rx* side used a horn antenna, with a gain of 22.5 dBi and a beamwidth at − 3 dB of 13° in azimuth and 10° in elevation.

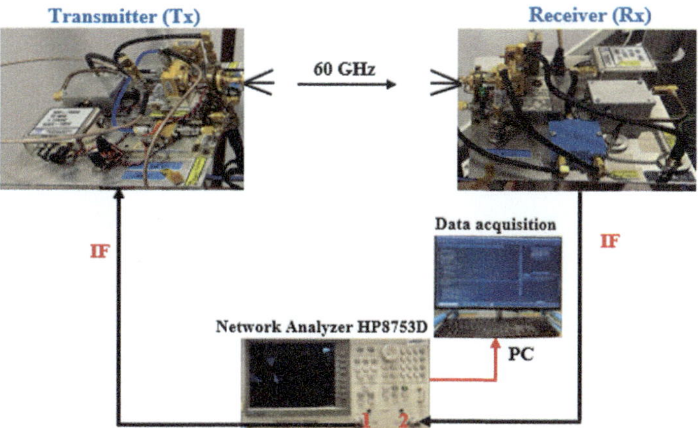

Fig. 10.1 Measurement system

Table 10.1 Parameters of the measurement system

f_c (GHz)	60
B (GHz)	2
IF (GHz)	3.5
Δf (MHz)	5
N	401
P_t (dBm)	0

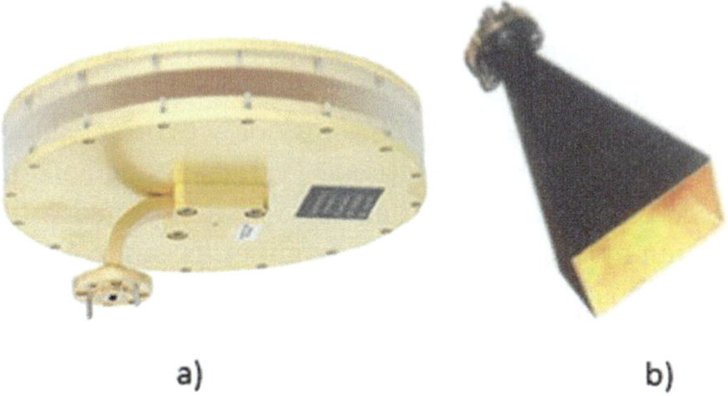

Fig. 10.2 60 GHz antennas: **a** omnidirectional and **b** horn

Environment and Measurement Scenario

As shown in Fig. 10.3, the measurement environment is an *L*-shaped corridor consisting of two parts *A* and *B* whose dimensions are $3.69 \times 2 \times 3$ m^3 and $4.7 \times 1.62 \times 3$ m^3, respectively [10].

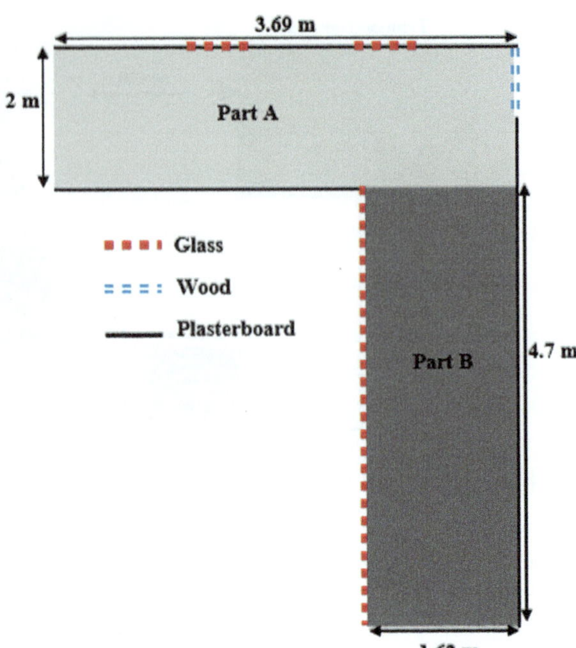

Fig. 10.3 *L*-shaped corridor

The considered corridor has the walls made of plasterboard. The environment was maintained time-invariant, i.e. without any movement. Figure 10.4 describes the considered environment. A metal plate (*AB*), with a length of 98.2 cm and a width of 59.5 cm, was used. The panel was placed firstly on a table in vertical position. In this case, the centre M of the panel was at a height of $h = 1.37$ m from the ground. This height has been set for *Tx* and *Rx* antennas. Moreover, the *Rx* horn antenna, fixed at a distance $L_R = 3.69$ m from the end of the corridor, was oriented towards the centre M of the reflector panel. The omnidirectional *Tx* transmitting antenna was placed in 16 positions spaced 25 cm apart. Measurements were repeated with a panel placed in horizontal position by keeping the same height of its centre M.

Geometric methods are used in [11] to calculate the optimal angles and finally properly orient the panel and the *Rx* antenna.

Measurement and Simulation Results

For these measurements, we have eliminated the frequency response of the cables by calibration. Then, the frequency response of RF blocks was obtained by back-to-back measurement. The RF frequency response and antenna gains were removed from measured data. We thus obtain the frequency response of the propagation channel. Its module was averaged in linear scale over the 401 frequency points in order to

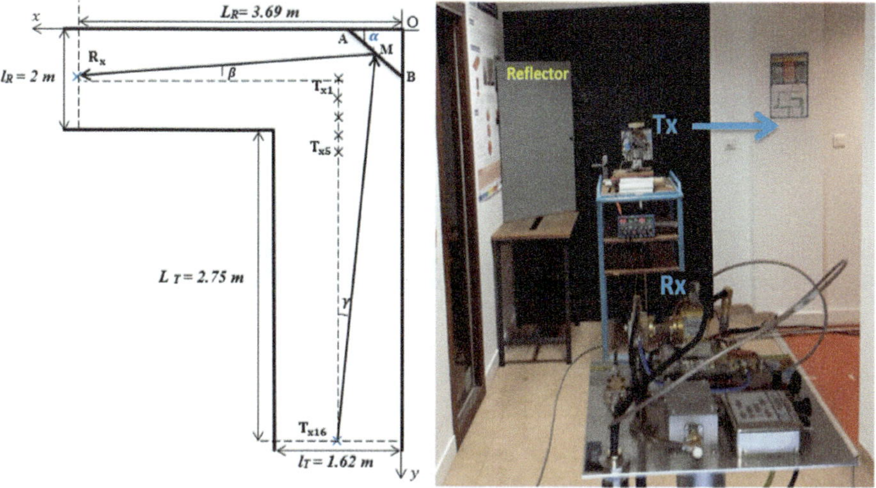

Fig. 10.4 *L*-shaped corridor measurement environment

obtain the average path loss for a given *Tx* position. Figure 10.5 shows the obtained path loss (PL) for all *Tx* positions.

This figure allows a comparison between the path loss made when the reflector is placed vertically and horizontally, with the same orientation and the same height of its centre *M*.

Overall, we find almost the same results when there is no reflector (Fig. 10.5) [11]. In this case, there are some differences that may be due to inaccuracies in the positions of the antennas. This difference can also be explained by the fact that the measurements (*V*) were carried out during one day and the measurements (*H*) during another day, which means that the back-to-back measurements were perhaps not identical. By using the reflector, it is possible to observe a PL decrease up to

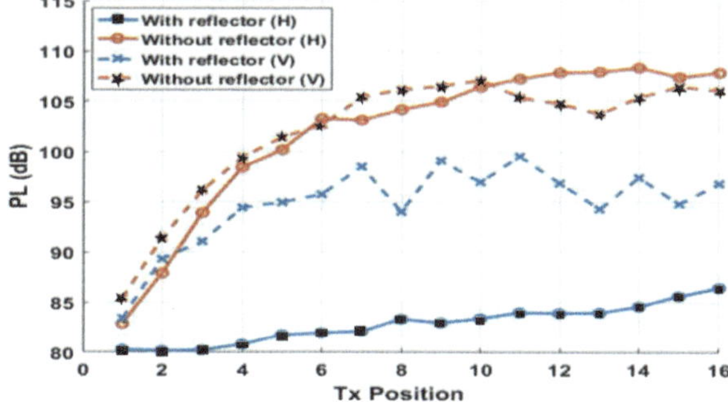

Fig. 10.5 Path loss versus *Tx* positions

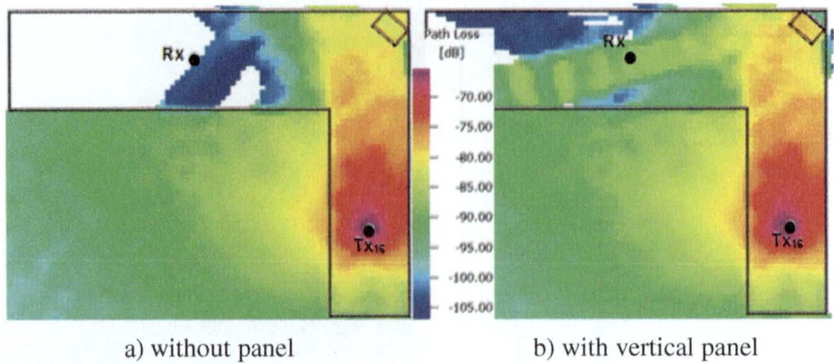

Fig. 10.6 Simulated path loss in *L*-shaped corridor

24 dB in the horizontal case (*H*) and a decrease to 12 dB considering the vertical case (*V*). Thus, in the horizontal case, the reflected energy is greater.

The same scenario was analysed by simulation carried out with the commercial software WinProp. Figure 10.6 gives the PL distribution for the position *Tx*16 with and without the vertical panel.

For the farthest position *Tx16* (Fig. 10.6a), without the use of the metal panel, the path loss (PL) is approximately 105 dB. When the metal panel is used, the PL decreases to around 95 dB (Fig. 10.6). These simulation results confirm that the use of the metal panel placed in vertical position can lead to a PL reduction of approximately 10 dB, which is close to the result obtained by measurement (Fig. 10.5). This improves radio coverage in this NLOS configuration.

10.2.2 *Use of a Passive Reflector Array in a* T-*Shaped Corridor*

In this indoor environment, a new measurement campaign was carried out using a groove cell reflector array.

The measurement system used is the same as previously, except that the VNA measures S21 parameter on 801 points over the same frequency band.

An antenna positioning system with a rotation accuracy of 0.02° was used (Fig. 10.7). Its role is to perform rotations of the Rx antenna in precise steps for a study of the angle of arrival (AoA) of the wave in azimuth.

The VNA and the antenna positioner are controlled by a computer via appropriate interfaces, which make it possible to perform antenna rotations using a programme developed at IETR–INSA and to record the measurement data. Figure 10.8 shows the radiation pattern of the horn antennas used for this measurement campaign. They have the same characteristics as the horn antenna used for the previous campaign.

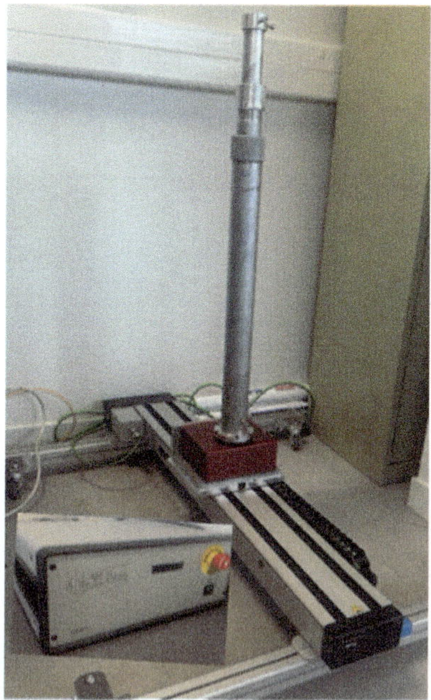

Fig. 10.7 Antenna positioner

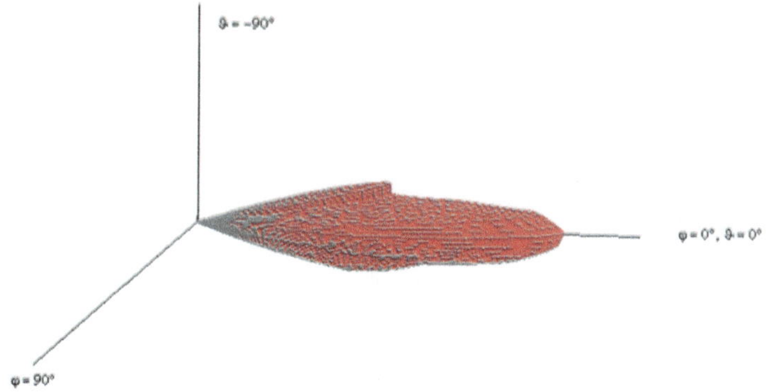

Fig. 10.8 Horn antenna radiation pattern

Environment and Measurement Scenario

The measurement environment is described in Fig. 10.9. The walls of the corridors are made of plasterboard. They have a height of 2.7 m. The doors are made of wood. Their dimensions are identical: 2 m height and 0.9 m width.

As previously mentioned, the two identical horn antennas were used during this measurement campaign. The reflector array was centred on the axis of the short

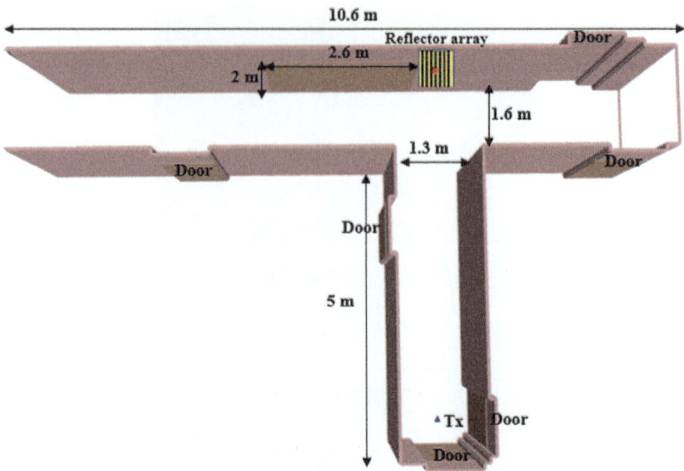

Fig. 10.9 *T*-corridor measurement environment

corridor (Fig. 10.10). The *Tx* antenna was placed in front of the array at a distance of 5.6 m. The *Rx* antenna was placed on the axis of the other part of the corridor, perpendicular to the transmitter-reflector axis, in 14 positions spaced 1 m apart. The array is placed 21.5 cm from the wall and its centre has been set at the height of the *Tx* and *Rx* antennas: 1.7 m.

For each position of the *Rx* antenna, a rotation is carried out over 360° in steps of 6°, which allows us to measure the S21 over the 60 angles of arrival (AoA) and to

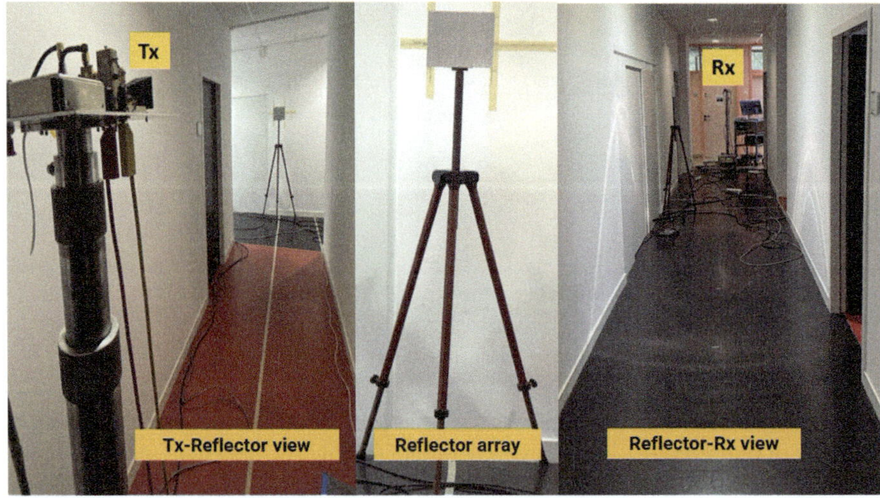

Fig. 10.10 General view of *T*-corridor measurement environment

identify the direction that corresponds to the maximum received power. During this measurement campaign, two scenarios were studied successively, with and without a reflector array.

Design of the Reflecting Panel

The reflecting panel is designed assuming an impinging plane wave under normal incidence, with x polarization. Referring to Fig. 10.11, the wave vector for the incident wave ($\overrightarrow{k}_{\text{inc}}$) can be expressed by

$$\vec{k}_{\text{inc}} = -k_0\vec{e}_z, \tag{10.1}$$

where k_0 is the free space wave number.

The goal is to steer the reflected wave in the $\pm x$ directions, as shown in Fig. 10.11, where $\pm\vec{k}_{\text{ref}}$ is the wave vector for reflected waves. More precisely, the reflected power is split into two symmetrical beams flowing along the panel surface in the ($\theta = 90°$, $\phi = 0°$) and ($\theta = 90°$, $\phi = 180°$) directions.

The dotted area shows two consecutive cells as detailed in Fig. 10.12. In order to do so, a phase gradient has to be applied over the reflected panel [12]. In practice, the gradient is produced by discretizing the panel into elementary cells, each of the cells reflecting the incident wave with a desired phase shift. The principle is basically the same as the one used in reflectarray design [13], except that here the illuminating feed is not attached to the reflecting panel but located at $z = +\infty$. Also, as the beams

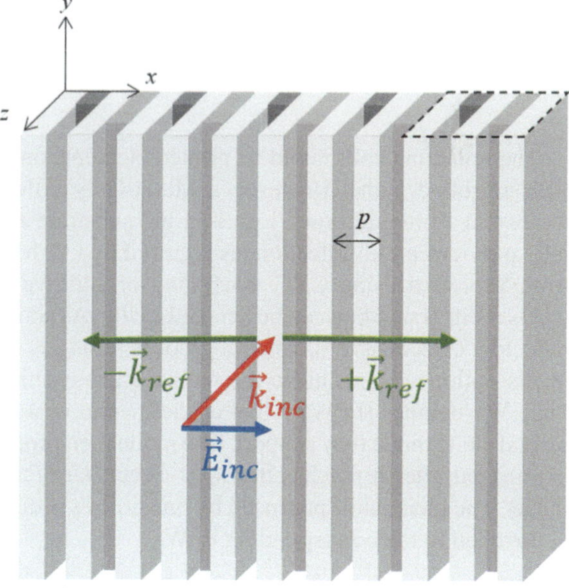

Fig. 10.11 Geometry and principle of the reflecting panel

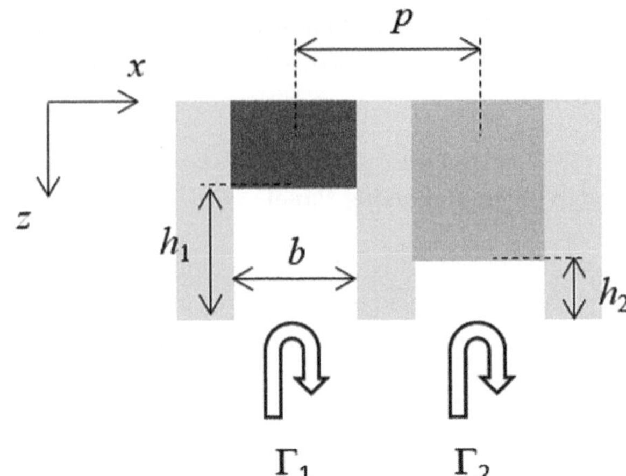

Fig. 10.12 Two consecutive cells and associated reflection coefficients (*Note* The different grey colours represent the same metal material and are just used for an easier readability of Fig. 10.11)

here have to be steered in (x, z) plane only, no phase variation along the y direction is required. Then, the discretization is only done along the x axis, with inter-element spacing p separating two consecutive cells.

Now, simple antenna array theory shows that the two required reflected beams can be produced by using opposite phase shifts (typically 0° and 180°) on consecutive cells, provided p is set to $\lambda_0/2$, λ_0 being the free space wavelength corresponding to the central frequency ($f_0 = 60$ GHz). Referring to Fig. 10.12, the reflection coefficients on consecutive cells (for instance Γ_1 on cell 1 and Γ_2 on cell 2) must comply with:

$$\arg(\Gamma_1) - \arg(\Gamma_2) = \pm 180^\circ, \tag{10.2}$$

$$|\Gamma_1| = |\Gamma_2| = 1. \tag{10.3}$$

The reflecting cells could be printed elements on top of a dielectric layer backed with a ground plane. However, a Metal-Only reflector is preferred here in order to prevent from additional losses in the substrate at 60 GHz and because it naturally provides a full reflection, as required by (3). Moreover, the fabrication is quite straightforward using CNC machining or additive manufacturing, and it directly yields a stiff frame that can be handled easily. As shown in Figs. 10.11 and 10.12, the reflecting cells consist of vertical grooves (along y) in the metal panel whose depths (h_i) is optimized to achieve the required phase shifts. The grooves act as Parallel Plate Waveguides (PPW). Most of the incoming wave is converted into the fundamental TEM mode they support. The mode then propagates along $-z$ in the PPW and is reflected after it reaches the short-circuit termination (at distance h_i from PPW input). The phase shift produced by one groove is directly controlled by the distance h_i travelled in the corresponding PPW.

Table 10.2 Geometry of the two different cells used in the panel

Cell index (i)	p (mm)	b (mm)	h_i (mm)
1	2.5	2	2.3
2	2.5	2	0.48

Here, only two different phase shifts and subsequently only two different heights (h_1 and h_2) are needed. The derivation of the dimensions of the grooves has been explained in [5] and will not be repeated here for the sake of concision. Table 10.2 summarizes the geometry of the two optimized cells. The panel is then fabricated by alternating cells with heights h_1 and h_2.

The fabricated panel is a 20 cm × 20 cm ($40\lambda_0 \times 40\lambda_0$) square made from aluminium ($\sigma = 37.8 \times 10^6$ S/m). It involves 80 vertical grooves. A larger panel would intercept more power and produce narrower reflected beams. Indeed, the size can be extended simply with no additional design steps, and it has to be chosen depending on the specific application. In addition, for more advanced scenarios, it could be possible to steer also the beam in the (y, z) plane, as shown for more complex Metal-Only panels in [14, 15].

The grating consists of 80 groove cells of size 20 cm × 20 cm (Fig. 10.15) spaced half a wavelength apart. The grating is capable of directing the reflection of the beam towards the lateral directions with a theta angle of approximately ± 75° from the normal incidence. The grating has horizontal polarization, and the beam width is equal to 10°. Figure 10.13 shows a picture of the fabricated panel.

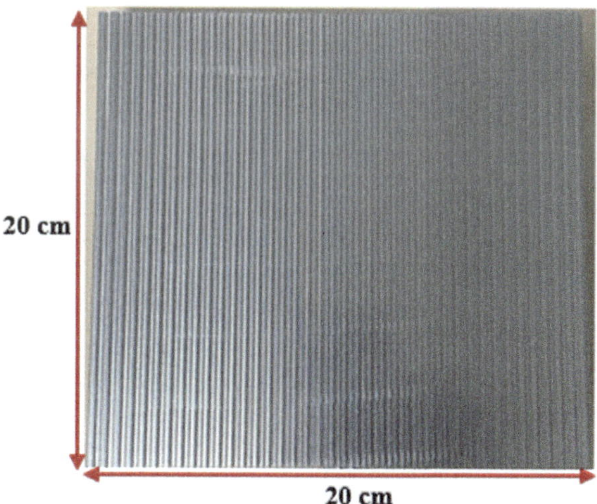

Fig. 10.13 Fabricated metal panel

Measurement Results

The results obtained by measurement are shown in Fig. 10.14, which represents the power received (P_r) for each *Rx* position and for the different angles of arrival (AoA), with (red curve) and without reflector (blue curve).

*P*0 is the position of *Rx* in the *Tx*-reflector alignment, which corresponds to the intersection of the central axes of the two corridors. At position *P*0, where the two

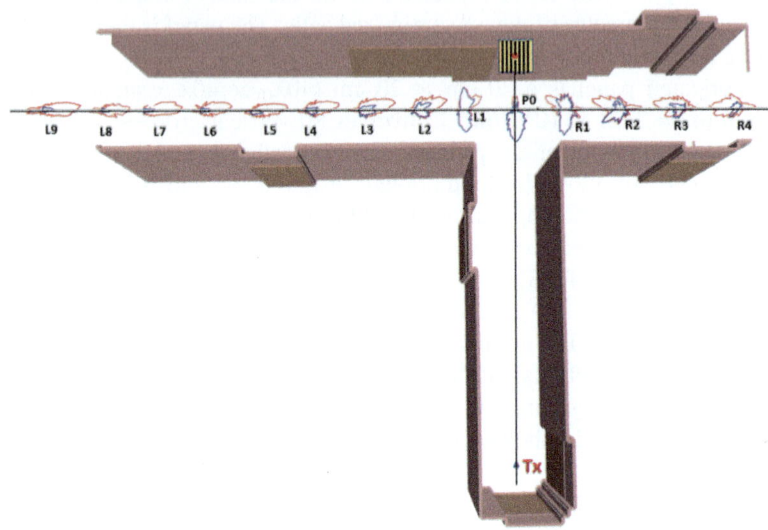

Fig. 10.14 Distribution of the angular power received for each position of Rx

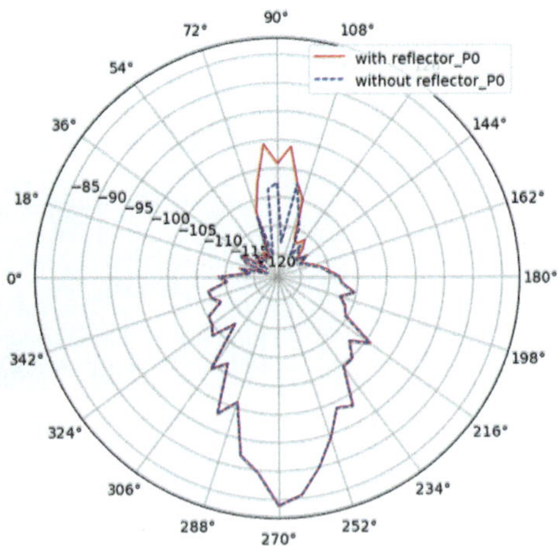

Fig. 10.15 Angular power received for position *P*0

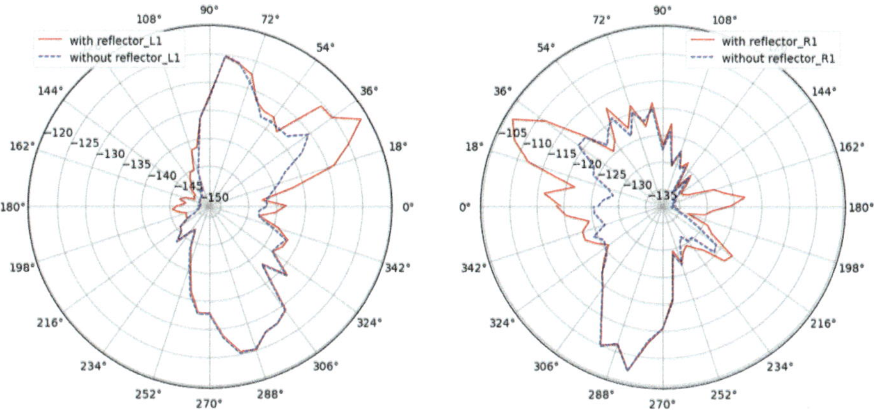

Fig. 10.16 Angular received power for $R1$ and $L1$

antennas are in LOS, the maximum received power is concentrated in the main lobe and is equal to – 83.8 dBm (Fig. 10.15). This value is close to that calculated for free space (– 82.18 dBm).

The positions $R1$–$R4$ are right-hand Rx positions and $L1$–$L9$ are left-hand positions. For displacements of Rx on positions $R1$ and $L1$, the distribution of the received power clearly shows the diffraction effect at the edges of the corridor. Figure 10.16 shows in detail the P_r versus the angle of arrival (AoA), for positions $R1$ and $L1$.

These results show that for the most distant positions, the largest P_r is obtained when the Rx antenna is oriented towards the reflector array. Using the reflector increases the received power by approximately 20 dB for position $L3$. The reflector array provides a significant gain in the link budget. This gain is maintained up to an $Rx - P0$ distance of 9 m, with an angle of arrival of 6° relative to the central axis of the corridor (Fig. 10.17).

These results also show that the maximum received power is practically maintained from position $L2$ to position $L9$, i.e. over a distance of 8 m (Fig. 10.14). Figures 10.18 and 10.19 show the maximum power received (Max_P_r) for the right and left positions. It is clear that the maximum power received is at position $P0$, which is equal to – 83.8 dBm. Figure 10.19 clearly shows that the reflector array provides a gain of around 20 dB from position $L3$.

This means that over this distance there is no need to update the beamforming, which can maintain the same dominant trajectory. Finally, this result confirms that the use of a reflector array is an effective solution to improve the radio coverage of indoor mmW off-line-of-sight (NLOS) links.

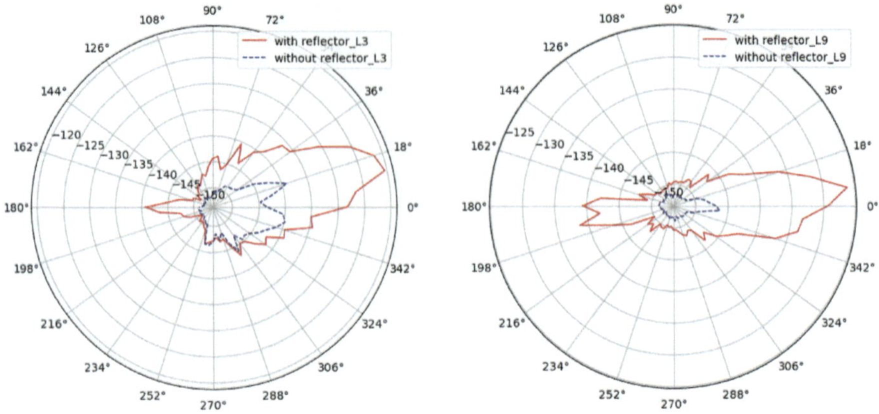

Fig. 10.17 Power received as a function of the angle of arrival at $L3$ and $L9$

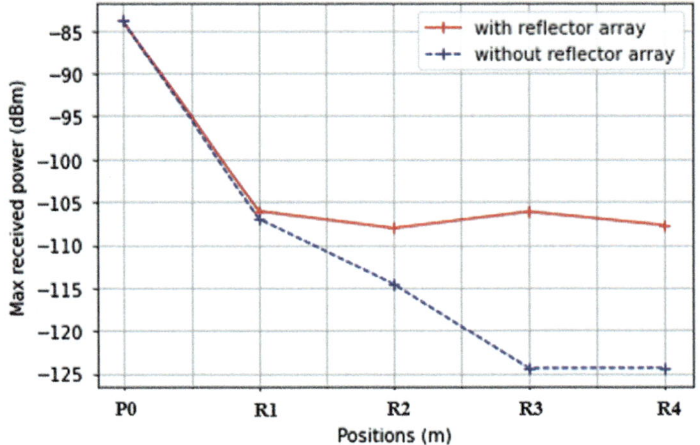

Fig. 10.18 Max_P_r obtained for right-hand positions

10.3 Impact of Blocking by the Human Body at 60 GHz

This section presents broadband radio propagation measurements conducted at 60 GHz inside a meeting room. The measurement system used is the same as previously.

10.3.1 Measurement Environment

The environment selected for this measurement is a time-invariant (no movement) meeting room whose dimensions are 6.5 $\times$ 2.2 $\times$ 2.5 m^3. Figure 10.20 shows

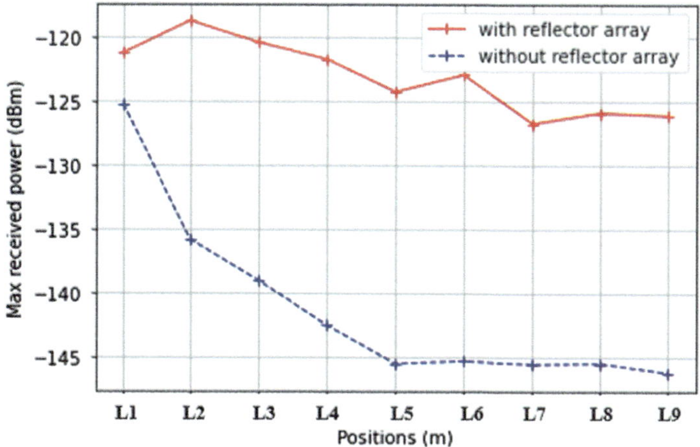

Fig. 10.19 Max_P_r obtained for left-hand positions

a photo of the measurement environment. In this room, there are two whiteboards with the same dimensions: 2 m × 1.2 m. The door remained closed throughout the measurement campaign.

In this room, the presence of the chairs and the table does not modify the measurement results, because of the use of very directional antennas. This room also has a wooden door with the height of 2.3 m and the width of 0.96 m (Fig. 10.21).

There was no movement during these measurements. The same horn antennas were used on *Tx* and *Rx* sides.

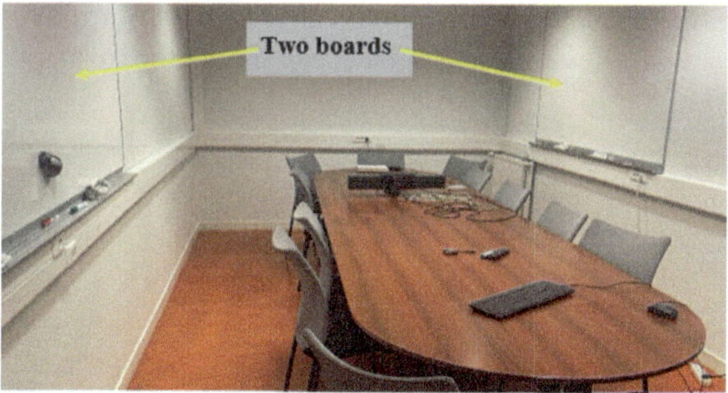

Fig. 10.20 Photo of the measurement environment

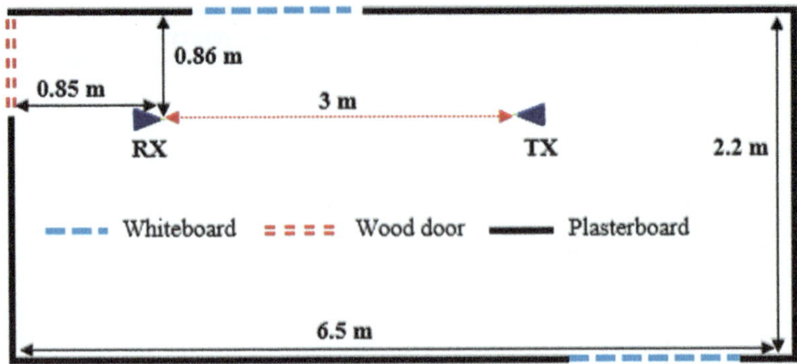

Fig. 10.21 Measurement environment in the meeting room

10.3.2 Measurement Scenario

The measurements were conducted in a meeting room of the IETR–INSA Rennes. The two antennas were put on supports at the same height of 1.71 m and separated by a distance of 3 m. The height of the two antennas was chosen so that it corresponds to the middle of the whiteboard. The receiving antenna was placed 0.85 m from the door. Its support is fixed on the positioner, which allows the rotation of the antenna. The whiteboard is at a distance of 1.35 m from the door (Fig. 10.22). The two antennas are equidistant from the centre of the board and are placed 0.86 m from the wall of the board. A human blocker, whose width at the shoulders, thickness and height are respectively 45 cm, 13 cm and 1.72 m, was considered.

An antenna positioning system was used. Its role is to perform antenna rotations by precise steps for a study of the angle of arrival (AoA) of the wave in azimuth. The positioner was used with a mast placed on the motor to support the receiving antenna.

Fig. 10.22 Measurement scenario in the meeting room

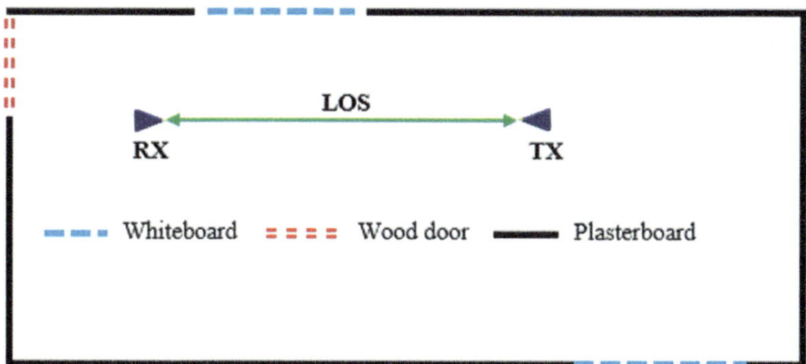

Fig. 10.23 LOS case

For this measurement campaign, three different scenarios are considered:

- Case where the antennas are in line-of-sight (LOS) and oriented towards each other.

Figure 10.23 shows the case where the two antennas are in LOS configuration, oriented towards each other.

- Case where the direct path is blocked.

In this case, the blocker was placed in the middle of the *Tx–Rx* distance and oriented towards the receiver, without any movement. The height of the antennas corresponds to the centre of the blocker's chest, as shown in Fig. 10.24. For this, the blocker was placed on a support, so that the centre of his chest was at a height of 1.71 m. The comparison with the previous LOS case makes it possible to estimate the attenuation introduced by the human body.

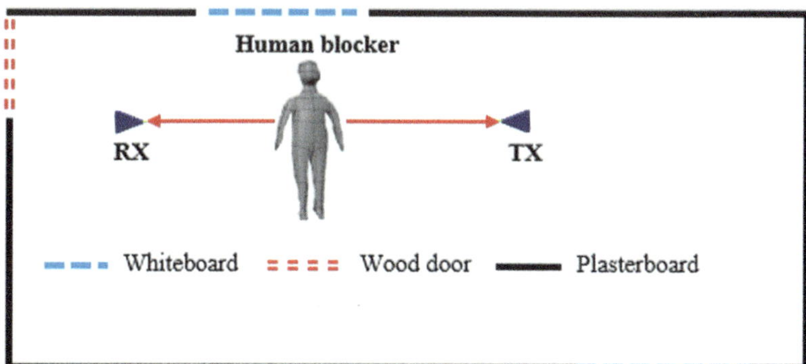

Fig. 10.24 Blocking the direct path

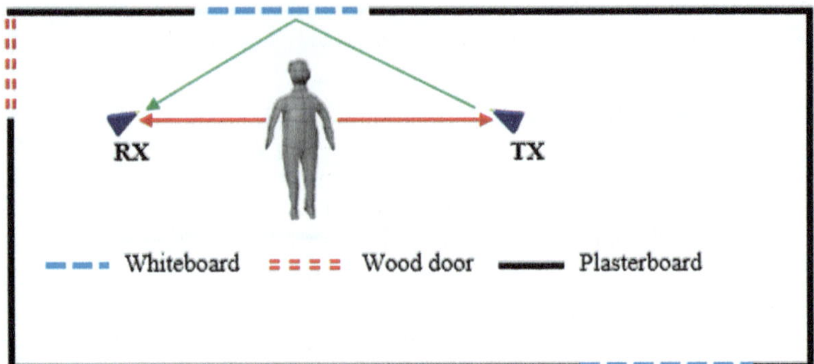

Fig. 10.25 Case with blocking and depointing of the antennas

- Case with blocking the direct path and depointing the antennas by 30° towards the whiteboard.

In this case, the two horn antennas have an azimuth pointing angle of 30° with respect to the *Tx*–*Rx* line. With this arrangement, the two antennas were oriented towards the centre of the whiteboard (Fig. 10.25), which allows us to obtain an optimal reflection. In this configuration, the position and height of the blocker and the antennas have not changed.

Moreover, in these three cases, the receiving antenna rotates 360° in azimuth using the positioner, with a step of 6°. This choice makes it possible to evaluate the channel power delay profile (PDP) on these 60 rotation positions for a study of the paths with the delays and the angles of arrival (AoA).

10.3.3 Measurement Results

By data post processing, it is possible to represent the power received (in dBm), with and without blocking, for each orientation of the Rx antenna (Fig. 10.26). We also calculated the impulse response of the propagation channel, in order to study the different paths with their delays.

The calculation of the maximum clearance radius of the first Fresnel ellipsoid, at a distance of 1.5 m, gives 0.06 m, which is much lower than the distance that separates the *Tx*–*Rx* line from the whiteboard and also from other furniture and equipment in the room. We can therefore consider that we are close to free space conditions, in the case where the two antennas are in direct visibility and without the human blocker. The calculation of the free space PL gives 77.54 dB. The PL obtained by measurement in the case of direct visibility (i.e. an AoA of 0°) and without blocking is 77.2 dB (Fig. 10.26), which is very close to the value previously calculated in free space condition. With the presence of a human blocker, and taking into account the distances adopted and the beam width of the antennas, it is checked that the main

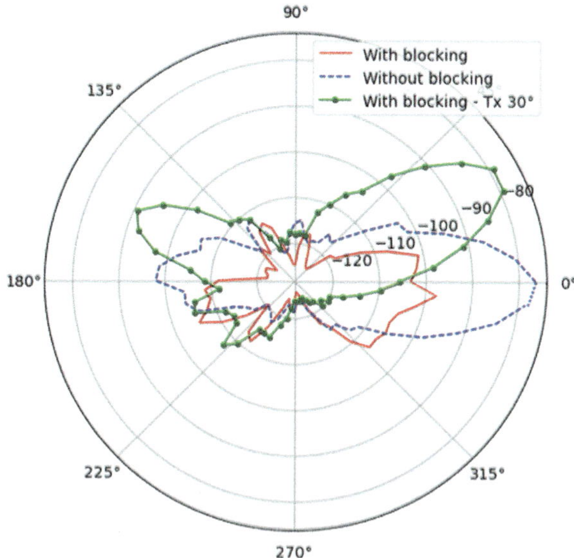

Fig. 10.26 Angular power received (in dBm)

radiation lobe is completely masked by the thorax of the blocker. In this case, when the antennas are pointed towards each other, a path loss of 103 dB is obtained. These results allow us to estimate the losses introduced by the blocker at around 25.8 dB, which is not far from the results presented in [8] at 60 GHz. The path loss is higher and equal to 105.6 dB in the case where only the transmitting antenna is offset by 30° towards the whiteboard and it becomes equal to 79 dB for the same offset of 30° of the two antennas. This shows that with the blocker, depointing the transmit antenna alone results in an increase in path loss of about 2.6 dB and a decrease of 24 dB if both antennas are depointed 30° compared to the blocked path without depointing. So, the reflected path causes a loss of only 1.8 dB compared to the direct path. This low path loss value can be justified by a very small difference in distance between the direct and reflected paths (0.46 m) and by the fact that the reflection is almost perfect with the board considered to be metallic. Indeed, a calculation of the difference in losses due to the only difference in paths gives 1.25 dB, which makes it possible to estimate the losses by reflection at approximately 0.55 dB.

On the other hand, the losses introduced by human blockage can be estimated by calculating the channel impulse response on a 2 GHz band. This makes it possible to obtain the relative power delay profile (PDP) for the different scenarios considered. Figure 10.27 shows the relative PDP, i.e. the PDP normalized to the maximum value computed over all PDPs. The largest value (0 dB) is obtained, as expected, in the case where the two antennas are in LOS (blue curve).

In this case, the direct path arrives with a delay of 10 ns, which corresponds to the *Tx–Rx* distance of 3 m. With the human blocker, we obtain a relative power value of about − 24.4 dB with a delay of 10.5 ns (red curve). This delay corresponds to

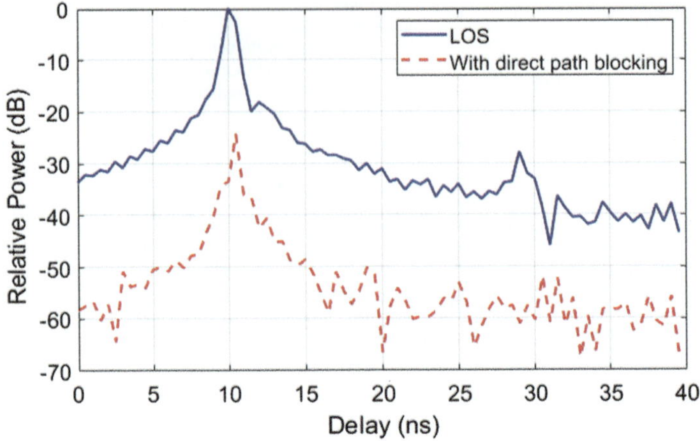

Fig. 10.27 Relative PDP with and without human blocking

a distance of 3.15 m, i.e. a path difference with respect to the direct path of 0.15 m. Thus, the human blocking attenuation of 24.4 dB, obtained by evaluating the relative received power, gives a value close to that calculated with the path loss, which is equal to 25.8 dB.

Figure 10.28 gives a comparison between the strongest paths in the LOS case and the case where the *Tx* antenna is oriented 30° towards the middle of the reflective board. This maximum corresponds to the path reflected by the whiteboard. We obtain a relative power value of about − 25.3 dB with a delay of 11.5 ns. Compared with the blocked direct path case, a small relative power increase of 0.8 dB is observed. This can be justified by the fact that we used directional antennas and that the depointing of one of the antennas can allow the signal to bypass the blocker.

In the last case, both antennas are pointed towards the centre of the whiteboard, with a 30° offset. This configuration seeks to exploit an alternative reflected path that could replace the direct path when the latter is blocked. Figure 10.29 shows the relative PDP of this configuration.

In the case where the two antennas are pointed towards the centre of the whiteboard, a reflected path is obtained with a delay of 11.5 ns and a relative power of approximately − 0.84 dB.

By comparing the received level relative to the direct path without blocking, with that received by the reflected path with blocking and depointing, there is a slight decrease of 0.84 dB for a distance difference of 0.45 m. These results are comparable to those obtained with the path loss calculation, where the reflected path gives additional losses of 1.8 dB compared to the direct path, with a distance difference of 0.46 m. This small difference in path loss can be justified by a very small difference between the lengths of the direct and the reflected paths, and by the fact that the reflection is almost perfect with the whiteboard considered metallic.

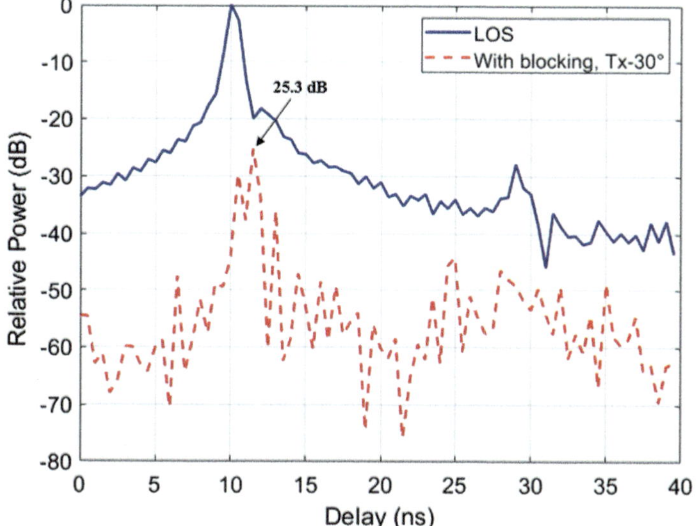

Fig. 10.28 Relative PDP with blocking and *Tx* offset by 30°

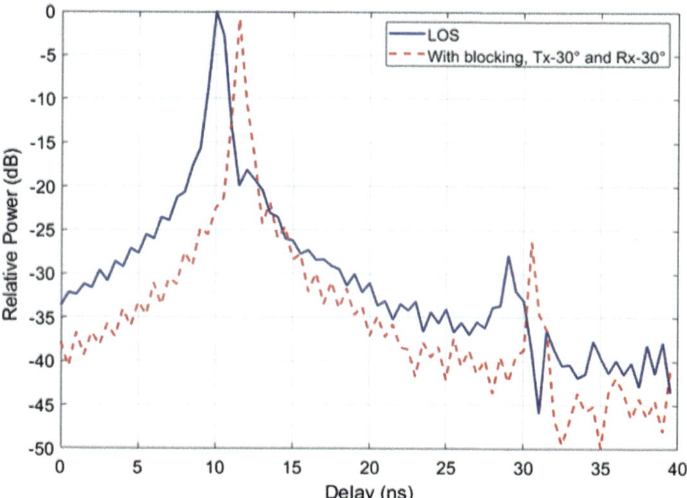

Fig. 10.29 Relative PDP with blocking and depointing of the two antennas by 30°

Therefore, in blocking situations, in order to limit outage times of the radio link, a reflected path can be used as a replacement for the direct path.

10.4 Conclusions

In this chapter, we first studied by measurement the influence of a rectangular metal reflector panel and a reflector antenna array on the radio coverage at 60 GHz in NLOS conditions. Measurements were carried out in an *L*-shaped corridor with and without the reflector. The results obtained show that with the reflector placed horizontally, then vertically, the received power increases with approximately 24 dB and 12 dB, respectively. These results show that the use of the metal panel increases the received power in this NLOS environment.

Then, we studied the use of a reflector antenna array in the *T*-shaped corridor. These results show that for the most distant positions, the maximum received power is obtained when the *Rx* horn antenna is oriented towards the direction of the reflector array. The reflector array provides a significant gain in the link budget of around 20 dB for the *L*3 position in NLOS. These results also show that the same received power and AOA are maintained over a distance of 8 m, so there is no need to update the beamforming. These results confirm that the use of a reflector array is an effective solution to improve radio coverage of indoor mmW out-of-line-of-sight (NLOS) links.

Finally, the results of broadband measurements on the blocking effect by the human body on an indoor radio link at 60 GHz were studied. These results show that the presence of a human blocker introduces additional losses, generally between 24 and 26 dB. On the other hand, the calculation of the channel impulse response makes it possible to highlight the multipath phenomenon and leads to obtaining the PDPs for the different measurement configurations. In particular, in the presence of a reflection on a metallic surface, the reflected path can be used when the direct path is blocked, using both transmit and receive beamforming.

However, in order to achieve a more precise modelling of the mmW propagation channel, further measurement campaigns remain necessary to be performed in other environments with less favourable reflection conditions, and using other types of antennas and reflective panels. Other parameters can also be studied such as the location of the blocker, the distance between the antennas or between the *Tx*–*Rx* line and the reflector and the height of the antennas.

Acknowledgements This work was financially supported by the French National Research Agency as part of the ANR MESANGES project (ANR-20-CE25-0016).

References

1. Han, S., Shin, K.G.: Enhancing wireless performance using reflectors. In: IEEE INFOCOM 2017—IEEE Conference on Computer Communications, 2017, pp. 1–9
2. Tan, X., Sun, Z., Jornet, J.M., Pados, D.: Increasing indoor spectrum sharing capacity using smart reflect-array. In: 2016 IEEE International Conference on Communications (ICC), 2016, pp. 1–6

3. Khawaja, W., Ozdemir, O., Yapici, Y., Erden, F., Guvenc, I.: Coverage enhancement for NLOS mmWave links using passive reflectors. IEEE Open J. Commun. Soc. **1**, 263–281 (2020)
4. Bennai, M., Talbi, L., Le Bel, J., Hettak K.: Medium range backhaul feasibility under NLOS conditions at 60 GHz. In: Global Symposium on Millimeter-Waves (GSMM), pp. 1–3. IEEE (2015)
5. Wang, D., Gillard, R., Loison, R.: A 60 GHz passive repeater array with quasi-endfire radiation based on metal groove unit-cells. Int. J. Microw. Wirel. Technol. **8**(3), 431–436 (2016)
6. Hiranandani, S., Mohadikar, S., Khawaja, W., Ozdemir, O., I. Guvenc, Matolak, D.: Effect of passive reflectors on the coverage of IEEE 802.11 ad mmWave systems. In: 2018 IEEE 88th Vehicular Technology Conference (VTC-Fall), pp. 1–6. IEEE (2018)
7. Sato, K., Manabe, T.: Estimation of propagation-path visible for indoor wireless LAN systems under shadowing condition by human bodies. In: Proceedings of 48th IEEE Vehicular Technology Conference, VTC, 21 May 1998, Ottawa, ON, Canada (1998)
8. Collonge, S., Zaharia, G., El Zein, G.: Influence of the human activity on wide-band characteristics of the 60 GHz indoor radio channel. IEEE Trans. Wirel. Commun. **3**(6), 2396–2406 (2004)
9. Collonge, S., Zaharia, G., El Zein, G.: Wideband and dynamic characterization of the 60 GHz indoor radio propagation futures home WLAN architectures. Annales des Télécomm. **58**, 417–447 (2003)
10. Dieng, M., El Hajj, M., Zaharia, G., El Zein, G.: Improvement of indoor radio coverage at 60 GHz in NLOS configuration. In: IEEE Conference on Antenna Measurements & Applications (CAMA'21), 15–17 Nov 2021, Antibes Juan-les-Pins, France (2021)
11. El Hajj, M., Dieng, M., Zaharia, G., El Zein, G.: Enhancement indoor mmWave coverage using passive reflector for NLOS scenario. In: EuCAP 2022, Europe's Flagship Conference on Antennas and Propagation, 27 Mar 2022, Madrid
12. Yu, N., Genevet, P., Kats, M.A., Aieta, F., Tetienne, J.-P., Capasso, F., Gaburro, Z.: Light propagation with phase discontinuities: generalized laws of reflection and refraction. Science **334**(6054), 333–337 (2011)
13. Nayeri, P., Yang, F., Elsherbeni, A.Z.: Reflectarray Antennas: Theory, Designs and Applications. Wiley (2018)
14. Palomares-Caballero, A., Molero, C., Padilla, P., Garcia-Vigueras, M., Gillard, R.: Wideband metal-only reflectarray for controlling orthogonal polarizations. IEEE TAP **71**(3), 2247–2258 (2023)
15. An, Z., Makdissy, T., Vigueras, M.G., Vaudreuil, S., Gillard, R.: A metal-only reflectarray made of 3D phoenix cells. In: EuCAP 2022, 27 Mar–1 Apr 2022. Madrid (Spain)

Chapter 11
Design and Analysis of Conformal Millimeter Wave Antenna for Next Generation Wireless Communication

K. C. Raja Rajeshwari, Tanvir Islam, N. Saravanakumar, and Sudipta Das

11.1 Introduction

Over the past few years, wireless technology has advanced significantly, passing over various periods. According to the aforementioned viewpoint, the connection to the wireless portion will be highly packed to enable upcoming mobile communications over few Gbps data capacity and very low latency. In this situation, mm-wave waves can cover the need for capacity and was successfully tested in numerous circumstances for accessibility or backhaul transmission. However, being able to be used concurrently for accessibility and backhaul links continues to be an unsolved problem for research. The real-life execution of these type of circumstances to meet the specifications and requirements of the cellular telephone sector is another issue that has not yet been resolved [1]. These features call for speeds for transmitting data of a number of megabits every operating hour. Subsequently is necessary to find novel approaches because the volume of information being transmitted is enormous and exceeds the capability of current systems of communication.

The millimeter radio (mmWave) frequency is especially interesting in this sense since mmWave networks have the ability to attain extremely high data rates. The harshness of the mmWave gearbox channel, which is exacerbated in a vehicle context, limits this capability, though [2] (Fig. 11.1).

K. C. Raja Rajeshwari (✉) · N. Saravanakumar
Department of ECE, Dr. Mahalingam College of Engineering and Technology, Pollachi, Tamil Nadu, India
e-mail: rajeswari.kc93@gmail.com

T. Islam
Department of Electrical and Computer Engineering, University of Houston, Houston, TX 77204, USA

S. Das
Department of Electronics and Communication Engineering, IMPS College of Engineering and Technology, West Bengal, Malda, India

M. El Ghzaoui et al. (eds.), *Next Generation Wireless Communication*, Signals and Communication Technology, https://doi.org/10.1007/978-3-031-56144-3_11

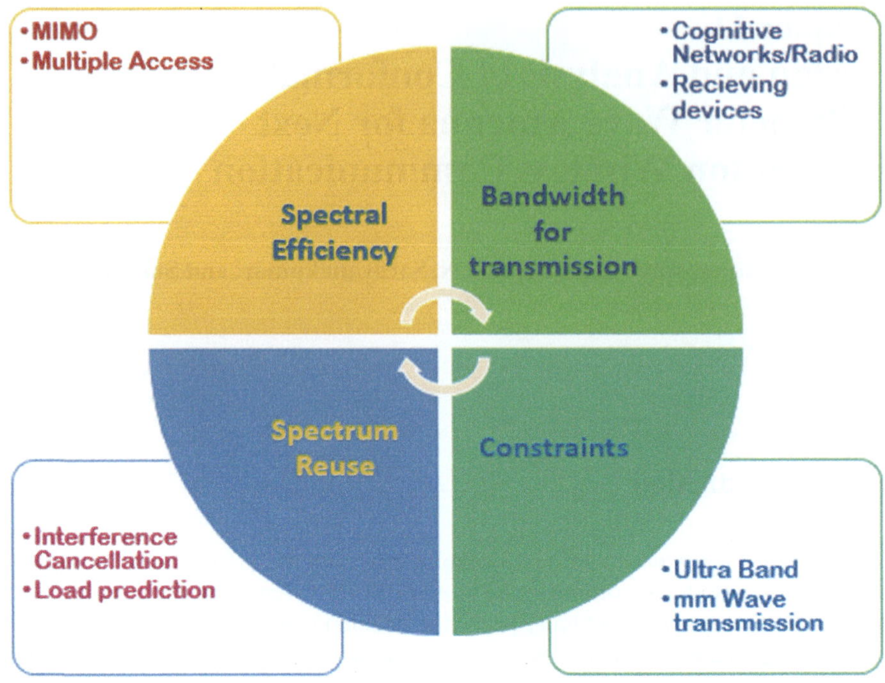

Fig. 11.1 Next generation wireless communication constraints

The difficulties that must be overcome with the goal to enable mmWave automobile situations are examined in this research. Greater frequency ranges, such as millimeter-wave (mm-Wave) bands of radiation, are being considered as applicants for possible next-generation mobile devices owing to the lack of spectrum availability beneath 6 GHz bands that and the requirement for greater speeds for data. This is because the significantly more major resources might be used in order to boost the ability and facilitate individuals to benefit from multiple states higher data rates. Yet, the increasing channel loss caused by the shorter frequency range and environment circumstances is a significant issue for mm-wave frequencies communications [3].

Numerous technical and practical difficulties, in addition to possibilities, must be solved before contemporary wireless communications can use pulses in the > 0.1 THz frequency range. This evaluation will concentrate on the following fundamental groups. The initial channels are the GHz band. This involves the novel obstacles associated on terahertz frequencies wavelength vibrations because they travel about transmitting to the receiving device in mostly earthbound networks of communication. Whereas the waves produced by terahertz have a number of similarities to RF (radio-frequency) waves, their respective positions shorter duration impacts beaming orientation, dispersion, and the antenna characteristics. Furthermore, the reflection, transmission, and penetration of materials, particularly the environment, vary greatly. Additionally, we will look at high-frequency devices. In recent years, a number of

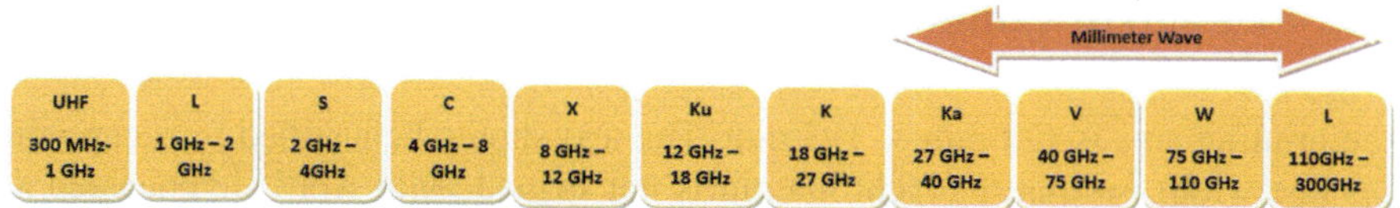

Fig. 11.2 Frequency band of millimeter wave communication

MIMO mobile antenna concepts were put forth. Yet, each of these configurations utilize a small number of big antenna components that may require up a significant amount of major boarder area or use a single frequency band working frequency. Several antennas designed for next generation mobile devices unveiled in 2017 only support one frequency band. Lastly, we look at space-based high-frequency possibilities because these uses could be distinctively appropriate for terahertz (THz) interactions and because they require internal remedies that can be prevalent to both outdoors and space mechanisms, in both environments likely imposing bigger constraints.

In a nutshell, terahertz telecommunications from space might be substantial replacements for subsequent earthly networks. Laser aiming safeguards, tremors, connection rigidity, and electrical consumption are among the important technical difficulties. The frequency band of millimeter wave communication is given as Fig. 11.2.

A rapid-gain set of antenna technology featuring wideband spectrum efficiency, appropriate pairing, steady radiating features, excellent beam analyzing capacity, suitable sidelobe level (SLL), and an intense transmitting beam is obtained to solve the aforementioned issues. The planned antenna encompasses the 5G bands in many nations, including the 28 GHz band in the United States of America and the 26 GHz spectrum band across Europe [3–5]. The 5G bandwidth is divided into two main categories: Sub6 GHz, which utilizes frequency below 6 GHz, and millimeter-wave area, which uses frequency beyond 24 GHz and higher. When trying to compensate for the substantial the attenuation and transmission losses encountered in the licensed mm-wave frequencies at 28, 37, and 39 GHz, powerful antenna arrays are necessary [6]. Greater quantities of transmitting components having low-loss supplying connections are present in these kinds of antenna arrangements [7]. In a rectangular spiral ring structure comprising spiral elements with numerous able-bodied waves, it has been shown as an alternative to traditional loop spiraling array. Within the impacts of spacing quantity and location, these have been examined in splitting rings resonance devices with just a single rectangular design with increasing number of rings. The properties of band spacing employing split ring arrangements as a function of their gap thickness relative to the resonating frequency are discovered. Lastly, the impact of the divided-ring alignment alongside regard to electromagnetic fields on spectral resonant structures were discussed [8].

11.2 Design of Conformal Antenna Array

The printing area is 122 mm × 193 mm. The text should be justified to occupy the full line width, so that the right margin is not ragged, with words hyphenated as appropriate. Please fill pages so that the length of the text is no less than 180 mm, if possible. The antenna's structure was created by merging two a microstrip and spiraling designs, with much thought given to the underlying substance, the size of each rectangle strip, the separation from the twisted loop, and the total number of twists of the spiral that were used. Algorithms developed using machine learning are additionally employed to increase effectiveness and precision. Artificial intelligence (AI) was recently investigated for modelling an antenna suggestion focusing on reliability criteria, but necessitates a large amount of learning information to increase correctness.

The array layout for the proposed conformal antenna is shown in Fig. 11.3. In this, the element 1 is designed as microstrip patch antenna, and element 2 is constructed as spiral loop antenna. The loop of the antenna element is varied and the dimension has been varied for getting maximum output [3, 7]. The spiral loop top view is shown in Fig. 11.4 and is fixed as element 2 of this proposed conformal array. There exist reports of rectangular and square spiral resonators. The spherical and multiple wire topology is realized on an electrical substrate in this research. Figure 11.4 depicts both frontal and top views of a single cell, wherein the delicate elements of the spiral and lines are constructed of copper with 17 m that with dielectric by $r = 3.84$. As illustrated in Fig. 11.4, every wire occupies the middle of an elongated arm and provides a receptive reaction corresponding to an electrically tiny capacitive dipole charged with inductive effect [9]. In this work, we explore a rectangular spiral-shaped antenna and assess its possibilities for usage in next generation communication. The metal conductor patchwork was spiraled on a high-permittivity insulating substrate that served to minimize the overall dimension of the antennas. To change the antenna resistance to the appropriate worth, the antennas feature a U-like loop construction.

Managing the form and duration of the U-like loop allows you to simply alter the antenna resistance. Because the spiral may be viewed as a tweaked dipole, initially an overall spiral length which was roughly half a wavelength was employed with no U-shaped loop after which it was carefully modified to create resonance at the desired frequency that was used in the very first phase. The U-like loop then fuses to the spiraling structure, as illustrated in Fig. 11.4. The minimal spiral arrangement has electrical impedance with a low immunity to radiation and conductive reactance; thus, the connected U-shaped looping process serves as an impedance development to the device, enhancing the resistivity and shunt inductive power, cancelling the capacitance part of the antenna's impedance and achieving connecting with the input impedance. As the frequency at which the resonance occurs and antennas susceptibility are mostly determined by the spiraling and U-shaped looping dimensions, accordingly, just a few repetitions are required [10] (Fig. 11.5).

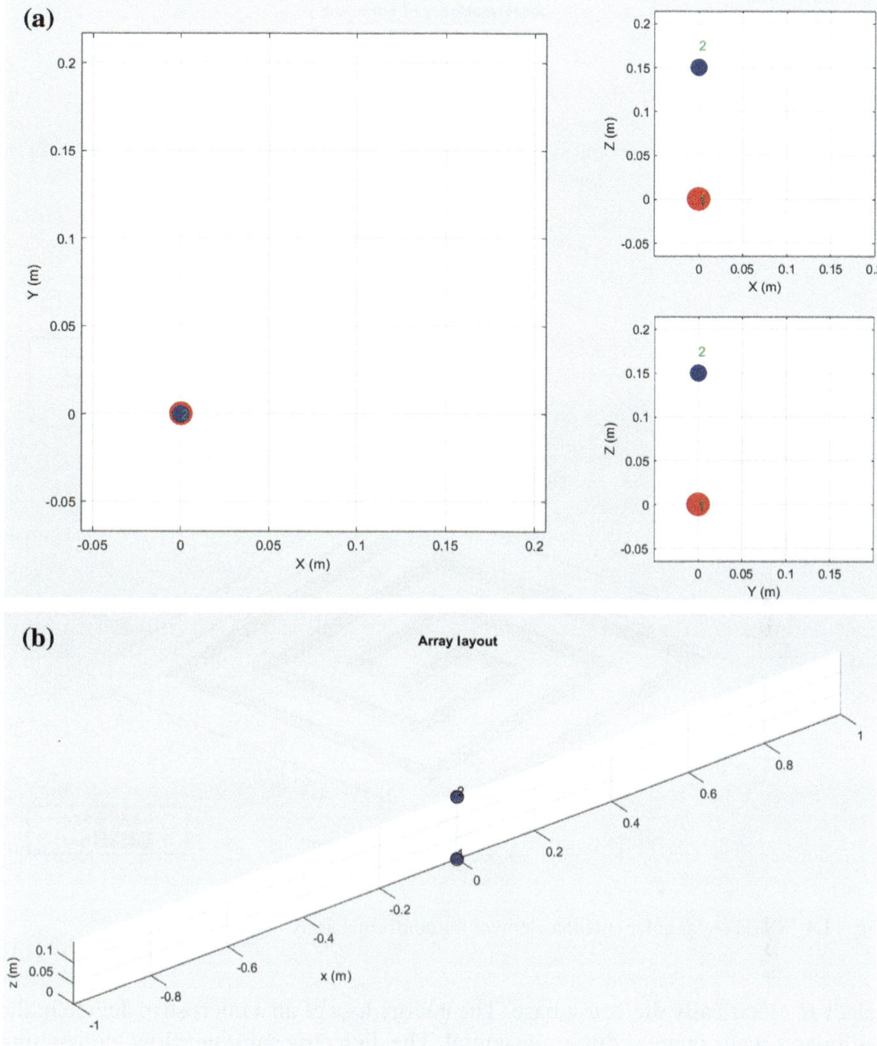

Fig. 11.3 **a** Array layout for the proposed conformal antenna, **b** lateral view

The antennas are often positioned over the transmitter body to obtain the necessary electromagnetic functionality. The conformal antenna had to be developed to do this. Conformal antennas may be established on the transmitter surface, that won't affect the carrier's internal structure and conserve storage. It can be positioned anyplace on the transmitter surfaces. Conformal antennas are often microstrip, horn, spiral, or fissure antennae. Numerous benefits of the small strip antenna include its small stature, tiny terms of size, compact size, and simplicity of integration with other carriers [11]. As a result, it is more suited for conformal antennas. To begin with, we

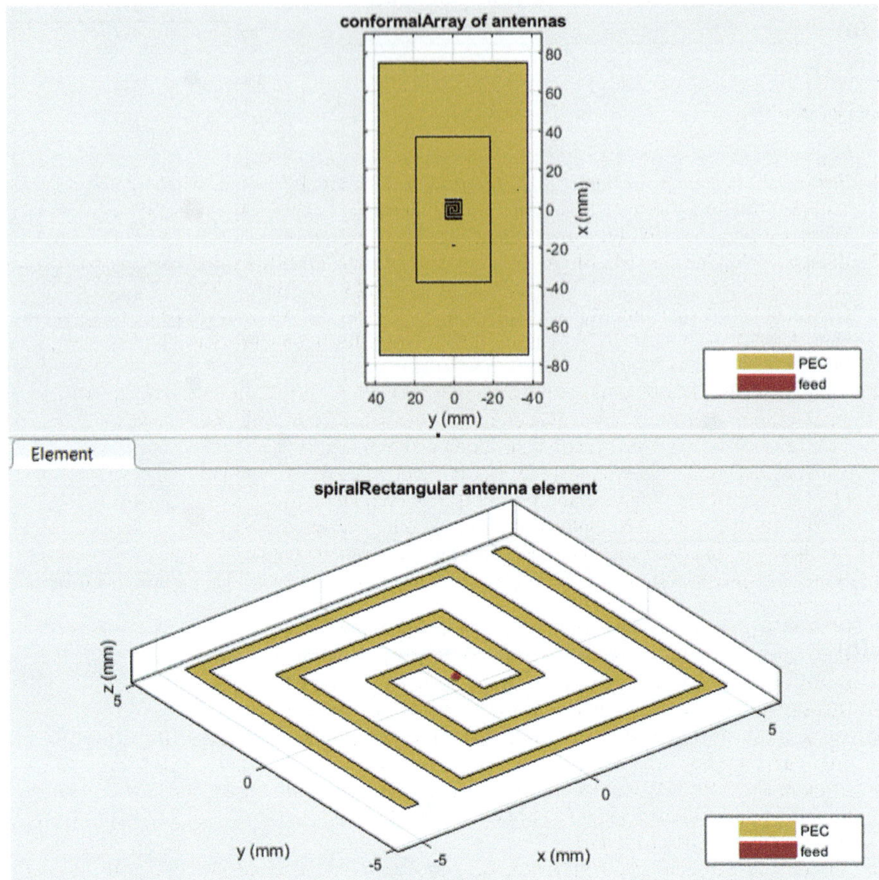

Fig. 11.4 Spiral rectangular antenna element for conformal array

select an electrically dielectric base. The energy loss of an a microstrip device in the millimeter-wave range is quite substantial. The dielectric substance loss, conducting material loss, and radiative loss are the three types of losses. To decrease the loss of dielectric, small loss tangential insulating substrates are frequently utilized.

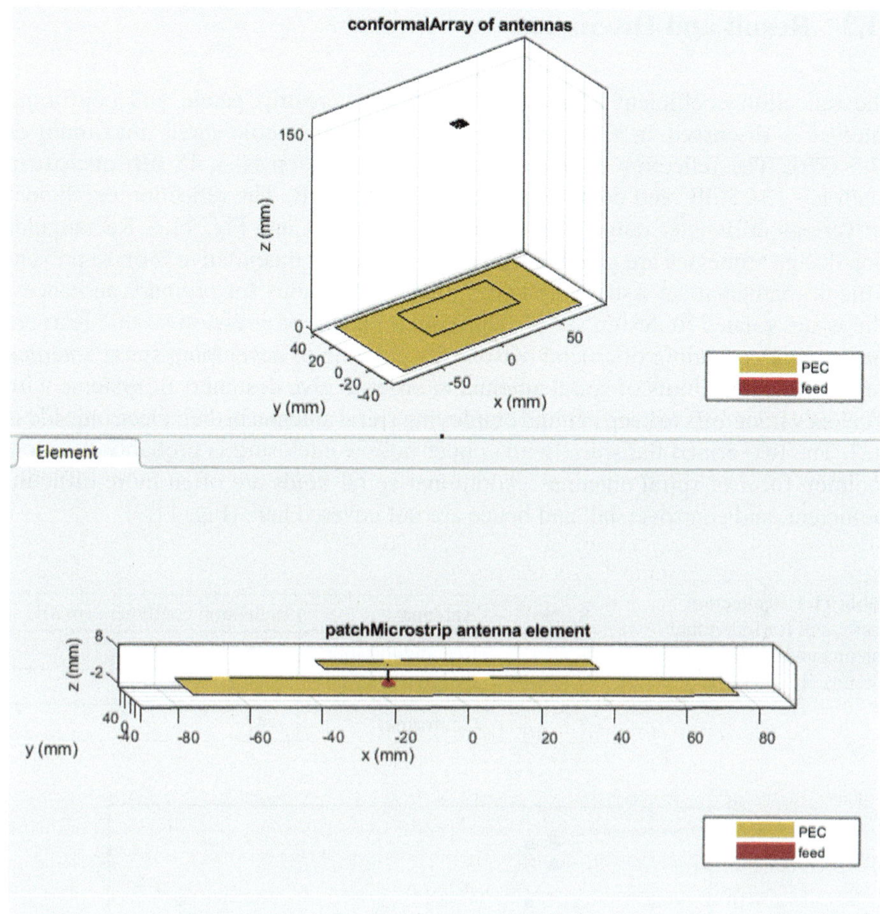

Fig. 11.5 Patch microstrip antenna element for conformal array

The complete loss of the tiny strip remains unchanged with its particular resistance while the constant for dielectric is minimal. Since the dielectric strength of the substrates is large, the electrical loss of the small strip changes fast with its special resistance. Thicker substrates enhance energy losses, and surface waves are more severe. A lower height suppresses the more powerful category more strongly and reduces electromagnetic leakage [5]. Furthermore, a thinner foundation with considerable elasticity is used for conformal antenna.

11.3 Result and Discussion

The reflection coefficient of rectangular loop, microstrip patch, and conformal antenna is discussed in this section. The reflection coefficient is maximum at 37.5 GHz. The reflection coefficient of rectangular loop is – 45 dB, microstrip patch is – 34.5 dB, and conformal antenna is – 47 dB. The reflection coefficient for various millimeter range is discussed in Table 11.1 and Fig. 11.6. Rectangular loop design strategies are given in subjective instead of quantitative form to prevent difficult mathematical issues that could prove superfluous for ordinary audiences. The issues related to design are presented in a logical progression so that learners may get understanding of crucial reasons for utilizing or developing spiral antenna. Furthermore, the limits of spiral antenna in order to give designers of systems with efficiency trade-offs to keep in mind employing spiral antenna in their electronic ideas [12]. The two-armed flat spiral with copper hollow enclosure is probably the most common form of spiral antennas. Additional spiral kinds are often more difficult, inefficient, and controversial, and hence are not covered here (Fig. 11.7).

Table 11.1 Reflection coefficient for individual antenna and conformal antenna at 37.5 GHz

S. No.	Antenna	Reflection coefficient (in dB)
1	Rectangular loop	– 45
2	Microstrip patch	– 34.5
3	Conformal	– 47

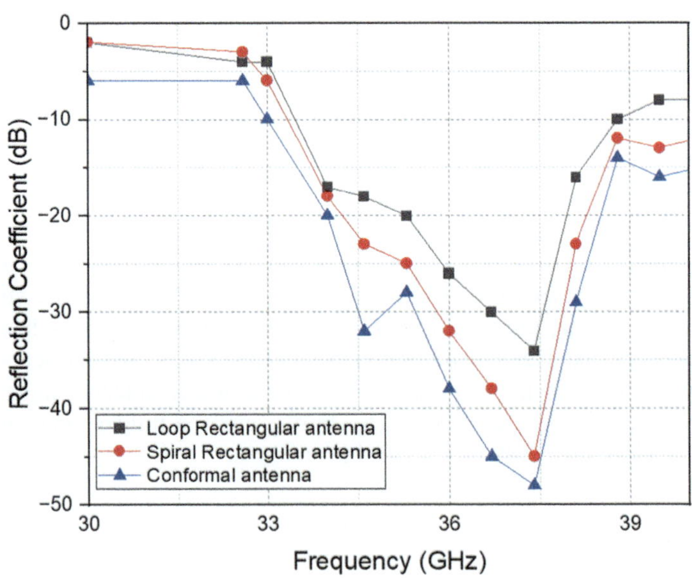

Fig. 11.6 Reflection coefficient of conformal antenna elements

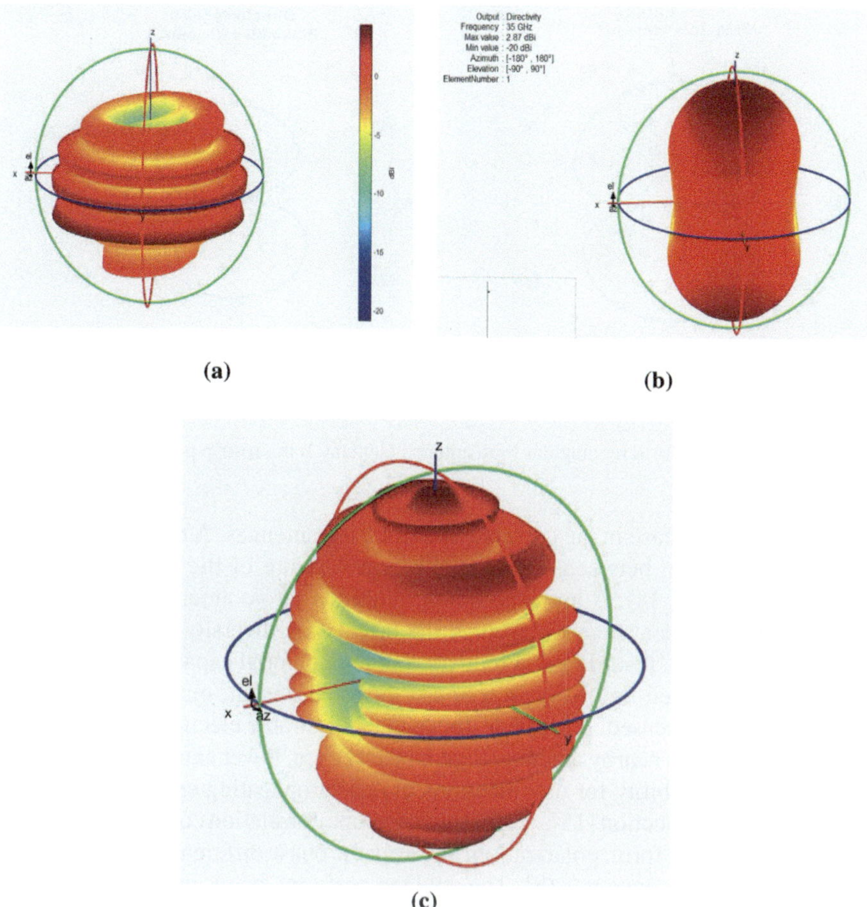

Fig. 11.7 Radiation pattern for **a** Rectangular loop antenna element, **b** microstrip patch antenna element, **c** conformal antenna

The mathematical calculation of directivity for a conformal array is more involved, requiring taking into account the locations of every array element with regard to each other and with the surrounding wavelength. The maximum directivity of a conformal rectangular array with non-isotropic components may be determined using the general ratio of effective aperture to directivity. The directivity of rectangular loop antenna element and microstrip patch antenna element is shown in Fig. 11.8.

The envelope correlation coefficient (ECC) indicates the degree to which the electromagnetic radiation characteristics of two antennas [10, 13]. So, if a single antenna was totally horizontally polarized whereas the other was entirely vertically polarized, the corresponding relationship between the two antennas would be 0. Similarly, if one antenna exclusively radiated energy into the atmosphere and the other solely emitted energy into the ground, both would have an ECC of 0 [2]. The reason for

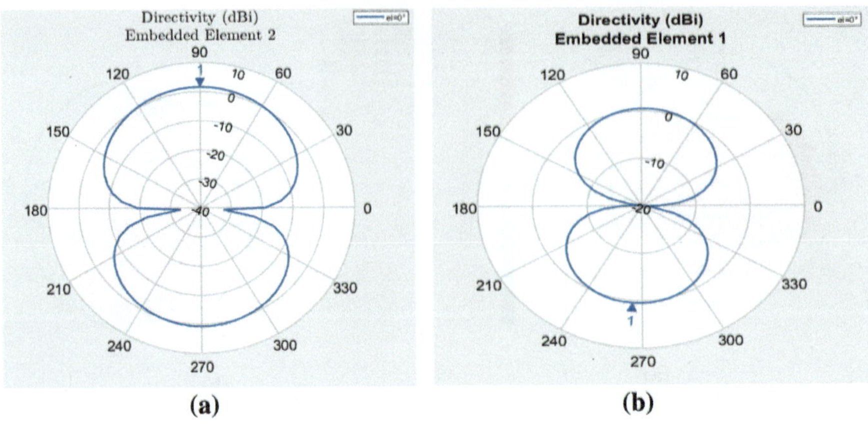

Fig. 11.8 Directivity for **a** rectangular loop antenna element, **b** microstrip patch antenna element

this is due to a decrease in the correlation among the antennas. A conformal antenna system's correlation between two antennas is a measure of the closeness of their acquired signals [6, 14]. A high level of correlations across antennas might restrict the capability of the system to leverage the orientation diversity provided by each antenna component, resulting in a reduction in the potential capability gain. Whenever antennae spaced tightly or in a non-orthogonal design, mutual coupling and correlations are increased. It happens due to the undesirable electromagnetic contact caused by the arrays nearby antenna. Signal disturbance, lower antenna performance, and higher susceptibility for modifications in their propagating environment can all come from this connection [15]. As a result, envelope correlation coefficient considers the radiation pattern form, polarization, and even the phase difference of the fields that exist across the two antennas [8]. The antenna performs better in the *xz*-plane with an omnidirectional-like design. The resulting design has great diagonal coverage, which makes it appropriate for a variety of telecommunications and detection uses [16]. The spatially correlated properties of conformal antenna employed in wireless networks are discussed in this part of the article. As a result, we provide a technique for determining the best conformal array shape for maximizing a mobile communication the system's ECC. This approach creates an individual goal function that will be used in the future to solve the resultant optimization issue using particle swarm optimization (PSO). The correlation graph for two antenna elements are given in Fig. 11.9.

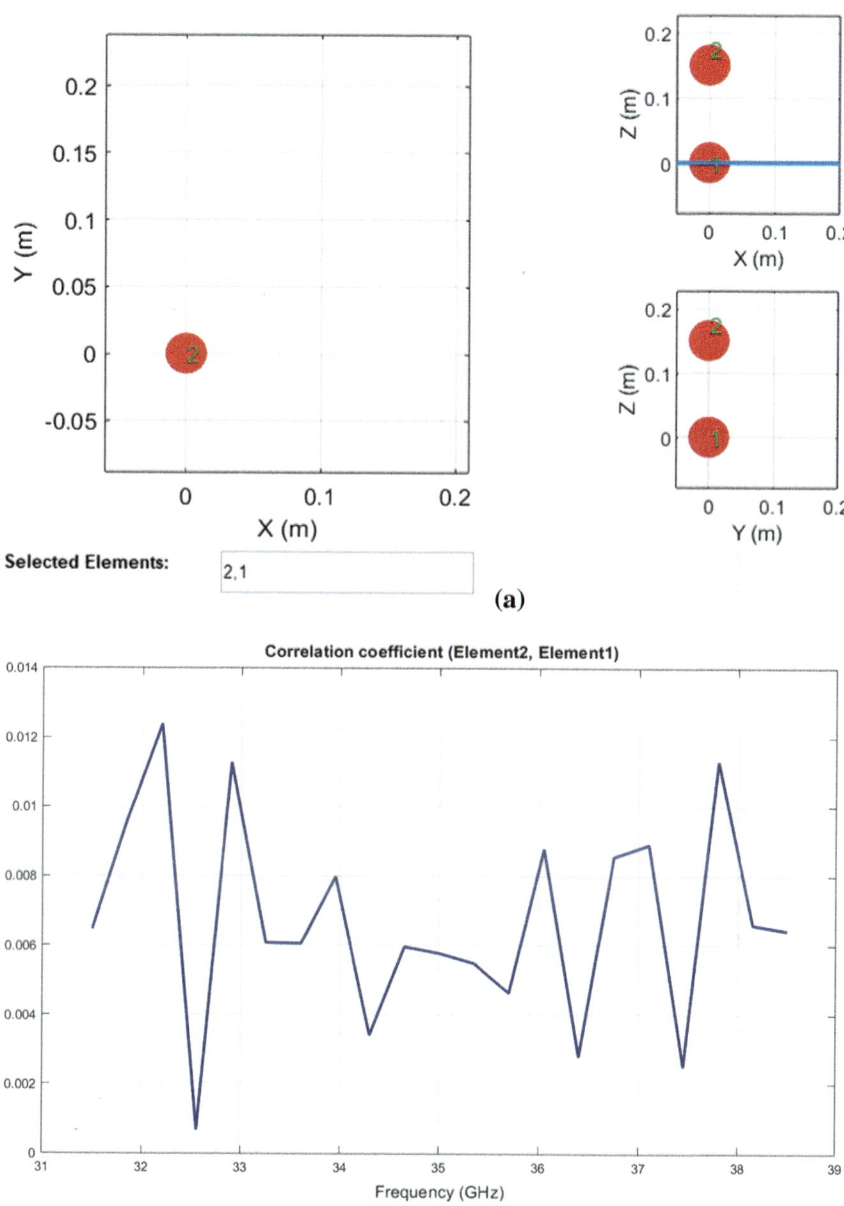

Fig. 11.9 Correlation coefficient **a** visualization, **b** for element 2 and element 1

11.4 Conclusion

We constructed and studied a compact conformal millimeter wave antenna for prospective next- generation applications in this paper. The resonance frequencies of the proposed antenna range from 32.5 to 39.5 GHz. The maximum reflection coefficient obtained of – 45 dB for a rectangular loop, – 34.5 dB for a microstrip patch, and – 47 dB for a conformal antenna. The analyzed conformal antenna features provide a good stable reflection coefficient in connection to the antennas construction across the operating band.

References

1. Khan, S., Farasat, M., Naseem, U., et al.: Performance evaluation of next-generation wireless (5G) UAV relay. Wireless Pers. Commun. **113**, 945–960 (2020). https://doi.org/10.1007/s11277-020-07261-x
2. Giordani, M., Zanella, A., Zorzi, M.: Millimeter wave communication in vehicular networks: challenges and opportunities. In: 2017 6th International Conference on Modern Circuits and Systems Technologies (MOCAST), Thessaloniki, Greece, pp. 1–6. https://doi.org/10.1109/MOCAST.2017.7937682
3. Khalily, M., Tafazolli, R., Xiao, P., Kishk, A.A.: Broadband mm-Wave microstrip array antenna with improved radiation characteristics for different 5G applications. IEEE Trans. Antennas Propag. **66**(9), 4641–4647 (2018). https://doi.org/10.1109/TAP.2018.2845451
4. OFCOM Notes: Update on 5G Spectrum in the U.K., Feb 2017 [online] Available https://www.ofcom.org.uk/data/assets/pdf_file/0021/97023/5G-update-08022017.pdf
5. Yalavarthi, U.D., Rukmini, M.S.S., Madhav, B.T.P.: A compact conformal printed dipole antenna for 5G based vehicular communication applications. Progr. Electromagn. Res. C **85**, 191–208.https://doi.org/10.2528/PIERC18041906
6. Ud Din, I., Alibakhshikenari, M., Virdee, B.S., Jayanthi, R.K.R., Ullah, S., Khan, S., See, C.H., Golunski, L., Koziel, S.: Frequency-selective surface-based MIMO antenna array for 5G millimeter-wave applications. Sensors **23**, 7009 (2023). https://doi.org/10.3390/s23157009
7. Munir, M.E., Al Harbi, A.G., Kiani, S.H., Marey, M., Parchin, N.O., Khan, J., Mostafa, H., Iqbal, J., Khan, M.A., See, C.H., et al.: A new mm-wave antenna array with wideband characteristics for next generation communication systems. Electronics **11**, 1560 (2022). https://doi.org/10.3390/electronics11101560
8. Elwi, T., Abbas, Z., Noori, M., Al-Naiemy, Y., Salih, E., Hamed, M.: Conformal antenna array for MIMO applications. J. Electromagn. Anal. Appl. **6**, 43–50 (2014). https://doi.org/10.4236/jemaa.2014.64007
9. Wang, Q., Mu, N., Wang, L.L., Safavi-Naeini, S., Liu, J.P.: 5G MIMO conformal microstrip antenna design. Wirel. Commun. Mob. Comput. **2017**, 11 (2017). https://doi.org/10.1155/2017/7616825
10. Rajarajeshwari, K.C., Poornima, T., Gokul Anand, K.R., Kumari, S.V.: SIMO array characterized THz antenna resonating at multiband ultra high frequency range for 6G wireless applications. In: Das, S., Nella, A., Patel, S.K. (eds.) Terahertz Devices, Circuits and Systems. Springer, Singapore (2022). https://doi.org/10.1007/978-981-19-4105-4_7
11. Aghoutane, B., Meskini, N., El Ghzaoui, M., Faylali, H.E.: Millimeter-wave microstrip antenna array design for future 5G cellular applications. In: 2018 International Conference on Electronics, Control, Optimization and Computer Science, ICECOCS 2018, 2018, 8610507
12. Rajarajeshwari, K.C., Sathiyapriya, T., Priya, E.L.D.: Surrogate optimization-assisted dual-band THz inverted-F coplanar graphene antenna. In: Patel, S.K., Taya, S.A., Das, S., Vasu

Babu, K. (eds.) Recent Advances in Graphene Nanophotonics. Advanced Structured Materials, vol. 190. Springer, Cham (2023). https://doi.org/10.1007/978-3-031-28942-2_11
13. Mohamed, H.A., Edries, M., Abdelghany, M.A., Ibrahim, A.A.: Millimeter-wave antenna with gain improvement utilizing reflection FSS for 5G networks. IEEE Access **10**, 73601–73609 (2022)
14. Berhab, S., Annou, A., Chebbara, F.: Reconfigurable low-profile antenna-based metamaterial for on/off body communications. Handbook Res. Emerg. Des. Appl. Microw. Millimeter Wave Circ. 166–200 (2023). https://doi.org/10.4018/978-1-6684-5955-3.ch007
15. Jeong, M.J., Hussain, N., Park, J.W., Park, S.G., Rhee, S.Y., Kim, N.: Millimeter-wave microstrip patch antenna using vertically coupled split ring metaplate for gain enhancement. Microw. Opt. Technol. Lett. **61**, 2360–2365 (2019)
16. Lakrit, S., Das, S., Alami, A.E., Barad, D., Mohapatra, S.: A compact UWB monopole patch antenna with reconfigurable Band-notched characteristics for Wi-MAX and WLAN applications. AEU Int. J. Electron. Commun. **105**, 106–115 (2019). ISSN 1434-8411. https://doi.org/10.1016/j.aeue.2019.04.001

Chapter 12
H-Shaped Resonators for UWB Bandpass Filter for 5G Applications

Kaoutar El Bakkar, Mohammed El Ghzaoui, and Ali El Alami

12.1 Introduction

The upcoming era of fifth-generation (5G) communication is poised to significantly influence the development of wireless communication systems [1–4]. These systems employ wireless technologies such as radio waves, microwaves, infrared, or other wireless mediums to facilitate the transmission and reception of data between devices [5–8]. Given the crucial role of filters in these systems, they are utilized in various applications such as satellites, mobile phone systems, and radar systems.

The use of high frequencies in 5G aims to meet the increasing demand for high data rates and improved performance, although this is accompanied by challenges such as reduced range, sensitivity to physical obstacles, and atmospheric interference, which can decrease the quality of communication. In this context, UWB filters play a crucial role in managing wideband signals associated with 5G. Their ability to transmit and receive signals across an extensive range of frequencies, particularly between 3.1 and 10.6 GHz, is essential to ensure efficient data transmission and minimize interference.

This paper introduces an UWB bandpass filter crafted to meet the demands of 5G applications. The suggested filter possesses a streamlined form, with dimensions of 40×59.52 mm^2, and employs a substrate made of Rogers RT/duroid 5880.

In the second section, presented a comprehensive introduction to filters and their diverse parameters. After that in Sect. 12.3, we formulated the design of a basic bandpass filter using a rectangular resonator. This design was then modified in Sect. 12.4 to broaden the bandwidth. In Sect. 12.5, the performance of the bandpass filter was enhanced by adding an H-shaped in the middle of the last resonator.

K. El Bakkar (✉) · M. El Ghzaoui
Sidi Mohamed Ben Abdellah University, Fez, Morocco
e-mail: kaoutar.elbakkar@usmba.ac.ma

A. El Alami
Moulay Ismail University, Meknes, Morocco

M. El Ghzaoui et al. (eds.), *Next Generation Wireless Communication*, Signals and Communication Technology, https://doi.org/10.1007/978-3-031-56144-3_12

The proposed UWB bandpass filter exhibits a large band, with bandwidth range (25.75–29.7 GHz), a return loss > 15 dB, and an insertion loss < 2.8 dB.

All design and simulations are performed using the Ansys HFSS software.

12.2 Overview of Filters

Filters are two-port devices widely used in various telecommunications systems and RF/microwave applications. Their role is of fundamental importance, consisting of eliminating all unwanted components of a useful signal, whether they originate from internal or external sources [7–10]. They characterized by:

- *Bandwidth* (*Passband*): This is the range of frequencies between which a signal at the input passes through to the output.
- *Attenuated Bandwidth* (*Stopband*): This is the range of frequencies where the amplitude of a signal is attenuated, so that it does not appear at the output.
- *Cutoff Frequency*: The frequency at which a filter begins to attenuate a signal.
- *Input and Output Impedance*: The impedance that the filter presents to the source and the load. Good impedance matching can be crucial in certain applications.

Microwave filters are widely used in wireless communications, radar systems, medical devices, measurement equipment, and other applications that require signal processing at high frequencies. To design and understand the behavior of these microwave filters, we use *S*-parameters (or scattering parameters), also known as scattering parameters [11]. These parameters provide detailed information on how signals are reflected and transmitted through the device.

The *S*-parameters are typically expressed in matrix form, where S_{ij} represents the contribution of wave j to port i. To illustrate, for a two-port device, the *S*-matrix would be formulated as follows:

$$S = \begin{bmatrix} S_{11} & S_{12} \\ S_{21} & S_{22} \end{bmatrix}$$

On the other hand, evaluating the performance of a filter can be achieved using two crucial formulas defined by the *S*-parameters. The first one pertains to insertion loss IL, while the second one concerns return loss RL, which also indicates the presence or absence of a reflected wave such as

$$\begin{aligned} \text{IL} &= -20\text{Log}\left|S_{ij}\right|\text{d}B \quad \text{where } i; j = 1; 2 \text{ and } i \neq j \\ \text{RL} &= 20\text{Log}|S_{ii}|\text{d}B \quad \text{where } i = 1; 2 \end{aligned}$$

The Voltage Standing Wave Ratio (VSWR) is another parameter used to analyze imperfections and the efficiency of transmission in a two-port network, instead of using return loss. It is determined by the following equation:

$$\text{VSWR} = \frac{1 + |S_{ii}|}{1 - |S_{ii}|} \quad \text{where } i = 1; 2$$

To design filters suitable for high-frequency signals, several high-frequency filtering technologies are utilized. Here are some commonly used high-frequency filtering technologies:

- Waveguide Filters: Waveguide filters direct and control electromagnetic waves using waveguides, providing excellent selectivity and a wide bandwidth.
- Surface Acoustic Wave (SAW) Filters:

SAW filters utilize surface acoustic waves to filter signals at high frequencies, commonly used in telecommunications applications and mobile devices.

- Planar filters:

Planar filters are created on a flat substrate, typically using printed circuit board manufacturing techniques. This approach is widely used for its ease of integration and compact footprint.

The microstrip technology is a type of planar technology. It provides various advantages for filters:

- Enables the design of compact and lightweight filters through the use of conductive traces on flat substrates. This provides significant advantages in applications where space and weight constraints are crucial.
- Adjustable Bandwidth: By modifying the geometry of the conductive trace, the bandwidth of microstrip filters can be altered.
- Microstrip filters can be configured to have minimal insertion loss.
- The production of microstrip filters is generally carried out using standard printed circuit board manufacturing processes, thereby simplifying the production process.

According to the frequency selection, microwave filters are classified into four families:

- Low-pass filters, which transmit signals with frequencies lower than a certain frequency f_c, called the cutoff frequency.
- High-pass filters, which transmit signals with frequencies higher than the cutoff frequency.
- Bandpass filters, which transmit signals with frequencies between two limiting frequencies, f_{c1} and f_{c2}.
- Band-stop filters diminish frequencies within the range of f_{c1} and f_{c2} (undesirable band) while permitting the passage of other frequencies. This filter type is effective in eliminating and rejecting noises or interferences characterized by specific, well-defined frequencies.

Fig. 12.1 represents the various families of filters.

Among these filters, the bandpass filter (BPF) is an essential component of the communication system, commonly used in radiofrequency to attenuate interferences

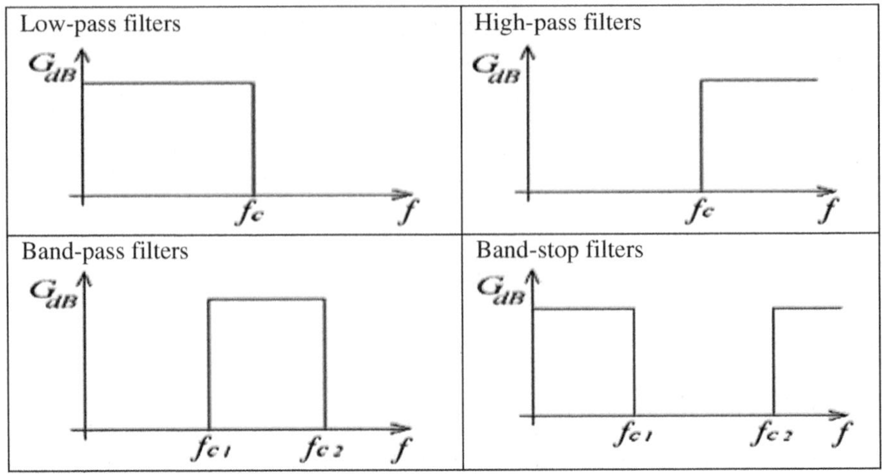

Fig. 12.1 Various families of filters

caused by noise that can hinder the transmission of information. Therefore, there is an urgent requirement to design a BPF for the radiofrequency interface used in 5G communication systems that is characterized by its compactness and low power consumption, while providing superior performance at a reduced cost. This is the goal of designing filters using metamaterials [12–18].

Since 2002, the FCC in the USA has authorized the commercial use of Ultra-Wideband (UWB) technology, leading to extensive research in this field. UWB systems operate within the frequency range of 3.1–10.6 GHz, requiring minimal power while delivering high data rates. Notably resistant to multipath interference, UWB can effortlessly penetrate obstacles and walls.

Ultra-Wideband bandpass filter an electronic device designed to allow the transmission of signals within a very broad range of frequencies while attenuating frequencies outside this specific range. This filter stands out for its capacity to allow signals across an exceptionally broad frequency range, making it well-suited for tasks demanding the sending or receiving of information across a wide spectrum of frequencies. The design of an Ultra-Wideband bandpass filter involves balancing various parameters, such as bandwidth and selectivity, to achieve optimal performance across the desired frequency range. These filters are commonly used in wireless communications, radar systems, and other applications where data transmission occurs over a very wide frequency band.

Various approaches for the bandpass filters with Ultra-Wideband (UWB) are introduced in the literature [19–23]. In [19], a simple Ultra-Wideband (UWB) bandpass filter utilizing a folded stepped impedance resonator (SIR) introduced. Thomas George, Lethakumary, improved the higher-frequency attenuation band of this filter by adding an L-shaped defected microstrip structure (LDMS). This filter achieved an excellent upper rejection band of 14.2 GHz.

Reference [20] introduces a microstrip Ultra-Wideband filter featuring feed lines on the substrate's upper surface and a rectangular-shaped defected ground structure at the bottom. The filter, measuring $26.5 \times 7.54 \times 1.6\ \text{mm}^3$, provides an impressive passband spanning from 2.4 to 14.3 GHz. It demonstrates a notable reduction in return loss at − 21.24 dB and an insertion loss of − 0.7 dB at 6.9 GHz.

12.3 The First Bandpass Filter

We started the work using the initial structure depicted in Fig. 12.2. This configuration consists of a rectangle connected to two transmission lines of different widths and shapes. The filter is manufactured on a substrate of "Rogers RT/duroid 5880 (tm)" with a dielectric constant of 2.2, a thickness of 0.5375 mm, and a loss tangent of 0.0009. This substrate is positioned on a ground plane of identical dimensions. The structure of this filter has an area of $40 \times 59.52\ \text{mm}^2$. Table 12.1 shows a list of the appropriate dimensions for the first bandpass filter.

Figure 12.3 illustrates the simulation outcomes for the *S*-parameters of the initial bandpass filter. The transmission range of this filter spans from 25.75 to 27.79 GHz, exhibiting an insertion loss of −3 dB and a return loss of − 12.5 dB.

The initial filter's Voltage Standing Wave Ratio (VSWR) functions as a metric for assessing its effectiveness in efficiently transferring power from the source to the load, while minimizing reflected power. VSWR plays a pivotal role in evaluating the filter's performance and quality, especially concerning impedance matching [23]. A low VSWR indicates successful impedance matching with minimal signal reflection,

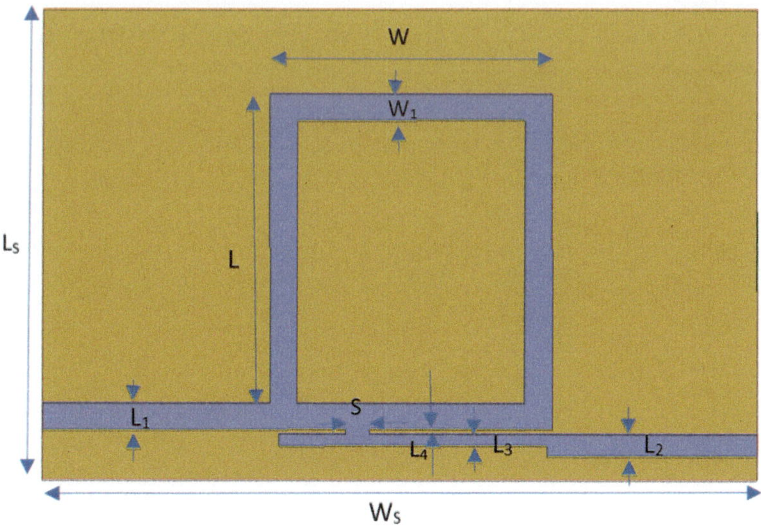

Fig. 12.2 Structure of the first bandpass filter

Table 12.1 Dimensions for the first bandpass filter

Parameters	Value (mm)	Parameters	Value (mm)
W_s	59.52	L_1	2.27
L_s	40	L_2	1.9
W	23.49	L_3	1
L	26.27	L_4	0.4
W_1	2.27	S	2

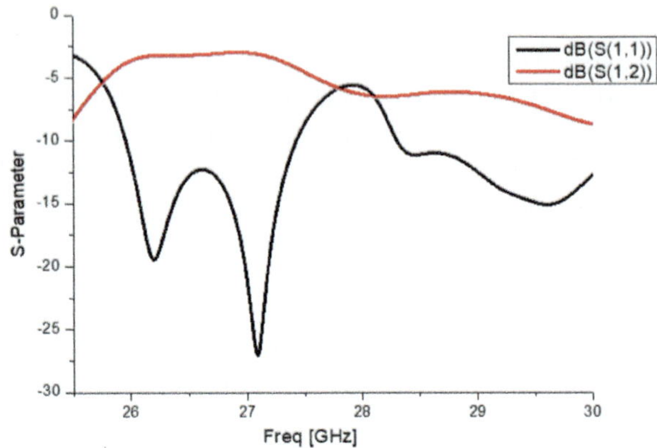

Fig. 12.3 Simulated S-parameters of the first bandpass filter

while a high VSWR indicates insufficient impedance matching and increased signal loss due to reflections.

Figure 12.4 shows the evolution of VSWR as a function of frequency for the first bandpass filter.

The VSWR value is less than 2 at the resonance frequency (26.6 GHz), suggesting a favorable impedance match and effective signal transmission.

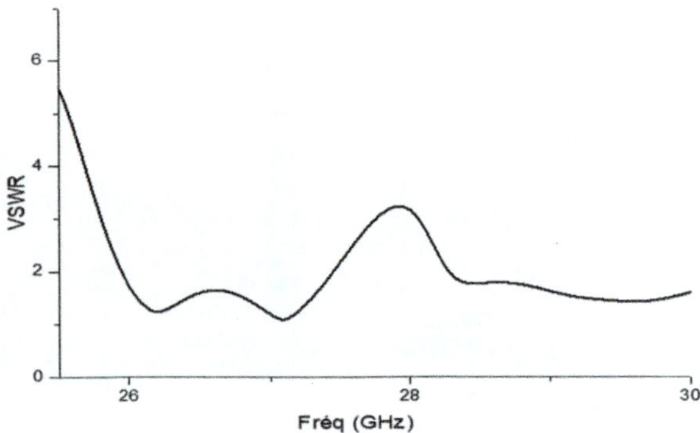

Fig. 12.4 VSWR of the first bandpass filter versus frequency

12.4 The Second Bandpass Filter

We keep the structure of the previous bandpass filter, and we make a modification to the rectangle, as illustrated in Fig. 12.5, in order to widen the bandwidth, such as $L_5 = 1.24$ mm and $W_2 = 1.27$ mm.

The results of a new simulation for S_{11} and S_{12} can be seen in Fig. 12.6, demonstrating the impact of the latest modification on the filter's bandwidth.

The plot reveals that the second bandpass filter exhibits an Ultra-Wideband [25.6; 29.9 GHz] with an insertion loss of − 3.75 dB and return loss − 14 dB.

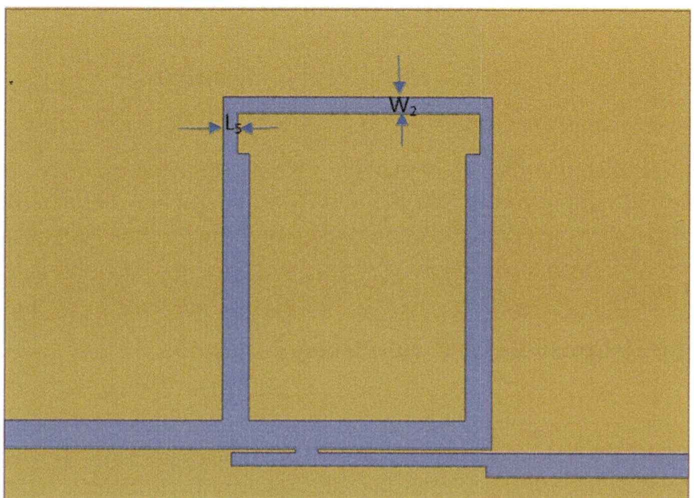

Fig. 12.5 Structure of the second bandpass filter

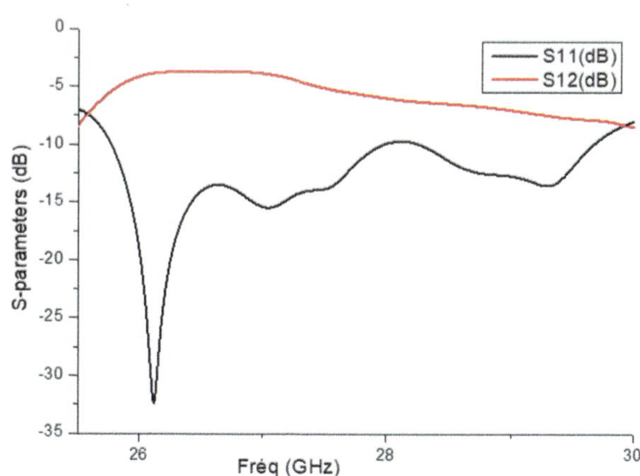

Fig. 12.6 Simulated *S*-parameters of the second bandpass filter

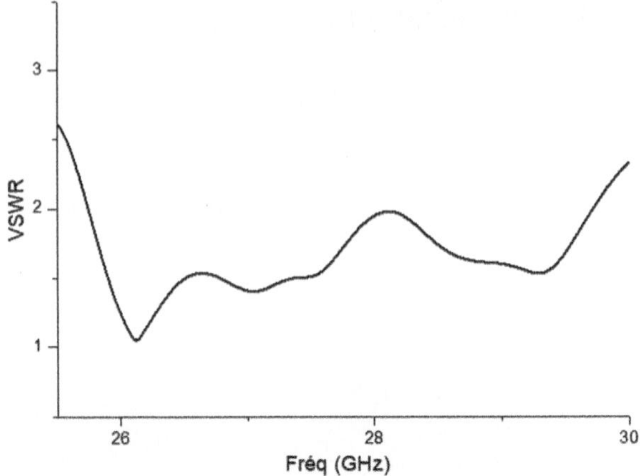

Fig. 12.7 VSWR of the second bandpass filter versus frequency

Figure 12.7 illustrates the variation of the VSWR for the second filter as a function of frequency. It is noteworthy that the VSWR value remains below 2 despite the modifications, ensuring efficient signal transmission for the second bandpass filter.

12.5 The Proposed UWB Bandpass Filter

Insertion loss exerts a pivotal role in the design and analysis of communication systems, particularly in applications like fiber optics and radio frequency (RF) systems. A lower insertion loss value signifies superior performance, indicating that minimal signal power is dissipated as it traverses through the device or system. This parameter is of utmost importance for maintaining signal integrity and ensuring efficient signal transmission within the specified system.

To achieve an UWB bandpass filter with low insertion loss, the structure of the filter depicted in Fig. 12.8 is implemented. This proposed filter formed by coupling between the previous resonator and an *H*-shaped resonator, with the dimensions of the *H*-shaped resonator detailed in Table 12.2.

The simulation result of the reflection coefficient and transmission coefficient for the proposed UWB bandpass filter is illustrated in Fig. 12.9.

The suggested UWB bandpass filter features an insertion loss lower than 2.8 dB, signifying minimal signal attenuation, and a return loss below − 15 dB, demonstrating outstanding signal reflection properties. Furthermore, the proposed filter demonstrates a bandwidth range of [25.75–29.7 GHz].

Figure 12.10 presents the effect of the *H*-shaped on the VSWR value.

The value of the Voltage Standing Wave Ratio (VSWR) approaches 1 at the resonance frequency, indicating a satisfactory impedance match and efficient signal transmission.

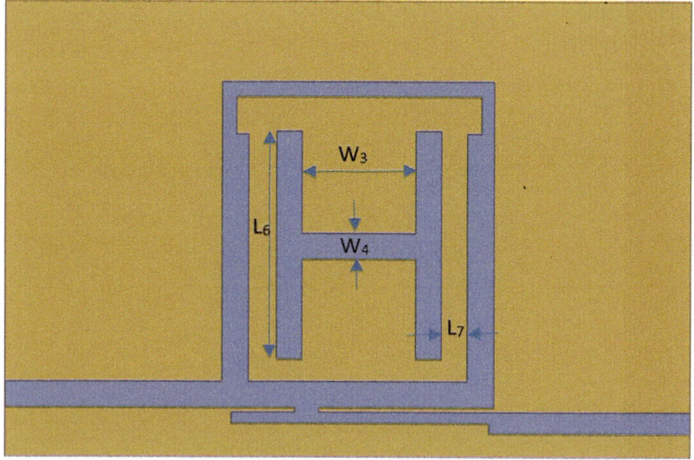

Fig. 12.8 Structure of the proposed UWB bandpass filter

Table 12.2 Dimensions of *H*-shaped

Parameters	Value (mm)
W_3	9.8
W_4	2.2
L_6	20
L_7	2.275

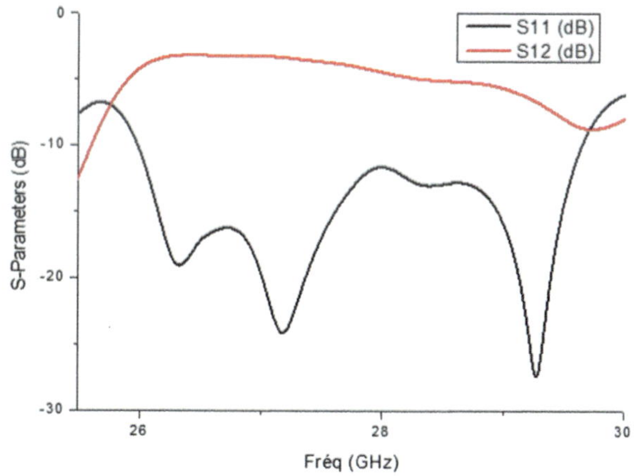

Fig. 12.9 Simulated S-parameters of the suggested UWB bandpass filter

Figure 12.11 provides a comparison of the S_{11} and S_{12} responses among the three filters, and Table 12.3 summarizes their results.

According to Fig. 12.11 and Table 12.3, we observe that the level of signal reflection at the input is minimal for our proposed filter. This suggests that our filter

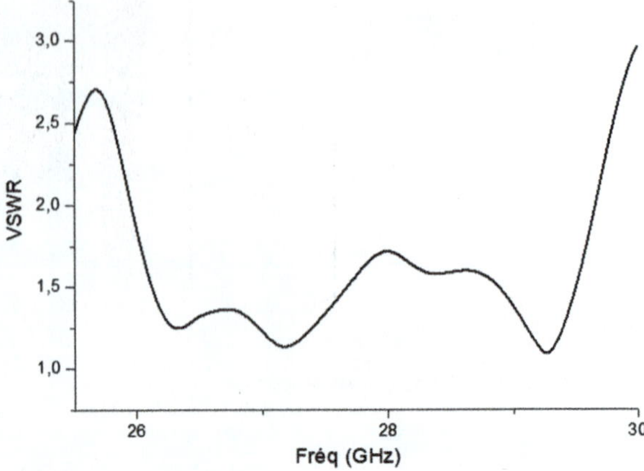

Fig. 12.10 VSWR of the suggested UWB bandpass filter versus frequency

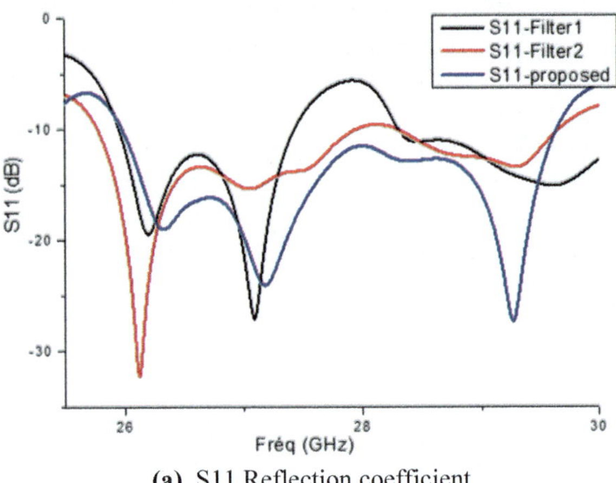

(a). S11 Reflection coefficient

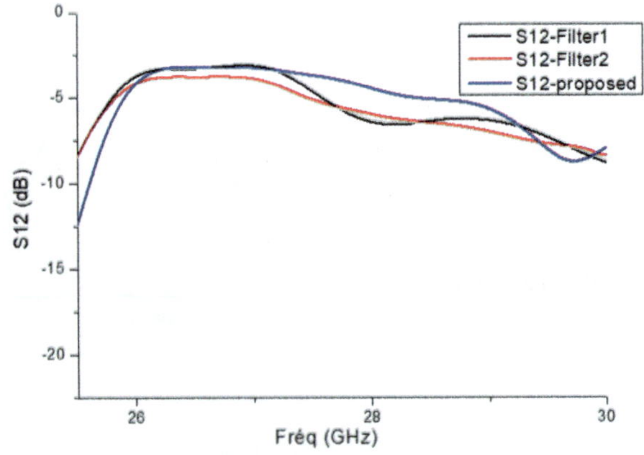

(b). S12 Transmission coefficient.

Fig. 12.11 Simulated *S*-parameters of the three filters

Table 12.3 Comparison of the simulation results of the three filters

Filter	IL (dB)	RL (dB)	BW range (GHz)
Filter 1	− 3	− 12.5	[25.75–27.79]
Filter 2	− 3.75	− 14	[25.6–29.9]
Proposed filter	≥ 2.8	≤ 15	[25.75–29.7]

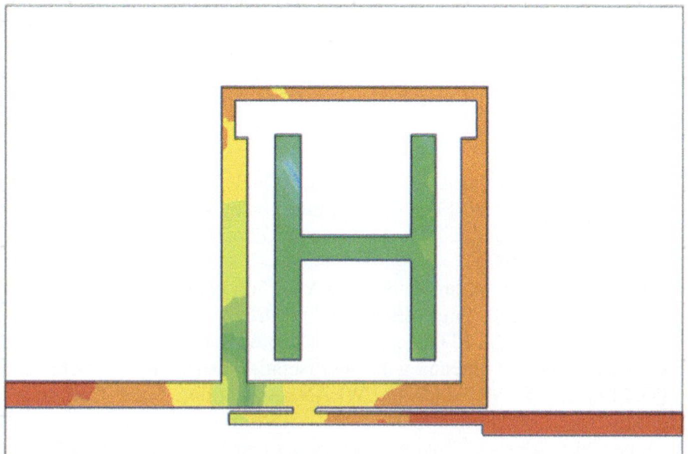

Fig. 12.12 Simulated surface current distributions

demonstrates good impedance matching and high efficiency in transferring the signal through the filter. Thus, the high transmission coefficient S_{12} of the proposed filter, compared to other filters, attests to an efficient signal transmission through this filter within its bandwidth.

A surface current distribution is an important electromagnetic characteristic that can influence the behavior of a filter, particularly concerning its performance in transmitting and rejecting signals at different frequencies. Figure 12.12 depicts the surface current distribution of the proposed UWB bandpass filter.

The diagram illustrates efficient signal transmission within the proposed filter's bandwidth while attenuating unwanted signals.

12.6 Conclusion

In this paper, an *H*-shaped resonator for UWB bandpass filter for 5G is presented.

This filter composed of a combination of two resonators: a rectangular resonator and another in *H*-shaped.

Based on the outcomes of simulations, the suggested UWB bandpass filter delivers a bandwidth of 3.95 GHz. It exhibits an insertion loss below 2.8 dB, a return loss exceeding 15 dB, and VSWR values approaching 1.

Due to its performance and wide frequency range [25.75–29.7] GHz, the filter becomes applicable to various applications such as high-speed wireless communication and radar systems. These performance characteristics make the proposed UWB bandpass filter well-suited for integration into 5G networks.

References

1. Oughton, E.J., Lehr, W., Katsaros, K., Selinis, I., Bubley, D., Kusuma, J.: Revisiting wireless internet connectivity: 5G vs Wi-Fi 6. Telecommun. Policy **45**(5), 102127 (2021)
2. Yang, K., He, C., Fang, J., Cui, X., Sun, H., Yang, Y., Zuo, C.: Advanced RF filters for wireless communications. Chip (2023)
3. Mestoui, J., El Ghzaoui, M.: A survey of NOMA for 5G: implementation schemes and energy efficiency. Lect. Notes Electr. Eng. **745**, 949–959 (2022)
4. Babu, K.V., Das, S., Ali, S.S., El Ghzaoui, M., Madhav, B.T.P., K. Patel, S.: Broadband sub-6 GHz flower-shaped MIMO antenna with high isolation using theory of characteristic mode analysis (TCMA) for 5G NR bands and WLAN applications. Int. J. Commun. Syst. **36**(6), e5442 (2023)
5. Jarry, P., Beneat, J.: Advanced Design Techniques and Realizations of Microwave and RF Filters. Wiley (2007)
6. White, J.F.: High Frequency Techniques: An Introduction to RF and Microwave Engineering. Artech House (2004)
7. Kiouach, F., Aghoutane, B., El Ghzaoui, M.: Novel microstrip bandpass filter for 5G mm-wave wireless communications, e-Prime—advances in electrical engineering. Electron. Energy **6**, 100357 (2023)
8. Ghazaoui, Y., Ghzaoui, M.E., Bri, S., Alami, A.E., Kumari, S.V., Das, S.: A flower-shaped quad-port dual-band MIMO antenna of 29/39 GHz millimeter-wave for 5G applications | Чотирипортова дводіапазонна MIMO-антена у формі квітки 29/39 ГГц міліметрової хвилі для додатків 5G. J. Nano Electron. Phys. **14**(3), 03023 (2022)
9. Elaage, S., Ghzaoui, M.E., Mrani, N., Das, S.: Optimum GMSK based transceiver model for cellular IoT networks. Simul. Modell. Pract. Theory **125**, 102756A (2023)
10. Potelon, B.: Étude et conception de filtres hyperfréquences hybrides planaires volumiques, Laboratoire d'Electronique et des Systèmes de Télécommunications (LEST-UMR CNRS 6165), France (2010)
11. Aghoutane, B., Meskini, N., El Ghzaoui, M., Faylali, H.E.: Millimeter-wave microstrip antenna array design for future 5g cellular applications. In: 2018 International Conference on Electronics, Control, Optimization and Computer Science, ICECOCS 2018, 2018, 8610507
12. Chen, J.H., Liao, S.S.: Design of compact printed 2.4 GHz band-pass filter using LC resonator. In: 2016 International Conference on Advanced Materials for Science and Engineering (ICAMSE), IEEE (2016)
13. Xiao, Z., Xu, Q., Li, C.: Metamaterial bandpass filter based on three-dimensional structure. J. Electron. Mater. (2021)
14. Çınar, A., Bicer, S.: Band-stop filter design based on split ring resonators loaded on the microstrip transmission line for GSM-900 and 2.4 GHz ISM band. Int. Adv. Res. Eng. J. (2020)
15. Ali, W.A.E., Hamdalla, M.Z.M.: Compact triple band-stop filter using novel epsilon-shaped metamaterial with lumped capacitor. J. Instrum. **13**(04), P04007 (2018)

16. Revathi, G., Nivetha, M., Roshini, K., Nishanthi, T.S.: Design of microstrip bandpass filter with DGS for WLAN applications. OSR J. Electron. Commun. Eng. **47** (2017)
17. Rahman, M., Park, J.-D.: A compact tri-band bandpass filter using two stub-loaded dual mode resonators. Progr. Electromag. Res. M **64**, 201 (2018)
18. Slimani, A., Das, S., Ali, W.A.E., El Alami, A., Bennani, S.D., Jorio, M.: Second order microstrip bandpass filter design based on square resonator for 5G sub-6 GHz band. JINST **17**, P07002 (2022)
19. George, T., Lethakumary, B.: High frequency rejection using L shaped defected microstrip structure in ultra wideband bandpass filter. Mater. Today Proc. **25**, 265–268 (2020)
20. Jose, D.S., Kumar, A.S., Shanmugantham, T.: A band pass coupled line filter with DGS for ultra-wide band application. Procedia Comput. Sci. **171**, 561–567 (2020)
21. Wu, X.H., Chu, Q.X., Tian, X.K., Ouyang, X.: UWB quintuple-mode, bandpass filter with sharp roll-off and super-wide upper stopband. IEEE Microwave Wirel. Comput. Lett., 661–663 (2011)
22. Song, K., Xue, Q.: Compact ultra-wideband (UWB) bandpass filter with multiple notch bands. IEEE Microwave Wirel. Comput. Lett., 447–449 (2010)
23. Wu, H.W., Chen, Y.-F.: Ultra wideband bandpass filter with dual-notched bands using stub-loaded rectangular ring multi-mode resonator. Microelectron. J. **43**, 257–262 (2012)

Chapter 13
Spatially Correlated Channels Investigation: Estimation and Hardening in Millimeter-Wave Massive MIMO Systems

Jamal Amadid, Asma Khabba, Zakaria El Ouadi, Lahcen Sellak, and Abdelouhab Zeroual

13.1 Introduction

In the ever-evolving landscape of wireless communication systems, the pursuit of higher data rates, improved spectral efficiency, and enhanced connectivity experiences has led to the emergence of massive MIMO (M-MIMO) technology as a pivotal enabler for next-generation networks [1–4]. M-MIMO represents a groundbreaking paradigm shift in the design and deployment of wireless communication systems, promising the potential to significantly enhance the quality of service for an ever-increasing number of connected devices and users [5]. At the heart of M-MIMO lies the intricate process of channel estimation (CE) [6–9], a fundamental aspect that underpins the successful transmission and reception of data across a multitude of antennas. CE within the context of M-MIMO poses unique challenges and opportunities [3], driven by the sheer scale of antennas employed, dynamic channel conditions, and the imperative to mitigate interference effectively. It is this very challenge that forms the core focus of this paper: The comprehensive exploration of CE techniques tailored specifically for the M-MIMO framework [4, 10]. CE in M-MIMO is far from a conventional endeavor. Unlike traditional MIMO systems, where a relatively modest number of antennas are deployed, M-MIMO leverages an extensive array of antennas at both the transmitter and receiver ends. This multitude of antennas holds the promise of substantially increasing spectral efficiency and

J. Amadid (✉)
Electronics Department, Higher Institute of Engineering and Business (ISGA), Marrakesh, Morocco
e-mail: m.Jamal.amadid@gmail.com

A. Khabba · Z. E. Ouadi · A. Zeroual
Instrumentation, Signals and Physical Systems (I2SP) Group, Faculty of Sciences Semlalia, Cadi Ayyad University, Marrakesh, Morocco

L. Sellak
Smart Systems and Applications (SSA) Group, National School of Applied Sciences of Marrakesh, Cadi Ayyad University, Marrakesh, Morocco

M. El Ghzaoui et al. (eds.), *Next Generation Wireless Communication*, Signals and Communication Technology, https://doi.org/10.1007/978-3-031-56144-3_13

enhancing system capacity. However, the inherent complexity of estimating channel parameters across these numerous antennas necessitates the development of novel and efficient estimation techniques [10–12].
Modern wireless communication networks are at the cusp of a transformative revolution, driven by the promise of Millimeter-wave (mmWave) technology [13–15] and M-MIMO systems. These advancements hold the key to unlocking unprecedented data rates and network capacity, revolutionizing the way we connect and communicate. However, realizing this potential comes with its own set of challenges, chief among them being the accurate estimation of channel parameters in spatially correlated channel environments. In this era of mmWave M-MIMO, where the number of antennas can scale into the hundreds or even thousands, the need for robust and efficient CE techniques becomes paramount. This paper will unravel the intricacies of CE in M-MIMO, shedding light on the various challenges, innovations, and methodologies that drive this critical aspect of modern wireless communication. From exploiting spatial diversity to mitigating pilot contamination (PC) [16–18] and adapting to the unique characteristics of spatially correlated channels. The spatial characteristics of the channel, known as Spatial Directions (SDs), play a crucial role in determining how signals propagate from transmitters to receivers. Due to the inherent variability in many scenarios, it is more likely that certain SDs will experience a stronger signal reception compared to others [1]. This variation arises from the irregular radiation patterns, making it impossible to achieve uniform received power across all SDs. Consequently, some SDs are more prone to receiving higher power levels than others. This correlation, referred to as spatial correlation (SC), implies a strong correlation between the signal's standard deviation and the average channel gain. This SC phenomenon becomes particularly significant when dealing with M-MIMO technology, as it aids in accurately determining user positions, especially in cases where a substantial number of users exhibit distinct SC behavior.

Within the body of existing literature, numerous models have been introduced with the aim of depicting spatial channel correlation. These include the exponential correlation (EC) model [19–21] and the local scattering model. The latter encompasses three distinct distribution types for deviations, specifically Gaussian [1, 3, 22–28], uniform [25, 29–31], and Laplace [11, 32, 33, Sect. 7.4.2]. It is worth noting that the EC model, although widely used, does not constitute a physical model. In certain investigations, efforts have been undertaken to depict it as an approximation well-suited for a uniform linear configuration (ULC).

Conversely, the deviation uniformly distributed is frequently termed the "one-ring (OR) model" due to its assumption that scatterers encompass the user in a circular manner. Moreover, the Gaussian scattering model is widely employed as a framework to characterize spatial correlation of channels. This model utilizes a set of parameters to regulate the degree of correlation among antennas within an array, primarily serving to provide a numerical representation of spatial correlation among antennas in a ULC [22]. The objective behind introducing VD within 2D antenna arrays is to harness and manage the anticipated strong channel gain and accuracy for efficient utilization in 5th generation wireless communication networks, particularly when employing a substantial number of antennas (NoA). However, it's important to acknowledge that while 2D antenna arrays offer advantages, they also present challenges, notably the need for a more complex three-dimensional (3D) channel model description. Among the designs that closely resemble the 2D antenna layout, rectangular configurations (RC) [1] and cylindrical configurations [4, 34, 35] are the most common [23, 36–38]. In practical terms, both URC and UCYC configurations offer several advantages for M-MIMO systems. Their small form factor proves highly beneficial in cases featuring spatially correlated (ScD) channels, given that the array design plays a substantial role in determining the SC results [39]. Furthermore, URC can be deployed on building facades, while UCYC is typically employed above buildings. Notably, the UCYC configuration possesses a shape that enables signal reception from all directions.

13.2 Related Works

We acknowledge the significance of comprehending the influence of SC on CE within the context of M-MIMO systems, given that SC reflects real-world propagation conditions. Consequently, this subject has garnered substantial attention from numerous research studies. In references [34, 35], the authors demonstrate that the power received by individual antennas in a network fluctuates randomly, leading to disparate gains from each antenna and uneven spatial correlation of the channel among antennas. In [1], the authors examine channel SC and its impact on CE and spectral efficiency (SE) employing the OR model, concluding that the OR model leads to low-rank channel matrices (CMs). In [19], the authors evaluate and analyze system outcomes, specifically CE and PC, using the MMSE estimator under scenarios involving ScD channels [40]. Additionally, the authors characterize SC using the EC model for ULC. Nevertheless, this work does not delve into URC or UCYC and their potential influence on SC since the EC model is an approximation model primarily tailored for ULCs. In reference [5], the employment of the EC model is seen in describing spatial correlation, considering variations in large-scale fading (LSF) across the array. In addition, this model is utilized to tackle the PC issue. Furthermore, it is shown that the PC problem diminishes as the NoA approaches infinity. Nonetheless, it's essential to note that this model serves as an approximation tailored for ULCs and specifically focuses on this array configuration. Uniform circular configuration (UCC) and their implications on SC are not covered. Reference [22] provides an approximate expression for the multi-path channel in the ULC configuration using the Gaussian local scattering model. However, this work exclusively addresses ULC and does not encompass for UCC arrangement. Furthermore, the discussion primarily revolves around SE, with no treatment of the CE problem. Recently, researchers have turned their attention to an alternative network known as the cell-free network [41–46], which offers distinct advantages over traditional M-MIMO networks. Numerous works have explored ScD channels within cell-free M-MIMO networks, as evident in references [47–50]. Conversely, several studies have contributed to wireless communication MIMO networks using various techniques, particularly in ScD channel scenarios [51–54]. Notably, the incorporation of M-MIMO technology is absent from these investigations.

13.2.1 Work Organization

To provide a structured overview of our study, we organize the main sections as follows:
In Sect. 13.3, we delve into the system model, focusing on an in-depth examination of the pilot sequence (PS) phase. Section 13.4 is dedicated to the investigation of the CE phase, where, we present the expression for the Normalized Mean Square Error (NMSE) associated with the MMSE estimator. Our exploration of the impact of SC on channel hardening (CH) is detailed in Sect. 13.5. Section 13.6 introduces the spatial channel correlation model utilized in our study. Section 13.7 presents numerical results, offering empirical evidence to validate our theoretical expressions. Finally, in Sect. 13.8, we provide a concise summary of our research conclusions, highlighting the key insights and contributions of our study.

13.3 System Model

In wireless communication systems, pilot-based estimation stands out as a pivotal technique. It hinges on the transmission of pilots by users, serving as reference signals for CE at each base station (BS). The temporal division of the coherence block, which denotes a period where the channels remain relatively constant, is determined by the time-division duplex (TDD) phase [36, 55]. This division typically comprises three segments: The sequence begins with uplink (UL) pilots, followed by UL data, and then transitions to downlink data. In the context of pilot-based CE, particular emphasis is placed on the UL pilot phase within each time-frequency slot. During this phase, users send their pilot signals to the base stations assigned to them, each equipped with a ULC comprising M antennas. In multi-cell M-MIMO systems, the PS allocated to a user in one cell is duplicated across all remaining cells, constrained by the coherence block limitations (i.e., limited number of samples) [56]. Consequently, a pilot reuse strategy is essential to serve all users network-wide, giving rise to a common interference phenomenon known as PC. For the scope of this study, we assume orthogonal pilots for users within each cell, thus circumventing intra-cell interference concerns. Given the limitations imposed by the coherence block duration (τ_c), pilots from one cell are reused in all other cells. Furthermore, our examination incorporates Rayleigh fading to model Spatially Correlated Channels (ScD). Here, $\mathbf{g}_{jlk} \in \mathbb{C}^{M\times 1}$ represents the channel linking the kth user in the lth cell to the jth BS. The channel is characterized as $\mathbf{g}_{jlk} \sim (0_M, \mathbf{C}_{jlk})$, where $\mathbf{C}_{jlk}$ denotes the CM capturing LSF effects, encompassing factors like path-loss as well as shadows fading. Alternatively, fast-scale fading is assumed to adhere to a complex

Gaussian distribution. Moreover, channel correlation is mathematically expressed through elements of $\mathbf{C}_{jlk}$ that do not have zero values off the main diagonal, assuming perfect knowledge of the CM.

13.3.1 Pilots Sequence Phase

During this stage, we establish the PS length as τ for each user. In each cell, users employ unique and mutually orthogonal PSs, effectively preventing interference within the cell. Moreover, we suppose $\tau = K$ to guarantee the elimination of intra-cellular interference, with K represents the number of users within each cell. Additionally, we adopt the strategy of reusing the segment of pilot signals utilized in each cell across all the remaining cells [57, 58]. This is due to the limitations associated with the flat coherence block duration τ_c, resulting in a frequency reuse factor of 1. Consequently, we can express the de-spread vector $\mathbf{Z}_j \in \mathbb{C}^{M\times\tau}$ at the jth BS in the following manner:

$$\mathbf{Z}_j = \sqrt{\eta}\sum_{l=1}^{L}\sum_{k=1}^{K}\mathbf{g}_{jlk}\psi_k^T + B_j. \tag{13.1}$$

In this context, η denotes the pilot transmission strength, ψ_k represents the kth user's PS, and B_j characterizes the normalized noise, with each component being independently and identically distributed as $\mathcal{N}_{\mathbb{C}}(0, 1)$. Furthermore, Eq. (13.1) can be reframed without considering the impact of pilots, resulting in a signal that depends solely on the channel and noise. Accordingly, the subsequent calculation assumes the scenario where the jth BS is tasked with estimating the ith user's channel in the lth BS.

$$z_{jli} = Z_j\psi_i^* = \sqrt{\eta}\sum_{l'=1}^{L}\mathbf{g}_{jl'i} + b_{ji}. \tag{13.2}$$

In this context, z_{jli} is a vector with dimensions $M \times 1$, and b_{ji}, represented asbreak $(B_j \times \psi_i^*) \in \mathbb{C}^{M\times\tau}$, follows a normal distribution with parameters $\mathcal{N}_{\mathbb{C}}(0, \mathbf{I}_M)$. It's worth emphasizing that η is equal to $\tau \times \rho_{uplink}$, where ρ_{uplink} represents the normalized power for UL transmission. Consequently, enhancing the PS length yields substantial improvements in system outcomes, given that the UL transmission power (η) exhibits a linear dependence on the length of PS.

13.4 Channel Estimation

The MMSE estimator falls under the realm of Bayesian estimation techniquesbreak [12, 28, 55, 56, 59]. As per existing literature, the Bayesian-MMSE estimator excels in terms of channel estimate, particularly when foreknowledge of channel statistics is accessible. A comprehensive methodology is outlined in [40], where a meticulously detail for the MMSE estimator is given. Furthermore, as outlined in [60], the channel linking the ith user in the lth cell to the jth BS is characterized by

$$\hat{\mathbf{g}}_{jli} = \sqrt{\eta}\mathbf{C}_{jli}\Theta_{ji}^{-1}z_{jli} \sim \mathcal{N}_{\mathbb{C}}(0, \Xi_{jli}). \tag{13.3}$$

In this context, Θ_{ji} is used to represent the variance of the vector z_{jli}, denoting the expected value of $\{z_{jli}z_{jli}^H\}$. Similarly, Ξ_{jli} is employed to signify the estimated vector' variance $\hat{\mathbf{g}}_{jli}$, representing $\mathbb{E}\{\hat{\mathbf{g}}_{jli}\hat{\mathbf{g}}_{jli}^H\}$. Furthermore, we introduce $\tilde{\mathbf{g}}_{jli}$ to represent the disparity between the actual channel and the estimated channel, effectively characterizing the estimation error. In other words, $\tilde{\mathbf{g}}_{jli}$ is a representation of the estimation error, and it follows a normal distribution with parameters $\mathcal{N}_{\mathbb{C}} \sim (0, \Omega_{jli})$, where Ω_{jli} is calculated as $\mathbf{C}_{jli} - \Xi_{jli}$.
It's important to note that the estimation channel error and estimated vectors are mutually orthogonal due to the principles underlying the B-MMSE estimator. An essential observation is that users sharing the same pilot exhibit correlated estimated vectors. As a result, we introduce the average-antenna correlation (AAC) coefficient metric to assess the level of correlation between a pair of users. The AAC coefficient metric depends on both the target channel $\mathbf{g}_{jji}$ and the interfering channel $\mathbf{g}_{jli}$, and it is formulated as follows [1]:

$$\begin{aligned} \upsilon_{jk,li} &= \frac{\mathbb{E}\{\hat{\mathbf{g}}_{jli}^H\hat{\mathbf{g}}_{jjk}\}}{\sqrt{\mathbb{E}\{\|\hat{\mathbf{g}}_{jli}\|^2\}\mathbb{E}\{\|\hat{\mathbf{g}}_{jjk}\|^2\}}} \\ \upsilon_{jk,li} &= \frac{tr(\mathbf{C}_{jli}\mathbf{C}_{jjk}\Xi_{jli})}{tr(\mathbf{C}_{jli}\mathbf{C}_{jli}\Xi_{jli})tr(\mathbf{C}_{jli}\mathbf{C}_{jjk}\Xi_{jli})} \end{aligned}. \tag{13.4}$$

In line with Eq. (13.4), when the parameter $\upsilon_{jk,li}$ approaches 0, the interfering user's contribution to interference remains minimal compared to the target user. Conversely, when $\upsilon_{jk,li}$ tends toward 1, the interfering user significantly elevates the interference level in comparison with the target user. Consequently, the level of correlation plays a pivotal role in determining the quality of CE. Furthermore, the accuracy of CE is assessed and scrutinized using a widely recognized metric, namely the NMSE, defined as

$$\iota_{jli}^{\mathrm{NMSE}} = \frac{\mathbb{E}\{\|\mathbf{g}_{jli} - \hat{\mathbf{g}}_{jli}\|^2\}}{\mathbb{E}\{\|\mathbf{g}_{jli}\|^2\}} = \frac{\mathbb{E}\{\|\tilde{\mathbf{g}}_{jli}\|^2\}}{\mathbb{E}\{\|\mathbf{g}_{jli}\|^2\}} = \frac{tr(\Omega)_{jli}}{tr(\mathbf{C})_{jli}}. \tag{13.5}$$

By employing Eq. (13.5), it becomes possible to assess the precision of the estimation within the studied system. Furthermore, the NMSE falls within a scale spanning from 0 to 1. In instances where $\iota_{jli}^{\mathrm{NMSE}}$ equals 0, it signifies the optimal scenario, wherein the estimated channel vector perfectly aligns with the actual channel vector. Conversely, when $\iota_{jli}^{\mathrm{NMSE}}$ equals 1, it denotes the most unfavorable scenario, characterized by the highest degree of estimation error.

13.5 Channel Hardening

In this section, we examine the influence of SC over the CH [1, 3]. Moreover, any M-MIMO network with a substantial NoA at the BS can leverage this feature, which offers multiple advantages to M-MIMO systems [1]. It's important to note that CH is closely linked to the channel propagation, which, in turn, depends on the SC as the underlying cause of SC. In their study [1], Björnson et al. examine SC within ULC configurations and its effect on CH. Their research demonstrates that heightened SC leads to a significant reduction in CH. Additionally, smaller angular spread deviation (ASD) values are associated with increased SC, while larger ASD values result in higher CH variability. Notably, this phenomenon appears to be less advantageous for UPC as simulations have verified that UPC tends to exhibit significantly higher SC when compared to ULC. While SC typically aids in separating user channels, the reverse is true for CH. On the other hand, the consideration of diversity concept and utilization of concatenated received antennas effectively mitigates SSF effect, where diversity concept encompasses scenarios with unrelated fading, and the deployment of large concatenated antennas at the receiver helps suppress SSF variations. This, in turn, leads to deterministic signal behavior, characteristic of CH. Moreover, in the presence of a large NoA, this pattern is anticipated to yield greater efficiency, offering numerous enhancements for M-MIMO systems, including SSF mitigation and advanced beam-forming techniques. Consequently, the next equation defines CH for the kth user in the jth cell communicating with the jth BS.

$$\frac{\|\mathbf{g}_{jjk}\|^2}{\mathbb{E}\{\|\mathbf{g}_{jjk}\|^2\}} \xrightarrow[M\to\infty]{a.s} 1. \tag{13.6}$$

This equation summarizes the core concept of CH, representing the relationship between channel gain and the average channel gain. As the NoA approaches infinity, these values tend to converge and become equal. It's worth noting that this expression can be applied to any channel $\mathbf{g}_{jjk}$ [1]. Furthermore, to assess how a specific channel approaches CH requirements, a more efficient approach, as opposed to Eq. (13.6), can be employed, which is represented by the next equation, regardless of asymptotic analysis

$$\gamma_{jjk} = \mathbb{V}\left\{\frac{\|\mathbf{g}_{jjk}\|^2}{\mathbb{E}\{\|\mathbf{g}_{jjk}\|^2\}}\right\} = \frac{\mathbb{V}\{\|\mathbf{g}_{jjk}\|^2\}}{(\mathbb{E}\{\|\mathbf{g}_{jjk}\|^2\})^2} = \frac{tr(\mathbf{C}_{jjk})^2}{(tr(\mathbf{C}_{jjk}))^2}. \tag{13.7}$$

In line with Eq. (13.7), as the variance approaches zero, the channel experiences a greater degree of hardening. Both the numerator and denominator in this equation rely on the CMs $\mathbf{C}_{jjk}$. This equation underscores the influence of SC on CH, as it directly impacts the eigenvalues.

13.6 Spatial Correlation Matrix

In this dedicated section, our foremost goal is to present a comprehensive analysis of the covariance matrices (CMs) that serve as critical descriptor of spatial channel correlation. Our intention is to offer a detailed and thorough explanation of these CMs, shedding light on their fundamental characteristics and properties. Furthermore, we will not limit our examination to a single scenario but rather extend our investigation to encompass both ULC and UCC. By doing so, we aspire to provide a holistic understanding of the spatial channel correlation in various scenarios, thus contributing to a more comprehensive body of knowledge in the field of wireless communication systems.

13.6.1 Spatial Correlation Matrix Over ULC

In this section, our focus is directed toward the utilization of GLS model [1] in the context of a multi-scatterers scenario, where the number of scatterers exceeds one. Within this framework, we aim to characterize the SC of the channel within the ULC [3, 4, 10, 28, 59]. This model is governed by the interplay of two key parameters, namely the N-AoA and the ASD, where the ASD dictates the deviation from this nominal angle [60–62]. It's worth noting that in the context of a multi-scattering scenario, the number of scatterers becomes a significant factor influencing system's outcomes. Consequently, we can mathematically express the (e, p)th component as

$$[\mathbf{C}]_{e,p}^{ulc} = \frac{\beta}{S} \sum_{s=1}^{S} e^{j\pi(e-p)\sin(\chi_s)} e^{-\frac{\sigma_\chi^2}{2}(\pi(e-p)\cos(\chi_s))^2}. \tag{13.8}$$

Within this context, the symbol β corresponds to the coefficients characterizing the LSF. The term χ_s, which spans the range $\chi_s \sim \mathcal{U}[\chi - 40°,\ \chi + 40°]$, signifies the N-AoA. The variable S is employed to denote the presence of multiple scattering clusters (ScC) (i.e., $S > 1$). Our modeling assumption involves the consideration of Gaussian-distributed multi-path elements within each cluster, exhibiting scattering around the nominal AoA. The standard deviation for these deviations from the nominal AoA is represented by σ_χ. To maintain simplicity, the subscripts for the variables in Eq. (13.8) are omitted. However, it's essential to emphasize that the model described in Eq. (13.8) is exclusively suitable for the ULC configuration, where the AoA primarily pertains to azimuth angles. Notably, the significant variations in the eigenvalues of $\mathbf{C}$ contribute to a more pronounced SC.
As we address a Millimeter-wave M-MIMO system, thus, the channel gain is expressed as [63, 64]

$$\beta = \alpha + 10\kappa \log_{10}(d) + \xi \quad \text{[dB]}. \tag{13.9}$$

Here, d stand for the user-BS distance and $\xi \sim \mathcal{N}_{\mathcal{C}}(0, \sigma)$ [63].

13.6.2 Spatial Correlation Matrix Over UCC

In this section, we delve into the proposed UCC configuration and introduce a GLMS tailored specifically for UCC. We also provide a theoretical explanation of this model. The SC matrices in the UCC setup are determined by the azimuth and elevation angles assigned to each user, which differentiates our approach from existing channel models. Our proposed method stands out for its simplicity compared to the complex models discussed in previous literature. The key aim is to formulate a GLMS model that delineates spatial correlation within the UCC arrangement. In essence, we aim to establish a covariance matrix $\mathbf{C} \in \mathcal{C}^{M\times M}$ to represent the UCC setup, considering non-line-of-sight channel paths connecting users to BSs equipped with UCC. We consider the assumption that the angle of elevation is fixed at $\pi/2$, resulting in the received signal being a composite of multiple multi-path elements. These elements propagate as waves, entering the array at specific angles, and are mathematically expressed in the received vector

$$\Lambda_s(\mathring{\chi}, \phi) = g_s \begin{bmatrix} e^{j2\pi\delta \sin(\phi)\cos(\mathring{\chi}_s - \theta_1)} \\ e^{j2\pi\delta \sin(\phi)\cos(\mathring{\chi}_s - \theta_2)} \\ \vdots \\ e^{j2\pi\delta \sin(\phi)\cos(\mathring{\chi}_s - \theta_m)} \end{bmatrix}. \tag{13.10}$$

As mentioned earlier, we have set the elevation angle to $\pi/2$, making $\sin(\phi = \frac{\pi}{2})$ equal to 1. Consequently, Eq. (13.10) can be expressed in the following manner:

$$\Lambda_s\left(\mathring{\chi}\right) = \Lambda_s(\mathring{\chi}, \phi = \frac{\pi}{2}\Big) = g_s \begin{bmatrix} e^{j2\pi\delta \cos(\mathring{\chi}_s - \theta_1)} \\ e^{j2\pi\delta \cos(\mathring{\chi}_s - \theta_2)} \\ \vdots \\ e^{j2\pi\delta \cos(\mathring{\chi}_s - \theta_j)} \end{bmatrix}. \tag{13.11}$$

With

$$\theta_j = 2\pi \frac{(j-1)}{M}, \qquad j = 1, \ldots, M. \tag{13.12}$$

The summation of all the paths within the vector $\Lambda_s(\mathring{\chi})$, denoted as P_{path}, results in the channel $\mathbf{g}$, which can be mathematically represented as follows:

$$\mathbf{g} = \sum_{s=1}^{P_{\text{path}}} \Lambda_s(\mathring{\chi}), \tag{13.13}$$

where $g_s \sim (0, \mathbb{E}\{|g_s|^2\})$ and $\beta = \sum_{s=1}^{P_{\text{path}}} \mathbb{E}\{|g_s|^2\}$. Therefore, by applying the central limit theorem [40] to the multi-dimensional scenario, we can represent the channel h as follows:

Hence, utilizing the central limit theorem [40] in the multi-dimensional context allows us to express the channel $\mathbf{g}$ in the following manner:

$$\mathbf{g} \to \mathcal{N}_{\mathcal{C}}(0,\ \mathbf{C}), \qquad P_{\text{path}} \to \infty. \tag{13.14}$$

With $\mathbf{C} = \mathbb{E}\{\sum_s \Lambda_s \Lambda_s^H\}$. Hence, we can describe the $(e,\ p)$th component of $\mathbf{C}$ as follows:

$$\begin{aligned}[\mathbf{C}]_{e,p}^{\text{ucc}} &= \sum_{s=1}^{P_{\text{path}}} \mathbb{E}\{|g_s|^2\}\mathbb{E}\{e^{j2\pi\delta\cos(\mathring{\chi}_s-\theta_e)}e^{-j2\pi\delta\cos(\mathring{\chi}_s-\theta_p)}\} \\ &= \beta\int e^{j2\pi\delta[\cos(\mathring{\chi}-\theta_e)-\cos(\mathring{\chi}-\theta_p)]} f(\mathring{\chi})d\mathring{\chi}\end{aligned}. \tag{13.15}$$

In this context, β represents the LSF coefficient, while $\mathring{\chi}$ signifies the angle of arrival for a specific spot in the multi-path component. Furthermore, θ_e and θ_p are used to specify the angular positions of the antennas within the UCC arrangement. It's important to note that we make an assumption that there are no ScC surrounding the BS; instead, all the ScC are considered to be positioned around the users. As a result, we define $\mathring{\chi}$ as the AoA, which is derived from a deterministic angle along with a variation around it, and this can be expressed as follows:

$$\mathring{\chi} = \chi + \omega. \tag{13.16}$$

Hence, we can describe the $(e,\ p)$th component of $\mathbf{C}$ as follows:

$$[\mathbf{C}]_{e,p}^{\text{ucc}} = \beta\int_{-\infty}^{+\infty} e^{j2\pi\delta[\cos(\chi-\theta_e+\omega)-\cos(\chi-\theta_p+\omega)]}\frac{1}{\sqrt{2\pi\sigma_\chi}}e^{\frac{-\omega^2}{2\sigma_\chi^2}}d\omega. \tag{13.17}$$

One can set

$$\Psi = \cos(\chi - \theta_e + \omega) - \cos(\chi - \theta_p + \omega). \tag{13.18}$$

Thus,

$$\begin{aligned}\cos(\chi - \theta_l + \omega) &= \cos(\chi - \theta_l)\cos(\omega) - \sin(\chi - \theta_l)\sin(\omega) \quad l \in (e,\ p) \\ &= \cos(\chi - \theta_l) - \sin(\chi - \theta_l)\omega\end{aligned}. \tag{13.19}$$

Therefore, by utilizing Eq. (13.19), we can express Eq. (13.18) as follows:

$$\begin{aligned}\Psi &= \cos(\chi - \theta_e + \omega) - \cos(\chi - \theta_p + \omega) \\ &= \cos(\chi - \theta_e) - \sin(\chi - \theta_e)\omega - \cos(\chi - \theta_p) + \sin(\chi - \theta_p)\omega \\ &= [\cos(\chi - \theta_e) - \cos(\chi - \theta_p)] - [\sin(\chi - \theta_e) - \sin(\chi - \theta_p)]\omega \\ \Psi &= E_{(e,p)} - R_{(e,p)}\omega\end{aligned}. \tag{13.20}$$

With

$$\begin{aligned}E_{(e,p)} &= [\cos(\chi - \theta_e) - \cos(\chi - \theta_p)], \\ R_{(e,p)} &= [\sin(\chi - \theta_e) - \sin(\chi - \theta_p)].\end{aligned} \tag{13.21}$$

After providing an approximate expression for Eq. (13.19), we proceed to calculate the closed formula for the suggested UCC as

$$\begin{aligned}[\mathbf{C}]^{\text{ucc}}_{e,p} &= \beta \int_{-\infty}^{+\infty} e^{2\pi\delta[E_{(e,p)} - R_{(e,p)}\omega]} \frac{1}{\sqrt{2\pi}\sigma_\chi} e^{\frac{-\omega^2}{2\sigma_\chi^2}} d\omega \\ &= \beta e^{2\pi\delta E_{(e,p)}} \int_{-\infty}^{+\infty} \frac{1}{\sqrt{2\pi}\sigma_\chi} e^{-2\pi\delta R_{(e,p)}\omega + \frac{-\omega^2}{2\sigma_\chi^2}} d\omega \\ &= \beta e^{2\pi\delta E_{(e,p)}} \int_{-\infty}^{+\infty} \frac{1}{\sqrt{2\pi}\sigma_\chi} e^{\frac{-[4\sigma_\chi^2\pi\delta R_{(e,p)}\omega + \omega^2]}{2\sigma_\chi^2}} d\omega \\ &= \beta e^{2\pi\delta E_{(e,p)}} \int_{-\infty}^{+\infty} \frac{1}{\sqrt{2\pi}\sigma_\chi} e^{\frac{-[(2\pi\delta\sigma_\chi^2 R_{(e,p)} + \omega)^2 - (2\pi\delta\sigma_\chi^2 R_{(e,p)})^2]}{2\sigma_\chi^2}} d\omega \\ &= \beta e^{2\pi\delta E_{(e,p)}} e^{\frac{(2\pi\delta\sigma_\chi^2 R_{(e,p)})^2}{2\sigma_\chi^2}} \underbrace{\int_{-\infty}^{+\infty} \frac{1}{\sqrt{2\pi}\sigma_\chi} e^{\frac{-[(2\pi\delta\sigma_\chi^2 R_{(e,p)} + \omega)^2]}{2\sigma_\chi^2}} d\omega}_{=1} \\ [\mathbf{R}]^{\text{ucc}}_{e,p} &= \beta e^{2\pi\delta E_{(e,p)}} e^{\frac{\sigma_\chi^2}{2}(2\pi\delta R_{(e,p)})^2}\end{aligned}. \tag{13.22}$$

Accordingly, in a multi-scatterer scenario, the above equation can be rewritten as follows:

$$[\mathbf{C}]^{\text{ucc}}_{e,p} = \frac{\beta}{S} \sum_{s=1}^{S} e^{2\pi\delta E^s_{(e,p)}} e^{\frac{\sigma_\chi^2}{2}(2\pi\delta R^s_{(e,p)})^2}. \tag{13.23}$$

With

$$\begin{aligned}E^s_{(e,p)} &= [\cos(\chi_s - \theta_e) - \cos(\chi_s - \theta_p)], \\ R^s_{(e,p)} &= [\sin(\chi_s - \theta_e) - \sin(\chi_s - \theta_p)].\end{aligned} \tag{13.24}$$

13.7 Simulation Results

In Fig. 13.1, we present a plot illustrating the ANE, created with a ULC configuration, versus the eigenvalues of **C**. This analysis is conducted for various values of σ_χ, specifically $\sigma_\chi = 0°, 5°, 10°$, and $15°$. It is important to note that the NoA at the BS in this scenario is $M = 100$. The comparison is made between two circumstances: the correlated (CS) and unCS (i.e., Rayleigh fading). The graph reveals several key insights. In the CS, employing the GLMS model, approximately 30% of the eigenvalues exhibit higher values than those in the unCS. However, the remaining eigenvalues generated in the CS are notably smaller than their unCS. As anticipated, when ASD values are large, the spatial channel correlation weakens, and vice versa. In simpler terms, as ASD increases, the term $e^{-\frac{\sigma_\chi^2}{2}(\pi(e-p)\cos(\chi_s))^2}$ (as seen in Eq. (13.8) tends toward 0. This term is significant for off-diagonal CM elements since, for $e = p$, it equals 1. Consequently, as the off-diagonal components of the matrix decrease, its rank increases. For instance, when $\sigma_\chi = 15°$, a mild degree of of spatial channel correlation is generated. Conversely, in the situation of $\sigma_\chi = 0°$, a substantial degree of spatial channel correlation is observed. Furthermore, as σ_χ approaches infinity, the scatterers become uniformly distributed across the range of $-\pi$ to π. Nevertheless, this doesn't result in a completely unCS because of the high resolution of the ULC arrangement in specific angular directions when compared to others.

In Fig. 13.2, we present a plot illustrating the ANE, created with a UCC configuration, versus the eigenvalues of **C**. This analysis is conducted for various values of σ_χ, specifically $\sigma_\chi = 0°, 5°, 10°$, and $15°$, while $\sigma_\phi = \pi/2$. It is important to note that the NoA at the BS in this scenario is $M = 100$. The comparison is made between two scenarios: the correlated scenario (CS) and the uncorrelated scenario (unCS) (i.e., Rayleigh fading). The graph reveals several key insights. In the CS, employing the GLMS model, approximately 20% of the eigenvalues exhibit higher values than those in the unCS. However, the remaining eigenvalues generated in the CS are notably smaller than their unCS. As anticipated, when ASD values are large, the spatial channel correlation weakens, and vice versa. In simpler terms, as ASD increases, the term $e^{-\frac{\sigma_\chi^2}{2}(\pi(e-p)\cos(\chi_s))^2}$ (as seen in Eq. (13.8) tends toward 0. This term is significant for off-diagonal CM elements since, for $e = p$, it equals 1. Consequently, as the off-diagonal components of the matrix decrease, its rank increases. For instance, when $\sigma_\chi = 15°$ and $\sigma_\phi = \pi/2$, a mild degree of spatial channel correlation is generated. Conversely, in the situation of $\sigma_\chi = 0°$, a substantial degree of spatial channel correlation is observed. Furthermore, as σ_χ and σ_ϕ still equal to $\pi/2$ approaches infinity, the scatterers become uniformly distributed across the range of $-\pi$ to π. However, this does not lead to a perfectly unCS due to the ULC arrangement's high resolution in specific angular directions when compared to others.

To comprehensively assess these impacts, we present a comparative analysis of the UCC and ULC arrangements in relation to unCS Rayleigh fading, which serves as our reference situation, as depicted in Fig. 13.3. Figure 13.3 provides valuable insights into the influence of ASD values on the CH phenomenon. Notably, within

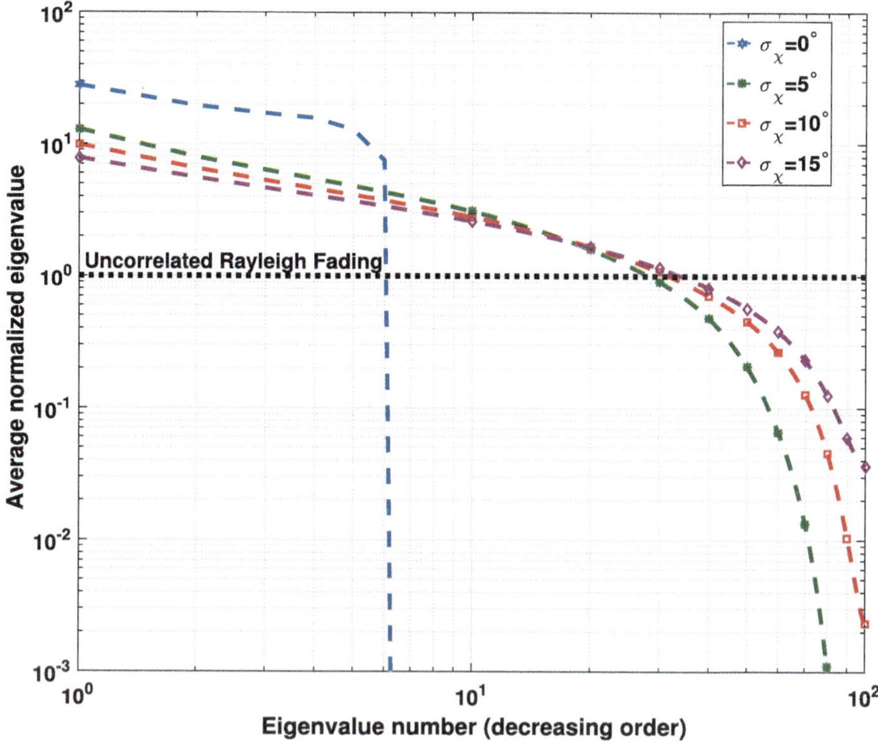

Fig. 13.1 Mean normalized eigenvalues (MNE) in descending order versus eigenvalue number for ULC

both considered arrangements (ULC and UCC), the ASD values exert a significant influence on the CH. Specifically, smaller ASD values result in higher variance, as they contribute to heightened SC of the channel. Furthermore, it's evident that the UCC arrangement exhibits a slightly more pronounced hardening effect compared to the ULC arrangement. The ASD values play a crucial role in determining the required NoA needed to achieve CH since they impact the rank of the SC matrix. It's essential to emphasize that CH cannot be attained without a substantial level of SC. In Fig. 13.3, although both configurations (ULC and UCC) ultimately converge toward a comparable upper limit, the UCC attains this limit at a swifter pace than the ULC scenario

Another perspective to consider is how SC influences CH by setting a threshold variance value of 10^{-2} as the target. This threshold is considered appropriate to demonstrate the benefits of CH. Furthermore, this threshold is predetermined based on the unCS, where the variance is represented only in terms of M, as detailed in [1]. For instance, when ASD $= 15°$, the required NoA (M) to achieve a variance of 0.01 in each of the 3 discussed cases is as follows: $M = 10^2$ for unCS, approxi-

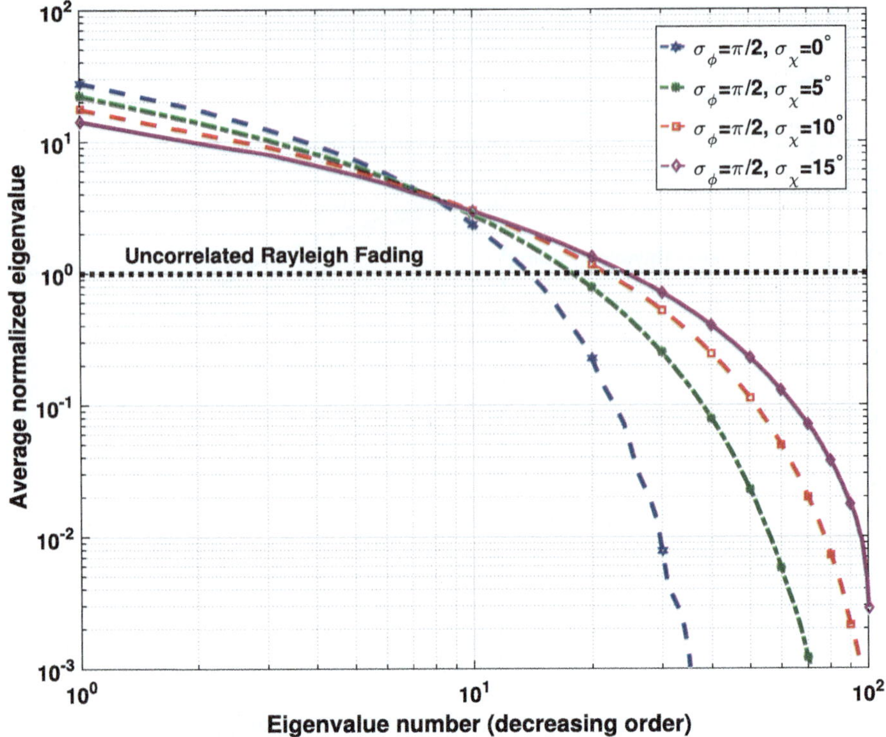

Fig. 13.2 MNE in descending order versus eigenvalue number for UCC

mately $M = 155$ for ULC, and approximately $M = 315$ for UCC. Conversely, when ASD $= 5°$, only $M = 100$ is needed for unCS, approximately $M = 275$ for ULC, and approximately $M = 660$ for UCC in the respective cases.

13.8 Conclusion

In this comprehensive study, we conducted a thorough evaluation, analysis, and investigation into spatial channel correlation utilizing the GLMS model, with a particular emphasis on its impact on the CE process for Millimeter-wave M-MIMO system. Furthermore, we extended the application of the GLMS model across various antenna arrangements, recognizing its versatility in constructing spatial channel correlations that closely emulate real-world scenarios. It is worth noting that the GLMS model is most suitable for the ULC configuration. Our investigations encompassed a detailed analysis of both CE and CH for two distinct configurations: the ULC and UCC arrangements. Additionally, we introduced a novel GLMS model tailored for the UCC configuration and supported its theoretical foundation with a rigorous proof. In

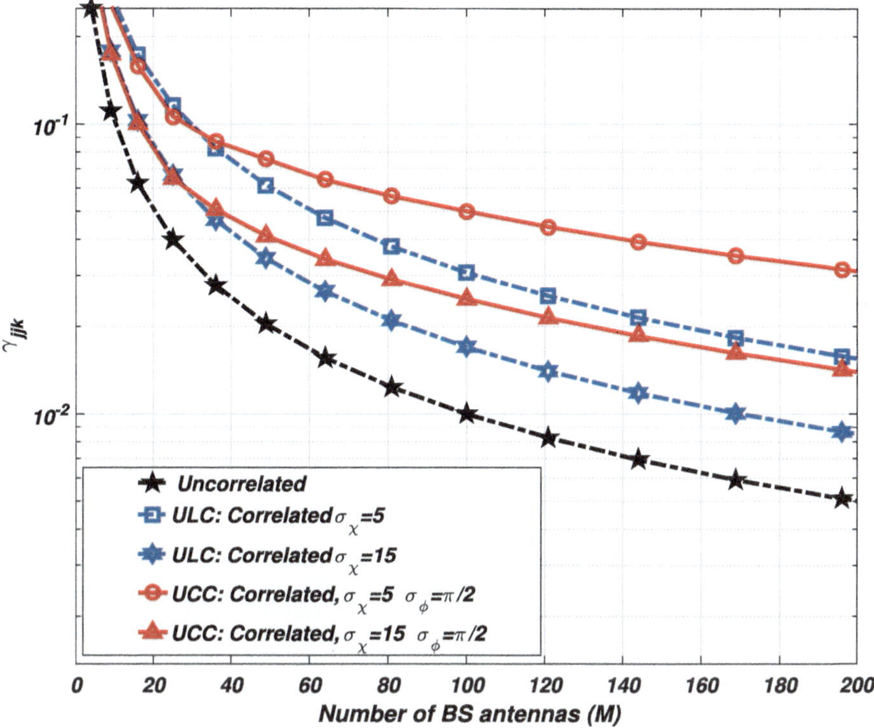

Fig. 13.3 Variance γ_{jjk} in function of M

summary, our findings revealed that the introduction of spatial channel correlation using the GLMS model can potentially compromise CH, with the UCC exhibiting a more pronounced variance gain in comparison with the ULC configuration. Furthermore, our research highlights the significant influence of the GLMS model on CH. Moreover, we underscored the positive impact of spatial channel correlation on CE quality, where higher levels of SC resulted in more accurate channel estimates. Furthermore, our investigations indicated that the proposed GLMS model tailored for the UCC configuration surpasses the ULC configuration in terms of CE quality, as evidenced by smaller NMSE values. This performance advantage can be attributed to the UCC configuration's more compact design, which fosters a higher level of spatial channel correlation, ultimately enhancing the CE process.

References

1. Björnson, E., Hoydis, J., Sanguinetti, L.: Massive MIMO networks: spectral, energy, and hardware efficiency. Found. Trends Signal Process. **11**(3–4), 154–655 (2017). https://doi.org/10.1561/2000000093
2. Marzetta, T.L.: Fundamentals of Massive MIMO. Cambridge University Press (2016)
3. Amadid, J., Boulouird, M., Riadi, A.: Channel estimation in massive MIMO-based wireless network using spatially correlated channel-based three-dimensional array. Telecommun. Syst. **79**(3), 323–340 (2022)
4. Amadid, J., Zeroual, A.: An efficient hybrid strategy for uplink channel estimation in massive MIMO systems based on spatially correlated channels. Int. J. Commun. Syst. **36**(9), e5478 (2023)
5. Björnson, E., Hoydis, J., Sanguinetti, L.: Massive MIMO has unlimited capacity. IEEE Trans. Wirel Commun. **17**(1), 574–590 (2017). https://doi.org/10.1109/TWC.2017.2768423
6. Verenzuela, D., Björnson, E., Wang, X., Arnold, M., ten Brink, S.: Massive-MIMO iterative channel estimation and decoding (MICED) in the uplink. IEEE Trans. Commun. **68**(2), 854–870 (2019)
7. Shariati, N., Björnson, E., Bengtsson, M., Debbah, Mérouane.: Low-complexity polynomial channel estimation in large-scale MIMO with arbitrary statistics. IEEE J. Sel. Top. Signal Process. **8**(5), 815–830 (2014)
8. Arslan, H., Bottomley, G.E.: Channel estimation in narrowband wireless communication systems. Wirel. Commun. Mob. Comput. **1**(2), 201–219 (2001)
9. Oyerinde, O.O., Mneney, S.H.: Review of channel estimation for wireless communication systems. IETE Techn. Rev. **29**(4) (2012)
10. Amadid, J., Khabba, A., El Ouadi, Z., Zeroual, A.: Impacts of three-dimensional array over channel estimation for cell-free massive MIMO considering spatially correlated channels. Trans. Emerg. Telecommun. Technol. **33**(12), e4638 (2022)
11. Molisch, A.F.: Wireless Communications, vol. 34. Wiley (2012)
12. Amadid, J., Boulouird, M., Belhabib, A., Zeroual, A.: On channel estimation for Rician fading with the phase-shift in cell-free massive MIMO system. Wirel. Pers. Commun., 1–21 (2021)
13. Busari, S.A., Huq, K.M.S., Mumtaz, S., Dai, L., Rodriguez, J.: Millimeter-wave massive MIMO communication for future wireless systems: a survey. IEEE Commun. Surv. Tutor. **20**(2), 836–869 (2017)
14. Alkhateeb, A.: Deepmimo: A Generic Deep Learning Dataset for Millimeter Wave and Massive MIMO Applications (2019). arXiv preprint arXiv:1902.06435
15. Lee Swindlehurst, A., Ayanoglu, E., Heydari, P., Capolino, F.: Millimeter-wave massive MIMO: the next wireless revolution? IEEE Commun. Mag. **52**(9), 56–62 (2014)
16. Elijah, O., Leow, C.Y., Rahman, T.A., Nunoo, S., Iliya, S.Z.: A comprehensive survey of pilot contamination in massive MIMO-5G system. IEEE Commun. Surv. Tutor. **18**(2), 905–923 (2015)
17. Jose, J., Ashikhmin, A., Marzetta, T.L., Vishwanath, S.: Pilot contamination and precoding in multi-cell TDD systems. IEEE Trans. Wirel. Commun. **10**(8), 2640–2651 (2011)
18. Jose, J., Ashikhmin, A., Marzetta, T.L., Vishwanath, S.: Pilot contamination problem in multi-cell TDD systems. In: 2009 IEEE International Symposium on Information Theory, pp. 2184–2188. IEEE (2009)
19. Albdran, S., Alshammari, A., Ahad, M.A.R., Matin, M.: Effect of exponential correlation model on channel estimation for massive MIMO. In: 2016 19th International Conference on Computer and Information Technology (ICCIT), pp. 80–83. IEEE (2016). https://doi.org/10.1109/ICCITECHN
20. Kim, H., Choi, J.: Channel estimation for spatially/temporally correlated massive MIMO systems with one-bit ADCS. EURASIP J. Wirel. Commun. Netw. **2019**(1), 1–15 (2019). https://doi.org/10.1186/s13638-019-1587

21. Temiz, M., Alsusa, E., Danoon, L.: Impact of imperfect channel estimation and antenna correlation on quantised massive multiple-input multiple-output systems. IET Commun. **13**(9), 1262–1270 (2019). https://doi.org/10.1049/iet-com.2018.5860
22. Özdogan, Ö., Björnson, E., Larsson, E.G.: Massive MIMO with spatially correlated Rician fading channels. IEEE Trans. Commun. **67**(5), 3234–3250 (2019). https://doi.org/10.1109/TCOMM.2019.2893221
23. Adachi, F., Feeney, M.T., Parsons, J.D., Williamson, A.G.: Crosscorrelation between the envelopes of 900 MHz signals received at a mobile radio base station site. IEE Proc. F-Commun. Radar Signal Process. **133**, 506–512 (1986). https://doi.org/10.1049/ip-f-1.1986.0083
24. Trump, T., Ottersten, B.: Estimation of nominal direction of arrival and angular spread using an array of sensors. Signal Process. **50**(1–2), 57–69 (1996). https://doi.org/10.1016/0165-1684(96)00003-5
25. Yin, H., Gesbert, D., Filippou, M., Liu, Yingzhuang: A coordinated approach to channel estimation in large-scale multiple-antenna systems. IEEE J. Sel. Areas Commun. **31**(2), 264–273 (2013). https://doi.org/10.1109/JSAC.2013.130214
26. Zetterberg, P., Ottersten, B.: The spectrum efficiency of a base station antenna array system for spatially selective transmission. IEEE Trans. Vehic. Technol. **44**(3), 651–660 (1995). https://doi.org/10.1109/VETEC.1994.345349
27. Bengtsson, M.: Antenna Array Signal Processing for High Rank Data Models. Ph.D. thesis, Signaler, sensorer och system (2000)
28. Amadid, J., Belhabib, A., Zeroual, A.: On channel estimation in cell-free massive MIMO for spatially correlated channels with correlated shadowing under rician fading. Int. J. Commun. Syst. **35**(1), e5011 (2022)
29. Adhikary, A., Nam, J., Ahn, J.-Y., Caire, Giuseppe: Joint spatial division and multiplexing-the large-scale array regime. IEEE Trans. Inf. Theor. **59**(10), 6441–6463 (2013). https://doi.org/10.1109/TIT.2013.2269476
30. Salz, J., Winters, J.H.: Effect of fading correlation on adaptive arrays in digital mobile radio. IEEE Trans. Vehic. Technol. **43**(4), 1049–1057 (1994). https://doi.org/10.1109/25.330168
31. Shiu,D.-S., Foschini, G.J., Gans, M.J., Kahn, J.M.: Fading correlation and its effect on the capacity of multielement antenna systems. IEEE Trans. Commun. **48**(3), 502–513 (2000). https://doi.org/10.1109/26.837052
32. Jiang, Z., Molisch, A.F., Caire, G., Niu, Z.: Achievable rates of FDD massive MIMO systems with spatial channel correlation. IEEE Trans. Wirel. Commun. **14**(5), 2868–2882 (2015). https://doi.org/10.1109/TWC.2015.2396058
33. Pedersen, K.I., Mogensen, P.E., Fleury, B.H.: Power azimuth spectrum in outdoor environments. Electron. Lett. **33**(18), 1583–1584 (1997). https://doi.org/10.1049/el:19971029
34. Gao, X., Edfors, O., Tufvesson, F., Larsson, E.G.: Massive MIMO in real propagation environments: do all antennas contribute equally? IEEE Trans. Commun. **63**(11), 3917–3928 (2015). https://doi.org/10.1109/TCOMM.2015.2462350
35. Gao, X., Edfors, O., Rusek, F., Tufvesson, F.: Massive MIMO performance evaluation based on measured propagation data. IEEE Trans. Wirel. Commun. **14**(7), 3899–3911 (2015). https://doi.org/10.1109/TWC.2015.2414413
36. Ngo, H.Q., Larsson, E.G.: No downlink pilots are needed in TDD massive MIMO. IEEE Trans. Wirel. Commun. **16**(5), 2921–2935 (2017). https://doi.org/10.1109/TWC.2017.2672540
37. Zhou, J., Sasaki, S., Muramatsu, S., Kikuchi, H., Onozato, Y.: Spatial correlation functions for a circular antenna array and their applications in wireless communication systems. IEICE Trans. Fund. Electron. Commun. Comput. Sci. **86**(7), 1716–1723 (2003). https://doi.org/10.1109/GLOCOM.2003.1258410
38. Chang, A.-C., Jen, C.-W.: Subspace-based techniques for 2-D DOA estimation with uniform circular array under local scattering. J. Chin. Inst. Eng. **29**(4), 663–673 (2006). https://doi.org/10.1080/02533839.2006.9671162
39. Forenza, A., Love, D.J., Heath, R.W.: Simplified spatial correlation models for clustered MIMO channels with different array configurations. IEEE Trans. Vehic. Technol. **56**(4), 1924–1934 (2007). https://doi.org/10.1109/TVT.2007.897212

40. Kay, S.M.: Fundamentals of Statistical Signal Processing: Estimation Theory. Prentice-Hall, Inc. (1993)
41. Ngo, H.Q., Ashikhmin, A., Yang, H., Larsson, E.G., Marzetta, T.L.: Cell-free massive MIMO versus small cells. IEEE Trans. Wirel. Commun. **16**(3), 1834–1850 (2017)
42. Interdonato, G., Björnson, E., Ngo, H.Q., Frenger, P., Larsson, E.G.: Ubiquitous cell-free massive MIMO communications. EURASIP J. Wirel. Commun. Netw. **2019**(1), 1–13 (2019)
43. Ngo, H.Q., Ashikhmin, A., Yang, H., Larsson, E.G., Marzetta, T.L.: Cell-free massive MIMO: uniformly great service for everyone. In: 2015 IEEE 16th International Workshop on Signal Processing Advances in Wireless Communications (SPAWC), pp. 201–205. IEEE (2015)
44. Zhou, S., Zhao, M., Xibin, X., Wang, J., Yao, Y.: Distributed wireless communication system: a new architecture for future public wireless access. IEEE Commun. Mag. **41**(3), 108–113 (2003)
45. Özdogan, Ö., Björnson, E., Zhang, J.: Performance of cell-free massive MIMO with Rician fading and phase shifts. IEEE Trans. Wirel. Commun. **18**(11), 5299–5315 (2019)
46. Interdonato, G., Ngo, H.Q., Larsson, E.G., Frenger, P.: On the performance of cell-free massive mimo with short-term power constraints. In: 2016 IEEE 21st International Workshop on Computer Aided Modelling and Design of Communication Links and Networks (CAMAD), pp. 225–230. IEEE (2016)
47. Wang, Z., Zhang, J., Björnson, E., Ai, B.: Uplink performance of cell-free massive MIMO over spatially correlated Rician fading channels. IEEE Commun. Lett. **25**(4), 1348–1352 (2020). https://doi.org/10.1109/LCOMM.2020.3041899
48. Björnson, E., Sanguinetti, L.: Making cell-free massive MIMO competitive with MMSE processing and centralized implementation. IEEE Trans. Wirel. Commun. **19**(1), 77–90 (2019). https://doi.org/10.1109/TWC.2019.2941478
49. Zheng, J., Zhang, J., Björnson, E., Ai, B.: Impact of channel aging on cell-free massive MIMO over spatially correlated channels. IEEE Trans. Wirel. Commun. (2021). https://doi.org/10.1109/TWC.2021.3074421
50. Femenias, G., Riera-Palou, F., Álvarez-Polegre, A., García-Armada, A.: Short-term power constrained cell-free massive-MIMO over spatially correlated Ricean fading. IEEE Trans. Vehic. Technol. **69**(12), 15200–15215 (2020). https://doi.org/10.1109/TVT.2020.3037009
51. Loyka, S.L.: Channel capacity of MIMO architecture using the exponential correlation matrix. IEEE Commun. Lett. **5**(9), 369–371 (2001). https://doi.org/10.1109/4234.951380
52. Shin, H., Win, M.Z., Lee, J.H., Chiani, M.: On the capacity of doubly correlated MIMO channels. IEEE Trans. Wirel. Commun. **5**(8), 2253–2265 (2006). https://doi.org/10.1109/TWC.2006.1687741
53. Byers, G.J., Takawira, F.: Spatially and temporally correlated MIMO channels: modeling and capacity analysis. IEEE Trans. Vehic. Technol. **53**(3), 634–643 (2004). https://doi.org/10.1109/TVT.2004.825766
54. Pätzold, M., Hogstad, B.O.: A space-time channel simulator for MIMO channels based on the geometrical one-ring scattering model. Wirel. Commun. Mob. Comput. **4**(7), 727–737 (2004). https://doi.org/10.1109/VETECF.2004.1399949
55. Amadid, J., Khabba, A., Ouadi, Z.E., Benedress, L.G., Wakrim, L., Zeroual, A.: The influence of phase-shifted line-of-sight component with correlated-shadows rician fading over the performance of cell-free massive MIMO. In: 2023 IEEE 3rd International Maghreb Meeting of the Conference on Sciences and Techniques of Automatic Control and Computer Engineering (MI-STA), pp. 353–358. IEEE (2023)
56. Amadid, J., Khabba, A., Ouadi, Z.E., Wakrim, L., Zeroual, A.: Channel gain in micro-urban environment assisted intelligent reflecting surface. In: 2023 IEEE 3rd International Maghreb Meeting of the Conference on Sciences and Techniques of Automatic Control and Computer Engineering (MI-STA), pp. 595–598. IEEE (2023)
57. Amadid, J., Ouadi, Z.E., Wakrim, L., Khabba, A., Zeroual, A.: Pilot sequence-based channel estimation in massive MIMO wireless communication networks under strong pilot contamination. In: 2022 International Conference on Decision Aid Sciences and Applications (DASA), pp. 1577–1582. IEEE (2022)

58. Amadid, J., Khabba, A., Ouadi, Z.E., Zeroual, A.: Performance evaluation of pilot-based channel estimation for conventional, cell-free, and small-cell massive MIMO networks. In: International Conference on Digital Technologies and Applications, pp. 341–350. Springer (2022)
59. Amadid, J., Belhabib, A., Khabba, A., Zeroual, A., Hassani, M.M.: On channel estimation and spectral efficiency for cell-free massive MIMO with multi-antenna access points considering spatially correlated channels. Trans. Emerg. Telecommun. Technol. **33**(5), e4438 (2022)
60. Amadid, J., Belhabib, A., Khabba, A., Ouadi, Z.E., Zeroual, A.: Channel estimation evaluation for a massive MIMO system considering spatially correlated channels in an urban network. E3S Web Conf. **351**, 01055 (2022)
61. Amadid, J., Khabba, A., Ouadi, Z.E., Wakrim, L., Zeroual, A.: Influence of spatial correlation over the channel estimation in 6G cell-free massive MIMO network. In: 2022 IEEE 2nd International Maghreb Meeting of the Conference on Sciences and Techniques of Automatic Control and Computer Engineering (MI-STA), pp. 436–440. IEEE (2022)
62. Amadid, J., Khabba, A., Wakrim, L., Ouadi, Z.E., Zeroual, A.: Performance of massive MIMO systems considering three dimensional array-based spatially correlated channel. In: 2022 International Conference on Decision Aid Sciences and Applications (DASA), pp. 1583–1588. IEEE (2022)
63. Akdeniz, M.R., Liu, Y., Samimi, M.K., Sun, S., Rangan, S., Rappaport, T.S., Erkip, E.: Millimeter wave channel modeling and cellular capacity evaluation. IEEE J. Sel. Areas Commun. **32**(6), 1164–1179 (2014)
64. Alkhateeb, A., Ayach, O.E., Leus, G., Heath, R.W.: Channel estimation and hybrid precoding for millimeter wave cellular systems. IEEE J. Sel. Top. Signal Process. **8**(5), 831–846 (2014)

Part II
Terahertz Technology and Its Applications

Chapter 14
THz Antennas: Applications and Challenges—A Review

Ouafae Elalaouy, Mohammed El Ghzaoui, and Jaouad Foshi

14.1 Introduction

Over the decades, wireless communication technology has undergone a continual evolution to address the growing requirements for enhanced specifications. The primary limitations of the existing wireless communication system include insufficient bandwidth, suboptimal spectrum utilization, interference issues, and lower data rates [1, 2]. The terahertz (THz) band has garnered considerable attention as a viable solution to tackle these issues. Existing within a range of the electromagnetic spectrum extending from 0.1 to 10 terahertz, the THz band, also known as the far-infrared region, positions itself between the microwave and infrared bands. Due to its distinctive qualities, this frequency range is appealing for a wide range of scientific and technological purposes. Historically, it has been referred to as the "terahertz gap" due to challenges associated with generating and detecting signals within this range. The diverse range of applications for the terahertz spectrum stems from its distinctive characteristics [2, 3]. Notably, they possess the ability to effortlessly traverse everyday materials like clothing, paper, and plastics without undergoing significant weakening. Additionally, the THz band stands out due to its shorter wavelength compared to microwaves, enabling the attainment of high-resolution communication. Lastly, THz waves operate at an exceptionally low power level, enabling a non-ionizing detection process. Owing to these features, terahertz (THz) waves are effectively employed in imaging, and sensing, particularly in scenarios requiring non-destructive penetration of common materials. Moreover, THz waves contribute significantly to advancements in wireless communication such as antenna technology. Indeed, the scarcity of available spectrum resources prompted the design

O. Elalaouy (✉) · J. Foshi
Faculty of Sciences and Techniques, University of Moulay Ismail, Errachidia, Morocco
e-mail: ou.elalaouy@edu.umi.ac.ma

M. El Ghzaoui
Faculty of Sciences Dhar El Mahraz, Sidi Mohamed Ben Abdellah University, Fes, Morocco

M. El Ghzaoui et al. (eds.), *Next Generation Wireless Communication*, Signals and Communication Technology, https://doi.org/10.1007/978-3-031-56144-3_14

of antennas at higher frequency bands [3, 4]. The key advantage lies in the significantly expanded bandwidth provided by THz antennas in this range compared to traditional antennas. THz antennas are integral elements in THz wireless communication systems, and their efficacy plays a pivotal role in determining the operating bandwidth and antenna gain, thus exerting a direct impact on the overall quality of the system. Critical parameters including data transmission rate, system image processing, and the operating range of detecting systems are all correlated with the attributes of THz antennas. THz antennas are designed and engineered to address the spectrum resources escalating scarcity [5, 6]. The anticipation of enhancing data transmission rates is tied to their broadband advantages. However, a thorough examination reveals that incorporating THz waves into wireless communication brings about specific challenges that require thorough consideration. Within the domain of terahertz (THz) wireless communication, a key obstacle arises due to air attenuation, mainly caused by water vapor absorption. A critical solution proposed to address this challenge involves the strategic use of high-gain antennas in specific frequency windows is crucial [7–9]. Moreover, beyond 300 GHz frequencies, the notable issue of free space path loss becomes more pronounced. The recommended solution emphasizes the implementation of high-gain antenna arrays. Through the utilization of arrays, the communication system seeks to compensate for inherent losses in free space transmission, promoting efficient signal propagation at these high frequencies. Some frequency bands, such 0.45, 0.55, and 0.76 THz, are more attenuated due to unfavorable air conditions [10, 11]. Given the significant propagation losses in THz wireless communication systems, beamforming techniques become essential. Using beamforming techniques in conjunction with Multiple-Input Multiple-Output (MIMO) antennas is a realistic approach [12]. Due to the narrow beamwidth array, the MIMO antennas effectively reduce interference and offer a tactical way to address the problems associated with propagation losses in the THz band. Therefore, in this work, we have discussed the design of THz single, array, and MIMO antennas.

14.2 Single THz Antenna

Optimal material selection is a crucial aspect in the design of high-performance terahertz (THz) microstrip antennas. Factors such as resonance frequency substrate characteristics stand as primary considerations in material choice. The distinctive structure of the microstrip patch antenna detector adds complexity, necessitating consideration not only of individual material parameters but also of the compatibility across all structural planes. Among these, the radiating patch plane assumes paramount importance in microstrip patch antenna configurations. For THz frequencies, copper is commonly employed for the metal patch in THz antennas. However, copper exhibits notable energy loss and reduced radiation efficiency at THz frequencies due to limitations in mobility and conductivity. In contrast, carbon-based materials have demonstrated satisfactory performance in the THz regime, presenting a viable alternative for designing efficient THz antennas. Another material commonly employed in terahertz (THz) antennas is graphene, a monolayer of graphite renowned for its exceptional electrical conductivity. Graphene outperforms traditional materials such as copper and carbon nanotubes in performance. Its thin, flexible, highly mobile, and transparent properties make it a preferred choice for THz antenna designs, contributing to enhanced capabilities in the realm of electrical conductivity and flexibility compared to conventional counterparts. Additionally, choosing the right dielectric substrate for terahertz (THz) antennas involves evaluating various materials, each with its set of pros and cons. Quartz stands out for its low loss tangent and high transparency in the THz range, yet it lacks flexibility. Polyimide stands as a frequently employed material owing to its blend of exceptional features, including resilience to elevated temperatures and robust resistance to vibrations. While providing flexibility, it is worth noting that polyimide may exhibit a comparatively higher loss tangent. Rogers RT/duroid provides a commercially available option with a low loss tangent, though it may have limited transparency. Polyethylene is low cost with a low loss tangent but is less flexible. Liquid Crystal Polymer (LCP) combines a low loss tangent with good flexibility, but it has limited transparency in certain THz frequency ranges [13, 14].

Fractal geometry proves highly beneficial in the realm of antenna engineering. Therefore, the authors in [15] introduce a fractal floral radiator with a partial ground that is etched on a 400 × 600 μm^2 polyamide substrate. The suggested floral radiator's operating band, which is between 0.7 and 4.8 THz, is enhanced by the usage of the partial ground. Additionally, it achieves a maximum gain of 12.5 dBi. In [16], the authors propose a departure from the conventional substrate-integrated waveguide (SIW) by introducing a simplified SIW on-chip antenna to enhance performance. This innovative approach not only addresses the complexity and bulkiness issues but also enables seamless integration into CMOS technology. The suggested antenna boasts a band which is between 38 and 0.420 THz, achieving a maximum gain of 6.5 dBi and a radiation efficiency of 87.9% at 0.4 THz. A terahertz patch antenna featuring a slotted ground, specifically designed for the terahertz frequency spectrum is introduced in [17]. The antenna designs undergo optimization through the Artificial Neural Networks (ANN) to identify the antenna's optimal substrate as well as identify the most favorable gain value by minimizing the error value. Compact dimensions measuring 264 × 400 μm^2 characterize the selected substrate, duroid 3210. The final suggested antenna has good performance metrics, with a gain of 11.97 dBi and a bandwidth that is adequate, spanning from 0.13 to 0.26 THz. The presented helical antenna in [18] is circularly polarized and reasonably priced. Noteworthy results include an operational band of 0.50992 THz ranging from to. The antenna's outcomes further reveal that the proposed antenna achieves impressive metrics, namely a realized gain of 11.8 dBi, and a remarkable radiation efficiency of 95.31% at 1 THz. Researchers leverage the graphene's attributes to design THz antennas. Graphene, made of a single layer of carbon atoms, is super thin and lightweight, helping reduce energy loss in antennas. It is excellent at conducting electricity, making the antenna work well. Graphene also allows the antenna to be tuned, made smaller, and flexible, making it versatile for different uses. It is much faster at conducting electricity than silicon and can handle high temperatures.

All these special qualities make graphene a crucial part of making antennas work effectively in the terahertz (THz) frequency [19, 20]. The graphene-based antenna presented in [21] integrates a triangular patch and ground (GND) shape, showcasing a remarkable bandwidth of 1.27 THz and exhibiting -33.22 dB in reflection coefficient. Additionally, it reaches 7.72 dBi for the gain, contributing to a broader range, superior signal quality, and directed radiation. Furthermore, the antenna demonstrates a maximum efficiency of 95.22%. Similarly, the graphene-based antenna reported in [22] provides a substantial operating band of 450 GHz, ranging from 0.57 to 1.02 THz. The noteworthy enhancement in the bandwidth is attributed to the modified ground. At its peak, the antenna achieves a radiation efficiency of 92% and a gain of 3.47 dBi. Another novel approach is introduced in [23] by integrating graphene, a nanomaterial, onto a silicon dioxide substrate's slotted bow-tie shape. The resulting configuration has a compact size measuring 40 $\times$ 40 μm^2 resonates in the triband within the THz region. According to simulation results, the bow-tie structure exhibits significant performance metrics, including a wide bandwidth that lies between 6 and 15 THz, an efficiency of 71%, and a peak gain of 17.529 dB. An innovative Tapered Vivaldi Antenna (TVA) with high gain, leveraging Graphene Nano Ribbon (GNR) for terahertz band applications is presented in [24]. To enhance radiation properties, the design integrates a combination of a metal structure with graphene and a quartz substrate extension approach. The antenna exhibits reconfigurability by altering the chemical potential. The outcomes reveal multiple performance metrics, among them an impedance range of 0.5–0.9 THz, a maximum gain of 12.31 dBi, a radiation efficiency of 83.63%, and a lowest return loss of − 58.68 dB. The structures of Photonic Band Gap (PBG) involve a systematic arrangement of hollow air cylinders within the dielectric substrate material, exhibiting diverse shapes like circular, hexagonal, triangular, and elliptical. The effectiveness of PBG in developing efficient patch antennas has been shown. Several studies were undertaken aimed at showing the antenna's capabilities, highlighting the influence of the orderly placement of cylindrical holes in the PBG material. In the subsequent study [25], the PBG crystal has been investigated and compared to a homogeneous substrate structure to assess its impact on the microstrip antenna performance. The designed antenna using this technique exhibits a bandwidth of 36.25 GHz within an operational frequency range of 0.6152–0.6514 THz, and achieves a noteworthy gain of 7.934 dBi. In another research [26], the authors proposed a PBG-based antenna made for THz applications. To improve the PBG-based substrate, the optimization uses a brand-new Segmented Objects Technique (SOT) in conjunction with Genetic Algorithm (GA). The optimized antenna, with dimensions of 118 $\times$ 118 μm^2, achieves a remarkable bandwidth of 133 GHz ranging from 255 to 388 GHz. It provides also a return loss of − 58.70 dB and a gain of 9.43 dBi. Optimization results showcase a substantial improvement of 3.74% in gain and 104.62% in bandwidth. In [27], an antenna having circular polarization is presented. This latter incorporates hexagonal-shaped metamaterial and is etched on a Substrate Integrated Waveguide (SIW). The antenna may operate as a lens and polarization converter thanks to the usage of metamaterial, increasing gain by 6–7 dBi. Additionally, the operational frequency range is 176.5–285 GHz, and the peak gain is 11.3 dBi. The work in [28] presents a graphene-based antenna with a superstrate configuration using a graphene metasurface and silicon dioxide (SiO_2) substrate. The inclusion of a superstrate layer enhances the radiation properties of the antenna, resulting in a gain of 8.91 dBi, an efficiency of 70.14%, an impressive return loss of 35.54 dB, and a substantial operational band of 1.4 THz. A circularly polarized (CP) antenna tailored for terahertz (THz) applications is introduced in [29]. The antenna's dimensions, compactly set at $29 \times 50 \times 4\ \mu m^3$, are achieved through the utilization of two distinct substrate materials. Besides, the proposed antenna stands out for its remarkable attributes, featuring a wide bandwidth ranging from 0.9 to 17.3 THz and a high gain of 12.8 dBi. Notably, the antenna exhibits circularly polarized radiations with an Axial Ratio (AR) value consistently below 3 dB throughout the frequency range of 13.6–17.3 THz. A nano-antenna design comprises a circular patch measuring $130 \times 130\ nm^2$, situated on a glass substrate is explored in [30]. A superstrate layer has been incorporated against environmental threats. The proposed configuration demonstrates resonance at 182 THz, featuring a frequency range from 160 to 207 THz. Furthermore, the antenna maintains a radiation efficiency of 96.54% across the entire operating band.

A circular polarized antenna holds significant utility in communication systems due to its unique and advantageous characteristics. Unlike linearly polarized antennas, circular polarized antennas can efficiently transmit and receive signals regardless of the orientation of the transmitting and receiving antennas. This omnidirectional capability makes them highly versatile in various communication applications. One key advantage of circular polarization is its resistance to signal degradation caused by changes in the relative orientation between transmitting and receiving antennas. In scenarios where the position or orientation of the communicating devices is unpredictable or subject to frequent changes, such as in mobile communications or satellite systems, circular polarized antennas excel in maintaining a stable and reliable connection. Circular polarization also aids in minimizing signal fading and multipath effects, common challenges in communication systems. This makes circular polarized antennas particularly well-suited for urban environments where signals may bounce off buildings and other structures. Moreover, circular polarization enhances the robustness of communication links in challenging environments, such as industrial settings or areas with high electromagnetic interference. The ability of circular polarized antennas to penetrate obstacles and provide consistent signal quality contributes to their effectiveness in these demanding scenarios. In satellite communication, circular polarized antennas are often preferred because they mitigate the signal loss that can occur when the satellite changes its orientation. This ensures a more consistent and stable communication link, especially in Earth observation or satellite broadcasting applications (Table 14.1).

Table 14.1 Table comparison of various single antenna

Refs.	Bandwidth (THz)	Size (μm^2)	Gain (dBi)	Efficiency (%)	Antenna type	Applications
Keshwala et al. [15]	0.56–6.6	400 × 600	12.56	–	Fractal flower-shaped antenna	High-speed THz applications
Kushwaha et al. [16]	0.38–0.42	490 × 450	6.5	87.9	SIW slot	High-speed communication and threat detection
Rajawat and Gupta [17]	0.13–0.26	264 × 400	11.97	–	Patch with slotted Ground	Applications in WBAN
Hajiyat et al. [18]	0.52–1.03	–	11.8	95.31	Metal helical antenna	Ultra-high-speed THz applications Wi-Fi, and medical applications
Islam et al. [21]	1.71–2.98	160 × 120	7.72	95.22	Triangular-based graphene patch	High-speed THz applications
Krishna et al. [22]	0.57–1.02	130 × 100	3.47	92	E-shaped antenna-based graphene	Ultra-speed in limited distance communication and sensing applications
Bansal et al. [23]	6–15	40 × 40	17.529	71	Slotted bow-tie antenna-based graphene	Wireless applications
Kushwaha [24]	0.5–0.9	590 × 1400	12.31	83.63	Configurable TVA-based graphene	Wireless communication, security, and medical imaging
Kushwaha et al. [25]	0.6152–0.6514	800 × 600	7.934	85.71	PBG-based antenna	Detection of explosive and material characterization applications
Khezzar et al. [26]	0.255–0.388	118 × 118	9.43	–	PBG-based antenna	Future wireless communications components and other THz and millimeter-wave applications
Ghzaoui et al. [27]	0.1765–0.285	11,000 × 10,800	11.3	–	MTM-based SIW antenna	Sub-THz 6G applications

(continued)

Table 14.1 (continued)

Refs.	Bandwidth (THz)	Size (μm^2)	Gain (dBi)	Efficiency (%)	Antenna type	Applications
Shubam et al. [28]	1.44–2.84	216 × 256	8.71	70.14	MTM-based antenna	Biomedical, satellite communication, high-quality video imaging,
Lchhab and El Ghzaoui [29]	0.9–17.3	21 × 37.5	12.8	–	Circular patch antenna	High-speed short-range indoor wireless communication, explosive detections, arms detection,
Aghoutane et al. [30]	0.160–0.207	130 × 130	–	96.54	Nano-scaled circular patch antenna	Optical communication, nano-photonics, and biosensing

14.3 Array THz Antenna

Antenna arrays in the THz frequency band are instrumental in addressing the challenges posed by the unique characteristics of THz waves. Their deployment facilitates improved communication reliability, extended range, and enhanced sensing capabilities, paving the way for advancements in wireless communication and high-resolution imaging technologies. Thus, in this section, different antenna arrays for THz are presented. A circularly polarized pattern antenna array is demonstrated in [31]. The designed antenna array covers the WM834 band. At 317 GHz, the antenna array attains a peak gain of 16.1 dBi and radiation efficiency of 97.5%. Furthermore, it has a wide circular polarization (CP) band lying between 289.1 and 317.38 GHz. An antenna array, tailored for the future iterations of cellular communication systems, has been meticulously designed in [32]. The array is constructed on a Rogers R03003 substrate measuring 90,000 × 85,000 μm^2. Each patch measures 1.349 × 0.70 mm^2. The reported gain is 21.1 dBi, and the efficiency is about 67.6%. A THz antenna array comprising 56 elements for miniaturization and reconfigurability purposes is introduced in [33]. Implemented on a PEN substrate, the array occupies 5000 × 5000 μm^2 within the 0.9–1.2 THz operating band. The proposed antenna array incorporates mixed materials, blending gold, and perovskite and measures 70 × 119 μm^2. The antenna demonstrates a gain of 11.3 dBi, accompanied by an impressive 83% radiating efficiency. Two antenna arrays, one with a 1 × 2 series feed and another with a 1 × 3 series feed, are presented in [34]. They are designed on a Rogers RT duroid 5880 substrates, with a size of 706.46 × 626.46 μm^2. The 1 × 2 array achieved a gain of 7.16 dBi, while the 1 × 3 array reported a gain of 11.2 dBi at 0.312 THz. A graphene-based patch antenna array with complementary split ring resonators (CSRR) was designed in [35] to enhance radiation performance through various design strategies, including a homogeneous substrate, photonic bandgap (PBG) substrate, and a superstrated dielectric grating. Utilizing substrates with distinct dielectric constants results in a return loss of 40.28 dB. Introducing a PBG substrate with specific dimensions resulted in a notable improvement, yielding a return loss of 56.57 dB. Further enhancement was observed by employing a superstrated grating, leading to the highest gain (15.5 dBi) and radiation efficiency (83.67%). The antenna demonstrated covers the frequency range 0.84–0.94 THz. Silicon (Si) complementary metal oxide semiconductor (CMOS) technology is integral to the development of cost-effective terahertz (THz) integrated circuits (ICs), given its affordability and seamless integration. Thus, Lee and Jeong [36] presents a CMOS antenna array with a Defected Ground Structure. The dimensions of the array are $920 \times 790\ \mu m^2$, operating 1.48–2.67 THz. The antenna array exhibits a measured gain of 8.9 dBi.

Expanding on the concept of an array of terahertz (THz) antennas introduces several exciting possibilities for advanced applications in wireless communication and sensing. A THz antenna array refers to a collection of multiple antennas working collaboratively to achieve specific objectives. So, extending the array configuration to terahertz antennas opens up new frontiers in wireless communication, imaging, and sensing. The unique characteristics of the terahertz spectrum, coupled with the collaborative capabilities of antenna arrays, position THz technology as a key player in the evolution of next-generation wireless systems (Table 14.2).

14.4 MIMO THz Antenna

The integration of MIMO technology in terahertz communication systems holds great potential for unlocking the capabilities of this high-frequency range [37]. Through spatial multiplexing, improved reliability, and advanced beamforming techniques, MIMO contributes to achieving higher data rates and robust communication links in THz applications, paving the way for innovative wireless communication and sensing technologies [38, 39].

A small footprint quad-port MIMO antenna is etched onto a Rogers RT/duroid 5880 substrate and is studied in [40]. The antenna, with dimensions measuring $67.5 \times 67.5\ \mu m^2$, operates efficiently within the frequency band of 1.5–10 THz, delivering an impressive gain of 11 dBi. The orthogonally placed radiating elements, coupled with the integration of L-stubs in the ground plane, ensure effective isolation exceeding 20 dB. The reported ECC value stands at 0.001, attesting to the antenna's robust performance. A compact microstrip feed antenna designed for wideband operation in the terahertz band is presented in [41]. The antenna features a circular shape and is remarkably small, measuring $480 \times 480\ \mu m^2$. It is constructed on a diffused quartz substrate coated with gold, characterized by a relative permittivity of the proposed configurationp3.50. This antenna is fabricated on a gold-plated diffused

Table 14.2 Table comparison of various antenna array

Refs.	Bandwidth (THz)	Size (μm^2)	Gain (dBi)	Efficiency (%)	Antenna type	Applications
Ansha et al. [31]	0.227–0.38	863.6 × 431.8	16.1	97.5	Array antenna using WR3 waveguide	Imaging, spectroscopy, remote sensing, 6G wireless communications
Alibakhshikenari et al. [32]	0.123	90,000 × 85,000	21.2	67	50 × 32 array depending on Wilkinson power divider	THz applications
Abohmra et al. [33]	0.9–1.2	5000 × 5000	11.3	83	57 elements antenna array	Future wireless communication systems
Nurfitri et al. [34]	0.312	706 × 626	7.6	–	1 × 2 array	Imaging
Kushwaha and Karuppanan [35]	0.84–0.94	1402 × 949.61	15.5	83.67	array-based on PBG and dielectric grating	Medical imaging and wireless communication
Lee and Jeong [36]	0.3	920 × 790	8.9	–	THz on-chip antenna array	Radar and wireless communications

quartz substrate with a relative permittivity of 3.50. The antenna exhibits an operational bandwidth spanning from 0.51 to 1.46 THz, encompassing 80.76% of the targeted frequency range. At its peak, the antenna achieves a gain of 10.16 dBi. Moreover, the radiation efficiency remains consistently high, surpassing 70% across the entire desired bandwidth. The presented two-port antenna in [42] is implemented onto a polyimide substrate with a defective ground structure. The antenna dimensions are compact, measuring 100 × 105 μm^2. To enhance port isolation, a T-shaped stub and an open-ended slot are strategically employed. These elements effectively modify capacitive and inductive reactance, leading to a reduction and interruption of surface currents. The dual use of the stub and slot contributes to the improvement of isolation, surpassing a value of 30. The antenna's operational bandwidth spans from 1.74 THz to 2.52 THz. In [43], a compact multiband terahertz (THz) MIMO antenna, measuring 80 × 100μm^2 is introduced. Fabricated on a gold-plated Arlon AD410 substrate with $\epsilon_r = 4.1$, the antenna has distinct resonant frequencies that are achieved through the incorporation of a metamaterial and Substrate Integrated Waveguide (SIW) technologies. The suggested configuration consists of two elements and has a low Envelope Correlation Coefficient (ECC) that is found to be less than 0.02. The antenna further exhibits peak gains of 6.82 dBi and maintains a radiation efficiency of approximately 51%. An ultra-broadband quad elements THz MIMO antenna, featuring a unique design with a defective ground plane, is presented in [44]. The reported MIMO is implemented on a polyamide substrate, covers 2 to 0.7 THz, and exhibits compact dimensions of 40 × 46 μm^2. Additionally, the proposed configuration achieves isolation exceeding 20 dB at the whole bandwidth owing to the use of a defective ground structure. The proposed THz MIMO demonstrates an Envelope Correlation Coefficient (ECC) of < 0.001. A small footprint and broadband quad MIMO antenna tailored for terahertz (THz) frequency applications is presented in [45]. The antenna size is 125 mm × 125 mm. To enhance performance, multiple types of fractal geometries are incorporated into the MIMO configuration. This design strategy aims to achieve an extended operating band and low mutual coupling. A satisfactory operating band lying between 0.72 and 10 THz is obtained. Furthermore, the antenna design achieves isolation exceeding 20 dB amid the antenna elements, and a peak gain of 8.2 dBi is attained. In [46] a MIMO antenna array is constructed using a tetradecagonal ring radiator and a semicircular Defected Ground Structure on a polyimide substrate. The antenna operates in the 0.1–15.1 THz band, with dimensions of 800 × 1170 μm^2. Spatial diversity achieves 30 dB isolation, and the Envelope correlation coefficient is measured at 10^{-5}. A dual-port MIMO antenna utilizing graphene etched on a substrate composed of SiO2 is introduced in [47]. Operating in the 1.76–1.87 THz band with a compact size of 60 × 40 µm2, the antenna achieves excellent isolation (> 25 dB) throughout the frequency range. The proposed design exhibits a gain of 4.45 dBi, and low Envelope Correlation Coefficient (< 0.01).

The implementation of Multiple-Input Multiple-Output (MIMO) antennas is a crucial aspect of 5G (fifth-generation) wireless communication systems, contributing significantly to their enhanced performance and capabilities. Expanding on the concept of MIMO antennas in the context of 5G involves considering various aspects that underscore their importance and the unique benefits they bring to the evolving landscape of wireless communication. The integration of MIMO antennas is a cornerstone of the 5G revolution, contributing to higher data rates, improved spectral efficiency, and enhanced reliability. As 5G networks continue to evolve, MIMO technology will play a central role in shaping the future of wireless communication by addressing the growing demand for higher performance and connectivity (Table 14.3).

14.5 THz Antenna's Challenges

Terahertz (THz) technology faces challenges such as atmospheric absorption limiting signal range, restricted penetration capabilities hindering applications requiring depth, susceptibility to scattering impacting performance in complex environments,

Table 14.3 Table comparison of various THz MIMO antenna types

Refs	No. of ports	Bandwidth (THz)	Size (μm^2)	Gain (dBi)	Isolation (dB)	ECC	Antenna type	Applications
Raj et al. [40]	Four	1.5–10	67.5 × 67.5	11	> 20	< 0.001	Four element MIMO antenna	Deep space network, aerospace applications
Saxena et al. [41]	Two	0.52–1.47	480 × 480	10.3	> 20	< 0.025	Metamaterial inspired MIMO antenna	High-speed RADAR, health care and astronomical radiometric
Kumar et al. [42]	Two	1.74–2.52	100 × 105	4.5	> 30	< 0.006	Two port MIMO antenna using unconnected ground	wireless communication
Saxena et al. [43]	Two		80 × 100	6.28	> 15	< 0.02	Multiband CSRR loaded MIMO antenna	Nano-communications and sensing
Kumar et al. [44]	Four	1.8–9.9	40 × 46	5.07	> 20	< 0.001	MIMO antenna with DGS	Wireless communication applications
Das et al. [45]	Four	0.72–10	125 × 125	8.2	> 20	< 0.02	Fractal-loaded MIMO antenna	THz applications
Singhal et al. [46]	Two	0.1–15.1	800 × 1170	–	> 30	$< 10^{-5}$	Ring-shaped MIMO	THz sensing, imaging screening, and communication over a limited distance
Varshney et al. [47]	Two	1.76–1.87	60 × 40	4.45	> 25	< 0.01	Graphene-based MIMO antenna	Future communication devices

regulatory constraints, and the need for efficient sources and detectors. Additionally, the complexity of THz system design, privacy concerns in security applications, and material interaction issues pose obstacles. Addressing these challenges requires ongoing research and technological advancements to improve THz device efficiency, develop novel materials, and overcome the limitations imposed by atmospheric conditions. Despite these obstacles, the unique properties of THz waves continue to drive innovation, with researchers actively working to unlock the full potential of THz technology in diverse applications.

The development and implementation of terahertz (THz) antennas, while holding immense potential for advanced applications, also pose several challenges that researchers and engineers are actively addressing. Exploring these challenges sheds light on the complexities involved in harnessing the terahertz frequency range and underscores the ongoing efforts to overcome these hurdles. One significant challenge in THz antennas is the high absorption and attenuation of THz signals by atmospheric gases and moisture. Addressing these absorption and attenuation issues is crucial for reliable long-distance communication and sensing applications. Another challenge is the materials availability; the availability of suitable materials for constructing THz antennas is a challenge. Many conventional materials used in lower-frequency antennas may not perform optimally in the THz range. Researchers are exploring new materials, such as metamaterials and quantum-confined structures, to design antennas with improved performance in the THz spectrum. Moreover, the wavelength of THz signals is short, posing challenges for the design and miniaturization of antennas. Developing compact yet efficient THz antennas for practical applications, especially in portable devices, requires innovative design approaches. Furthermore, fabricating THz antennas with precision is challenging due to the small scale and high frequencies involved. Researchers are exploring advanced fabrication techniques, including nanofabrication and additive manufacturing, to overcome these challenges and achieve high-performance THz antennas. Addition, terahertz signals are sensitive to obstacles and may experience significant signal degradation due to scattering and absorption. Developing efficient beamforming techniques to control and maintain signal integrity in various environments is an ongoing area of research. Integrating THz antennas with electronics poses challenges due to the limited availability of electronic components that operate seamlessly in the THz frequency range. Developing compatible and efficient electronic components for THz systems is crucial for the overall success of THz technology. The use of THz frequencies raises regulatory and safety concerns, particularly in terms of potential health risks and interference with existing communication systems. Addressing and mitigating these concerns is vital for the widespread adoption of THz technology. However, implementing THz technology can be expensive due to the specialized materials and fabrication techniques involved. Researchers are exploring cost-effective solutions and scalable manufacturing processes to make THz technology more accessible.

14.6 Conclusion

Terahertz (THz) band communication is poised to be pivotal in meeting the growing need for ultra-high data rates in upcoming wireless communication systems. Achieving these high data rates mandates the use of antennas with significant gain. However, designing a high-gain THz antenna array remains intricate due to the micron-scale size of radiating elements and the limited availability of dielectric materials. This research conducts a comprehensive analysis of various THz antenna designs. Initially, the study delves into the detailed description and findings of single THz antennas developed on diverse substrates. The paper proceeds to discuss the contemporary THz antenna designs, covering those employing graphene-based and Photonic Band Gap (PBG), to overcome limitations on gain. The research extends to THz Multiple-Input Multiple-Output (MIMO) antenna arrays onto diverse substrates. An extensive analysis is done, exploring single-port antennas, antenna arrays, and multiport antennas, while highlighting their distinctive practical purposes. Finally, these hurdles, encompassing system complexity, privacy concerns, and material interaction issues, necessitate ongoing research and technological advancements. Finally, the challenges facing the THz technology like system complexity, privacy concerns, and material interaction issues are presented. Overcoming these obstacles necessitates ongoing research and technological advancements. Nevertheless, the unique properties of THz waves drive innovation as researchers aim to maximize the technology's potential across diverse applications.

References

1. Kushwaha, R.K., Karuppanan, P., Kishore, N.: High-gain patch antenna design using PRS and ground plane reflector for THz band applications. Optik **232**, 166559 (2021)
2. El Ghzaoui, M., Mestoui, J., Hmamou, A., Elaage, S.: Performance analysis of multiband on–off keying pulse modulation with noncoherent receiver for THz applications. Microw. Opt. Technol. Lett. **64**(12), 2130–2135 (2022)
3. Kushwaha, R.K.: Reconfigurable GNR based elliptical dielectric resonator antenna for THz band applications. Results Opt. **10**, 100354 (2023)
4. Teng, F., Wan, J. and Liu, J. Review of terahertz antenna technology for science missions in space. In: IEEE Aerospace and Electronic Systems Magazine (2022)
5. He, Y., Chen, Y., Zhang, L., Wong, S.-W., Chen, Z.N.: An overview of terahertz antennas. China Commun. **17**(7), 124–165 (2020)
6. El Ghzaoui, M., Belkadid, J., Benbassou, A.: Performance analysis of CE-OFDM over terahertz bands. J. Inst. Eng. Ser. B **104**(5), 1147–1154 (2023)
7. Pant, R., Malviya, L.: THz antennas design, developments, challenges, and applications: a review. Int. J. Commun Syst **36**(8), e5474 (2023)
8. Elaage, S., Ghzaoui, M.E., Hmamou, A., Foshi, J., Mestoui, J.: MB-OOK transceiver design for terahertz wireless communication systems. Int. J. Syst. Control Commun. **12**(4), 309–326 (2021)
9. El Ghzaoui, M., Das, S.: Data transmission with terahertz communication systems. In: Emerging trends in terahertz solid-state physics and devices: sources, detectors, advanced materials, and light-matter interactions, pp. 121–141. Springer, Singapore (2020)

10. Maurya, N.K., Kumari, S., Pareek, P., Singh, L.: Graphene-based frequency agile isolation enhancement mechanism for MIMO antenna in terahertz regime. Nano Commun. Netw. **35**, 100436 (2023)
11. El Ghzaoui, M., Das, S., Lenka, T.R., Biswas, A.: Terahertz wireless communication components and system technologies. Springer, Singapore (2022)
12. Yadav, R., Gotra, S., Pandey, V., Kumar, S.: Graphene based two-port MIMO yagi-uda antenna for THz applications. Micro Nanostruct. **181**, 207616 (2023)
13. Muthukrishnan, K., Kamruzzaman, M., Lavadiya, S., Sorathiya, V.: Superlative split ring resonator shaped ultrawideband and high gain 1× 2 MIMO antenna for terahertz communication. Nano Commun. Netw. **36**, 100437 (2023)
14. Malhotra, I., Jha, K.R., Singh, G.: Terahertz antenna technology for imaging applications: a technical review. Int. J. Microw. Wirel. Technol. **10**(3), 271–290 (2018)
15. Keshwala, U., Rawat, S., Ray, K.: Design and analysis of eight petal flower shaped fractal antenna for THz applications. Optik **241**, 166942 (2021)
16. Kushwaha, R.K., Karuppanan, P., Dewang, R.K.: Design of a SIW on-chip antenna using 0.18-μm CMOS process technology at 0.4 THz. Optik **223**, 165509 (2020)
17. Rajawat, A., Gupta, S.H.: Design and optimization of THz antenna for onbody WBAN applications. Optik **223**, 165563 (2020)
18. Hajiyat, Z.R., Ismail, A., Sali, A., Hamidon, M.N.: Design and analysis of helical antenna for short-range ultra-high-speed THz wireless applications. Optik **243**, 167232 (2021)
19. Singh, G., Sandha, K.S., Kansal, A.: GA based optimized graphene antenna design for detection of explosives and drugs using THz spectroscopy. Micro Nanostruct. **179**, 207566 (2023)
20. Kiani, N., Hamedani, F.T., Rezaei, P.: Implementation of a reconfigurable miniaturized graphene-based SIW antenna for THz applications. Micro Nanostruct. **169**, 207365 (2022)
21. Islam, M.A., Alam, F., Rahman, M.A., Hossain, M.R., Roy, S.C.: Design and analysis the performance of triangular patch antenna for THz applications. Multidiscip. Sci. J. **6**(4), 2024043–2024043 (2024)
22. Krishna, M., Islam, T., Suguna, N., Kumari, S.V., Devi, R.D.H., Das, S.: A micro-scaled graphene-based wideband (0.57–1.02 THz) patch antenna for terahertz applications. Results Opt. **12**, 100501 (2023)
23. Bansal, G., Singh, A. and Bala, R. A triband slotted bow-tie wideband THz antenna design using graphene for wireless applications. Optik, 1852019), 1163–1171.
24. Kushwaha, R.K.: GNR based tapered Vivaldi antenna for THz band applications. Results in Optics **13**, 100574 (2023)
25. Kushwaha, R.K., Karuppanan, P., Malviya, L.: Design and analysis of novel microstrip patch antenna on photonic crystal in THz. Phys. B Condens. Matter **545**, 107–112 (2018)
26. Khezzar, D., Khedrouche, D., Denidni, T.A.: New design of a broadband PBG-based antenna for THz band applications. Photon. Nanostruct. Fund. Appl. **46**, 100947 (2021)
27. Ghzaoui, M.E., Belkadid, J., Benbassou, A.: Near zero index Metamaterial-based SIW antenna for 6G Sub-Terahertz applications. Results Opt. **12**, 100468 (2023)
28. Shubham, A., Samantaray, D., Ghosh, S.K., Dwivedi, S., Bhattacharyya, S.: Performance improvement of a graphene patch antenna using metasurface for THz applications. Optik **264**, 169412 (2022)
29. Lchhab, T., El Ghzaoui, M.: A circularly polarized wideband high gain antenna for THz wireless applications. Opt. Quant. Electron. **54**(12), 787 (2022)
30. Aghoutane, B., Ghzaoui, M.E., Kumari, S., Das, S., Faylali, H.E.: A circularly polarized super wideband transparent optical nanoantenna for advanced THz communication applications. Opt. Quant. Electron. **55**(3), 211 (2023)
31. Ansha, K., Abdulla, P.: Design of broadband circularly polarized THz antenna with stable radiation pattern for 6G communications. Optik **243**, 167397 (2021)
32. Alibakhshikenari, M., Virdee, B.S., Salekzamankhani, S., Aïssa, S., See, C.H., Soin, N., Fishlock, S.J., Althuwayb, A.A., Abd-Alhameed, R., Huynen, I., McLaughlin, J.A.: High-isolation antenna array using SIW and realized with a graphene layer for sub-terahertz wireless applications. Sci Rep. **11**(1), 1–4 (2021)

33. Abohmra, A., Abbas, H., Al-Hasan, M., Mabrouk, I.B., Alomainy, A., Imran, M.A., Abbasi, Q.H.: Terahertz antenna array based on a hybrid perovskite structure. IEEE Open J Antennas Propag. **1**, 464–471 (2020)
34. Nurfitri, I., Apriono, C.: Rectangular linear array microstrip antenna design for terahertz imaging. In: 2019 International Conference on Information and Communications Technology (ICOIACT). IEEE, pp. 719–722 (2019)
35. Kushwaha, R.K., Karuppanan, P.: Enhanced radiation characteristics of graphene-based patch antenna array employing photonic crystals and dielectric grating for THz applications. Optik **200**, 163422 (2020)
36. Lee, C., Jeong, J.: THz CMOS on-chip antenna array using defected ground structure. Electronics **9**(7), 1137 (2020)
37. Babu, K.V., Sree, G.N.J., Islam, T., Das, S., Ghzaoui, M.E., Saravanan, R.A.: Performance analysis of a photonic crystals embedded wideband (141–30 THz) fractal MIMO antenna over SiO_2 substrate for terahertz band applications. Silicon **15**, 7823–7836 (2023)
38. Babu, K.V., Das, S., Ali, S.S., El Ghzaoui, M., Madhav, B.T.P., Patel, K., S.: Broadband sub-6 GHz flower-shaped MIMO antenna with high isolation using theory of characteristic mode analysis (TCMA) for 5G NR bands and WLAN applications. Int. J. Commun. Syst. **36**(6), e5442 (2023)
39. Aghoutane, B., El Ghzaoui, M., Das, S., Ali, W., El Faylali, H.: A dual wideband high gain 2×2 multiple-input-multiple-output monopole antenna with an end-launch connector model for 5G millimeter-wave mobile applications. Int. J. RF Microwave Comput. Aided Eng. **32**(5), e23088 (2022)
40. Raj, U., Sharma, M.K., Singh, V., Javed, S., Sharma, A.: Easily extendable four port MIMO antenna with improved isolation and wide bandwidth for THz applications. Optik **247**, 167910 (2021)
41. Saxena, G., Chintakindi, S., Kasim, M.A., Maduri, P.K., Awasthi, Y., Kumar, S., Kansal, S., Jain, R., Sharma, M.K., Dewan, C.: Metasurface inspired wideband high isolation THz MIMO antenna for nano communication including 6G applications and liquid sensors. Nano Commun. Netw. **34**, 100421 (2022)
42. Kumar, A., Saxena, D., Jha, P., Sharma, N.: Compact two-port antenna with high isolation based on the defected ground for THz communication. Results Opt. **13**, 100522 (2023)
43. Saxena, G., Alam, M., Roy, M., Barnawi, A.B., Khan, T.Y., Yadava, R.L., Chintakindi, S., Jain, R., Singh, H., Awasthi, Y.K.: CSRR loaded Multiband THz MIMO Antenna for Nano-Communications and Bio-Sensing Applications. Nano Communication Networks **38**, 100481 (2023)
44. Kumar, P., Ali, T., Pathan, S., Shetty, N.K., Huchegowda, Y.B., Nanjappa, Y.: Design and analysis of ultra-wideband four-port MIMO antenna with DGS as decoupling structure for THz applications. Results Opt. **13**, 100573 (2023)
45. Das, S., Mitra, D., Chaudhuri, S.R.B.: Fractal loaded planar super wide band four element MIMO antenna for THz applications. Nano Commun. Netw. **30**, 100374 (2021)
46. Singhal, S.: Tetradecagonal ring shaped terahertz superwideband MIMO antenna. Optik **208**, 164066 (2020)
47. Varshney, G., Gotra, S., Pandey, V.S., Yaduvanshi, R.S.: Proximity-coupled two-port multi-input-multi-output graphene antenna with pattern diversity for THz applications. Nano Commun. Netw. **21**, 100246 (2019)

Chapter 15
Fractal MIMO Antenna Design for High-Frequency Terahertz Applications

K. Vasu Babu, Gorre Naga Jyothi Sree, Tanvir Islam, Sudipta Das, and Bhuma Anuradha

15.1 Introduction

In the rapidly advancing field of wireless communication, the requirement for higher data rates and spectral efficiencies over multi-band frequency ranges has greatly increased in recent years.

The numerous references [1–37] given below, covering the work of the current lead author and many others, demonstrate clearly how rapidly the field of THz antenna design is advancing, and the great potential offered by novel approaches and materials. The current paper describes a novel THz antenna structure having the following advantages:

- Compact structure with less complexity

K. Vasu Babu (✉)
Department of Electronics and Communication Engineering, BVRIT HYDERABAD College of Engineering for Women, Bachupally, Telangana, India
e-mail: vasubabuece@gmail.com

G. N. J. Sree
Department of Electronics and Communication Engineering, Vasireddy Venkatadri Institute of Technology, Guntur, Andhra Pradesh, India

T. Islam
Department of Electrical and Computer Engineering, University of Houston, Houston, TX 77204, USA

S. Das
Department of Electronics and Communication Engineering, IMPS College of Engineering and Technology, Malda, West Bengal, India

B. Anuradha
Department of Electronics and Communication Engineering, S. V. University, Tirupati, Andhra Pradesh, India

M. El Ghzaoui et al. (eds.), *Next Generation Wireless Communication*, Signals and Communication Technology, https://doi.org/10.1007/978-3-031-56144-3_15

- Improved isolation between elements within the chosen space
- Better MIMO parameters
- 40 dB isolation.

15.2 Antenna Geometry

Figure 15.1 shows the proposed geometry within a rectangle of dimensions 90 × 44 μm. The substrate is SiO_2 with relative permittivity 3.9 and thickness 1.5 nm. The patch design is a circular fractal structure in the shape of a cross made of gold 0.015 μm thick. Removing the cross-shaped fractal elements improves isolation and reduces coupling over the full ground plane. Table 15.1 shows the geometrical parameters of the corresponding structure.

15.3 Results Explanation

Figure 15.2a shows the S-parameter analysis of the proposed structure at the resonant frequencies of 1.62 and 3.13 THz. At 1.62 THz S_{11} and S_{21} are − 24 dB and − 40 dB respectively, and at 3.13 THz they are − 21 dB and − 102 dB. Figure 15.2b shows the impedance of the design as real and imaginary parts. At 1.62 THz the real and imaginary values are 152 ohms and − 26 Ω respectively, and at 3.13 THz they are 98 Ω and − 48 Ω. Figure 15.3 shows the co-polarization and cross-polarization radiation patterns at 1.62 and 3.13 THz. Figures 15.4a–d show the predicted surface current distributions of the proposed structure at 1.62 THz. Figure 15.4a indicates that in the front plane of Port 1 more current flows through the edges of the fractal structure, whilst Fig. 15.4b shows that more current flows through the rear of Port 1 ground plane. Figure 15.4c, d show a similar pattern for Port 2 and that more current flows at the bottom edges of the ground plane. Similarly Fig. 15.5a–d show the predicted currents at 3.13 THz. These display very similar behaviour to that at 1.62 THz, except that more current is predicted to flow at the feed points.

Figure 15.6a–d show the predicted electric and magnetic field distributions of the proposed structure at 1.62 and 3.13 THz. Figure 15.6a shows the E-field at 1.62 THz, for which a greater electric field is generated over the entire patch structure. Port 1 excitation generates a maximum field of around 2 v/m at the bottom of the feeder. Figure 15.6b shows the H-field at 1.62 THz. The distribution is similar to that of the electric field with a maximum value around 1.72 A/m. Figure 15.6c reveals a similar E-field distribution at 3.13 THz with a maximum value around 2.5 v/m, whilst Fig. 15.6d again shows the same pattern with a maximum value around 1.97 A/m at 3.13 THz.

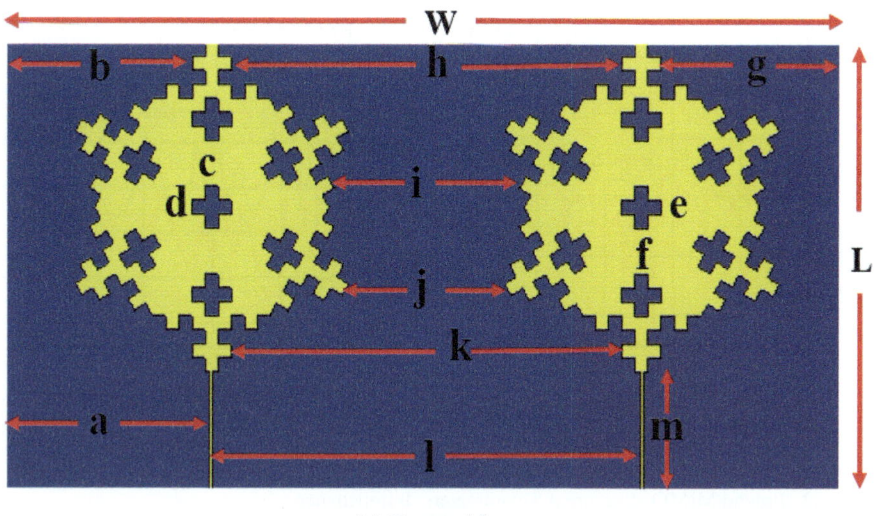

(a) Front side

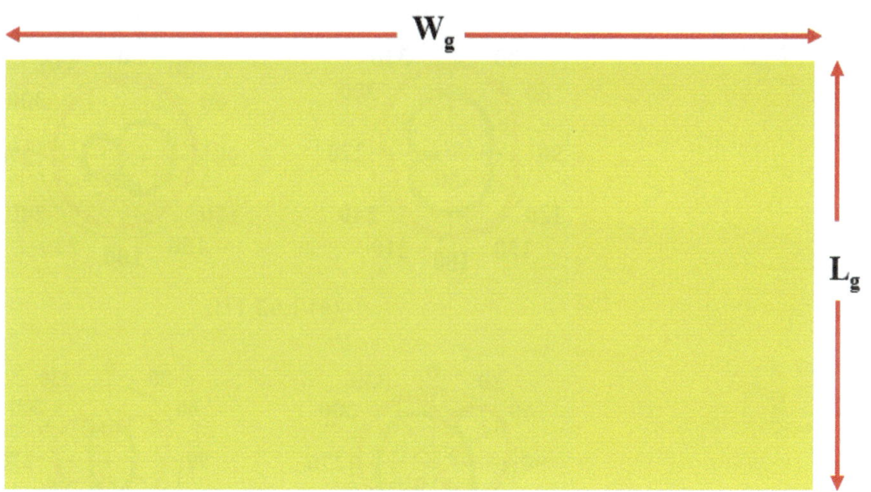

(b) Back side

Fig. 15.1 Fractal structure design

Table 15.1 Fractal MIMO structure designed parameters in micrometers

Parameter	W	L	a	b	c	d	e	f	g
Value	90	44	22	20	1.4	1.4	1.4	1.4	20
Parameter	h	i	j	k	l	m	W_g	L_g	ε_r
Value	50	40	20	30	46	12	90	44	3.9

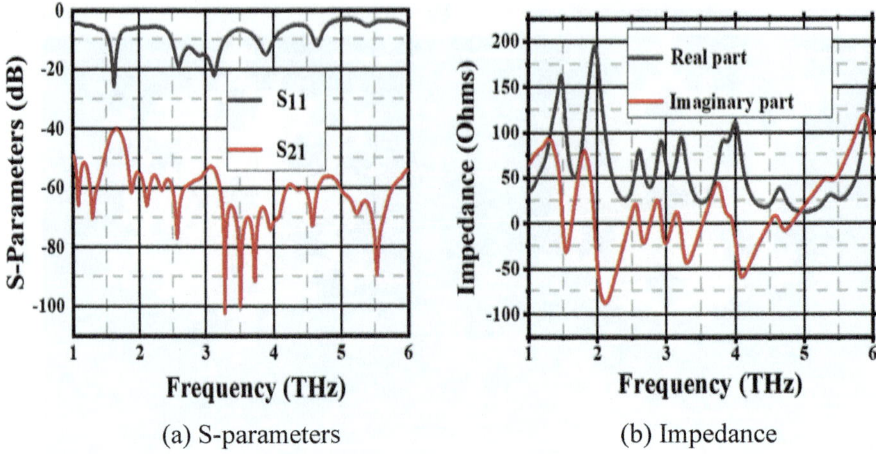

(a) S-parameters (b) Impedance

Fig. 15.2 Fractal MIMO structure **a** S-parameters, **b** impedance

Fig. 15.3 Radiation patterns at 1.62 and 3.13 THz

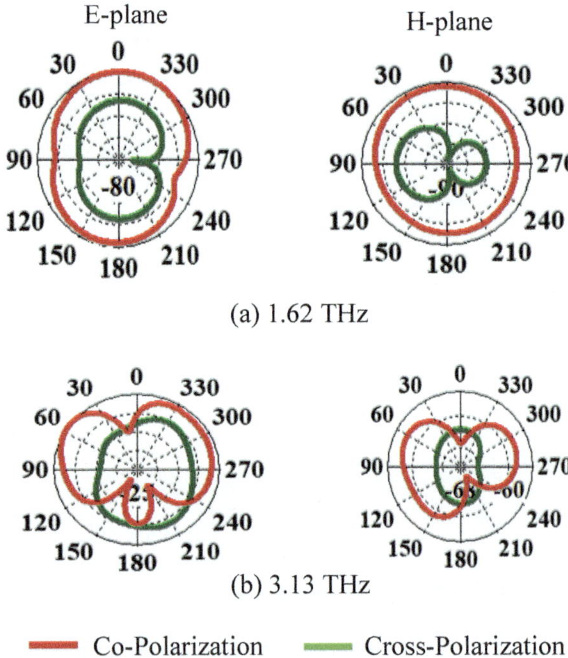

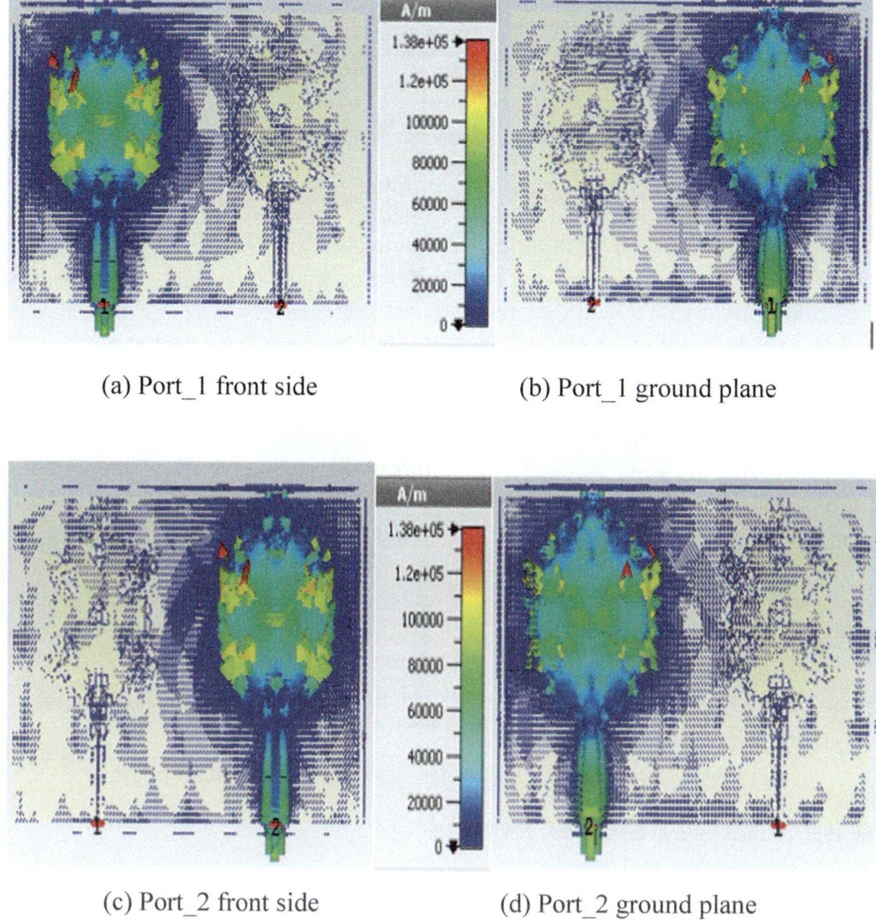

(a) Port_1 front side (b) Port_1 ground plane

(c) Port_2 front side (d) Port_2 ground plane

Fig. 15.4 Surface current distribution at 1.62 THz

15.4 MIMO Parameters

Figure 15.7a–e show the predicted multiple-input multiple-output (MIMO) parameters of the proposed structure, based upon expressions (15.1–15.9) given below. Figure 15.7a shows the envelope correlation coefficient (ECC) of the proposed structure as a function of frequency, which is within the acceptable limits of MIMO design at the resonant frequencies. The ECC was evaluated using expressions (15.1) and (15.2) with far-field and S-parameter analysis [28].

$$\rho_{ij} = \frac{\left|S_{11}^{*}S_{12} + S_{21}^{*}S_{22}\right|^{2}}{\left(1 - \left(|S_{11}|^{2} + |S_{21}|^{2}\right)\right)\left(1 - \left(|S_{22}|^{2} + |S_{12}|^{2}\right)\right)} \tag{15.1}$$

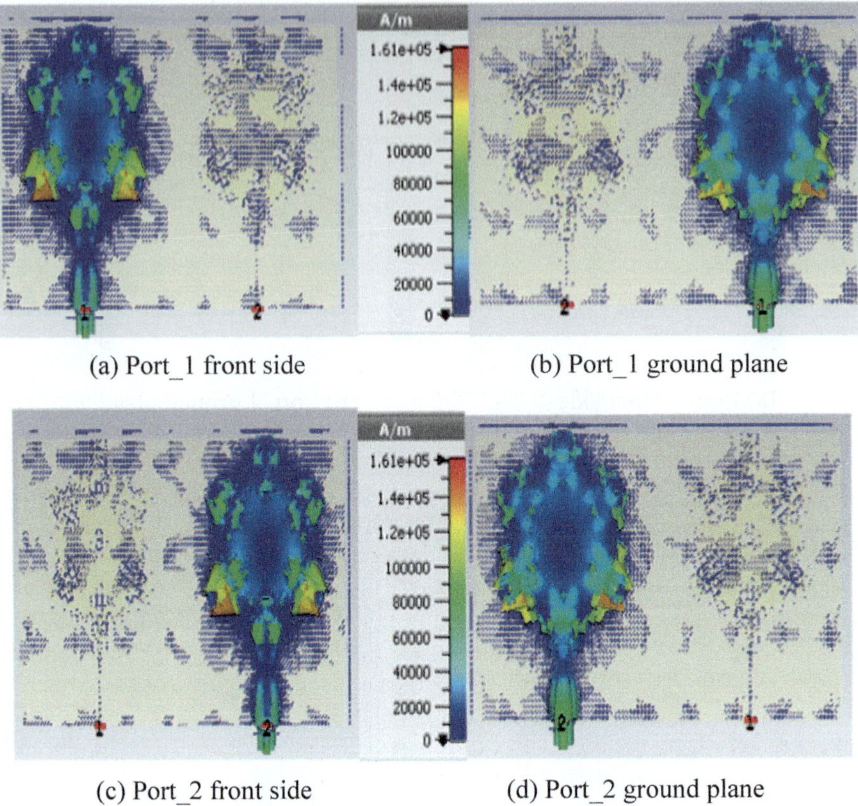

Fig. 15.5 Surface current distribution at 3.13 THz

$$\rho_{ij} = \left| \frac{\iint \left(E_{\theta_i}.E^*_{\theta_j} + E_{\varphi_i}.E^*_{\varphi_j} \right) d\Omega}{\iint \left(E_{\theta_i}.E^*_{\theta_i} + E_{\varphi_i}.E^*_{\varphi_i} \right) \mathrm{d}\Omega \iint \left(E_{\theta_j}.E^*_{\theta_j} + E_{\varphi_i}.E^*_{\varphi_j} \right) \mathrm{d}\Omega} \right|^2 \tag{15.2}$$

A diversity gain (DG) of 9.99 dB within the operating band has been calculated using expression (15.3) [32]. Figure 15.7b shows the diversity gain as a function of frequency.

$$\mathrm{DG} = 10\sqrt{1 - |\mathrm{ECC}|^2} \tag{15.3}$$

Figure 15.7c shows the predicted total active reflection coefficient (TARC) for the simulated fractal-loaded MIMO radiator, calculated using expressions (15.4) and (15.5) [30].

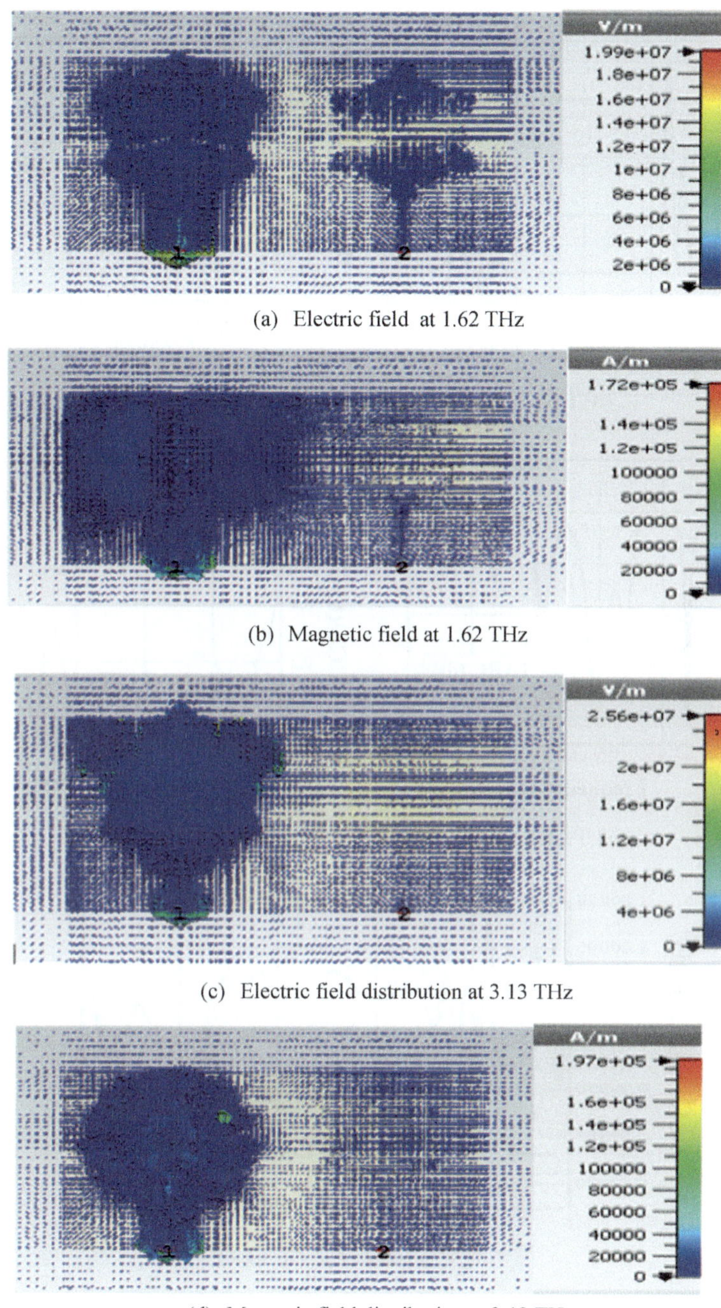

(a) Electric field at 1.62 THz

(b) Magnetic field at 1.62 THz

(c) Electric field distribution at 3.13 THz

(d) Magnetic field distribution at 3.13 THz

Fig. 15.6 Distribution of electric and magnetic fields

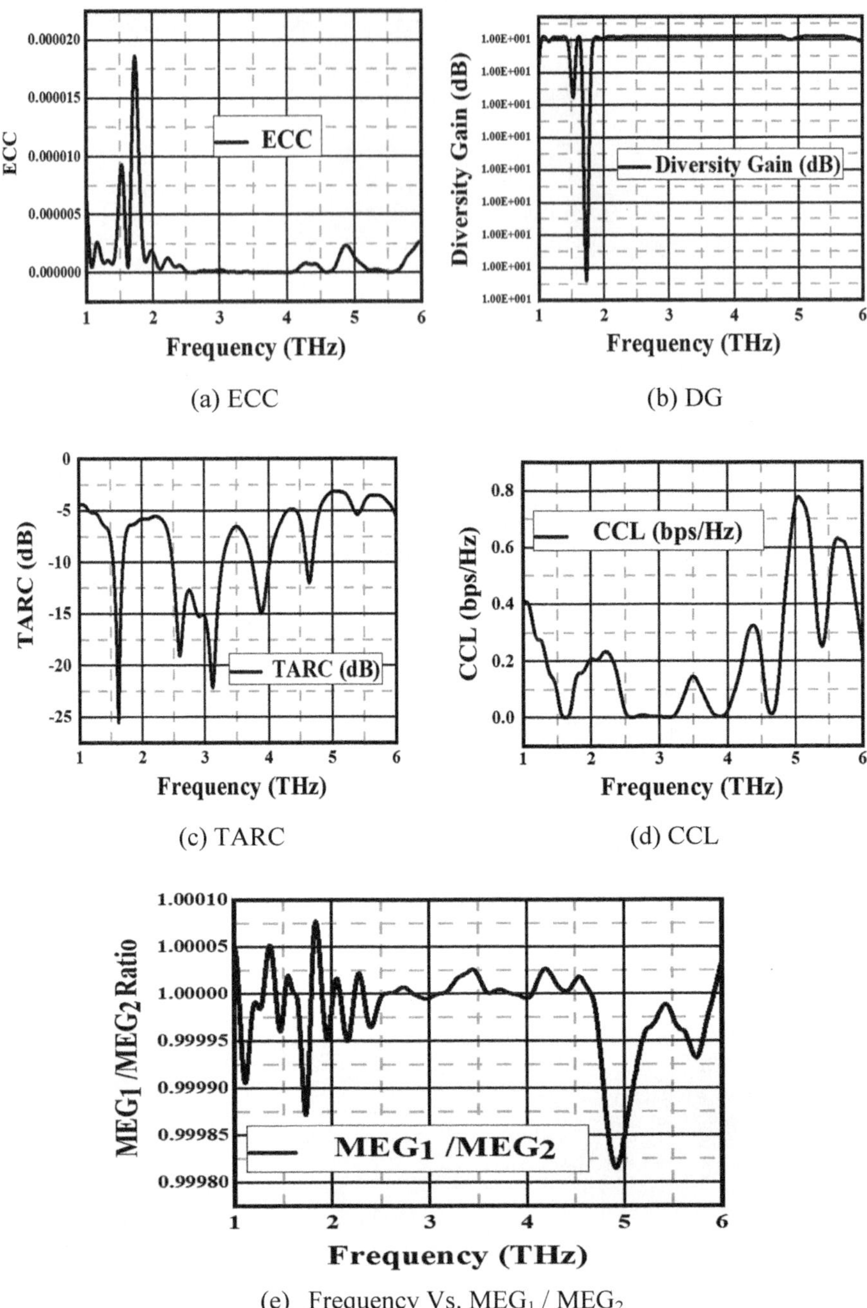

Fig. 15.7 MIMO parameters of fractal designed structure

$$\Gamma_a^t = \frac{\sqrt{\Sigma_j^M |b_j|^2}}{\sqrt{\Sigma_j^M |a_j|^2}} \tag{15.4}$$

$$\Gamma_a^t = \sqrt{\frac{\left|S_{11} + S_{12}e^{j\theta}\right|^2 + \left|S_{21} + S_{22}e^{j\theta}\right|^2}{2}} \tag{15.5}$$

Expressions (15.6–15.8) [27] were used to determine the channel capacity loss (CCL). Figure 15.7d shows it to be less than 0.028 bits/s/Hz over the working band.

$$C_{\text{loss}} = \log_2 \det\left(a^R\right) \tag{15.6}$$

$$a^R = \begin{pmatrix} \rho_{11} & \rho_{12} \\ \rho_{21} & \rho_{22} \end{pmatrix} \tag{15.7}$$

$$\rho_{ii} = 1 - \left(|S_{ii}|^2 + \left|S_{ij}\right|^2\right), \text{ and } \rho_{ij} = -\left(s_{ii}^* s_{ij} + s_{ij}^* s_{jj}\right), \text{ where } \quad i, j = 1 \text{ or } 2 \tag{15.8}$$

Figure 15.7e shows the expected mean effective gain (MEG) evaluated using expression (15.9) for the MEG performance of an N-port radiating system [35].

$$\text{MEG}_i = 0.5\eta_{i,\text{rad}} = 0.5\left[1 - \sum_{j=1}^{M} \left|S_{ij}\right|^2\right] \tag{15.9}$$

15.5 Conclusion

A 2-element fractal-type MIMO antenna structure for wireless applications in the THz range has been designed and its expected behaviour analyzed. It is a circular structure built from cross-shaped fractals over a full ground plane. The arrangement of the fractals yielded the required return loss and improved isolation among the patch elements. Isolation in excess of 40.0 dB is achieved at the dual-band resonant frequencies. The MIMO parameters MEG, CCL, DG, TARC and CCL are within threshold limits at the dual-band resonant frequencies. This antenna design could be used in real time imaging, sensing, human skin and water content region applications.

References

1. Babu, K.V., Anuradha, B.: Design of multi-band Minkowski MIMO antenna to reduce the mutual coupling. J. King Saud Univ.-Eng. Sci. **32**(1), 51–57 (2020)
2. Saraereh, O.A., et al.: Design and analysis of a novel antenna for THz wireless communication. Intell. Autom. Soft Comput. **31**(1), 607–619 (2022)
3. Fakhte, S., Taskhiri, M.M.: Graphene-enabled terahertz dielectric rod antenna with polarization reconfiguration. Opt. Quantum Electron. **55**(14), 1258 (2023)
4. Vasu Babu, K., et al.: A micro-scaled graphene-based tree-shaped wideband printed MIMO antenna for terahertz applications. J. Comput. Electron. **21**(1), 289–303 (2022)
5. Khaleel, S.A., Hamad, E.K.I., Saleh, M.B.: High-performance tri-band graphene plasmonic microstrip patch antenna using superstrate double-face metamaterial for THz communications. J. Electr. Eng. **73**(4), 226–236 (2022)
6. Kiani, N., Hamedani, F.T., Rezaei, P.: Designing of a circularly polarized reconfigurable graphene-based THz patch antenna with cross-shaped slot. Opt. Quant. Electron. **55**(4), 356 (2023)
7. Babu, K.V., Anuradha, B.: Design of MIMO antenna to interference inherent for ultra wide band systems using defected ground structure. Microw. Opt. Technol. Lett. **61**(12), 2698–2708 (2019)
8. Wu, Y.H., Yu, T., Shen, Z.X.: Two-dimensional carbon nanostructures: fundamental properties, synthesis, characterization, and potential applications. J. Appl. Phys. **108**(7) (2010)
9. Kiani, N., Hamedani, F.T., Rezaei, P.: Implementation of a reconfigurable miniaturized graphene-based SIW antenna for THz applications. Micro Nanostruct. **169**, 207365 (2022)
10. Babu, K.V., Anuradha, B.: Design of UWB MIMO antenna to reduce the mutual coupling using defected ground structure. Wirel. Pers. Commun. **118**(4), 3469–3484 (2021)
11. Babu, K.V., Anuradha, B.: Design of inverted L-shape & ohm symbol inserted MIMO antenna to reduce the mutual coupling. AEU-Int. J. Electron. Commun. **105**, 42–53 (2019)
12. Kiani, N., Afsahi, M.: Design and fabrication of a compact SIW diplexer in C-band. Iran. J. Electr. Electron. Eng. **15**(2), 189 (2019)
13. Deng, Q., et al.: Adjustable plasmonic multi-channel demultiplexer with graphene sheets and ring resonators. Plasmonics **14**, 993–998 (2019)
14. Babu, K.V., et al.: Design and optimization of micro-sized wideband fractal MIMO antenna based on characteristic analysis of graphene for terahertz applications. Opt. Quant. Electron. **54**(5), 281 (2022)
15. Asgari, S., Granpayeh, N., Fabritius, T.: Controllable terahertz cross-shaped three-dimensional graphene intrinsically chiral metastructure and its biosensing application. Opt. Commun. **474**, 126080 (2020)
16. Vasu Babu, K., Anuradha, B.: Design of Wang shape neutralization line antenna to reduce the mutual coupling in MIMO antennas. Anal. Integr. Circ. Signal Process. **101**(1), 67–76 (2019)
17. Luan, J., et al.: Design and optimization of a graphene modulator based on hybrid plasmonic waveguide with double low-index slots. Plasmonics **14**, 133–138 (2019)
18. Fakhte, S., Taskhiri, M.M.: Graphene-based terahertz antenna with polarization reconfiguration. Physica Scripta **98**(11), 115541 (2023)
19. Babu, K.V., et al.: Design and development of miniaturized MIMO antenna using parasitic elements and Machine learning (ML) technique for lower sub 6 GHz 5G applications. AEU-Int. J. Electron. Commun. **153**, 154281 (2022)
20. Babu, K.V., et al.: Compact dual-band design and analysis of half-circular U-shape MIMO radiator for wireless applications. Microsyst. Technol. **29**(4), 501–514 (2023)
21. Giddens, H., et al.: Mid-infrared reflect-array antenna with beam switching enabled by continuous graphene layer. IEEE Photon. Technol. Lett. **30**(8), 748–751 (2018)
22. Hanson, G.W.: Dyadic green's functions for an anisotropic, non-local model of biased graphene. IEEE Trans. Antennas Propag. **56**(3), 747–757 (2008)
23. Hanson, G.W.: Dyadic green's functions and guided surface waves for a surface conductivity model of graphene. J. Appl. Phys. **103**(6) (2008)

24. Babu, K.V., et al.: Deep learning assisted fractal slotted substrate MIMO antenna with characteristic mode analysis (CMA) for Sub-6 GHz n78 5G NR applications: design, optimization and experimental validation. Physica Scripta **98**(11), 115526 (2023)
25. Kiani, N., Hamedani, F.T., Rezaei, P.: Reconfigurable graphene-gold-based microstrip patch antenna: RHCP to LHCP. Micro Nanostruct. **175**, 207509 (2023)
26. Babu, K.V., et al.: Design and analysis of fractal-based THz antenna with co-axial feeding technique for wireless applications. In: Recent Advances in Graphene Nanophotonics, 351–358. Springer Nature Switzerland, Cham (2023)
27. Qin, X., et al.: A tunable THz dipole antenna based on graphene. In: 2016 IEEE MTT-S International Microwave Workshop Series on Advanced Materials and Processes for RF and THz Applications (IMWS-AMP). IEEE (2016)
28. Babu, K.V., Sree, G.N.J.: Design and circuit analysis approach of graphene-based compact metamaterial-absorber for terahertz range applications. Opt. Quant. Electron. **55**(9), 769 (2023)
29. Ezzulddin, S.K., Hasan, S.O., Ameen, M.M.: Microstrip patch antenna design, simulation and fabrication for 5G applications. Simul. Modell. Pract. Theor. **116**, 102497 (2022)
30. Vasu Babu, K., et al.: Design of monopole ground graphene disc-inserted THz antenna for future wireless systems. In: Recent Advances in Graphene Nanophotonics, 305–312. Springer Nature Switzerland, Cham (2023)
31. Kushwaha, R.K., Karuppanan, P., Malviya, L.D.: Design and analysis of novel microstrip patch antenna on photonic crystal in THz. Physica B Condens. Matter **545**, 107–112 (2018)
32. Shamim, S.M., et al.: Design and analysis of microstrip patch antenna with photonic band gap (PBG) structure for high-speed THz application. Opt. Quant. Electron. **55**(7), 618 (2023)
33. Bala, R., Marwaha, A.: Investigation of graphene based miniaturized terahertz antenna for novel substrate materials. Eng. Sci. Technol. Int. J. **19**(1), 531–537 (2016)
34. Babu, K.V., et al.: Design of graphene-based broadband metamaterial absorber with circuit analysis approach for terahertz region applications. Opt. Quant. Electron. **55**(13), 1188 (2023)
35. Shalini, M., Ganesh Madhan, M.: Design and analysis of a dual-polarized graphene based microstrip patch antenna for terahertz applications. Optik 194, 163050 (2019)
36. Rabbani, M.S., Ghafouri-Shiraz, H.: Liquid crystalline polymer substrate-based THz microstrip antenna arrays for medical applications. IEEE Antennas Wirel. Propag. Lett. **16**, 1533–1536 (2017)
37. Babu, K.V., et al.: Design and implementation of MIMO graphene patch antenna to improve isolation for THz applications. Microsyst. Technol., 1–11 (2023)

Chapter 16
Evaluating the Performance of a Transparent MIMO Nano-Antenna for Wireless Health: Addressing Terahertz Challenges in Integrated Medical Device Networks

Bilal Aghoutane, Hamid Bezzout, Fatima Kiouach, Houda Hiddar, Mohammed El Ghzaoui, and Hanan El Fayalali

16.1 Introduction

In the rapidly evolving realm of medical science, technological advancements have played an indispensable and pivotal role in reshaping the complex and intricate landscape of health care [1]. Over the course of recent years, we have observed and borne witness to significant and noteworthy strides in the field of medical science, particularly in the realm of integrating wireless transmission capabilities into medical devices [2]. This integration has heralded the arrival and onset of an era that can only be described as transformative, wherein patient care and health management have been revolutionized and transformed [3]. As medical devices become interconnected and interlinked within a vast network, they not only serve to alleviate and mitigate the burden that is placed upon healthcare professionals, but also serve to optimize and enhance time management and overall efficiency in the provision of patient care [4]. At the very core and foundation of this remarkable and astounding transformation lies the Wireless Body Area Network (WBAN) as illustrated in Fig. 16.1, which relies and depends upon cutting-edge and state-of-the-art Radio Frequency (RF) technology to facilitate and enable seamless and uninterrupted communication among medical

B. Aghoutane (✉) · H. Bezzout · H. E. Fayalali
Faculty of Sciences, Ibn Tofail University, Kenitra, Morocco
e-mail: bilal.aghoutane@gmail.com

F. Kiouach · M. E. Ghzaoui
Faculty of Sciences Dhar El Mahraz-Fes, Sidi Mohamed Ben Abdellah University, Fes, Morocco

H. Hiddar
Laboratory of Microbiology and Molecular Biology, Faculty of Sciences, BioBio Research Center, University Mohammed V, Rabat, Morocco

M. El Ghzaoui et al. (eds.), *Next Generation Wireless Communication*, Signals and Communication Technology, https://doi.org/10.1007/978-3-031-56144-3_16

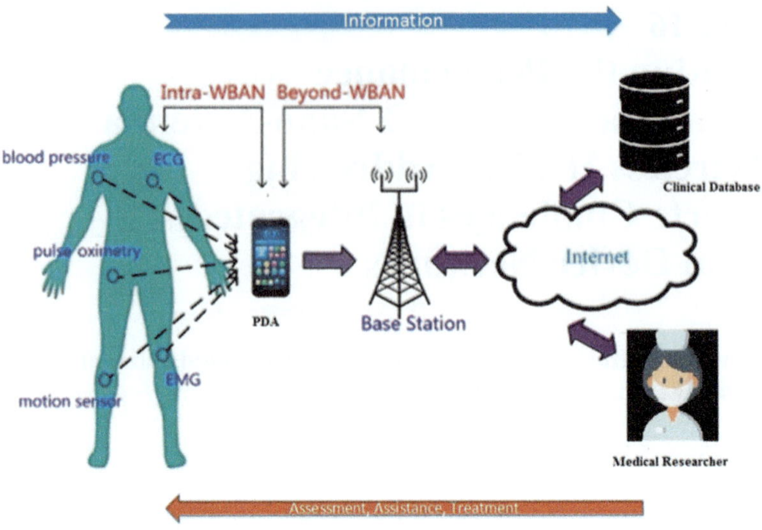

Fig. 16.1 Path of the data in the wireless body area network [6]

devices and central monitoring systems [5]. This groundbreaking and innovative technology offers and provides real-time health monitoring and patient-centered care, thereby revolutionizing and transforming the very nature and essence of health care as we know it.

However, the evolution and progress of wireless transmission technology has not been without its fair share of challenges and obstacles [6, 7]. One proposed and envisioned solution that holds great promise and potential is the use and utilization of the terahertz (THz) band [8], which spans and covers a wide and extensive frequency range from 0.1 to 10 THz as mentioned in Fig. 16.2. Nevertheless, the utilization and application are shown in Fig. 16.3, of this band presents and poses its own unique and distinct set of challenges and hurdles due to the drastic and significant changes and fluctuations in THz band characteristics that occur and manifest over varying distances.

These challenges include but are not limited to the occurrence and presence of very high propagation loss, which in turn places severe and stringent limitations and restrictions on the available bandwidth for longer distances, thus posing and presenting a significant and formidable hurdle and obstacle to the achievement and attainment of efficient and effective wireless communication.

In order to effectively and efficiently address and overcome these aforementioned challenges and obstacles, the utilization and implementation of THz antennas and antenna arrays have emerged as crucial and instrumental components and entities [9–13]. THz antennas play a pivotal and indispensable role in ensuring and guaranteeing the transmission and reception of signals within the THz band, but the increasing and escalating path loss that occurs and takes place over greater and longer distances necessitates and calls for the development and exploration of creative and innovative

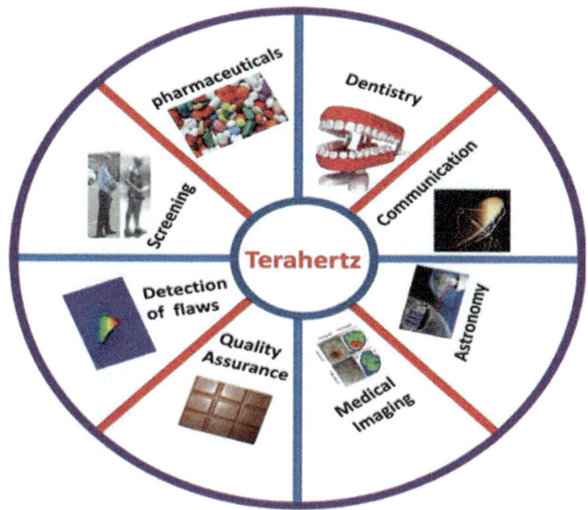

Fig. 16.2 Examples of application of terahertz radiation

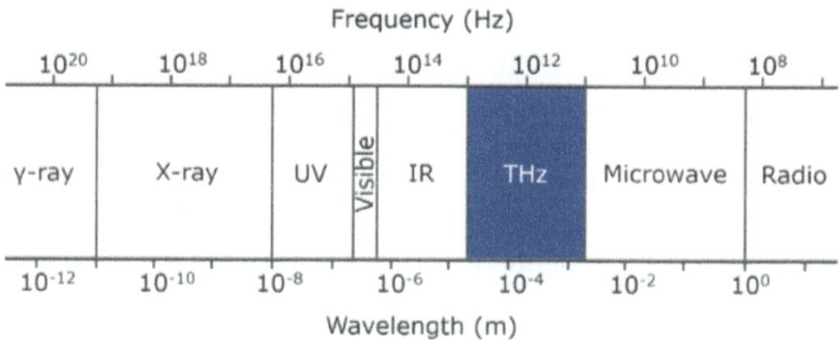

Fig. 16.3 THz band from 0.1 to 10 THz

solutions. One such approach and methodology that has gained traction and prominence involves the utilization and deployment of antenna arrays to implement and establish Multiple-Input Multiple-Output (MIMO) communication systems, which in turn enable and facilitate high-capacity and reliable communication. For instance, MIMO systems that incorporate and encompass 4×4 or 16 antennas in both the transmission and reception processes have become commonplace and standardized in existing and current wireless communication standards. The application and utilization of MIMO technology not only serves to alleviate and mitigate the challenges and hurdles posed by the THz band, but also opens up new and unexplored avenues and possibilities in the realm of wireless healthcare technology.

Our MIMO antenna system as illustrated in Fig. 16.4, which operates within the high-frequency range of 170 THz, constitutes a pioneering development in the realm of telemedicine and remote surgery. At this remarkable frequency, we have harnessed the potential of Multiple-Input Multiple-Output (MIMO) technology not only to facilitate secure and dependable communication, but also to significantly augment data throughput. The utilization of MIMO antenna technology at 170 THz is on the cusp of transforming the manner in which we approach remote medical procedures, offering a substantial enhancement in data transmission capabilities. Within the context of remote surgery and telemedicine, where instantaneous data exchange holds paramount importance, our innovative system employs AI-driven optimization, beamforming, interference mitigation, and dynamic resource allocation in order to ensure that medical practitioners can promptly access and share crucial information with precision. By substantially amplifying data throughput, our MIMO antenna system becomes an indispensable tool in enabling remote medical procedures, permitting seamless communication and the transfer of high-resolution medical data, thereby ultimately enhancing patient care and broadening the potential of telehealth services. The present study introduces an innovative and compact antenna design, which possesses dimensions of 130 × 260 × 112 nm. This antenna operates at an exceedingly high frequency of 170 THz and showcases an impressive bandwidth of 44,000 GHz. By utilizing the patch antenna configuration and these compact dimensions, a 2 × 1 Multiple-Input Multiple-Output (MIMO) antenna system has been developed. Furthermore, this system has been enhanced with a plexiglass housing, which measures 90 × 58 × 1.5 nm. The antenna and plexiglass housing both incorporate glass substrates, which are characterized by a relative permittivity of 5.5 and a mass density of 2500. Additionally, the circular patch elements are meticulously positioned at a distance of $\lambda/2$, thereby amplifying their resonant properties. This cutting-edge antenna technology demonstrates significant potential in the fields of telemedicine and remote surgery. Its compact size, high-frequency operation, and improved data throughput capabilities have the potential to revolutionize communication systems. Furthermore, this technology has the capability to facilitate seamless and high-resolution data exchange, thereby advancing the quality of healthcare services in remote or telemedical scenarios.

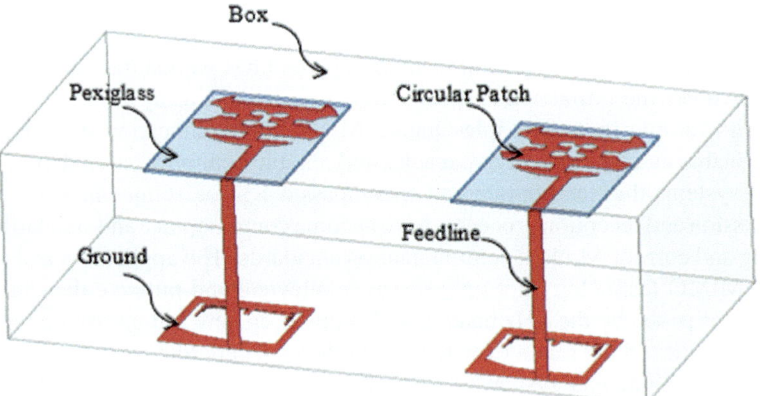

Fig. 16.4 Proposed MIMO nano-antenna

16.2 Related Works

In the landscape of high-frequency antenna technology, a wealth of innovative approaches has been explored in prior research. Studies such as the one outlined in reference [16] have delved into the use of graphene-based frequency selective surfaces, with the aim of reducing mutual coupling in plasmonic nano-antenna arrays, thereby striving for enhanced performance. Reference [17] has notably focused on the realization of ultra-massive Multiple-Input Multiple-Output (MIMO) systems, tailored for terahertz band communication, with the ultimate goal of achieving exceptional data throughput and reliability. Equally compelling is reference [18], which delves into the application of MIMO nano-antennas in optical frequency domains, underscoring their immense potential in high-frequency communication and photonics. Furthermore, reference [19] delves into channel modeling for ultra-massive MIMO communication systems operating in the terahertz band, taking into account mutual coupling effects that are pivotal for realistic performance evaluation. Reference [20] spotlights the design of silicon-based MIMO nano-dielectric resonator antennas, emphasizing their utility in photonics applications and the inherent advantages of polarization diversity. Similarly, reference [21] is dedicated to the design of nano-antennas tailored for terahertz (THz) communication, with a specific focus on material selection and strategies to augment their performance in THz-frequency applications. In reference [22], the work likely revolves around the design of MIMO antenna arrays, targeting millimeter-wave frequencies and catering to 5G communication terminals. Reference [23] explores the design of terahertz (THz) MIMO antennas inspired by metamaterials (MTM) and graphene, presumably engaging characteristic mode analysis and discussing potential applications in 6G and IoT scenarios. Reference [24] is likely dedicated to the design and practical operation of a graphene-based plasmonic nano-antenna array developed specifically for communication in the terahertz band. Furthermore, reference [25] appears to delve into the development of nano-antennas for advanced THz communication, shedding light on wideband transparent antennas and their diverse applications. Reference [26] could potentially explore reconfigurable nano-antennas, offering valuable insights into adaptability and performance optimization for nano-antennas within various photonic applications. Reference [27] is expected to present an alternative design approach for super wideband nano-antennas operating in the THz range, which aligns with our paper's emphasis on advanced THz communication. In reference [28], the study likely provides essential context for the utilization of MIMO technology in THz communication, complementing our paper's discussions regarding data throughput enhancement. Reference [29] is anticipated to discuss the performance of plasmonic nano-antennas, offering insights that can relate to our paper's exploration of materials and strategies for performance optimization. Reference [30] is expected to offer insights into the design of nano-antennas tailored for photonic applications, which can enrich the understanding of potential applications within our research domain. Reference [31] could present a design approach for triband optical nano-antennas, further expanding the horizon of multi-frequency considerations. Lastly, reference

[32] likely provides insights into multiarray antennas operating in the THz band, harmonizing well with our paper's emphasis on Multi-Input Multi-Output (MIMO) technology at an impressive frequency of 170 THz. Collectively, these referenced works enrich the understanding of advanced antenna designs, materials, and applications in the high-frequency domains, thus underlining the significance and potential impact of our paper in the context of telemedicine and remote surgery.

16.3 Antenna Characteristics

Antenna characteristics encompass a multitude of parameters and properties that define the behavior and performance of an antenna system. These characteristics include but are not limited to the antenna's radiation pattern, gain, directivity, bandwidth, polarization, impedance, and efficiency. Understanding and optimizing these characteristics is crucial for designing and implementing reliable and efficient antenna systems in various applications, such as wireless communication, radar systems, satellite communication, and broadcasting. The radiation pattern of an antenna describes the distribution of electromagnetic energy in space, indicating the direction and intensity of the radiated or received signals. The gain of an antenna represents the ability of the antenna to direct and concentrate the radiated or received energy in a specific direction, typically measured in decibels (dB). Directivity is a measure of how well an antenna focuses its energy in a particular direction and is closely related to the antenna's gain. The bandwidth of an antenna refers to the range of frequencies over which the antenna can operate effectively. It is a crucial parameter for determining the capacity and performance of wireless communication systems. The polarization of an antenna refers to the orientation of the electric field vector of the radiated or received signals. It can be linear, circular, or elliptical, and matching the polarization of the transmitting and receiving antennas is essential for efficient signal transmission and reception. Impedance represents the opposition that an antenna presents to the flow of current when a voltage is applied, and it plays a vital role in maximizing power transfer and minimizing signal loss. Lastly, antenna efficiency is a measure of how effectively an antenna converts electrical power into radiated or received electromagnetic energy, and it is influenced by various factors such as conductor losses, dielectric losses, and mismatch losses. By comprehensively studying and optimizing these antenna characteristics, engineers and researchers can develop antennas that meet the specific requirements and constraints of different applications, leading to improved performance and reliability in wireless communication and other fields. Antennas are essential components in the field of electromagnetic communication and wireless technology. They come in various types and designs, each with its own set of characteristics.

16.3.1 AI-Enhanced Antenna Radiation Pattern Optimization

Radiation patterns are a fundamental concept in antenna technology, delineating how antennas disperse electromagnetic energy in space. These patterns can be omnidirectional, directional, or exhibit bidirectional characteristics, depending on their intended purpose. Omnidirectional patterns offer uniform radiation in all directions, commonly used in applications like Wi-Fi routers and cell towers. In contrast, directional patterns concentrate radiation in a specific direction, suited for point-to-point communication and radar systems. AI enhances the understanding and optimization of these radiation patterns, employing methods like genetic algorithms and machine learning to design antennas that offer tailored performance. AI's adaptive beamforming capabilities enable dynamic adjustments for precise targeting and interference reduction, making it crucial for applications such as phased array radar systems and 5G communication. AI's predictive maintenance and cognitive radio systems leverage radiation patterns to ensure optimal antenna performance and spectrum efficiency, offering practical solutions for the evolving world of wireless communication.

In the realm of antenna technology, radiation patterns are a key factor in determining how electromagnetic energy is distributed in space. These patterns can be omnidirectional, directional, or bidirectional, depending on the antenna's purpose. Omnidirectional patterns provide uniform coverage in all directions, commonly found in Wi-Fi routers and cell towers, while directional patterns are used in point-to-point communication and radar systems. AI plays a vital role in understanding and optimizing radiation patterns by employing techniques like genetic algorithms and machine learning for antenna design. Adaptive beamforming, powered by AI, dynamically adjusts antenna properties to target specific areas or mitigate interference, crucial for applications such as phased array radar and 5G communication. AI's predictive maintenance and cognitive radio systems utilize radiation patterns to ensure reliable antenna performance and efficient spectrum utilization, making it a valuable tool in the ever-evolving field of wireless communication.

16.3.2 Enhancing Antenna Gain with Artificial Intelligence

Antenna gain, expressed in decibels (dBi), measures an antenna's ability to focus electromagnetic energy in a specific direction, compared to an ideal isotropic radiator. It plays a pivotal role in antenna technology, quantifying directional efficiency. Various types of gain, including directivity gain, front-to-back ratio, and effective isotropic radiated power (EIRP), are associated with antennas. The integration of intelligent artificial systems opens new possibilities for harnessing and optimizing antenna gain. AI-enhanced antenna gain is a game-changer in various applications and optimizations. Beamforming, empowered by AI, dynamically adjusts antenna parameters to steer gain patterns toward specific targets or nullify sources of interference, a critical advancement in 5G communication and phased array radar systems. AI also assists in designing antennas tailored to specific gain patterns using genetic algorithms and machine learning. Adaptive communication, enabled by AI, allows real-time gain pattern adjustments to adapt to environmental changes, improving mobile communication and wireless networks.

Intelligent beam management, a key application, leverages AI to analyze real-time data for optimizing gain patterns based on signal quality, interference levels, and user/device locations. Energy and spectrum efficiency are further benefits, as AI optimizes gain patterns to minimize power consumption and interference, ensuring energy-efficient communication networks. Moreover, AI-enabled predictive maintenance continuously monitors antenna performance, including gain characteristics, and schedules maintenance when deviations from ideal patterns are detected, ensuring optimal gain and system performance.

16.3.3 Antenna Selection Factors

Antennas are integral for effective communication, and they possess a range of critical attributes that impact their performance. The first set of attributes to consider includes frequency range, impedance, and polarization. Frequency range determines the signals an antenna can handle, while impedance, typically 50 or 75 Ω, plays a vital role in power transfer efficiency. Additionally, polarization—whether linear (horizontal or vertical) or circular—must align with the incoming signal for optimal reception. Bandwidth is another key factor, influencing the range of frequencies an antenna can efficiently work with.

The second set of attributes encompasses directivity, efficiency, and size/form factor. Directivity measures the focus of an antenna's radiation pattern, varying from highly directional to broad. Efficiency gauges how effectively the antenna converts input power into radiated energy. The antenna's size and form factor are essential for practical implementation, with options ranging from small chip antennas to large parabolic dishes.

Lastly, attributes like mounting and installation, environmental considerations, beamwidth, radiation efficiency, VSWR, and resonant frequency complete the considerations. Mounting and installation methods impact antenna performance, while the antenna's ability to withstand specific environmental conditions is crucial. Beamwidth defines the antenna's coverage area, and radiation efficiency ensures power is not lost as heat. VSWR reflects how well the antenna matches its transmission line, with lower values indicating a better match and less signal reflection. The resonant frequency is the frequency at which the antenna operates most efficiently, determined by its physical dimensions. These attributes collectively influence an antenna's performance, coverage, and suitability for its intended use, making them pivotal in the antenna selection process.

16.3.4 Key Antenna Parameters and Practical Design Guidelines

Efficiency in antennas, denoted as η, signifies how effectively they convert input power into radiated power. This efficiency factor accounts for losses inherent in the energy conversion process. Antennas can focus their power in specific directions, a characteristic known as directivity. Directivity (D) measures the concentration of radiation in a particular direction and is distinct from gain (G), which also considers losses. When efficiency is 100%, directivity and gain align. These factors are pivotal for wireless communication.

The efficiency of the antenna is determined by this process [33]:

$$\eta = \frac{P_r}{P_e}$$

$$\eta = \frac{R_r}{R_r + R_j}$$

With the directivity and gain obtained by

$$\begin{aligned} D(\theta, \phi) &= \frac{K(\theta, \phi)}{K_{\text{moy}}} = \frac{4\pi K(\theta, \phi)}{\langle P_t \rangle} \\ &= \frac{4\pi K_{\max} K_n(\theta, \phi)}{K_{\max} \Omega_n} \\ &= \frac{4\pi}{\Omega_a} K_n(\theta, \phi) \\ G(\theta, \phi) &= \varepsilon_r D(\theta, \phi) \end{aligned}$$

Radiation resistance, quantified by R_r, characterizes the energy associated with an antenna's emitted electromagnetic wave. It is proportional to the square of the antenna's intensity. Understanding radiation resistance is vital for comprehending an antenna's energy radiating efficiency. Additionally, radiation intensity, denoted as K, defines the power radiated by an antenna in a specific direction, remaining constant in relation to distance but varying with direction. This parameter is crucial for optimizing antenna design and performance. The resistance radiation given by the equation [34].

$$P_r = \frac{k^2}{32\pi^2} Z \frac{(Il)^2}{r^2} \int_0^{\pi} 2\pi r^2 \theta \sin\theta \mathrm{d}\theta$$

where

$$Z = \sqrt{\frac{\mu_0}{\varepsilon_0}} \approx 377\,\Omega$$

The outcome of the integration yields

$$P_r = \frac{k^2}{12\pi^2} Z(Il)^2$$

Let be again, by relating the size of the antenna to the wavelength l:

$$P_r = \frac{\pi}{3}\frac{l^2}{\lambda^2} Z l^2$$

$$P_r = R_r I_{\text{eff}} = R_r \frac{1}{2} I^2$$

$$R_r = \frac{2\pi}{3} Z \frac{l^2}{\lambda^2} \approx 800 \frac{l^2}{\lambda^2}$$

With intensity

$$K(\theta, \phi) = \langle P(r, \theta, \phi)\rangle r^2 = \frac{E^2(r, \theta, \phi)}{2\eta_0}$$

$$= \frac{E_\theta^2(r, \theta, \phi) + E_\phi^2(r, \theta, \phi)}{2\eta_0}$$

η_0 represents the intrinsic impedance of the vacuum, approximately equal to 120π. The total emitted power $\prec P_t \succ$ can be obtained by integrating $K(\theta, \phi)$ over the entire 4π steradians. This can be directly derived from the integral of the dot product of power density across a closed surface. By selecting a sphere as the closed surface, we maximize the dot product at every point because

$$\langle P\rangle = \langle P_r\rangle a_r \quad \text{and} \quad \mathrm{d}S = \mathrm{d}S a_r.$$

So,

$$\langle P_r\rangle = \oint_S \langle P(r, \theta, \phi)\rangle \mathrm{d}S$$

$$\mathrm{d}S = r^2 \sin(\theta)\mathrm{d}\theta\mathrm{d}\phi$$

And,

$$\langle P_t\rangle = \oint_{4\pi} K(\theta, \phi)\mathrm{d}\Omega$$

With,

$$\mathrm{d}\Omega = \sin(\Theta)\mathrm{d}\theta\mathrm{d}\phi$$

The reflection coefficient (S11) reveals how power reflects at the antenna's input, considering complex amplitudes *B* and *A*. It plays a significant role in achieving impedance matching for optimal antenna performance. Voltage Standing Wave Ratio (VSWR) assesses antenna matching by measuring the ratio between the maximum and minimum amplitudes of a standing wave formed due to signal reflection in the transmission line. These parameters are crucial for evaluating the efficiency and performance of antenna systems. Designing patch antennas necessitates careful consideration of substrate properties, antenna dimensions, and operating frequency. Parameters like substrate thickness, patch width, and length significantly influence the antenna's resonance frequency and size. Achieving impedance matching is crucial for efficient antenna operation. Proper design and calculations are fundamental to ensuring the antenna's optimal performance, with electromagnetic simulation tools aiding in fine-tuning. Designing a rectangular patch antenna entails a systematic procedure. Calculations involving substrate properties, antenna dimensions, and the operating frequency determine the antenna's size and characteristics. Substrate thickness, patch width, and effective wavelength calculations are key to achieving the desired performance. Further adjustments and fine-tuning are possible with electromagnetic simulation tools. The antenna's impedance and dimensions are essential factors influencing its efficiency in wireless communication.

16.4 MIMO Nano-Antenna

The book chapter introduces a novel ultra-compact antenna design with dimensions of 130 × 260 × 112 nm, resonating at an exceptionally high frequency of 170 THz and boasting a remarkable bandwidth of 44,000 GHz. Leveraging the patch antenna configuration and these compact dimensions, a 2 × 1 Multiple-Input Multiple-Output (MIMO) antenna system has been developed, further augmented with a plexiglass housing measuring 90 × 58 × 1.5 nm. The substrates utilized for both the antenna and plexiglass housing consist of glass, characterized by a relative permittivity of 5.5 and a mass density of 2500. Additionally, the circular patch elements are meticulously spaced at a distance of $\lambda/2$, enhancing their resonant properties. This cutting-edge antenna technology exhibits tremendous potential in the realms of telemedicine and remote surgery, where its compact size, high-frequency operation, and enhanced data throughput capabilities could revolutionize communication systems, enabling seamless, high-resolution data exchange, and ultimately advancing the quality of healthcare services in remote or telemedical scenarios.

16.5 Performance Analysis and Discussion

The optimized dimension of the single nano-antenna is demonstrated in [25]. The performance analysis of the ultra-compact MIMO antenna proposed operating at 170 THz presents a comprehensive evaluation of its functionality and efficacy. Key parameters, including data throughput, signal reliability, and bandwidth utilization, are thoroughly examined. The antenna's ability to resonate at this exceptionally high frequency, with a bandwidth of 44,000 GHz as shown in Fig. 16.5, is scrutinized to ensure its suitability for advanced applications, particularly in the realms of telemedicine and remote surgery. The analysis considers the impact of the antenna's compact dimensions, the use of a plexiglass housing, and the characteristics of the glass substrate material, such as its relative permittivity and mass density. Additionally, the performance analysis delves into the circular patch design and its spacing in relation to the wavelength, aiming to optimize the antenna's resonant properties. These assessments collectively serve to provide a comprehensive understanding of the antenna's capabilities, making it a valuable asset in the pursuit of high-frequency communication for critical medical applications. Figure 16.5 provides a critical insight into the performance of the suggested MIMO nano-antenna, specifically focusing on the reflection coefficient (S11). The resonance behavior is prominently evident, with the antenna demonstrating resonance at approximately 170 THz. The striking feature observed in the figure is the remarkable return loss, reaching an impressive − 35 dB. This low return loss indicates a high level of power transfer efficiency and minimal reflection of signals back to the source.

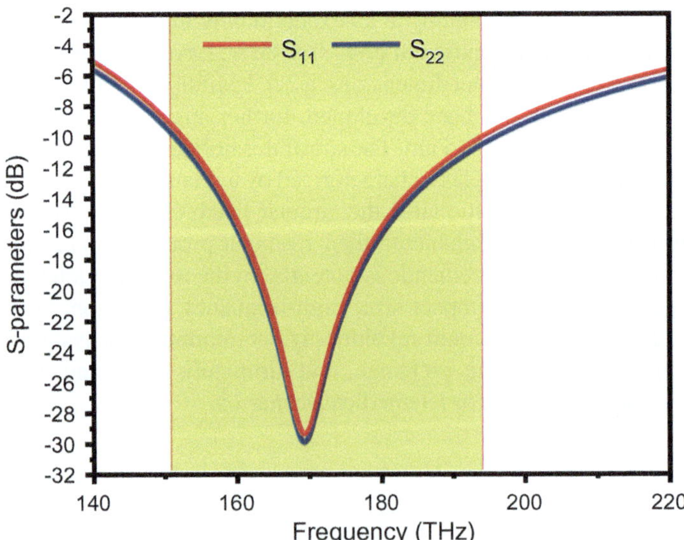

Fig. 16.5 S11 and S22 for the proposed MIMO antenna

Such an attractive return loss is indicative of the antenna's capability to effectively couple with the transmission line and surrounding environment, ensuring optimal energy transfer. The super wide bandwidth exhibited by the nano-antenna is a notable highlight. The bandwidth spans an extensive range from 151 to 195 THz for a − 10 dB threshold. This exceptional bandwidth, amounting to 44,000 GHz, is a distinctive feature attributed to the suggested MIMO nano-antenna. The ability to cover such a broad frequency range is pivotal, especially in applications demanding high data throughput, such as telemedicine and remote surgery.

The obtained bandwidth is a result of the nano-antenna's design characteristics and resonant properties. The suggested MIMO nano-antenna's ability to operate across this wide frequency band enhances its versatility and makes it well-suited for communication systems in the specified frequency range. The findings presented in Fig. 16.5 reinforce the efficacy of the proposed MIMO nano-antenna in achieving not only resonance at 170 THz but also a broad operational bandwidth, thereby positioning it as a promising candidate for advanced high-frequency communication applications.

In Fig. 16.6, the real and imaginary parts of the impedance are depicted, offering crucial insights into the antenna's performance characteristics. Notably, at the resonance frequency, the impedance is observed to be nearly 50 Ω. This observation is indicative of excellent impedance matching, a critical factor for maximizing power transfer and ensuring efficient communication system operation. The fact that the impedance is close to the standard 50 Ω signifies that the antenna is well-matched to the transmission line, minimizing signal reflections and optimizing the utilization of available power. The proximity of the real part of the impedance to 50 Ω attests to the antenna's resonance, aligning with its intended operating frequency at 170 THz. This favorable impedance matching is paramount, particularly in high-frequency applications such as telemedicine and remote surgery, where signal integrity and efficient power transfer are pivotal for reliable data transmission. The results presented in Fig. 16.6 underscore the antenna's robust design and its capability to maintain optimal impedance characteristics, thereby enhancing its overall performance in the specified frequency range. Figure 16.7 presents the Voltage Standing Wave Ratio (VSWR) of the antenna, a crucial metric in assessing the efficiency of the antenna system. VSWR is a measure of how well the antenna is matched to the transmission line and is indicative of the power reflections in the system. In the context of the presented figure, a VSWR value of 1.25 is observed, which signifies an excellent match between the antenna and the transmission line.

Figure 16.8, illustrating the E (electric) and H (magnetic) plane patterns for the transparent MIMO nano-antenna, provides valuable insights into the antenna's radiation characteristics across the frequency range from 151 to 195 THz. One notable observation from Fig. 16.8 is the directive nature of the antenna. The radiation patterns indicate a well-defined principal lobe, particularly toward $\pm 90°$ in the H-plane. This implies that the antenna exhibits a strong and focused radiation pattern along the horizontal axis, emphasizing its directional capabilities. The concentration of energy in the specified direction is indicative of the antenna's ability to transmit and receive signals with increased gain in a specific angular region. This directional behavior

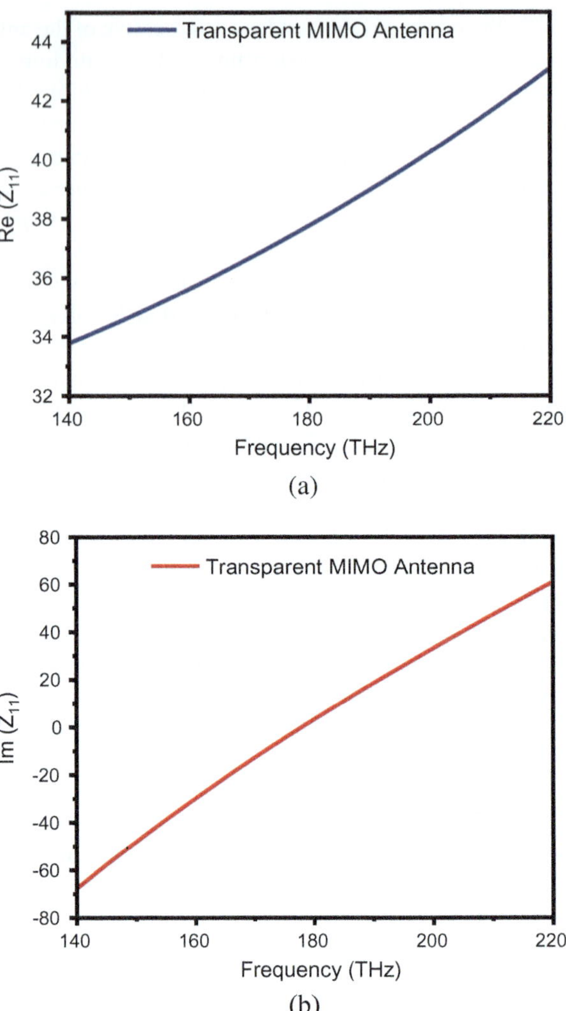

Fig. 16.6 **a** Re Z11 and **b** Im Z11 for transparent MIMO nano-antenna

enhances the antenna's performance, ensuring efficient and focused electromagnetic radiation in applications demanding precision and reliability.

Figure 16.9, displaying the normalized radiation patterns of the transparent MIMO nano-antenna at two distinct angles (a) $\Phi = 0^\circ$ and (b) $\Phi = 90^\circ$, succinctly affirms the antenna's directive characteristics. The principal lobe orientation toward $\pm$ 90° is clearly evident in both patterns, underscoring the antenna's ability to focus electromagnetic radiation in specific directions. This directive behavior is crucial for applications requiring precise signal transmission and reception. The consistent observation of a well-defined principal lobe in both radiation patterns reinforces the antenna's suitability for high-frequency communication within the frequency range under consideration. In practical terms, this directional property enhances the

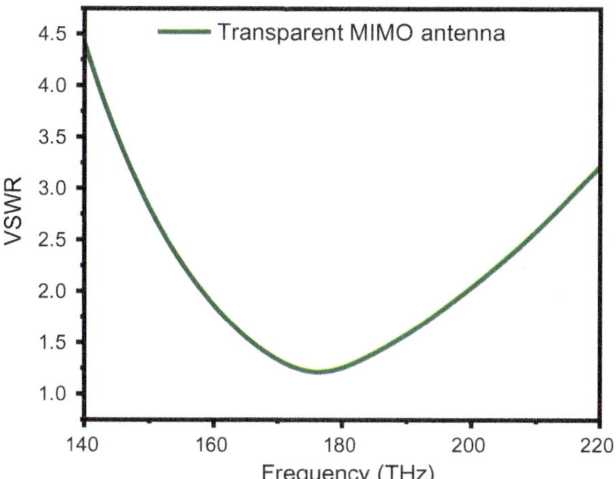

Fig. 16.7 VSWR for the transparent MIMO nano-antenna

antenna's effectiveness in scenarios such as telemedicine and remote surgery, where targeted and reliable signal propagation is paramount.

Figure 16.10 illustrates the radiation efficiency of the transparent MIMO nano-antenna throughout its operating band from 151 to 195 THz. Notably, the antenna exhibits a remarkably high radiation efficiency, consistently surpassing 99.80% across the entire frequency range. This outstanding efficiency underscores the antenna's effectiveness in converting electrical power into radiated energy, minimizing losses and maximizing the overall performance. The consistently high radiation efficiency is a positive indicator of the antenna's efficacy in maintaining signal integrity and optimizing the use of available power resources, making it a promising candidate for applications demanding reliable and efficient high-frequency communication.

In Fig. 16.11, the axial ratio values for the transparent MIMO nano-antenna are presented. Notably, the antenna consistently maintains an axial ratio of less than 3 dB. This observation underscores the antenna's ability to preserve polarization integrity, a critical factor in high-frequency communication. A low axial ratio is indicative of minimal signal distortion and polarization mismatch, emphasizing the antenna's suitability for applications where maintaining signal quality and consistency is paramount.

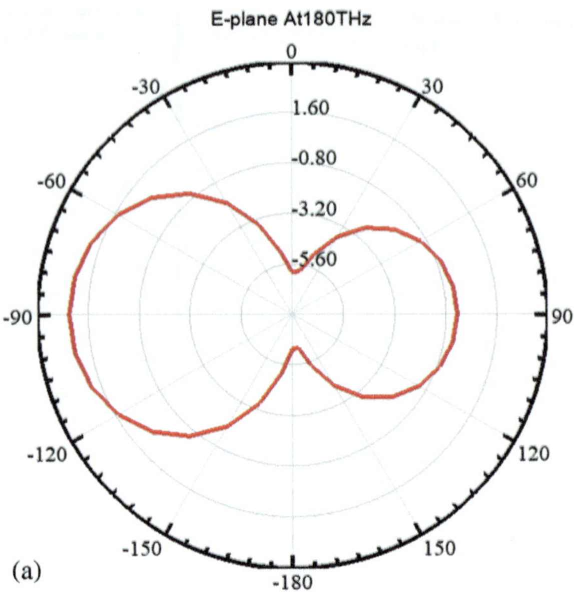

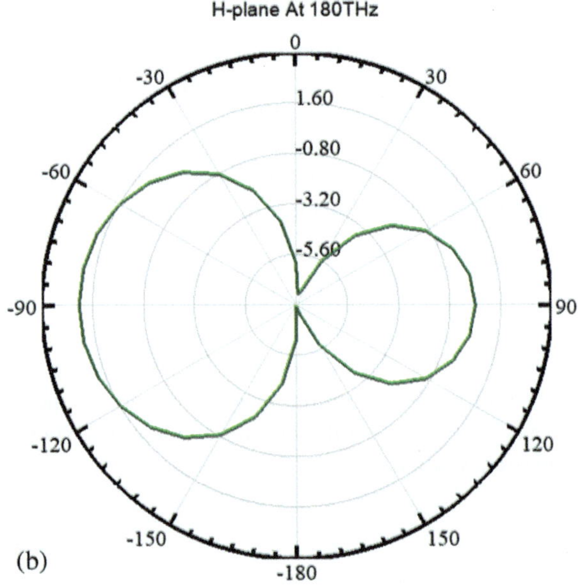

Fig. 16.8 **a** *E*-plane and **b** *H*-plane for the transparent MIMO nano-antenna

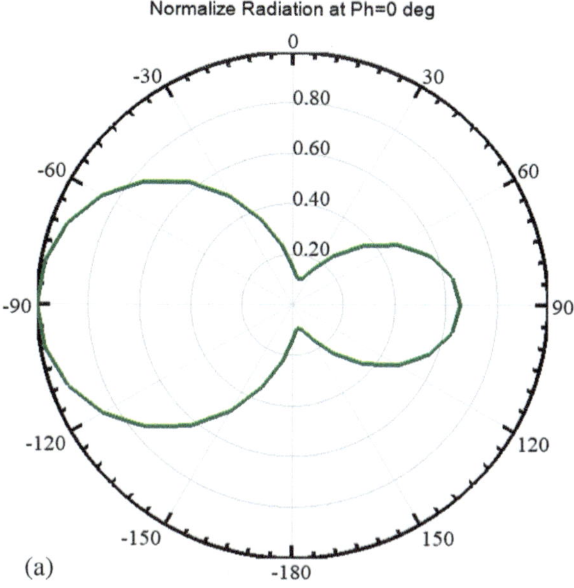

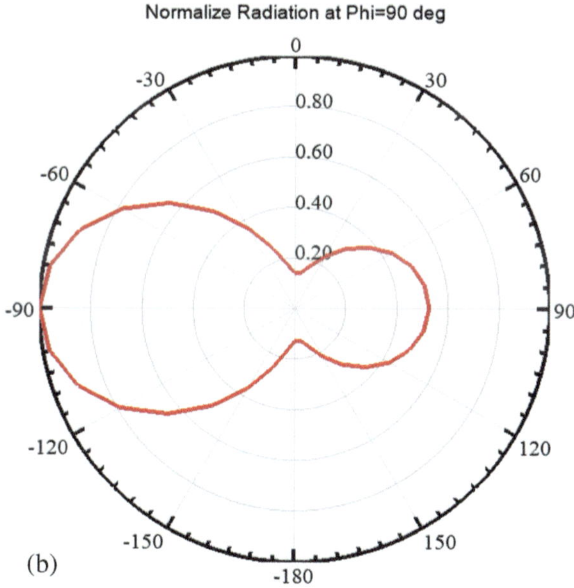

Fig. 16.9 Normalize radiation **a** at $\Phi = 0°$ and **b** at $\Phi = 90°$ for the transparent MIMO nano-antenna

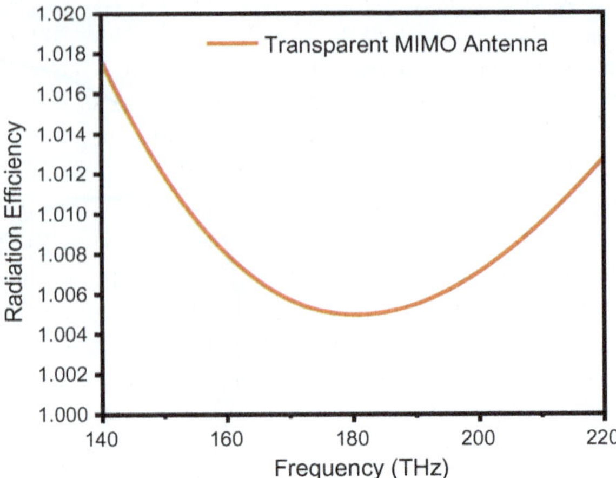

Fig. 16.10 Radiation efficiency for the transparent MIMO nano-antenna

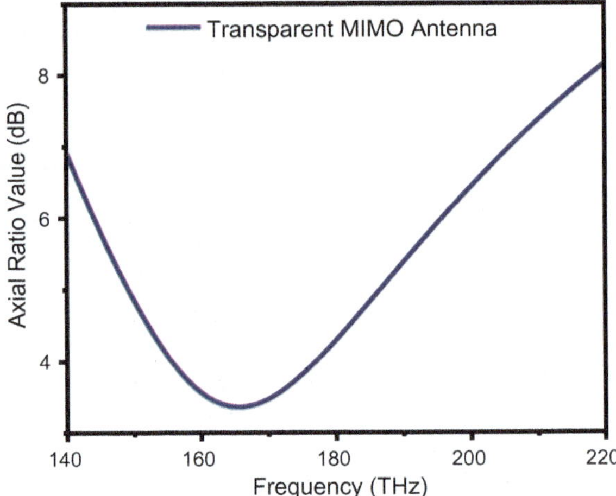

Fig. 16.11 Axial ratio value for the transparent MIMO nano-antenna

16.6 Conclusion

In conclusion, the proposed MIMO nano-antenna, operating at an impressive frequency range of 151–195 THz, demonstrates exceptional performance characteristics. Through extensive analysis, the antenna exhibits remarkable features such as a resonance frequency at 170 THz, a low return loss of – 35 dB, a super wide bandwidth of 44,000 GHz, and a VSWR close to 1.25. The radiation patterns illustrate the antenna's directive nature, with a principal lobe concentrated toward ± 90°. Moreover, the transparency of the antenna is highlighted, making it suitable for applications where optical transparency is crucial. The consistently high radiation efficiency, exceeding 99.80% throughout the operating band, underscores its efficiency in converting electrical power into radiated energy. Additionally, the low axial ratio of less than 3 dB emphasizes the antenna's ability to maintain polarization integrity. Collectively, these findings position the proposed MIMO nano-antenna as a promising candidate for advanced high-frequency communication applications, particularly in fields such as telemedicine and remote surgery, where reliability, efficiency, and precision are paramount.

References

1. Flores, M.A., Glusman, G., Brogaard, K., Price, N.D., Hood, L.: P4 medicine: how systems medicine will transform the healthcare sector and society. Pers. Med. **10**(6), 565–576 (2013). https://doi.org/10.2217/pme.13.57
2. Roco, M. C., Williams, R. S., Alivisatos, P. (eds.).: Nanotechnology Research Directions: IWGN Workshop Report: Vision for Nanotechnology in the Next Decade. Springer Science & Business Media (2000)
3. Wessel, L., Baiyere, A., Ologeanu-Taddei, R., Cha, J., Blegind-Jensen, T.: Unpacking the difference between digital transformation and IT-enabled organizational transformation. J. Assoc. Inf. Syst. **22**(1), 102–129 (2021)
4. Aileni, R.M., Suciu, G., Suciu, V., Pasca, S., Strungaru, R.: Health monitoring using wearable technologies and cognitive radio for IoT. In: Cognitive Radio, Mobile Communications and Wireless Networks, 143–165 (2019)
5. Lamiri, M., El Ghzaoui, M., Aghoutane, B.: Monopole patch antenna to generate and detect THz pulses. In: Advances in Terahertz Technology and Its Applications, 273–291 (2021)
6. Raihani, H., Benbassou, A., El Ghzaoui, M., Belkadid, J.: Performance evaluation of a passive UHF RFID tag antenna using the embedded T-Match structure. In: 2017 International Conference on Wireless Technologies, Embedded and Intelligent Systems (WITS), pp. 1–6. IEEE (2017)
7. Nadeem, A., Hussain, M.A., Owais, O., Salam, A., Iqbal, S., Ahsan, K.: Application specific study, analysis and classification of body area wireless sensor network applications. Comput. Netw. **83**, 363–380 (2015)
8. Aghoutane, B., Faylali, H.E., Das, S.: High gain of a canine MIMO antenna for terahertz applications. In: Terahertz Wireless Communication Components and System Technologies, pp. 139–152. Springer, Singapore (2022)
9. Babu, K.V., Sree, G.N.J., Islam, T., et al.: Performance analysis of a photonic crystals embedded wideband (1.41–3.0 THz) fractal MIMO antenna over SiO_2 substrate for terahertz band applications. Silicon (2023). https://doi.org/10.1007/s12633-023-02622-0
10. Ibili, H., et al.: Modeling plasmonic antennas for the millimeterwave & THz range. IEEE J. Sel. Top. Quant. Electron. **29**(5: Terahertz Photonics), 1–15 (2023). https://doi.org/10.1109/JSTQE.2023.3314696
11. Sun, T., Bai, Z., Li, Z., Liu, Y., Chen, Y., Xiong, F., Chen, L., Xu, Y., Zhang, F., Li, D., Li, J., Wen, L.: Generation of tunable terahertz waves from tailored versatile spintronic meta-antenna arrays. ACS Appl. Mater. Interf. **15**(19), 23888–23898 (2023)
12. Raihani, H., Benbassou, A., El Ghzaoui, M., Belkadid, J.: Gain enhancement of a looped slots patch antenna for passive UHF RFID tags applications in metallic supports. Int. J. Commun. Antenna Propag. IRECAP **9**, 172 (2019)
13. Aghoutane, B., El Ghzaoui, M., Das, S., Ali, W., El Faylali, H.: A dual wideband high gain 2×2 multiple-input-multiple-output monopole antenna with an end-launch connector model for 5G millimeter-wave mobile applications. Int. J. RF Microw. Comput. Aided Eng. **32**(5), e23088 (2022)
14. Singh, R., Varshney, G.: Isolation enhancement technique in a dual-band THz MIMO antenna with single radiator. Opt. Quant. Electron. **55**(6), 539 (2023)

15. Mestoui, J., El Ghzaoui, M.: A survey of NOMA for 5G: implementation schemes and energy efficiency. In: WITS 2020: Proceedings of the 6th International Conference on Wireless Technologies, Embedded, and Intelligent Systems, pp. 949–959. Springer, Singapore (2022)
16. Zhang, B., Jornet, J.M., Akyildiz, I.F., Wu, Z.P.: Mutual coupling reduction for ultra-dense multi-band plasmonic nano-antenna arrays using graphene-based frequency selective surface. IEEE Access **7**, 33214–33225 (2019). https://doi.org/10.1109/ACCESS.2019.2903493
17. Akyildiz, I.F., Jornet, J.M.: Realizing ultra-massive MIMO (1024 × 1024) communication in the (0.06–10) terahertz band. Nano Commun. Netw. **8**, 46–54 (2016). https://doi.org/10.1016/j.nancom.2016.02.001
18. Rubani, Q., Begh, G.R., Gupta, S.H.: Design and investigation of a MIMO nanoantenna operating at optical frequency. Optik **223**, 165481 (2020). https://doi.org/10.1016/j.ijleo.2020.165481
19. Li, S., Ai, G.: The channel modeling of ultra-massive MIMO terahertz-band communications in the presence of mutual coupling. IEICE Trans. Fund. Electron. Commun. Comput. Sci. (2023). https://doi.org/10.1587/transfun.2023eal2046
20. Gotra, S., Pandey, V.S.: Two-port silicon-based MIMO nano-dielectric resonator antenna with polarization diversity for photonics applications. Progr. Electromagn. Res. C **129**, 221–230 (2023). https://doi.org/10.2528/pierc22111103
21. Dash, S., Liaskos, C., Akyildiz, I.F., Pitsillides, A.: Nanoantennas design for THz communication. In: NanoCom'20: Proceedings of the 7th ACM International Conference on Nanoscale Computing and Communication (2020). https://doi.org/10.1145/3411295.3411312
22. Aghoutane, B., Meskini, N., Elghzaoui, M., El Faylali, H.: Millimeter-wave microstrip antenna array design for future 5G cellular applications. In: 2018 International Conference on Electronics, Control, Optimization and Computer Science (ICECOCS), pp. 1–4. IEEE (2018)
23. Khaleel, S.A., Hamad, E.K.I., Parchin, N.O., Saleh, M.A.: MTM-inspired graphene-based THZ MIMO antenna configurations using characteristic mode analysis for 6G/IoT applications. Electronics **11**(14), 2152 (2022). https://doi.org/10.3390/electronics11142152
24. Hmamou, A., Elaage, S., Foshi, J.: Spatial modeling of the transmission channel for THz radio communication system. Progr. Electr. Eng. Appl. Phys. **1**(2), 1–13 (2023). https://doi.org/10.5281/zenodo.8351686
25. Aghoutane, B., Ghzaoui, M.E., Kumari, S.V., et al.: A circularly polarized super wideband transparent optical nanoantenna for advanced THz communication applications. Opt. Quant. Electron. **55**, 211 (2023). https://doi.org/10.1007/s11082-022-04484-z
26. Sharmila, P., Gayathri, A., Priya, M.R., Movithasree, V., Nithya, P.: reconfigurable multiband nano-antenna design for photonic applications. Int. J. Health Sci. (IJHS), 5681–5692 (2022). https://doi.org/10.53730/ijhs.v6ns3.7211
27. Das, S., Lakrit, S., Krishna, C.M., et al.: A novel flower petal-shaped super wideband (439.36–557.59 THz) optical nano-antenna for terahertz (THz) wireless communication applications. Opt. Quant. Electron. **55**, 516 (2023). https://doi.org/10.1007/s11082-023-04802-z
28. Yuan, Y., et al.: A 3D geometry-based THz channel model for 6G ultra massive MIMO systems. IEEE Trans. Veh. Technol. **71**(3), 2251–2266 (2022). https://doi.org/10.1109/TVT.2022.3143500
29. Ezzulddin, S.K., Hasan, S.O., Ameen, M.M.: Performance analysis of plasmonic nano-antenna based on graphene with different dielectric substrate materials for optoelectronics application. Plasmonics (2023). https://doi.org/10.1007/s11468-023-02030-5
30. Ahmad, I., Ullah, S., Din, J.U., Ullah, S., Ullah, W., Habib, U., Khan, S., Anguera, J.: Maple-leaf shaped broadband optical nano-antenna with hybrid plasmonic feed for nano-photonic applications. Appl. Sci. **11**(19), 8893 (2021). https://doi.org/10.3390/app11198893
31. Sugumaran, S., Maurya, N.K., Gudivada, B.: Design and analysis of triband optical nanoantenna. In: 2022 IEEE 2nd Mysore Sub Section International Conference (MysuruCon), Mysuru, India, 2022, pp. 1–4. https://doi.org/10.1109/MysuruCon55714.2022.9972578
32. Chopra, K., Misra, S., Gupta, S.H., Rajawat, A.: Design and optimization of multiarray antenna operating in terahertz (THz) band for in-vivo nanonetworks. Optik **265**, 169475 (2022). https://doi.org/10.1016/j.ijleo.2022.169475

33. El Ghzaoui, M., Hmamou, A., Foshi, J., Mestoui, J.: Compensation of non-linear distortion effects in MIMO-OFDM systems using constant envelope OFDM for 5G applications. J. Circ. Syst. Comput. (2020). https://doi.org/10.1142/S0218126620502576
34. Ghazaoui, Y., El Alami, A., El Ghzaoui, M., Das, S., Baradand, D., Mohapatra, S.: Millimeter wave antenna with enhanced bandwidth for 5G wireless application. J. Instrum. **15** (2020). https://doi.org/10.1088/1748-0221/15/01/T01003

Chapter 17
Design of a Broadband (0.84–1.2 THz) Microstrip Patch Antenna Utilizing Graphene Material and Polyimide Substrate for Terahertz 6G Applications

Abdelaaziz El Ansari, Tanvir Islam, Sudipta Das, Najiba El Amrani El Idrissi, Hassan Yakkou, and Shobhit K. Khandare

17.1 Introduction

Over the last ten years, there has been a significant transformation in wireless communication systems. This transformation necessitates the development of high-speed communication systems to establish faster connections. However, the currently available microwave spectrum bands are fully assigned and being used by numerous wireless communication applications. This constraint has prompted researchers worldwide to explore alternative unassigned frequency bands. As a result, the accessible spectrum and the possibility of high-speed communication ranging up to 10 Gbits per second [1, 2] make the terahertz frequency range a viable choice for supporting a

A. El Ansari (✉) · N. E. A. El Idrissi
Signal, System and Component Laboratory, Sidi Mohamed Ben Abdellah University—FST Fez, Fes, Morocco
e-mail: abdelaaziz.elansari1@usmba.ac.ma

T. Islam
Department of Electrical and Computer Engineering, University of Houston, Houston, TX 77204, USA

S. Das
Department of Electronics and Communication Engineering, IMPS College of Engineering and Technology, Malda, India

H. Yakkou
Engineering Sciences Laboratory, Sidi Mohamed Ben Abdellah University—FP Taza, Fes, Morocco

S. K. Khandare
Vivekanand Education Society's Institute of Technology (VESIT), Mumbai, India

M. El Ghzaoui et al. (eds.), *Next Generation Wireless Communication*, Signals and Communication Technology, https://doi.org/10.1007/978-3-031-56144-3_17

variety of wireless applications. The frequencies of the terahertz band, which range from 0.1 to 10 THz, are between those of the microwave and optical bands [3].

The carbon crystallizes in many allotropic varieties such as diamond, C60 fullerene, and graphite. The graphite exhibits dual conductivity: weak vertical conductivity and high vertical horizontal conductivity. In 2010, a single layer of graphite atoms was successfully isolated and named graphene. Then, a ground-breaking experiment with graphene was realized and has shown that the graphene has remarkable properties such as perfect conductivity and exceptional strength. These remarkable properties have motivated many scientists and researchers worldwide to consider its application in some devices including filters, amplifiers, and microstrip antennas in THz band [4–10]. Furthermore, graphene is used in the THz band for a variety of technical applications, particularly wireless communication, due to its exceptional optical, mechanical, and electrical capabilities [11–15].

The utilization of graphene in wireless communication is primarily seen in the widespread adoption of graphene-based power dividers, filters, modulators, resonators, absorbers, waveguides, and phase shifters. Additionally, graphene-based antennas offer intriguing possibilities within the terahertz frequency range [9–22].

As the antenna holds a vital role in all communication systems, the absence of it renders any data exchange impossible [5, 6]. The microstrip antenna is the most popular antenna and has been used in the literature thanks to its advantages such as low cost, low weight, and ease of integration with printed circuit board (PCB) [7, 23–26]. Nevertheless, the patch antenna has a few shortcomings, including a small bandwidth, poor directivity, and low gain. In this work, we are going to challenge all these problems of narrow bandwidth using graphene as conductor for the patch and the gold as conductor for the ground plane in the terahertz band.

Therefore, in this paper, we will concentrate on the analysis and improvement of a patch antenna's performance in the THz band. After this introduction, this article will be outlined as follows: Section shows the presentation of graphene. The design of the suggested patch antenna-based graphene is covered in Sect. 17.3. Section 17.4 offers an analysis of comparisons. The simulation results and discussion are covered in Sect. 17.5. This article will be concluded in Sect. 17.6.

17.2 Presentation of Graphene

Carbon has four valance electrons and exhibits several allotropes, which means it can exist in various structural forms at the atomic level. The most well-known allotropes of carbon are:

i. **Fullerenes**: Fullerenes are carbon molecules that form spherical or ellipsoidal structures composed of hexagonal and pentagonal cages (see Fig. 17.1a). The most well-known fullerene is C60. Despite fullerenes are made of carbon atoms, they are generally considered to be poor conductors of electricity. They typically do not exhibit the same level of electrical conductivity as metals.

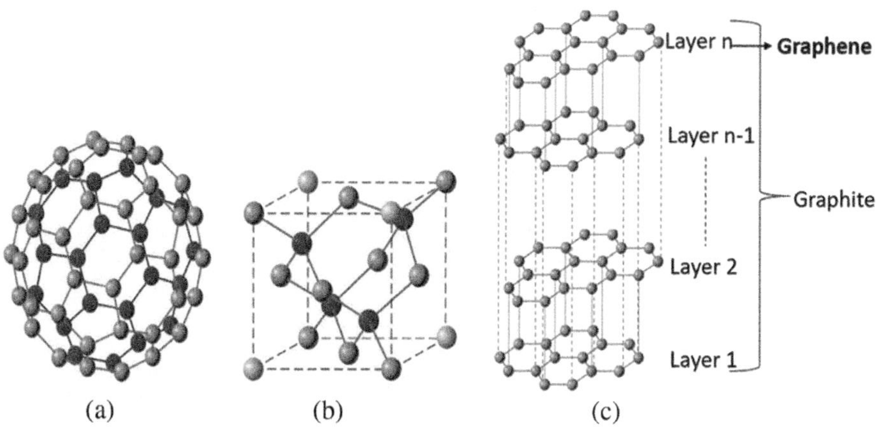

Fig. 17.1 Allotropic varieties of carbon **a** fullerene structure, **b** diamond structure, **c** graphite and graphene structure

ii. **Diamond**: Diamond is a crystalline form of carbon in which each carbon atom is bonded to four other carbon atoms in a three-dimensional tetrahedral structure. It is an extremely hard and transparent substance. Although the diamond is made of carbon atoms, its specific arrangement and bonding structure make it an electrical insulator rather than a conductor.

iii. **Graphite**: Graphite is a form of carbon where carbon atoms are arranged in flat hexagonal layers called graphene sheets. Van der Waals forces hold these layers together as they are piled on top of one another. Dual conductivity in graphite is demonstrated by its high vertical horizontal conductivity and weak vertical conductivity. Thus, it is a good electrical conductor.

iv. **Graphene**: A single layer of carbon graphite is known as graphene. It is a two-dimensional material possessing remarkable mechanical and electrical qualities. This carbon allotrope has since captured the attention of scientists and engineers due to its extraordinary properties. Graphene is renowned for its exceptional conductivity, mechanical strength, and flexibility, making it a promising candidate for a wide range of applications [20]. Its high electron mobility and thermal conductivity have spurred interest in electronic and optoelectronic devices, with potential applications in fields such as transparent conductive films, sensors, and even advanced materials for next-generation batteries. Furthermore, graphene's robustness and lightweight make it an ideal material for enhancing the mechanical properties of composites [22]. As researchers delve deeper into the possibilities of graphene, its unique characteristics continue to open new avenues for innovation across various industries, heralding a future where this nanomaterial plays a pivotal role in shaping technological advancements.

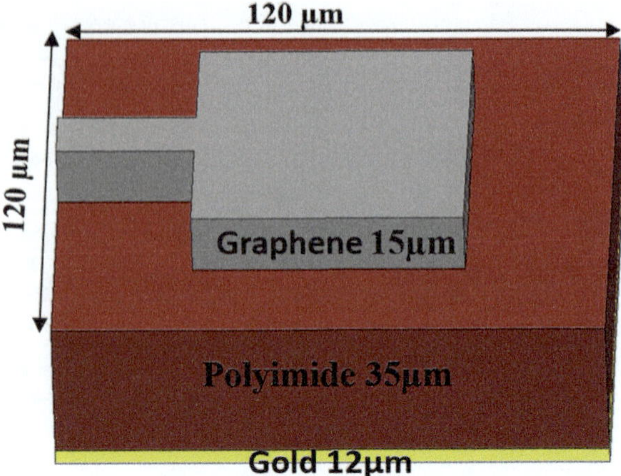

Fig. 17.2 Proposed graphene-based microstrip antenna

17.3 Design of the Proposed Graphene Patch Antenna

The structure of proposed microstrip antenna-based graphene is shown in Fig. 17.2. It contains a radiating element having a rectangular shape and feeding by a 50 Ω microstrip straight line. The radiation element and the feed line are made of graphene material with a high of 15 µm. The ground plan is made of gold with a high equal to 12 µm. The ground plan and the radiating element are printed on the substrate polyimide having a relative primitivity of 4.3, dielectric loss tangent of 0.004, and its high equals to 35 µm.

At first, the rectangular patch antenna was designed using the analytical equations existing in reference [6]. Next, a parametric sweep was done on the deep of the microstrip patch in order to obtain an optimal high. As results obtained the optimal high reached is 15 µm. Then, a parametric study was done on the deep of the ground plan in order to obtain an optimal high. As results obtained the optimal high of the ground plan is 12 µm.

17.4 Simulation Results and Discussion

The proposed antenna was designed and simulated using HFFS electromagnetic simulator. Its performance has been investigated in terms of reflection coefficient S_{11}, gain, directivity, co-polarization, and cross-polarization in the planes E and H, 3D radiation patterns, and surface current distribution as depicted in Figs. 17.3, 17.4, 17.5, 17.6, 17.7, 17.8 and 17.9. The return loss in dB versus frequency in terahertz is illustrated in Fig. 17.3. The suggested antenna has a wide 10 dB impedance bandwidth of 280 GHz spacing from 0.84 to 1.2 THz, as demonstrated by the return loss. Furthermore, the resonance frequency is 0.98 THz, and its minimum value is equal to

− 30 dB. This indicates that nearly all of the input power was transformed into electromagnetic waves via the patch antenna. The gain is one of the important radiating parameters that characterize. When the gain tends toward infinity, the antenna has a significant range in transmission and reception. Figure 17.4 plots the antenna gain in dB versus frequency in terahertz. According to this figure, it can be seen that all over the impedance bandwidth (from 0.84 to 1.2 THz), the grain is greater than 5 dB which means that the proposed antenna has a good gain. Moreover, it is clear that the highest gain is 7.5 dB at the resonance frequency 0.98 THz. Also, directivity is an important parameter to characterize the radiation pattern of an antenna. When the directivity is high, the antenna is more directive and concentrates energy in a specific direction. Figure 17.5 plots the directivity in dB versus frequency. According to this figure, it is clear that all over the impedance bandwidth (from 0.84 to 1.2 THz), the directivity is greater than 6 dB which means that the proposed is directional. Furthermore, the peak directivity is 7 dB at the resonance frequency of 0.98 THz. Figure 17.6 exhibits co-polarization versus cross-polarization in the E plane of the suggested antenna at the resonance frequency 0.98 THz. As depicted in this diagram, it can be seen that in the E plane (phi = 0°) when the colorization is maximal the cross-polarization is minimal. Figure 17.7 indicates co-polarization versus cross-polarization in the H plane of the suggested antenna at the resonance frequency 0.98 THz. Based on what is shown in this graph, it is clear that in the H plane (phi = 90°), both co-polarization and cross-polarization are almost overlapped which means that the proposed antenna has a linear polarization. Figure 17.8 shows the three-dimensional radiation pattern at the resonance frequency 0.98 THz of the proposed antenna. In accordance with this illustration, it is clear that the radiation patterns resemble almost omnidirectional in shape. Moreover, radiation patterns in terms of gain 7.5 dB and the maximum beam are directed to the ax normal to the antenna, and thus the proposed antenna has a directional radiation pattern. Further, the proposed antenna is evaluated using the surface current distribution at the resonant frequency 0.98 THz as shown in Fig. 17.9. As depicted in this diagram, it can be seen that the current density is totally distributed in whole surface of the patch with good values of Jsurf.

17.5 Comparison Analysis

Table 17.1 compares the performance study of the proposed rectangular antenna-based graphene with other published works in the terahertz range. This table makes it evident that the suggested antenna has the maximum gain and the widest impedance bandwidth.

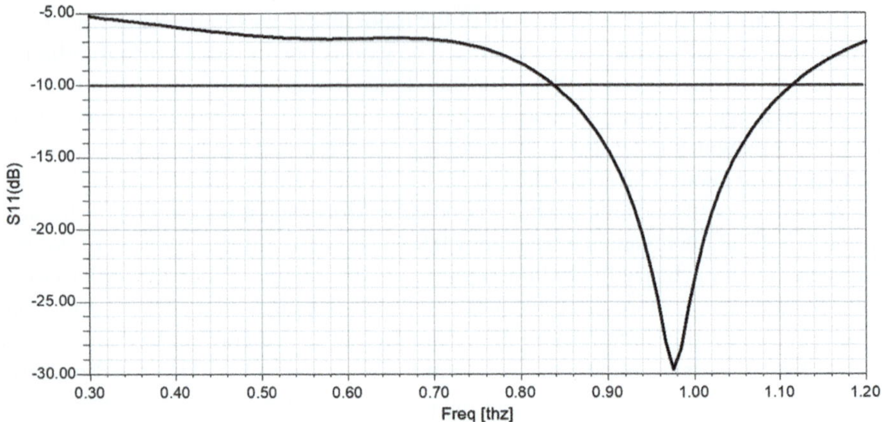

Fig.17.3 Reflection coefficient S_{11} in dB versus frequency in THz

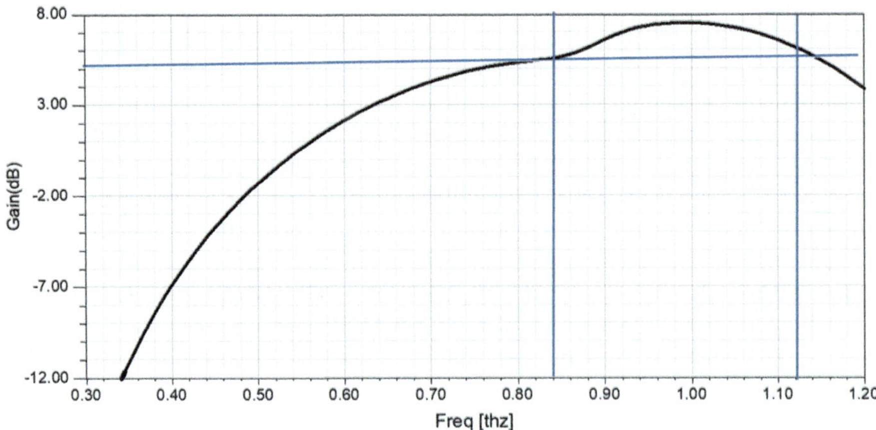

Fig. 17.4 Gain in dB versus frequency in THz

17.6 Conclusion

Within this paper, we have suggested a broadband graphene-based microstrip antenna base for THz 6G utilizations. The proposed antenna was designed using HFSS software. The obtained results have shown that it operates within a broadband 10 dB impedance bandwidth of band spacing from 0.84 to 1.12 THz. The offered antenna is well matched with a minimal insertion loss and has an optimal gain of 7.41 dB, optimal directivity of 7.48 dB, and a high efficiency of 98.95%. Furthermore, it offers directional radiation patterns. As a result, this antenna is highly suited for upcoming

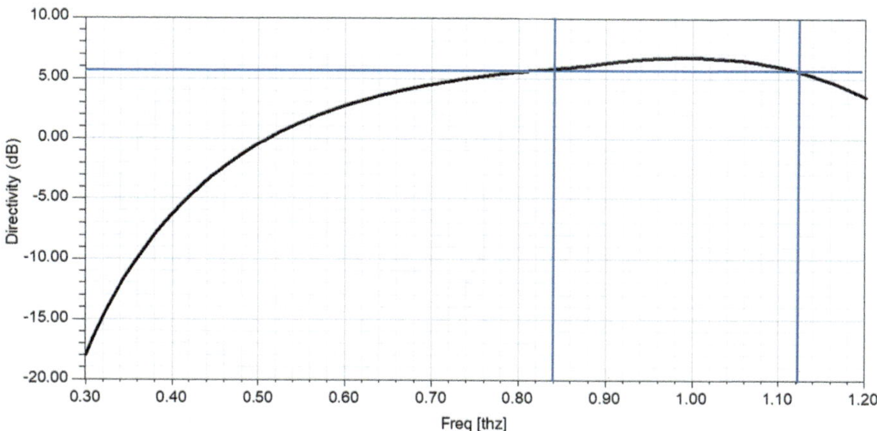

Fig. 17.5 Directly in dB versus frequency in THz

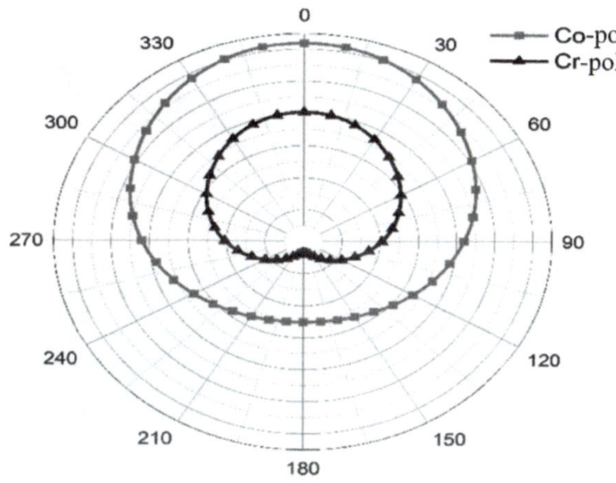

Fig. 17.6 Co-polarization versus cross-polarization in the E plane of the suggested antenna

terahertz band applications, particularly those involving high-speed wireless communication, medical imaging, material characterization in the THz band, and explosive and arm detection.

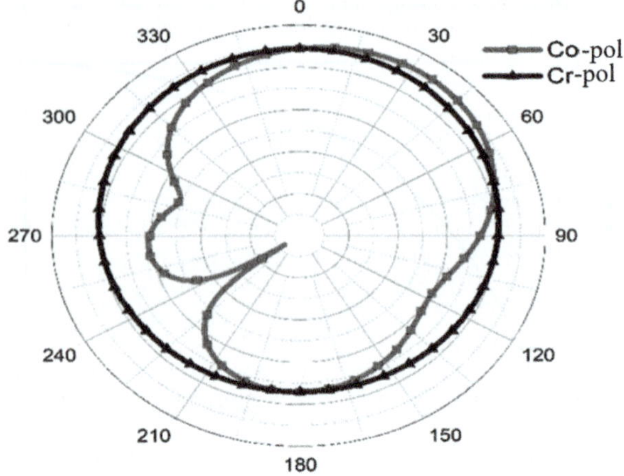

Fig. 17.7 Co-polarization versus cross-polarization in the H plane of the suggested antenna

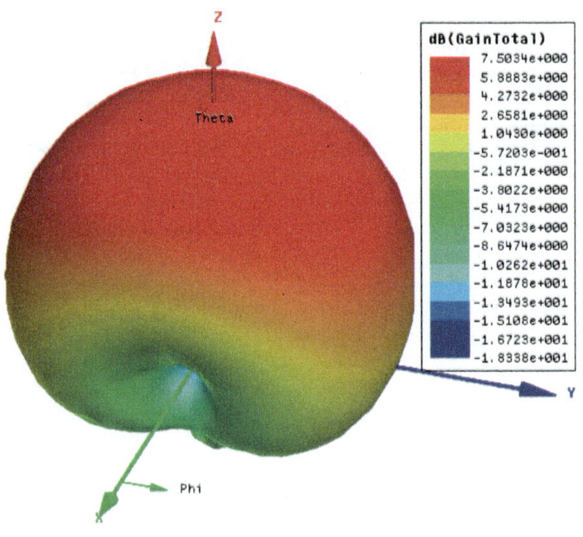

Fig. 17.8 3D radiation pattern

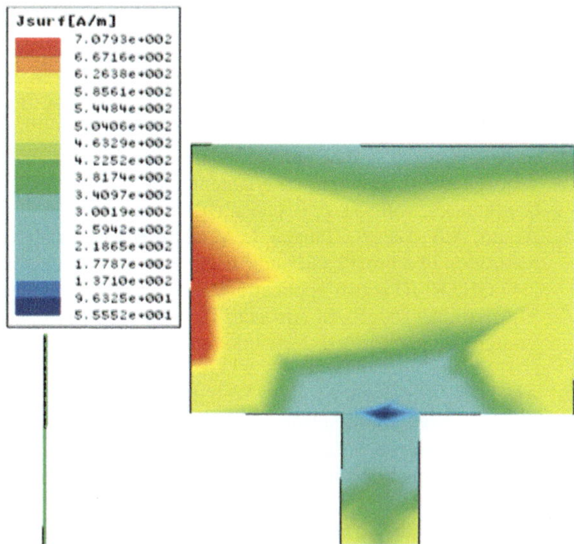

Fig. 17.9 Surface current distribution

Table 17.1 Comparative analysis with other relevant array designs

References	Impedance bandwidth (GHz)	Spacing frequency (THz)	Peak gain (dB)
[8]	50	0.725–0.775	5.71
[9]	10	0.296–0.306	4.39
[10]	20	0.73–0.75	6.395
[11]	6	0.849–0.855	2.5
[12]	20	0.68–0.70	6.793
[13]	26	0.688–0.714	5.234
[14]	58	0.582–0.64	–
[15]	88	1.053–1.141	3.56
[16]	40	0.65–0.69	5.22
[17]	17	0.303–0.320	4.4
[18]	67	[4.062–4.129]	3.81
[19]	60	–	5
[20]	269	[0.445–0.714]	5.7
Proposed work	280	[0.84–1.120]	7.5

References

1. Kushwaha, R.K., Karuppanan, P., Malviya, L.D.: Design and analysis of novel microstrip patch antenna on photonic crystal in THz. Phys. B: Conden. Matter **545**, 107–112 (2018)
2. Choudhury, B., Danana, B., Jha, R.M., et al.: PBG Based Terahertz Antenna for Aerospace Applications. Springer, Singapore (2016)

3. Ansari, A.E., Jayaprakasan, V., Duraisamy, K., Das, S., El-Arrouch, T., El Idrissi, N.E.A.: A wideband microstrip 1 × 2 array antenna fed by coupler for beam steering terahertz (THz) band applications. J. Nano-Electron. Phys. **15**(3), 03028 (2023)
4. Shamim, S.M., Uddin, M.S., Hasan, M.R., Samad, M.: J. Comput. Electron. **20**(1), 604–610 (2021)
5. El Ansari, A., Das, S., El Idrissi, N.E.A., El-Arrouch, T., Bendali, A.: Slot incorporated high gain printed RFID reader array antenna for 2.4 GHz ISM band applications. In: E3S Web of Conferences, vol. 351, p. 01056. EDP Sciences (2022)
6. Ansari, A.E., Das, S., Tabakh, I., Madhav, B.T.P., Bendali, A., El Idrissi, N.E.A.: Design and realization of a broadband multi-beam 1 × 2 array antenna based on 2 × 2 butler matrix for 2.45 GHz RFID reader applications. J. Circ. Syst. Comput. **31**(17), 2250305 (2022)
7. El Ansari, A., Das, S., El-Arrouch, T., El Idrissi, N.E.A.: Proceedings—2022 9th International Conference on Wireless Networks and Mobile Communication (WINCOM 2022), pp. 4–9 (2022)
8. Anand, S., Sriram Kumara, D., Wub, R.J., Chavali, M.: Graphene nanoribbon-based terahertz antenna on polyimide substrate. Optik **125**, 5546–5549 (2014)
9. Jha, K.R., Sharma, S.K.: Waveguide integrated microstrip patch antenna at THz frequency. In: 2014 IEEE Antennas and Propagation Society International Symposium (APSURSI), pp. 1851–1852 (2014)
10. Kushwaha, R.K., Karuppanan, P.: Investigation and design of microstrip patch antenna employed on PCs substrates in THz regime. Aust. J. Electr. Electron. Eng. **18**(2), 118–125 (2021)
11. Rubani, Q., Gupta, S.H., Pani, S., Kumar, A.: Design and analysis of a terahertz antenna for wireless body area networks. Optik **179**, 684–690 (2019)
12. Ullah, S., Ruan, C., Ul Haq, T., Zhang, X.: High performance THz patch antenna using photonic band gap and defected ground structure. J. Electromagn. Waves Appl. **33**(15), 1943-1954 (2019). https://doi.org/10.1080/09205071.2019.1654929
13. Paul, L.C., Islam, M.: Proposal of wide bandwidth and very miniaturized having dimension of μm range slotted patch THz microstrip antenna using PBG substrate and DGS. In: 20th International Conference on Computing and Information Technology (ICCIT), pp. 1–6 (2017)
14. Azarbar, A., Masouleh, M.S., Behbahani, A.K.: A new terahertz microstrip rectangular patch array antenna. Int. J. Electromagn. Appl. **4**, 25–29 (2014)
15. Sirmaci, Y.D., Akin, C.K., Sabah, C.: Fishnet based metamaterial loaded THz patch antenna. Opt. Quantum Electron. **48**(2), 168 (2016)
16. Dhillon, A.S., Mittal, D., Sidhu, E.: THz rectangular microstrip patch antenna employing polyimide substrate for video rate imaging and homeland defence applications. Optik **144**, 634–641 (2017)
17. Zhu, H., Li, X., Qi, Z., Xiao, J.: A 320 GHz octagonal shorted annular ring on-chip antenna array. IEEE Access **8**, 84282–84289 (2020). https://doi.org/10.1109/ACCESS.2020.2991868
18. Varshney, G.: Silicon **13**(6), 1907–1915 (2021)
19. Vijayalakshmi, K., Selvi, C.S.K., Sapna, B.: Opt. Quantum Electron. **53**(7), 1–13 (2021)
20. Krishna, C.M., Das, S., Nella, A., Lakrit, S., Madhav, B.T.P.: Plasmonics **16**(6), 2167–2177 (2021)
21. Khan, M.A.K., Ullah, M.I., Kabir, R., Alim, M.A.: Plasmonics **15**(6), 1719–1727 (2020)
22. Thakur, E., Jaglan, N., Gupta, S.D.: Wirel. Pers. Commun. **123**(1), 407–420 (2022)
23. El Ansari, A., Kabouri, L., Ahouzi, E.: Random attack on asymmetric cryptosystem based on phase-truncated Fourier transforms. In: 2014 International Conference on Next Generation Networks and Services (NGNS), pp. 65–68. IEEE (2014)

24. El Arrouch, T., El Idrissi, N.E.A., El Ansari, A.: Microstrip patch antenna using a parasitic mushroom for 5G application at 28 GHz. In: 2022 9th International Conference on Wireless Networks and Mobile Communications (WINCOM), pp. 1–6. IEEE (2022, Oct)
25. El Ansari, A., Das, S., El-Arrouch, T., El Idrissi, N.E.A.: A hybrid coupler integrated 1 × 4 printed array antenna with broadband and high performance for beamforming RFID reader. In: 2022 9th International Conference on Wireless Networks and Mobile Communications (WINCOM), pp. 1–6. IEEE (2022, Oct)
26. Ansari, A., Islam, T., Rama Rao, S.V., Saravanan, A., Das, S., El Idrissi, N.E.A.: A broadband microstrip 1 × 8 magic-T power divider for ISM band array antenna applications. J. Nano-Electron. Phys. **15**(3), 03003 (2023)

Chapter 18
Investigation on LP-OFDM for THz Application

Serghini Elaage, Mohammed El Ghzaoui, Nabil Mrani, and El Alami Ali

18.1 Introduction

The most relevant applications of terahertz (THz) frequencies are high-speed wireless telecommunications [1, 2]. Free space telecommunications currently use radio frequencies and the increasing need for throughput requires the use of wider bandwidths [3–7]. The transfer to higher frequencies would make it possible to achieve higher bit rates without complicating the modulation used. In addition, since the bands above 275 GHz are not allocated, large frequency bands are available. Concepts of new applications have multiplied in recent years. However, the lack of powerful, compact and reliable sources remains an obstacle to the deployment of this technology, which remains mostly confined to research laboratories [8–11]. This lack of sources is due to their power which decreases beyond 100 GHz. Indeed, the increase in the operating frequencies of electronic sources comes up against the critical influence of their passive elements (*R*, *L*, *C*). On the other hand, the increase in the emission wavelengths of optical sources comes up against strong attenuations of the emitted powers [11–14].

Fiber optic networks exhibit very low attenuation with distance. The current limit in frequency and power of these networks is dependent on the optoelectronic conversion of the transmitted signal. This conversion to terahertz frequencies [15], carried out by photomixing, is a possible solution to increase throughputs without requiring a complete change of current infrastructures.

In this chapter, we propose a new waveform for the THz band called LP-OFDM [16–18]. It is acquired by adding to the OFDM system a linear precoding function advantageously combining OFDM and the spread spectrum technique. In fact, linear

S. Elaage (✉) · N. Mrani · El Alami Ali
Moulay Ismail University, Meknes, Morocco
e-mail: s.elaage@edu.umi.ac.ma

M. El Ghzaoui
Sidi Mohamed Ben Abdellah University, Fez, Morocco

M. El Ghzaoui et al. (eds.), *Next Generation Wireless Communication*, Signals and Communication Technology, https://doi.org/10.1007/978-3-031-56144-3_18

precoding techniques (LPT) allow better use of the frequency diversity of the channel. It also offers greater granularity in the choice of flow rates, thus increasing the flexibility of the system. It is significant to emphasize that the addition of the linear precoding function comes with a minimal increase in the complexity of the system. We will then focus on the linear precoding technique on OFDM modulation (LP-OFDM) and the additional advantages presented by this LP-OFDM technique for our study in order to understand the interest of having recourse to this additional operation.

Based on the LPT, which, applied to OFDM systems, has brought a significant gain in throughput on the power lines. Within the framework of OFDM multicast systems, simulations carried out on THz channel show a significant gain contribution compared to the classic multicast method. The proposed scheme provides a better number of bits for each user and the gain offered in terms of number of bits increases significantly as the number of users increases from two to twelve users. The LP-LCG method offers the best performance.

18.2 The LP-OFDM System

18.2.1 Description of the LP-OFDM

The objective of this section is to present the LP-OFDM technique used in this study. The system, developed from the MB-OFDM system to which we have added the functions of linear precoding on transmission and linear precoding on reception, is shown in Fig. 18.1.

We will begin this description by introducing the expressions of the LP-OFDM signals. We will then focus on the new functions by specifying the modifications induced by the use of linear precoding concerning the reception techniques used as well as the system parameters.

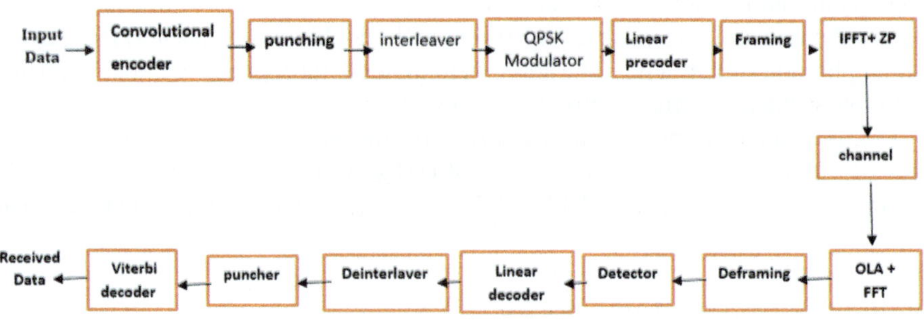

Fig. 18.1 Synoptic of the LP-OFDM communication chain studied

18.2.2 LP-OFDM Signal Expressions

Expression of the Transmitted Signal

We consider the emission of a frame of N_S successive LP-OFDM symbols. Assuming that the prefix cyclic is greater than the channel delay, the OLA operation restores orthogonality between the sub-carriers. Remember that the LP operation is performed before OFDM technique [19, 20]. The general equation making it possible to express the LP-OFDM signal at the output of the OFDM modulator can then be written [21]:

$$\underset{N\times N_S}{S} = \underset{N\times N}{F_N^H} \; \underset{N\times L}{D} \; \underset{L\times N_L}{C} \; \underset{N_L\times N_S}{X}, \tag{18.1}$$

where S is the symbols sent, each symbol composed of N samples in the time domain. C is the LP matrix applied to X which precodes the N_L complex symbols from the N_S LP-OFDM symbols to be transmitted on the L sub-carriers. Finally, D denotes the distribution matrix utilized to distribute the data on the frequency grid, so it is the matrix that defines the chip mapping. It is about a permutation matrix, that is to say that an element $d_{i,j}$ of this matrix is worth 1 if the element of the jth column of the matrix resulting from the product CX must be emitted on the ith sub-carrier. The distribution of the data can respond to a procedure of random distribution of the data such as the introduction of a frequency interleaving of the precoded data or else to a distribution policy adapted to the channel. Furthermore, to allow the insertion of the pilot, guard and/or zero sub-carriers after linear precoding, the matrix D includes lines of zeros at the locations provided for these sub-carriers.

The expression (18.1) of the matrix of symbols at the output of the OFDM transmitter applied to each sub-band can be reformulated and becomes:

$$S = F_N^H.D.\begin{bmatrix} C_1 & & & & 0 \\ & \ddots & & & \\ & & C_b & & \\ & & & \ddots & \\ 0 & & & & C_B \end{bmatrix} \begin{bmatrix} X_1 \\ \vdots \\ X_b \\ \vdots \\ X_B \end{bmatrix}. \tag{18.2}$$

Received Signal Expression

The LP-OFDM configuration studied combines OFDM and the precoding function by placing the chips of the symbols spread over the locations of an OFDM frame, e.g., on a sub-carrier k of an OFDM symbol i. It is therefore essentially the structure of the symbols transmitted on these locations which differentiates an LP-OFDM system, and more generally an MC-SS system, from an OFDM system. In the first case, it is the product of a data symbol and a spread symbol chip and in the second of a simple data symbol. The mathematical representation of the coefficient of the propagation channel on each location is therefore identical in both cases, and each symbol transmitted on a given location is received after OFDM demodulation.

The linear deprecoding operation which follows that of OFDM demodulation is carried out quite simply by applying the same precoding matrix C as on transmission (cf. Eq. 18.2). It makes it possible to recover the data symbols of an LP-OFDM symbol from the L chips of this same symbol which were transmitted on the locations of the OFDM frame defined by the chip mapping operation. The characteristics of L distortion coefficients $H_{k,i}$ which multiply the L chips of a spread symbol at the input of the precoding function will be linked to the choice of chip mapping. We will denote h_l ($l \in [1 \ldots L]$) the coefficient of channel $H_{k,i}$ which multiplies the l^{th}chip.

The LP-OFDM technique makes it possible to transmit on each spread symbol the data $x_{n,j}$ of a single user j ($n \in [1, \ldots, N_L]$). After the OFDM demodulation FFT operation and the linear precoding operation, we obtain the following expression for the estimate $\widehat{y}_{n,j}$ of the symbol $x_{n,j}$:

$$\widehat{y}_{n,j} = \underbrace{\sum_{l=1}^{L} C_{l.n}^{2}\, x_{n.j} h_l}_{\text{signal util}} + \underbrace{\sum_{\substack{l=1 \\ p \neq n}}^{N_L} x_{p.j} \sum_{l=1}^{L} C_{l.n}\, C_{l.p} h_l}_{\text{SI}} + \underbrace{\sum_{l=1}^{L} C_{l.n}\, n_{ll}}_{\text{AWGN}} . \tag{18.3}$$

18.2.3 The Choice of the Linear Precoding Matrix

Among the many families of possible spreading codes for the linear precoding matrix [22], we find the Walsh–Hadamard codes. These codes are produced by the Sylvester–Hadamard transformation matrix. These codes correspond exactly to the orthogonal rows or columns of this matrix whose coefficients are ± 1. The orthogonality property of this transformation matrix H_{THE} of dimensions $L \times L$ can be written in the following form:

$$H_L\, H_L^{\mathrm{T}} = L\, I_L, \tag{18.4}$$

where $H_L{}^{\mathrm{T}}$ is the transposed matrix of the Sylvester–Hadamard matrix and I_{THE} the identity matrix of dimensions $L \times L$. This definition shows that the rows or columns are mutually orthogonal. The fact of interchanging the rows or the columns does not therefore in any way affect the properties of such a matrix.

18.2.4 Receiving LP-OFDM Signals

Equalization of Received Signals

All the chips transmitted simultaneously on the same sub-carrier undergo the same distortions h_{the} introduced by the channel (Eq. 18.3). The equalization operation with a single coefficient g_{the} per sub-carrier therefore makes it possible to restore the data transmitted $x_{n,j}$ and to restore the orthogonality by making the term of SI tend toward 0. This is then referred to as single user detection given that the detector only requires knowledge of the spreading sequences of a single user whose data is to be restored. The term single user defines the fact that the code chips of a given sub-carrier have all undergone the same distortions in the channel. The expression of the signal obtained after equalization is:

$$\widehat{y}_{n,j} = x_{n.j} \sum_{l=1}^{L} C_{l.n}^{2} h_l\, g_l + \sum_{\substack{l=1 \\ p\neq n}}^{N_L} x_{p.j} \sum_{l=1}^{L} C_{l.n}\, C_{l.p} h_l\, g_l + \sum_{l=1}^{L} C_{l.n}\, n_l\, g_l. \tag{18.5}$$

This technique is considered to be optimal with respect to additive noise when the same information is transmitted simultaneously on two diversity branches. Thus in the absence of SI, that is to say when $N_L = 1$, the performance of this technique is the best in terms of BER since the processing of diversity is optimal. However, in the presence of interference, when $N_L > 1$, the performance decreases rapidly. In addition, the multiplication of the symbols received by $h_{\text{the}}{}^*$ has the effect of increasing the SI term and thus severely deteriorating the performance of the LP-OFDM system in reference to BER.

Calculation of Confidence Values

The expression of the received, despread and equalized signal in Eq. (18.5) is:

$$\widehat{y}_{n,j} = \overbrace{\underbrace{x_{n.j} \sum_{l=1}^{L} C_{l.n}^{2} h_l\, g_l}_{\vartheta_{n,j}}}^{\text{signal util}} + \overbrace{\sum_{\substack{l=1 \\ p\neq n}}^{N_L} x_{p.j} \underbrace{\sum_{l=1}^{L} C_{l.n}\, C_{l.p} h_l\, g_l}_{\varsigma_{p,n}}}^{\text{SI}} + \overbrace{\underbrace{\sum_{l=1}^{L} C_{l.n}\, n_l\, g_l}_{\eta_{n,j}}}^{\text{AWGN}}. \tag{18.6}$$

Complex equalized values $\widehat{y}_{n,j}$ are demodulated via the symbol-to-binary encoder in order to obtain the soft and real values, $\widehat{Z}_{n,j} = [\widehat{Z}^{1}_{n,j}, \ldots, \widehat{Z}^{\upsilon}_{n,j}, \ldots, \widehat{Z}^{m}_{n,j}]$ relating to the bits transmitted and equal to:

$$\widehat{Z}^{\upsilon}_{n,j} = x^{\upsilon}_{n,j}\upsilon'_{n,j} + \sum_{\substack{p=1\\ p\neq n}}^{N_L} x^{\upsilon}_{p,j}\varsigma'_{p,j} + \eta'_{n,j}. \tag{18.7}$$

The real terms $\vartheta'_{n,j}$, $\varsigma'_{n,j}$ and $\eta'_{n,j}$ correspond respectively to complex terms $\vartheta_{n,j}$, $\varsigma_{n,j}$ and $\eta_{n,j}$ after demapping. As for the MB-OFDM system, it is necessary to make the vector reliable. $\widehat{Z}^{\upsilon}_{n,j}$ to ensure effective channel decoding. The real SI term of variance σ^2_{SI} can be considered as an additive Gaussian noise with zero mean in the same way as the real noise term σ^2_{bruit}. Thus, with the definition of $\widehat{Z}^{\upsilon}_{n,j}$ of expression (*A*), we obtain the expression of the log likelihood ratio (LLR) for the vector $\widehat{Z}^{\upsilon}_{n,j}$:

$$l^{\upsilon}_n \approx \frac{2|\upsilon_{n,j}'|}{\sigma^2_{\text{SI}} + \sigma^2_{\text{bruit}}}\widehat{Z}^{\upsilon}_{n,j}. \tag{18.8}$$

Or $\widehat{Z}^{\upsilon}_{n,j}$ is multiplied by 2 instead of 4, because in (*B*) the variance is attributed to an actual noise value.

We now have to determine the expressions of the terms $|\vartheta'_{n,j}|$, $\sigma_{\text{SI}}{}^2$ and $\sigma_{\text{bruit}}{}^2$ in order to obtain the expression of l^{υ}_n. To do this, it is good to come back to certain aspects of the Walsh–Hadamard codes used. *p* taking its values in the alphabet $\left\{-\frac{1}{\sqrt{L}}, +\frac{1}{\sqrt{L}}\right\}$. Further assuming that the occurrence of having $x^{\upsilon}_{n,j}$ is equal to $\pm$ 1 is equally likely. $\left|\vartheta'_{n,j}\right|$ is then written:

$$\left|\vartheta'_{n,j}\right| = \frac{1}{L}\left|\sum_{l=1}^{L} h_l g_l\right|. \tag{18.9}$$

The variance of the SI is:

$$\sigma^2_{\text{SI}} = \frac{N_L - 1}{L}\left(\varepsilon\left[|h_l g_l|^2\right] - |\varepsilon|h_l g_l||^2\right). \tag{18.10}$$

And the noise variance is:

$$\sigma^2_{\text{bruit}} = \frac{\sigma^2}{2}\varepsilon\left[|g_l|^2\right]. \tag{18.11}$$

By replacing the expressions of Eqs. (18.10), (18.11) and (18.12) in (18.8), we obtain:

$$l^{\upsilon}_n \approx \frac{\frac{2}{L}\left|\sum_{l=1}^{L} h_l g_l\right|}{\frac{N_L-1}{L}\left(\varepsilon\left[|h_l g_l|^2\right] - |\varepsilon|h_l g_l||^2\right) + \frac{\sigma^2}{2}\varepsilon\left[|g_l|^2\right]}\widehat{Z}^{\upsilon}_{n,j}. \tag{18.12}$$

It can be seen that the expression of the LLR of an LP-OFDM system using Walsh–Hadamard codes is relatively complex to define and depends on the equalization technique employed. Generally, multiples expressions of $L^{v}{}_{n}$ can be obtained depending on the equalization technique used. Assuming that MMSE detection is used and that the chosen Walsh–Hadamard codes are relatively short, we can end up with a simplified expression of $L^{v}{}_{n}$.

18.2.5 Characteristics of an LP-OFDM Signal

Multi-carrier modulation originates from frequency multiplexing and is based on frequency parallelization of the information to be transmitted. The OFDM signal consists of sub-carriers forming an orthogonal vector base in frequency and with minimum spectral occupancy. We assume an OFDM signal adapted to the channel and the reception equation, in its digital and matrix form, is written

$$Y = HX + Z, \tag{18.13}$$

where X is the vector of symbols emitted and containing the information, H channel matrix, Z the noise and Y the received signal.

The reception equation is written:

$$Y = HCX + Z, \tag{18.14}$$

where C represent the precoding matrix made up of Hadamard matrices. Vector estimated symbols $\widehat{X}$ are written:

$$\widehat{X} = C^{\mathrm{T}} H^{-1} Y. \tag{18.15}$$

The advantage of this technique is to be able to group the sub-carriers of the OFDM modulation into subsets, called blocks, in order to increase the transmission rates. The sub-carriers of each frame are not certainly adjacent. For simplicity, we assume a fixed block size for all users and, in the case of unconstrained modulations, the total throughput achieved on a block S_b is,

$$\Re_{u,b} = L \log_2\left(1 + \frac{1}{\Gamma} \frac{L}{\sum_{n \in S_b} \frac{1}{|h_{u,n}|^2}} \frac{E}{N_0}\right), \tag{18.16}$$

where E represents the constraint of DSP, Γ the SNR margin, N_0 the noise and $|h_{u,n}|^2$ the amplitude of the channel. To maximize throughput $\Re_{u,b}$, it suffices to minimize the sum $\sum_{n \in S_b} \frac{1}{|h_{u,n}|^2}$ by choosing the sub-carriers with the best amplitudes $|h_{u,n}|^2$.

18.3 Formulation of the BER Maximization Problem

18.3.1 The Classic Method

The classic OFDM multicast method, LCG (low channel gain), (18.10), is a simple method which allows multicast information to be allocated while satisfying the quality of service of users. The throughput achieved in the CPL context is

$$\Re_n^{\mathrm{LCG}} = \min_u \log_2\left(1 + \frac{E}{\Gamma N_0}\left|h_{u,n}\right|^2\right), \tag{18.17}$$

and $\left|h_{u,n}\right|^2$ the amplitude of the channel on the sub-carrier n of the user u. The status of all users' channels is known.

Within the framework of the LP-OFDM system, the bit rate achieved in the multicast block S_b will be the lowest BER of the users on this Frame. This rate is:

$$\Re_b^{\mathrm{LP}} = \min_u \Re_{u,b} = \min_u L \log_2\left(1 + \frac{1}{\Gamma}\frac{L}{\sum_{n\in S_b}\frac{1}{\left|h_{u,n}\right|^2}}\frac{E}{N_0}\right) \tag{18.18}$$

Our goal is to maximize the throughput in multicast; it will therefore be a question of maximizing the lowest throughput of users on each block. For this, a strategy of grouping the sub-carriers in the different blocks which maximizes the lowest BER of the block is necessary. As,

$$\min_u \Re_{u,b} \Leftrightarrow \max \sum_{n\in S_b}\frac{1}{\left|h_{u,n}\right|^2} \tag{18.19}$$

and vice versa, the optimization problem is written:

$$\min_{S_b}\max_u \sum_{n\in S_b}\frac{1}{\left|h_{u,n}\right|^2}. \tag{18.20}$$

We are in the presence of a "min–max" problem which consists in finding the solution presenting the best bad rate among all the scenarios.

Combinatorial Solution

The problem (18.17) is a combinatorial optimization problem, and the basic method of its resolution is to test all the possibilities of grouping the sub-carriers into blocks and to choose the best case. It is clear that such an algorithm would be finished. But as soon as the number N of sub-carriers becomes large, it is easy to realize that it would take too much computing time for the computer to enumerate all the possibilities. The total number of possibilities is given by:

$$\frac{C_N^L \times C_{N-L}^L \times C_{N-2L}^L \times \cdots \times C_{2L}^L}{\left(\frac{N}{L}\right)!} = \frac{N!}{(L!)^{\left(\frac{N}{L}\right)}\left(\frac{N}{L}\right)!}, \tag{18.21}$$

where N denotes the total number of sub-carriers. N represents the total number of combinations of L among N, L among $N - L$ and so on. It is then divided by the total number of arrangement of the blocks. This solution gives the optimal throughput in LP-OFDM multicast but its implementation quickly becomes impractical when the number of sub-carriers increases.

Maximizing Throughput in a Single User Context

Based on LCG method, the BER can be considered as the rate calculated on an equivalent channel in the context of a single user. This channel is created from the combination of different user channels. For each sub-carrier index in the channel, the equivalent amplitude is the user amplitude with the lowest amplitude.

$$\left|h_n^{\text{eq}}\right|^2 = \min_u \left|h_{u,n}\right|^2. \tag{18.22}$$

The sub-carriers of the equivalent channel are thus arranged in decreasing order for the formation of the blocks. It was proved in [22] that the linear precoding component makes it possible to increase the OFDM BER in the case of a single user.

18.3.2 Improvement of the LP-LCG Method

By reasoning by block, we have:

$$\left|h_{u,n}\right|^2 \geq \left|h_n^{\text{eq}}\right|^2 \Leftrightarrow \max_u \sum_{n \in S_b} \frac{1}{\left|h_{u,n}\right|^2} \leq \sum_{n \in S_b} \frac{1}{\left|h_n^{\text{eq}}\right|^2}. \tag{18.23}$$

Therefore, it can be deduced that the "worst" user on S_b enhanced throughput than the LP-LCG.

18.4 Simulation Results and Discussion

18.4.1 Performance of the Proposed System

This section will analyze the performance of the suggested system over THz band. To do that, a channel model for THz frequencies must be given [23]. Generally, to model THz channel, we had to take into account to aspect: Line of sight (LOS) path and non-LOS paths. However, the LOS configuration is dominant in THz channel, which means the transmit power should be mostly used to transmit signal over the LOS path [24]. In this work, we will use a channel model that contains the two configurations. So that, the channel model proposed in [24] will be used here to determine the performance of the proposed LP-OFDM. The impulse channel model that will be used in this work is given by:

$$h(x_0, y_0, \tau) = \sum_{l=1}^{N_C} \sum_{k=1}^{R_l} \hat{\mu}_{k,l}(t) e^{j\lambda_{k,l}(t)} \delta\Big(\tau - T_l - \frac{1}{c}[(x - x_0)\cos(\vartheta_l + \varphi_{k,l}) + (y - y_0)\sin(\vartheta_l + \varphi_{k,l})]\Big), \quad (18.24)$$

where ϑ_l is the average arrival azimuth in the lth cluster. Here, we consider ϑ_l as uniformly distributed function in the interval $[0, 2\pi]$, $\varphi_{k,l}$ is the distribution of the arrival azimuth within the cluster, $\lambda_{k,l}(t)$ is the phase corresponds to the kth radius inside of the lth cluster, τ represents the arrival time, N_C is the number of clusters, R_l is the number of kth cluster and $T_l(t)$ the arrival time of the lth cluster and $\hat{\mu}_{k,l}(t)$ is given by:

$$\hat{\mu}_{k,l}(t) = \mu_{1,1}^2 \frac{c}{4\pi f d} \exp\Big(-\frac{1}{2}\gamma(f, T_k, p)d\Big) e^{-\frac{T_l - T_1}{\rho}} e^{-\frac{T_l - T_1}{\gamma}}, \quad (18.25)$$

where $\mu_{k,l}(t)$ represents the amplitude associated with the kth radius inside the lth cluster, $\gamma(f, T_k, p)$ is the molecular absorption coefficient [25], d is the distance that separates the transmitter from the receiver and ρ and γ are inter- and intra-cluster of the different radius respectively.

The channel transfer function $H(f)$ can be calculated easily by the discrete time Fourier transform (DTFT) of the channel impulse response given in (18.24). Figure 18.2 shows the channel transfer function of the proposed THz channel model versus frequency in the band (260–300 GHz).

Based on the channel presented in Fig. 18.2, we simulated the total path loss of the THz channel. The obtained results are given in Fig. 18.3. This figure indicates clearly that the THz channel is lousy channel, especially in some frequency such as 2.93 GHz. These results validate the discussion given above about the channel transfer function.

In Fig. 18.4, we depicted the bit error rate (BER) versus signal to noise (SNR) for LP-OFDM and OFDM. From Fig. 18.4, we observe that as the number of users

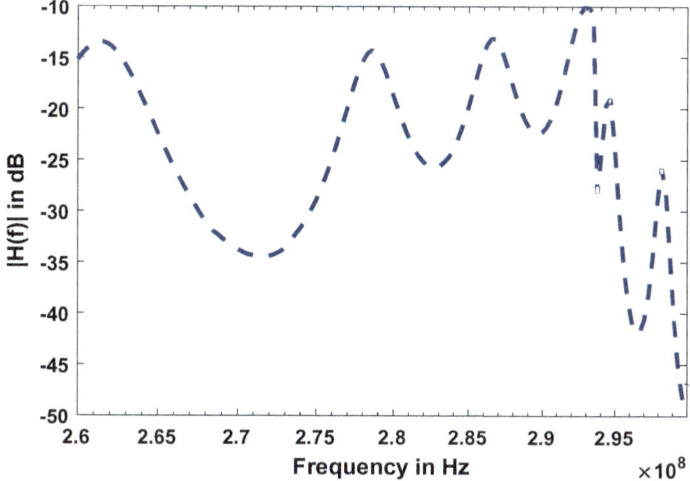

Fig. 18.2 Proposed THz channel transfer function versus frequency

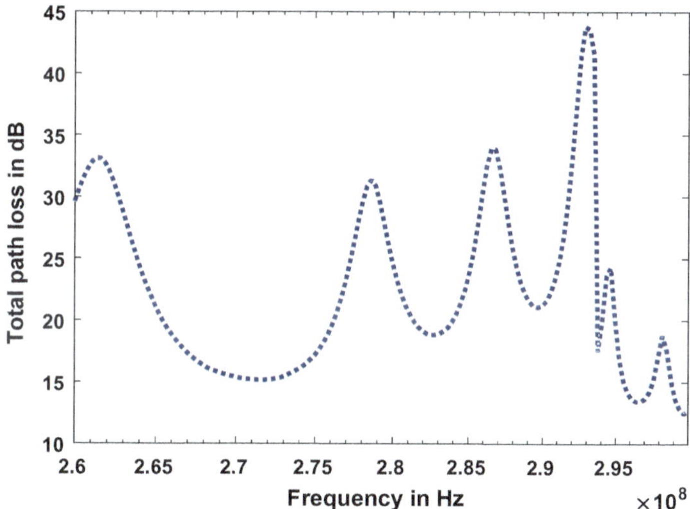

Fig. 18.3 Total path loss versus frequency in THz context

increases, the BER increases too. Besides, the BER decreases for higher value of SNR. For two users, we obtained a BER near to 10–3 when the SNR equal 9 for LP-OFDM, but for OFDM, we need a SNR of about 12 dB to obtain the same BER. For twelve users, we need a SNR of 13 dB to have a BER of about 10–3 for LP-OFDM; this BER can be obtained with a SNR of 17 dB for OFDM. As can be seen from Fig. 18.4, OFDM and LP-OFDM have the same performance when the number of

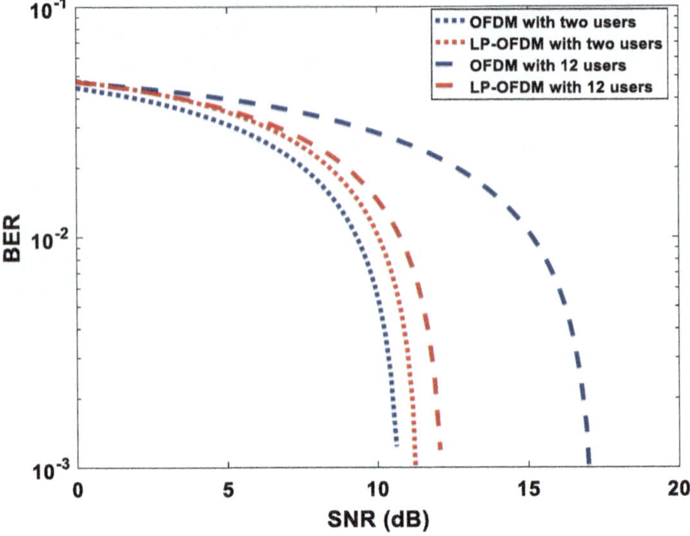

Fig. 18.4 BER versus SNR for different number of users

users equal to two. However, when the number of users is equal to twelve, LP-OFDM performs better than OFDM.

18.4.2 Comparison of the Different Solutions

In this part, we present the simulation results in the THz context for the different allocation methods in LP-OFDM multicast. The signal generated is $N = 2048$ sub-carriers, transmitted in the band of (260–300) THz. Synchronization and channel estimation are assumed to be perfect. The considered background noise, DSP and maximum number of bits per sub-carrier are respectively set at – 110 dBm, – 50 dBm, 10, for all users. First, we compare the allocation methods in LP-OFDM multicast to each other. The ratio of the average channel amplitudes of the users varies from 0 to 20 dB, and the sum of the average channel amplitudes is fixed at – 30 dB. Figure 18.5 gives the different numbers of bits allocated to each user. In order to avoid the sizing of the system necessary for the calculation of the bit rates, we compare the number of bits received by users. As predictable, the combinatorial solution outperforms the others, and the linear precoding component improves the number of allocated bits, obtained with the traditional LCG method. The LP-LCG method will be compared with the classical LCG method; the simulations will be carried out without the combinatorial method. Compared to the traditional LCG method, the LP-LCG method provided a complexity of multiplication by an orthogonal Hadamard matrix.

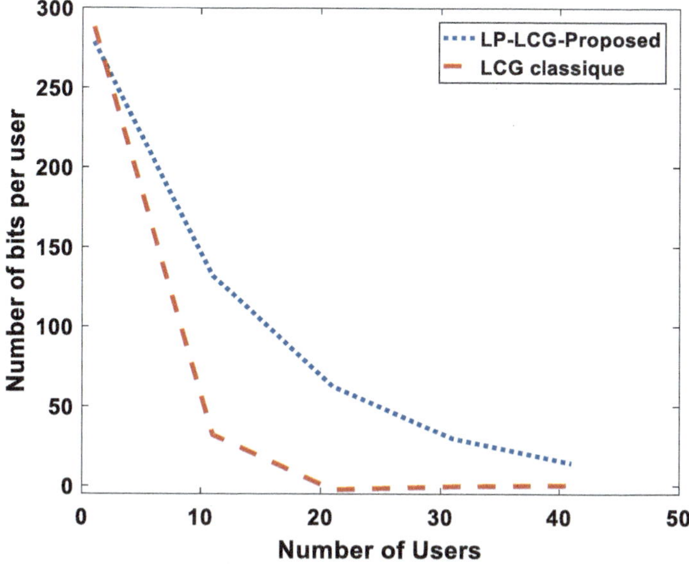

Fig. 18.5 Number of bits allocated to each user according to the number of users

As expected, the proposed scheme provides a better number of bits for each user and the gain offered in terms of number of bits increases considerably as the number of users rises from two to twelve users. The LP-LCG method offers the best performance.

18.5 Conclusion

In this chapter, we have described and characterized the LP-OFDM signal. Multi-carrier systems effectively overcome degradations introduced by the channel such as frequency selectivity. Thanks to the progress in the manufacture of digital circuits, the realization of the OFDM system becomes possible. Although the latter has been chosen as the standard physical layer for a variety of important systems, the theory, algorithms and techniques of its implementation remain the subjects of current interest. Wireless communications have experienced tremendous growth almost everywhere in the world in recent years. The public's enthusiasm for wireless amenities has spawned a multitude of offers across networks, terminals and services of all kinds. In this direction, we have drawn up a fairly detailed description of the multi-carrier transmission represented by the LP-OFDM modulation technique. The simulation results show the benefit of adding the linear precoding component to OFDM multicast systems, allowing considerable gains in transmission rates compared to the conventional OFDM multicast allocation method.

References

1. Babu, K.V., Sree, G.N.J., Islam, T., Das, S., El Ghzaoui, M., Agilesh Saravanan, R.: Performance analysis of a photonic crystals embedded wideband (1.41–3.0 THz) fractal MIMO antenna over SiO_2 substrate for terahertz band applications. Silicon (2023)
2. El Ghzaoui, M., Belkadid, J., Benbassou, A.: Performance analysis of CE-OFDM over terahertz bands. J. Inst. Eng. (India): Ser. B **104**(5), 1147–1154 (2023)
3. Koenig, S., Lopez-Diaz, D., Antes, J., et al.: Wireless sub-THz communication system with high data rate. Nature Photon. **7**, 977–981 (2013). https://doi.org/10.1038/nphoton.2013.275
4. Moldovan, A., Kisseleff, S., Akyildiz, I.F., Gerstacker, W.H.: Data rate maximization for terahertz communication systems using finite alphabets. In: 2016 IEEE International Conference on Communications (ICC), pp. 1–7 (2016). https://doi.org/10.1109/ICC.2016.7511275
5. Liebermeister, L., Nellen, S., Kohlhaas, R., et al.: Ultra-fast, high-bandwidth coherent cw THz spectrometer for non-destructive testing. J. Infrared Millimeter Terahertz Waves **40**, 288–296 (2019). https://doi.org/10.1007/s10762-018-0563-6
6. Mestoui, J., El Ghzaoui, M.: A survey of NOMA for 5G: implementation schemes and energy efficiency. Lecture Notes in Electrical Engineering, vol. 745, pp. 949–959 (2022)
7. Belkadid, J., Benbassou, A., El Ghzaoui, M.: PAPR reduction in CE-OFDM system for numerical transmission via PLC channel. Int. J. Commun. Antenna Propag. **3**(5), 267–272 (2013)
8. Castro-Camus, E., Koch, M., Kleine-Ostmann, T., et al.: On the reliability of power measurements in the terahertz band. Commun. Phys. **5**, 42 (2022). https://doi.org/10.1038/s42005-022-00817-2
9. Tekbıyık, K., Ekti, A.R., Kurt, G.K., Görçin, A.: Terahertz band communication systems: challenges, novelties and standardization efforts. Phys. Commun. **35**, SP-100700 (2019). https://doi.org/10.1016/j.phycom.2019.04.014
10. Gopalan, P., Sensale-Rodriguez, B.: 2D materials for terahertz modulation. Adv. Opt. Mater. **8**, 1900550 (2020). https://doi.org/10.1002/adom.201900550
11. El Ghzaoui, M., Mestoui, J., Hmamou, A., Serghini, E.: Performance analysis of multiband on–off keying pulse modulation with noncoherent receiver for THz applications. Microwave Opt. Technol. Lett. (2021). https://doi.org/10.1002/mop.33051
12. Sheikh, F., Zarifeh, N., Kaiser, T.: Terahertz band: channel modelling for short-range wireless communications in the spectral windows. IET Microwaves Antennas Propag. **10**, 1435–1444 (2016). https://doi.org/10.1049/iet-map.2016.0022
13. Kaltenecker, K.J., Kelleher, E.J.R., Zhou, B., et al.: Attenuation of THz beams: a "How to" tutorial. J. Infrared Millimeter Terahertz Waves **40**, 878–904 (2019). https://doi.org/10.1007/s10762-019-00608-x
14. Yilmaz, T., Akan, O.: Attenuation constant measurements of clear glass samples at the low terahertz band. Electron. Lett. **56**, 1423–1425 (2020). https://doi.org/10.1049/el.2020.1593
15. Elaage, S., El Ghzaoui, M., Hmamou, A., Foshi, J., Mestoui, J.: MB-OOK transceiver design for terahertz wireless communication systems. Int. J. Syst. Control Commun. **12**(4), 309–326 (2021). https://doi.org/10.1504/IJSCC.2021.118627
16. Wang, Z., Giannakis, G.B.: Linearly precoded or coded OFDM against wireless channel fades? In: 2001 IEEE Third Workshop on Signal Processing Advances in Wireless Communications (SPAWC'01). Workshop Proceedings (Cat. No.01EX471), pp. 267–270 (2001). https://doi.org/10.1109/SPAWC.2001.923899
17. Stephan, A., Guéguen, E., Crussière, M., et al.: Optimization of linear precoded OFDM for high-data-rate UWB systems. J. Wirel. Commun. Network **2008**, 317257 (2007). https://doi.org/10.1155/2008/317257
18. Maiga, A., Baudais, J.-Y., Hélard, J.-F.: Bit rate optimization with MMSE detector for multicast LP-OFDM systems. J. Electr. Comput. Eng. **2012** (2012). https://doi.org/10.1155/2012/232797
19. Mestoui, J., El Ghzaoui, M., Hmamou, A., Foshi, J.: BER performance improvement in CE-OFDM-CPM system using equalization techniques over frequency-selective channel. Procedia Comput. Sci. **151**, 1016–1021 (2019)

20. Mestoui, J., El Ghzaoui, M., Fattah, M., Hmamou, A., Foshi, J.: Performance analysis of CE-OFDM-CPM Modulation using MIMO system over wireless channels. J. Ambient Intell. Humanized Comput. **11**(10), 3937–3945 (2020)
21. Bouvet, P.J., Hélard, M., Le Nir, V.: Simple iterative receivers for MIMO LP-OFDM systems. Ann. Télécommun. **61**, 578–601 (2006). https://doi.org/10.1007/BF03219924
22. Kuo, C.C.J., Tsai, S.H., Tadjpour, L., Chang, Y.H.: Precoding techniques in multiple access channels. In: Precoding Techniques for Digital Communication Systems. Springer, Boston (2008)
23. El Ghzaoui, M., Das, S., Lenka, T.R., Biswas, A.: Terahertz Wireless Communication Components and System Technologies, pp. 1–310 (2022)
24. Aghoutane, B., El Ghzaoui, M., El Faylali, H.: Spatial characterization of propagation channels for terahertz band. SN Appl. Sci. **3**, 233 (2021). https://doi.org/10.1007/s42452-021-04262-8
25. Bharathi, M., Amsaveni, A., Sasikala, S.: Orthogonal frequency division multiplexing for improving the performance of terahertz channel. J. Infrared Millimeter Waves **38**(3) (2019). https://doi.org/10.11972/j.issn.1001-9014.2019.03.001

Chapter 19
Terahertz Technology and Its Importance in the Field of Biomedical Application: A Review

Krishanu Kundu and Narendra Nath Pathak

19.1 Introduction

Conventional wireless communication [2] facilitating technologies, including LTE, GSM and others, provide an extreme data transmission rate of 100 Mbps. In combination with 4G, IEEE 802.16e was installed in various nations, running at 2.5–2.7 GHz and having max data rate of 128 Mbps. IEEE 802.11 works on the 2.4 GHz band with max data rate of 150 Mbps, while IEEE 802.15 functions on the 2.4 GHz, 868 as well as 915 MHz bands with a peak transmission rate of 250 Kbps. Bluetooth 4.0 works at 2.4 GHz and 1 Mbps data rate. LoRaWAN [3] has a peak data throughput of 50 Kbps. Narrowband IoT (NB-IoT) operates in the 700–900 MHz frequency band and may be united with LTE, as approved by 3GPP. NB-IoT has a maximum possible data rate of 200 Kbps. THz frequencies are particularly vulnerable to water moisture absorption; thus, THz waves suffer a significant reflection loss in an NLoS environment. The scattering effect reduces as the wavelength associated with the transmitted waves decreases. However, atmospheric losses like as fog, rain, pollution, and so on have a greater effect on frequencies over 10 GHz. THz may be used to dramatically boost data rate. The THz spectrum consists mostly of frequencies spanning from 100 GHz to 10 THz, having data speeds varying between 10–160 Gbps having a transmission span of 10 m.

Novel transceiver along with physical layer architectures is required to increase spectral efficacy as well as data throughput throughout the THz range. The THz band [4] is placed between microwave as well as infrared frequencies on the regular

K. Kundu (✉)
Department of Electronics and Communication Engineering, G. L. Bajaj Institute of Technology & Management, Greater Noida, India
e-mail: krishanukundu08@gmail.com

N. N. Pathak
Department of Electronics and Communication Engineering, Dr. B. C. Roy Engineering College, Durgapur, India

M. El Ghzaoui et al. (eds.), *Next Generation Wireless Communication*, Signals and Communication Technology, https://doi.org/10.1007/978-3-031-56144-3_19

radio spectrum. THz communications cannot substitute laser as well as microwave communications. THz, on the contrary hand, possesses various unique properties that contribute to its superiority over laser as well as microwave. According to ITU guidelines, THz has the following important properties and characteristics: (a) The penetrating strength of radio signals over 275 GHz is exceptional for dielectric substances as well as non-polar liquids. THz's increased penetration power allows it to scan opaque objects. Because a THz wave has a longer wavelength, transmission loss caused by a THz wave in smoke/dust is quite minor. As a result, it might be used for imaging in smoky conditions like firefighting areas or deserts. (b) The attenuation caused by radio signals beyond 275 GHz is noteworthy, and it may be used for recognition and analysis in a variety of medical professions. THz can be employed to detect or identify cancerous cells by accessing data about the water content of tissues. (c) THz wave photon energy is represented in meV and is suggestively lesser than chemical bond energy. As a consequence, the ionization reaction induced by THz waves is generally modest, making it suitable for biological sample identification and human body checks. Because THz has a large absorption of water effect, it is more unlikely to infiltrate the human body, thus rendering it excellent for skin disease diagnostics. (d) THz vibrations give a variety of spectrum data, including physical and chemical features of materials. Organic substances have considerable absorption and dispersion properties in this band. THz may be used to identify the traits, properties, and compositions of substances for both chemical and physical inquiry using these spectrum attributes. (e) THz waves have a better spatial resolution than the microwave spectrum. Especially when compared with microwave imaging, wavelengths in the sub-mm range improve image resolution. (f) THz waves possess a high degree of directivity due to the significant absorption and reflection losses during such high frequencies, which confine communication to the directed LoS set-up because NLoS state faces spreading losses. Because of said properties, THz has been proposed for high-speed wireless communications. Today, terahertz technology is used across industries like security screening, medical and dental imaging and pharmaceuticals, etc.

19.2 Terahertz Technology: A Historical Perspective

Terahertz technology [5–8] refers to the section of the electromagnetic spectrum between around 0.1–10 THz. This range sits between microwave frequencies on the low end and infrared frequencies on the high end, thus sometimes referred to as the "terahertz gap". Research into terahertz radiation dates back to the nineteenth century, but major advancements have occurred from the 1980s onwards with the deployment of terahertz time-domain spectroscopy. Today, terahertz technology is employed across security screening, medical and dental imaging, wireless communications, pharmaceutical quality control, and more.

19.2.1 Early Research

In the 1800s, infrared radiation was discovered by William Herschel, spurring research into properties of light beyond the visible spectrum. In 1890s, Heinrich Rubens investigated far-infrared radiation and developed water vapour absorption spectra. In between 1900–1970, we can see steady progress in generating and detecting mm and sub-mm waves, the lower end of terahertz band. But it lacked detailed understanding.

19.2.2 Key Developments in 1980s–1990s

In the 1980s, THz-TDS was developed at Bell Labs and allowed detailed study of THz waves. Other advances in photoconductive THz antennas are: in 1985, the first photoconductive THz emitter was demonstrated at Bell Labs; in 1988, the first demonstration of free-space THz beams; in 1989, the first THz spectra collected and showed water vapour absorption lines; and in the 1990s, improved THz sources and components and first THz images in 1995.

19.2.3 Applications and Progress in 2000s–2010s

In the early 2000s, commercial THz-TDS systems were introduced. Applications grew in various fields. In 2002, the first passive THz image of a hand, for security screening was introduced. In 2004, THz cameras for security screening was introduced. In the 2010s, portable, compact THz systems emerged. THz for wireless communications researched.

19.2.4 Advancements in Terahertz Technology Post-2010

Major advancements in terahertz technology [9–11] from 2010 to 2023 include improved terahertz sources, detectors, and components for applications like imaging, spectroscopy, and communications. Key applications enabled by terahertz technology advances include security screening, medical imaging, material characterization, wireless communications, and astronomy. Notable contributors to terahertz technology include researchers at universities and companies developing terahertz devices and systems. Major progress has come from improved semiconductor and photonic devices. Hundreds of scientific papers have been published on terahertz technology from 2010 to 2023, with increasing focus on communications, imaging, spectroscopy, sources, and components. Future prospects for terahertz technology are very positive, with potential for new applications in 6G wireless networks, quantum sensing, manufacturing, and healthcare. However, challenges remain in improving power, efficiency, and costs.

19.2.5 *Major Advancements 2010–2023*

(a) Improved terahertz Sources: Semiconductor sources like quantum cascade lasers (QCLs) with higher power, efficiency and operating temperatures. Photonic sources like photo mixers and optical rectification with broader bandwidths. Electronic sources like oscillators and multipliers reaching up to several THz.
(b) Advanced terahertz detectors: Faster semiconductor detectors like Schottky diodes. More sensitive superconducting detectors. Large-scale detector arrays for imaging.
(c) Terahertz components: Low-loss waveguides and antennas. High-power amplifiers > 1 THz. Integrated terahertz circuits and systems-on-chip.
(d) Enabled applications: Security screening of concealed objects. Medical imaging for cancer detection. Material characterization through spectroscopy. High-speed wireless communications. Astronomy and cosmology studies.
(e) Future prospects: 6G wireless networks using terahertz frequencies. Faster wireless data rates > 100 Gbps. Quantum sensing and computing applications. Improved medical imaging and healthcare. More sensitive material characterization. Higher resolution terahertz astronomy.
(f) Challenges: Increasing terahertz power and efficiency. Reducing costs of terahertz systems. Overcoming high terahertz attenuation in atmosphere. Enabling terahertz arrays with thousands of pixels. Developing integrated terahertz circuits and packaging.

19.3 Biomedical Applications of Terahertz Technology

Terahertz technology [12–15] refers to the use of electromagnetic radiation, which lies in between microwave as well as infrared radiation. Often referred to as T-rays or T-waves, terahertz radiation exhibits properties between electronics and photonics. The terahertz band is also called the sub-millimetre band since the wavelengths range from 1 to 0.1 mm (Fig. 19.1).

Terahertz waves can penetrate many non-conducting materials like paper, plastic, cardboard, wood, and textiles. This enables imaging of concealed objects. Unlike X-rays, terahertz radiation is non-ionizing and does not damage human tissue, making it safe for medical use. Many materials have unique terahertz spectral signatures that act as fingerprints for precise identification. Terahertz frequencies can enable ultra-high bandwidth wireless communication, exceeding 100 Gbps speeds. The

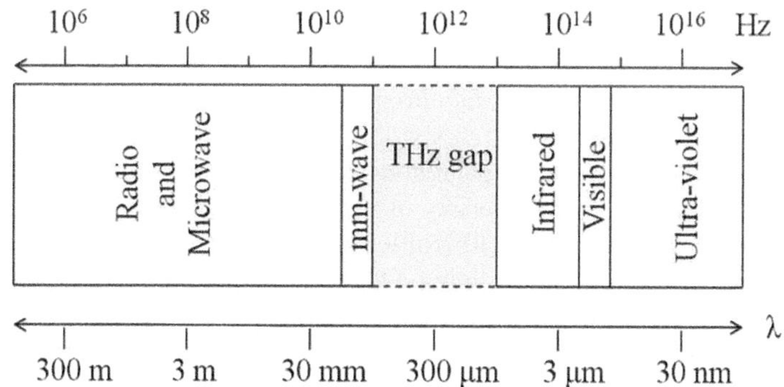

Fig. 19.1 THz band spectrum

short terahertz wavelengths allow for sub-millimetre resolution imaging. Terahertz imaging can provide elemental composition contrast for chemical mapping. Recent advances in THz instrumentation have enabled growing biomedical applications in imaging and spectroscopy.

19.3.1 Terahertz Imaging

Terahertz imaging, a cutting-edge technique that uses electromagnetic waves in the terahertz frequency range, is transforming a variety of sectors. Terahertz imaging provides new insights into the structure and composition of things concealed from the human eye due to its unique ability to penetrate non-conductive materials such as plastics, ceramics, and clothes. Terahertz imaging uses frequencies between microwave as well as infrared rays. Terahertz waves, commonly known as T-rays, have wavelengths ranging from 0.1 to 1 mm and may interact with materials in a unique way. Unlike X-rays, which may be dangerous and must be used with caution, terahertz waves are non-ionizing and hence safe for both humans and the environment. This technique makes use of the concepts of terahertz wave reflection, transmission, and absorption. Terahertz waves can be reflected, transmitted, or absorbed when they collide with an object, depending on the structure and composition of the material. Terahertz imaging systems may gather precise information about an object's internal structure, surface qualities, and chemical composition as a result of these interactions. Security is one of the most popular uses of terahertz imaging. When it comes to identifying non-metallic threats and concealed objects, traditional security screening systems such as metal detectors and X-ray scanners have limits. Terahertz imaging overcomes these constraints by producing detailed pictures that reveal hidden things such as hidden weapons, explosives, and illegal materials. Terahertz imaging technology is increasingly being used in airports and other high-risk places to improve security and assure individual safety. Terahertz imaging is essential in quality control and non-destructive testing in industrial applications. Terahertz waves are useful for checking a wide range of items, including semiconductors, medicines, and composite materials, due to their ability to penetrate non-conductive materials. Manufacturers can discover faults, monitor production processes, and verify product quality by analysing terahertz pictures without hurting or destroying the materials being evaluated. Terahertz imaging's non-destructive nature saves time, money, and enhances overall efficiency in a variety of sectors. As terahertz imaging advances, researchers are resolving present difficulties and exploring new boundaries for this technique. One of the difficulties being addressed is increasing the resolution and imaging speed of terahertz devices. Researchers want to obtain greater resolution and quicker picture collection by improving hardware components and creating sophisticated algorithms, allowing real-time imaging for time-critical applications. Another area of interest is the creation of terahertz imaging devices that can function in hostile conditions. Terahertz imaging may be used in sectors such as aerospace, automotive, and oil and gas by designing strong equipment that can survive harsh temperatures, humidity, and other demanding environments. Terahertz imaging is proven to be a game changer for non-invasive medical diagnostics in the field of healthcare. Doctors can detect skin cancer, assess dental health, and spot prospective tumours with amazing precision by using terahertz radiation. Terahertz waves may penetrate the skin's outer layers, enabling deep tissue investigation without the need for intrusive treatments.

Terahertz Imaging for Cancer Detection

THz imaging [16–19] provides contrast between normal and diseased tissue due to variances in water content as well as tissue structure. Skin cancer margins can be delineated in THz images, aiding surgical planning. Dynamic THz imaging showed changes in skin hydration during scar healing. Breast cancer imaging ex vivo using THz pulses distinguished tumours from normal tissue based on higher cell density and water content. Oral cancer imaging at – 20 °C enhanced contrast between tumours and healthy tissue by reducing water content. Colon cancer tissues showed ~ 23% higher THz absorption than normal tissues, but more research is needed for statistical significance. THz imaging utilizes the differences in THz absorption by water to differentiate between tissues (Fig. 19.2).

Cancerous tissues tend to have higher water content compared to normal tissues. This provides natural contrast in THz images. During a heart attack, cell death leads to breakdown of cell membranes, allowing influx of water into infarcted tissue. This leads to increased THz absorption. THz imaging can detect these water content changes to identify infarcted heart tissue. THz technology is a promising new technique for heart attack detection due to its ability to detect morphological and molecular changes during infarction. With further research to address limitations, it can become a useful addition to the clinical toolbox for rapid heart attack diagnosis.

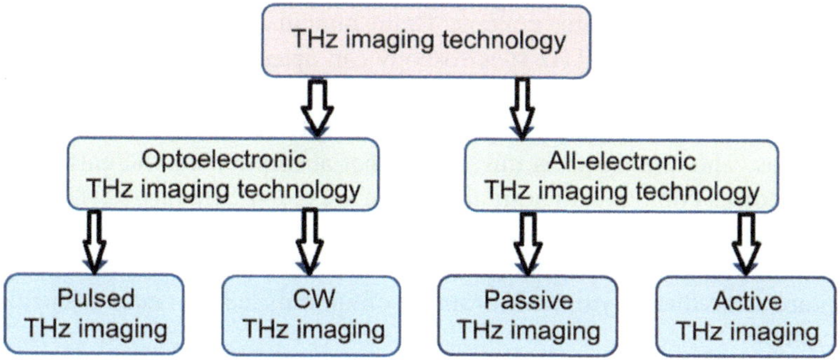

Fig. 19.2 THz imaging technology types

19.3.2 Advantages of THz for Heart Attack Detection

- Non-invasive and non-ionizing. Safe for patients.
- Fast acquisition times. Enables real-time imaging.
- Provides both morphological (imaging) and molecular (spectroscopy) information.
- High sensitivity and contrast for detecting water content changes in infarcted tissue.
- Can complement existing diagnostic methods like ECG, cardiac MRI.

THz spectroscopy detects unique spectral signatures of molecules in the THz range. It can detect conformational changes in proteins like C-reactive protein during a heart attack. Spectral signatures provide molecular information to complement morphological imaging.

19.3.3 Terahertz Spectroscopy for Disease Diagnosis

- THz spectroscopy can identify spectral fingerprints of biological molecules for disease diagnosis.
- DNA methylation in cancers produces distinct THz absorption signatures, enabling quantification of DNA methylation levels.
- In vivo spectroscopy of skin nevi revealed a strong distinction between normal as well as dysplastic lesions in THz dielectric characteristics.
- THz spectroscopy has been used to sense blood glucose levels in vivo, working towards non-invasive glucose monitoring.

THz imaging [20–22] is being researched for detecting brain tumours and defining their margins more clearly to guide surgical resections. The key advantages are: THz radiation is non-ionizing and does not damage tissue like X-rays. THz waves have high sensitivity to water content. Brain tumours contain more water due to angiogenesis and oedema. THz spectroscopy can detect unique spectral signatures of brain tumour tissue. Proof-of-concept studies on rat glioma models have shown the capability of THz imaging to discriminate in between normal as well as tumour brain regions. The tumour areas unveiled higher absorption coefficients along with refractive indices compared to normal tissue in the 0.1–4 THz range. The contrast is accredited to the greater cell density as well as water content inside the tumour. The high nucleic acid content in cancerous cells increases refractive index. The leaky vasculature and reduced lymphatic drainage of tumours lead to oedema and elevated water content.

19.3.4 Advantages of Terahertz Technology in Brain Tumour Detection

The key advantages of THz imaging for brain tumour detection are:

- **Non-invasive and non-ionizing**: Safe for repeated patient scans.
- **High sensitivity to water content**: Enables tumour discrimination based on oedema.
- **Provides morphological and functional information**: Absorption coefficients and refractive indices can detect tumours.
- **Spectral fingerprints of tissue**: Tumour has distinct THz signatures compared to normal brain.
- **Real-time imaging**: Enables intraoperative tumour margin assessment.

19.4 Importance of Terahertz Technology in Biomedicine

Biomedicine is essential for disease diagnosis, treatment, and prevention. Biomedicine has revolutionized healthcare by delivering more accurate diagnostic tools, effective medicines, and preventative methods via the use of scientific knowledge and technology advances. The creation of tailored medicines is one of the most important contributions of biomedicine to healthcare. Researchers have been able to build medicines that selectively attack cancer cells while causing minimal harm to healthy cells. For many patients, this method has greatly improved therapy results while reducing adverse effects. Over the years, biomedicine has seen significant discoveries and breakthroughs that have revolutionized the medical environment. The completion of the Human Genome Project in 2003 was one of the most momentous breakthroughs. This massive project mapped the whole human genome, giving scientists with a blueprint of our genetic make-up. This development has paved the way for personalized medicine, which tailors treatments and interventions to an individual's genetic profile. The discovery of gene editing methods such as CRISPR-Cas9 is another major milestone in biology. This ground-breaking method enables scientists to accurately modify DNA sequences, potentially correcting genetic mutations and treating hereditary illnesses. The applications of gene editing in biomedicine are numerous, ranging from correcting hereditary illnesses to better understanding human biology. One of the most attractive features of biomedicine is the potential for personalized medicine to revolutionize healthcare. To personalize treatments and interventions, personalized medicine considers an individual's genetic composition, lifestyle, and environmental variables. Healthcare practitioners may give more accurate and effective care if they understand the unique qualities of each patient. This method not only improves treatment outcomes, but it also lowers the chance of unpleasant responses and unneeded therapies. Biomedicine's future has limitless potential. We should expect many more remarkable discoveries and advancements in the sector as technology advances. The future of biomedicine, from artificial intelligence and machine learning to nanotechnology and regenerative medicine, has the potential to alter healthcare. Despite the difficulties, the future of terahertz technology in biomedicine appears bright. Researchers are continually attempting to solve the limits of terahertz technologies and enhance their capabilities.

One area of emphasis is the advancement of improved terahertz imaging systems. Scientists hope to create higher-quality photos with greater detail by enhancing the resolution and sensitivity of terahertz detectors. This will allow for more precise diagnosis and improved treatment planning. Another area of study is the creation of terahertz waveguides and fibres. These technologies attempt to improve terahertz wave propagation, allowing for greater transmission lengths and deeper penetration into biological tissues. This breakthrough broadens the potential uses of terahertz technology in biology. Additionally, efforts are being undertaken to enhance the computer methods and data processing techniques utilized in terahertz imaging and spectroscopy. Researchers can extract valuable information from terahertz data and increase the accuracy of terahertz measurements by inventing more efficient algorithms.

Terahertz radiation can excite low-frequency vibrations of biomolecules, providing spectral signatures for identification and characterization. Terahertz spectroscopy enables label-free detection of blood cells, cancer cells, and bacteria with high accuracy by exploiting their distinct "terahertz fingerprints". Key techniques like FTS, photo mixing spectrometer, and THz-TDS facilitate spectroscopic analysis of biological samples by measuring amplitude and phase information. Terahertz technology holds promise for developing rapid, non-invasive biosensors for point-of-care diagnosis of diseases (Fig. 19.3).

Terahertz imaging allows visualization of tissues and discrimination between normal and pathological conditions with high contrast. Techniques like TPI and CW THz imaging enable 2D and 3D terahertz imaging for medical applications. TPI provides depth resolution and 3D visualization by measuring time-of-flight information. CW THz imaging [23, 24] offers simpler instrumentation and faster acquisition through continuous-wave illumination. Ongoing research aims to advance in vivo terahertz imaging in clinical settings. Key challenges include improving imaging resolution, increasing penetration depth in tissues, and assessing biological

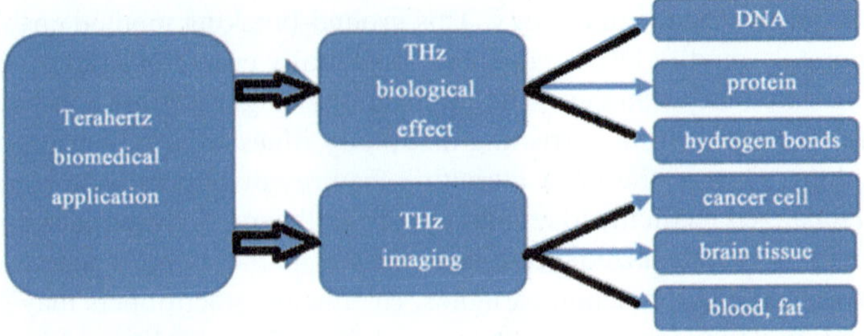

Fig. 19.3 THz in biomedical application

effects of terahertz radiation. Emerging techniques like near-field imaging and higher frequency excitation aim to enhance resolution. Higher power sources and detectors are being developed to enable deeper tissue imaging. More biological studies are needed to establish safety standards for clinical use. The future outlook is promising for terahertz biosensors and in vivo imaging applications with technological advancements. In summary, terahertz technology plays an important role in biomedicine owing to its ability to probe biomolecular interactions and enable tissue discrimination. With continued research to overcome existing challenges, terahertz-based techniques hold enormous potential to transform diagnostic and imaging applications in the clinical setting.

19.5 Case Studies: Terahertz Technology in Biomedical Research

The uses of THz characterization methods along with their biological impacts in biomedicine from 2010 to 2018 are reviewed by Gong et al. [25]. The five sections that make up this review of THz characterization techniques are: amino acid plus polypeptide, DNA, protein, cancer discovery, and supplementary applications. The study of THz's biological impacts focuses mostly on how it affects both cells and tissue of organisms. Additionally, it provides a basic overview of the chemometric techniques, which are mostly used in multivariate analysis and data preparation.

It is possible to use THz absorption and/or reflection during cancer diagnostics because cancer cells have a water content that is much greater than that of healthy cells and since these techniques are more sensitive to water content. Recently created very sensitive plasmonic TeraFET detectors and TeraFET arrays used in standard Si CMOS technology may detect small amounts of biological samples and scan enormous regions of biological tissues having a high sensitivity and nanometre scale resolution [26].

Terahertz imaging along with spectroscopy has showcased ability to differentiate tissue categories of expunged breast cancer tumours. Mostly terahertz technology uses 0.1–4 THz to detect cancerous tissue. In work by El-Shenawee et al. [27], pictures of breast tumours expunged from human along with animal models were reviewed.

Despite the overwhelming craving to adopt THz imaging as a benign and non-invasive tool early cancer diagnosis, a substantial quantity of interstitial water present in living tissue might make image interpretation challenging. As a result, THz is ineffective for cancer discovery, particularly in deep-seated tumours. When it comes to boosting the precision and efficacy of THz imaging in preoperative as well as intraoperative assessment, contrast chemicals could prove to be a game changer. Recent advancements in THz advancements and the affordability of THz imaging employing nanoparticle contrast agents might be able to overcome inadequate in vivo sensitivity and suggestively increase image contrast. Such advancements allow THz imaging to be used as a novel diagnostic device in cancer diagnosis [28].

Study by Wang [29] examined the viability of identifying breast illness using THz imaging methods. Discussions were held about breast tissue's dielectric characteristics, THz imaging and spectroscopic methods, THz radiation sources, and THz imaging for the diagnosis of breast cancer. Viability of enhancing THz picture resolution with machine learning classifiers was also covered in this research. It has been suggested that THz imaging might be used to image breast tumours during operations.

An overview of terahertz (THz) technology improvements towards optimized, compact THz devices that are miniaturized is provided by Gezimati and Singh [30]. The creation and detection of THz radiation, THz methods, and their biological applications for cancer diagnosis were main areas of interest of said study. Future research is anticipated by recognizing THz technology's problems and suggesting that THz systems be integrated with AI as well as IoT.

THz endoscopes are being developed by researchers employing photoconductive antennas that must operate at higher voltage. Doradla et al. [31] describe the design and crucial steps of developing a single-channel terahertz endoscopic system.

In order to capture ex vivo pictures of recent human colonic excisions, Doradla et al. [32] presented a reflecting continuous-wave THz imaging scheme. Using a polarization-specific detection method, reflection measurements of colorectal tissues in 5-mm-thick slices were obtained. At 584 GHz frequency, there is a clear distinction between healthy and tumorous tissues.

Terahertz (THz) technology has found applications in astronomy, earth research, and flexible radars along with the identification of hidden weapons. THz technology's potential killer applications include 5G Wi-Fi along with the IoT. All of these potential uses might be further increased by plasmonic TeraFETs with feature sizes as tiny as 3 nm, which are similar to Si CMOS [33].

The THz spectroscopy and picture data preparation, reconstruction techniques, and qualitative and quantitative analysis are the technological and major hurdles of machine learning approaches. An extensive evaluation of current relevant efforts using THz detection, imaging, and machine learning approaches is provided by Jiang et al. [34].

Terahertz waves are particularly useful for identifying cellular as well as biological macromolecules. Water has a high terahertz absorption, which has created a bottleneck in the identification of cells using terahertz technology. Shi et al. [35] provide a series of THz single measurement systems based upon the tilt wavefront of grating pulse method.

Work by Cherkasova et al. [36] focuses at glioma molecular markers in brain tissues and bodily fluids, explains how they develop, and provides established techniques of examination. The most important optical characteristics of glioma markers throughout THz range are also highlighted. At THz frequencies, novel metamaterial-based technologies for molecular marker detection are addressed. A variety of machine learning techniques are being investigated in order to improve the sensitivity and differentiation of healthy and cancer tissues utilizing THz equipment. Kumar et al. [37] present the design and procedures involved in building a terahertz antenna operating at 257.04 THz suitable for in vivo lung cancer diagnostics. The proposed antenna aids in the identification and separation of distinct phases of lung cancer.

Colorectal cancer is associated with chronic colitis. Terahertz technology, thanks to its non-invasive and unique spectroscopic capabilities, has the potential to aid in disease diagnostics. The goal of this study is intended to look at the utilization of terahertz technology in colitis-associated cancer utilizing a mouse model [38].

THz scanning reflectometry, THz 3D imaging, as well as THz-TDS were used to detect and compare features in basal cell carcinoma (BCC)-positive human skin biopsy samples to healthy skin samples. Healthy skin samples displayed a normal cellular sequence, while BCC skin samples did not, according to the 3D images [39].

A circular disc metasurface with a single as well as array of rectangular slots is proposed as a non-ionized approach in the THz spectrum suitable for cancer detection at an early stage, by Tabatabaeian [40]. The rectangular slot is used as the primary model for 0.86 THz to concentrate energy and provide a higher Q-factor. The second phase includes employing slot arrays, and the results show that the Q-factor improves from 1331.8 in the first construction to 1898.5 in the second structure.

Oh et al. demonstrated THz reflection imaging could discriminate between normal and tumour brain tissue in rat models [41]. Ji et al. showed THz reflectance imaging could identify the border between glioma and normal brain tissue [42]. Yamaguchi et al. [43] found the tumour regions had higher refractive indices and absorption coefficients compared to normal tissue. Meng et al. [44] reported differences in the complex refractive index between paraffin-embedded gliomas and normal brain tissue. Emerging techniques include contrast agents, tissue preparation methods, THz tomography, THz endoscopy, and integration with ultrasound imaging to improve imaging depth and resolution [45, 46].

19.6 Challenges and Limitations of Terahertz Technology in Biomedical Application

Biomedical applications hold great promise for improving healthcare through earlier diagnosis, more targeted treatments, and better monitoring. However, realizing this potential faces several key challenges. We can segregate these challenges across five areas: technical, ethical, regulatory, cost, and integration. Understanding and addressing these challenges is critical for advancing biomedical innovation in a responsible manner.

19.6.1 Technical Challenges

Several technical challenges arise in developing and applying biomedical technologies:

- **Complex systems integration**: Biomedical systems must integrate hardware, software, imaging, sensors, robotics, and analytical components into complex architectures. Ensuring seamless integration and interoperability is difficult.
- **Data volume and variety**: Biomedical data comes in many formats from diverse sources. Managing and analysing massive, heterogeneous datasets requires scalable storage, processing power, and analytics.
- **Robust models and algorithms**: Developing robust, accurate and transparent analytical models requires high-quality training data. However, biases can be introduced during data collection and processing.
- **Data security**: Securing biomedical data against breaches requires strong access controls, encryption, and monitoring. Attackers seek health data for identity theft and fraud.
- **Privacy protection**: De-identifying biomedical data for research must balance privacy risks and data utility. But re-identification of anonymized data may still be possible.

19.6.2 Ethical Challenges

Several ethical issues arise with biomedical applications:

- **Informed consent**: Patients may not fully understand how their data is used. Dynamic consent processes that allow ongoing control are being explored.
- **Equitable access**: Disparities exist in access to advanced biomedical treatments. Cost, geography, bias and other factors contribute to this.
- **Algorithmic bias**: Biomedical algorithms can perpetuate existing biases in data. Ongoing audits are needed to assess fairness.

- **Dual use**: Biomedical advances like gene editing have dual use potential if misapplied. Governance processes should evaluate risks.
- **Oversight**: Independent oversight is needed to assess emerging biomedical applications against ethical principles of beneficence, autonomy and justice.

19.6.3 Regulatory Challenges

Several key regulatory challenges exist:

- **Pace of innovation**: Regulatory processes struggle to keep up with rapid biomedical innovation. More agile, adaptive approaches are needed.
- **Safety vs access**: Regulators must balance patient safety with timely access to beneficial treatments, especially for serious conditions.
- **Global coordination**: Regulatory frameworks differ across countries, making it difficult to launch treatments globally and share data.
- **Accountability**: With complex biomedical systems, accountability for adverse events can be unclear. Improved traceability is needed.
- **Standards**: Biomedical research lacks standardized processes and protocols, hindering reproducibility and translation.
- **Transparency**: Details of clinical trials and regulatory submissions are often opaque. Greater transparency would build trust.

19.6.4 Cost Challenges

Several cost-related challenges exist:

- High development costs: Bringing a new biomedical product to market can cost billions of dollars over many years. Most candidates fail to reach approval.
- Reimbursement: Innovative products may not fit existing reimbursement models. New value-based schemes are needed to incentivize adoption.
- Sustainable funding: Public research funding is limited. Private investment tends to focus on later stages, creating a "valley of death".
- Manufacturing: Scaling complex personalized biologics like cell therapies is costly. Improving processes could expand access.

19.6.5 Integration Challenges

Integrating biomedical applications into healthcare systems also poses challenges:

- **Clinical workflows**: New biomedical technologies can disrupt established clinical workflows. Change management helps drive adoption.

- **Training**: Healthcare workers require extensive training on new biomedical systems. Shortages of skilled staff are a constraint.
- **Interoperability**: Biomedical data must be integrated into electronic health records. Standards are lacking.
- **Infrastructure**: Many healthcare settings lack the infrastructure to support advanced biomedical technologies.

19.7 Conclusion

Terahertz technology has advanced significantly since 2010, leading to new applications. Key innovations were in improved semiconductor, photonic and electronic terahertz sources, detectors, and components. Research expanded rapidly in areas like communications, imaging, and spectroscopy. Prospects are positive for terahertz technology in 6G networks, healthcare, manufacturing, and science. Ongoing challenges include power, efficiency, cost, attenuation, and integration that need to be addressed. The current paper provides a thorough understanding of Terahertz technology including its applicability in numerous biological applications.

References

1. Siegel, P.H.: Terahertz technology. IEEE Trans. Microw. Theory Tech. **50**(3), 910–928 (2002)
2. Tse, D., Viswanath, P.: Fundamentals of Wireless Communication. Cambridge University Press, Cambridge (2005)
3. Haxhibeqiri, J., De Poorter, E., Moerman, I., Hoebeke, J.: A survey of LoRaWAN for IoT: from technology to application. Sensors **18**(11), 3995 (2018)
4. El Ghzaoui, M., Das, S.: Data transmission with terahertz communication systems. In: Emerging Trends in Terahertz Solid-State Physics and Devices: Sources, Detectors, Advanced Materials, and Light-matter Interactions, pp. 121–141 (2020)
5. Siegel, P.H.: Terahertz technology in biology and medicine. IEEE Trans. Microw. Theory Tech. **52**(10), 2438–2447 (2004)
6. Hangyo, M.: Development and future prospects of terahertz technology. Jpn. J. Appl. Phys. **54**(12), 120101 (2015)
7. Nagatsuma, T.: Terahertz technologies: present and future. IEICE Electron. Expr. **8**(14), 1127–1142 (2011)
8. Kemp, M.C., Taday, P.F., Cole, B.E., Cluff, J.A., Fitzgerald, A.J., Tribe, W.R.: Security applications of terahertz technology. In: Terahertz for Military and Security Applications, vol. 5070, pp. 44–52. SPIE (2003)
9. Hosako, I., Sekine, N., Patrashin, M., Saito, S., Fukunaga, K., Kasai, Y., Baron, P., et al.: At the dawn of a new era in terahertz technology. Proc. IEEE **95**(8), 1611–1623 (2007)
10. Rostami, A., Rasooli, H., Baghban, H.: Terahertz Technology: Fundamentals and Applications, vol. 77. Springer Science & Business Media, Berlin (2010)
11. El Ghzaoui, M., Das, S., Lenka, T.R., Biswas, A.: Terahertz Wireless Communication Components and System Technologies. Terahertz Wireless Communication Components and System Technologies, pp. 1–310 (2022)
12. Shur, M.: Terahertz technology: devices and applications. In: Proceedings of the 31st European Solid-State Circuits Conference, 2005. ESSCIRC 2005, pp. 13–21. IEEE (2005)

13. Arnone, D.D., Ciesla, C.M., Corchia, A., Egusa, S., Pepper, M., Martyn Chamberlain, J., Bezant, C., Linfield, E.H., Clothier, R., Khammo, N.: Applications of terahertz (THz) technology to medical imaging. In: Terahertz Spectroscopy and Applications II, vol. 3828, pp. 209–219. SPIE (1999)
14. Elaage, S., et al.: MB-OOK transceiver design for terahertz wireless communication systems. Int. J. Syst. Control Commun. **12**(4), 309–326 (2021)
15. Tribe, W.R., Newnham, D.A., Taday, P.F., Kemp, M.C.: Hidden object detection: security applications of terahertz technology. In: Terahertz and Gigahertz Electronics and Photonics III, vol. 5354, pp. 168–176. SPIE (2004)
16. Mittleman, D.M., Gupta, M., Neelamani, R., Baraniuk, R.G., Rudd, J.V., Koch, M.: Recent advances in terahertz imaging. Appl. Phys. B **68**, 1085–1094 (1999)
17. Mittleman, D.M.: Twenty years of terahertz imaging. Opt. Expr. **26**(8), 9417–9431 (2018)
18. Valušis, G., Lisauskas, A., Yuan, H., Knap, W., Roskos, H.G.: Roadmap of terahertz imaging 2021. Sensors **21**(12), 4092 (2021)
19. Jansen, C., Wietzke, S., Peters, O., Scheller, M., Vieweg, N., Salhi, M., Krumbholz, N., Jördens, C., Hochrein, T., Koch, M.: Terahertz imaging: applications and perspectives. Appl. Opt. **49**(19), E48–E57 (2010)
20. Guerboukha, H., Nallappan, K., Skorobogatiy, M.: Toward real-time terahertz imaging. Adv. Opt. Photon. **10**(4), 843–938 (2018)
21. Castro-Camus, E., Koch, M., Mittleman, D.M.: Recent advances in terahertz imaging: 1999 to 2021. Appl. Phys. B **128**(1), 12 (2022)
22. Chan, W.L., Deibel, J., Mittleman, D.M.: Imaging with terahertz radiation. Rep. Prog. Phys. **70**(8), 1325 (2007)
23. Mittleman, D.: Terahertz imaging. In: Sensing with Terahertz Radiation, pp. 117–153. Springer, Berlin (2003)
24. Yu, C., Fan, S., Sun, Y., Pickwell-MacPherson, E.: The potential of terahertz imaging for cancer diagnosis: a review of investigations to date. Quant. Imaging Med. Surg. **2**(1), 33 (2012)
25. Gong, A., Qiu, Y., Chen, X., Zhao, Z., Xia, L., Shao, Y.: Biomedical applications of terahertz technology. Appl. Spectrosc. Rev. **55**(5), 418–438 (2020)
26. Shur, M., Liu, X.: Biomedical applications of terahertz technology. Adv. Terahertz Biomed. Imaging Spectro. **11975**, 1197502 (2022)
27. El-Shenawee, M., Vohra, N., Bowman, T., Bailey, K.: Cancer detection in excised breast tumors using terahertz imaging and spectroscopy. Biomed. Spectro. Imaging **8**(1–2), 1–9 (2019)
28. Sadeghi, A., Hossein Naghavi, S.M., Mozafari, M., Afshari, E.: Nanoscale biomaterials for terahertz imaging: a non-invasive approach for early cancer detection. Transl. Oncol. **27**, 101565 (2023)
29. Wang, L.: Terahertz imaging for breast cancer detection. Sensors **21**(19), 6465 (2021)
30. Gezimati, M., Singh, G.: Advances in terahertz instrumentation and technology for cancer applications. In: 2022 47th International Conference on Infrared, Millimeter and Terahertz Waves (IRMMW-THz), pp. 1–2. IEEE (2022)
31. Doradla, P., Alavi, K., Joseph, C.S., Giles, R.H.: Development of terahertz endoscopic system for cancer detection. In: Terahertz, RF, Millimeter, and Submillimeter-Wave Technology and Applications IX, vol. 9747, pp. 63–74. SPIE (2016)
32. Doradla, P., Alavi, K., Joseph, C., Giles, R.: Detection of colon cancer by continuous-wave terahertz polarization imaging technique. J. Biomed. Opt. **18**(9), 090504–090504 (2013)
33. Shur, M.S.: Terahertz plasmonic technology. IEEE Sens. J. **21**(11), 12752–12763 (2021). https://doi.org/10.1109/JSEN.2020.3022809
34. Jiang, Y., et al.: Machine learning and application in terahertz technology: a review on achievements and future challenges. IEEE Access **10**, 53761–53776 (2022). https://doi.org/10.1109/ACCESS.2022.3174595
35. Shi, W., Wang, Y., Hou, L., Ma, C., Yang, L., Dong, C., Wang, Z., et al.: Detection of living cervical cancer cells by transient terahertz spectroscopy. J. Biophoton. **14**(1), e202000237 (2021)

36. Cherkasova, O., Peng, Y., Konnikova, M., Kistenev, Y., Shi, C., Vrazhnov, D., Shevelev, O., Zavjalov, E., Kuznetsov, S., Shkurinov, A.: Diagnosis of glioma molecular markers by terahertz technologies. Photonics **8**(1), 22 (2021)
37. Kumar, M., Goel, S., Rajawat, A., Gupta, S.H.: Design of optical antenna operating at terahertz frequency for in-vivo cancer detection. Optik **216**, 164910 (2020)
38. Chen, Y., Li, D., Liu, Y., Hu, L., Qi, Y., Huang, Y., Zhang, Z., et al.: Optimal frequency determination for terahertz technology-based detection of colitis-related cancer in mice. J. Biophoton. e202300193 (2023)
39. Rahman, A., Rahman, A.K., Rao, B.: Early detection of skin cancer via terahertz spectral profiling and 3D imaging. Biosens. Bioelectron. **82**, 64–70 (2016)
40. Tabatabaeian, Z.S.: Developing THz metasurface with array rectangular slot with high Q-factor for early skin cancer detection. Optik **264**, 169400 (2022)
41. Oh, S.J., Kang, J., Maeng, I., Suh, J.-S., Huh, Y.-M., Haam, S., Son, J.-H.: Nanoparticle-enabled terahertz imaging for cancer diagnosis. Opt. Expr. **22**(19), 22720–22726 (2014). https://doi.org/10.1364/OE.22.022720
42. Ji, X., Huang, L., Wang, H., Zhu, X., Zhao, Z.: Terahertz reflectance imaging for glioma identification during neurosurgery. Biomed. Opt. Expr. **8**(1), 238–248 (2016). https://doi.org/10.1364/BOE.8.000238
43. Yamaguchi, S., Fukushi, Y., Kubota, O., Itsuji, T., Ouchi, T., Yamamoto, S.: Brain tumor imaging of rat fresh tissue using terahertz spectroscopy. Sci. Rep. **6**(1), 1–6 (2016). https://doi.org/10.1038/srep30124
44. Meng, K., Chen, T., Chen, Y., Li, Z., Li, X., Wu, X.: Terahertz pulsed spectroscopy of paraffin-embedded brain glioma. J. Biomed. Opt. **19**(7), 077001 (2014). https://doi.org/10.1117/1.JBO.19.7.077001
45. Yamaguchi, S., Fukushi, Y., Kubota, O., Itsuji, T., Yamamoto, S., Ouchi, T.: Effective enhancement of refraction contrast for terahertz imaging of fresh tissue without refractive index matching. Biomed. Opt. Expr. **7**(10), 4241–4249 (2016). https://doi.org/10.1364/BOE.7.004241
46. Ji, X., Huang, L., Wang, H., Zhu, X., Zhao, Z., Li, W.: Terahertz dual-mode imaging through an extremely narrow channel for brain tumor surgical resection. Biomed. Opt. Expr. **8**(9), 4273–4283 (2017). https://doi.org/10.1364/BOE.8.004273

Chapter 20
Design and Analysis of Frequency Selective Surface for Gain Enhancement in Terahertz Applications

K. Vasu Babu, Gorre Naga Jyothi Sree, Tanvir Islam, Sudipta Das, and Bhuma Anuradha

20.1 Introduction

In the current era, THz frequencies' requirement is rigorously explored in the field of wireless systems which provide enormous bandwidth and high data rates for the demand of unlicensed frequency spectrum and massive data traffic. A single layered magnetoelectric dipole radiator designed for broad band for new 5G systems [1], tree-shaped THz micro-scaled design [2], for broadband gain enhancement-printed antenna resonant structure with H-shaped [3], micro-sized structure MIMO antenna using graphene material optimized [4], for 5G systems designed broad-band improved FSS filter [5], a photonic crystal embedded fractal structure THz

K. Vasu Babu (✉)
Department of Electronics and Communication Engineering, BVRIT HYDERABAD College of Engineering for Women, Bachupally, Telangana, India
e-mail: vasubabuece@gmail.com

G. N. J. Sree
Department of Electronics and Communication Engineering, Vasireddy Venkatadri Institute of Technology, Guntur, Andhra Pradesh, India

T. Islam
Department of Electrical and Computer Engineering, University of Houston, Houston, TX 77204, USA

S. Das
Department of Electronics and Communication Engineering, IMPS College of Engineering and Technology, Malda, West Bengal, India

B. Anuradha
Department of Electronics and Communication Engineering, S V University, Tirupati, Andhra Pradesh, India

M. El Ghzaoui et al. (eds.), *Next Generation Wireless Communication*, Signals and Communication Technology, https://doi.org/10.1007/978-3-031-56144-3_20

band antenna designed [6], applications, challenges, research directions and technologies 6G wireless systems [7], opportunities beyond 5G system THz communications [8], U-shaped radiator for wireless system designed the dual-band compact structure [9], fractal slotted deep learning MIMO radiator for 5G NR applications designed and experimental validation [10], bandpass FSS design in THz band [11], fractal-based co-axial feeding THz antenna analyzed and designed [12], for THz sensing designed and enhanced FSS structure [13], graphene constituted metamaterial absorber designed with circuit analysis approach [14], a microstructure photothermal FSS direct wiring designed at THz frequency [15], a disk structure inserted monopole ground plane THz design for wireless systems [16], a BPF for THz system on FSS [17], a CPW feeding fractal MIMO THz designed system using parasitic elements analyzed and designed [18], an optimum planar FSS structure with square loop using various dielectric substrates [19], a broadband metamaterial absorber for THz region designed the graphene constituted explained [20], for isolation improvement designed and implemented MIMO patch antenna with graphene material [21], a fractal MIMO antenna for sub-6-GHz regions analyzed and designed [22], in the region of graphene nanophononics applications designed advanced and recent structures implementations [23], a dual-band MIMO structure antenna simulated and designed for 5G regions and radar [24] and a Minkowski structure dual-band MIMO radiator for better spacing of antenna elements [25].

20.2 Unit Cell and Frequency Selective Surface Geometry

Meta surfaces are employed for the manipulation of polarization, amplitude, and phase of terahertz waves which have more limitations in the calculation of radiation efficiency. These problems can be encountered by losses in material constituting the various devices. Metallic-type resonators related to terahertz suffer from ohmic losses. With the help of spacers and dielectric substrates loss tangent and high permittivity also reduce efficiency and bandwidth. The designed unit cell with gold coated of 35 nm structure of cross-I-shaped with curved ends of semicircular oriented by principal diagonal using the polyimide substrate having thickness of 3 μm and permittivity of 3.5. The unit cell dimensions are $38 \times 38\ \mu m^2$. Figure 20.1 shows the geometry of unit cell. Figure 20.2 shows the surface current distribution of the designed unit cell. The maximum number of current flows through I-shaped structure, and minimum amount of current flows through the end of the patches. Figure 20.3 shows the E-field distribution of proposed unit cell representation with a value of 1.12 V/m. Figure 20.4 shows the H-field distribution of proposed unit cell representation with a value of 47,701 A/m.

Figure 20.5 shows the patch design L-slots inserted patch design using the substrate of silicon dioxide with monopole ground plane, and it resonates at 3.8 THz. Silicon dioxide loss tangent 0.0004 and permittivity 3.75 are considered in the current design. The dimensions of the current design are 50 μm $\times$ 50 μm, and patch dimensions are 30 μm $\times$ 30 μm with monopole ground structure for better

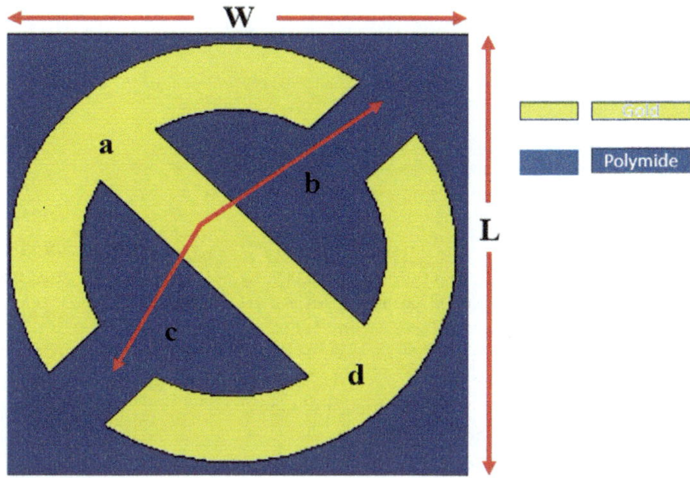

Fig. 20.1 Unit cell representation

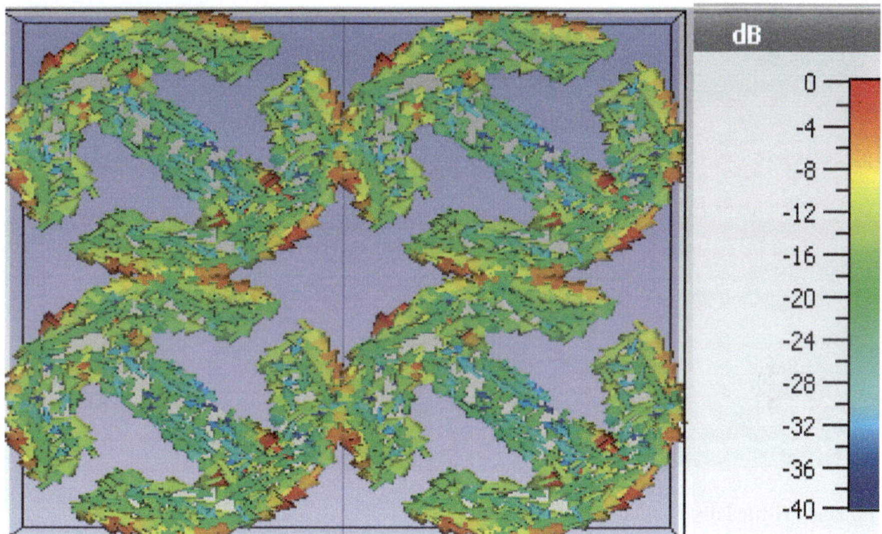

Fig. 20.2 Surface current distribution of unit cell

return loss. At the resonant frequency of 3.81 THz for patch design, its S_{11} is − 36. 58 dB which is shown in Fig. 20.6. Figure 20.7 shows the FSS design combination of unit cell and patch arrangement. In the frequency selective surface unit cell array and patch are placed 8 μm away. Figure 20.8 shows the return loss of FSS design is resonate at 5.25 THz varying the wider bandwidth of 2.41–6.66 THz with S_{11} of − 24.53 dB. Figure 20.9 shows the surface distribution of FSS designed structure

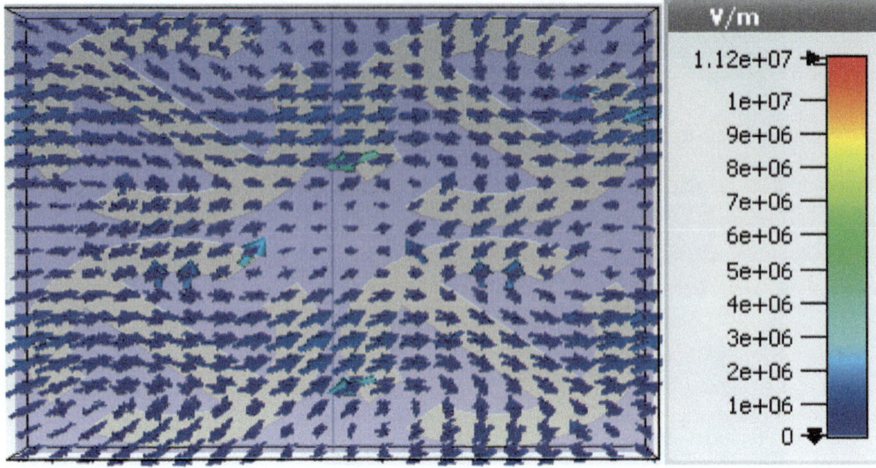

Fig. 20.3 E-filed distribution of unit cell

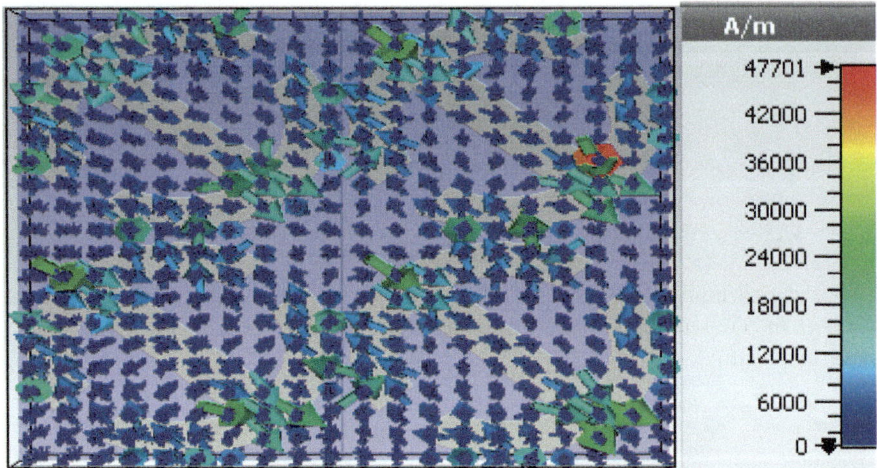

Fig. 20.4 H-filed distribution of unit cell

more current obtained at the center of the patch and designed unit cell representation of I-shaped structure. Figure 20.10 shows E-filed representation of FSS designed structure more electric filed at the center of the designed structure and less amount of electric filed generation at the edges of the patch design. Figure 20.11 shows H-field representation of FSS designed structure with more magnetic field at the center of the designed structure and less amount of magnetic field generation at the edges of the patch design.

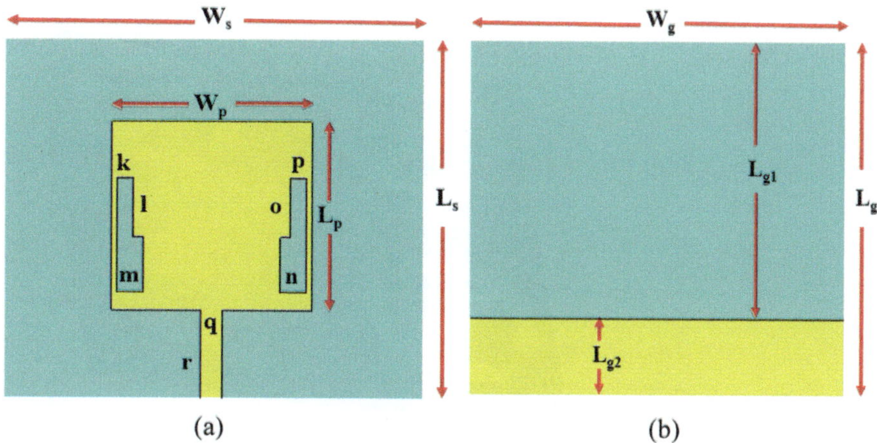

Fig. 20.5 Microstrip patch design

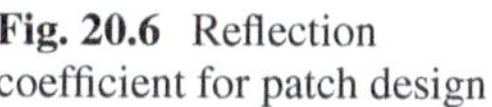
Fig. 20.6 Reflection coefficient for patch design

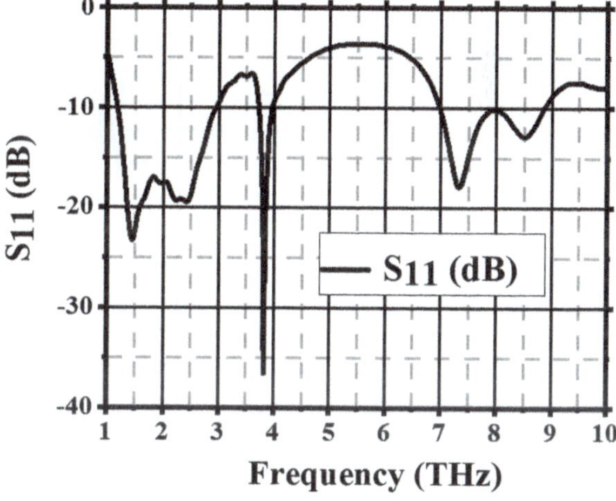

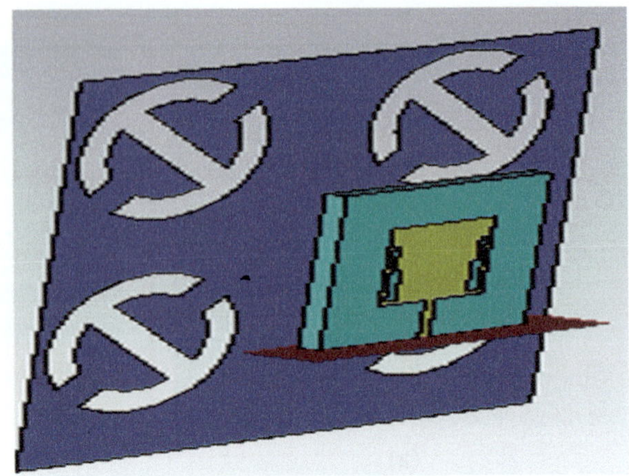

Fig. 20.7 FSS design with patch antenna

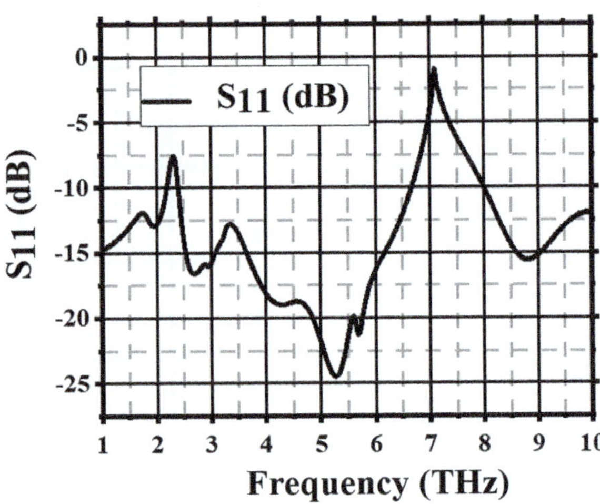

Fig. 20.8 Reflection coefficient of FSS designed structure

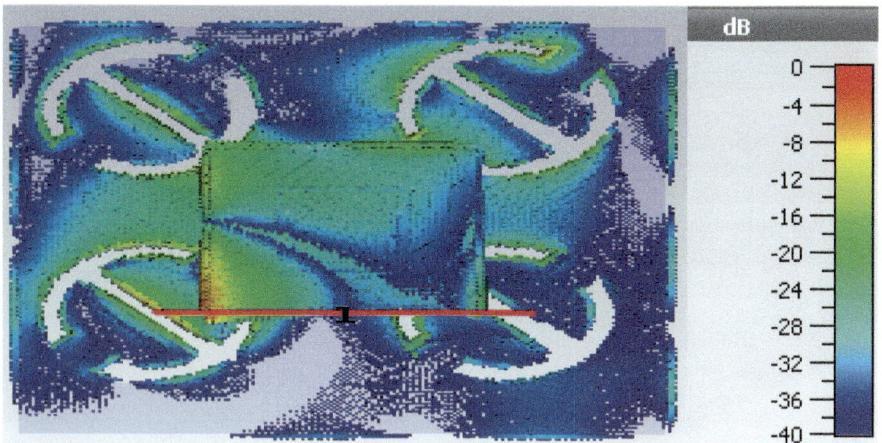

Fig. 20.9 Surface current of FSS designed structure

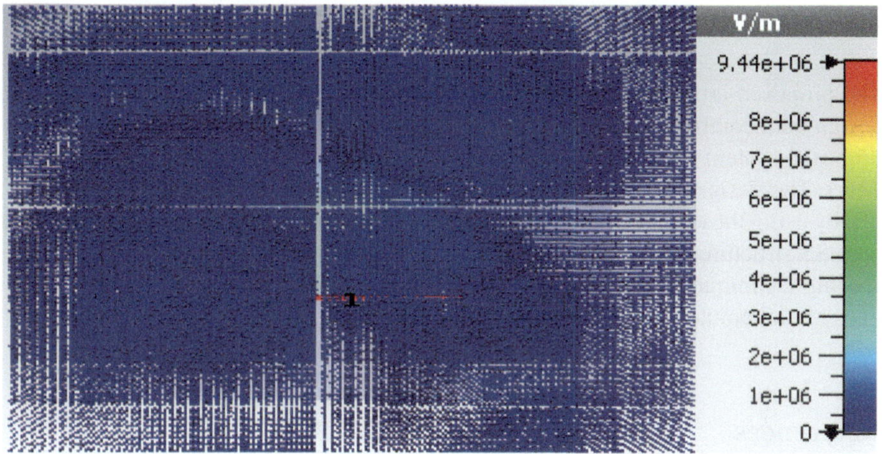

Fig. 20.10 E-filed distribution of FSS designed structure

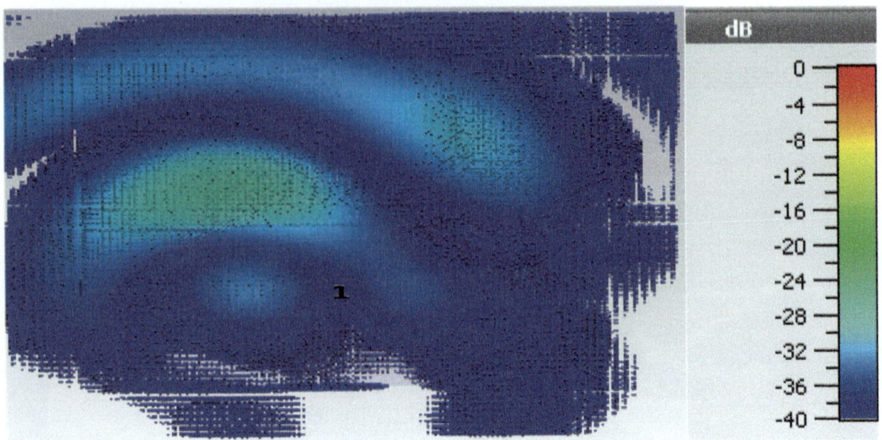

Fig. 20.11 H-filed distribution of FSS designed structure

20.3 Conclusion

The proposed article provides a low-cost highly gain enhancement method of FSS design with combination of FSS structure with patch that has been investigated. The proposed system provided the wide impedance bandwidth of 2.41–6.66 THz (4.25 THz). The designed structure obtained the greater improvement in the gain of 9.7 dBi by using the technique of FSS structure and placing the patch antenna away from the FSS structure at a distance of 8 μm has been observed. With the help of spatial feeding technique and planar structure employment, the designed FSS structure is used in the remote wireless communications.

References

1. Zeng, J., Luk, K.-M.: Single-layered broadband magnetoelectric dipole antenna for new 5G application. IEEE Antennas Wirel. Propag. Lett. **18**(5), 911–915 (2019)
2. Vasu Babu, K., et al.: A micro-scaled graphene-based tree-shaped wideband printed MIMO antenna for terahertz applications. J. Comput. Electron. **21**(1), 289–303 (2022)
3. Wang, H., et al.: Gain enhancement for broadband vertical planar printed antenna with H-shaped resonator structures. IEEE Trans. Antennas Propag. **62**(8), 4411–4415 (2014)
4. Vasu Babu, K., et al.: Design and optimization of micro-sized wideband fractal MIMO antenna based on characteristic analysis of graphene for terahertz applications. Opt. Quantum Electron. **54**(5), 281 (2022)
5. Wu, T.-K.L: Improved broadband bandpass FSS filters for 5G applications. In: 2018 IEEE International Symposium on Antennas and Propagation & USNC/URSI National Radio Science Meeting. IEEE (2018)
6. Vasu Babu, K., et al.: Performance analysis of a photonic crystals embedded wideband (1.41–3.0 THz) fractal MIMO antenna over SiO_2 substrate for terahertz band applications. Silicon 1–14 (2023)

7. Chowdhury, M.Z., et al.: 6G wireless communication systems: applications, requirements, technologies, challenges, and research directions. IEEE Open J. Commun. Soc. **1**, 957–975 (2020)
8. Elayan, H., et al.: Terahertz communication: the opportunities of wireless technology beyond 5G. In: 2018 International Conference on Advanced Communication Technologies and Networking (CommNet). IEEE (2018)
9. Vasu Babu, K., et al.: Compact dual-band design and analysis of half-circular U-shape MIMO radiator for wireless applications. Microsyst. Technol. **29**(4), 501–514 (2023)
10. Vasu Babu, K., et al.: Deep learning assisted fractal slotted substrate MIMO antenna with characteristic mode analysis (CMA) for sub-6 GHz n78 5 G NR applications: design, optimization and experimental validation. Phys. Scr. **98**(11), 115526 (2023)
11. Chen, B.J., Chan, C.H.: High-selectivity bandpass frequency-selective surface in terahertz band. IEEE Trans. Terahertz Sci. Technol. **6**(2), 284–291 (2016)
12. Vasu Babu, K., et al.: Design and analysis of fractal-based THz antenna with co-axial feeding technique for wireless applications. In: Recent Advances in Graphene Nanophotonics, pp. 351–358. Springer Nature, Cham (2023)
13. El-Rayes, S., et al.: Enhancing the selectivity of frequency selective surfaces for terahertz sensing applications. In: 2015 8th UK, Europe, China Millimeter Waves and THz Technology Workshop (UCMMT). IEEE (2015)
14. Vasu Babu, K., Sree, G.N.J.: Design and circuit analysis approach of graphene-based compact metamaterial-absorber for terahertz range applications. Opt. Quantum Electron. **55**(9), 769 (2023)
15. Chen, X., et al.: Photothermal direct writing of metallic microstructure for frequency selective surface at terahertz frequencies. In: 2012 International Workshop on Metamaterials (Meta). IEEE (2012)
16. Vasu Babu, K., et al.: Design of monopole ground graphene disc-inserted THz antenna for future wireless systems. In: Recent Advances in Graphene Nanophotonics. Springer Nature, Cham, pp. 305–312 (2023)
17. Li, J.-S., Li, Y., Zhang, L.:Terahertz bandpass filter based on frequency selective surface. IEEE Photon. Technol. Lett. **30**(3), 238–241 (2017)
18. Vasu Babu, K., et al.: Design and analysis of a CPW-fed fractal MIMO THz antenna using an array of parasitic elements. In: Terahertz Devices, Circuits and Systems: Materials, Methods and Applications, pp. 53–60. Springer Nature, Singapore (2022)
19. Seman, F.C., Khalid, N.K.: Design strategy for optimum planar square loop FSS with different dielectric substrates. In: Theory and Applications of Applied Electromagnetics: APPEIC 2014, pp. 87–94. Springer International Publishing, Cham (2015)
20. Vasu Babu, K., et al.: Design of graphene-based broadband metamaterial absorber with circuit analysis approach for terahertz region applications. Opt. Quantum Electron. **55**(13), 1188 (2023)
21. Vasu Babu, K., et al.:Design and implementation of MIMO graphene patch antenna to improve isolation for THz applications. Microsyst. Technol. 1–11 (2023)
22. Vasu Babu, K., et al.: Design and analysis of fractal type MIMO radiator for the applications of sub 6-GHz 5G systems. In: Microelectronics, Circuits and Systems: Select Proceedings of Micro2021, pp. 243–250. Springer Nature, Singapore (2023)
23. Patel, S.K., et al. (eds.): Recent Advances in Graphene Nanophotonics. Springer, Berlin (2023)
24. Vasu Babu, K., Kokkirigadda, S., Das, S.: Design and simulation of dual-band MIMO antenna for radar and sub-6-GHz 5G applications. In: Futuristic Communication and Network Technologies: Select Proceedings of VICFCNT 2020. Springer, Singapore (2022)
25. Vasu Babu, K., Anuradha, B.: A dual-band Minkowski-shaped MIMO antenna to reduce the mutual coupling. In: Optical and Wireless Technologies: Proceedings of OWT 2018. Springer, Singapore (2020)

Chapter 21
Terahertz Waves in Biomedicine: Pioneering Imaging and Sensing for Healthcare Revolution

Maitri Mohanty, Premansu Sekhara Rath, Ambarish G. Mohapatra, Anita Mohanty, and Sasmita Nayak

21.1 Introduction

The Terahertz frequency range, spanning from 0.1 to 10 THz, represents a remarkable frontier in biomedical applications, offering unique advantages for a wide range of uses. Unlike ionizing radiation such as X-rays or gamma rays, terahertz waves are non-ionizing, making them ideal for studying biological materials without causing damage. Their sensitivity to molecular vibrations and water content allows for in-depth analysis of biological cells and tissues, enabling the detection of tumor margins, early disease diagnosis, and non-invasive monitoring of wound healing processes. Figure 21.1 represents the advantages of terahertz waves over other forms of radiation in biomedical applications. The subsequent section describes the terahertz frequency range in biomedical applications.

M. Mohanty · P. S. Rath
Department of Computer Science and Engineering, GIET University, Gunupur, Odisha, India
e-mail: maitri.mohanty@giet.edu

P. S. Rath
e-mail: premansusekhararath@giet.edu

A. G. Mohapatra (✉) · A. Mohanty
Department of Electronics Engineering, Silicon University, Silicon Institute of Technology, Bhubaneswar, Odisha, India
e-mail: ambarish.mohapatra@gmail.com

A. Mohanty
e-mail: anitamohanty776@gmail.com

S. Nayak
Department of Mechanical Engineering, Government College of Engineering, Kalahandi, Odisha, India
e-mail: snayak@gcekbpatna.ac.in

M. El Ghzaoui et al. (eds.), *Next Generation Wireless Communication*, Signals and Communication Technology, https://doi.org/10.1007/978-3-031-56144-3_21

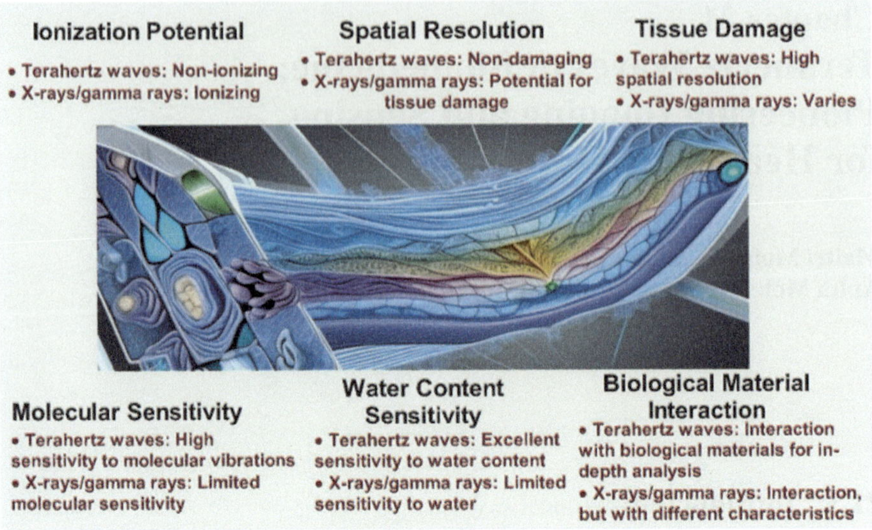

Fig. 21.1 Comparative analysis of biomedical imaging modalities

21.1.1 The Terahertz Frequency Range in Biomedical Application

The electromagnetic spectrum is a broad band of radiation with distinct qualities and possible uses in each section. Between microwaves as well as infrared radiation in this spectrum is the terahertz (THz) frequency range, which spans from 0.1 to 10 THz or wavelengths between 30 μm as well as 3 mm [1, 2]. This range's unique qualities—its non-ionizing nature and remarkable material penetrating capabilities—offer a commitment to revolutionary biomedical applications [3]. The terahertz wave frequency range holds the promise of permeating everyday materials such as plastics, clothing, and ceramics [4]. This distinct feature creates a wealth of new opportunities for imaging and sensing applications. Terahertz waves cause partial absorption in biological cells' water molecules when they interact with them. Figure 21.2 depicts a wide view of the representation of biomedical applications using terahertz radiation. It is feasible to use this absorption phenomenon to differentiate between sick as well as healthy cells [5]. Terahertz waves have unique properties that make them attractive for many applications in biomedicine, imaging, spectroscopy, communication, and technology [6, 7], as will be discussed below.

- Because terahertz waves are non-ionizing and have a remarkable sensitivity to molecular vibrations and water content, they are particularly well-suited for the investigation of biological materials. Unlike X-rays or gamma rays, these waves

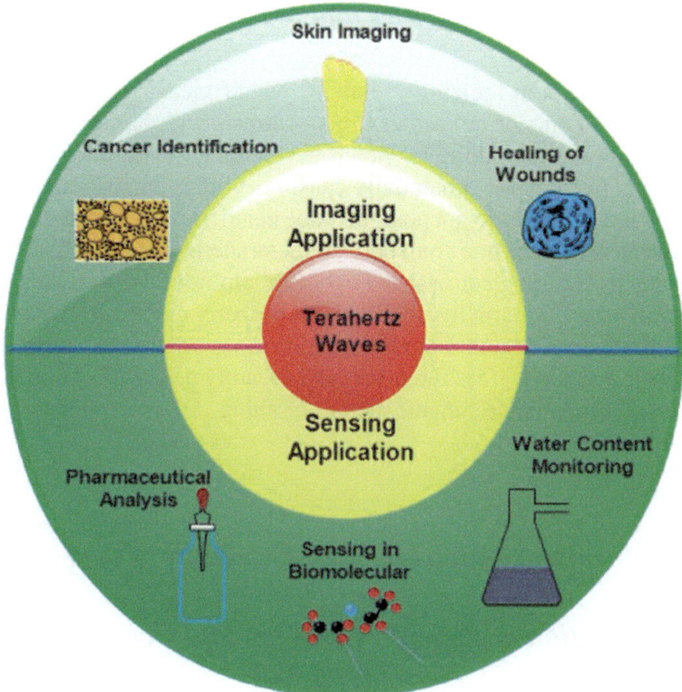

Fig. 21.2 Representation of biomedical application using terahertz waves

allow the structural and chemical characteristics of biological cells to be investigated without inflicting harm [8]. However, gamma or X-ray radiation can cause damage to living cells.

- The capacity to create three-dimensional, high-resolution photographs of biological cells and tissues facilitates non-invasive wound healing process monitoring, early illness diagnosis, and tumor margin detection [9].
- Researchers utilize terahertz spectroscopy to analyze the composition of biological tissue samples, offering valuable insights into the molecular makeup of these materials. It is useful for identifying and quantifying Biomolecules, such as proteins, nucleic acids, and lipids, contributing to advances in diagnostics and pharmaceutical research [10].
- Terahertz technology is employed in security applications, such as airport scanners, for the non-invasive detection of concealed weapons or explosives. This has potential applications in healthcare for non-invasive screening of patients and visitors in a clinical setting [11].
- Terahertz waves hold promise for high-speed wireless communication and data transfer due to their high frequency. Researchers are exploring THz technology for future applications in ultra-fast wireless networks and data transmission [12].

21.1.2 Non-ionizing Attributes and Material Penetration

Terahertz waves are non-ionizing radiation, which means they lack the energy required to ionize atoms or molecules. This contrasts with higher-energy radiation types like X-rays and gamma rays, which have enough energy to remove tightly, bound electrons from atoms, potentially causing cellular damage and DNA mutations [13]. Terahertz waves do not pose the same health risks associated with ionizing radiation. Because of their non-ionizing nature, terahertz waves are considered safe for interacting with biological tissues [14]. They do not have sufficient energy to break chemical bonds or harm living cells. This property makes terahertz waves suitable for medical imaging and diagnostic applications where the safety of patients and operators is a concern [15]. Terahertz waves readily penetrate common materials found in our everyday environment, such as clothing, ceramics, paper, plastic, and biological tissues.

21.2 Interaction of Materials with Terahertz Waves

21.2.1 Penetration of Plastics, Clothing, and Ceramics

- Penetration of Plastics:
  One major advantage of terahertz radiation in industrial and production environments is its ability to penetrate different types of polymeric materials. Plastic is frequently used in packaging materials, and in sectors like food and medicine, monitoring quality control depends on the integrity of the packaging. Plastic containers and packaging may be successfully penetrated by terahertz radiation, enabling a non-intrusive inspection of the contents without causing damage to the material. This non-destructive testing method aids in the preservation of items' quality and safety by identifying flaws, impurities, or anomalies [16].
- Penetration of Clothing:
  Terahertz waves can easily pass through clothing made of common materials, which is an amazing property with great potential for security applications. For example, terahertz imaging systems are used by airport security to find weapons or other hidden objects under clothes. Through the use of terahertz radiation in non-invasive body scans and the analysis of reflected signals, security officers may quickly find irregularities. Terahertz imaging thus turns out to be a useful instrument for non-intrusive security enhancement and privacy preservation [16].
- Penetration of Ceramics:
  Ceramics are widely used in crucial applications where structural integrity is of the highest significance. Terahertz waves are a good choice for evaluating ceramics because they can penetrate these materials to a limited extent. Terahertz non-destructive testing can be used to find hidden flaws, cracks, or structural weaknesses in ceramic components in industries including construction and aerospace.

In order to ensure the dependability and safety of vital systems, this procedure is essential [17]. Overall, the ability of terahertz waves to penetrate clothing, plastics, and ceramics underscores the enhancement of security, improving quality control, or ensuring structural integrity, and their versatility in diverse applications.

21.2.2 Water Absorption and Contrast in Biological Tissues

Terahertz waves, operating in the terahertz (THz) frequency range, interact uniquely with biological tissues due to their sensitivity to water absorption and the resulting contrast they provide. This section explores how terahertz waves interact with water in biological tissues for various biomedical applications.

- Water Absorption in Biological Tissues:
  One of the distinctive features of terahertz waves is their interaction with water molecules. In the THz frequency range, water molecules exhibit significant absorption, primarily due to the rotational and vibrational transitions of water's molecular structure. This absorption property makes terahertz waves highly sensitive to the water content within biological tissues [18].
- Contrast Enhancement in Imaging:
  Terahertz imaging in biomedical applications benefits from the strong water absorption characteristics of THz waves. Since biological tissues contain varying water content, terahertz imaging can provide excellent contrast between different tissue types [19]. This property is particularly valuable in medical imaging scenarios where distinguishing healthy tissue from diseased or abnormal tissue is essential.

21.3 Terahertz in Imaging Applications

In medical imaging, terahertz waves have shown promise in detecting and characterizing various conditions, such as skin diseases, cancerous tissues, and burns shown in Fig. 21.3. This imaging system can generate detailed images that highlight variations in water content, enabling healthcare professionals to identify anomalies in tissues without the need for ionizing radiation or invasive procedures [20]. This non-invasive and non-destructive imaging modality has the potential to improve early disease detection and treatment planning.

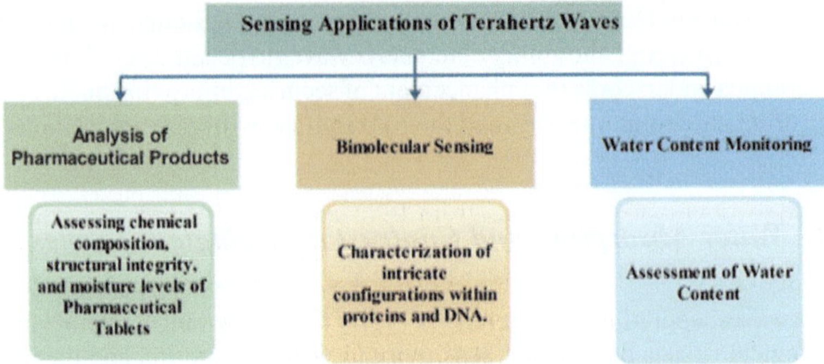

Fig. 21.3 Representation of imaging applications using terahertz waves

21.3.1 Cancer Identification

(a) Contrasting Terahertz Absorption Profiles in Healthy and Cancerous Tissues
Terahertz waves occupy a unique swath within the electromagnetic spectrum, rendering them exceptionally sensitive to the vibrational and rotational behaviors of molecules, particularly water molecules, within biological tissues. The presence of water molecules and their hydrogen bonding with other molecules constitutes a major contributor to terahertz absorption within tissues. Divergent terahertz absorption profiles characterize healthy and cancerous tissues, primarily stemming from disparities in their molecular makeup and water content. Changes in cell density, tissue architecture, also the appearance of pathological states in malignant tissues, are the reasons for these differences. In the terahertz absorption spectrum, healthy tissues typically show well-defined absorption peaks that correlate to the vibrational modes of water and other biomolecules. These peaks outline the expected characteristics of a typical terahertz absorption profile and act as a point of reference. Alternatively, altered terahertz absorption profiles are often observed in malignant tissues. The anatomical changes linked to cancer, including increased cell density, variations in water content, and the existence of aberrant proteins, because the typical absorption patterns found in healthy tissues to diverge. Terahertz spectroscopy can be used to identify as well as measure these anomalies [21].

(b) Non-invasive Early Cancer Detection
Research on non-invasive procedures is gaining traction due to the importance of early detection of cancer in improving treatment outcomes and patient survival rates. Non-ionizing terahertz waves are one of these cutting-edge methods that have shown promise as an early cancer diagnosis aid. These waves, which operate in the terahertz frequency range, provide special benefits. They can be safely used for repeated tests on biological tissues because they are non-ionizing and do not pose a risk of ionization or damage to cells. Furthermore, through tracking changes in water distribution, terahertz waves can disclose

pathological diseases, including cancer, due to their exceptional sensitivity to water content and water-related features within tissues [22].

21.3.2 Skin Imaging

Terahertz waves are essential to the realm of biomedical applications because of their special capacity to safely and non-invasively penetrate epidermal layers.

(a) Penetration of epidermal Layers by Terahertz Waves
Terahertz waves are absorbed by the epidermis, but their ability to cooperate with water molecules makes them sensitive to changes in the moisture content as well as the composition of the skin, making them valuable for skin imaging as well as analysis. This characteristic helps in the early identification of skin conditions like melanoma [23].

(b) Imaging Subsurface Features
Terahertz waves are the best way to diagnose various health issues such as sweat ducts. This is related to skin issues and these are very much necessary for controlling body temperature. Terahertz radiation, when absorbed by the skin, exposes precise information about the number, location, and activity of sweat ducts. Diseases like hyperhidrosis and anhidrosis can thus be more easily detected as well as monitored [23]. Terahertz waves are also very good at assessing skin hydration, giving real-time information required for dermatological and cosmetic applications, customized skincare routines, and wound healing tracking. This provides medical personnel with the knowledge necessary to make informed decisions about patient care.

21.3.3 Wound Healing Assessment

Since wound healing is a dynamic and intricate process, precise monitoring is essential to giving patients the best therapy possible and helping them make educated treatment choices. Terahertz radiation has shown to be a beneficial tool in this area due to its unique ability to assess the water content of wounds. The interaction of terahertz pulses and water molecules allows for a precise measurement of the affected area's hydration condition.

(a) Assessment of Water Content
When wounds heal, changes in the moisture content and composition of the tissue are visible. Healthcare personnel can considerably benefit from terahertz imaging's capacity to detect these changes in real time and obtain critical information on wound healing progress. It is feasible to discover fluctuations in the moisture content of wounds with this technology, which acts as a helpful gauge of the general health of the healing tissue and the efficiency of treatment. These waves help determine the water content as well as the composition and

form of the tissue. This comprehensive strategy advances our knowledge of wound healing and makes it easier to identify any problems early in the healing process [24]. Terahertz waves are now considered an essential tool in the field of wound care due to their ability to provide accurate, non-invasive, and real-time wound assessments. This has led to better patient outcomes and more successful treatment for the sufferer.

(b) Wound Infections Detection without Dressing Removal
An essential part of managing wound care is monitoring and early diagnosis of wound infections. In the past, this procedure frequently required removing wound coverings, which can be uncomfortable and interfere with the healing process. But new cutting-edge technology known as terahertz waves makes it possible to detect wound infections non-invasively and very effectively—all without removing bandages. Terahertz waves interact differently with different substances, such as healthy tissue versus sick tissue, and can penetrate a wide range of materials, including bandages and wound dressings. These waves display distinct properties, such as modifications in absorption and reflection patterns, when they come into contact with an infection within a wound. Healthcare personnel can detect the existence of infection behind the dressing by using this imaging system's ability to record and analyze the unique signals [24]. This cutting-edge technology offers the following several substantial benefits in the field of wound care.

- Firstly, it reduces patient discomfort and the chance of further contamination by doing away with the need to disturb the wound and dressing.
- Secondly, it makes early infection diagnosis possible, allowing for prompt intervention and modifications to the treatment plan.
- Thirdly, the great degree of precision provided by terahertz wave-based detection reduces the possibility of false positives or negatives.
- Lastly, because terahertz imaging is real-time, medical professionals can closely track how infections develop over time and make sure the right steps are taken to encourage healing and avoid problems.

The utilization of terahertz waves for detecting wound infections without necessitating the removal of dressings signifies a noteworthy advancement in the field of wound care. Terahertz wave technology is poised to revolutionize the way healthcare professionals diagnose and address wound infections, ultimately leading to more efficient and patient-friendly wound care protocols.

21.4 Terahertz Sensing Applications

Terahertz sensing applications in pharmaceutical analysis represent a rapidly growing field that offers valuable insights into the quality, composition, and characteristics of pharmaceutical products.

21.4.1 Pharmaceutical Analysis

Terahertz spectroscopy, with its ability to probe the unique spectral fingerprints of molecules and materials in the terahertz frequency range (typically 0.1–10 THz), has found diverse applications in the pharmaceutical industry [8] that are described below.

(a) Non-destructive Analysis of Pharmaceutical Tablets

- Tablet Content Uniformity
  Terahertz spectroscopy is employed to assess the uniform distribution of active pharmaceutical ingredients (APIs) within tablets. Variations in API concentration can lead to inconsistent drug delivery and reduced effectiveness. This sensing can quickly determine if the API is evenly distributed throughout the tablet, helping ensure product quality.
- Polymorph Analysis
  Different polymorphic forms of a drug material may provide unique solubility and bioavailability characteristics, which could impact the medicine's efficacy. In the pharmaceutical business, terahertz spectroscopy plays a crucial role in differentiating between various polymorphic forms, which helps to improve formulation and guarantee quality control.
- Moisture Content Determination
  The stability and shelf life of medicinal medicines can be impacted by their moisture content. Terahertz sensing is a useful technique for moisture control in pharmaceutical manufacturing because it can determine moisture levels precisely without requiring sample pre-treatment.
- API-Excipient Interaction
  Tablet formulation interactions between APIs and excipients are investigated using Terahertz spectroscopy. This information is essential for guaranteeing ingredient compatibility and optimizing medicine formulation.
- Detection of Counterfeit Drugs
  Terahertz imaging can help identify counterfeit pharmaceuticals by comparing the Terahertz spectra of genuine and counterfeit tablets. Counterfeit drugs may have different compositions or structures that can be detected through spectral analysis.
- Research and Development
  In pharmaceutical research and development, terahertz sensing aids in understanding the physicochemical properties of drug compounds, formulation development, and the optimization of drug delivery systems. It provides critical data for the development of innovative pharmaceutical products.

(b) Determining Chemical Composition, Structural Integrity, and Moisture Content
Determining chemical composition, structural integrity, and moisture content are critical aspects of material analysis across various industries, including pharmaceuticals, food production, construction, and more. Accurate and precise measurements in these areas are essential for quality control, safety, and

performance optimization. In this article, we will delve into the significance and methods for determining chemical composition, structural integrity, and moisture content in materials [8].

- Determining Chemical Composition Significance:
  Chemical composition analysis is essential in sectors where the qualities and usefulness of materials are directly affected by their composition. For instance, knowing the chemical composition guarantees that active pharmaceutical ingredients (APIs) are present in the right amounts in the pharmaceutical industry, while in the food industry, it confirms nutritional content and compliance with regulations, which are delineated by the various methods that follow
  Methods:
  I. Spectroscopy: To identify and measure the chemical elements and compounds in a material, methods such as infrared (IR), Raman, and X-ray fluorescence (XRF) spectroscopy are frequently employed. Every technique has special powers; for example, it can analyze elements (XRF) or organic molecules (IR).
  II. Mass Spectrometry: This technique determines and measures the ions' mass-to-charge ratios in a sample. It is very helpful for analyzing complicated mixes and identifying unknown substances.
  III. Chromatography: To isolate and measure distinct chemical components within a mixture, liquid chromatography (LC) and gas chromatography (GC) are used. Environmental and pharmaceutical analysis frequently uses these methods.
- Determining Structural Integrity Significance:
  The evaluation of structural integrity guarantees that the materials and components are safe and able to tolerate variations in temperature, mechanical stress, and other environmental conditions. Several techniques can be used to evaluate this [25]. In sectors like aerospace, automotive, and civil engineering, this is essential.
  Methods:
  I. Non-Destructive Testing (NDT): NDT methods assess the internal structure of materials and components without causing harm. Examples of NDT methods include magnetic particle testing, radiography testing, and ultrasonic testing. For instance, sound waves are used in ultrasonic testing to find fractures, cavities, and other flaws.
  II. Microscopy: High-resolution images of a material's microstructure are provided by scanning electron microscopy (SEM) and transmission electron microscopy (TEM), which enables the microscopic identification of flaws and evaluation of structural integrity.
  III. Mechanical Testing: A material's elasticity, strength, and other mechanical qualities can be ascertained by mechanical testing techniques like

tensile, compression, and hardness testing. These examinations evaluate a material's structural soundness and appropriateness for a certain use.

- Determining Moisture Content Significance:
  The analysis of moisture content plays a pivotal role in industries such as agriculture, food production, pharmaceuticals, and construction. Excessive moisture can jeopardize product quality, safety, and longevity. Various methods, as outlined in reference [25], are employed to determine moisture levels.
  Methods:
  I. Gravimetric Method: In this traditional method, a material is weighed both before and after it is dried at a specific temperature in an oven. The weight loss is used to compute the moisture content.
  II. Infrared Moisture Analyzers: These devices measure the absorption of water molecules in a sample using infrared radiation to determine the moisture content. They offer non-destructive measurements quickly.
  III. Capacitance and Resistance Sensors: These sensors track how moisture affects electrical characteristics like capacitance or resistance. They are typically used in industrial settings to determine the moisture content of commodities like grains or powders, or in agriculture to monitor the moisture content of the soil.
  IV. Karl Fischer II Titration: This technique was created expressly to measure low moisture levels precisely. Water and a reagent are involved in a chemical reaction, and the amount of reagent used determines the moisture content.

The methods chosen will depend on the requirements and characteristics of the material being analyzed. Accuracy, precision, and non-destructive capabilities will be prioritized where necessary.

21.4.2 Bimolecular Sensing

In the realm of bimolecular sensing, Biomolecules are essential, particularly for terahertz (THz) sensing applications. This novel method provides novel insights into the structural and functional characteristics of biomolecules, paving the way for advances in several disciplines, such as materials science, biology, and medicine.

(a) Probing Vibration Modes of Biomolecules
Examining the vibration modes of biomolecules is crucial in the field of biomolecular sensing. Terahertz sensing is a state-of-the-art technique that interacts with and assesses materials using electromagnetic waves in the terahertz frequency range (often 0.1–10 THz). The non-destructive nature of this approach and its capacity to yield important insights into molecular structures, chemical compositions, and molecular interactions has led to its recent surge in popularity. Terahertz sensing presents the following benefits when used with biomolecules.

I. Because terahertz spectroscopy is so sensitive to molecular vibrations, it can identify even minute modifications to the conformation or structure of molecules.
II. Terahertz waves are perfect for examining biomolecules in watery environments, including biological tissues or solutions, because they are sensitive to the amount of water present.
III. Comprehensive structural data on biomolecules, such as the identification of hydrogen bonds, secondary structures, as well as conformational changes, can be obtained by terahertz sensing.

Using the following methods and applications is part of investigating the vibrational modes of biomolecules in terahertz sensing:

I. Terahertz spectroscopy can be used to identify the unique vibrational modes of biomolecules such as proteins, nucleic acids, and carbohydrates. These modes correlate to torsional, bending, and stretching vibrations of chemical bonds.
II. Two instances of distinct secondary protein structures that can be identified with terahertz spectroscopy are alpha-helices and beta-sheets. Folding or denaturation can be used to determine changes in secondary structure.
III. When biomolecules interact with ligands, substrates, or other substances, terahertz sensors can be used to measure molecular vibrations.
IV. Terahertz spectroscopy can identify structural changes in biomolecules. This can be used to monitor structural alterations brought on by variations in temperature, the environment, or biological processes.
V. Terahertz sensing has applications in biomolecular analysis such as medical diagnostics, biomarker research, and pharmaceutical research. It can be used, for instance, to monitor the stability of biopharmaceuticals or spot changes in biomolecules that are connected to diseases.

In summary, terahertz sensing applications provide a potent and non-invasive means of exploring biological molecules' vibration modes and gaining additional insight into their structural and functional characteristics. There will continue to be potential for groundbreaking scientific and medical discoveries as long as this technology develops.

(b) Identification of Complex Structures in Proteins and DNA

Applications involving terahertz (THz) detection depend on the ability to identify complex structures in DNA and proteins. In this session, we discuss how THz sensing improves the recognition of complex structures in proteins and DNA, emphasizing the importance and practicality of this technology.

(c) The Importance of Determining Complex Structures in DNA and Proteins
The activities of proteins and DNA are closely linked to their distinct three-dimensional structures. Identifying complex structures is essential for deciphering how biomolecules perform crucial biological processes such as enzymatic reactions, gene expression, and signal transduction [25]. This is given below:

I. Disease Mechanisms: Misfolded or structurally altered proteins and DNA are often associated with diseases, including neurodegenerative disorders and cancer. Identifying complex structures in biomolecules helps in understanding disease mechanisms and developing targeted therapies.
II. Drug Development: In pharmaceutical research, identifying complex structures is vital for designing drugs that interact specifically with certain protein targets or DNA sequences. It aids in rational drug design and the optimization of therapeutic compounds.
III. Secondary Structure: THz spectroscopy can differentiate between various secondary structures in proteins, such as alpha-helices, beta-sheets, and random coils. The identification of these structural elements is crucial for understanding protein folding and stability.
IV. Tertiary Structure: The general three-dimensional configuration of atoms within a protein can be provided by THz sensing. It can recognize intricate tertiary structures as well as conformational changes in proteins brought on by ligand binding or external factors. Quaternary Structure: Terahertz spectroscopy is useful in determining the quaternary structure of protein complexes when proteins consist of many subunits. It determines this by exposing the interactions and interfaces between the subunits.
V. Dynamic Changes: Protein dynamics and structural changes are susceptible to THz spectroscopy. It provides information on processes like allosteric modulation and protein–ligand interactions by monitoring variations in protein structure as they happen.
VI. DNA Double Helix: When it comes to identifying DNA's unique double helix structure, THz spectroscopy is an admirable choice.

The fields of biology, medicine, and materials science can significantly advance through the recognition of intricate protein and DNA structures using THz sensing.

21.4.3 Monitoring Water Content

Terahertz sensors are becoming more and more popular tools in a variety of industries for water content assessment. They are an excellent tool for monitoring and managing water content in a variety of materials and applications because of their high sensitivity, non-destructive nature, and adaptability. This helps to advance scientific research, resource management, and product quality.

(a) Water Content Assessment by Terahertz Sensors

Terahertz sensors work on the principle of absorption of water molecules and react to terahertz radiation in a certain way. They operate using the following procedures [26]:

- Generation of Terahertz: THz sources and lasers are examples of specialized equipment used to produce terahertz radiation. After that, the material being studied is exposed to this radiation.
- Transmission or Reflection: When terahertz waves pass through or reflect off the material, they interact with the water molecules present. These interactions lead to absorption and scattering phenomena.
- Detection: A detector records terahertz radiation following its interaction with the material. Alterations in the radiation's characteristics, including amplitude and phase, provide insights into the sample's water content.
- Analysis of Data: To precisely quantify the water content, the collected data is analyzed using mathematical models or algorithms. Water usually has distinctive peaks in the Terahertz absorption spectrum.

Applications of Terahertz Sensors for Water Content Assessment.

I. Agriculture: To optimize crop management and irrigation, farmers can monitor soil moisture levels with the use of terahertz sensors. This preserves water resources and raises agricultural production.
II. Food Industry: In the food processing industry, terahertz sensors assess moisture content in products such as grains, fruits, and baked goods, ensuring product quality, freshness, and compliance with food safety regulations.
III. Pharmaceuticals: By measuring the amount of water present in pharmaceutical formulations, these sensors help producers keep drugs stable and high-quality throughout manufacturing and storage.
IV. Materials Science: Terahertz sensors assess water content in construction materials, polymers, and composites. This information is essential for quality control and ensuring material integrity.
V. Environmental Monitoring: Researchers use terahertz sensors to study soil moisture, aiding in ecological and hydrological studies, weather forecasting, and natural resource management.

Terahertz sensors are appropriate for quality control in the manufacturing process because they don't damage the substance being tested. Even at low concentrations; they have a great sensitivity to water content, enabling precise readings. Because of their adaptability, these sensors can be used in a wide range of materials and applications. Since measurements are frequently made, quick monitoring and decision-making can be done in real time. Because these sensors do not need to come into close contact with the material, there is less chance of contamination, and the integrity of the material is preserved.

(b) Diverse Applications: Biological Tissues, Food Products, and Building Materials

Terahertz sensing is a cutting-edge technology that has profound diverse applications across a wide range of domains. It utilizes terahertz waves and explores the diverse applications of terahertz sensing in biological tissues, food products, and building materials [27].

Terahertz Sensing in Biological Tissues:

Terahertz sensing has become a cornerstone in the medical field for non-invasive imaging and diagnostics. Its ability to penetrate biological tissues without ionizing radiation ensures its safety in healthcare [28]. Some significant applications encompass:

- It is useful for evaluating dental restorations, detecting cavities, and assessing the health of teeth.
- Terahertz imaging can help diagnose skin malignancies early by revealing properties of the subsurface tissue.
- During cancer surgery, it can help surgeons detect the margins of tumors, which lowers the possibility of leaving cancerous tissue behind.

Terahertz Sensing in Food Products:

Terahertz sensing is advantageous for quality control, safety, and inspection in the food business. It has multiple applications [29], including:

- Terahertz sensors can evaluate moisture levels in food products, assisting in quality control and the prediction of shelf life.
- It can detect objects, such as plastic, glass, or metals, in packaged food items.
- Access to the freshness of vegetables, fruits, and other perishable goods can be done by analyzing their water content.

Terahertz Sensing in Building Materials: Terahertz sensing is a burgeoning technology in the construction and civil engineering industries, providing the following applications [30]:

- Terahertz radiation may examine structural elements without inflicting harm and can identify cavities, cracks, and water intrusion.
- These sensors are employed to gauge the coatings' thickness on different kinds of materials, guaranteeing their quality and longevity.
- It aids in evaluating the concrete curing process, thus optimizing construction schedules.

This sensor technology is adaptable and has a wide range of uses in industries including construction, food processing, healthcare, security, and more. Its exceptional capacity to offer non-destructive, non-invasive insights into materials and objects has the potential to transform several industries and raise the standard and safety of goods and services.

21.5 Challenges and Ongoing Developments

This section examines current advancements and challenges in the realm of terahertz waves [31].

21.5.1 Resolution, Signal-to-Noise Ratio, and Instrumentation Complexity

Achieving excellent resolution and signal-to-noise ratios while controlling the complexity of instrumentation is one of the main problems in terahertz technology. These difficulties result from terahertz radiation' comparatively lengthy wavelengths. Researchers are continuously working on:

- Enhance clarity is done by cutting-edge sensor technologies.
- Improving signal-to-noise ratios using advanced signal processing algorithms.
- The advancement of terahertz instrumentation that is more portable and intuitive.

21.5.2 Advancements in Terahertz Time-Domain Imaging (THz-TDI)

Terahertz time-domain imaging (THz-TDI) is a potent imaging and spectroscopy technique. Its speed and sensitivity have been improved recently, which is important for applications in areas like materials characterization, security screening, and medical diagnostics. Ongoing efforts aim to:

- Enhance the speed of scanning for real-time imaging.
- Detect subtle material variations by increasing sensitivity.
- Optimize THz-TDI systems for various applications.

21.5.3 Exploration of Terahertz-Computed Tomography (THz-CT)

One prospective imaging method that can produce remarkably detailed 3D reconstructions of things is terahertz-computed tomography (THz-CT). The complexity of data processing and scan duration are challenges. Researchers are focusing on:

- Data acquisition faster methods.
- Rapid 3D reconstruction using advanced algorithms.
- Integrating THz-CT with additional imaging modalities to enable a thorough investigation.

21.5.4 *Therapeutic Potential: Selective Heating and Cellular Modulation*

Terahertz waves have demonstrated potential for use in therapeutic applications, such as cellular regulation and selective tissue heating. Challenges in this area include optimizing safety and efficacy. Ongoing developments include:

- Investigating the safety of terahertz exposure.
- Developing terahertz-based therapies for conditions like cancer treatment and wound healing.
- Understanding the biological effects of terahertz radiation at the cellular level.

21.5.5 *Advancements in Terahertz Sources and Detectors: Toward Miniaturization*

Miniaturization of terahertz sources and detectors is essential for enabling portable and integrated terahertz systems. Ongoing developments include:

- Compact terahertz sources using technologies like quantum cascade lasers and frequency combs.
- Miniature detectors based on nanomaterials and innovative sensor designs.
- Integration of terahertz components into consumer devices for applications such as non-invasive medical diagnostics and security screening.

As research and innovation in this field continue to progress, we can expect even more exciting applications to emerge.

21.6 Future Prospects

The future of terahertz technology is marked by remarkable prospects, particularly in the field of healthcare. Portable terahertz systems are expected to revolutionize point-of-care diagnostics and telemedicine, offering transformative implications for healthcare, and enhancing diagnostics, and treatments [31].

Portable Terahertz Systems for Point-of-Care Diagnostics and Telemedicine:

- Miniaturization: Future terahertz systems will continue to shrink in size, making them handheld and easily portable. This portability is a game-changer, as healthcare professionals can carry advanced diagnostic tools to remote locations or directly to the patient.
- Imaging in real-time: Terahertz technologies of the future will be able to deliver imaging in real-time.

- Increasing Access: Telemedicine facilitated by terahertz technology can help remove geographical barriers and make healthcare more accessible in remote or impoverished areas.

Transformative Diagnostics and Healthcare Treatments:

- Non-invasive Monitoring: Without the need for uncomfortable blood tests, patients with chronic disorders like diabetes or cardiovascular diseases can be continuously monitored thanks to the non-invasive nature of the technology.
- Personalized Treatment Plans: By offering more precise information about the patient's health, terahertz diagnostics will enable individualized treatment programs.
- Therapeutic Applications: In addition to diagnostics, terahertz technology may find use in therapeutic domains like medication transport, cancer treatment, and non-invasive cellular manipulation.

Terahertz portable devices have a bright future ahead of them in healthcare, with revolutionary effects on telemedicine, diagnosis, and therapy. These devices will give patients and healthcare professionals quick, accurate, non-invasive diagnostic tools that may be used in a variety of environments. This technology has the potential to completely transform healthcare as it develops, guaranteeing everyone has fair access to high-quality treatment regardless of geographic location.

21.7 Conclusion

Portable terahertz devices hold great promise for the future of healthcare since they have the potential to revolutionize telemedicine, diagnosis, and treatment methods. These small, multipurpose gadgets have the power to revolutionize the medical industry. A major improvement in patient care is anticipated with the miniaturization of terahertz technology, which when paired with real-time imaging and quick analysis, promises early disease identification and individualized treatment programs. One of the key advantages of terahertz technology is its non-invasive nature, eliminating the need for invasive procedures and enhancing patient comfort. This opens the door to non-invasive therapeutic applications, offering a patient-centric approach to healthcare. Terahertz capabilities will allow telemedicine, which will democratize healthcare. Expanded access to medical services and remote consultations are made possible by it, especially in underprivileged areas. It is essential for tracking chronic illnesses and offers contactless diagnostics in an emergency, enhancing the quality and accessibility of healthcare. It is an effective tool for boosting patient outcomes, expanding healthcare accessibility, and providing a patient-centered approach to medical care because of its non-invasiveness and adaptability.

References

1. Panwar, K., Singh, A., Kumar, A., Kim, H.: Terahertz imaging system for biomedical applications: current status. System **28**, 44 (2013)
2. Zhang, X., Hu, M., Zhao, X., Zhou, J., Zhang, Z., Zhou, J., et al.: Biomedical applications of terahertz near-field imaging. In: 2021 46th International Conference on Infrared, Millimeter and Terahertz Waves (IRMMW-THz), pp. 1–2, Aug 2021
3. Choudhury, B., Menon, A., Jha, R.M.: Active Terahertz Metamaterial for Biomedical Applications, pp. 1–41. Springer, Singapore (2016)
4. Yadav, A., Gorodetsky, A.: Biomedical applications of terahertz radiation. In: Advanced Photonics Methods for Biomedical Applications, vol. 126, p. 67 (2023)
5. Yin, X., Ng, B.W.H., Abbott, D.: Terahertz Imaging for Biomedical Applications: Pattern Recognition and Tomographic Reconstruction. Springer Science & Business Media, Berlin (2012)
6. El Ghzaoui, M., Mestoui, J., Hmamou, A., Elaage, S.: Performance analysis of multiband on–off keying pulse modulation with noncoherent receiver for THz applications. Microwave Opt. Technol. Lett. **64**(12), 2130–2135 (2022)
7. Lchhab, T., El Ghzaoui, M.: A circularly polarized wideband high gain antenna for THz wireless applications. Opt. Quant. Electron. **54**(12), 787 (2022)
8. Samanta, D., Karthikeyan, M.P., Agarwal, D., Biswas, A., Acharyya, A., Banerjee, A.: Trends in terahertz biomedical applications. In: Generation, Detection and Processing of Terahertz Signals, pp. 285–299 (2022)
9. Sun, Q., He, Y., Liu, K., Fan, S., Parrott, E.P., Pickwell-MacPherson, E.: Recent advances in terahertz technology for biomedical applications. Quant. Imaging Med. Surg. **7**(3), 345 (2017)
10. Wan, M., Healy, J.J., Sheridan, J.T.: Terahertz phase imaging and biomedical applications. Opt. Laser Technol. **122**, 105859 (2020)
11. Sirkeli, V.: Recent advances in Terahertz technology for security and biomedical applications. In: Abordări inter/transdisciplinare în predarea ştiinţelor reale (concept STEAM), vol. 2, pp. 82–88 (2021)
12. Yeo, W.G.: Terahertz spectroscopic characterization and imaging for biomedical applications. The Ohio State University (2015)
13. Sizov, F.F.: Infrared and terahertz in biomedicine. Semicond. Phys. Quantum Electron. Optoelectron. **20**(3), 273–283 (2017)
14. Son, J.H., Oh, S.J., Cheon, H.: Potential clinical applications of terahertz radiation. J. Appl. Phys. **125**(19) (2019)
15. Fan, S., He, Y., Ung, B.S., Pickwell-MacPherson, E.: The growth of biomedical terahertz research. J. Phys. D Appl. Phys. **47**(37), 374009 (2014)
16. Tabata, H.: Application of terahertz wave technology in the biomedical field. IEEE Trans. Terahertz Sci. Technol. **5**(6), 1146–1153 (2015)
17. Koul, S.K., Kaurav, P.: Sub-Terahertz Sensing Technology for Biomedical Applications. Springer Nature, Berlin (2022)
18. Siegel, P.H.: Terahertz technology in biology and medicine. IEEE Trans. Microw. Theory Tech. **52**(10), 2438–2447 (2004)
19. Zhang, Y., Wang, C., Huai, B., Wang, S., Zhang, Y., Wang, D., et al.: Continuous-wave THz imaging for biomedical samples. Appl. Sci. **11**(1), 71 (2020)
20. Karmakar, J., Samanta, D., Banerjee, A., Karthikeyan, M.P.: Terahertz image processing: a boon to the imaging technology. In: International Conference on Advances in Data Science and Computing Technologies, pp. 277–283, June 2022
21. Bowman, T., Chavez, T., Khan, K., Wu, J., Chakraborty, A., Rajaram, N., et al.: Pulsed terahertz imaging of breast cancer in freshly excised murine tumors. J. Biomed. Opt. **23**(2), 026004–026004 (2018)
22. Gezimati, M., Singh, G.: Open research challenges and opportunities in terahertz imaging and sensing for cancer detection. In: 2023 First International Conference on Microwave, Antenna and Communication (MAC), pp. 1–6, Mar 2023

23. Banerjee, A., Vajandar, S., Basu, T.: Prospects in medical applications of terahertz waves. In: Terahertz Biomedical and Healthcare Technologies, pp. 225–239 (2020)
24. Yang, X., Zhao, X., Yang, K., Liu, Y., Liu, Y., Fu, W., et al.: Biomedical applications of terahertz spectroscopy and imaging. Trends Biotechnol. **34**(10), 810–824 (2016)
25. Ajito, K.: Terahertz spectroscopy for pharmaceutical and biomedical applications. IEEE Trans. Terahertz Sci. Technol. **5**(6), 1140–1145 (2015)
26. Chen, X., Lindley-Hatcher, H., Stantchev, R.I., Wang, J., Li, K., Hernandez Serrano, A., et al.: Terahertz (THz) biophotonics technology: instrumentation, techniques, and biomedical applications. Chem. Phys. Rev. **3**(1) (2022)
27. Yan, Z., Ying, Y., Zhang, H., Yu, H.: Research progress of terahertz wave technology in food inspection. Terahertz Phys. Devices Syst. **6373**, 142–151 (2006)
28. Tewari, P., Taylor, Z.D., Bennett, D., Singh, R.S., Culjat, M.O., Kealey, C.P., et al.: Terahertz imaging of biological tissues. In: Medicine Meets Virtual Reality, vol. 18, pp. 653–657 (2011)
29. Afsah-Hejri, L., Hajeb, P., Ara, P., Ehsani, R.J.: A comprehensive review on food applications of terahertz spectroscopy and imaging. Compr. Rev. Food Sci. Food Saf. **18**(5), 1563–1621 (2019)
30. Jansen, C., Piesiewicz, R., Mittleman, D., Kurner, T., Koch, M.: The impact of reflections from stratified building materials on the wave propagation in future indoor terahertz communication systems. IEEE Trans. Antennas Propag. **56**(5), 1413–1419 (2008)
31. Humphreys, K., Loughran, J.P., Gradziel, M., Lanigan, W., Ward, T., Murphy, J.A., et al.: Medical applications of terahertz imaging: a review of current technology and potential applications in biomedical engineering. In: The 26th Annual International Conference of the IEEE Engineering in Medicine and Biology Society, vol. 1, pp. 1302–1305, Sept 2004

Chapter 22
The Analog Design of a Voltage Generator in 180 nm CMOS Technology

Hanae Mejdoub, Mohammed El Ghzaoui, Qjidaa Hassan, and Rachid El Alami

22.1 Introduction

RFID is a wireless technology that harnesses the power of radio waves to enable the identification and tracking of objects or individuals. It has emerged as a prominent solution in various industries, offering immense potential in enhancing efficiency and streamlining operations. At its core, a typical RFID system comprises three fundamental components: a reader or scanner, an antenna, and a tag or transponder [1, 2]. The reader serves as the central unit that emits radio waves, which are subsequently intercepted by the antenna. Acting as an intermediary, the antenna then relays the received signal to the tag. The tag, equipped with a microchip for data storage and an antenna to capture the radio wave signal emitted by the reader, plays a pivotal role in responding to the reader's signal. It accomplishes this by transmitting a distinct identifier or other pertinent information back to the reader, enabling efficient and reliable communication [3–6]. The applications of RFID technology span diverse domains, showcasing its versatility and widespread adoption. From inventory tracking and supply chain management to livestock monitoring, access control in restricted areas, and contactless payment systems, RFID has revolutionized various aspects of modern-day operations [7, 8]. One of the key drivers behind the popularity of RFID lies in its remarkable ability to automatically and accurately identify objects without the need for direct line of sight or physical contact. This inherent capability has contributed to increased efficiency, reduced errors, and enhanced security across numerous industries [9].

Many RFID tags are presented in literature [6–13]. Authors of Ref. [10] explored the creation of an integrated circuit (IC) for ultrahigh-frequency (UHF) band passive RFID tags, adept at accommodating both generation-2 (Gen-2) RFID mode and noticeable RFID mode. For both Gen-2 and visible RFID modes, power is primarily

H. Mejdoub (✉) · M. El Ghzaoui · Q. Hassan · R. El Alami
Faculty of Sciences Dhar El Mahraz, Sidi Mohamed Ben Abdellah University, Fez, Morocco
e-mail: hanae.mejdoub@usmba.ac.ma

M. El Ghzaoui et al. (eds.), *Next Generation Wireless Communication*, Signals and Communication Technology, https://doi.org/10.1007/978-3-031-56144-3_22

provided by the RF source. In Ref. [11], the authors detail the creation of a sensor tag detecting wireless temperature and capable of functioning in both fully passive and semi-passive modes. The latter mode is supported by the inclusion of a flexible thermoelectric generator (TEG). The wireless sensor tag is crafted with an UHF RFID integrated circuit (IC) of the EPC C1G2/ISO 18000-6C type, linked to a low-power microcontroller unit (MCU). This MCU is responsible for sampling and collecting temperature data from a digital temperature sensor. In [12], authors discussed the development of an RFID reader to tackle the issues related to expenses, energy consumption, stability, and dependability in extensive Internet of Things (IoT) implementations. The SoC integrates various functions and supports multiple communication protocols, making it a potentially valuable component in RFID-based systems across various industries and applications. The authors emphasize in their work that conventional RFID reader modules integrate various chips onto a single printed circuit board (PCB), which includes a microcontroller and a high-frequency (HF) RFID reader IC. This approach faces challenges related to cost, power consumption, stability, and reliability. Authors of Ref. [13] presented an RFID sensor device that is capable of simultaneously collecting data on various environmental parameters, including temperature, lightning intensity, relative humidity, and pressure. The architecture of the proposed RFID sensor device emphasizes low-power consumption. This is essential for ensuring the device's long-term operation without the need for frequent battery replacements or recharging. The RFID sensor presented in [11] is built upon passive ultra-high frequency (UHF) RFID technology. In [14], authors focused on the design of a CMOS active rectifier for wirelessly powered biomedical applications. The main goal of the research presented in [14] is to design a CMOS active rectifier that achieves high-power conversion efficiency (PCE) specifically tailored for wirelessly powered biomedical applications. Authors of this paper used voltage mode switching to reduce the static power consumption of the control circuit. This is important for minimizing power wastage in the rectifier. Research presented in [15] introduces the design of a VDD generator using CMOS technology with a 180 nm process node, operating at a frequency of 900 MHz. The primary objective of the work presented in [15] is to produce a stable output voltage of 1 V, which is crucial for applications in passive ultra-high frequency (UHF) RFID tags. Authors of Ref. [16] presented an UHF RFID reader receiver with a focus on low-noise performance and fast settling time. The key innovation lies in the use of digitally controlled leakage cancelation methods in both passive and active configurations, along with optimizing the Tx to Rx isolation ratio and utilizing a CMOS 55 nm process. This technology has the potential to enhance the overall performance and reliability of UHF RFID reader systems. The work of [17] highlights a breakthrough in RFID technology by demonstrating a highly compact and digital RFID tag chip that is compatible with the EPC Gen-2 standard. The chip's small size, sensitivity, and reliance on digital IP blocks make it a promising candidate for various RFID applications where space constraints, sensitivity, and power efficiency are critical considerations.

In the context of this particular paper, our objective revolves around the analog design and layout of an RFID tag utilizing the advanced features of the Cadence software. By delving into the intricacies of the UHF RFID tag design, we aim to explore the potential for optimizing performance, enhancing reliability, and further advancing the capabilities of RFID technology in this specific frequency range. Through meticulous design and layout methodologies, we strive to develop a robust and efficient UHF RFID tag solution that aligns with industry standards and best practices. In fact, RFID is facilitating the identification and tracking of objects or individuals over a distance. This involves the transmission of a signal from the reader via electromagnetic waves to the tag. Subsequently, the antenna of the tag receives these waves, powering the chip, which interprets the information and transmits it back to the reader. This paper delves into the analysis of the analog design and layout of an ultra-high-frequency (UHF) RFID tag, utilizing the advanced features of Cadence software. The focus is on an extensive examination of the circuit simulation process, emphasizing the optimization of key parameters. Specifically, the paper presents the analog design of a VDD voltage generator for a radio frequency identification system, with a particular emphasis on designing a passive RFID tag in the UHF band at a frequency of 900 MHz, using CMOS 180 nm technology. The goal is to explore the nuances of UHF RFID tag design to optimize performance, enhance reliability, and advance RFID technology capabilities within this specific frequency range. By employing meticulous design and layout methodologies, the objective is to create a robust and efficient UHF RFID tag solution that aligns with industry standards and best practices. This work contributes to the ongoing efforts to push the boundaries of RFID technology and its applications in various domains.

22.2 Architecture

An RFID chip comprises two primary components that engage in communication with one another: an analog block, also known as the radio frequency (RF) part, and a digital block [18].

The analog block within the RFID chip (Fig. 22.1) is subdivided into several distinct sub-blocks, each serving a specific function. These include the voltage generator VDD, the power-on reset mechanism, the oscillator, the demodulator, and the modulator. Conversely, the digital block of the RFID chip encompasses a memory block, an encoding/decoding block, and a response generation block. An RFID chip is composed of two fundamental components that interact to facilitate communication. The first component is the analog block, often referred to as the radio frequency (RF) part. This analog block plays a pivotal role in managing the wireless communication process within the RFID system. It includes elements such as antennas, radio frequency circuitry, and signal processing mechanisms. The RF part is responsible for receiving signals from an RFID reader, transmitting data, and ensuring effective communication between the chip and the reader. This critical aspect of the RFID chip enables the exchange of information and the seamless functioning of the entire RFID system.

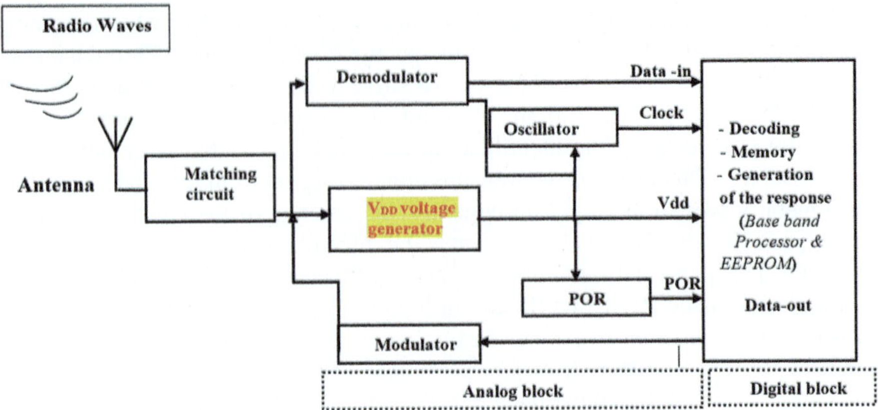

Fig. 22.1 Blocks of the RFID tag

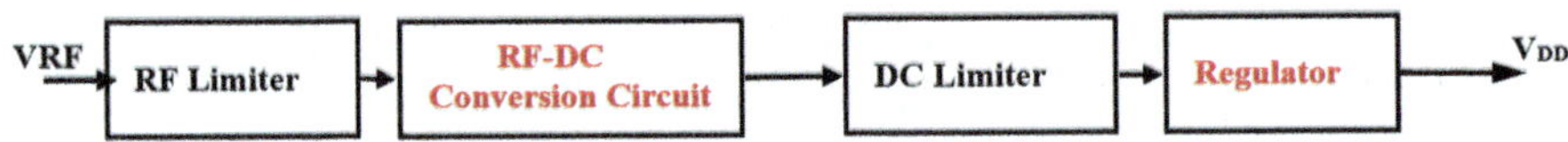

Fig. 22.2 VDD voltage generator block

Our focus lies specifically on the voltage regulator and the rectifier components within the VDD voltage generator (Fig. 22.2). The rectifier is comprised of the following sub-blocks.

In electronic systems, particularly in integrated circuits (ICs), the VDD voltage generator block and regulator are crucial components responsible for managing the supply voltage. The VDD voltage generator block is designed to generate a stable and controlled supply voltage (often denoted as VDD) required for the proper operation of the integrated circuit. This block typically includes circuits that convert an input voltage or power source into a voltage level suitable for the IC. The goal is to ensure a reliable and constant supply voltage, which is essential for the functionality and performance of the electronic components within the IC. Besides, the regulator is a specific component within the VDD voltage generator block. Its primary function is to regulate and maintain a consistent output voltage regardless of variations in the input voltage or load conditions. There are different types of voltage regulators, such as linear regulators and switching regulators. Linear regulators reduce the voltage to the desired level through a controlled linear element, while switching regulators use a switching element to regulate the voltage efficiently. The VDD voltage generator block and regulator work together to provide a stable and regulated supply voltage to the integrated circuit. This is crucial for ensuring proper functionality, preventing damage to electronic components, and optimizing the performance of the overall system. The voltage generator creates the necessary voltage, while the regulator ensures that this voltage remains within specified limits, contributing to the reliability and longevity of electronic devices.

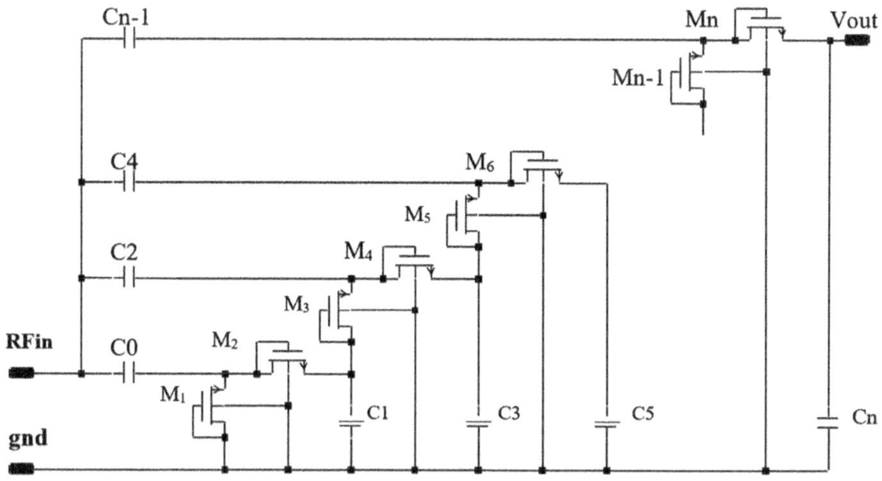

Fig. 22.3 Rectifier circuit based on MOS transistor-based charge pump: N-stage

22.2.1 Rectifier

The rectifier in Fig. 22.3 plays a crucial role in the RFID system by transforming the input RF wave, acquired through the tag antenna, into a direct current (DC) voltage. This DC voltage serves as the primary power source for the entire tag chip. Conventional rectifier circuits rely on the utilization of the Dickson charge pump topology, which employs MOS transistors operating in the diode-connected mode. The Dickson charge pump is a widely adopted methodology in the realm of integrated circuits, specifically in the context of CMOS-based designs [19]. MOSFETs, serving as the prevailing logic gates in integrated circuits, play a crucial role in realizing the diodes within the Dickson assembly by employing suitable connections.

22.2.2 Voltage Regulator

A voltage regulator which is presented in Fig. 22.5 is an electronic circuit designed to provide a stable, low-noise output voltage with minimal difference between the input and output voltages. However, the voltage obtained after rectification exhibits a significant drawback: as the current drawn by the load increases, the ripple increases due to excessive discharge of the output capacitor, resulting in a decrease in the average rectified voltage [20]. To address this issue, the implementation of a low-power voltage regulator is proposed to ensure a constant voltage across the load terminals, regardless of the varying load current. Figure 22.4 depicts the schematic diagram utilized for our RFID tag implementation.

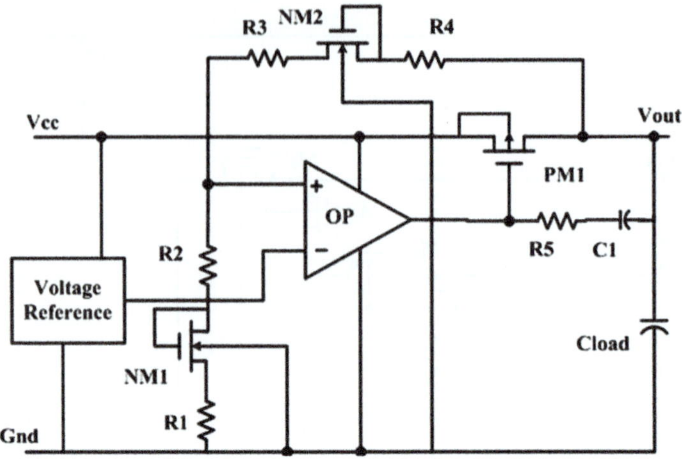

Fig. 22.4 Circuit of the voltage regulator

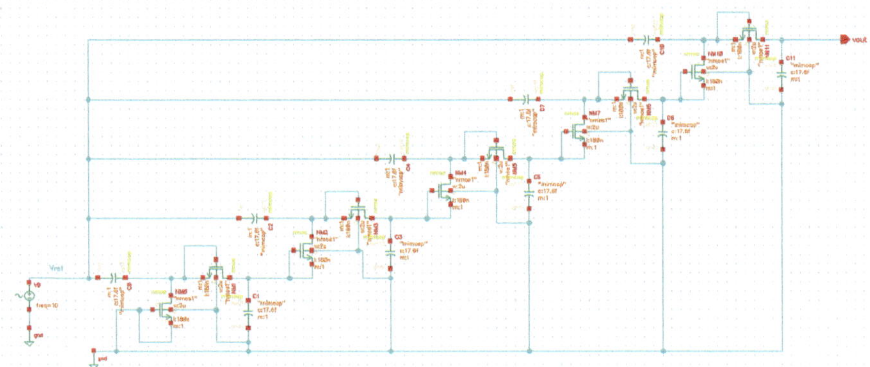

Fig. 22.5 Rectifier circuit on Cadence

22.3 Results and Discussions

In the subsequent sections, we will provide an in-depth examination of the circuit simulation process, with a primary focus on achieving comprehensive simulations of the entire circuit while optimizing key parameters. Notably, the spectrum simulator integrated within the Cadence Virtuoso platform has emerged as an indispensable tool for designers in their endeavors to anticipate and validate the electrical characteristics of a circuit. By leveraging the topological description and component specifications, this sophisticated simulator enables designers to predict and verify the expected performance of the circuit accurately. This sophisticated simulator plays a pivotal role by leveraging the topological description and component specifications provided by designers. By doing so, it empowers designers to make accurate predictions and verify the expected performance of the circuit. The spectrum simulator within Cadence Virtuoso serves as a virtual testing ground, allowing designers to assess the behavior of the circuit under various conditions and configurations. This predictive capability is invaluable in the design phase, enabling designers to fine-tune parameters, identify potential issues, and optimize the circuit for optimal performance before physical implementation.

To facilitate our simulation and layout endeavors, a range of specialized tools were employed. These tools played a pivotal role in conducting thorough simulations and efficiently optimizing the circuit's parameters:

Cadence Virtuoso software, Library: gpdk180, Technology: 180 nm, Types of analysis: DC, AC and TR.

22.3.1 *Voltage Rectifier Circuit*

A rectifier circuit is commonly used to convert alternating current (AC) into direct current (DC). It is a fundamental component in power supplies and various electronics. The most basic rectifier circuit is a single-diode rectifier, often referred to as a half-wave rectifier. Figure 22.5 shows the proposed schematic of the rectifier circuit.

Transient analysis is a computational technique employed to determine the time-domain responses of a circuit, specifically the variations in voltage or current as a function of time, in response to a predefined set of excitations. This analysis provides valuable insights into the dynamic behavior of the circuit, highlighting instantaneous reactions and transient phenomena that occur during state transitions. Figure 22.6 shows the transient analysis results of proposed rectifier.

Layout is very essential when designing a rectifier circuit, especially if you are working on an integrated circuit (IC) or a printed circuit board (PCB). Proper layout

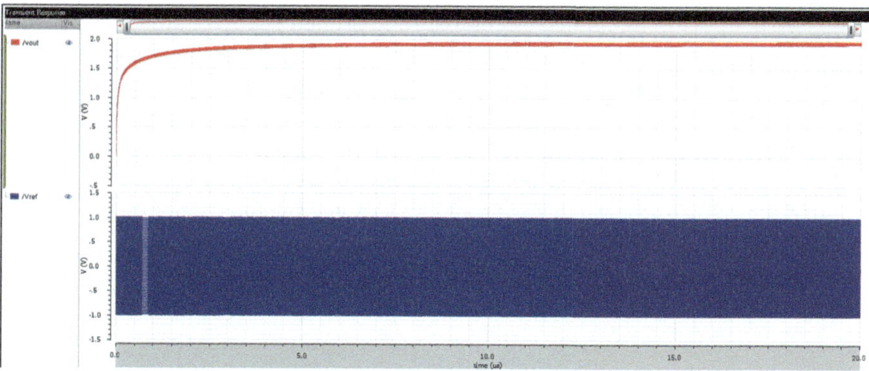

Fig. 22.6 Transient analysis result for the Rectifier

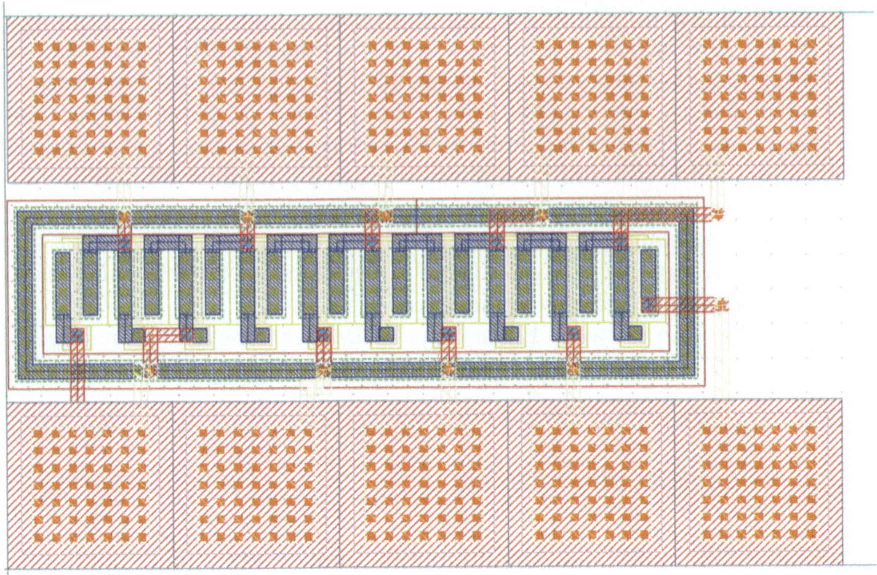

Fig. 22.7 Rectifier layout

can significantly impact the performance, efficiency, and reliability of the rectifier circuit. We carefully placed the components in the circuit to minimize parasitic capacitance and inductance. Ensure that components are positioned to minimize signal path lengths and reduce the risk of interference. We implemented a solid ground plane to provide a low-resistance return path for the current. This helps reduce noise and improve the efficiency of the rectifier. We placed decoupling capacitors close to the rectifier circuit to filter out high-frequency noise and ensure stable operation. Proper placement of these capacitors is critical. The proposed layout of the rectifier is shown in Fig. 22.7.

A Design Rule Checker (DRC) is a software tool used in the field of electronics design, particularly in integrated circuit (IC) and printed circuit board (PCB) design. Its main purpose is to analyze and verify whether a design adheres to the specified manufacturing rules and constraints. These rules ensure that the resulting design is manufacturable and functional without any critical errors or violations that could lead to performance issues, shorts, opens, or other manufacturing defects. Using a DRC tool is crucial in ensuring that a design is manufacturable and meets the desired performance specifications. Figure 22.8 shows that our design for the rectifier contains no error.

The next step will be the design of the voltage regulator circuit.

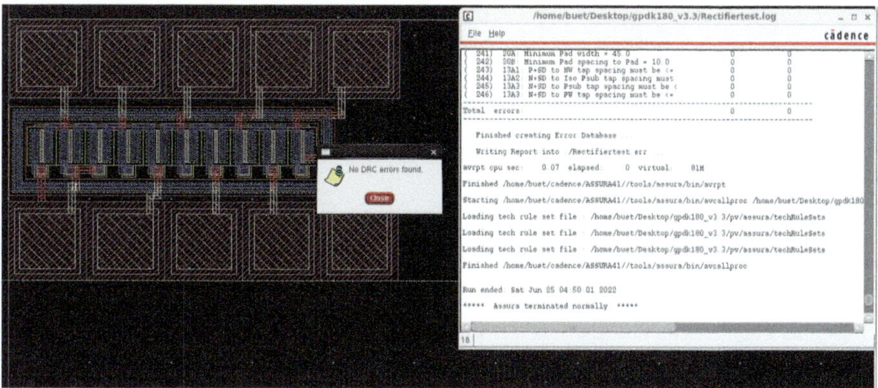

Fig. 22.8 Design rule checker for the Rectifier

22.3.2 *Voltage Regulator Circuit*

Voltage regulator circuits are indeed essential in radio frequency identification (RFID) systems. RFID devices, such as RFID tags and readers, often require stable and regulated power supplies to operate reliably. In fact, RFID chips and components typically have specific voltage requirements for proper operation. Voltage regulators ensure that the supply voltage remains within a specified range, preventing voltage fluctuations or spikes that could damage or disrupt the RFID components. In addition, voltage regulators help filter out noise and voltage fluctuations from the power source, providing clean and stable power to the RFID components. This is crucial for maintaining signal integrity and preventing data corruption in RFID communication. Moreover, in battery-powered RFID devices, such as RFID tags, voltage regulators help maximize battery life by ensuring that the device operates at the most efficient voltage levels. They prevent the device from operating at too high or too low a voltage, which can lead to increased power consumption. RFID readers and tags may operate at different voltage levels or have varying voltage tolerances. Voltage regulators can adapt the input voltage to match the requirements of the specific RFID components, ensuring compatibility and preventing damage due to voltage mismatches. When designing the proposed circuit, we selected and designed the voltage regulation circuit to meet the specific voltage requirements and power consumption of the RFID components. The voltage regulator is made up of the GAP Strip plus the operational amplifier (Fig. 22.9). Figure 22.10 shows transient analysis simulations of the regulator. Figure 22.11 shows the layout of the regulator.

A bandgap voltage reference (BGR) is a stable and accurate voltage source based on semiconductor characteristics. It is typically designed to generate a constant and precise reference voltage regardless of temperature and power supply variations. By using an operational amplifier to regulate an output voltage with a bandgap voltage reference, you can create a precise and stable voltage regulator.

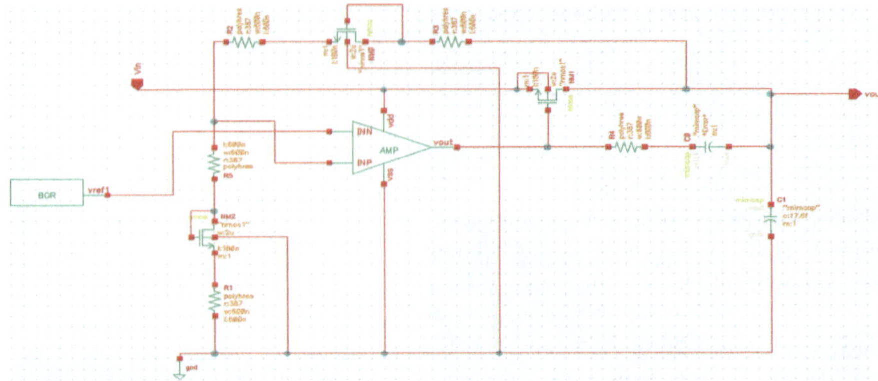

Fig. 22.9 Regulator circuit on cadence

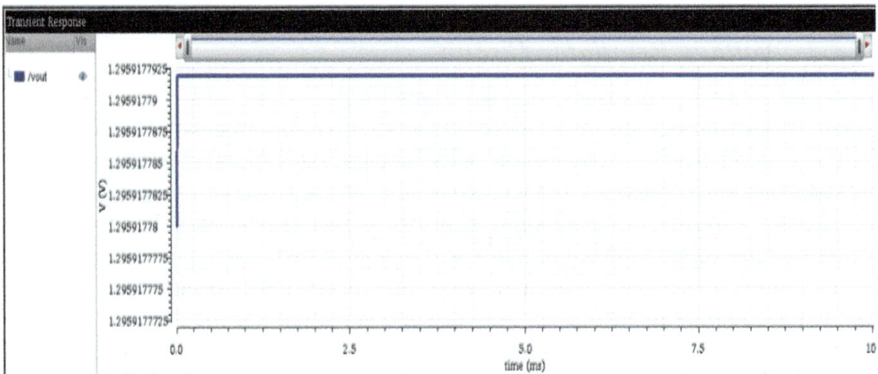

Fig. 22.10 Voltage regulator transient analysis

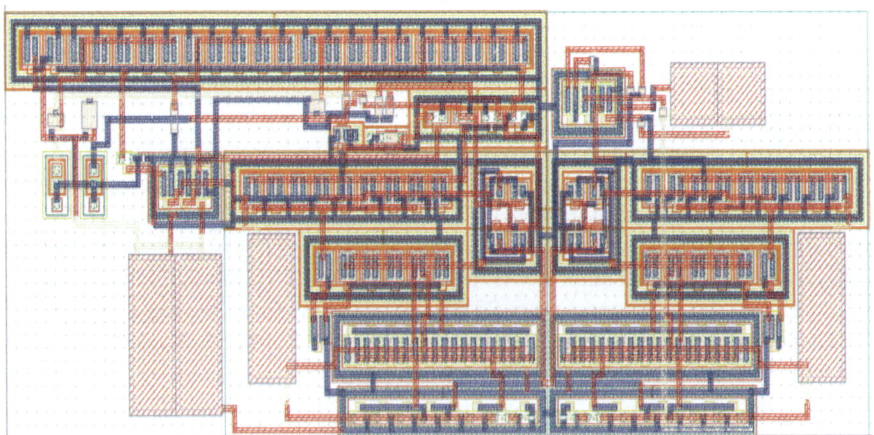

Fig. 22.11 Voltage regulator layout

The operational amplifier can be configured as a negative feedback amplifier to adjust the output voltage based on the BGR reference voltage. This maintains the output voltage at a constant level despite variations in input voltage or load conditions. Figure 22.12 shows the BGR circuit and Fig. 22.13 shows its simulations results. Figure 22.14 shows the layout of the BGR.

As discussed before, the Design Rule Checker is a crucial tool in the design and manufacturing of electronic circuits, whether you are working on integrated circuits (ICs), printed circuit boards (PCBs), or other electronic systems. Figure 22.15 shows the DRC of layout presented in Fig. 22.14.

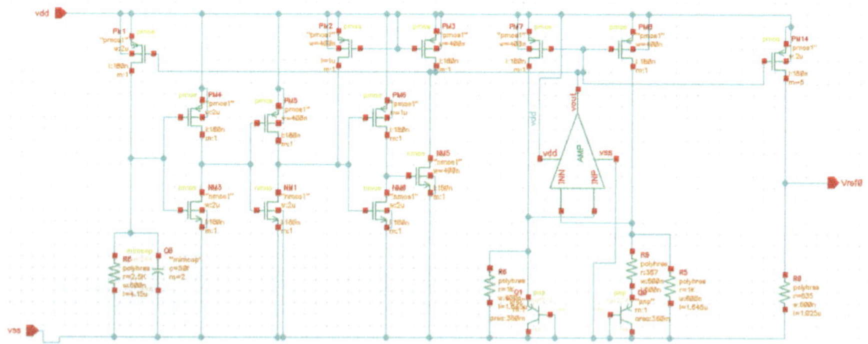

Fig. 22.12 BGR circuit on cadence

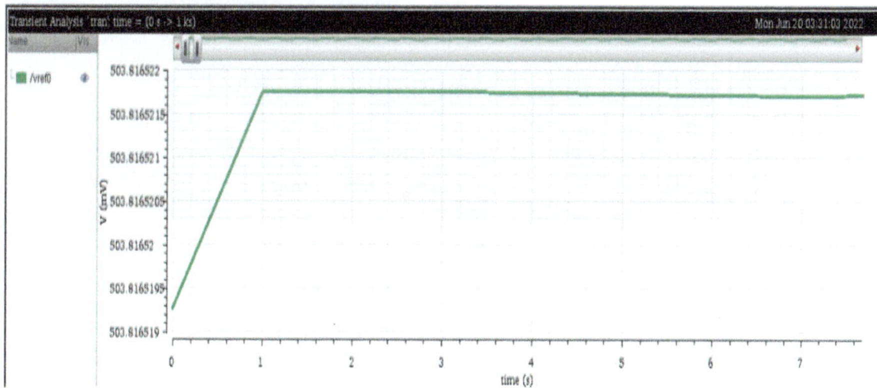

Fig. 22.13 Transient analysis result for BGR

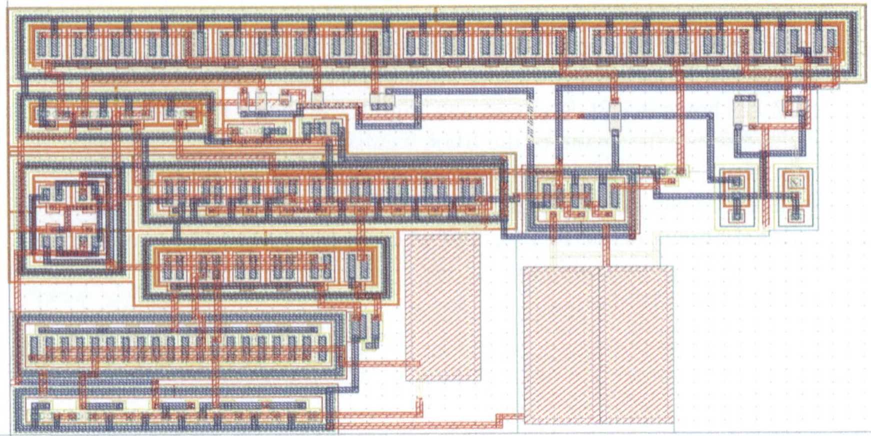

Fig. 22.14 BGR layout

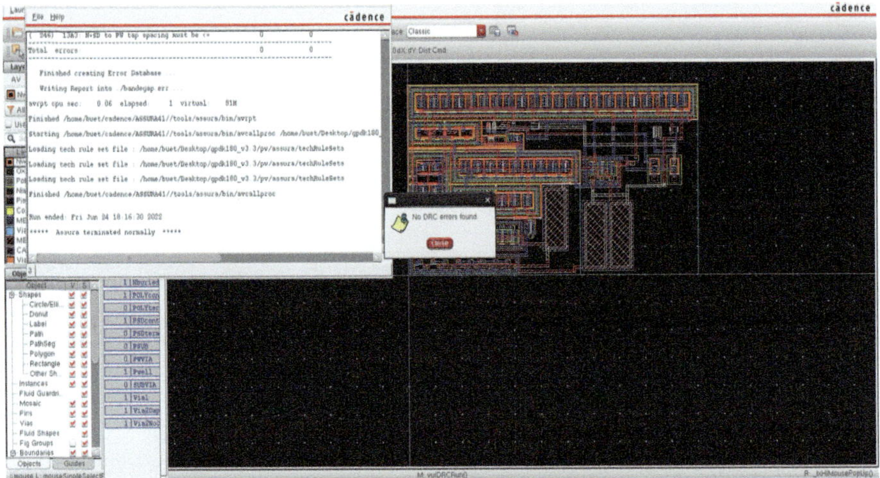

Fig. 22.15 Design rule checker for BGR

22.3.3 *Rectifier and Regulator Circuit*

A rectifier and regulator circuit is a combination of two essential components used in RFID tag. The rectifier converts alternating current (AC) into direct current (DC), and the regulator ensures that the DC voltage remains stable within a specified range. This combination is used to provide clean and regulated power to electronic devices. The rectifier circuit converts AC voltage into DC voltage by allowing current to flow in only one direction. After rectification, the DC voltage may still have some fluctuations and ripple. A regulator circuit is used to stabilize and maintain the output voltage at a constant level. In this circuit, the voltage regulator ensures that V_{out} remains stable, regardless of variations in V_{in} or load current. Figure 22.16 shows the rectifier and regulator circuit while Fig. 22.17 shows its results. Figure 22.18 shows the layout of the rectifier and regulator circuit.

RFID system faced some challenges. First, interference with other signal used the same frequency. UHF signals may face interference from metals or liquids, impacting performance. The cost is another challenge which should be solved. Implementing UHF RFID systems can initially be more expensive than lower-frequency alternatives. The RFID technology is an anti-collision technology. UHF RFID systems often incorporate anti-collision technology, allowing multiple tags to be read simultaneously. This is crucial for scenarios where numerous tags are present in the reading field. Regulatory compliance: Compliance with regional regulations and standards is important when deploying UHF RFID systems, as frequency allocations and power limits can vary. The integration of RFID technology with IoT and Industry 4.0 is the most challenges issue in the revolution of RFID system. UHF RFID plays a vital role

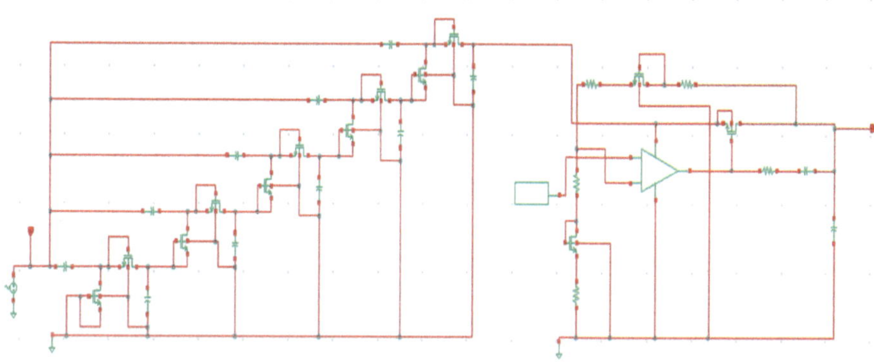

Fig. 22.16 TOP circuit on Cadence (rectifier and regulator)

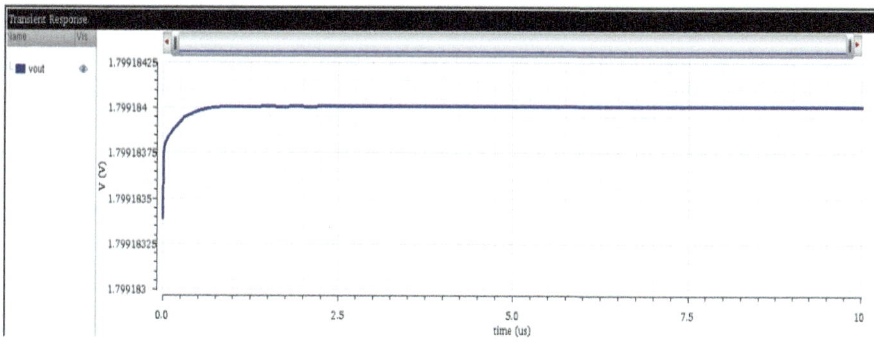

Fig. 22.17 Transient analysis result of TOP (rectifier and regulator)

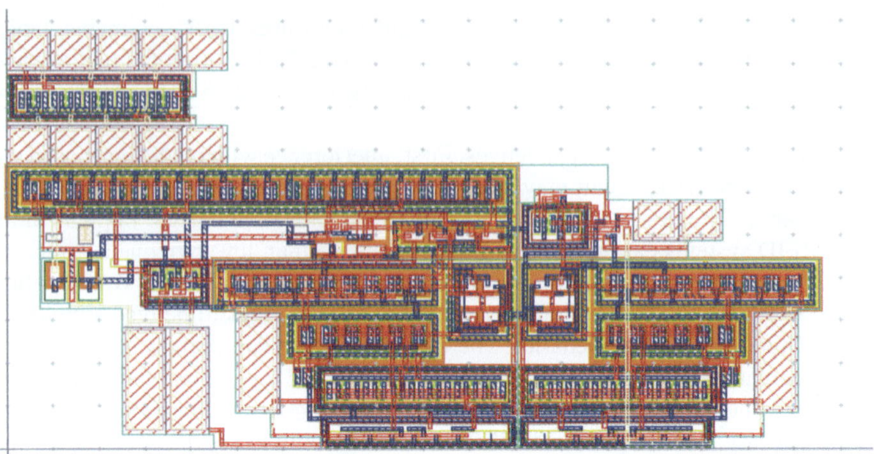

Fig. 22.18 Layout of TOP (rectifier and regulator)

in the Internet of Things (IoT) and Industry 4.0, enabling seamless connectivity and real-time data exchange in smart and automated environments.

22.4 Conclusion

In this paper, we presented a VDD voltage generator for a passive UHF RFID tag, consisting of an RF limiter, a CMOS rectifier, a DC limiter, and a regulator. Theoretical analyses and simulations of the rectifier and regulator circuits used for design optimization are presented. The circuit demonstrates the ability to generate stable DC voltages independent of input power as a power source for a passive UHF RFID tag. The VDD Voltage Generator for Passive UHF RFID tag comprises an RF limiter. The RF limiter likely serves to protect the subsequent circuitry from high-power RF signals, preventing damage and ensuring stable operation. The CMOS rectifier is crucial for converting the RF signal received by the tag's antenna into a DC voltage. This is a common method used in passive RFID tags to harvest energy from incoming RF waves. The DC limiter likely plays a role in stabilizing the DC voltage output, preventing excessive fluctuations and ensuring a consistent power supply. The regulator is responsible for maintaining a stable and regulated output voltage, ensuring that the RFID tag operates reliably across varying input power conditions. The paper includes theoretical analyses and simulations of the rectifier and regulator circuits. This is crucial for understanding the performance characteristics, efficiency, and limitations of the VDD voltage generator. Design optimization, as presented in the paper, involves refining the circuit based on theoretical insights and simulation results to achieve desired outcomes. A significant achievement of the presented circuit is its capability to generate stable DC voltages independently of input power. This is particularly important for passive UHF RFID tags, as they rely on harvested energy from incoming RF signals for their operation.

The presented VDD voltage generator contributes to the development of efficient and reliable passive UHF RFID tags. Stable DC voltage generation enhances the performance and longevity of these tags in real-world applications. The insights and findings from this paper could pave the way for further advancements in passive UHF RFID technology. Future research may explore enhancements to the VDD voltage generator for even more efficient and robust performance.

References

1. Finkenzeller, K.: RFID Handbook: Fundamentals and Applications in Contactless Smart Cards and Identification. Wiley (2003)
2. Raihani, H., Benbassou, A., Ghzaoui, M.E., Belkadid, J.: Gain enhancement of a looped slots patch antenna for passive UHF RFID tags applications in metallic supports. Int. J. Commun. Antenna Propag. **9**(3), 172–181 (2019)
3. Karthaus, U., Fischer, M.: Fully integrated passive UHF RFID transponder IC with minimum RF input power. IEEE J. Solid-State Circ. **38**(10) (2003)
4. Raihani, H., Benbassou, A., El Ghzaoui, M., Belkadid, J.: Novel miniaturized folded line antenna for passive UHF RFID tags. Int. J. Commun. Antenna Propag. **8**(2), 95–102 (2018)
5. Raihani, H., Benbassou, A., El Ghzaoui, M., Belkadid, J.: Performance evaluation of a passive UHF RFID tag antenna using the embedded T-Match structure. In: 2017 International Conference on Wireless Technologies, Embedded and Intelligent Systems, WITS 2017, p. 7934636 (2017)
6. Rongsawat, K., Thanachayanout, A.: Ultra low power analog front-end for UHF RFID transponder. IEEE ISCIT (2006)
7. Ashry, A., Sharaf, K.: Ultra low power UHF RFID tag in 0.13um CMOS. IEEE ICM, December (2007)
8. El Alami, A., Ghazaoui, Y., Das, S., Bennani, S.D., El Ghzaoui, M.: Design and simulation of RFID array antenna 2 × 1 for detection system of objects or living things in motion. Procedia Comput. Sci. **151**, 1010–1015 (2019)
9. Dickson, J.F.: On-chip high-voltage age generation in NMOS integrated circuits using an improved voltage multiplier technique. IEEE J. Solid State Circ. (1976)
10. Din, A.U., Lee, J.-H., Hieu, N.X., Lee, J.-W.: Dual-mode RFID tag IC supporting Gen-2 and visible RFID modes using a process-compensating self-calibrating clock generator. IEEE Trans. Industr. Electron. **67**(1), 569–580 (2020). https://doi.org/10.1109/TIE.2019.2896278
11. Solar, H., Beriain, A., Rezola, A., del Rio, D., Berenguer, R.: A 22-m operation range semi-passive UHF RFID sensor tag with flexible thermoelectric energy harvester. IEEE Sens. J. **22**(20), 19797–19808 (2022), https://doi.org/10.1109/JSEN.2022.3202634
12. Wang, D.-M., Hu, J.-G., Wu, J.: A fully integrated low-cost HF multistandard RFID reader SoC and module for IoT applications. IEEE Internet Things J. **9**(19), 19201–19213 (2022). https://doi.org/10.1109/JIOT.2022.3164919
13. Rennane, A., Benmahmoud, F., Tayeb Cherif, A., Touhami, R., Tedjini, S.: Design of autonomous multi-sensing passive UHF RFID tag for greenhouse monitoring, Sens. Actuators A: Phys. **331**, 112922 (2021). https://doi.org/10.1016/j.sna.2021.112922
14. Banerjee, A., Bhattacharyya, T.K., Nag, S.: High efficiency CMOS active rectifier with adaptive delay compensation. Microelectron. J. **112**, 105052 (2021). https://doi.org/10.1016/j.mejo.2021.105052
15. Menssouri, Z., Mrabet, Z., Zenkouar, L., Qjidaa, H., El Khadiri, K.: Design of VDD generator circuit for a passive UHF RFID tag in 180 nm CMOS. In: 2018 6th International Conference

on Multimedia Computing and Systems (ICMCS), Rabat, Morocco, pp. 1–6 (2018). https://doi.org/10.1109/ICMCS.2018.8525861
16. Kim, S., Choi, K.-S., Kim, K.-M., Ko, J., Kim, J., Lee, S.-G.: A low-noise and fast-settling UHF RFID receiver with digitally controlled leakage cancellation. IEEE Trans. Circ. Syst. II Expr. Briefs **68**(8), 2810–2814 (2021). https://doi.org/10.1109/TCSII.2021.3066644
17. Bhanushali, K., Zhao, W., Pitts, W.S., Franzon, P.D.: A 125 μm $\times$ 245 μm mainly digital UHF EPC Gen2 compatible RFID tag in 55 nm CMOS process. IEEE J. Radio Freq. Identif. **5**(3), 317–323 (2021). https://doi.org/10.1109/JRFID.2021.3087448
18. Najafi, V., Jenabi, M., et al.: A dual mode EPC Gen 2 UHF RFID transponder in 0.18um CMOS. IEEE Electron. Circ. Syst. (2008)
19. Umeda, T., Yoshida, H., et al.: A 950-MHz rectifier circuit for sensor network tags with 10-m distance. IEEE J. Solid State Circ. **41**(1) (2006)
20. Vita, G., Iannaccone, G.: Ultra low power RF section of a passive microwave RFID transponder in 0.35 un BiCMOS. IEEE ISCAS (2005)

Chapter 23
Satellite and Terrestrial Mobile Integration-Potential and Open Issues for 5G and Terahertz Communication

Pia Sarkar, Arijit Saha, Amit Banerjee, and Vedatrayee Chakraborty

23.1 Introduction

Nowadays, the operators are looking for new frequency bands for upcoming wireless communication. To mitigate the frequency scarcity problem, we have adopted new techniques like frequency reuse, multiple-input multiple-output (MIMO) antenna, precoding to increase channel capacity. Fifth generation (5G) system demands high-quality signal, high security, massive device connectivity and low latency [1] to provide expected data rate up to up to 10 Gigabit per second (Gbps) and latency of 1–10 milli second (ms). Third-Generation Partnership Project (3GPP) accelerated the use of non-terrestrial networks (NTN) for the growth in communication [2]. If satellite systems can be integrated with terrestrial communication, 5G and beyond connection can be accessed anytime and anywhere. Satellite communication turns into backup link in place of terrestrial link which can expand the coverage area. It can be a cost-friendly solution to provide high-quality signal in the area where ground network fails to support connection. Satellite interfaces are deployed in earth ground station by which terrestrial network can get internet connectivity over satellite network. The hybrid communication of space and ground network can bring an evolution in upcoming terahertz communication. Satellite communication has opened up a new research area to explore. Instead of facilitating lots of aspects, there are some questions to be solved. The integration process of these two separate networks is

P. Sarkar (✉) · V. Chakraborty
B. P. Poddar Institute of Management and Technology, Kolkata, India
e-mail: pia.sarkar@bppimt.ac.in

A. Saha
Dumdum Motijheel Rabindra Mahavidyalaya, Kolkata, India

A. Banerjee
Physics Department, Microsystem Design-Integration Lab, Bidhan Chandra College, Asansol, India

M. El Ghzaoui et al. (eds.), *Next Generation Wireless Communication*, Signals and Communication Technology, https://doi.org/10.1007/978-3-031-56144-3_23

not simple. Satellite shows high latency and high bit error rate (BER). Satellite communication is not compatible with the general techniques used in terrestrial communication. The interconnection with LAN and Wi-Max [3, 4] results in progress of the communication. The interconnection of satellite and mobile communication can provide better capacity and throughput that will be useful in advanced 5G and 6G communication. This has provided the motivation of our study.

In our study, we have explained the integration process, key technologies, the challenges and solutions of the problems have been discussed. Besides, we have emphasized on the security of communication in our work. Here, the novelty of the work lies.

A comprehensive literature survey has been summarized in this study. In Sect. 23.2, existing work has been mentioned. In Sect. 23.3, non-terrestrial network has been discussed. In Sect. 23.4, the importance of hybrid network has been explained. The integration process has been provided in Sect. 23.5. The open issues are discussed in Sect. 23.6. In Sect. 23.7, key technologies have been presented. The use cases have been mentioned in Sect. 23.8. Discussions on results are presented in Sect. 23.9. The future scope has been discussed in Sect. 23.10. The study has been concluded in Sect. 23.11.

23.2 Related Work

Satellite communication has explored new area to develop data communication and networking. The latest research and published work of integrated satellite communication have been discussed in this paper.

Kota et al. provided how the QoS is related to each layer of Internet Protocol (IP) network for Satcom [5]. Akyildiz et al. summarized architecture, algorithm and protocol of interplanetary network [6]. Wang et al. proposed the protocol for reliable data transport. The classification, operation, design and performance comparison among these protocols for space internet have been discussed [7]. Zafar et al. compared between different handoff and mobility management approaches for satellite communication. Gaber et al. summarized the potential of satellite communication as a global network infrastructure [8]. This literature presented advanced technology to integrate 5G with satellite communication for better QoS. The new mobility management scheme for getting uninterrupted service during satellite handover process has been proposed. The use cases and technical challenges are also summarized. Giordani et al. demonstrated the establishment of satellite network for upcoming 6G while 5G network is still developing. The requirement and limitations toward service coverage, huge traffic load and continuous service in 6G have been mentioned. The challenge of integration process and design trade-off has been identified in this survey [9]. Giordani et al. described the key technologies of using non-terrestrial network (NTN) for 6G network to provide high capacity. They discussed the technical challenges of satellite network regarding using mmWave frequency. The performances of different network configurations are evaluated in this study

[10]. Lin et al. gave an overview of new technology that 3GPP explored to make 5G a commercial reality. The ideas for solving the difficulties have been provided [11]. Kapovits et al. presented extension of terrestrial communication to satellite communication as backhaul. It overviewed the energy-efficient and robust satellite network as an alternative for covering globally, ensuring connection in isolated areas [12]. Tirmizi et al. overviewed the design of hybrid network and the process to improve the architecture, technique and challenges for 6G communication [13]. Wang et al. discussed a new method to use high frequency to enhance the capacity in integrated space-ground communication [14]. Sheng et al. explained how artificial intelligence (AI) can be applied for increasing the capacity of 6G communication [15]. Xue et al. proposed the method to select proper relay and to share the spectrum for facilitating the hybrid terrestrial and non-terrestrial communication [16]. Fang et al. discussed the necessity of integration of 5G communication with satellite communication for effective 6G communication. They mentioned the challenges of integration process due to mismatch of technology between space and ground networks [17]. Wang et al. presented the architecture for communication in ocean. They suggested to deploy base station through unmanned aerial vehicles for supporting 6G communication in ocean [18]. Wang et al. summarized the need of converging base connection with satellite communication to enhance capacity and quality of service [19]. Zhu et al. proposed new network design of Medium Earth Orbit (MEO) satellite and terrestrial communication and analyzed the efficiency of the new architecture [20]. Dicandia et al. mentioned the application of antenna used in space and ground to increase the coverage for hybrid communication [21]. Höyhtyä et al. discussed the motivation of integration of satellite and ground communication for safe communication of ships [22]. Yan et al. proposed a method for interference mitigation of space and ground-integrated communication [23].

There are many differences between satellite communication (Satcom) and terrestrial communication (Tercom) [24, 25]. In satellite communication as the signal has to traverse larger transmission distance, propagation loss and transmission delay increase. The scatterer becomes less as signal transmission path is direct. Rain and other atmospheric conditions [26–28] result in more attenuation. Due to these inherent characteristics, LEO and MEO Satcom cannot apply the existing routing approach and mobility management of Tercom. These facts demand rapid change in network topology.

23.3 Non-terrestrial Networks (NTN)

Satellite is a wide-flying object with transponder that receives signals to a certain range of frequency and uses another frequency range to transmit. A signal is reflected by satellite from one point to another. There are limited number of orbits such as nearly circular shaped for Low Earth Orbit (LEO), Medium Earth Orbit (MEO) and Geostationary Earth Orbit (GEO). Satellites provide transmission delay due to large altitudes. Relative speed of LEO, MEO, GEO and HEO are high, low, stationary and varying. The signal faces significant delay and attenuation in case of GEO satellite. But as it is continuously visible from terrestrial terminal, its coverage is more. Signal quality is better for LEO satellites. But for stationary communication, GEO satellites are used.

Antenna types and sizes are different for different altitude, smaller footprint (spot beam) or wider footprint (Large beams). Antenna size is inversely proportional to wavelength. Exclusive opportunities in satellite communication are large coverage, robust network, multicasting and broadcasting capabilities. However, many propagation challenges have to be solved, viz. atmospheric condition, delay, path loss, indoor and mobility management.

NTN communication operates through air/space-borne vehicle. 3GPP first introduced the possibility of satellite integration in Release-15 (Rel-15), the deployment scenarios and parameters in Rel-16 and Rel-17 and enhanced specification in Rel-18 and Rel-19.

23.3.1 General Architecture

It consists of

(i) a terrestrial terminal: User equipment (UE) can be either terrestrial or satellite-mobile system.
(ii) an aerial/space station: Filtering, demodulation/decoding, modulation/coding can be done.
(iii) a service link/feeder link: It connects satellite with UE or with terrestrial gateway.
(iv) a gateway through a feeder link for connecting the NTN with the core network.
(v) Inter-satellite links (ISL): This connection transmits data through propagating light using RF or free space optical (FSO) communication.

The architecture of integrated satellite and terrestrial connection has been depicted in [29]. Satellites are connected with satellite gateways via satellite links. The terrestrial network consists of multi-access edge computing (MEC), user equipment (UE), 5G network, long-term evolution (LTE) and wireless fidelity (WiFi), etc. To support 5G network in remote areas, satellite network should be integrated with terrestrial network.

23.3.2 Unmanned Aerial Vehicle (UAV)

It is flexible and remains at few hundred meters. During natural disaster, it can provide wireless broadband connectivity. Terrestrial infrastructures are fixed, whereas deployment of UAV can be done on demand and is energy efficient.

23.3.3 High Altitude Platform (HAP)

It remains in stratosphere around 20 km. It can be deployed easily and coverage is high. There are two ways by which satellite can cooperate with HAP.

Backhauling: It is a downlink communication between satellite and ground receiver and HAP is intermediate element. At first, signal traverses from satellite to HAP. High-bandwidth optical links should be used to recover atmospheric attenuation. Then signal transmits from HAP to ground receiver. As this distance is shorter than satellite altitude, the link budget is cost-effective using smaller antenna.

Trunking: Users directly connect in regional coverage areas using HAP and use satellites for inter-coverage communications.

Challenging issues [30] in HAP: (i) Lack of sufficient amount of daylight hours due to high altitude. (ii) High-speed wind can displace the HAP from their operating area. Battery lifetime is reduced due to low temperature. Geographical coverage, fast deployment, reconfigurable and low propagation delay are the benefits of HAP [31].

23.4 Necessity of Integration

Long-term evaluation advanced (LTE-A) has been overburdened to handle enormous data with increasing wireless device user and Internet of Things (IoT) applications. Various communication mediums can be selected considering the requirement of bandwidth, capacity and cost, etc. to overcome this congestion. The importance of satellite-terrestrial network integration lies for 5G reliable service [32].

Non-satellite network plays an important role as backup service with high capacity and large coverage. It enables the connection where terrestrial network is not feasible in remote and rural areas, or out of order due to natural disaster. Integration of non-terrestrial network with terrestrial network can facilitate many opportunities in 5G and beyond communication. Current network cannot provide sufficient broadband coverage in remote or rural areas. Lack of reliability of future wireless applications is still there in countries that are technologically advanced. The ubiquity and high scalability of integrated communication is a feasible alternative for enhancing coverage and energy efficiency. This robust network can operate during natural disaster using backhaul. In some remote areas, deployment of terrestrial network is economically challenging. Global coverage of integrated network can extend the communication service at affordable cost in isolated and rural areas.

Three main services have been defined by 3GPP in Technical Report 22.822 (TR22.822) and further introduced in [33]. Three major services are given in the following [34].

Service continuity: Terrestrial networks cannot assure continuous service in urban areas. Lack of reliable coverage in some areas can be overcome by this category for providing continuous connection to users in cars, trains, airborne platforms, maritime vessels.

Service ubiquity: Terrestrial service is vulnerable in some isolated areas. Furthermore, natural disaster can destruct this service. This category can provide 5G service where terrestrial services are interrupted.

Service scalability: Satellite network is capable for multicasting and data broadcasting. This *service* can support 5G services.

5G is an evolution of wireless communication that introduces advanced application NTN has been approved for 5G after two years, and work item (WI) has been started from January 2020.

Satellite applications for 5G uses are as follows:

23.4.1 Use in Enhanced Mobile Broadband (eMBB)

The use has been summarized in [35]. It optimizes global 5G network service using backhaul.

23.4.2 Use for Massive Machine-Type Communication (mMTC)

In massive connection, traffic load is less but its effect in network load is significant. Satellites can be used to reduce the load using backhaul. It continues the services in isolated areas where terrestrial network connection is not possible. Low complex and cheap devices (sensors/actuators) are used in Internet of Things (IoT). This use case has been divided into two categories depending on satellite applications or distribution of IoT sensors on earth.

Wide-area IoT services: Satellites are used for energy, transport and agricultural applications.

Local area IoT services: After collecting local data, it is reported to central server by IoT devices. Applications like advanced metering or on-board service to a vessel, a truck or a train are included in this category.

23.4.3 Application in Ultra-reliable Low Latency Communication (uRLLC)

This is applied for reliable communication where latency is low (lower than 1 ms) such as autonomous driving, remote surgery, industry automation, etc. Low latency at the user side is achieved by wide-area broadcasting of satellites. Satellites can broadcast the car software and traffic updates in autonomous driving.

23.5 How the Integration is Done?

Mobile communication system uses backhaul that is GEO satellite based [36]. Satellite services are provided via S1 interface between EPC and evolved Node B (eNodeB) and Ethernet. For satellite access, one 120 centimetre (cm) antenna, two Ka antennas and VSAT hub are used.

As C and Ku-band becomes saturated, Ka band can be used as bandwidth is higher than L and S bands. It can provide broadband services, avoid interferences, and reduce cost with enhancing capacity. Though high-frequency usage is a critical issue, some atmospheric perturbations affect link quality QoS.

23.6 Challenges

NTN has been developed through standardization. Besides several open challenges like interface issues and proper protocol designs need to be solved. The signal has to cover longer distance in satellite communication. Propagation delay and Doppler effect increases in satellite communication. There are other challenges regarding mobility management and movement of base stations.

23.6.1 Use in eMBB

As available power, transmitting and receiving antenna gain, link budget should be estimated in satellite communication. Beamforming is introduced to tackle these propagation issues. Satellite can be assumed as a relay for two-way transmission. Signals from earth stations are amplified and reflected back from satellite by beamforming and other technology. To alleviate interference and spectrum constraints, a digital communication method with phased-array beamforming is proposed to produce interference immune signal and use the spectrum efficiently. Satellite signals can less penetrate the building. So proper satellite-terrestrial network coordination should be done.

23.6.2 Spectrum Scarcity

The key enabler eMBB in 5G requires high-speed data rate and large capacity. To avail these facilities in limited spectrum millimeter Wave (mmWave), frequency should be utilized by terrestrial mobile communication. 1–40 GHz frequency range is used for satellite communication. Specific frequency band has been allocated for 5G mobile communication in World Radio Communication Conference 2019 (WRC-19). Cognitive radio (CR) with machine learning is an optimized spectrum utilization strategy.

23.6.3 Satellite Propagation Delay

Latency depends on the factors of distance and elevation angle. Propagation delay is less for LEO than MEO and GEO as the altitude for LEO is less. Forward error correction (FEC) adds parity bits with information bits for error correction. If receiver detects no error, it sends positive acknowledgment and negative acknowledgment for error detection. Random access is a proposed solution of delay problem depending on UE complexity.

23.6.4 Doppler Effect

The Doppler effect results in change of frequency due to satellite motion. A Doppler mitigation method is introduced in [11].

23.6.5 Security

Security is the most challenging issue for terrestrial and satellite communication. Individually satellite and terrestrial communication performs independently. For the integration, different technologies should be applied. Besides, maintaining safe and reliable communication is difficult. For data transmission, stable connection should be formed and authentication should be done. Due to mobility of satellite, the connection can be unstable. For secured data transmission, encryption is effective technology that should be incorporated. For data encryption and computation, sufficient resources and storage should be necessary. But there is shortage of storage and resources in satellite. Challenges and solutions regarding security have been elaborately presented in [37].

There can be different types of attacks in routing technology [38]. The routing network can be accessed and program can be altered by attacker. Sometimes, they can take the control of routing network. For maintaining confidentiality, some algorithms have been discussed in [39]. For hybrid connection, if the key enabler software function is applied in satellite, then it can hamper the security because some attacker can alter the program. To facilitate terrestrial and satellite hybrid connection, the use of these two key enablers has been discussed in [40]. If the handover can change rapidly, then security can be a great concern. This happens due to movement of mobile, due to mobility of airplane from the coverage of LEO Earth Orbit (LEO) to Geostationary Earth Orbit (GEO). During frequent handoff, secured connection should be ensured.

23.7 Air Interface Key Enablers

For getting proper performance, the interfacing of technologies should be tackled. The key technologies have been summarized below.

23.7.1 Channel Modeling

For proper system design, channel and propagation characteristics are key factors. It depends on the system design frequency. Satellite channel propagation characteristics at mmWave frequencies have been proposed in 3GPP [41]. But statistics (space–time correlation), Doppler, fading and multipath effect are not specified in 3GPP. A proper model of space-air-ground channel needs to be designed. Channel variations are summarized in [42].

Common channel modeling depending on long-term components, dynamic channel factors and lack of scatterers near satellite antennas are described as below.

Fixed satellite: The operating frequency range for next-generation satellites will be above 10 GHz. Above this, atmospheric fading effect is severe in Ku and Ka bands which are categorized as:

Long-term channel effects: Precipitation attenuation, gaseous and cloud absorption, scintillation of troposphere, signal depolarization effects play a key role to this category involving first-order statistics [42].

Dynamic channel effects: Rain effect is a key constituent that changes the properties of AWGN channel. The channel fading between the satellites or user terminals gets correlated due to absence of scatterer near satellite transmitting antenna and longer distance between user terminal and satellite.

Mobile satellites: Non-LoS multipath is included with LoS due to mobility. Channel characteristics are different from fixed satellite. A multi-state Markov model in which narrowband and broadband channel modeling is discussed has been presented in [42]. Each state can decide channel parameters. Three-state Markov model (narrowband) with statistics is proposed in ITU Recommendation P.681 [43]. If shadow strength increases, the attenuation also increases. New satellite-mobile channel at 2.2 GHz is modeled in [44]. User terminal (UT) speed determines channel state statistics. For multipath propagation, ITU multitap model is approached for wideband satellite [45]. For short duration, the channel properties are quasi-stationary and for long duration are stationary stochastic processes.

23.7.2 Antennas

Passive reflector antennas with focused beam: It can offer broadband service. If multibeam antennas (MBAs) transmit narrower beam into a geographical area, the gain and spectral efficiency increases by frequency reusing. A single feed produces each beam in SFPB antenna. It increases the gain and decreases the side lobe level. Carrier to interference ratio becomes suitable. Contiguous coverage is achieved by three or four reflectors. The relative electrical aperture of horn antenna is large. Antenna efficiency increases, and side lobes are reduced. Beam separation is large for small displacement of feed horn antenna. Another reflector produces the beam that is interleaved in these.

Alternatively, a cluster of feeds forms each beam in MFPB. Single feed may produce multiple adjacent beams. Adjacent feed cluster overlaps to get contiguous coverage by one or two main reflectors. The network for beamforming is complex and aperture efficiency is low. It is not always preferred in passive antenna as operating frequency bandwidth gives rise to a challenge.

For GEO satellite, these two architectures are commonly used. Power allocation will be flexible by using multiport amplifier. More directive beams can be produced if large reflectors can be deployed. Feeds form number of beams that are decoupled by active antenna array. The power is shared between beams and coverage is flexible applying beam technology.

Active antenna arrays: It is different from passive antenna in which radiating elements are integrated with amplifier and radiating signal gets spatially distributed amplifications. The peak power level decreases improving reliability. The transition from traditional vacuum (traveling wave tube) to antenna array is suitable choice to reduce the power of amplifier.

Direct radiating array (DRA) is chosen for large field view in LEO and MEO. Throughput and gain increase with electrical size of antenna. Reflector architecture is selected in GEO. The selection of antenna used in GEO remains open. Antenna gain is related to aperture size.

Antennas in space and the ground segment: LEO/MEO satellite distributes the traffic spatially. This approach aligns with a steerable beam approach adopted in MEO constellation by O3B. For limited capacity of on-board smaller platform, antenna architectures are described in [46]. Flat antennas that can steer the beam, can apply

Satellite on the Move (SOTM) and can connect non-GEO platform are deployed in ground terminals. Two-axis positioner tracks azimuth and elevation using two narrow-width arrays. Interference due to broad beam width is a problem of this architecture. In avionic industry, the deployment is mentioned [47]. Flat panel antenna that is mechanically steerable can mitigate this problem. Nematic liquid crystals meet the performance objective and reduce the cost.

23.7.3 Ultra-high Throughput Satellites (UHTS)

Digital payloads: Connection should be reliable and flexible and throughput should be high to meet increased traffic demand. Multibeam systems architecture, transition to high frequency, novel precoding technique and resource management have been proposed in [48]. In this proposal, on-board processing (OBP) method considers space-based assets. On-board satellite processing digital signal is proposed in [49].

Digital transparent processors (DTPs): It processes the sampled signal. No demodulation or decoding is performed.

Regenerative processing: Waveform digitization, demodulation and decoding produce digital baseband data that is operated by regenerative processing. Cost is high and flexibility is limited.

Continuous monitoring and degradation identification technique for on-board amplifier still have to be developed. Spread spectrum-based method is considered in [50].

Precoding/MU-MIMO: In multibeam architectures, antenna arrays have spatial degree of freedom and enhanced throughput. Spatial multiplexing reuses the available spectrum. Multi-user can interfere in multibeam system and degrades the operation. Multiple antennas can prevent this problem.

Linear precoding (or beamforming) becomes a demanding research area. Recently, broadband multibeam Satcom standards extend to precoding technique mitigating interference at Satcom gateway. Data streams are multiplied with precoder matrix in satellite gateway. The satellite shifts the frequency routing the signal to antenna array. Then multiple beams serve precoded data and cover large geographical area. Same frequency is reused by another beam while transmitting in downlink.

The problem of this technique lies in complexity because of large sized precoding matrix to arrange several hundreds of beams. This limitation requires new research attention.

Symbol-level precoding exploits both channel knowledge and data information (DI). It does not only mitigate interference like conventional precoding but forms constructive interference to each user. Gateway consumes low power improving quality of service (QoS). Several precoding schemes have been included to provide robust signal in nonlinear effects.

Non-orthogonal multiple access (NOMA): It is attracting technology in 5G and beyond. Same time and frequency resources are shared among multiple terminals and spectrum efficiency increases. Receiver operates Successive Interference Cancellation (SIC) to remove co-channel interference. NOMA performs superior than orthogonal multiple access (OMA).

Terrestrial NOMA is not applicable to satellite NOMA because they have different characteristics. In satellite communication, the same path loss is achieved by users in a beam. Implementation of grouping of users is difficult. Due to huge number of beams, the system is complex. Compared with cellular systems, satellite communication faces new constraints like long propagation delay, power supply limitation and signal distortion.

23.7.4 Data Support

Satellite can benefit IOT services by low power wide-area networks (LPWANs) that use different technologies. The signal of narrowband is modulated with differential binary phase shift keying (DBPSK). Chirp spread spectrum signal (CSS) is utilized by long range (LoRa). As the satellite channel experiences large delay and high Doppler effects, it is tough to collect huge data from IOT devices over satellite link. The LoRa for LEO satellite and new modulation strategy Folded Chirp-rate Shift Keying (FCrSK) are analyzed to make the signal Doppler-immune.

For conventional modem, the data rate is up to 500 MHz. To support Ka band, a modem should be designed for large spectrum about 1.5 GHz. The hardware can limit the frequency selectivity and large differences of minimum and maximum frequency result in impairment.

23.7.5 Others

Optical communications: The transition from radio frequency (RF) to optical frequencies can meet the demand of large bandwidth. Rain and cloud attenuation severity in high frequency makes this communication challenging. Though it facilitates as secured, low power requirement and low-cost communication as few gateways are needed to get high throughput.

Fading is caused due to cloud coverage and turbulence. These can be limited by following two techniques.

Micro-diversity: Multiple optical ground stations are placed by hundreds of miles to combat cloud blockage.

Macro-diversity: This transmitter diversity technique places multiple apertures at a longer distance than turbulence coherence length.

Satellite swarms and synchronization: They play an important role in low-cost miniaturized electronics.

They should be synchronized regarding phase, frequency and time of clock.

Deep space communications: The challenge rises for the large distance (about 2 million km away) of satellite from earth reducing SNR. Power generation in spacecraft from sun is limited leading to main challenge in case of downlink. A supplement approach for power generation is radioisotope thermoelectric generator (RTG).

Transmitting/receiving sites known as deep space network is specified by space agency using large sizes of 35 and 70 m diameter dish antenna.

23.8 Applications

The hybrid communication brings a new research direction for industrialist and academician. Satellite communication can facilitate the 5G and 6G communication in remote areas. Satellite can provide broadband connection in rural areas where terrestrial networks cannot be established due to unfavorable situations. If data transmission rate for both satellite and terrestrial communication becomes compatible with each other, then user can access both networks simultaneously. If this integration can be implemented properly, then signal coverage and signal quality will be improved. This hybrid communication can facilitate 6G connection in moving vehicle, Internet of Technology (IoT) devices, navigation and remote sensing etc.

23.9 Results and Discussions

The capacity for the altitudes of LEO (300 km), MEO (10,000 km) and GEO (36,000 km) satellites has been analyzed with frequencies in MATLAB 21. 6, 28, 70 and 150 GHz frequencies are selected at $\alpha = 10°$ elevation angle. Antenna gain is taken as high value of Grx = 50 dB to make up large attenuation in high frequencies.

In Fig. 23.1, the capacities for dense urban region have been shown at $\alpha = 10°$ elevation angle for LEO, MEO and GEO satellites. At 6 GHz, the capacities are near about 1, 0.2 and 0.12 Gbps. For LEO, satellite capacity reaches to maximum value of 12.5 Gbps at 70 GHz. As attenuation increases for long distances, capacity deteriorates for high altitudes of MEO and GEO satellites. Maximum capacities are of 0.7 and 0.12 Gbps at 28 GHz for MEO and at 6 GHz for GEO satellites, respectively.

Shannon capacity for rural area has been depicted with frequencies in Fig. 23.2. The elevation angle has been taken as $\alpha = 10°$ for the distances of LEO, MEO and GEO satellites. The capacities are at 2, 0.5 and 0.2 Gbps at 6 GHz. For LEO satellite, altitude maximum capacity is of 25 Gbps at 70 GHz. Maximum capacity reaches to 4.5 and 1.7 Gbps at 28 GHz for the distances of MEO and GEO satellites.

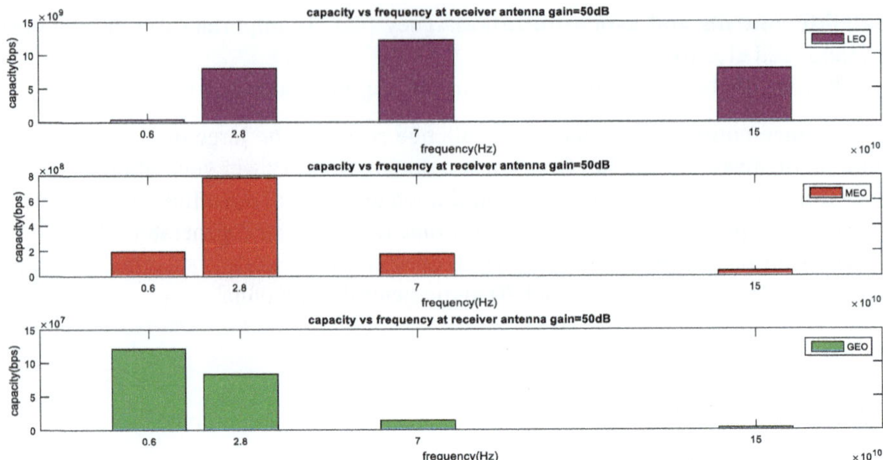

Fig. 23.1 Shannon capacity for different altitudes at different frequencies, varying the satellite altitude h. $\alpha = 10°$, dense urban scenario, Grx = 50 dB

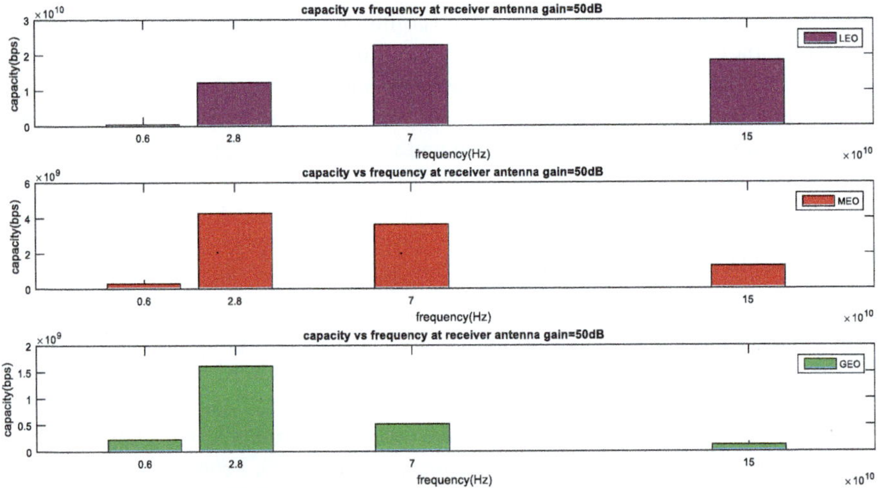

Fig. 23.2 Shannon capacity for different altitudes at different frequencies, varying the satellite altitude h. $\alpha = 10°$, rural scenario, Grx = 50 dB

It can be explained that growth in capacity has been resulted in rural areas due to less blockage in rural areas than dense urban region. Capacity becomes better for less altitude of LEO satellites. As the distance increases for MEO and GEO satellites, capacity becomes poor. Besides, severe loss occurs for long distances of MEO and GEO satellites in high frequencies. If the MEO and GEO satellites perform in lower frequencies than LEO satellites, the capacity can be improved.

23.10 Future Direction

High latency for long distance, higher bit error rate, key encryption, software updation and storage for computation are the challenges for satellite. Terrestrial connection also can hamper in remote, rural area and the area effected by natural calamity. If these two networks can be connected, the dual network can bring huge progress in wireless communication. Artificial intelligence (AI) can be implemented for autonomous integrated communication [51, 52]. Throughput of satellite-terrestrial network can be improved by applying hybrid NOMA/OMA technique [53]. In future user equipment, ground station will be directly connected with satellite to establish huge coverage and broadcasting services with improved quality of signal. More instigation should be performed in integration procedure to implement 6G communication with enhanced capacity.

23.11 Conclusion

Satellite communication can be considered as new horizon paving the way of 6G communication. Latency and path loss at millimeter Wave (mmWave) frequencies open up new research direction. At low elevation angle, received signal severely attenuates due to atmospheric and tropospheric absorption. Elevation angle is an important parameter for proper set-up of satellite systems at mmWave frequency.

Various challenges have been discussed in this present work regarding latency and coverage constraints. Still the establishment of space stations and the interconnection of space-ground networks require to be more explored.

There are various questions to be answered related to deployment of satellite stations, practical integration of terrestrial and non-terrestrial communication.

Capacity and geographical coverage can be improved using highly directional satellite systems. The signal quality can be better using satellite communication. Through this integration process, the advantages of both terrestrial and non-terrestrial communication can be provided for 5G and 6G communication.

References

1. Gupta, A., Jha, R.K.: A survey of 5G network: architecture and emerging technologies. IEEE Access **3**, 1206–1232 (2015)
2. 3GPP: Solutions for NR to support non-terrestrial networks (NTN). 3rd Generation Partnership Project (3GPP), Technical Specification (TS) 38.821, Jan 2020, version 16.0.0 [Online]. Available: https://portal.3gpp.org/desktopmodules/Specifications/SpecificationDetails.aspx?specificationId=3525
3. Ansari, A., Dutta, S., Tseytlin, M.: S-WiMAX: adaptation of IEEE 802.16 e for mobile satellite services. IEEE Commun. Mag. **47**(6), 150–155 (2009)

4. Giuliano, R., Luglio, M., Mazzenga, F.: Interoperability between WiMAXand broadband mobile space networks [topics in radio communications]. IEEE Commun. Mag. **46**(3), 50–57 (2008)
5. Kota, S., Marchese, M.: Quality of service for satellite IP networks: a survey. Int. J. Satell. Commun. Network. **21**(4–5), 303–349 (2003)
6. Akyildiz, I.F., Akan, Ö.B., Chen, C., Fang, J., Su, W.: InterPlaNetary internet: state-of-the-art and research challenges. Comput. Netw. **43**(2), 75–112 (2003)
7. Wang, R., Taleb, T., Jamalipour, A., Sun, B.: Protocols for reliable data transport in space internet. IEEE Commun. Surv. Tutorials **11**(2), 21–32 (2009)
8. Gaber, A., ElBahaay, M.A., Mohamed, A.M., Zaki, M.M., Abdo, A.S., AbdelBaki, N.: 5G and satellite network convergence: survey for opportunities, challenges and enabler technologies. In: 2020 2nd Novel Intelligent and Leading Emerging Sciences Conference (NILES), pp. 366–373. IEEE (2020)
9. Giordani, M., Zorzi, M.: Satellite communication at millimeter waves: a key enabler of the 6G era. In: 2020 International Conference on Computing, Networking and Communications (ICNC), pp. 383–388. IEEE (2020, Feb)
10. Giordani, M., Zorzi, M.: Non-terrestrial networks in the 6G era: challenges and opportunities. IEEE Network **35**(2), 244–251 (2020)
11. Lin, X., Hofström, B., Wang, Y.P.E., Masini, G., Maattanen, H.L., Rydén, H., Sedin, J., Stattin, M., Liberg, O., Euler, S., Muruganathan, S.: 5G new radio evolution meets satellite communications: opportunities, challenges, and solutions. In: 5G and Beyond: Fundamentals and Standards, pp. 517–531 (2021)
12. Kapovits, A., Corici, M.I., Gheorghe-Pop, I.D., Gavras, A., Burkhardt, F., Schlichter, T., Covaci, S.: Satellite communications integration with terrestrial networks. China Commun. **15**(8), 22–38 (2018)
13. Tirmizi, S.B.R., Chen, Y., Lakshminarayana, S., Feng, W., Khuwaja, A.A.: Hybrid satellite–terrestrial networks toward 6G: key technologies and open issues. Sensors **22**(21), 8544 (2022)
14. Wang, L., Tang, W., Fan, Y., Yan, T., Jiang, L.: Performance analysis of PDMA-based hybrid satellite–terrestrial networks with decode-and-forward relaying. Phys. Commun. **52**, 101628 (2022)
15. Sheng, M., Zhou, D., Bai, W., Liu, J., Li, H., Shi, Y., Li, J.: Coverage enhancement for 6G satellite-terrestrial integrated networks: performance metrics, constellation configuration and resource allocation. Sci. China Inf. Sci. **66**(3), 130303 (2023)
16. Xue, G., Yang, M., Guo, Q., Yuan, S.: Performance of spectrum sharing in hybrid satellite terrestrial network with opportunistic relay selection. Wirel. Commun. Mob. Comput. (2022)
17. Fang, X., Feng, W., Wei, T., Chen, Y., Ge, N., Wang, C.X.: 5G embraces satellites for 6G ubiquitous IoT: basic models for integrated satellite terrestrial networks. IEEE Internet Things J. **8**(18), 14399–14417 (2021)
18. Wang, Y., Feng, W., Wang, J., Quek, T.Q.: Hybrid satellite-UAV-terrestrial networks for 6G ubiquitous coverage: a maritime communications perspective. IEEE J. Sel. Areas Commun. **39**(11), 3475–3490 (2021)
19. Wang, P., Zhang, J., Zhang, X., Yan, Z., Evans, B.G., Wang, W.: Convergence of satellite and terrestrial networks: a comprehensive survey. IEEE Access **8**, 5550–5588 (2019)
20. Zhu, X., Jiang, C.: Creating efficient integrated satellite-terrestrial networks in the 6G era. IEEE Wirel. Commun. **29**(4), 154–160 (2022)
21. Dicandia, F.A., Fonseca, N.J., Bacco, M., Mugnaini, S., Genovesi, S.: Space-air-ground integrated 6G wireless communication networks: a review of antenna technologies and application scenarios. Sensors **22**(9), 3136 (2022)
22. Höyhtyä, M., Martio, J.: Integrated satellite–terrestrial connectivity for autonomous ships: survey and future research directions. Remote Sens. **12**(15), 2507 (2020)
23. Yan, S., Cao, X., Liu, Z., Liu, X.: Interference management in 6G space and terrestrial integrated networks: challenges and approaches. Intell. Converged Networks **1**(3), 271–280 (2020)
24. Burkhart, R.M.: Tutorial: propagation phenomena and terrestrial interference in satellite television transmission. SMPTE J. **98**(9), 658–663 (1989)

25. Stone, W.R.: Mobile, Terrestrial, and Satellite Propagation Modeling (1999)
26. Kanellopoulos, S.A., Panagopoulos, A.D., Kourogiorgas, C.I., Kanellopoulos, J.D.: Satellite and terrestrial links rain attenuation time series generator for heavy rain climatic regions. IEEE Trans. Antennas Propag. **61**(6), 3396–3399 (2013)
27. Timothy, K.I., Ong, J.T., Choo, E.B.: Raindrop size distribution using method of moments for terrestrial and satellite communication applications in Singapore. IEEE Trans. Antennas Propag. **50**(10), 1420–1424 (2002)
28. Gusler, L.T., Hogg, D.C.: Some calculations on coupling between satellite communications and terrestrial radio-relay systems due to scattering by rain. Bell Syst. Tech. J. **49**(7), 1491–1511 (1970)
29. Höyhtyä, M., Ojanperä, T., Mäkelä, J., Ruponen, S., Järvensivu, P.: Integrated 5G satellite-terrestrial systems: use cases for road safety and autonomous ships. In: Proceedings of the 23rd Ka and Broadband Communications Conference, Trieste, Italy, pp. 16–19 (2017)
30. Wibisono, M.A., Priatna, S., Juhana, T., Rachmana, N.: On the design and development of flying BTS system using balloon for remote area communication. In: 2017 11th International Conference on Telecommunication Systems Services and Applications (TSSA), pp. 1–5. IEEE (2017, Oct)
31. Haps-teleo: High-altitude pseudo-satellites for telecommunication and complementary space applications. Technical Report, July 2018 [Online]. Available: https://nebula.esa.int/sites/default/files/nebstudy/2470/C4000118800ExS.pdf
32. Jia, M., Gu, X., Guo, Q., Xiang, W., Zhang, N.: Broadband hybrid satellite-terrestrial communication systems based on cognitive radio toward 5G. IEEE Wirel. Commun. **23**(6), 96–106 (2016)
33. ITU-R: IMT vision framework and overall objectives of the future deployment of IMT for 2020 and beyond [Online]. Available: https://www.itu.int/dmspubrec/itu-r/rec/m/R-REC-M.2083-0-201509-I!!PDF-E.pdf
34. 3GPP: Technical specification group radio access network; study on new radio (NR) to support non terrestrial networks; (release 15). 3rd Generation Partnership Project (3GPP), Technical Report (TR) 38.811, 09-2019, version 15.2.0
35. Liolis, K., Geurtz, A., Sperber, R., Schulz, D., Watts, S., Poziopoulou, G., Evans, B., Wang, N., Vidal, O., Tiomela Jou, B., Fitch, M.: Use cases and scenarios of 5G integrated satellite-terrestrial networks for enhanced mobile broadband: the SaT5G approach. Int. J. Satell. Commun. Network. **37**(2), 91–112 (2019)
36. Zeydan, E., Turk, Y.: On the impact of satellite communications over mobile networks: an experimental analysis. IEEE Trans. Veh. Technol. **68**(11), 11146–11157 (2019)
37. Ahmad, I., Suomalainen, J., Porambage, P., Gurtov, A., Huusko, J., Höyhtyä, M.: Security of satellite-terrestrial communications: challenges and potential solutions. IEEE Access **10**, 96038–96052 (2022)
38. Yan, Y., Han, G., Xu, H.: A survey on secure routing protocols for satellite network. J. Netw. Comput. Appl. **145**, 102415 (2019)
39. Liu, Z., Rong, J., Jiang, Y., Zhang, L.: Satellite network security routing technology based on deep learning and trust management. Sensors **23**(20), 8474 (2023)
40. Bertaux, L., Medjiah, S., Berthou, P., Abdellatif, S., Hakiri, A., Gelard, P., Planchou, F., Bruyere, M.: Software defined networking and virtualization for broadband satellite networks. IEEE Commun. Mag. **53**(3), 54–60 (2015)
41. 3GPP: Solutions for NR to support non-terrestrial networks (NTN), TR 38.821 (release 16), 2020. Differences Between SatComs and TerComs: (satcom 2011)
42. Arapoglou, P.D., Liolis, K., Bertinelli, M., Panagopoulos, A., Cottis, P., De Gaudenzi, R.: MIMO over satellite: a review. IEEE Commun. Surv. Tutorials **13**(1), 27–51 (2010)
43. I.-R. P.681-6: Propagation data required for the design of earth space land mobile telecommunication systems (2003)
44. Prieto-Cerdeira, R., Perez-Fontan, F., Burzigotti, P., Bolea-Alamañac, A., Sanchez-Lago, I.: Versatile two-state land mobile satellite channel model with first application to DVB-SH analysis. Int. J. Satell. Commun. Network. **28**(5–6), 291–315 (2010)

45. I.-R. M.1225: Guidelines for evaluation of radio transmission technologies for IMT-2000
46. Toso, G., Angeletti, P., Mangenot, C.: Multibeam antennas based on phased arrays: an overview on recent ESA developments. In: The 8th European conference on antennas and propagation (EuCAP 2014), pp. 178–181. IEEE (2014, Apr)
47. Gao, S., Rahmat-Samii, Y., Hodges, R.E., Yang, X.X.: Advanced antennas for small satellites. Proc. IEEE **106**(3), 391–403 (2018)
48. Rao, A., Chatterjee, A., Payne, J., Trujillo, J.: Multi-beam/multiband aeronautical antenna: opportunities and challenges. In: 29th ESA Antenna Workshop (2018, Oct)
49. Perez-Neira, A.I., Vazquez, M.A., Shankar, M.B., Maleki, S., Chatzinotas, S.: Signal processing for high-throughput satellites: challenges in new interference-limited scenarios. IEEE Signal Process. Mag. **36**(4), 112–131 (2019)
50. Angeletti, P., De Gaudenzi, R., Lisi, M.: From "bent pipes" to "software defined payloads": evolution and trends of satellite communications systems. In: 26th International Communications Satellite Systems Conference (ICSSC) (2008, June)
51. Kato, N., Fadlullah, Z.M., Tang, F., Mao, B., Tani, S., Okamura, A., Liu, J.: Optimizing space-air-ground integrated networks by artificial intelligence. IEEE Wirel. Commun. **26**(4), 140–147 (2019)
52. Hu, J., Zhang, H., Song, L., Schober, R., Poor, H.V.: Cooperative internet of UAVs: distributed trajectory design by multi-agent deep reinforcement learning. IEEE Trans. Commun. **68**(11), 6807–6821 (2020)
53. Clerckx, B., Mao, Y., Schober, R., Poor, H.V.: Rate-splitting unifying SDMA, OMA, NOMA, and multicasting in MISO broadcast channel: a simple two-user rate analysis. IEEE Wirel. Commun. Lett. **9**(3), 349–353 (2019)

Chapter 24
Design of an Image Frequency Rejection Mixer for the Terahertz Band

Abdeladim El Krouk, Abdelhafid Es-saqy, Mohammed Fattah, Said Mazer, Mahmoud Mehdi, Moulhime El Bekkali, and Catherine Algani

24.1 Receiver System Architectures

Current research into the sixth generation of wireless networks (6G) is attracting significant interest, not least because of the rapid development of high-speed wireless communication systems using millimeter and terahertz waves. The terahertz (THz) frequency band, which extends from 0.1 to 10 THz, is attracting attention both scientifically and for its potential applications in fields such as medical imaging, genetic research, identification of biochemical materials and security monitoring [1–3].

The utilization of this frequency band enables the attainment of expanded bandwidth and increased communication rates [4]. 3G networks predominantly operate within the 900 MHz and 2.1 GHz frequency bands [5]. 4G networks can use frequencies up to 2.5 GHz [6–8], while the 28 and 39 GHz bands are increasingly exploited by 5G networks [9–14]. The development of future wireless applications beyond 100 GHz is scheduled for the 6G between 2025 and 2035 [1].

The mixer is considered the core module of the wireless communication system, and its performance directly determines the advantages and disadvantages of the wireless communication system.

A. El Krouk (✉) · M. Fattah
EST, Moulay Ismail University, Meknes, Morocco
e-mail: Abdeladim.elkrouk@usmba.ac.ma

A. Es-saqy · S. Mazer · M. El Bekkali
AIDSES Laboratory, Sidi Mohamed Ben Abdellah University, Fez, Morocco

M. Mehdi
CRITC, Faculty of Sciences and Fine Arts, AUL University, Beirut, Lebanon

C. Algani
CNRS, Gustave Eiffel University, Le Cnam, Paris, France

M. El Ghzaoui et al. (eds.), *Next Generation Wireless Communication*, Signals and Communication Technology, https://doi.org/10.1007/978-3-031-56144-3_24

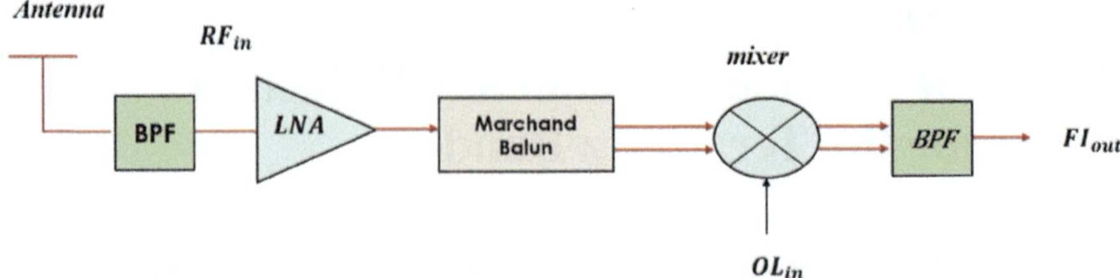

Fig. 24.1 Basic simplified architecture of a superheterodyne receiver system

Channel selection filtering is complicated at high carrier frequencies. We need to transpose the desired channel to a much lower center frequency to enable channel selection filtering with a reasonable Q factor.

A bandpass filter is placed directly after the receiving antenna to select the desired channel [15, 16]. Then, the low-noise amplifier (LNA) is used to increase the amplitude of the radio frequency signal. The sub-harmonic mixer (SHM) is emerging as a relevant choice for frequency conversion in transceivers operating in the microwave and millimeter ranges. It allows two signals to be multiplied in time: the RF signal received by the antenna and the LO signal from a local oscillator. to generate the IF signal, as shown in Fig. 24.1.

Unfortunately, most high-frequency SHM sub-harmonic mixers above 100 GHz have considerable conversion losses. This imposes problems of noise figure (NF), conversion gain and linearity within the low-noise amplifier, as well as the following elements of the receiver chain [9, 17, 18].

In recent years, a considerable number of studies concentrated on optimizing the performance of millimeter-wave mixers, particularly in the frequency band above 100 GHz, and more specifically at 140 GHz [19–26]. A down-conversion mixer is described in [19], using 130 nm commercial technology, simply balanced. Simulation results indicate a conversion gain of 2.6 dB in the 100–140 GHz frequency band, with an LO power of 5 dBM. In [20], a sub-harmonic mixer (SHM) based on a 140 GHz Schottky diode is presented, using high-precision MEMS micromachining technology. It has a conversion loss of around 8.2 dB.

This chapter presents a down-conversion frequency mixer offering high-conversion gain and wideband transconductance, designed to operate in a frequency range from 135 to 145 GHz. This study aims to compare the performance of this topology beyond 100 GHz using MMIC technology from the UMS foundry. Two separate techniques and approaches were implemented to enhance the mixer's gain, bandwidth, and linearity characteristics.

This chapter is structured as shown below: Sect. 24.2 outlines the mixer's operational principle and performance parameters. Section 24.3 focuses on the architecture of the proposed mixer, including the two techniques, charge injection and shunt series, as well as the simulation results. These approaches are then examined and compared in Sect. 24.4. A final conclusion is formulated in the last part.

24.2 Theoretical Study of the Mixer

24.2.1 Operating Principle

The mixer is an electronic circuit that multiplies two signals in order to transpose the frequency of an input signal. The output signal frequency may result from the sum or difference of the frequencies of the two input signals. On reception, the frequency of the wanted signal is transposed to the intermediate frequency (IF). On transmission, the IF signal frequency is transposed to the radio frequency RF. In this way, a distinction is made between two mixer operating modes: high conversion and low conversion.

Figure 24.2 shows a functional description of a low-conversion mixer. It consists of three input/output ports, as shown in the figure:

- The input signal radio frequency (RF) corresponds to the signal to which we want to apply frequency conversion.
- The input signal local oscillator (LO) comes from local oscillator.
- The intermediate frequency (IF) output signal for the low-frequency signal resulting from this mixing.

The down-conversion mixer transposes an RF frequency from the antenna to the IF frequency. Its role is to lower the frequency of the wanted signal. Some mixers are optimized to operate with either an up-converter or a down-converter, while others can operate in both modes with the same performance.

Theoretically, the mathematical formulas for RF, LO and IF signals are given by the following equations:

$$V_{\mathrm{RF}}(t) = A\cos(\omega_{\mathrm{RF}}t) \tag{24.1}$$

$$V_{\mathrm{LO}}(t) = B\cos(\omega_{\mathrm{LO}}t) \tag{24.2}$$

$$V_{\mathrm{IF}}(t) = V_{\mathrm{LO}}(t) * V_{\mathrm{RF}}(t). \tag{24.3}$$

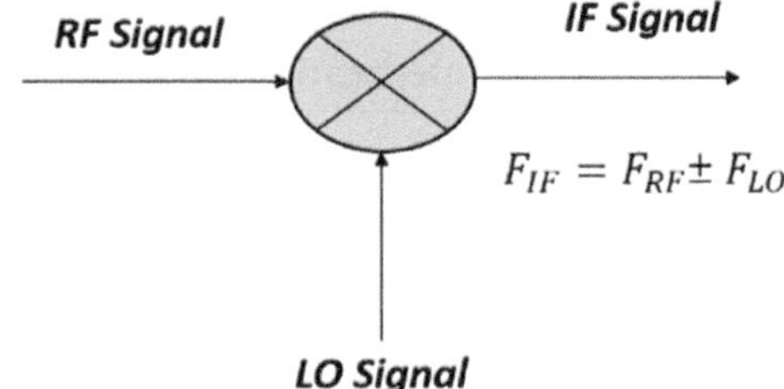

Fig. 24.2 Low-conversion mixer operation

The mixer generates an IF output whose frequency is equal to the addition and difference of the RF and LO input signals, expressed by the following relationships:

$$V_{IF}(t) = \frac{AB}{2}[\cos(\omega_{RF} - \omega_{LO})t + \cos(\omega_{RF} + \omega_{LO})t]. \tag{24.4}$$

We can see that the IF output signal from the product comprises two sinusoidal signals whose frequencies are the difference and sum of the input frequencies. In the case of the receiver, only the difference in frequency is of interest.

24.2.2 *Mixer Performance*

a. **Conversion gain**

Conversion gain represents the efficiency of converting the input RF signal to the desired IF frequency. In the case of a receiver, it is defined as the ratio between the IF output power and the input RF signal power.

The conversion gain, CG, is expressed by:

$$CG = \frac{P_{IF_{output}}}{P_{RF_{input}}} \tag{24.5}$$

$$CG_{dB} = 10\log\left(\frac{P_{IF_{output}}}{P_{RF_{input}}}\right) = P_{IF_{output}}\,(dBm) - P_{RF_{input}}\,(dBm), \tag{24.6}$$

with

P_{RF}: is the power of the RF input signal available at the RF frequency.

P_{IF}: is the IF output signal power at the IF frequency.

b. **Compression point at 1 dB**

The 1 dB compression point assesses the 1 dB change in conversion gain concerning the power applied to the mixer. The graph in Fig. 24.3 illustrates the fluctuation in IF output power for a low-conversion mixer relative to the RF input power.

c. **Third-order interception point (IP3)**

The third-order intercept point (IP3) characterizes the distortion of the nonlinear system; it is determined by a test using two signals of the same amplitude and very close in RF frequency and is separated by a small frequency deviation Δk. Intermodulation occurs when two signals of neighboring frequencies RF_1 and RF_2 are applied to the RF access, with

$$RF_1 = RF + \Delta k/2 \tag{24.7}$$

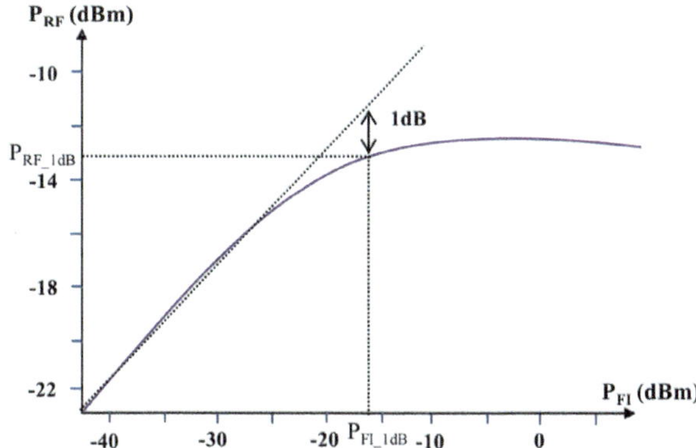

Fig. 24.3 Single-dB compression in a high-conversion mixer

$$\mathrm{RF}_2 = \mathrm{RF} - \Delta k/2. \tag{24.8}$$

Third-order intermodulation concerns the frequencies:

$$2F_{\mathrm{RF}_1} - F_{\mathrm{RF}_2} - F_{\mathrm{LO}}\,\mathrm{et}\,2F_{\mathrm{RF}_2} - F_{\mathrm{RF}_1} - F_{\mathrm{LO}}. \tag{24.9}$$

It is calculated using the following relationship: $P_{\mathrm{IIP3}} = P1\,(\mathrm{dBm}) + \frac{\Delta P}{2}$, where P is the difference between one of the test lines RF_1 or RF_2 and the third-order intermodulation product line $2\mathrm{RF}_1$–RF_2 or $2\mathrm{RF}_2$–RF_1.

d. **Isolation between access points**

Mixer isolation is an essential parameter for overall transceiver performance. Signals sent and received on different gates can end up on different gates in unwanted ways. This is quantified by isolation. These parameters define the level of signal leakage between port pairs, i.e., RF to LO, LO to IF and RF to IF, as shown in Fig. 24.4.

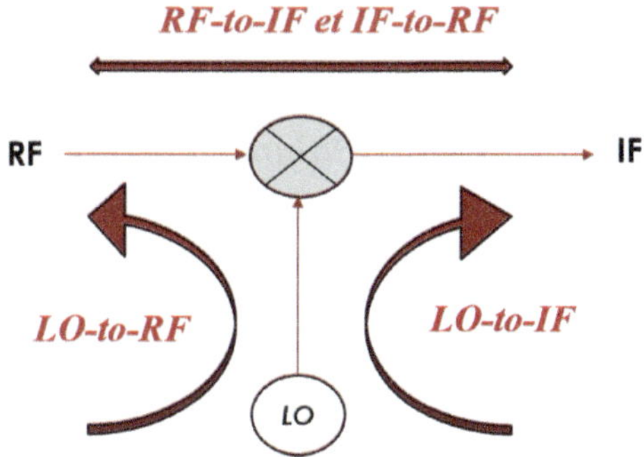

Fig. 24.4 Main insulation of a low-conversion mixer

e. **Noise in the mixer**

It is a parameter that evaluates the performance of a device in terms of the noise it generates. A mixer's noise figure is a measure of the degradation in signal quality when mixed at a different frequency.

There are two categories of noise factors applicable to mixers:

- Single-sideband noise figure (FSSB)

$$F_{ssb} = 10 \log\left(\frac{S/N_{in}}{S/N_{out}}\right). \tag{24.10}$$

- Double sideband noise figure (DSB)

$$F_{dsb} = F_{ssb} - 3\,\text{db}. \tag{24.11}$$

f. **Image frequency**

Two cases can be distinguished for the mixer frame rate:

- In the case of up-conversion, the image frequency of the upper sideband (USB) is written as follows:

$$F_{img} = 2 * F_{LO} + F_{RF} \tag{24.12}$$

- In the case of down-converted frequencies, the frequency of the upper image of the LSB lower sideband is written as follows:

$$F_{img} = 2 * F_{LO} - F_{RF}. \tag{24.13}$$

To avoid the emergence of these undesired signals at the output, it is advisable to incorporate an image rejection filter at the mixer's input to mitigate this effect or by utilizing the phase opposition method [27, 28].

24.2.3 Simply Balanced Mixer (SBM) Architecture

The architecture of the simply balanced mixer (SBM) used in this work is illustrated in Fig. 24.5. The chosen architecture uses two devices. The local oscillator is accessed in differential mode, while the radio frequency is accessed in single-ended mode, which improves isolation between the RF and LO signals. Transistor M1 is a voltage-to-current converter, receiving the RF signal and evenly distributing the current (I_s) between coupled sources M2 and M3. Consequently, the RF signal modulates the drain-source current of M1, and the switching action of M2 and M3 amplifies this modulation through interaction with the LOLO signal from a local oscillator [12]. The voltage across the transistors M2 and M3 drains represents the resulting IF output signal.

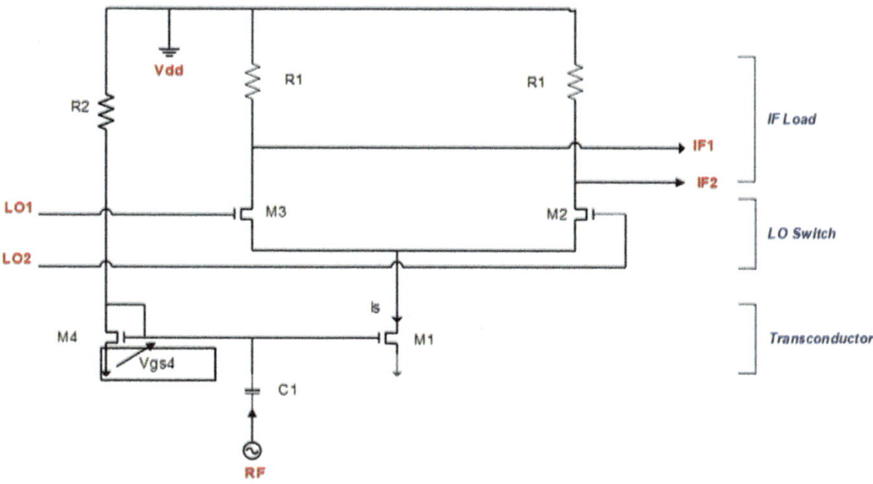

Fig. 24.5 Single-balanced mixer architecture

In brief, our circuit consists of three sections:

- **RF stage**: This stage functions as a transconducting stage, converting a voltage signal into a current signal and providing sufficient gain.
- **LO stage**: This stage is a switching mechanism that converts the RF current from the RF section into an IF current. It achieves this by modulating the RF current with the voltage driven by the local oscillator (LO).
- **IF stage**: Acting as the mixer load, this section facilitates the conversion of IF current to an IF output voltage.

The output current (I_{out}) of the circuit, which is controlled by the LO local oscillator signal, is determined by the following relationship:

$$I_{\text{out}} = I_S(t)\,(\text{t}) \cdot \text{sign}[\text{LO}(t)] \tag{24.14}$$

with

$$\text{sign}[\text{LO}(t)] = \frac{4}{\pi}\left\{\cos(w_{\text{lo}}t) - \frac{1}{3}(\cos(3w_{lo}t)) + \frac{1}{5}(\cos(5w_{lo}t)) + \cdots\right\} \tag{24.15}$$

is the Fourier transform of the signal LO and we have:

$$I_S(t) = g_m V_{\text{GS1}} + g_m V_{\text{RF}} \cos(w_{\text{RF}}t) \tag{24.16}$$

where g_m is the transconductance of transistor M1.

V_{GS1} is the bias voltage of gate M1,

We then obtain:

$$I_{out}(t) = \{g_m V_{GS1} + g_m V_{RF}\cos(w_{RF}t)\}$$
$$\frac{4}{\pi}\left\{\cos(w_{lo}t) - \frac{1}{3}(\cos(3w_{lo}t)) + \frac{1}{5}(\cos(5w_{lo}t)) + \cdots\right\} \quad (24.17)$$

With

$$V_{out(t)} = R1 \cdot I_{out}(t) \quad (24.18)$$

$$V_{out(t)} = \{g_m R1 V_{GS1} + g_m R1 V_{RF}\cos(w_{RF}t)\}$$
$$\frac{4}{\pi}\{\cos(w_{lo}t) - \frac{R1}{3}(\cos(3w_{lo}t)) + \frac{R1}{5}(\cos(5w_{lo}t)) + \cdots\} \quad (24.19)$$

The following relationship gives the conversion gain:

$$CG = \frac{|V_{out}(t)|\,\grave{a}(w_{RF} - w_{lo})}{|V_{RF}(t)|\grave{a}\; w_{RF}} \quad CG = \frac{2}{\pi} g_m R_1 \quad (24.20)$$

The advantages of such a topology are:

- Good isolation between LO and RF ports.
- High-conversion gain.
- Better common-mode rejection.
- Elimination of a greater number of spurious responses,

However, the disadvantage of this structure is that circuit complexity is greater.

24.3 Design of a pHEMT Mixer for the THz Band

24.3.1 Mixer Bias Circuit

For the design of a 140 GHz mixer, a 2 μm × 25 μm GaAs pHEMT transistor is accurately modeled, as the circuit's performance is highly dependent on bias conditions. The primary objective is to select the appropriate transistor bias parameters to ensure that the amplifier stage functions in the saturation zone while the switching stage function is approaching the pinch region. The mixer transconductance bias voltage and the switching tube pinch-off voltage can be taken from the $I = f(V)$ curve in Fig. 24.6. To obtain a good transconductance value Gm and moderate DC current consumption, the threshold voltage of this transistor is around – 0.8 V with VGS = – 0.3 V and VDS = 0.2 V.

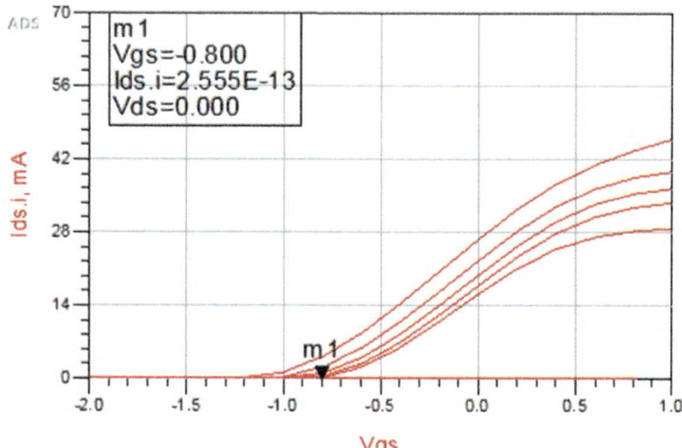

Fig. 24.6 Threshold voltage curve for 0.15 μ GaAs pHEMT technology

The bias points for the quadruple switching transistors (M3–M4), the full-current transistor (M9), and the current mirror (M10) are shown in Figs. 24.7, 24.8, and 24.9, respectively.

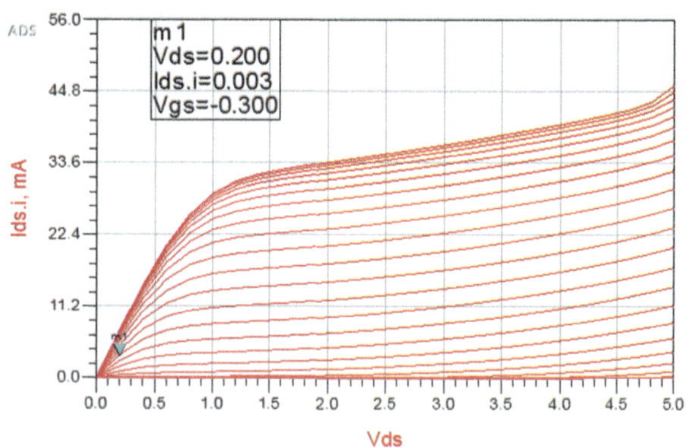

Fig. 24.7 Polarization point of the differential pair

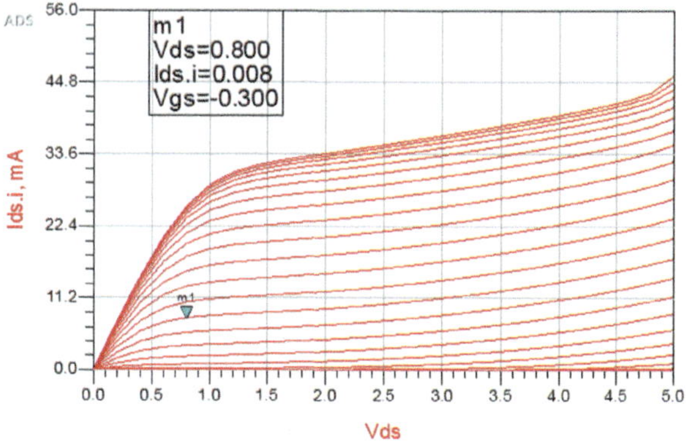

Fig. 24.8 Bias point for the total current of transistor M9

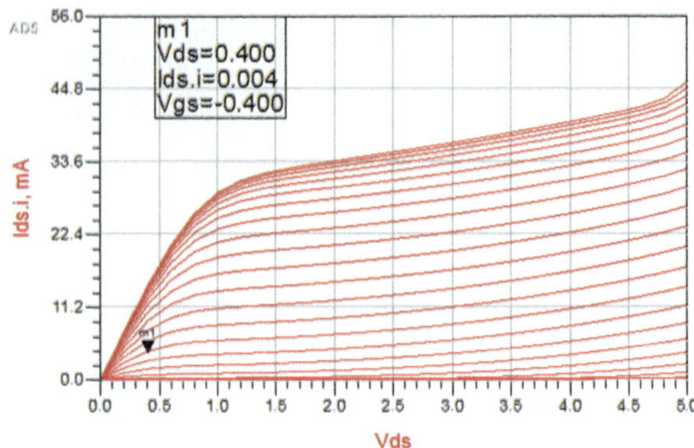

Fig. 24.9 Bias point of the M10 transistors for the current mirror (VGS = VDS)

Static characteristics of the PHEMT PH15NHF transistor for a gate width Wu = 50 μm.

24.3.2 Improving SBM Performance with Charge Injection

Electrical Circuit Analysis

Figure 24.10 shows the proposed mixer architecture using the charge injection method [29], which is a single-balance down-converter mixer in the 0.14 THz frequency band. Our mixer's RF and LO frequencies are set at 140 and 139 GHz, respectively, giving an intermediate frequency of 1 GHz. This IF value aligns with current wireless communication standards, which generally operate around 1 GHz [30, 31].

The principle of the charge injection technique is to augment the current in the transconductance stage by adding a 198 Ω R3 resistor to supply current to the M1 transconductance amplifier, which can make M2–M3 best matched to the switching tube pairs and reduce the voltage drop across the load resistor R1 = R2. This method improved the conversion gain CG, reduced the circuit's NF noise, and improved all other mixer design specifications. The balun converts the local oscillator LO signal from a single-ended to a double-ended differential signal to obtain a balanced mixing function.

The conversion gain, expressed by the relation:

$$\mathrm{CG} = \frac{|V_{\mathrm{out}}(\mathrm{t})|}{V_{\mathrm{RF}}(\mathrm{t})} = \frac{2}{\pi} g_{\mathrm{m}} R_{\mathrm{c}}, \tag{24.21}$$

where g_{m} is the transconductance and R_{c} is the circuit's load resistance.

We add two buffer stages, consisting of two common-source transistors, M5 and M6, each with four fingers and a width of 70 μm. This enables wideband matching,

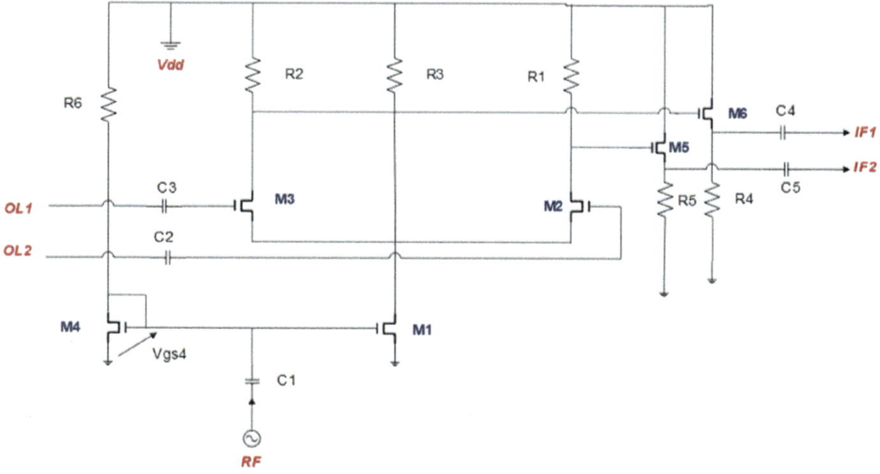

Fig. 24.10 Mixer architecture with charge injection

Table 24.1 Different parameter values used in our circuit

Parameters	Values
M2–M3	51 μm
M1	58 μm
M4	70 μm
M5–M6	70 μm
C2–C3	3.8 pF
C4–C5	12 pF
C1	4 pF
R1–R2	800 Ω
R3	93 Ω
R4–R5	53 Ω
R6	600 Ω
Vdd	3 V

improving RFRF to IFIF isolation at the output ports. Two series resistors (R4/R5) are also positioned at the sources of transistors M5 and M6 to avoid damaging the transistors at high current levels. The current mirror circuit is the typical current source configuration. It is a simple structure consisting of two transistors M1, M4, and a resistor R6. It is designed primarily to maintain the drain current at a constant value.

Table 24.1 presents the different parameter values of the simply balanced mixer using the charge injection technique.

Simulation and Results

a. Conversion gain

Setting the frequency IF = 1 GHz, Fig. 24.11 indicates that the conversion gain (C_G) curve exceeds a value of 25 dB when the injected RF power, varying from −50 to 0 dBm, is < − 25 dBM and is superior to 5 when the RF power is between −1 and − 25 dBM. The curve of the conversion gain (C_G) as a function of the LO local oscillator power, which varies from − 20 dBM to 20 dBM, reaches a maximum value of 25 dB in the power range from 3 to 9 dBM, as illustrated in Fig. 24.12.

C_G is found to be supperior to a value of 24 dB when power LO is between 0 dBM and 10 dBM. The evolution of conversion gain as a function of LO frequency in GHz is illustrated in Fig. 24.13. We can see that the C_G is over 15 dB in the LO frequency band from 135 to 145 GHz.

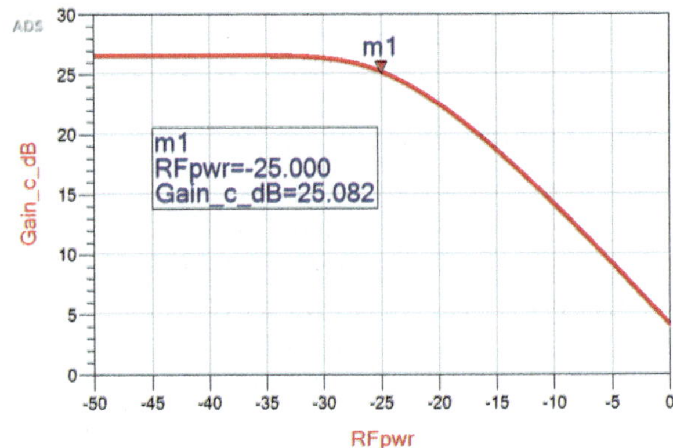

Fig. 24.11 Evolution of conversion gain as a function of injected RF power in dBm

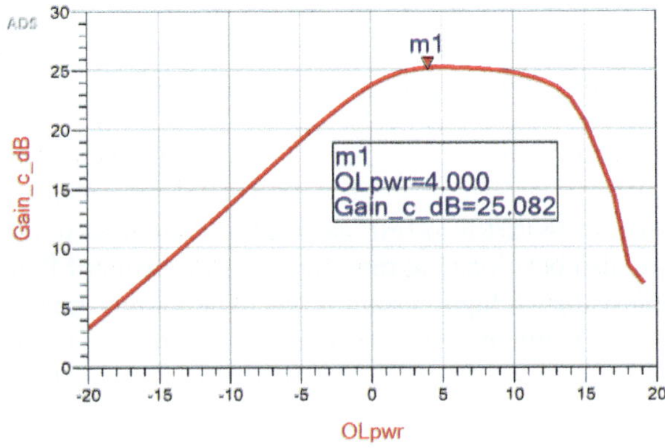

Fig. 24.12 Simulated conversion gain as a function of input LO power in dBm

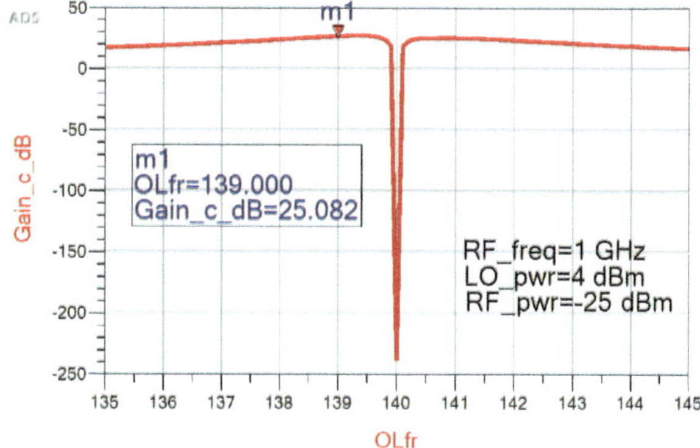

Fig. 24.13 Evolution of conversion gain with input LO frequency

These results show that the LO and RF input power of 4 dBm and − 25 dBm, respectively, are optimal values for the best conversion gain.

b. **The noise figure of the circuit**

Figure 24.14 presents the two noise figures simulated for the double sideband (NF_{dsb}) and the single sideband (NF_{ssb}), with the LO power of the local oscillator.

The minimum values of these noise factors are 5.38 and 6.45 dB, respectively, for an LO input power of 5 dBm.

c. **Compression point at 1 dB**

The evolution of IF output power as a function of RF input power is shown in Fig. 24.15. The 1 dB compression point (P1dB) is achieved at an injected RF power of − 26 dBm, corresponding to an IF power of − 0.5 dBm. It indicates a 1 dB deviation of the conversion gain from the RF power applied to the circuit.

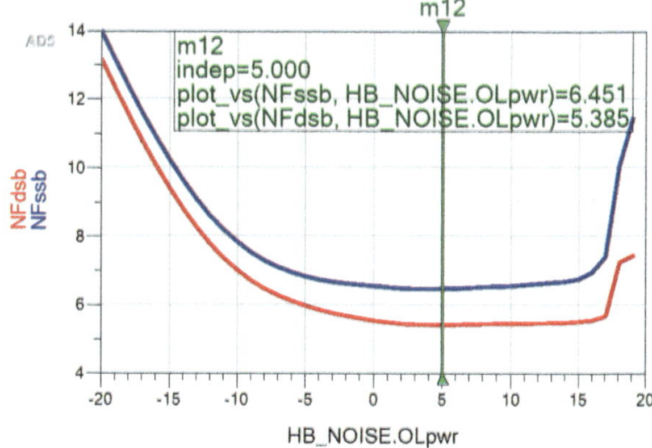

Fig. 24.14 DSB and SSB mixer noise factors

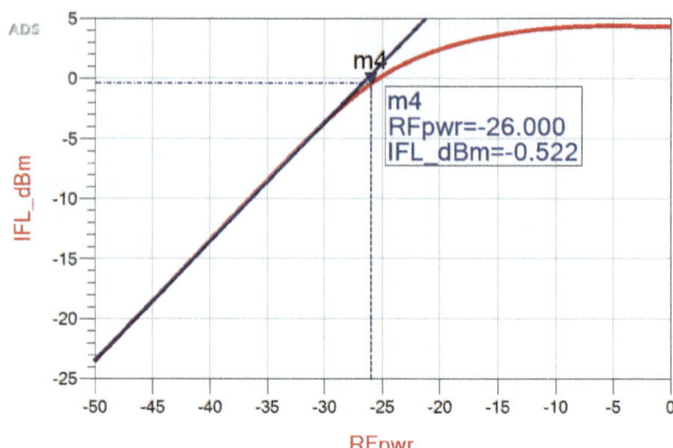

Fig. 24.15 1 dB compression point (P1dB)

d. **Image frequency rejection**

Figure 24.16 shows the image frequency rejection rate as a function of the input LO power. A rejection of the order of – 36.5 dB is obtained for an injected LO power equal to 5 dBm.

Figure 24.17 shows the spectra of our circuit's output signal, providing a clear visualization of image frequency rejection. Figure 24.18 presents a simulation of the VIF output voltage as a function of time.

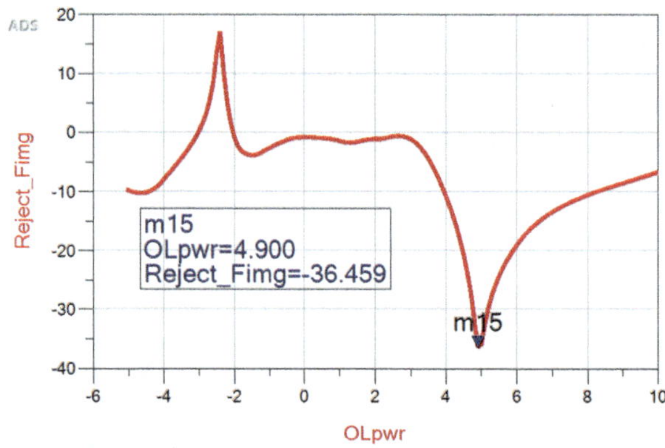

Fig. 24.16 Simulation of frequency rejection Image

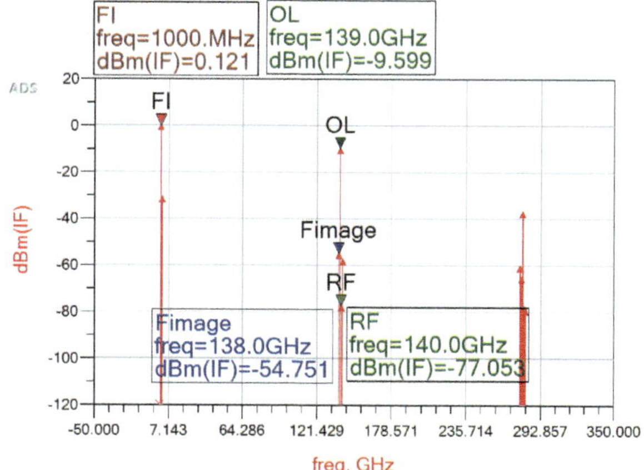

Fig. 24.17 Mixer output spectrum with charge injection

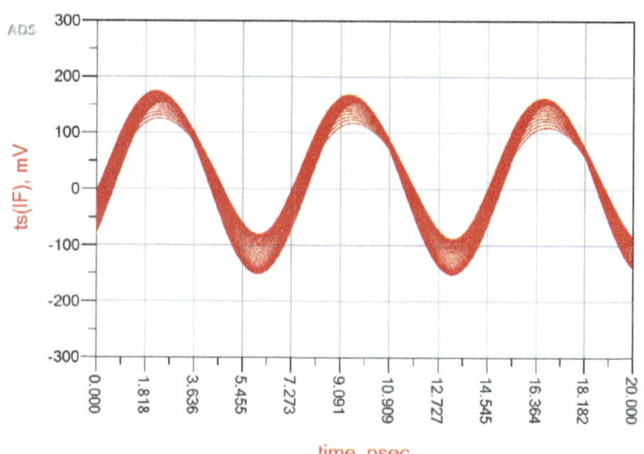

Fig. 24.18 Variation of output signal versus time

24.3.3 SBM Performance with a Serial Shunt Point

Electrical Circuit Analysis

In this section, using Advanced Design System (ADS) software, we designed a down-conversion mixer at the same 0.14 THz radio frequency (RF) as the one used in the previous section but with a different technique.

We find the tip series shunt technique among the techniques and architectures proposed to improve the performance of the simply balanced mixer [32]. Figure 24.19 illustrates the electrical schematic of a simply balanced mixer using the tip series shunt technique.

This technique uses L2 and L1 inductances instead of resistors, allowing a zero into the magnitude response, thus compensating for the bandwidth error. In addition, L3 and L4 inductances are used for capacitive division.

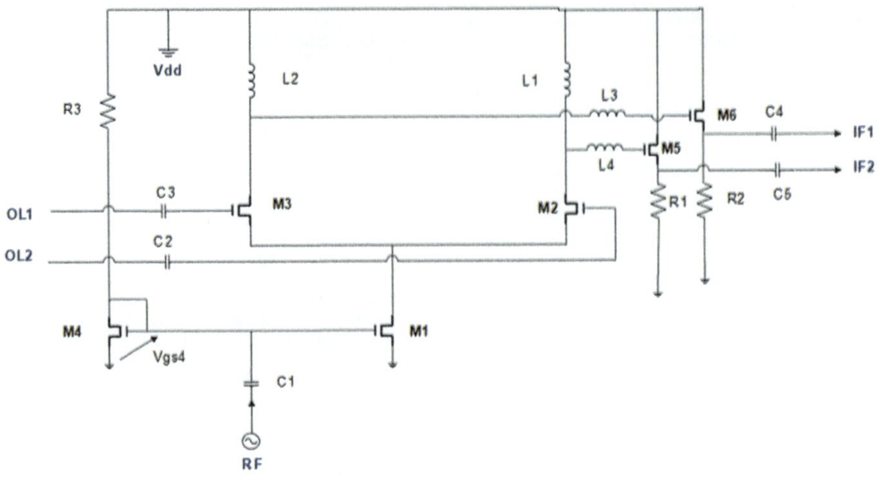

Fig. 24.19 Simply balanced mixer with the tip in shunt series

Simulation and Results

a. Conversion gain

The diagram in Fig. 24.20 illustrates the evolution of C_G in dB as a function of injected LO power. A maximum conversion gain of around 14.5 dB is achieved with an LO power of 9 dBm.

Figure 24.21 illustrates the evolution of the C_G in dB as a function of the injected RF power. It achieves a maximum conversion gain of around 14.5 dB at an RF power of – 25 dBm after setting the injected LO power at 9 dBm.

Dynamic analysis of the single balanced system enables us to obtain optimum values for the LO and RF powers injected. These allow us to maximize the mixer's

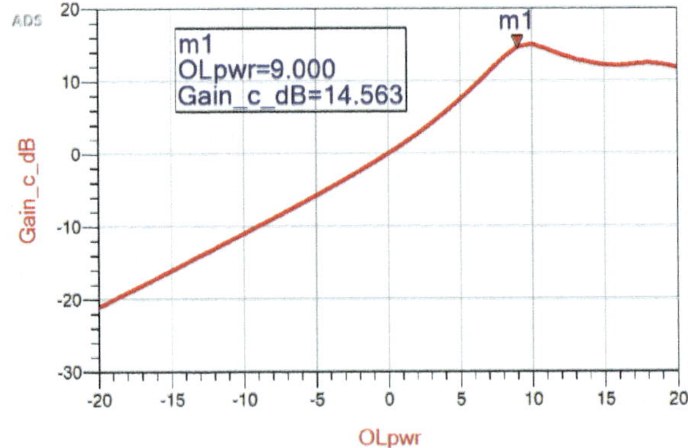

Fig. 24.20 Choosing the optimum LO power input for the circuit

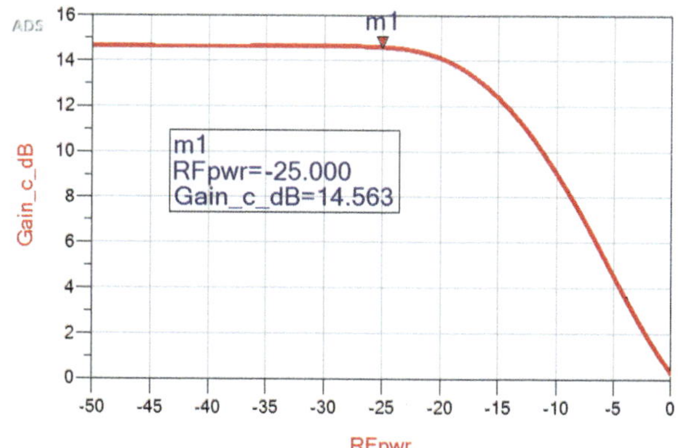

Fig. 24.21 Choosing the optimum RF power input for the circuit

conversion gain, minimize our circuit's noise figure, and choose a good linearity point.

b. **The 1 dB compression point**

Figure 24.22 shows the variation in IF_{pwr} output as a function of the injected LO_{pwr}. The 1 dB compression point is reached at an injected RF_{pwr} of − 4 dBm, corresponding to an IF output power of − 18.5 dBm. This is achieved with RF power ($RF_{pwr} = -25$ dBm) and LO power ($LO_{pwr} = 9$ dBm).

c. **The noise figure of the circuit**

Figure 24.23 shows the evolution of noise factors at double sideband (DSB) and single sideband (SSB). The minimum values observed for NF_{ssb} and NF_{dsb} are around 6.39 and 8.85 dB, respectively, obtained at LO input power of 9 dBm.

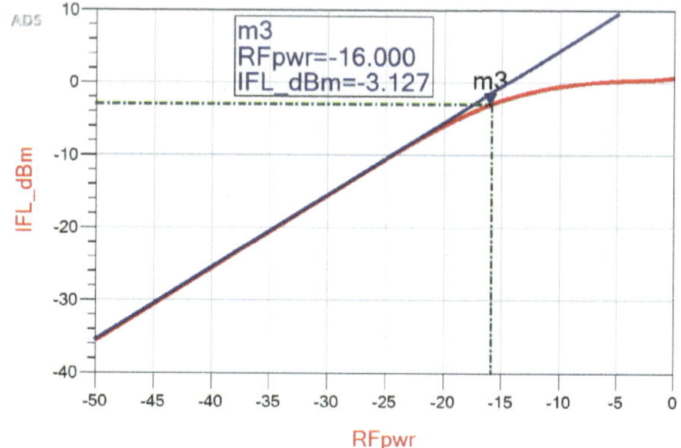

Fig. 24.22 Determination of compression point at 1 dB

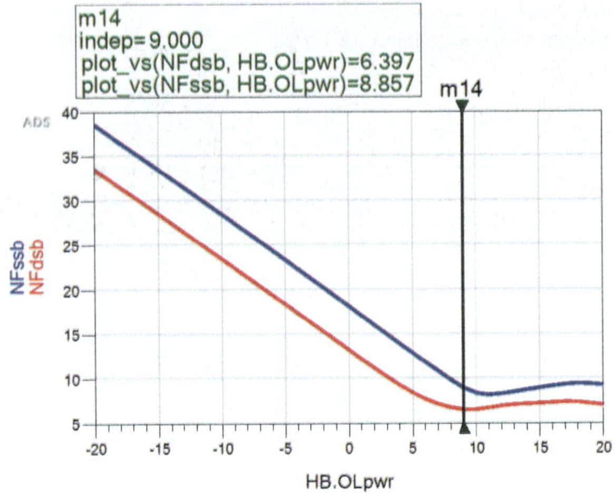

Fig. 24.23 Variation of NF_{ssb} and NF_{dsb} noise figures as a function of injected LO_{pwr}

d. **Image frequency rejection**

Figure 24.24 presents the spectrum of the output signal, allowing us to better understand the notion of image frequency rejection.

The image frequency rejection ratio is calculated as the difference in dB between the RF_{pwr} input and the image output power of the mixer.

$$\text{Reject-Image (dB)} = P_{RF}(\text{output}) - P_{\text{image}}(\text{output}).$$

With $P_{RF} = -25$ dBm.

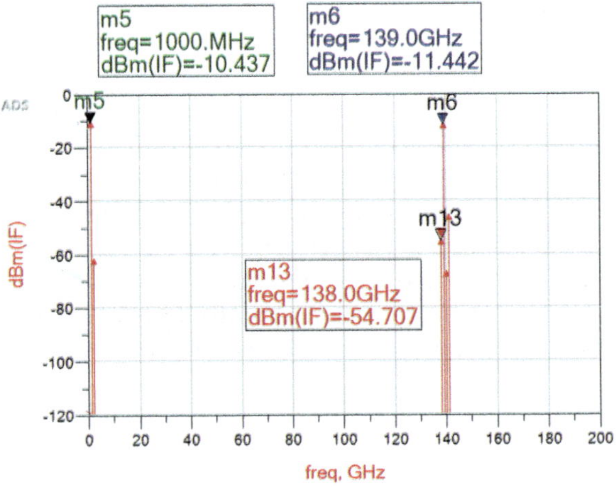

Fig. 24.24 Mixer output spectrum

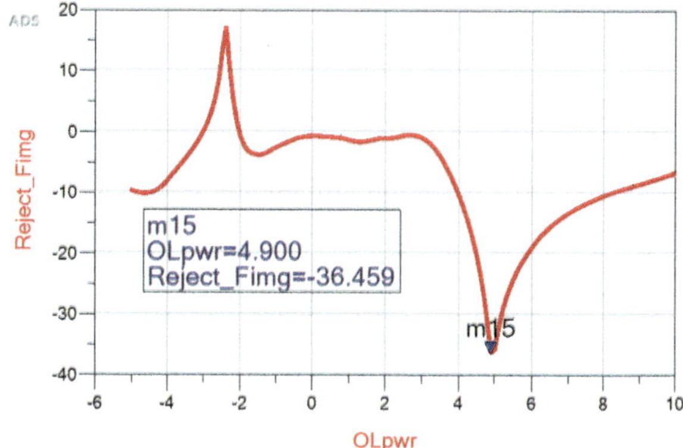

Fig. 24.25 Image frequency rejection as a function of LO_{pwr}

The evolution of the image frequency rejection as a function of the input power of the local oscillator LO is shown in Fig. 24.25, attaining a value of − 33.5 dB at a power of 9 dBm.

24.4 Comparison of Results

The results of the comparative analysis of the two techniques studied in this chapter are presented in Table 24.2.

Simulation of the simply balanced mixer using the charge injection method demonstrates a higher conversion gain than that obtained using the shunt series clipping technique at 140 GHz. In addition, the me-mixer using charge injection has a lower noise figure and local oscillator (LO) power than the shunt series clipping technique.

Table 24.2 Comparison of results obtained using the ADS tool

Parameter	Gilbert mixer with load injection	Gilbert mixer with RCR Shunt peaking
Technology	PH15 process in MMIC technology	PH15 process in MMIC technology
f (GHz)	140	140
Noise figure DSB (dB)	5.3	6.39
Conversion gain (dB)	25	14.5
P1dB	−26	−16
Image frequency rejection	−36.5	−33.5
LO power dBm	4	9

24.5 Conclusion

This chapter implements a simply balanced active down-conversion mixer based on the GaAs PH15NHF transistor of the Ph15 process of MMIC technology in the 140 GHz frequency band. Simulation results, carried out using the ADS tool, show that the proposed mixer architecture, with the charge injection technique, presents a very high C_G conversion gain (15 $\pm$ 25) dB over a wide bandwidth in the 135–145 GHz frequency band with a low LO_{pwr} of 4 dBm. In addition, the noise figure of 5.3 dB is significantly lower than that of other mixers in this paper.

References

1. Rappaport, T.S., et al.: Wireless communications and applications above 100 GHz: opportunities and challenges for 6g and beyond. IEEE Access **7**, 78729–78757 (2019). https://doi.org/10.1109/ACCESS.2019.2921522
2. Vazquez, P.R.: 6G wireless communication links operating at frequencies beyond 200 GHz: an analysis of their performance and main limitations (2022). https://doi.org/10.25926/TVMC-Z778
3. Shan, G., Zhang, M., Huang, W.: A model for THz silicon nanotube transistor. In: 2010 3rd IEEE International Nanoelectronics Conference (INEC), vol. 1(1), pp.181–182 (2010)
4. I. F. Akyildiz, J. M. Jornet, C. Han.: Terahertz band: next frontier for wireless communications. Phys. Commun. **12**, 16–32 (2014). https://doi.org/10.1016/j.phycom.2014.01.006
5. Abdellaoui, M., Fattah, M.: Characterization of ultra wide band indoor propagation. In: 7th Mediterranean Congress of Telecommunications (CMT). IEEE (2019). https://doi.org/10.1109/CMT.2019.8931367
6. Naik, B.V., Nath, D., Sharma, R.: Design and analysis of rectangular slot microstrip patch antenna for millimetrewave communication and its SAR evaluation. Int. J. Antennas (JANT) **7**(1), 15–23 (2021). https://doi.org/10.5121/jant.2021.7102
7. Hussain, N., Awan, W.A., Ali, W., Naqvi, S.I., Zaidi, A., Le Tu, T.: Compact wideband patch antenna and its MIMO configuration for 28 GHz applications. Int. J. Electron. Commun. (2021). https://doi.org/10.1016/j.aeue.2021.153612
8. Daghouj, D., Fattah, M., Mazer, S., Balboul, Y., El Bekkali, M.: UWB waveform for automotive short range radar. Int. J. Eng. Appl. **8**(4), 158–164 (2020). https://doi.org/10.15866/irea.v8i4.18997
9. Tsai, K.J., Yang, H., Huang, T., Wang, H.: A 30–100 GHz wideband sub-harmonic active mixer in 90 nm CMOS technology. IEEE Microw. Compon. Lett. **18**(8), 554–556 (2008)
10. Niu, Y., Li, D., Jin, Su, L., Vasilakos, A.V.: A survey of millimeter wave communications (mmWave) for 5G: opportunities and challenges. Wireless Netw. **21**(8), 2657–2676 (2015)
11. Es-Saqy, A., Abata, M., Mazer, S., Fattah, M., Mehdi, M., El Bekkali, M., Algani, C.: Very low phase noise voltage controlled oscillator for 5G mm-wave communication systems. In: 2020 1st International Conference on Innovative Research in Applied Science, Engineering and Technology (IRASET), Meknes, Morocco, pp. 1–4 (2020)
12. Es-Saqy, A., Abata, M., Mehdi, M., Mazer, S., Fattah, M., El Bekkali, M., Algani, C.: 28 GHz balanced pHEMT VCO with low phase noise and high output power performance for 5G mm-wave systems. Int. J. Electr. Comput. Eng. (IJECE) **10**(5), 4623–4630 (2020)
13. Es-saqy, A. et al.: A 5G mm-wave compact voltage-controlled oscillator in 0.25 μm pHEMT technology. Int. J. Electr. Comput. Eng. (IJECE) **11**(2), 1036–1042 (2021). https://doi.org/10.11591/ijece.v11i2.pp1036-1042

14. Mamane, A., Fattah, M., El Ghazi, M., Balboul, Y., El Bekkali, M., Mazer, S.: The impact of scheduling algorithms for real-time traffic in the 5G femto-cells network. In: 9th International Symposium on Signal, Image, Video and Communications, ISIVC 2018, pp. 147–1512 (2018). https://doi.org/10.1109/ISIVC.2018.8709175
15. Salah-Eddine, D., Imane, H., Mohammed, F., Younes, B., Said, M., Moulhime, E.L.: Design of a microstrip antenna patch with a rectangular slot for 5G applications operating at 28 GHz. J. TELKOMNIKA Telecommun. Comput. Electron. Control. **20**(3), 527–536 (2022). https://doi.org/10.12928/TELKOMNIKA.v20i3.23159
16. Salah-Eddine, D., Imane, H., Mohammed, F., Younes, B., Said, M., Moulhime, E.L.: Study and design of printed rectangular microstrip antenna arrays at an operating frequency of 27.5 GHz for 5G applications. J. Nano-Electron. Phys. **13**(6), 06035 (2021). https://doi.org/10.21272/JNEP.13(6).06035
17. Parveg, D., Varonen, M., Kangaslahti, P., Safaripour, A., Hajimiri, A., Tikka, T., Gaier, T., Halonen, K.A.: CMOS I/Q subharmonic mixer for millimeter-wave atmospheric remote sensing. IEEE Microw. Wirel. Compon. Lett. **26**(4), 285–287 (2016)
18. Yan, Y., Bao, M., Gunnarsson, S.E., Vassilev, V., Zirath, H.: A 110–170-GHz multi-mode transconductance mixer in 250 nm In: DHBT
19. Jun, J., Deng, X.: 140 GHz single balance fundamental mixer d sign based on CPWG GaAs IC.pdf. J. Terahertz Sci. Electron. Inf. Technol. (2018)
20. Shan, G., Zhao, X., Zhu Shan, H. et al.: Critical components in 0.14 THz communication syst.pdf
21. Fuse, M., Sekiya, H., Otani, A.: High-dynamic-range measurement of millimeter-wave amplifier using 140-GHz fundamental mixer. In: Proceedings of APMC 2012, Kaohsiung, Taiwan (2012)
22. Koch, S., Guthoerl, M., Kallfass, I.: A 120–145 GHz heterodyne receiver chipset utilizing the 140 GHz atmospheric window for passive millimeter-wave imaging applications. IEEE J. Solid-State Circ. **45**(10), 1961–1967 (2010)
23. Laskin, E., Chevalier, P., Sautreuil, B.: 140-GHz double-sideband transceiver with amplitude and frequency modulation operating over a few meters. In: Conference Paper (2009). https://doi.org/10.1109/BIPOL.2009.5314245
24. Parveg, D., Varonen, M., Kärkkäinen, M.: Wideband millimeter-wave active and passive mixers in 28 nm. In: Bulk CMOS Technology Proceedings of the 10th European Microwave Integrated Circuits Conference, pp. 116–117, Paris, France (2015)
25. Seyedhosseinzadeh, N., Nabavi, A.: A 100–140 GHz SiGe-BiCMOS sub-harmonic down-converter mixer. In: Proceedings of the 12th European Microwave Integrated Circuits Conference, pp. 17–20. Nuremberg, Germany (2017)
26. El Krouk, A., Es-Saqy, A., Fattah, M.: Gilbert cell down-conversion mixer for THz wireless communication. Artif. Intell. Smart Environ. **635**, 475–480 (2023). https://doi.org/10.1007/978-3-031-26254-8_68
27. Varun, D., Mazumdar, T.: Design and development of a novel architecture for multistage RF downconversion with improved image rejection andnon-linearity corrections for 1–10 GHz range. SASTECH J. **10**(1), 43–51 (2011)
28. Chouchane, T., Sawan, M.: A 5 GHz CMOS RF mixer in 0.18 μm CMOS technology. In: IEEE CCECE 2003, Montreal, pp. 1905–1908 (2003)
29. Lee, S.-G., Choi, J.-K.: Current-reuse bleeding mixer. Electron. Lett. **36**(8), 696 (2000)
30. Coffing, D., Main, E.: Effects of offsets on bipolar integrated circuit mixer even-order distortion terms. IEEE Trans. Microw. Theory Techniq. **49**(1), 23–30 (2001)
31. Proakis, J.G., Salehi, M.: Digital communications (2008)
32. Kavivarman, J., Partibane, B., Vaithianathan, V.: Design of an improved gilbert mixer for 5G applications using 45nm CMOS technology. In: Proceedings of the Fifth International Conference on Communication and Electronics Systems (ICCES 2020). IEEE (2020)

Chapter 25
Spatially Efficient MIMO Antenna Design Using Photonic Crystal and Polyimide Substrate

K. Vasu Babu, Gorre Naga Jyothi Sree, Tanvir Islam, Sudipta Das, and Bhuma Anuradha

25.1 Introduction

For the next generation of wireless communication areas require the higher transmission of data rates transitioning carrier frequencies to THz band is viable solution. The bandwidth availability is an important factor for establishing and increasing communication data rate links. The comparison with lower band system over THz band channels has distinct type of characteristics such as extra molecular loss and higher amount of transmission path loss due to absorption of system. A type of trapezoidal structure patch antenna placed on the photonic crystal type substrate designed THz regions [1], using the ohm symbol and L-structure designed MIMO system [2], design and investigation of patch antenna on PCs substrates for the THz regime

K. V. Babu (✉)
Department of Electronics and Communication Engineering, BVRIT HYDERABAD College of Engineering for Women, Bachupally, Telangana, India
e-mail: vasubabuece@gmail.com

G. N. J. Sree
Department of Electronics and Communication Engineering, Vasireddy Venkatadri Institute of Technology, Guntur, Andhra Pradesh, India

T. Islam
Department of Electrical and Computer Engineering, University of Houston, Houston, TX 77204, USA

S. Das
Department of Electronics and Communication Engineering, IMPS College of Engineering and Technology, Malda, West Bengal, India

B. Anuradha
Department of Electronics and Communication Engineering, S V University, Tirupati, Andhra Pradesh, India

M. El Ghzaoui et al. (eds.), *Next Generation Wireless Communication*, Signals and Communication Technology, https://doi.org/10.1007/978-3-031-56144-3_25

[3], miniaturized and analyzed the multi-frequency antenna with required characteristics for WLAN systems [4], graphene-based metamaterial compact absorber for THz regime [5], MIMO graphene patch radiator for isolation improvement [6], based on the synthesized PBG structure and BPSO designed and analyzed [7], disk inserted THz system [8], fractal-based MIMO system for 5G [9], novel array patch antenna on the photonic crystal designed [10], various technique of designs implemented for photonic crystals with MIMO designs [11–20], graphene-based UWB patch antenna for THz region investigations [21], using air-silicon photonic crystal THz patch antenna with enhanced flexible operation [22], using nanostructures, photonics designed the novel broadband design [23], using DGS designed multi-band structure[24], UWB MIMO system designed using DGS [25], the techniques related to design of THz regions explained with different compact structures [26–30], performance analysis related to fractal MIMO structure using SiO_2 [31], without and with using of superstrate designed THz antenna [32], a rhombus shaped micro-sized THz antenna designed for short-range applications [33], penta-band compact multiple antenna system designed [34], using graphene material fractal wide-band MIMO system based on the characteristic analysis [35], a tree-shape micro-scaled printed MIMO radiator for THz regime [36], recent development of graphene material nanophotonics [37] and half-circular dual-band compact MIMO system designed for wireless areas [38].

25.2 Antenna Geometry

In this section, developed and designed photonic crystal-based MIMO antenna using polyimide substrate with dimensions of 1660 μm × 600 μm. The patch consists of square and considered as circular arcs are added at the three sides of the patch for the improvement of return loss and reflection parameters using the material as graphene. Beneath the patch, the full ground plane is used and the material considered as copper. In the proposed design, it consists of square shape structure with circular arcs are introduced and placing the same patch at suitable distance for obtain the better isolation value at the resonant frequency. The spacing among designed structure is considered as 340 μm and also considered the inner arc space among the both patches are considered as 380 μm. Figure 25.1a shows the arrangement of photonic crystal MIMO antenna design with front plane consists of patch, photonic crystals and substrate material. Figure 25.1b shows the arrangement of backplane of the proposed design with copper as ground material. Due to the arrangement front and back planes pf the current designed got S_{11} and S_{21} are − 27.13 dB and − 35.11 dB (Table 25.1).

The two different ways of modes propagation related to EM waves inserted inside photonic crystal material substrate is represented by transverse field of electric type (TE) and the transverse type of magnetic type (TM) represent modes of system. The expressions for mathematical analysis of EM waves considered [32].

For the case of TE related modes:

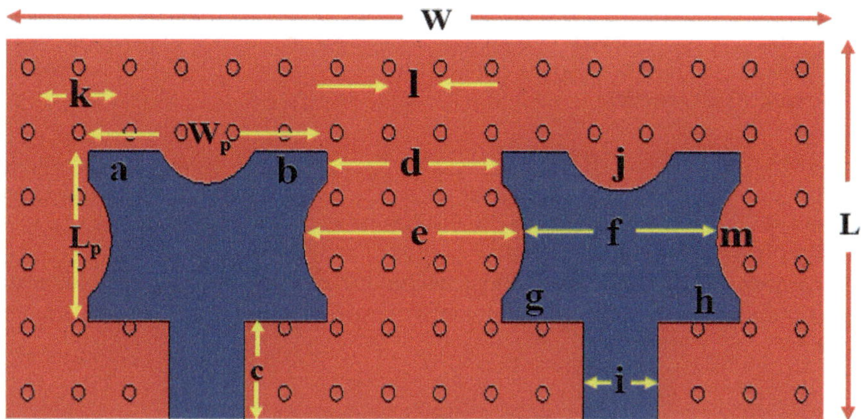

(a) Front-plane of photonic crystal

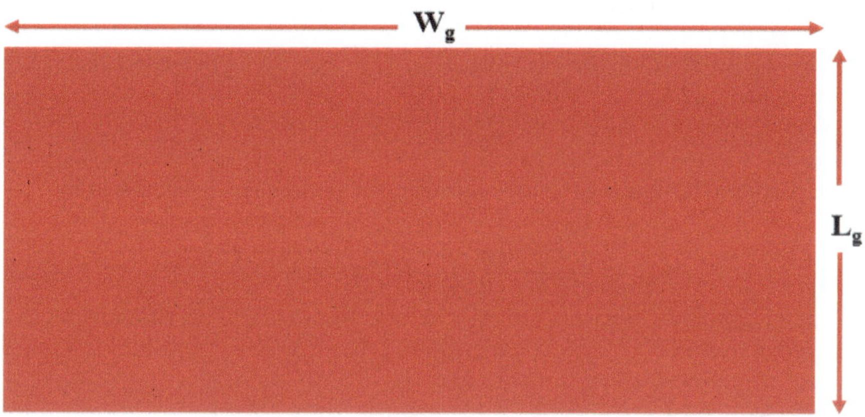

(b) Back-plane of photonic crystal

Fig. 25.1 Photonic crystal MIMO antenna design

Table 25.1 Geometrical parameters of photonic crystal design (μm)

Unit	W	L	l	W_p	k	a	b	c	d
Value	1660	600	123	322	16	166	166	150	340
Unit	e	f	g	L_p	h	i	j	W_g	L_g
Value	380	300	166	258	166	166	184	1660	600

$$\frac{\partial}{\partial x}\left(n^{-2}(x, y)\frac{\partial H_z}{\partial x}\right) = +\frac{\partial}{\partial y}\left(n^{-2}(x, y)\frac{\partial H_z}{\partial y}\right) + k^2 H_z = 0 \quad (25.1)$$

For the case of TM modes:

$$\frac{\partial^2 E_z}{\partial x^2} + \frac{\partial^2 E_z}{\partial y^2} + k^2 n^2(x, y) E_z = 0 \quad (25.2)$$

Here k represents the Wave number.

According to use of Floquet–Bloch technique realized by [28] and calculate magnetic component or electric component fields are represented by the system (H_z or E_z)

$$H_z = \sum_{m=-\infty}^{+\infty} \sum_{n=-\infty}^{+\infty} \mathrm{e}^{j\vec{\beta}\cdot\vec{r}}\left(c_{m,n} \cdot \mathrm{e}^{i\left(m\vec{b_1}+n\vec{b_2}\right)\cdot\vec{r}}\right) \quad (25.3)$$

The plane wave refractive index (η_{ref}) is evaluated by

$$\eta_{\mathrm{eff}}(x, y) = \left[\mathrm{e}^{j\vec{\beta}\cdot\vec{r}} \sum_{m=-\infty}^{+\infty} \sum_{n=-\infty}^{+\infty} f_{m,n} \mathrm{e}^{i\left(m\vec{b_1}+n\vec{b_2}\right)\cdot\vec{r}}\right]^{\pm\frac{1}{2}} \quad (25.4)$$

The proposed designed structure antenna resonant frequency calculated by

$$f_{m,n} = \frac{n_1^2 - n_0^2}{\sqrt{m^2 + n^2}} \cdot \frac{r}{a} \cdot J_1\left(2\pi\sqrt{m^2 + n^2}\frac{r}{a}\right) \quad (25.5)$$

$$f_{0,0} = n_0^2 + \left(n_1^2 - n_0^2\right)\pi\left(\frac{r}{a}\right)^2 \quad (25.6)$$

$$f_{\mathrm{r}} = \frac{12.7}{\left(4n + \frac{g}{6} + f\right)} \quad (25.7)$$

$$\varepsilon_{\mathrm{eff}} = \frac{\varepsilon_{\mathrm{r}} + 1}{2} \quad (25.8)$$

Here, the values are $n = 15.0$ µm, $g = m/6.0 = 6.0$ µm and $f = (d-t) = (6.0\text{–}3.5)$ µm $= 2.50$ µm.

25.3 Results Analysis

Figure 25.2 shows analysis parameters related photonic crystal MIMO structure. Figure 25.2a shows impedance versus S-parameters and its S_{11} bandwidth ≤ -10 dB using CST software among 0.60–0.70 THz exhibiting reasonable agreement of propose design which can be used for THz regime regions. At the resonant frequency due to the parametric variations and finalized the design dimensions obtained the S_{11} and S_{21} are − 27.13 dB and − 35.11 dB. Figure 25.2b shows the frequency versus impedance graph for both real and imaginary parts of the design resonated at 0.636 THz are 55 ohms and − 5 ohms, respectively. Figure 25.2c shows frequency versus VSWR of the design and identified that at resonate frequency which is ≤ 2.

Figure 25.3 shows radiation patterns of current designed structure of port_1 and port_2 at 0.636 THz. The designed MIMO antenna exhibits the nearly dipole like

Fig. 25.2 Photonic crystal MIMO design structure

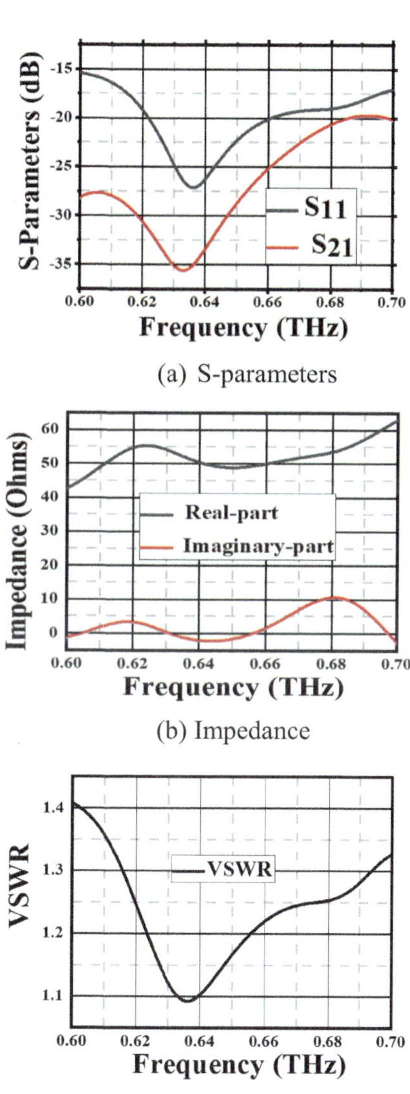

(a) S-parameters

(b) Impedance

(c) VSWR

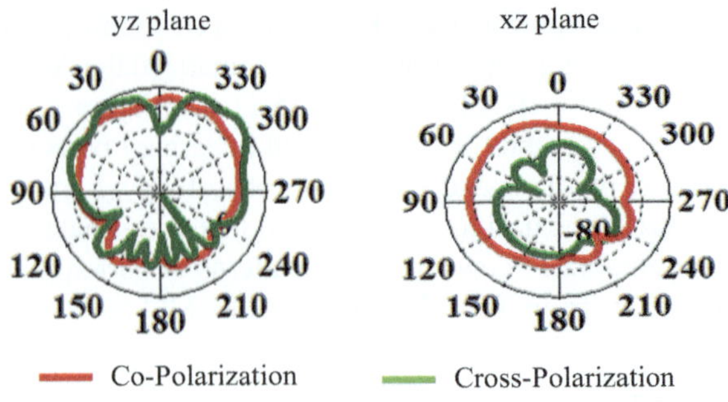

Fig. 25.3 Radiation patterns at 0.636 THz

structure at frequency in both the planes of *xz*-plane of $\Phi = 90°$ versus θ as well as $\Phi = 0°$ versus θ. At the resonant frequency identified that complementary radiation patterns are identified in the plane of *yz* for ports 1 and 2. Figure 25.4 shows surface current distribution of current design structure. Figure 25.4a shows excitation of front side plane port_1 and most radiation intensive around whole antenna structure expressing its fundamental frequency. Figure 25.4b shows excitation of back side plane port_1 and most radiation intensive around the full ground back side of the patch and at feeding terminal. Figure 25.4c shows excitation of front side plane port_2 and most radiation intensive around whole antenna structure and edges of the patch structure expressing its fundamental frequency. Figure 25.4d shows excitation of back side plane port_2 and most radiation intensive around the full ground back side of the patch and at feeding terminal. Figure 25.5 shows distribution of E-filed and H-field proposed designed structure. Figure 25.5a shows excitation of front side plane port_1 E-filed distribution and more amount of electric filed is generated at the circular arcs of the square patch with a maximum *E*-filed of 7.89 V/m. Figure 25.5b shows excitation of front side plane port_2 E-filed distribution and more amount of electric filed is generated at the circular arcs of the square patch with a maximum E-filed of 8.39 V/m. Figure 25.5c shows excitation of front side plane port_1 H-filed distribution and more amount magnetic field generated at circular arcs of square patch with a maximum H-filed of 2713 A/m. Figure 25.5b shows excitation of front side plane port_1 H-filed distribution and more amount magnetic field generated at circular arcs of square patch with a maximum H-filed of 2724 A/m.

25.4 Analysis of Diversity Parameters

In this section explained the study of metrics that involve to vital assessing performance related to photonic crystal MIMO structure parameters are DG, MEG, CCL, TARC, and ECC. Figures 25.6a shows ECC of current design structure. For design of MIMO system ECC has smaller than 0.50. Using the far-field analysis obtained the radiation pattern using relation Eq. (25.9) as [12]

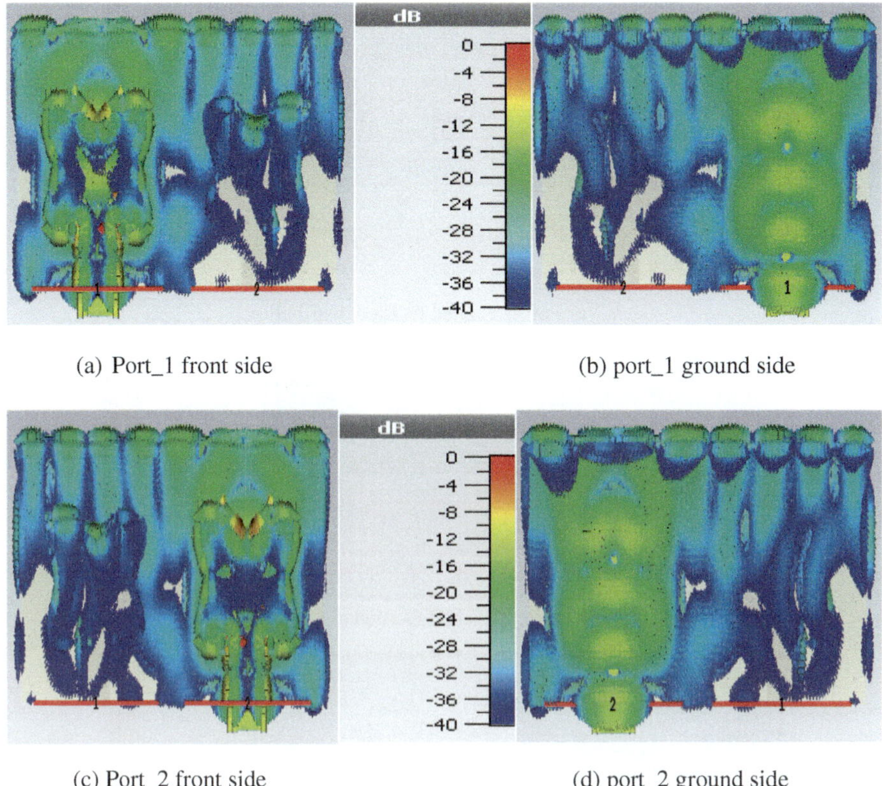

(a) Port_1 front side

(b) port_1 ground side

(c) Port_2 front side

(d) port_2 ground side

Fig. 25.4 Surface current distribution at 0.636 THz

$$\rho_{ij} = \left| \frac{\iint \left(E_{\theta_i} \cdot E_{\theta_j}^{*} + E_{\varphi_i} \cdot E_{\varphi_j}^{*} \right) \mathrm{d}\Omega}{\iint \left(E_{\theta_i} \cdot E_{\theta_i}^{*} + E_{\varphi_i} \cdot E_{\varphi_i}^{*} \right) \mathrm{d}\Omega \iint \left(E_{\theta_j} \cdot E_{\theta_j}^{*} + E_{\varphi_j} \cdot E_{\varphi_j}^{*} \right) \mathrm{d}\Omega} \right|^{2} \tag{25.9}$$

For 2-elelment MIMO system ECC is evaluated using Eq. (25.10) with the help of S-parameter analysis as [34]

$$\rho_{ij} = \frac{\left| S_{11}^{*} S_{12} + S_{21}^{*} S_{22} \right|^{2}}{\left(1 - \left(|S_{11}|^{2} + |S_{21}|^{2}\right)\right)\left(1 - \left(|S_{22}|^{2} + |S_{12}|^{2}\right)\right)} \tag{25.10}$$

Figure 25.6b shows frequency versus DG of the current design structure. When ECC = 0 and its diversity gain becomes 10 dB which indicates that reliability of the designed structure is increases for a radio link. The diversity gain is evaluated using the empirical relation (25.11) as [29]

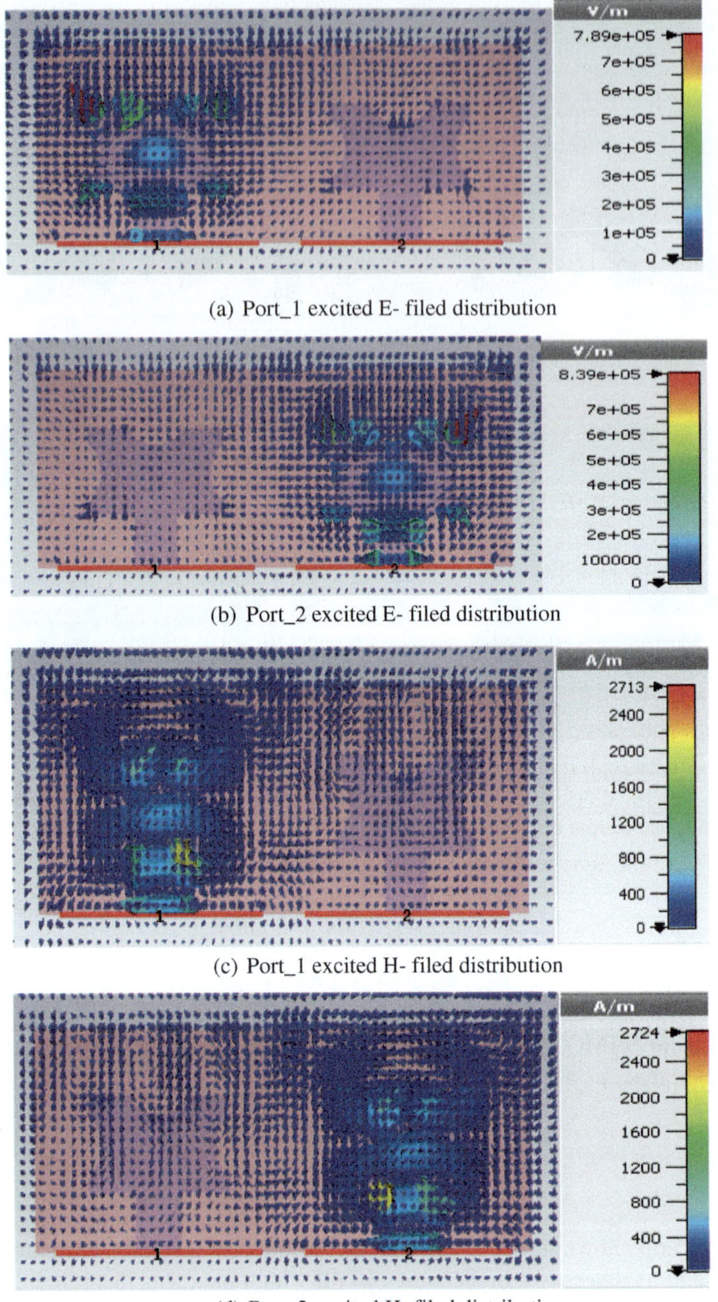

(a) Port_1 excited E- filed distribution

(b) Port_2 excited E- filed distribution

(c) Port_1 excited H- filed distribution

(d) Port_2 excited H- filed distribution

Fig. 25.5 E-field and H-filed distributions at 0.636 THz

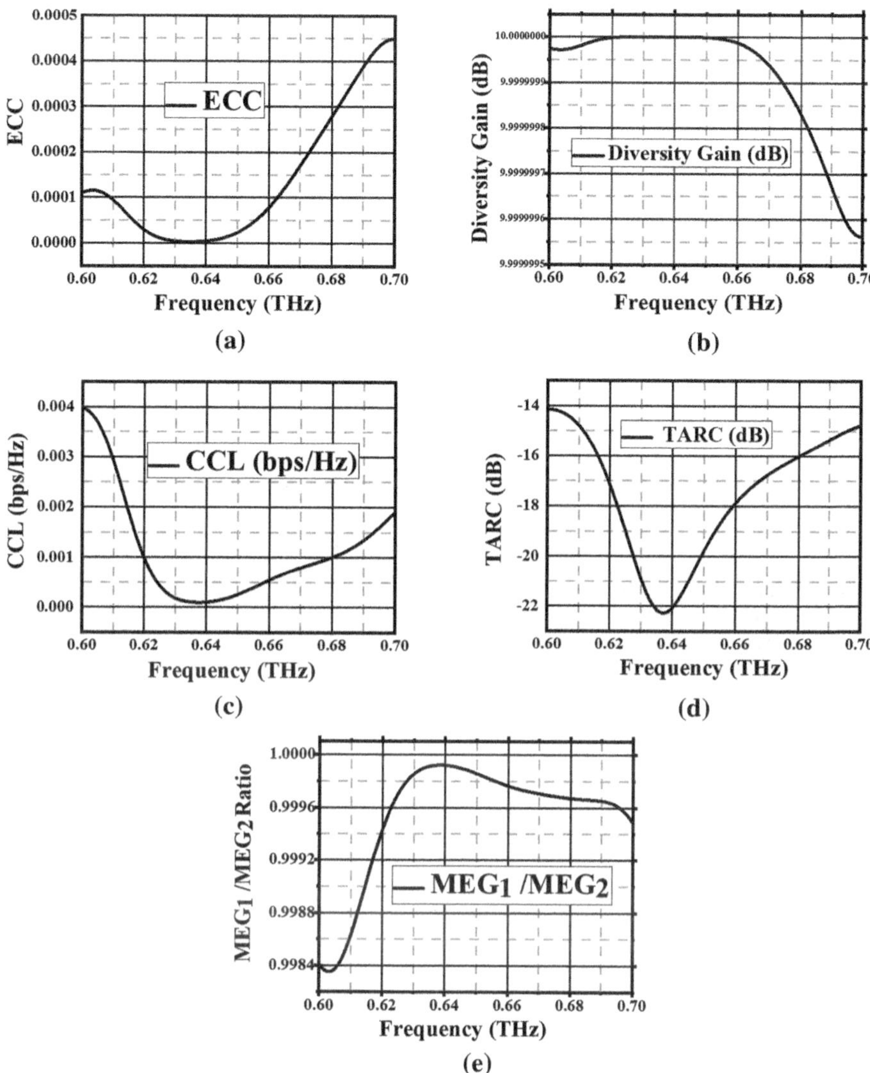

Fig. 25.6 MIMO Parameter analysis of the current designed structure **a** ECC, **b** DG, **c** CCL, **d** TARC, **e** $\text{MEG}_1/\text{MEG}_2$

$$\text{DG} = 10\sqrt{1 - |\text{ECC}|^2} \tag{25.11}$$

Figure 25.6c shows CCL of the designed structure system. The CCL of MIMO system is evaluated using the relation Eq. (25.12) as [18]

$$C_{\text{loss}} = -\log_2 \det\left(a^R\right), \tag{25.12}$$

where $a^R = \begin{pmatrix} \rho_{11} & \rho_{12} \\ \rho_{21} & \rho_{22} \end{pmatrix}$ and $\rho_{ii} = 1 - \left(|S_{ii}|^2 + |S_{ij}|^2\right)$, and $\rho_{ij} = -\left(s_{ii}^* s_{ij} + s_{ij}^* s_{jj}\right)$, where $i, j = 1$ or 2.

Figure 25.6d shows TARC of proposed designed structure which can evaluated by Eqs. (25.13) and (25.14) as [16]

$$\Gamma_a^t = \frac{\sqrt{\Sigma_j^M |b_j|^2}}{\sqrt{\Sigma_j^M |a_j|^2}}, \tag{25.13}$$

where a_j and b_j represents the incident and reflected waves.

$$\Gamma_a^t = \sqrt{\frac{|S_{11} + S_{12}e^{j\theta}|^2 + |S_{21} + S_{22}e^{j\theta}|^2}{2}} \tag{25.14}$$

Another circular metric in MIMO analysis is MEG which can be evaluated by power received combinations of isotropic and diversity antennas. Figure 25.6d shows MEG of proposed designed structure. At the same power level of the design requires the difference among the any two MEG values to be near zero which can be evaluated using the relation Eq. (25.15) as [25]

$$\text{MEG}_i = 0.5\eta_{i,\text{rad}} = 0.5\left[1 - \sum_{j=1}^{M} |S_{ij}|^2\right]. \tag{25.15}$$

25.5 Conclusion

A 2-port photonic crystal-based MIMO system was presented as for THz regime systems. The designed structure resonates at 0.636 THz for considered the patch with square patch and circular arcs which have been produced S_{11} and S_{21} are − 27.13 dB and − 35.11 dB. The proposed system radiation patterns exhibited the symmetrical patterns for the ports of 1 and 2 exploited the challenging scenario with THz communication systems. The obtained results related to the system showed that designed system of circular arc shapes due to the excessive length of the system optimized using photonic crystal using polyimide substrate achieved channel capacity loss have 0.105 bits/s/Hz in operating frequency based on the homogenous and periodic photonic crystals. By varying the various geometrical parameters there is an interesting improvement in channel capacity loss information, mean effective gain system and total active reflection coefficient.

References

1. Singh, A., Singh, S.: A trapezoidal microstrip patch antenna on photonic crystal substrate for high-speed THz applications. Photon. Nanostruct. Fund. Appl. **14**, 52–62 (2015)
2. Babu, K.V., Anuradha, B.: Design of inverted L-shape and ohm symbol inserted MIMO antenna to reduce the mutual coupling. AEU-Int. J. Electron. Commun. **105**, 42–53 (2019)
3. Kushwaha, R.K., Karuppanan, P.: Investigation and design of microstrip patch antenna employed on PCs substrates in THz regime. Aust. J. Electr. Electron. Eng. **18**(2), 118–125 (2021)
4. Das, S., et al.: Analysis of a miniaturized modified multifrequency printed antenna with broadband characteristics for WLAN application. In: Advances in Power Systems and Energy Management: Select Proceedings of ETAEERE 2020. Springer Singapore (2021)
5. Babu, K.V., Sree, G.N.J.: Design and circuit analysis approach of graphene-based compact metamaterial-absorber for terahertz range applications. Opt. Quant. Electron. **55**(9), 769 (2023)
6. Babu, K.V., et al.: Design and implementation of MIMO graphene patch antenna to improve isolation for THz applications. Microsyst. Technol. **29**, 1–11 (2023)
7. Temmar, M.N.E., et al.: Analysis and design of a terahertz microstrip antenna based on a synthesized photonic bandgap substrate using BPSO. J. Comput. Electron. **18**(1), 231–240 (2019)
8. Vasu Babu, K., et al.: Design of monopole ground graphene disc-inserted THz antenna for future wireless systems. In: Recent Advances in Graphene Nanophotonics, pp. 305–312. Springer Nature Switzerland, Cham (2023)
9. Vasu Babu, K., et al.: Design and analysis of fractal type MIMO radiator for the applications of sub 6-GHz 5G systems. In: Microelectronics, Circuits and Systems: Select Proceedings of Micro2021. Springer Nature Singapore, Singapore, pp. 243–250 (2023)
10. Benlakehal, M.E., et al.: Design and analysis of novel microstrip patch antenna array based on photonic crystal in THz. Opt. Quant. Electron. **54**(5), 1–162 (2022)
11. Benlakehal, M.E., Hocini, A., Khedrouche, D., et al.: Design and analysis of a 1 × 2 microstrip patch antenna array based on photonic crystals with a graphene load in THz. J. Opt. (2022). https://doi.org/10.1007/s12596-022-01006-8
12. Babu, K.V., et al.: Deep learning assisted fractal slotted substrate MIMO antenna with characteristic mode analysis (CMA) for Sub-6 GHz n78 5 G NR applications: design, optimization and experimental validation. Phys. Script. **98**(11), 115526 (2023)
13. Babu, K.V., Anuradha, B.: Design of MIMO antenna to interference inherent for ultra wide band systems using defected ground structure. Microw. Opt. Technol. Lett. **61**(12), 2698–2708 (2019)
14. Vasu Babu, K., Anuradha, B.: Design of nine-shaped MIMO antenna using parasitic elements to reduce mutual coupling. In: Optical and Wireless Technologies: Proceedings of OWT 2019. Springer Singapore (2020)
15. Hocini, A., Temmar, M.N., Khedrouche, D., Zamani, M.: Novel approach for the design and analysis of a terahertz microstrip patch antenna based on photonic crystals. Photon. Nanostruct. Fundam. Appl. **36**, 100723 (2019)
16. Goyal, R., Vishwakarma, D.K.: Design of a graphene-based patch antenna on glass substrate for high-speed terahertz communications. Microw. Opt. Technol. Lett. **60**(7), 1594–1600 (2018)
17. Babu, K.V., et al.: Design and development of miniaturized MIMO antenna using parasitic elements and Machine learning (ML) technique for lower sub 6 GHz 5G applications. AEU-Int. J. Electron. Commun. **153**, 154281 (2022)
18. Ananda, S., Sriram Kumara, D., Wub, R.J., Chavali, M.: Graphene nanoribbon-based terahertz antenna on polyimide substrate. Optik **125**, 5546–5549 (2014)
19. Vasu Babu, K., et al. Design and analysis of a CPW-fed fractal MIMO THz antenna using an array of parasitic elements. In: Terahertz Devices, Circuits and Systems: Materials, Methods and Applications, pp. 53–60. Springer Nature Singapore, Singapore (2022)

20. Babu, K.V., et al.: Design and analysis of fractal-based THz antenna with co-axial feeding technique for wireless applications. In: Recent Advances in Graphene Nanophotonics, pp. 351–358. Springer Nature Switzerland, Cham (2023)
21. Shamim, S.M., Das, S., Hossain, M.A., et al.: Investigations on graphene-based ultra-wideband (UWB) microstrip patch antennas for terahertz (THz) applications. Plasmonics **16**, 1623–1631 (2021). https://doi.org/10.1007/s11468-021-01423-8
22. EddineTemmar, M.N., et al.: Enhanced flexible terahertz microstrip antenna based on modified silicon-air photonic crystal. Optik **217**, 164897 (2020)
23. Khezzar, D., Khedrouche, D., Denidni, T.A.: New design of a broadband PBG-based antenna for THz band applications. Photonics Nanostruct. Fund. Appl. **46**, 100947 (2021)
24. Babu, K.V., Anuradha, B.: Design and analysis of multi-band circle shape MIMO antenna using defected ground structure to reduce mutual coupling. Int. J. Ultra Wideband Commun. Syst. **4**(1), 32–40 (2019)
25. Babu, K.V., Anuradha, B.: Design of UWB MIMO antenna to reduce the mutual coupling using defected ground structure. Wirel. Personal Commun. **118**(4), 3469–3484 (2021)
26. Mahmud, R.H.: Terahertz microstrip patch antennas for the surveillance applications. Kurdistan J. Appl. Res. (KJAR) **5**, 17–27 (2020). https://doi.org/10.24017/science.2020.1.2
27. Babu, K.V., et al.: Design of graphene-based broadband metamaterial absorber with circuit analysis approach for Terahertz Region applications. Opt. Quant. Electron. **55**(13), 1188 (2023)
28. Azarbar, A., Masouleh, M.S., Behbahani, A.K.: A new terahertz microstrip rectangular patch array antenna. Int. J. Electromagn. Appl. **4**, 25–29 (2014)
29. Babu, K.V., et al.: Broadband sub-6 GHz flower-shaped MIMO antenna with high isolation using theory of characteristic mode analysis (TCMA) for 5G NR bands and WLAN applications. Int. J. Commun. Syst. **36**(6), e5442 (2023)
30. Shamim, S.M., Uddin, M.S., Hasan, M., et al.: Design and implementation of miniaturized wideband microstrip patch antenna for high-speed terahertz applications. J. Comput. Electron. **20**, 604–610 (2021). https://doi.org/10.1007/s10825-020-01587-2
31. Babu, K.V., et al.: Performance analysis of a photonic crystals embedded wideband (1.41–3.0 THz) fractal MIMO antenna over SiO_2 substrate for terahertz band applications. Silicon **15**(18), 7823–7836 (2023)
32. Younssi, M., Jaoujal, A., Yaccoub, M.D., El Moussaoui, A., Aknin, N.: Study of a microstrip antenna with and without superstrate for terahertz frequency. Int. J. Innov. Appl. Stud. **2**(4), 369–371 (2013)
33. Krishna, C.M., Das, S., Nella, A., et al.: A Micro-sized rhombus-shaped THz antenna for high-speed short-range wireless communication applications. Plasmonics **16**, 2167–2177 (2021). https://doi.org/10.1007/s11468-021-01472-z
34. Das, S., et al. A compact penta-band printed monopole antenna for multiple wireless communication systems. In: Futuristic Communication and Network Technologies: Select Proceedings of VICFCNT 2020. Springer Singapore (2022)
35. Babu, K.V., Das, S., Sree, G.N.J., et al.: Design and optimization of micro-sized wideband fractal MIMO antenna based on characteristic analysis of graphene for terahertz applications. Opt. Quant. Electron. **54**, 281 (2022). https://doi.org/10.1007/s11082-022-03671-2
36. Vasu Babu, K., et al.: A micro-scaled graphene-based tree-shaped wideband printed MIMO antenna for terahertz applications. J. Comput. Electron. **21**(1), 289–303 (2022)
37. Patel, S.K., et al. (eds.): Recent Advances in Graphene Nanophotonics. Springer, Cham (2023)
38. Babu, K.V., et al.: Compact dual-band design and analysis of half-circular U-shape MIMO radiator for wireless applications. Microsyst. Technol. **29**(4), 501–514 (2023)

Chapter 26
Modeling Terahertz Patch Antenna Using the Wave Concept Iterative Process in Frequency Range 6–15 THz

Hamid Bezzout, Bilal Aghoutane, Mohammed El Ghzaoui, and Hanan El Faylali

26.1 Introduction

The electromagnetic spectrum includes various frequencies that have a broad range of potential uses. In particular, the terahertz band, which extends from approximately 0.1–30 THz, is a relatively unexplored area with considerable potential [1–3]. Terahertz (THz) frequency study has become a focus of electromagnetic research in recent years. The terahertz range, which lies in the range of microwaves and infrared rays, has a number of interesting characteristics that open up a wide range of applications, including improved communication systems [4], sensing applications [5, 6], imaging technologies [7], biomedical engineering [8], and more. As researchers have explored the specificity of these frequency range applications, it has become essential to design and optimize antennas for efficient operation in the THz frequency band.

Although the study of electromagnetic waves has progressed remarkably well, the THz band presents both challenges and opportunities [11, 12]. As it is limited by atmospheric absorption and the complexities of material interactions, innovative solutions are required to exploit the full potential of the terahertz spectrum. Challenges and gaps persist in the effective design and use of antennas in the terahertz range. As technology advances and applications of THz frequencies develop, antennas capable of efficiently capturing and transmitting signals in this spectrum are growing. Filling this gap requires a thorough understanding of the antenna design principles, advanced simulation techniques, and efficient numerical methodologies. To address these challenges, the present study explores and analyzes the patch antenna

H. Bezzout (✉) · B. Aghoutane · H. El Faylali
Department of Physics and Computer Sciences, Ibn Tofail University, Kenitra, Morocco
e-mail: hamid.bezzout@uit.ac.ma

M. El Ghzaoui
Faculty of Sciences Dhar El Mahraz-Fes, Sidi Mohamed Ben Abdellah University, Fes, Morocco

M. El Ghzaoui et al. (eds.), *Next Generation Wireless Communication*, Signals and Communication Technology, https://doi.org/10.1007/978-3-031-56144-3_26

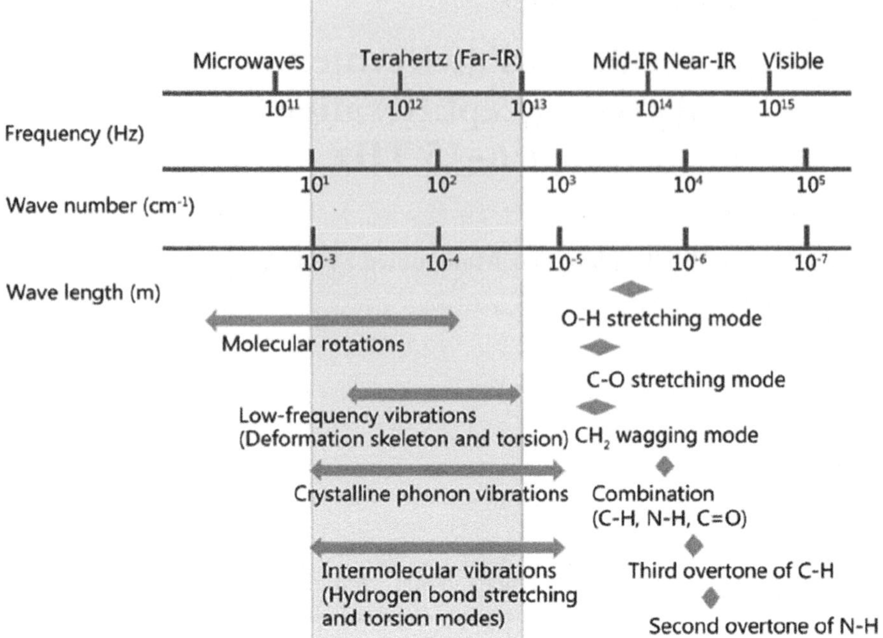

Fig. 26.1 Terahertz band of electromagnetic spectrum [9, 10]

[13] specifically adapted to the terahertz band. The central problem studied is the efficient capture and radiation of signals in this difficult frequency range.

Involving the wave concept iterative process (WCIP), the transmission line matrix (TLM) [14, 15], the finite element method (FEM) [16], and the finite difference time domain (FDTD) [17]. The WCIP is faster than other numerical methods, as it allows 2D discretization, which saves both memory and execution time. In this work, the application of the (WCIP) offers considerable computational advantages. In particular, it is simple to program using MATLAB and can analyze a variety of patch shapes. Its computational efficiency is ensured using the FFT (fast Fourier transform) two-dimensional algorithm [18]. This approach ensures simplicity, flexibility, and considerable time savings compared with other numerical methods.

The design and optimization of a patch antenna for effective terahertz band is the first goal of this work. The second is an investigation of the electromagnetic properties of the terahertz band, which includes S11 parameter analysis and current distribution visualization. The study tackles the difficulties associated with terahertz detection and communication, offering basic information for the creation of effective terahertz devices. Accuracy is maintained while increasing computational efficiency through the utilization of a discretization technique and WCIP. The paper is organized into three sections. In Sect. 26.2, the problem formulation for designing a patch antenna using WCIP is provided. In Sect. 26.3, the results and discussion are examined, along with electric field analysis and current distribution visualization. The bandwidth of 2

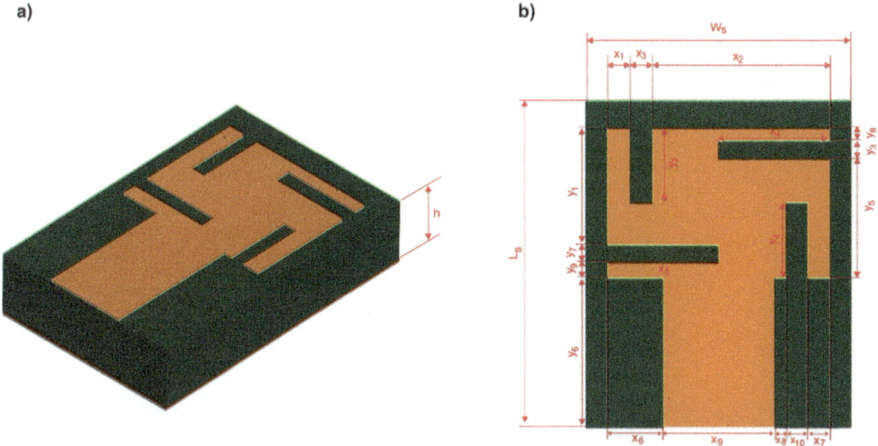

Fig. 26.2 Detailed configuration of the proposed antenna. a) 3D structure showing overall design, b) 2D structure highlighting key antenna parameters

THz is determined at the resonant frequency of 9.91 THz by calculating the parameter $|S_{11}|$. Additional computed parameters are the electric field, VSWR, admittance, and impedance (Sect. 26.4). The study concludes with ideas for more research.

26.2 Problem Formulation

26.2.1 Patch Antenna Geometry

Figure 26.2 illustrates the proposed patch antenna for terahertz applications. Figure 26.2a depicts the antenna's three-dimensional configuration, revealing the substrate's thickness, while Fig. 26.2b displays the antenna's complete design, including all of its characteristics, the values of which are given in Table 26.1. The substrate material, Roger/Duroid, has an $\varepsilon_r = 10.2$ characterization. The dimensions of the wave guiding region into which the antenna is put are $47.9 \times 37.7 \times 5\ \mu\text{m}^3$.

26.2.2 Theory of Wave Concept Iterative Method

The method (WCIP) was introduced on the basis of the formulation of problems for antennas as transverse waves. These two transverse waves are composed of electric and magnetic fields [19, 20]. As Fig. 26.3 shows, incident waves are linked to reflected waves by means of the following equation:

Table 26.1 Dimensions of the patch antenna

Parameter	Value (μm)	Parameter	Value (μm)
x_1	3.2	y_1	17.2
x_2	25.3	y_2	11
x_3	3.2	y_3	2.6
x_4	15.8	y_4	6.3
x_5	15.8	y_5	17.5
x_6	8	y_6	22
x_7	3.1	y_7	2.6
x_8	1.6	y_8	1.8
x_9	15.8	y_9	2.1
x_{10}	3.2	W_s	37.7
h	5	L_s	47.9

$$\vec{\mathbf{A}}_i = \frac{1}{2\sqrt{Z_{0i}}}\left[\vec{\mathbf{E}}_i + Z_{0i}\left(\vec{\mathbf{H}}_i \times \mathbf{n}_i\right)\right] \tag{26.1}$$

$$\vec{\mathbf{B}}_i = \frac{1}{2\sqrt{Z_{0i}}}\left[\vec{\mathbf{E}}_i - Z_{0i}\left(\vec{\mathbf{H}}_i \times \vec{\mathbf{n}}_i\right)\right], \tag{26.2}$$

where $\vec{\mathbf{A}}_i$ and are the incident and reflected transverse waves, respectively, through medium i. Z_{0i} represents the characteristic impedance of the medium i, $\vec{\mathbf{E}}_i$ is the electric field, $\vec{\mathbf{H}}_i$ is the magnetic field, and $\vec{\mathbf{n}}_i$ are the normal vectors to the medium i. The following equation defines the surface current density:

$$\vec{\mathbf{J}}_i = \vec{\mathbf{H}}_i \times \mathbf{n}_i \tag{26.3}$$

The following equations come from replacing the term in Eqs. (26.1) and (26.2) by the surface current density:

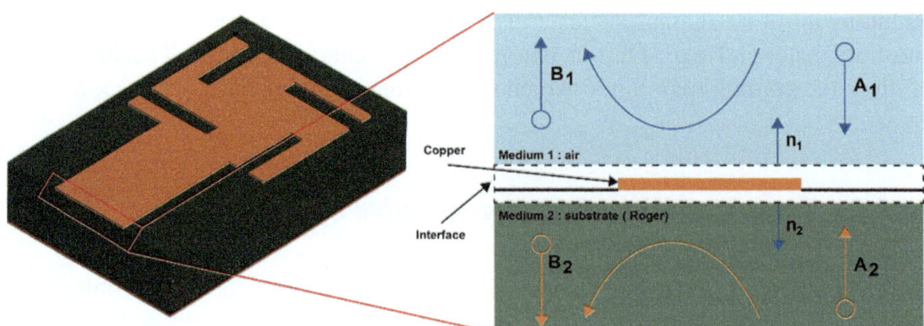

Fig. 26.3 A patch antenna's incident and reflected waves in the WCIP algorithm

$$\vec{\mathbf{A}}_i = \frac{1}{2\sqrt{Z_{0i}}}\left[\vec{\mathbf{E}}_i + Z_{0i}\vec{\mathbf{J}}_i\right] \tag{26.4}$$

$$\vec{\mathbf{B}}_i = \frac{1}{2\sqrt{Z_{0i}}}\left[\vec{\mathbf{E}}_i - Z_{0i}\vec{\mathbf{J}}_i\right] \tag{26.5}$$

Based on Eqs. (26.4) and (26.5), we derive the equations for electric field and current density in terms of incident and reflected waves $\vec{\mathbf{A}}_i$ and $\vec{\mathbf{B}}_i$:

$$\vec{\mathbf{E}}_i = \sqrt{Z_{0i}}\left[\vec{\mathbf{A}}_i + \vec{\mathbf{B}}_i\right] \tag{26.6}$$

$$\vec{\mathbf{J}}_i = \frac{1}{\sqrt{Z_{0i}}}\left[\vec{\mathbf{A}}_i - \vec{\mathbf{B}}_i\right] \tag{26.7}$$

The current density must be equal to zero for continuity of the electric field at the interface between media 1 (air) and 2 (Roger/Duroid). As a result, there must be an equal electric field on both sides of the interface. The following equations define the boundary conditions for the WCIP method:

$$\vec{\mathbf{J}}_{\text{tot}} = \vec{\mathbf{J}}_1 + \vec{\mathbf{J}}_2 \tag{26.8}$$

$$\vec{\mathbf{E}}_1 = \vec{\mathbf{E}}_2 \tag{26.9}$$

Concerning the boundary conditions outlined in Eqs. (26.8) and (26.9), in the scenario of a perfect conductor (metallic medium), the electric field is uniform and null $\vec{\mathbf{E}}_1 = \vec{\mathbf{E}}_2 = \vec{0}$ while the cumulative current density is $\vec{\mathbf{J}}_1 + \vec{\mathbf{J}}_2 = \vec{\mathbf{J}}_{\text{tot}}$. In contrast, for a dielectric medium, the electric fields are uniform $\vec{\mathbf{E}}_1 = \vec{\mathbf{E}}_2 = \vec{\mathbf{E}}$, and the total current density is $\vec{\mathbf{J}}_{\text{tot}} = 0$.

26.3 Results and Discussion

We have designed the proposed patch antenna according to electromagnetic analysis approach of the WCIP method. Firstly, the discretization of the antenna is done using 80×80 cell in the two directions x and y that is shown in Fig. 26.4. After 1000 iteration, the parameter current density distribution is obtained in Fig. 26.5. The behavior of electric field is show in Fig. 26.6.

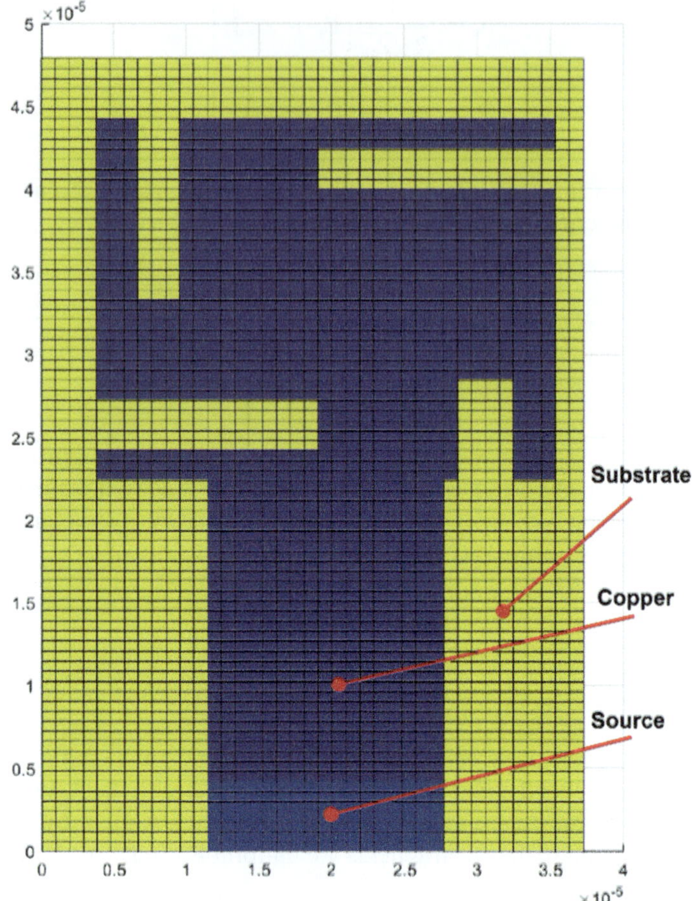

Fig. 26.4 The patch antenna's discretized design

26.3.1 Analysis of the Patch Antenna

The spatial distributions of electric field and current density, as observed in the simulation results, align consistently with the conditions stipulated in Eqs. (26.8) and (26.9). This agreement underlines the validity of the simulated electromagnetic characteristics to the theoretical framework described in the governing equations.

We analyzed the $|S_{11}|$ parameter to determine the antenna's frequency band width, and the results show that the resonant frequency is 9.91 THz (Fig. 26.7). A frequency band width of 2 THz was found for this resonance. As shown in Fig. 26.8, several calculation experiments were carried out to optimize the number of iterations for the best findings. The progression that has been noticed suggests

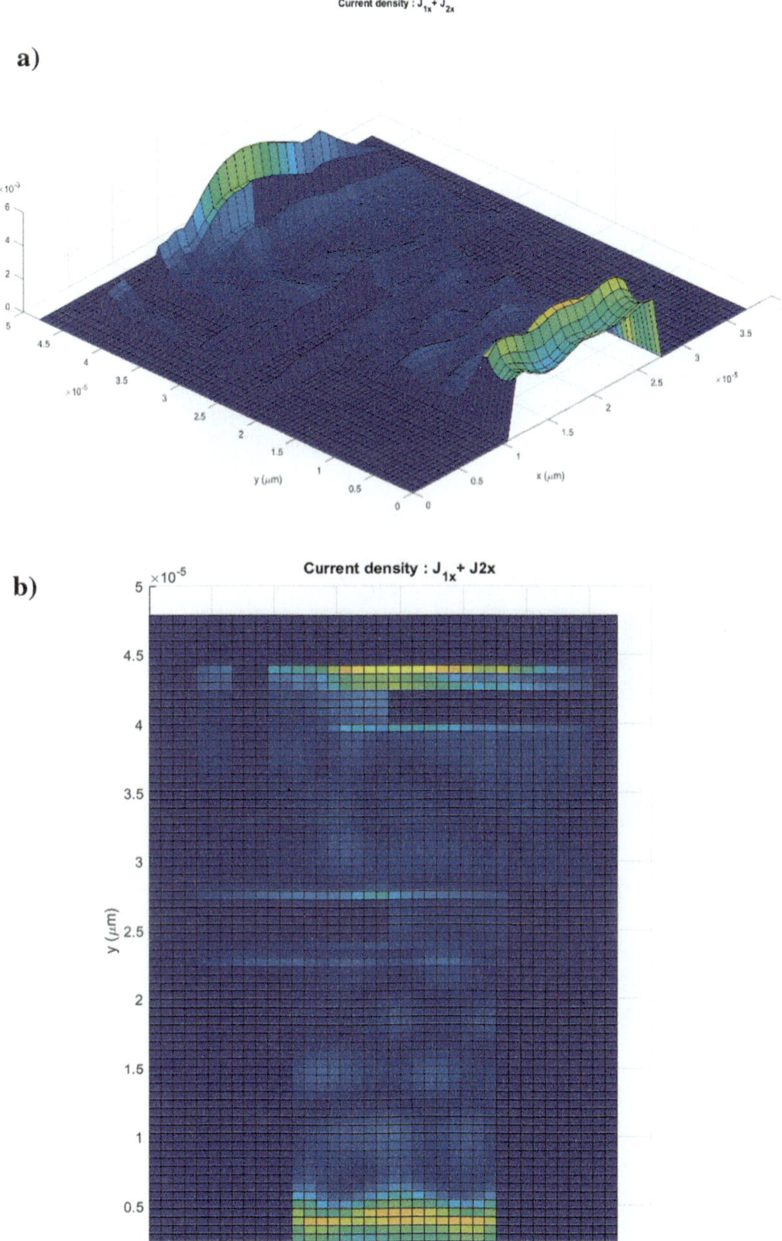

Fig. 26.5 The 3D and a 2D visualization of the patch antenna's current distribution

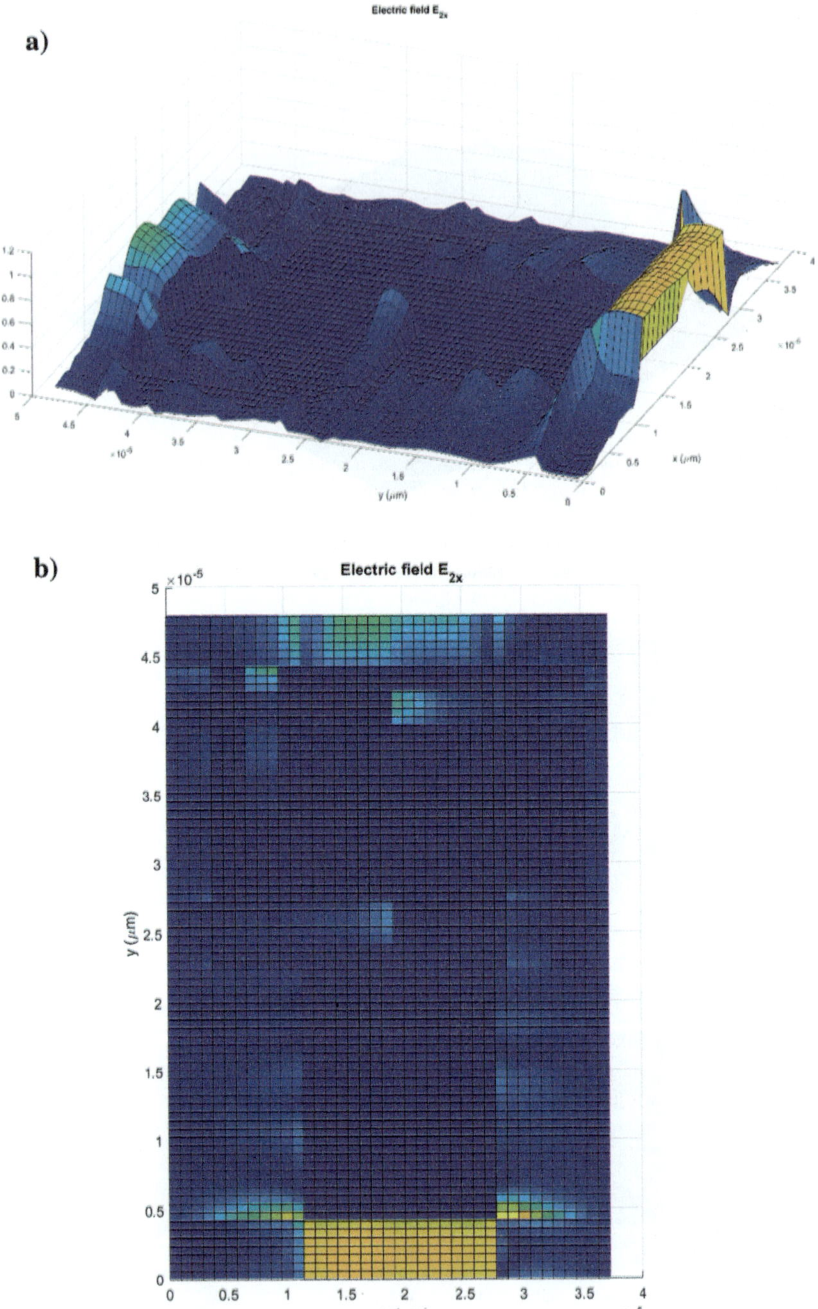

Fig. 26.6 The patch antenna's electric field: **a** in three dimensions and **b** in two dimensions

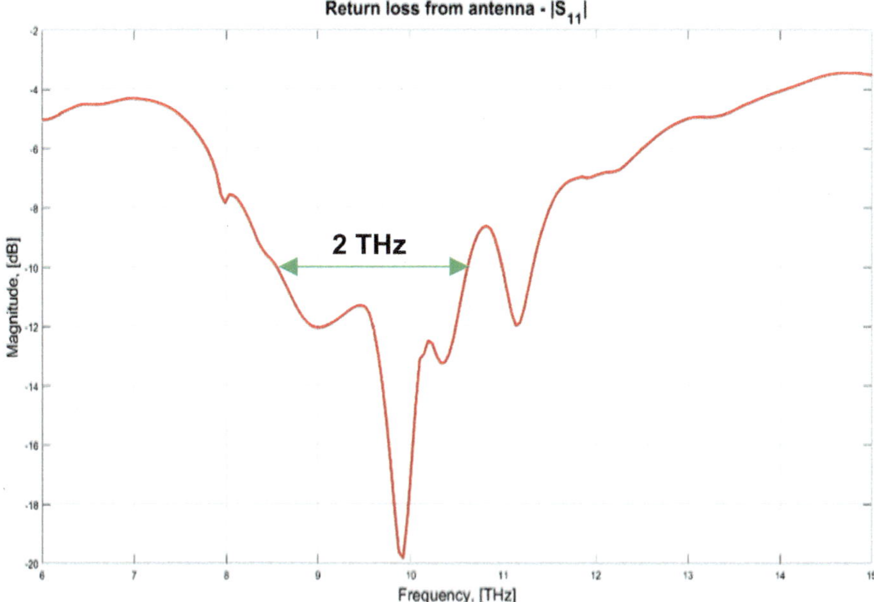

Fig. 26.7 The suggested patch antenna's return loss characteristics

that increasing the number of iterations improves the accuracy of the results, indicating the antenna's good performance. To confirm the efficiency of the suggested antenna design, two further parameters were calculated: Voltage Standing Wave Ratio (VSWR) (Fig. 26.9), impedance (Fig. 26.10) and admittance (Fig. 26.11).

Computer platform running on Intel Core i7 processor, embedded Intel graphics, Windows operating system, and MATLAB software was used to generate these results.

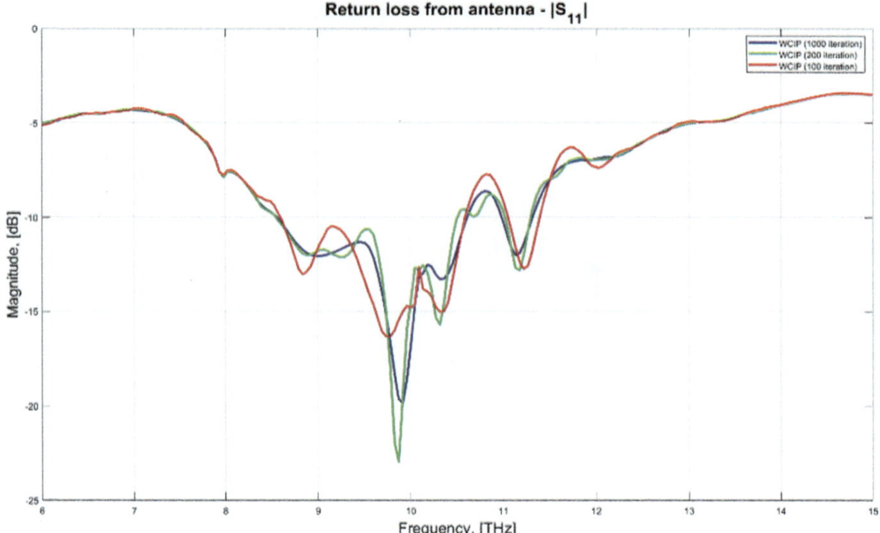

Fig. 26.8 The characteristics of return loss obtained during 100, 200, and 1000 iterations

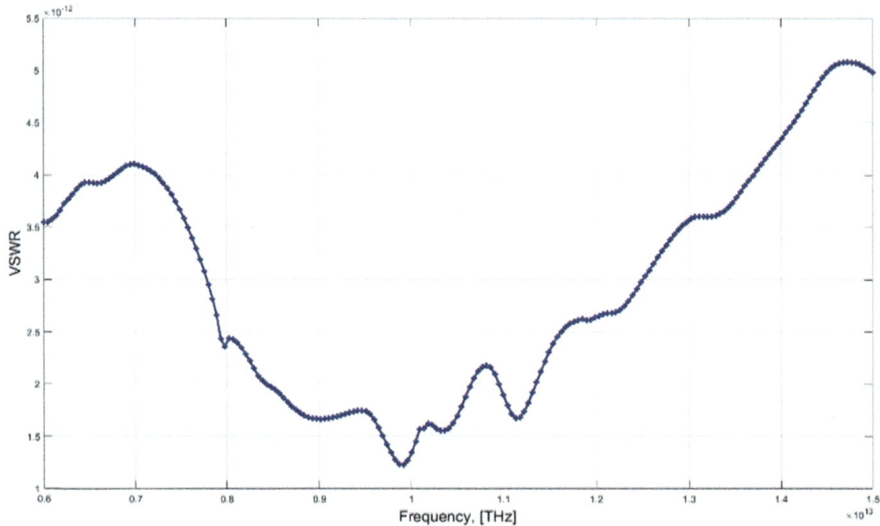

Fig. 26.9 Voltage standing wave ratio of the suggested patch antenna

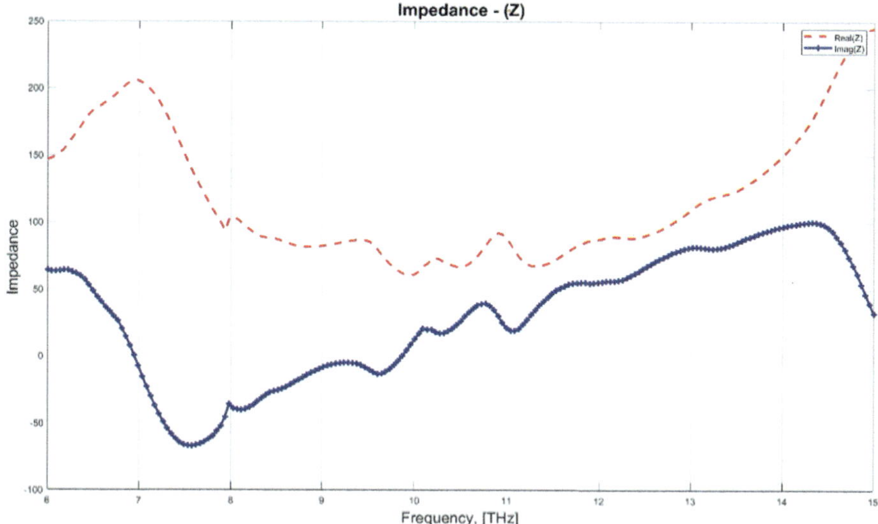

Fig. 26.10 Frequency-dependent impedance variation offered by the patch antenna

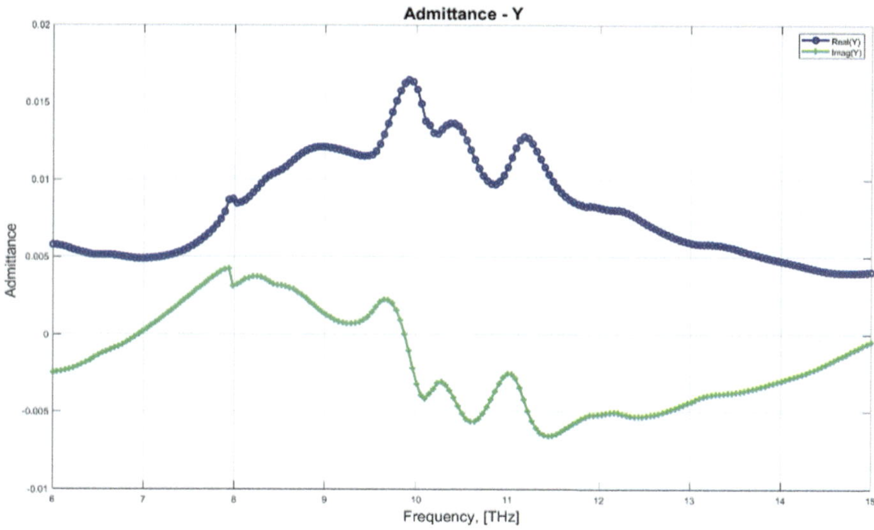

Fig. 26.11 Variation of the proposed patch antenna's impedance with frequency

26.4 Conclusion

Finally, this work provided an in-depth examination of the use of the WCIP method to design and analyze a terahertz (THz) patch antenna. Comprehensive simulations in conjunction with the discretization of the antenna through the WCIP framework have yielded significant understanding into its electromagnetic behavior. Together, the computed parameters—which include $|S_{11}|$, electric field distribution, current density, impedance, and admittance—confirm the suggested antenna design's accuracy and efficiency. The antenna's potential for use in THz transmission and sensing is further highlighted by the observed resonance frequency of 9.91 THz and its 2 THz bandwidth. The accuracy and reliability of the WCIP approach are demonstrated by the iterative optimization process, which is indicated by the $|S_{11}|$ parameter as the number of iterations increases.

References

1. Pawar, A.Y., Sonawane, D.D., Erande, K.B., Derle, D.V.: Terahertz technology and its applications. Drug Invention Today **5**(2), 157–163 (2013). https://doi.org/10.1016/j.dit.2013.03.009
2. Elaage, S., et al.: MB-OOK transceiver design for terahertz wireless communication systems. Int. J. Syst. Control Commun. **12**(4), 309–326 (2021)
3. El Ghzaoui, M., Mestoui, J., Hmamou, A., Elaage, S.: Performance analysis of multiband on–off keying pulse modulation with noncoherent receiver for THz applications. Microwave Opt. Technol. Lett. **64**(12), 2130–2135 (2022)
4. Tekbıyık, K., Ekti, A.R., Kurt, G.K., Görçin, A.: Terahertz band communication systems: challenges, novelties and standardization efforts. Phys. Commun. **35**, 100700 (2019). https://doi.org/10.1016/j.phycom.2019.04.014
5. Al-Douseri, F.M., Chen, Y., Zhang, X.-C.: THz wave sensing for petroleum industrial applications. Int. J. Infrared Milli Waves **27**(4), 481–503 (2006). https://doi.org/10.1007/s10762-006-9102-y
6. Hillger, P., Grzyb, J., Jain, R., Pfeiffer, U.R.: Terahertz imaging and sensing applications with silicon-based technologies. IEEE Trans. THz Sci. Technol. **9**(1), 1–19 (2019). https://doi.org/10.1109/TTHZ.2018.2884852
7. Davies, A.G., Linfield, E.H., Johnston, M.B.: The development of terahertz sources and their applications. Phys. Med. Biol. **47**(21), 3679–3689 (2002). https://doi.org/10.1088/0031-9155/47/21/302
8. Pickwell, E., Wallace, V.P.: Biomedical applications of terahertz technology. J. Phys. D: Appl. Phys. **39**(17), R301–R310 (2006). https://doi.org/10.1088/0022-3727/39/17/R01
9. Sun, L., Zhao, L., Peng, R.-Y.: Research progress in the effects of terahertz waves on biomacromolecules. Military Med. Res. **8**(1), 28 (2021). https://doi.org/10.1186/s40779-021-00321-8
10. El Ghzaoui, M., Das, S., Lenka, T.R., Biswas, A.: Terahertz wireless communication components and system technologies, pp. 1–310. Springer, Singapore (2022)
11. Lchhab, T., El Ghzaoui, M.: A circularly polarized wideband high gain antenna for THz wireless applications. Opt. Quant. Electron. **54**(12), 787 (2022)
12. El Ghzaoui, M., Das, S.: Data transmission with terahertz communication systems. In: Emerging Trends in Terahertz Solid-State Physics and Devices: Sources, Detectors, Advanced Materials, and Light-matter Interactions, pp. 121–141. Springer, Singapore (2020)
13. Khan, M.U., Sharawi, M.S., Mittra, R.: Microstrip patch antenna miniaturisation techniques: a review. IET Microwaves Antennas Propag. **9**(9), 913–922 (2015). https://doi.org/10.1049/iet-map.2014.0602
14. Attalhaoui, A., Bezzout, H., Habibi, M., El Faylali, H.: New TLM formulations for Debye medium. Optik **207**, 164436 (2020). https://doi.org/10.1016/j.ijleo.2020.164436

15. Attalhaoui, A., Bezzout, H., Habibi, M., El Faylali, H.: A New ADE-TLM for Lorentz Dispersive Medium. J. Russ Laser Res **42**(2), 2 (2021). https://doi.org/10.1007/s10946-021-09956-3
16. Adnet, N., Bruant, I., Pablo, F., Proslier, L.: The FEM–BIM approach using a mixed hexahedral finite element to model the electromagnetic and mechanical behavior of radiative microstrip antennas. Eng. Anal. Bound. Elem. **38**, 17–30 (2014). https://doi.org/10.1016/j.enganabound.2013.09.012
17. Gao, S.-C., Li, L.-W., Leong, M.-S., Yeo, T.-S.: Analysis of an H-shaped patch antenna by using the FDTD method. PIER **34**, 165–187 (2001). https://doi.org/10.2528/PIER01060501
18. Baudrand, H.: Application of wave concept iterative procedure in planar circuits. Recent Res. Dev. Microwave Theory Tech. **1**, 187–197 (1999)
19. El Hadri, D., Zugari, A., Zakriti, A.: Conception and characterisation of a novel 5_Shaped antenna for 5G application using WCIP method. Aust. J. Electr. Electron. Eng. **18**(2), 69–79 (2021). https://doi.org/10.1080/1448837X.2021.1916182
20. Ammar, N., Bdour, T., Aguili, T., Baudrand, H.: Investigation of electromagnetic scattering by arbitrarily shaped structures using the wave concept iterative process. J. Microw. Optoelectron. Electromagn. Appl. (JMOe) **7**(1), 26–43 (2008)

Chapter 27
Study and Design of a Printed Microstrip Antenna in the Terahertz Band

Salah-Eddine Didi, Imane Halkhams, Abdelhafid Es-Saqy, Mohammed Fattah, Said Mazer, and Moulhime El Bekkali

27.1 Introduction

Today, the THz frequency band (between 100 GHz and 10 THz) plays an increasingly important role in telecommunications systems, not least because of the ever-increasing need for higher data transmission rates, which means higher data transfer rates [1]. In the field of communications, there are many different types of applications, including remote sensing, biological detection, chip-to-chip or chip-on-chip communication, and body area networks (BANs) [1–3]. In fact, in THz frequency bands, many obstacles arise, including the presence of atmospheric molecules and the possibility of the signal travelling with or without line-of-sight. For a lower THz frequency band, 0.1–1.5 THz represents approximately one absorption wavelength at THz frequencies. A high-frequency (THz) antenna requires excellent directivity characteristics to reduce the path risk, irrespective of the THz signal path attenuation phenomenon. Wireless networks are also expected to reach around 75 billion consumers and users from now until around the year 2025, hence the research opportunity. "Millimeter" and "THz" antennas have been published for many advanced uses in remote communications [4–11]. The substrate is a photonic crystal, and the patch shape is trapezoidal, giving a bandwidth of 0.88–1.62 THz [12–14]

Scientific and technological advances in wireless communication are as rapid as ever. They have revolutionized our daily lives and accelerated the development of telecommunications technologies from the first generation of standard communications (1G) to the fifth generation of the Internet of Things (5G). Since a new mobile

S.-E. Didi (✉) · A. Es-Saqy · S. Mazer · M. E. Bekkali
IASSE Laboratory, Sidi Mohamed Ben Abdellah University, Fez, Morocco
e-mail: salaheddine.didi@usmba.ac.ma

I. Halkhams
LSEED Laboratory, UPF, Fez, Morocco

M. Fattah
IMAGE Laboratory, Moulay Ismail University, Meknes, Morocco

M. El Ghzaoui et al. (eds.), *Next Generation Wireless Communication*, Signals and Communication Technology, https://doi.org/10.1007/978-3-031-56144-3_27

communications system is developed every ten years, a 6th generation (6G) commercial mobile network should be ready around 2030 [15–17]. 6G technology will be associated with various application areas, including autonomous driving, robotics, and many other technological applications. This is made possible by using digital twin technology and an automatic control system, which will reduce the need for human intervention. A very low latency (0.1–1 ms) is also required to eliminate the barriers between the real world and the computer domain [18, 19].

In the context of scientific research, it is also important to focus on the development of wireless technologies worldwide. Over the years, the technological development of modern systems has become increasingly complex, requiring small, inexpensive, high-gain, portable, compatible or high-performance structures, in other words, small, compact antennas such as patch antennas. In this case, our study aims to design a new antenna for the THz frequency band, which will operate in the 300 GHz band, which is compatible with THz wireless communication systems. This antenna is also a candidate for THz technology applications, including 6G technology, THz imaging, spectroscopy, digital holography, and ptychography [20]. The applications of 6G technology also include the possibility of full connection to satellites, autonomous vehicles, and more. In addition, it is necessary for the behavioral and neurological processes linked to sensory integration to be triggered and for THz antennas to be able to provide all the data obtained through these five senses [21]. The THz imaging system developed recently has applications in imaging systems using pulsed and continuous THz imaging [22]. THz antennas are also useful for cell detection and bacterial identification in THz spectroscopy. Similarly, pulsed THz spectroscopy can detect cancer by identifying the characteristics of cancer spectral lines [23].

Terahertz radiation is an intermediate spectral range between microwaves and infrared. Its applications, long marginal or non-existent, are now booming. Among these applications, the most important are spectroscopy and imaging for pollutant detection, non-destructive testing, and medical diagnostics. These applications generally use Terahertz radiation propagating in free space, requiring transmitters and receivers fitted with antennas [24, 25]. The antennas generally used at Terahertz frequencies are planar, typically of the dipole (Hertz dipole) type, monolithically built on the substrate in the active elements employed within them for radiation creation as well as radiation detecting are integrated, and featuring a hemispherical silicon lens arranged on the rear face of the substrate. These devices have several drawbacks. Firstly, producing the silicon lens and positioning it on the substrate with a precision of the order of a micron about the antenna is very delicate and costly.

On the other hand, using a coupling lens is essential to prevent the radiation, mainly emitted in the semiconductor substrate, from being trapped in the latter. In fact, despite such a lens, only around 21% of the power emitted by a typical planar antenna is radiated into free space. The rest is trapped and absorbed by the substrate. In addition, the Hertz dipole, the most widely used planar antenna, has a low efficiency and is highly frequency-dependent. In addition, broadband antennas are ideal for precise pulse spectroscopy measurements in the terahertz range. On the other hand, they are inefficient and unsuitable for permanent use at very low power levels.

The design of high-gain antennas at terahertz frequencies suffers from numerous constraints. These include physical dimensions, which must be kept to a minimum. The challenge also lies in the choice of antenna construction, compatible with material-based techniques. Several techniques are available in the literature to facilitate the design and construction of terahertz antennas. In [25], researchers used the silicon micromachining technique to design and manufacture rectangular traveling-wave waveguide antennas at 23–245 GHz.

Similarly, the metal layer technique was used to manufacture a corrugated rectangular waveguide at frequencies of 130–180 GHz [26]. Researchers have also designed a reconfigurable loop antenna operating at terahertz frequencies using the graphene-metal approach [27]. In [28], the authors present the design of a microstrip patch antenna operating at frequencies from 700 to 850 GHz. One of the unique properties of the microstrip patch presented is its adaptation to the multilayer structure. They studied the effects of parameters related to the dimensions of the microstrip patch on electrical performance. For 300 GHz, high-gain Fabry–Perot cavity antennas are available, providing adequate bandwidths for 6G wireless applications. These elements can also be installed in seven-way metal structures, thus providing a horn-like element with frequency-selective surfaces [29]. In addition, the literature cited in references [30, 31] also offers studies of equivalent designs. This MIMO antenna offers the possibility of producing an extended bandwidth from 0.72 to 10.0 THz and an isolation of over 20 dB [32].

In this chapter, we'll be working on THz components and, more specifically, on the design of a patch antenna for the 300 GHz operating frequency. The work is organized as follows: We begin with an introduction. In the second part, we will present a general study of THz, including the properties, characteristics, and typical applications of terahertz waves at around 300 GHz. Then, in a third section, we will study and design a patch antenna in the 300 GHz frequency band. In this section, we will present the definition of an antenna, the structure of an antenna, the operating principle, the characteristics of an antenna and the design of a printed antenna operating at 300 GHz. Finally, we conclude our work.

27.2 Properties, Characteristics and Typical Applications of Terahertz Waves

Terahertz wireless communications are based on conventional wireless communications, which range from microwave and millimeter waves to terahertz waves. They also have several features in common with laser microwave communications. Terahertz communications will not replace microwave and laser communications. However, terahertz communications offer unique advantages that most microwave and laser communications lack.

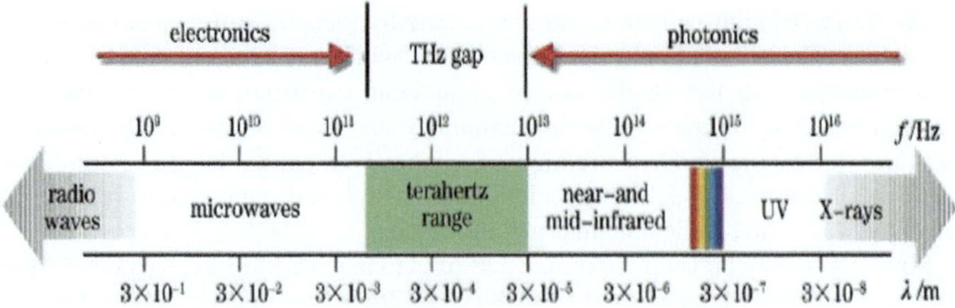

Fig. 27.1 Representation of the terahertz frequency spectrum in the radio band

27.2.1 Definition

The "terahertz" (THz, 1 THz = 1012 Hz) frequency range extends from around 100 GHz to 30 THz, i.e., at wavelengths between 0.01 and 3 mm. Historically known as far-infrared, it is now also known as T-rays. It lies in the electromagnetic spectrum between infrared (the optical domain) and microwaves (the radio domain).

The band below 100 GHz is generally defined as radioelectric, while frequencies above 30 THz are generally defined as infrared, but these boundaries are not standardized, as this is only a change of language or technology, not of nature.

27.2.2 General Description of the Frequency Range Above 275 GHz

For frequencies above the 275 GHz threshold, this essentially represents an important part of the "terahertz" domain. In other words, sub-millimeter waves refer to a frequency band between 0.1 and 10 THz. In other words, wavelengths between 0.003 and 3 mm, as shown in Fig. 27.1.

27.2.3 Characteristics of the Frequency Range Above 275 GHz

The specific characteristics of the frequency band above 275 GHz clearly distinguish it from all other radio frequency bands. The main characteristics of these bands are as follows:

a. **High permittivity**

Radio signals above 275 GHz easily penetrate dielectric materials and non-polar liquids. They are, therefore, ideal for imaging both materials and non-transparent objects and for non-destructive testing, safety checks, and general quality control.

Moreover, the wavelength of these signals is greater than dust as the particles of dust and grime in the air, so propagation loss due to dust or smoke is very low. They could, therefore, provide excellent opportunities for imaging in the presence of heavy smoke, for example, during fire-fighting operations or large quantities of windblown dust, as may be the case in the desert.

b. **Rapid attenuation in water**

Indeed, at frequencies above 275 GHz, many radio signals are strongly damped when present in water, and this can find many potential uses in the healthcare sector. For example, the content of water is very different from that of natural structures, so cancerous structures can be located by analyzing the water content of tissues.

c. **Defense and security**

- THz imaging is already used in airports for security checks to detect concealed weapons and is an alternative to microwave or X-ray systems.
- THz spectroscopy can detect explosive materials, particularly homemade explosives or illicit chemicals.
- The high attenuation of THz waves when propagating in free air due to water vapor in the atmosphere reduces their range to values in the kilometer range. Nevertheless, this short range can be an asset, making it more difficult for the enemy to intercept confidential communications between various armed units, particularly warships.

d. **High spatial resolution**

Frequencies in the 275 GHz + band offer better spatial resolution than the microwave band. This band offers better visual image quality due to its shorter wavelength than the microwave band.

e. **Short wavelength and good directivity**

Terahertz waves have a higher frequency than microwave waves, which could be exploited to carry more information per unit of time. Because they can transmit signals of shorter wavelengths and have good directivity, the terahertz wave range holds great promise for certain applications in wireless communications.

27.2.4 Main Applications of Terahertz Waves

Terahertz is used in radio astronomy, planetary radiometry, and meteorology like infrared. Potential applications are numerous in this field, and the first tests have been successfully carried out. In fact, this radiation's low-energy, non-ionizing aspect opens up a wide range of possibilities based on its particular spectroscopic properties, particularly in medicine and security.

Other potential applications for terahertz include high-speed telecommunications, wireless networks, radar, environmental monitoring, biomedical testing, characterization of materials and devices, detection of gases or pollutants, counter-terrorism, astronomical observation, and more.

Recently, terahertz waves have been used for airport security (for example, at Moscow's Domodedovo International Airport or in many American airports). Passengers enter a cylinder, and a moving part "scans" them with such waves. As these waves are not blocked by clothing (which contains no water or metals and is therefore transparent to this radiation), the passenger can be seen as if truly undressed. The advantage over conventional gates is that checks are much faster (no need to remove shoes or perform palpation). The potential drawback, which is not without controversy today, is that passenger privacy is compromised.

Terahertz technology can penetrate many types that are not conductors, namely fabrics, paper, cardboard, wood, plastic, foam, ceramics and even walls. In both layer-by-layer forms, they can uncover and detail the contents hidden within obstacles. As a result, they can image, recognize or analyze a whole range of terahertz fingerprints. Moreover, thanks to their specific characteristics, THz waves are particularly interesting in various applications. Indeed, these waves can be considered natural sources of energy. They are, therefore, not harmful and cannot harm the well-being of people or animals when they pass through objects.

The development of research into terahertz waves enables us to make even better use of their exceptional characteristics. Today, the terahertz wave range is mainly used for radio astronomy observations, but developing high-power terahertz radiation sources opens the way to many other applications. The main possible applications are as follows:

a. **Radio astronomy applications**

Numerous sources of scientific data are available on all aspects of celestial and interstellar radiation. Most of this data comes from frequencies above 275 GHz, with a much lower noise level. For this reason, fundamental research into these frequencies has been undertaken with a view to their application to astronomical science. In addition to the infrared radio telescope and the Hubble Space Telescope, there is a terahertz radio telescope to study the complex physics associated with stellar clouds within the galaxy, creating the world's largest far-infrared facility.

b. **Application to molecular detection**

Indeed, everything that makes up matter has a dynamic component. The structures of the molecules that make up objects can evolve rapidly since any movement produces radiation. This electromagnetic radiation has its specific vibrational frequency, known as a "fingerprint". The most common molecular "fingerprints" are in the infrared range, exceeding 275 GHz. Terahertz laser technology can detect radiation caused by light vibrations of molecules that cannot be detected using infrared rays.

c. **Application to security checks**

Since most molecular rotational levels of explosives and drugs are in the terahertz range, terahertz wave spectroscopy could be used to carry out security checks on the human body to detect the presence of explosives, drugs, biological macromolecules, weapons, and other contraband. Unlike existing techniques using X-rays and ultrasound imaging, terahertz wave spectroscopy and imaging can determine the shape of objects and their material properties by comparing measured terahertz radiation with that of known hazards stored in a library. Moreover, because of their very low energy, terahertz waves are unlikely to cause harmful ionization of biological tissues. They are, therefore, free from the shortcomings of X-rays, which are potentially harmful to the human body and whose use in metal detectors does not allow non-metallic objects to be detected. Terahertz technology, therefore, offers interesting application prospects in the field of security controls.

d. **Application to biomedicine**

Radiofrequency signals greater than 275 GHz are readily accepted and absorbed into polar molecular structures such as water molecules and oxygen, and these molecules exhibit varying degrees of absorption spectra. It is possible to diagnose early lesions due to skin cancer or other lesions affecting superficial tissues using these spectral lines and imaging techniques. In surgery, terahertz wave imaging systems are often used to monitor real-time cancer excision. Furthermore, we can use terahertz temporal and dimensional spectroscopy (THz-TDS) to study organic macromolecules whose vibrational and rotational energetics are in the terahertz range to guide drug production and medical research.

As part of non-ionizing radiation, THz waves are less harmful to living tissues than X-rays, making them particularly suitable for medical imaging, particularly for the detection of skin or breast cancer, for identification dental caries, and for assessing the severity of wounds and burns. In addition, all the more so since THz waves can pass through non-conductive dielectric media, such as medical dressings.

e. **Wireless communications applications**

The range above 275 GHz lies at the boundary between optics and electronics. It combines the characteristics of microwave and lightwave communications and has its own specific features. Firstly, the rapid development of communications makes it difficult to satisfy the need for high-speed, broadband wireless communications using conventional microwave communications. In contrast, the terahertz wave range, which enables high-speed data transmission and offers a large available bandwidth, could become the main medium for wireless communications. Furthermore, light waves suffer significant propagation attenuation in dust, walls, plastic, fabric, and other non-metallic, non-polar substances. Terahertz waves can penetrate these substances at low attenuation, making them highly penetrable in difficult environments. However, the range above 275 GHz also has its drawbacks, the most damaging of which is the ease with which it can be absorbed by polar molecules in the atmosphere, resulting in relatively high atmospheric attenuation, particularly in rainy weather. Given these characteristics, it was decided that this range would be used primarily for future interplanetary communications, short-range mobile broadband, ground communications, and communications in harsh environments, for example, in dry, smoky weather conditions or on battlefields.

f. **Radar applications**

Terahertz waves offer potential applications for radar, target recognition, precision guidance, and rockets. By taking advantage of the good directivity and energy concentration of terahertz waves, it is possible to develop high-resolution and low-angle tracking radars. The ability to image through materials can be used to detect objects hidden under cover or by smoke. The low level of attenuation due to fog and smoke should enable a navigation system capable of operating in all weather and enable guided landing of aircraft in foggy conditions. Last but not least, the terahertz band is wide compared to other wavebands. For example, it offers a wider range of frequencies than the band used today for stealth technologies so that an ultra-broadband radar using terahertz waves as a radiation source would be capable of capturing the image of stealth aircraft.

g. **Telecommunications**

The high frequency of terahertz waves makes it possible to envisage the development of very high-speed telecommunication systems, particularly in developing 5G and future 6G technologies. However, the absence of high-performance waveguides and the high absorption of ambient air limit this field of application to very short-distance transmissions. Work and demonstrations are underway to use a terahertz electromagnetic wave in a room inside a building to carry information at a rate of several tens of gigabits per second (Gbit/s). The idea, for example, is to use the signal brought into the room by an optical fiber to modulate a terahertz source that radiates throughout the room via an antenna. Each multimedia device in the room (TV, computer, tablet, etc.) receives this signal via a receiving antenna. As a result, these devices can be directly fed by a signal of several tens of gigabits per second. The attenuation of terahertz waves by the atmosphere and the room's walls is an advantage since the signal remains confined and cannot be detected from outside. While such high data rates are currently almost useless, the advent of future TV standards, such as super high definition (SHDTV), will require such performance (typically 2 Gbit/s per TV channel). In addition to the entertainment aspect of this technology, professional applications for super high definition are also envisaged, for example, in long-distance medical operations via video.

h. **Culture and history**

Since terahertz waves do not alter the objects they pass through, one of the key applications of terahertz technology is to inspect samples without destroying them. This makes it possible to inspect 100% of components or assembled products, whether packaged or not, to detect manufacturing defects and ensure quality control.

In many industries, such as manufacturing, pharmaceuticals, and food processing, terahertz technology detects production errors before they impact the production chain, thereby reducing the costs associated with non-quality and delivering better products to customers. Destructive testing could be avoided in a wide range of situations by taking advantage of the penetrating power of THz waves.

Terahertz technology has a wide range of applications in a multitude of industries. The harmlessness and layer-by-layer penetration of terahertz waves makes them a non-destructive inspection and non-contact detection tool of proven value in various applications.

27.2.5 Terahertz Wireless Communications

Research into developing high-speed wireless communications systems using the frequency band above 275 GHz is very active. Part of this research involves developing new types of wireless communications systems capable of interfacing with 40 and 100 Gbit/s Ethernet systems.

Communication links using terahertz technologies, with their high transmission capacity but high propagation loss, can be used as last-mile access links. Research and development organizations have conducted several trials using frequencies above 275 GHz.

When considering terahertz communications use cases, the following specific points should be borne in mind: Use of an ultra-wide frequency band, significant reduction in size for antennas as well as devices, significant increase in directivity as well as free-space degradation (the wavelength is more than five times shorter than in the 60 GHz band, and even though it is at least 25 times greater, the free-space degradation is compensated for by high antenna gains), development of manufacturing technologies including oscillators, power amplifiers, and beam steering antennas.

27.2.6 Properties

Terahertz radiation is highly penetrating. They are capable of penetrating many insulating materials (for example, fabrics, especially paper, clothing, wood, cardboard and plastics). Moreover, because they are both nonionizing and of low energetic (1 THz is equivalent to an energetic photon of 4.1 meV, which is considerably less than its thermal triggering energy at room temperature), they are relatively harmless.

The high water absorption at THz frequencies indicates a strong interaction between biological samples and THz waves. Indeed, these waves set polar water molecules into vibration/rotation and excite intermolecular low-energy bonds (hydrogen bonding) within water and proteins. THz spectroscopy has numerous applications in the biological field, including studying protein hydration and conformation, DNA hybridization, and detecting certain cancer cells (abnormally rich in water).

27.2.7 Terahertz Electromagnetic-Wave Detectors

In the electromagnetic spectrum, the terahertz (THz) range, also known as infrared (FIR), lies between infrared and microwaves. Typically, it extends from wavelengths of around 30 μm to 3 mm, in other words, from around 100 GHz to 10 THz in terms of frequency, photons with energies of between 0.4 and 40 meV. This spectral position is at the root of numerous difficulties in developing high-performance sources and detectors and, therefore, in carrying out studies at THz frequencies and in developing applications that are nevertheless promising. In the case of detectors, we are looking for compact, easy-to-use, low-cost devices with high sensitivity and dynamic range, and the ability to manufacture detector arrays for imaging.

We need to go back to the physical basis of electromagnetic radiation detection to understand the difficulties involved in designing and producing such high-performance detectors. This radiation, absorbed by the material illuminated in the detector, causes either a rise in the temperature of the material, transitions between the energy levels of the material's atoms/molecules, or the transfer of each photon's energy to the material's free charges. In the case of detector heating, THz beams are not very powerful in most application studies, and the temperature rise is minimal. In the case of the transfer of THz photon energy to the free or bound charges of the illuminated material, this energy is less than or of the order of the energy of the thermal quantum (24 meV) at room temperature. This prevents "optical" detectors, usually semiconductor-based, from operating efficiently in the THz range since they require an empty conduction band (or excited level) and a populated valence band (or fundamental level).

For this reason, detectors used in the visible and infrared range lose their effectiveness when THz frequencies are reached. On the microwave side, receivers are based on the principle that free electrons in the metal forming the receiving antenna are accelerated by the Coulomb force induced by the electromagnetic field coupled to the antenna. An electronic device reads the resulting electric current. These receiving systems lose their efficiency at THz frequencies due to the lower efficiency of electronic components and resistors and parasitic capacitances that limit their bandwidth.

Several terahertz radiation sources include free-electron lasers, nonlinear crystals pumped by short laser pulses (optical rectification), quantum-well photo emitters, and photoconductive antennas triggered by ultrashort laser pulses. There are also several terahertz radiation detectors, including bolometers, nonlinear crystals (electro-optical effect), quantum-well photodetectors, and photoconductive antennas. The references give a more exhaustive list of terahertz radiation sources and detectors.

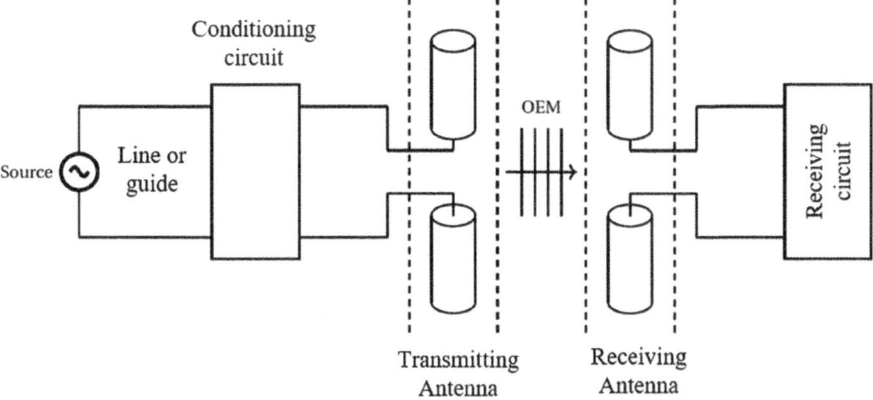

Fig. 27.2 Representation of a transmitting and receiving system

27.3 Printed Antenna for THz Frequencies

27.3.1 What is an Antenna?

The antenna is the part of a transmission chain that receives or transmits an electromagnetic signal. In most cases, an antenna is a metal-dielectric component, acting as the interface between guided and free-space propagation (see Fig. 27.2, for an example of a transmission and reception chain). Another way of putting it is that it enables impedance matching between the guided wave and the desired free-space propagation mode.

Following the work of MAXWELL [33], which gave rise to Maxwell's equations, Hertz first demonstrated electromagnetic wave propagation in 1886. A great deal of progress was made after the Second World War, but the design of antennas was based partly on empirical principles. It was the advent of numerical calculation that enabled real progress. Today, antennas come in all shapes and sizes and can be divided into 4 categories:

- wire antennas;
- radiating aperture antennas;
- Planar antennas (patch antennas, etc.).

Antenna arrays are a group of antennas acting as a single, more directive antenna than the antenna alone.

27.3.2 The Printed or Patch Antenna

Antennas can be classified into several categories, as follows:

- Wire antennas: dipole, loop, spiral.
- Aperture antennas: horn, slot, reflector antenna.
- Printed antennas: patch, printed dipole, spiral.

In this chapter, we focus on the third category, printed antennas. The concept of these antennas first appeared in the 1950s, but they were not put to use until the 1970s. Patch antennas are planar radiating elements. This concept has several advantages: low weight, simple manufacturing technology, and easy integration into electronic circuits occupying a reduced volume and for different surface types, but the main advantage lies in their low manufacturing cost.

a. **Microstrip antenna structure**

A microstrip antenna generally consists of a ground plane and one or more layers of substrate, which may have either equal or different permittivity ($\mathcal{E}_r$), on the surface of which there is a thin radiating element of any geometry (rectangular, square, circular, elliptical, slotted, or more elaborate shapes). Several antenna excitation techniques can produce radiation patterns with either linear or circular polarization. Figure 27.3 shows the simplest structure of a rectangular patch antenna.

Dielectric substrates generally have a low permittivity ($\mathcal{E}_r < 3$) to facilitate and promote radiation while avoiding the confinement of electromagnetic fields in the cavity between the radiating element and the ground plane [34, 35].

b. **Operating principle**

The antenna's behavior in its original configuration is controlled by the feeding technique used, and to better understand the operating principle of the microstrip antenna, it is necessary to understand electromagnetic fields, particularly the near field. When the transmission line is excited by an RF (radio frequency) source, an electromagnetic wave guided between the feed line and the ground plane will propagate to the radiating element. A charge distribution is then established above and below the radiating element. The patch–substrate–ground plane assembly shown

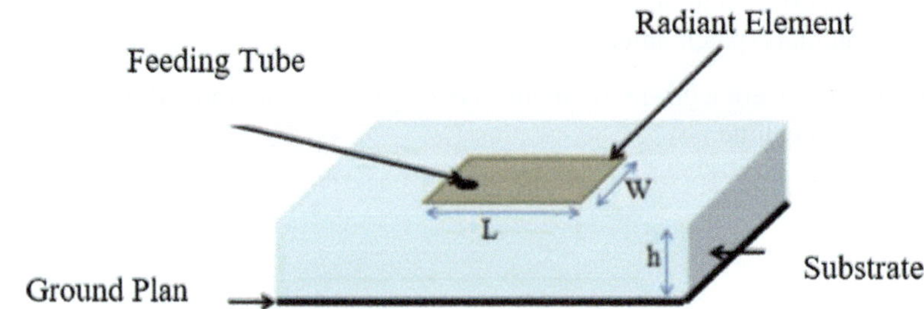

Fig. 27.3 Structure of a patch antenna

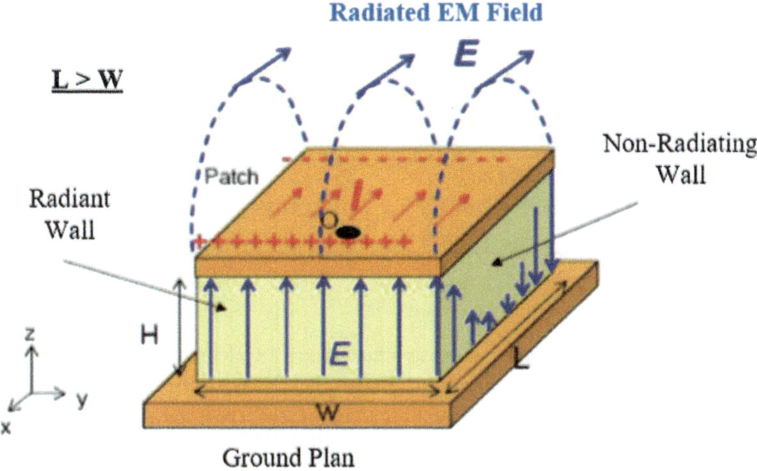

Fig. 27.4 Radiation from a rectangular patch antenna

in Fig. 27.4 can be likened to a closed cavity, bounded by electrical walls at the top by, the radiating element at the bottom by the ground plane, and a magnetic side wall. The antenna will resonate according to a set of modal frequencies called TMmn modes [36].

The field created between the patch's edges and the ground plane will overflow and help generate the radiated electromagnetic field. The field generated by the edges along the length "L" of the patch being maximum and in phase opposition will tend to add up and generate radiation inscribed in the YZ plane. These two edges are called "radiating edges".

c. **Analysis methods**

Two types of methods can be used to analyze printed antennas. Analytical methods are based on a physical approach to the phenomenon, where simplifying hypotheses lead to fairly simple equations, but in return, approximate results are obtained. These methods are generally based on the equivalent magnetic currents along the edges of the patch.

Full-wave numerical methods give more accurate results at the expense of losing the meaning of the physical phenomenon. They require the use of powerful numerical algorithms and a longer computing time than analytical methods. They are generally based on obtaining electrical current distributions on the patch and on the ground plane.

d. **Important antenna parameters**

i. **Far field and near field**

If we look at the field radiated by an antenna, we can distinguish three different zones depending on the distance from the antenna (see Fig. 27.5). These three zones are as follows and depend on the antenna's dimension D:

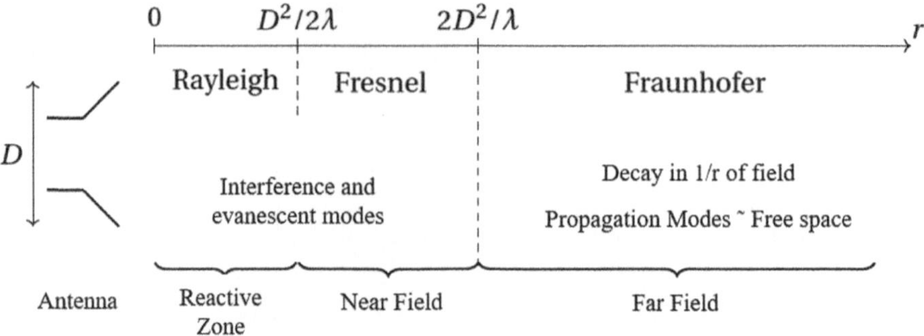

Fig. 27.5 Near-field and far-field positions as a function of wavelength and antenna dimensions *D*

- the Fraunhofer zone, at a distance greater than $2D^2/\lambda$, corresponding to the far field;
- the Fresnel zone, between $D^2/2\lambda$, and $2D^2/\lambda$, corresponding to the near field;
- the Rayleigh zone for distances less than $D^2/2\lambda$,, corresponding to the reactive zone.

The reactive zone or Rayleigh space corresponds to the field set-up by the sources. In this space, many modes, most of them evanescent, are superimposed. This is followed by the Fresnel zone, corresponding to the near field. In this zone, we mainly see interference and the disappearance of evanescent modes.

On the other hand, the far field is made up solely of propagative modes and corresponds to what interests us in antennas. Antennas are used for long-distance, far-field transmissions.

ii. **Directivity, gain, and radiation pattern**
a. **Directivity**

The directivity of the antenna in the horizontal plane is an important characteristic when choosing an antenna.

An equidirectional or omnidirectional antenna radiates equally in all directions in the horizontal plane.

A directional antenna has one or two lobes that are clearly larger than the others, known as main lobes. It also has secondary lobes, which we try to minimize. The narrower the main lobe, the more directive the antenna. If the radio station being received is not always in the same direction, it may be necessary to orientate the antenna by turning it with a motor.

Some satellite tracking antennas are steerable in azimuth (direction in the horizontal plane) and elevation (height above the horizon). Direction finding uses directional antennas to determine the direction of a transmitter. Radars use antennas with highly directional transmission and reception patterns.

b. **Gain**

The gain of an antenna compared to an isotropic antenna is what characterizes the main lobe. It is due to the fact that the energy is focused in one direction, just as the light energy from a candle can be concentrated by a converging mirror or lens. It is expressed in dBi (decibels relative to the isotropic antenna). For an antenna, the mirror can be a reflecting element (plane or parabolic screen), while a directing element (in a yagi antenna, for example) plays the role of the lens. Antenna measurements are carried out in free space or in an anechoic chamber.

A low-gain direction can be used to eliminate a troublesome signal (on reception), or to avoid radiating into a region where there may be interference from other transmitters.

The far field of an antenna is characterized by its decay in $1/r$, where r is the distance between the antenna and the point of view. The electric field can be written as:

$$\overrightarrow{E}(\vec{r}) = \vec{F}(\vec{u}_r)\frac{e^{-jkr}}{r}e^{jwt} \tag{27.1}$$

$$\overrightarrow{H}(\vec{r}) = \frac{1}{\eta}\vec{u}_r \times \vec{E}(\vec{r}) \tag{27.2}$$

The part $\vec{F}(\vec{u}_r)$ characterizes the antenna radiation, and η is the wave impedance (for propagation in air, $\eta = \eta_0$). We can also write the Poynting vector (in Wm^{-2}).

$$\overrightarrow{P}(\vec{r}) = \frac{1}{2\eta}\frac{\left\|\overrightarrow{F(\vec{u}_r)}\right\|^2}{r^2}\vec{u}_r \tag{27.3}$$

The amplitude $\left\|\vec{F}(\vec{u}_r)\right\|$ radiation pattern is called the magnitude, and the power radiation pattern is called the magnitude $\left\|\vec{F}(\vec{u}_r)\right\|^2$.

Another quantity of interest for antennas is directivity. It is defined as the ratio between the Poynting vector of the antenna divided by the Poynting vector of the isotropic antenna (see Fig. 27.6).

Wr is defined as the total power radiated by the antenna.

$$W_r = \int_{\theta=0}^{\pi}\int_{\phi=0}^{2\pi} p_i r^2 \sin(\theta) d\phi d\theta \tag{27.4}$$

The power density radiated by the isotropic antenna is then:

$$p_i = \frac{W_r}{4\pi r^2} \tag{27.5}$$

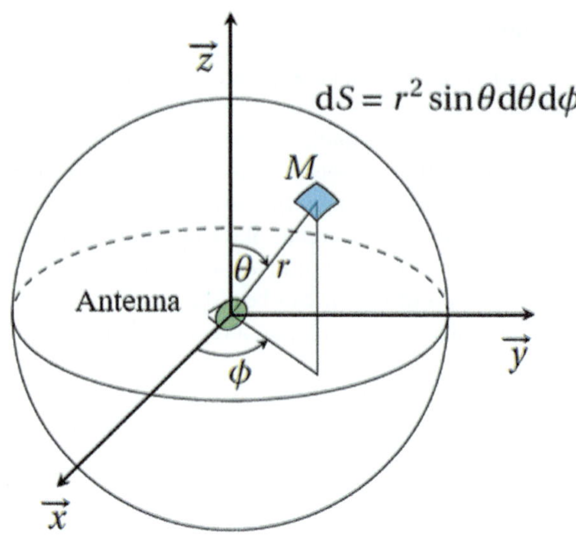

Fig. 27.6 Isotropic antenna and spherical coordinates

The directivity thus becomes:

$$D(\overrightarrow{u_r}) = \frac{\left\| \vec{P}(\vec{r}) \right\|}{W_r/4\pi r^2} = \frac{4\pi}{2\eta W_r} \left\| \overrightarrow{F}(\vec{u}_r) \right\|^2 \quad (27.6)$$

Finally, the gain $G(\vec{u}_r)$ of an antenna can be defined as the product of its directivity and its efficiency ξ:

$$G(\vec{u}_r) = \xi D(\vec{u}_r) \quad (27.7)$$

It is common practice to plot gain as a function of spherical coordinate angles, then normalize it with respect to the antenna's principal direction (the direction in which the antenna's gain is maximum). In practice, the antenna's power radiation pattern is measured and normalized by the power maximum. The maximum gain can then be obtained by comparison with an antenna of known gain or by integrating the measured pattern.

a. **Frequency bandwidth**

The bandwidth refers to a frequency interval within which an antenna's operation and specific characteristics meet specified standards. This frequency band can be considered the same way as those in which the antenna's characteristics are within an acceptable range of those of the center frequency. Generally speaking, antennas with a reflection coefficient of less than − 10 dB over the entire bandwidth are required in the field of high-frequency wireless communication.

The bandwidth characterizing the antenna can be defined as either absolute bandwidth (ABW) or fractional bandwidth (FBW). The factors of fH and fL represent successively the top and then the bottom of the antenna bandwidth, while the value of ABW represents rather the difference between these two values, and that of FBW, rather the difference between the center frequency and the frequency difference, as shown in Eqs. 27.8 and 27.9, respectively.

$$\mathrm{ABW} = f_H - f_L \tag{27.8}$$

$$\mathrm{FBW} = 2(f_H - F_L/f_H + f_L) \tag{27.9}$$

The bandwidth (BW) for broadband antennas (LB) can also be expressed as the ratio of the highest to the lowest frequency where antenna performance is acceptable, as given in Eq. 27.10.

$$f_H/F_L \tag{27.10}$$

b. **Efficiency**

The performance of an antenna depends essentially on the quality of the material, particularly concerning ohmic losses in the dielectric and the effects of reflections at the input terminals. In addition, reflection performance, or efficiency in impedance mismatch, is mainly related to parameter S_{11} (Γ). This $\mathbf{e_r}$ parameter is determined by the following:

$$e_r = 1 - |\Gamma|^2 \tag{27.11}$$

An antenna does not transmit all the power the feed supplies during operation. The radiation efficiency of an antenna eray is defined as the ratio between the power radiated by the antenna Pray and the power accepted by the antenna Pa. On the other hand, the total efficiency of an antenna etot is defined by the ratio between the total power radiated by the Pray antenna and the power supplied at its Pin input, which includes the power dissipated in the antenna and linked to its mismatch. Radiation efficiency can, therefore, be expressed as follows:

$$e_{\mathrm{rad}} = \frac{p_{\mathrm{rad}}}{p_{\mathrm{in}}} \tag{27.12}$$

The overall efficiency of an antenna is obtained from the combination of reflection and radiation efficiencies. This is essentially the result of both radiation and reflection. In practice, it is possible to find an overall efficiency in the 60–90% range, but many commercially available antenna structures generally do not exceed 50–60% due to the use of low-cost, lossy dielectric materials [37].

c. **Polarization**

Polarization is a property of light. Light is an electromagnetic wave traveling in a vacuum at speed $c = 299{,}792{,}458$ m/s. It comprises an orthogonal electric field (generally noted as E) and a magnetic field (noted as B). Maxwell's equations link the E and B fields: Knowing one is enough to know the other. So, for simplicity's sake—and this is also the convention chosen in polarimetry—we reason only in terms of the E field.

An electromagnetic plane wave is defined by its direction of propagation. The plane perpendicular to the direction of propagation is called the wave plane. In the wave plane, the E field (and the B field) evolves. At each instant, the E field has a different amplitude and direction in the wave plane. Put another way, if the wave propagates towards the observer, the latter will see the E field form different patterns in the wave plane as it evolves. This defines the wave's polarization.

Antenna polarization is characteristic of an antenna that reflects its adaptation to the orientation of the electric field of the received wave. In practice, polarization can be horizontal, vertical, or circular, as shown in Fig. 27.7 [38].

d. **Telecommunications equation or Friis formula**

An equation relates the power transmitted by one antenna to the power received by a second antenna at a distance r from the first (see illustration in Fig. 27.8). However, we assume that the receiving antenna is in the far field of the transmitting antenna and that the propagation medium is homogeneous, linear, isotropic, and reciprocal.

We are the power emitted by the transmitting antenna and $p_e(\vec{r})$ is the energy density emitted by the transmitting antenna. Similarly, Wr and $p_e(\vec{r})$ are the power and energy density the receiving antenna receives. Ge is the gain of the transmitting antenna, and G_r is the gain of the receiving antenna. We then have the following relationships:

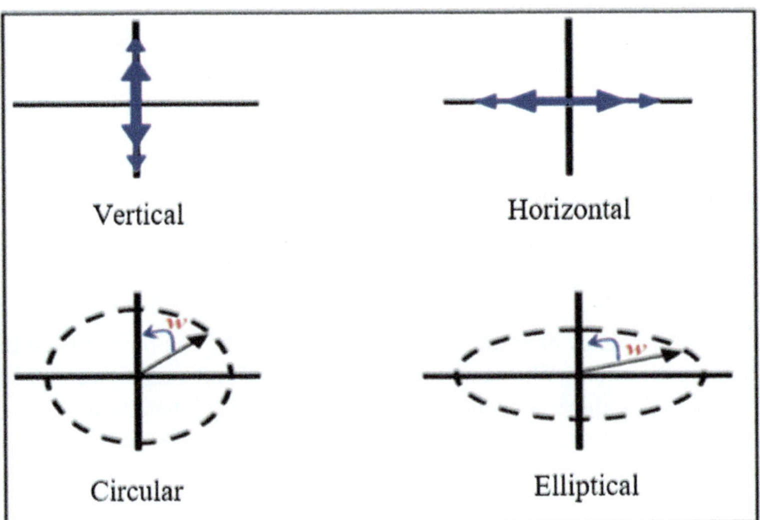

Fig. 27.7 Polarization of the electromagnetic wave

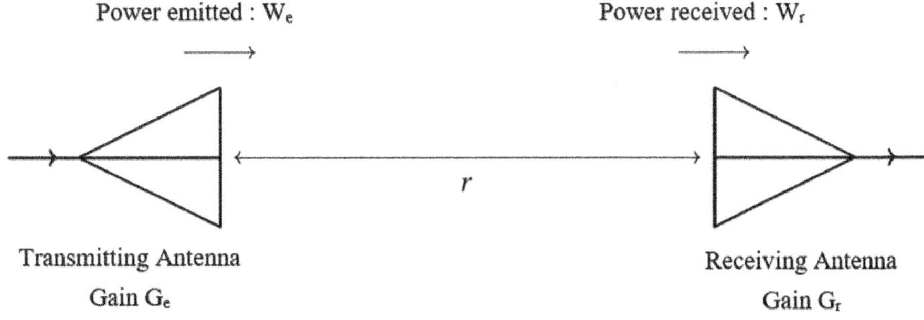

Fig. 27.8 Transmitting and receiving antenna geometry

$$p_r(\vec{r}) = p_e(\vec{r}) = p_i G_e = \frac{W_e}{4\pi r^2} G_e \tag{27.13}$$

The effective radiating surface Σ_r is defined as the ratio between the power received by the receiving antenna and the power density of the incident wave. $p_r(\vec{u}_r)$:

$$\Sigma_r = \frac{W_r}{p_r(\vec{u}_r)} = \frac{\lambda^2}{4\pi} G_r \tag{27.14}$$

Or:

$$p_r(\vec{u}_r) = \frac{W_r 4\pi}{G_r \lambda^2} \tag{27.15}$$

Equality between emitted and received energy densities links Eqs. (27.14) and (27.15) to give the telecommunications equation.

$$W_r = W_e G_e G_r \left(\frac{\lambda}{4\pi r} \right)^2 \tag{27.16}$$

The preceding relationship is better known as the Friis formula, which involves the quantities of the telecommunications equation in decibels:

$$w_r = w_e + G_{e\mathrm{dB}} + G_{r\mathrm{dB}} - A_{\mathrm{dB}}, \tag{27.17}$$

where W_r and W_e are in dB or dBm and $A_{\mathrm{dB}} = 20 \log(\frac{4\pi r}{\lambda})$ is the attenuation of the wave in free space due to diffraction.

27.3.3 Printed Antenna Design Operates at 300 GHz

General Method for Designing a Patch Antenna

a. **Cahier des charges**

A methodology has been established for the design of the rectangular patch antenna. This consists of:

- Choose a simulation software (HFSS);
- Calculate the geometrical parameters of the patch antenna to be designed at the desired frequency;
- Adapt and optimize the designed antenna, then study the influence of the antenna's geometrical parameters on its performance;
- Change the type of feed to see its impact on performance.

b. **Calculation of antenna geometrical parameters**

When designing microstrip patch antennas, it is necessary to take the following steps: Firstly, we must choose the precise substrate required for the patch. For this study, our preferred choice of substrate is an FR4-type substrate with permittivity $\varepsilon_r = 4.4$ and its height ($h = 20\ \mu\text{m}$). The fundamental characteristics of the microstrip antenna are influenced mainly by the height of the substrate, whose formula for maximum determination has been derived from the relationships described in [39, 40]. Next, the "patch" dimensions, i.e., its width and length, are determined. In a further step, the dimensions of the ground plane are determined, and finally, the power source is chosen. In this study, we selected a microstrip line power supply, which required the formulas described in [41, 42] at 300 GHz. We applied the formulas (27.18–27.27), which enabled us to calculate the parameters in Table 27.1, that is, the general dimensions of the proposed antenna. This antenna is illustrated in Fig. 27.9. In this work, we used the ground plane shape deformation technique to optimize the size and thus improve the bandwidth, which reaches a wide bandwidth of 24.12 GHz.

$$E_s \leq \frac{0.3c}{2\pi f_{\text{res}} \sqrt{\varepsilon_r}} \tag{27.18}$$

$$W_p = \frac{c}{2 f_{\text{res}}} \sqrt{\frac{2}{\varepsilon_r + 1}} = \frac{\lambda}{2} \sqrt{\frac{2}{\varepsilon_r + 1}} \tag{27.19}$$

$$\varepsilon_{\text{eff}} = \frac{\varepsilon_r + 1}{2} + \frac{\varepsilon_r + 1}{2} \left(1 + 12 \frac{E_s}{W_p}\right)^{\frac{-1}{2}} \tag{27.20}$$

$$L_{\text{eff}} = \frac{c}{2 f_{\text{res}}} \varepsilon_{\text{eff}}^{-\frac{1}{2}} \tag{27.21}$$

$$\Delta L_p = 0.412\frac{(\varepsilon_{\text{eff}} + 0.3)\left(\frac{W_p}{E_s} + 0.264\right)}{(\varepsilon_{\text{eff}} - 0.258)\left(\frac{W_p}{E_s} + 0.8\right)} \tag{27.22}$$

$$L_p = L_{\text{eff}} - 2\Delta L_p = \frac{c}{2 f_{\text{res}}\sqrt{\varepsilon_{\text{eff}}}} - 2\Delta L_p \tag{27.23}$$

$$W_g = W_p + 6E_s \text{ and } L_g = L_p + 6E_s \tag{27.24}$$

$$W_{\text{fed}} = \frac{2h}{\pi}\left[B - 1 - \ln(2B - 1) + \frac{\varepsilon_r - 1}{2\varepsilon_r}\left(\ln(B - 1) + 0.39 - \frac{0.61}{\varepsilon_r}\right)\right] \tag{27.25}$$

$$B = \frac{60\pi^2}{Z_C\sqrt{\varepsilon_r}} \tag{27.26}$$

$$L_{\text{fed}} = 3h \tag{27.27}$$

$$|S_{11}|^2 = \frac{|Z_e - Z_c|^2}{|Z_e + Z_c|^2} \tag{27.28}$$

E_s	generally corresponds to the thickness of the substrate.
C	indicates the speed of light such that
f_{res}	means the resonance frequency
ε_r	represents the dielectric constant
W_P	represents the patch width
L_p	measures patch length
ε_{eff}	corresponds to the effective dielectric constant
ΔL_p	corresponds to the patch length extension
L_g	corresponds to the length of the ground plane
W_g	corresponds to the ground plane width
W_{fed}	corresponds to the width of the feed line
L_{fed}	corresponds to feeder length
Z_c	50 Ω
return loss	S_{11}

27.4 Results of the Simulation and Discussion

a. **Simulation software HFSS**

HFSS (high-frequency structure simulator) is a commercial software package that calculates electromagnetic fields in the frequency domain, enabling us to analyze

Table 27.1 The parameters of the proposed antenna

Parameters	Values (μm)
L_g	286.4
W_g	380.22
h	20
W_p	304.29
L_p	230.22
L_{fed}	75
W_{fed}	10
Y_0	10
Y_1	5

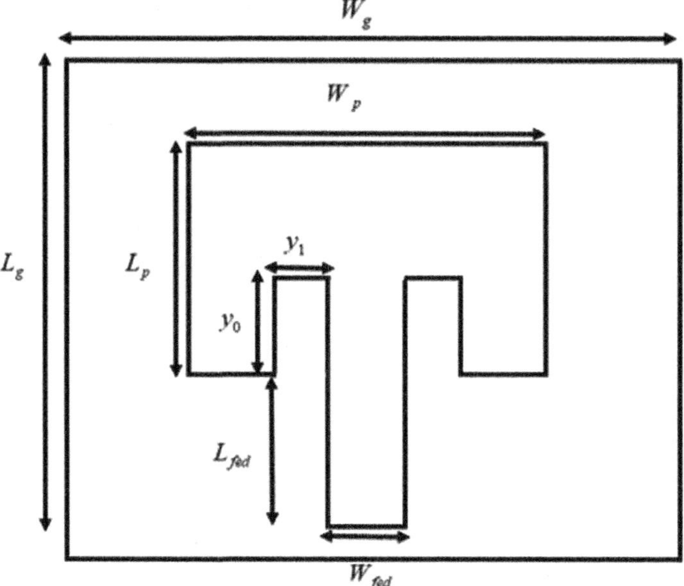

Fig. 27.9 Geometry of this proposed antenna

a structure's electromagnetic behavior. To analyze this behavior in detail, software provides post-processing interpretation tools. It performs electromagnetic modeling by solving Maxwell's equations using the finite element method. The latter is based on a geometric description of the structure in the form of a mesh. One of the strengths of this method is that it divides space into small homogeneous elements of practically any size [43].

An HFSS project is a folder containing one or more models called a design. Each model contains a geometric structure, its boundary conditions and the materials used, as well as electromagnetic field solutions and post-processing interpretations.

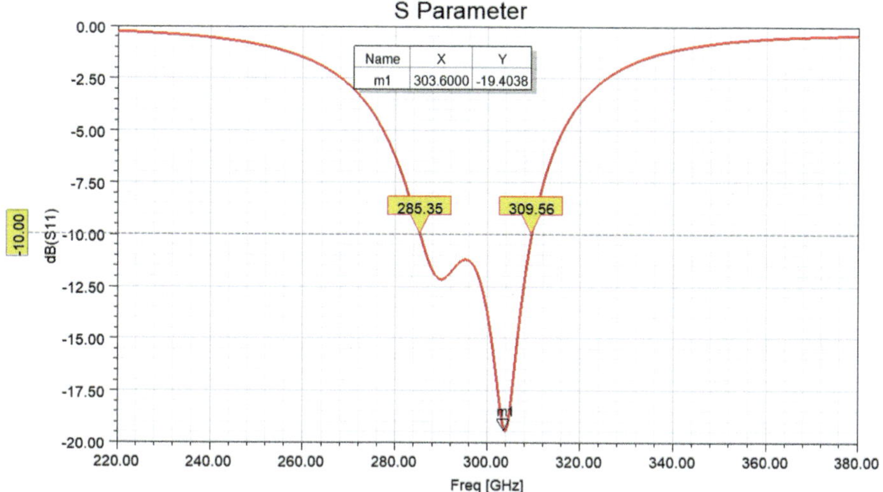

Fig. 27.10 Illustrated representation of the S_{11} curve

All these components can be accessed via the project manager window in the HFSS environment.

b. **Antenna design simulations**

Figure 27.10 shows the simulation curve for the reflection coefficient assigned to the proposed antenna. To check the antenna's compatibility with the load or feed, we use the reflection coefficient S11. This coefficient reaches − 19.4 dB, as shown in Fig. 27.10. From this curve, we also determined the bandwidth interval (from 285.35 to 309.56 GHz), which means the bandwidth is 24.21 GHz.

Figure 27.11 shows the standing wave ratio simulation curve assigned to the proposed antenna. This ratio and standing wave ratio express the quality of the antenna's adaptation to a coaxial or two-wire transmission line. Impedance matching is an electrical technique used to optimize the transfer of electrical power between a transmitter (source) and an electrical receiver (load) and to optimize the transmission of telecommunication signals. In the case of coaxial cable or coaxial line, this is a transmission line or asymmetrical link used at low or high frequencies, consisting of a cable with two conductors (central and outer), the outer conductor generally providing the shielding. Similarly, a two-wire line is a transmission line made up of two parallel wires separated by an insulator.

Figures 27.12 and 27.13 show the radiation characteristics, i.e., the gain and directivity assigned to the proposed antenna. This radiation pattern shows the lobes in three dimensions, in the horizontal or vertical plane, including the largest lobe. In this way, the antenna's radiation pattern can be used to evaluate its directivity.

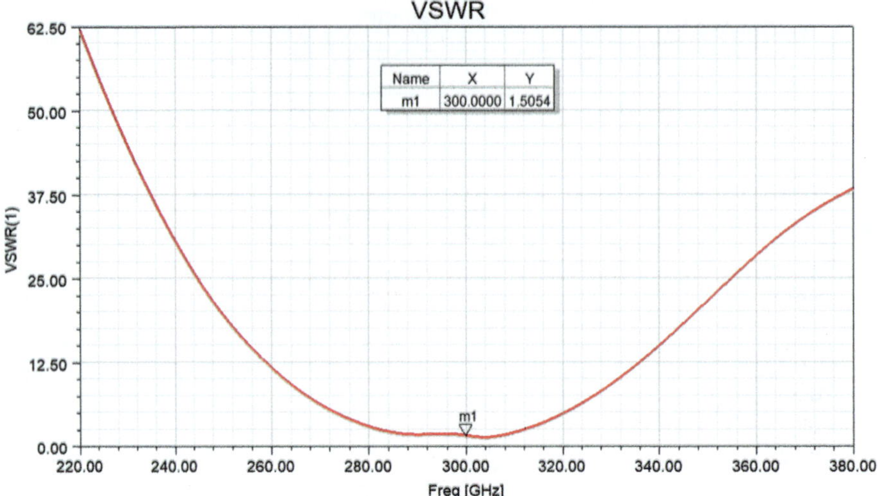

Fig. 27.11 Illustrated representation of the VSWR curve

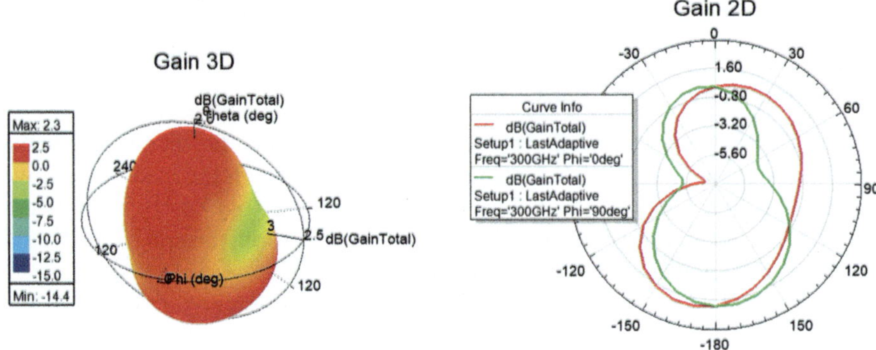

Fig. 27.12 Explicit representation of gain diagrams in 3D and 2D

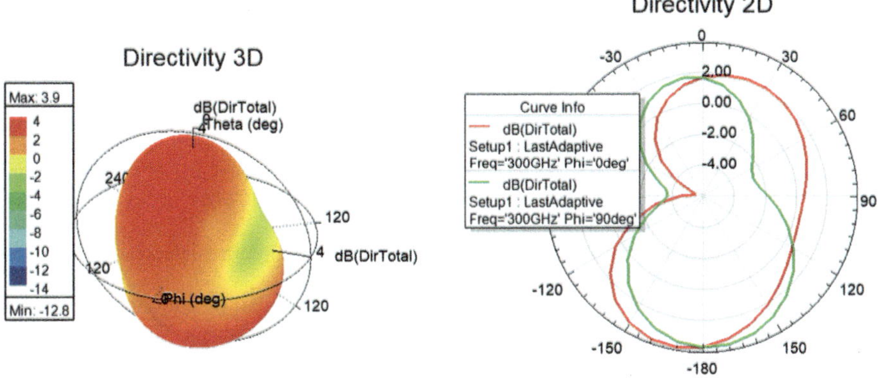

Fig. 27.13 Explicit representation of directivity diagrams in 3D and 2D

27.5 Conclusion

This chapter presents a detailed study of THz technology components. We have created a new component, called a "printed antenna". This antenna offers important features in a small, compact size, particularly in terms of bandwidth. It is applicable to the terahertz range. We will continue to improve its other characteristics, such as gain and directivity.

THz waves are a future field, with great potential for industrial applications and a fast-growing market. However, there are still several technological hurdles to overcome. Progress regarding device performance and manufacturing costs is still needed to enable large-scale industrialization.

We believe that the new possibilities offered by the THz spectrum will become a driving force for 6G technology. The demonstration could be an important step in exploring the feasibility of using the THz spectrum for 6G wireless communications. Exploiting the terahertz frequency bands could enable 6G to deliver multi-terabit-per-second data transfers, reaching up to 50 times faster than those promised by 5G, with 1–2 ms latencies.

References

1. Singh, A., Singh, S.: A trapezoidal microstrip patch antenna on photonic crystal substrate for high speed THz applications. Photon. Nanostruct. Fundam. Appl. **14**, 52–62 (2015)
2. Singhal, S.: Compact wideband antenna for 60 GHz millimeter wave applications. Microw. Rev. **27**(1), 23–27 (2021)
3. Es-saqy, A., et al.: A 5G mm-wave compact voltage-controlled oscillator in 0.25 μm pHEMT technology. Int. J. Electr. Comput. Eng. **11**(2), 1036–1042 (2021). https://doi.org/10.11591/ijece.v11i2.pp1036-1042
4. Vahdati, A., Parandin, F.: Antenna patch design using a photonic crystal substrate at a frequency of 1.6 THz. Wireless Pers. Commun. **109**(4), 2213–2219 (2019)

5. Dashti, M., Carey, J. D.: Graphene microstrip patch ultrawide band antennas for THz communications. Adv. Funct. Mater. **28**(11) (2018) (Art. no. 1705925)
6. Bie, S., Pu, S.: Array design of 300 GHz dual-band microstrip antenna based on dual-surfaced multiple split-ring resonators. Sensors **21**(14), 1–13 (2021)
7. Semkin, V.: Beam switching conformal antenna array for mm-wave communications. IEEE Antennas Wireless Propag. Lett. **15**, 28–31 (2015)
8. Babu, K.V., Das, S., Ali, S.S., et al.: Broadband sub-6 GHz flower-shaped MIMO antenna with high isolation using theory of characteristic mode analysis (TCMA) for 5G NR bands and WLAN applications. Int. J. Commun. Syst. **36**(6), e5442 (2023)
9. Lchhab, T., El Ghzaoui, M.: A circularly polarized wideband high gain antenna for THz wireless applications. Opt. Quant. Electron. **54**(12), 787 (2022)
10. Ullah, S., Ruan, C., Sadiq, M.S., Haq, T.U., He, W.: Microstrip system on-chip circular polarized (CP) slotted antenna for THz communication application J. Electromagn. Waves Appl. **34**(8), 1029–1038 (2020)
11. El Alami, A., Ghazaoui, Y., Das, S., Bennani, S.D., El Ghzaoui, M.: Design and simulation of RFID array antenna 2x1 for detection system of objects or living things in motion. Proc. Comput. Sci. **151**, 1010–1015 (2019)
12. Yan, J., Wang, H., Yin, J., Yu, C., Hong, W.: Planar series-fed antenna array for 77 GHz automotive radar. In: Proceedings of 6th Asia–Pacific Conference on Antennas Propagation (APCAP), pp. 1–3 (2017)
13. Es-saqy, A., et al.: Very low phase noise voltage controlled oscillator for 5G mm-wave communication systems. In: 1st International Conference on Innovative Research in Applied Science, Engineering and Technology. IRASET (2020). https://doi.org/10.1109/IRASET48871.2020.9092005
14. Mamane, A., Fattah, M., el Ghazi, M., Balboul, Y., el Bekkali, M., Mazer, S.: The impact of scheduling algorithms for real-time traffic in the 5G femto-cells network. In: 9th International Symposium on Signal, Image, Video and Communications, pp. 147–1512. ISIVC (2018). https://doi.org/10.1109/ISIVC.2018.8709175
15. Mohanty, A., Sahu, S.: A micro 4-port THz MIMO antenna for nano communication networks. Photon. Nanostruct. Fundam. Appl. **53**, 1–14 (2023)
16. Salah-Eddine, D., Imane, H., Mohammed, F., Younes, B., Said, M., Moulhime, E.: Design of a microstrip antenna two-slot for fifth generation applications operating at 27.5GHz. In: Springer International Conference on Digital Technologies and Applications(ICDTA), Lecture Notes in Networks and Systems, Fes, Morocco, vol. 211, pp. 1081–1089 (2021)
17. Abdellaoui, M., Fattah, M.: Characterization of ultra wide band indoor propagation. In: 7th Mediterranean Congress of Telecommunications (CMT). IEEE (2019)
18. Muthukrishnan, K., Kamruzzaman, M.M., Lavadiya, S., Sorathiya, V.: Superlative split ring resonator shaped ultrawideband and high gain 1×2 MIMO antenna array for terahertz communication. Int. J. Electron. Commun. **33**, 1–18 (2023)
19. Abdelhafid, E., Maryam, A., Mohammed, F., Said, M., Moulhime, E., Catherine, A.: High rejection self-oscillating up-conversion mixer for fifth-generation communicationsi. Int. J. Electr. Comput. Eng. **13**(5), 4979 (2023)
20. Anam, Z., Rizwan, Q., Muhammad, S., Ferhat, S., Mehmood, N., Haris, A.K., Muhammad, U.: Current advances in imaging spectroscopy and its state-of-the-art applications. Exp. Syst. Appl. **238**(Part E), 122172 (2024) (Elsevier)
21. Bi, Q.: The proximity radio access network for 5G and 6G. IEEE Commun. Mag. **60**(1), 67–73 (2022)
22. Rashmi, P., Leeladhar, M.: THz antennas design, developments, challenges, and applications: a review. Int. J. Commun. Syst. 5474 (2023). https://doi.org/10.1002/dac.5474

23. Vasu Babu, K., Das, S., Varshney, G., et al.: A micro-scaled graphene-based tree-shaped wide-band printed MIMO antenna for terahertz applications. J. Comput. Electron **21**, 289–303 (2022). https://doi.org/10.1007/s10825-021-01831-3
24. Didi, S.E., Halkhams, I., Mohammed, F., Balboul, Y., Mazer, S., Bekkali, M.E.L.: Study and design of the terahertz antenna array. In: Das, S., Nella, A., Patel, S.K. (eds.) Terahertz Devices, Circuits and Systems. Springer, Singapore. https://doi.org/10.1007/978-981-19-4105-4_13
25. Daghouj, D., Fattah, M., Mazer, S., Balboul, Y., El Bekkali, M.: UWB waveform for automotive short range radar. Int. J. Eng. Appl. **8**(4), 158–164 (2020). https://doi.org/10.15866/irea.v8i4.18997
26. Sarabandi, K., Jam, A., Vahidpour, M., East, J.: A novel frequency beam-steering antenna array for submillimeter-wave applications. IEEE Trans. Terahertz Sci. Technol. **8**, 654–665 (2018)
27. Ranzani, L., Kuester, D.K., Vanhille, J., Boryssenko, A., Grossman, E., Popović, Z.: G-band micro-fabricated frequency-steered arrays with 2°GHz beam steering. IEEE Trans. Terahertz Sci. Technol. 3, 566–573 (2013)
28. Rashad, H.M.: Terahertz microstrip patch antennas for the surveillance applications. Kurdistan J. Appl. Res. **5**(1), 1–13 (2020)
29. Didi, S.E., et al.: Study and design of the microstrip patch antenna operating at 120 GHz. In: El Ghzaoui, M., Das, S., Lenka, T.R., Biswas, A. (eds.) Terahertz Wireless Communication Components and System Technologies. Springer, Singapore (2022). https://doi.org/10.1007/978-981-16-9182-9_12
30. Aqlan, B., Himdi, M., Vettikalladi, H., Le-Coq, L.: A 300-GHz lowcost high-gain fully metallic Fabry-Pérot cavity antenna for 6G terahertz wireless communications. Sci. Rep. **11**(1), 7703 (2021)
31. Anveshkumar, N., Gandhi, A.S.: Lumped equivalent models of narrowband antennas and isolation enhancement in a three antennas system. Radioengineering **27**(3), 646–653 (2018)
32. Nella, A., Gandhi, A.: Typical approach of modeling lumped equivalents for narrowband antennas. Int. J. Microw. Opt. Technol. **14**(5), 298–305 (2019)
33. Das, S., Mitra, D., Chaudhuri, S.R.B.: Fractal loaded planar super wide band four element MIMO antenna for THz applications. Nano Commun. Netw. **30** (2021) (Art. no. 100374)
34. Salah-Eddine, D., Imane, H., Mohammed, F., Younes, B., Said, M., Moulhime, E.L.: Study and design of a 5G millimeter band patch antenna with a resonant frequency of 60 GHz. J. Nano Electron. Phys. **15**(2), 02015-1–02015-6 (2023)
35. El Krouk, A., Es-Saqy, A., Fattah, M., Mazer, S., El Bekkali, M., Mehdi, M.: Gilbert cell down-conversion mixer for THz wireless communication. In: Farhaoui, Y., Rocha, A., Brahmia, Z., Bhushab, B. (eds.) Artificial Intelligence and Smart Environment. ICAISE 2022. Lecture Notes in Networks and Systems, vol. 635. Springer, Cham (2023)
36. Ouadiaa, B., Abdelati, R.: Etude paramétrique des antennes patchs rectangulaire alimentées par proximité. Mediterranean Telecommun. J. **1**, 2–3 (2016)
37. Kushwaha, R.K., Karuppanan, P., Malviya, L.: Design and analysis of novel microstrip patch antenna on photonic crystal in THz. Phys. B Condens. Matter **545**, 107–112 (2018)
38. Abdelhafid, E. et al.: Simple inter-stage impedance matching technique for 5G mm-wave systems. In: Farhaoui, Y., Rocha, A., Brahmia, Z., Bhushan, B. (eds.) Artificial Intelligence and Smart Environment. ICAISE 2022. Lecture Notes in Networks and Systems, vol. 635. Springer, Cham (2023)
39. Salah-Eddine, D., Imane, H., Abdelhafid, E., Mohammed, F., Younes, B., Said, M., Moulhime, E.L.: New microstrip patch antenna array design at 28 GHz millimeter-wave for fifth-generation application. IJECE **13**(4), 4184–4193 (2023)
40. Didi, S.-E., Halkhams, I., Fattah, M., Balboul, Y., Mazer, S., El Bekkali, M.: Study and design of printed rectangular microstrip antenna arrays at an operating frequency of 27.5 GHz for 5G applications. J. Nano Electron. Phys. **13**(6) (2021). https://doi.org/10.21272/jnep.13(6).06035

41. Didi, S.-E., Halkhams, I., Fattah, M., Younes, B., Mazer, S., El Bekkali, M.: Design of a microstrip antenna patch with a rectangular slot for 5G applications operating at 28 GHz. J. Telecommun. Comput. Electron. Control **20**(3), 527–536 (2022). https://doi.org/10.12928/TELKOMNIKA.v20i3.23159
42. Hasan, S.O., Ezzulddin, S.K., Hammd, O.S., Mahmud, R.H.: Design and performance analysis of rectangular microstrip patch antennas using different feeding techniques for 5G applications. Int. J. Electr. Comput. Eng. Syst. **14**(8), 833–841 (2023)
43. Didi, S.-E., Halkhams, I., Fattah, M., Younes, B., Mazer, S., El Bekkali, M.: Study and design of a 28/38 GHz bi-band MIMO antenna array element for 5G. In: Farhaoui, Y., Rocha, A., Brahmia, Z., Bhushab, B. (eds.) Artificial Intelligence and Smart Environment. ICAISE 2022. Lecture Notes in Networks and Systems, vol. 635. Springer, Cham. https://doi.org/10.1007/978-3-031-26254-8_107

Part III
Advances in Materials for Optical Spectrum Applications

Chapter 28
Graphene-Based Nanomaterials for Photosensitive Spectrum Applications: An Inclusive Review

Hari Shankar Biswas

28.1 Introduction

Graphene is one of the most remarkable materials among the newly discovered and produced nanostructured carbon species. The word "graphene" was first devised in 1987, to define single layer honeycomb carbon atoms within a larger compound. Initially, the concept of graphene dates back to 1947, but it was widely believed that its 2D structure was thermodynamically and physically unstable, making it merely a theoretical material [1–5].

However, in 2004, Andre Geim and his team developed the field by successfully creating monolayer fragments of graphene. They established innovative techniques to separate individual layers from graphite crystals. This innovative technique, known as the "tape scotch technique" or micro-mechanical cleavage, involved using tape to gradually peel the layers until a single, uniformly distributed sheet remained. This sheet was then transferred onto a silica wafer (100) and examined under a microscope to determine its thickness. This discovery unlocked new avenues for the study of monolayers of graphene and various types of 2D crystal structures [6].

The success of Geim's team in isolating graphene disproved the previous belief that 2D materials could not exist due to the assumption that a single atomic layer would naturally roll or fold due to its lowest potential energy. Despite the challenges, imperfections such as ripples were observed on the monolayer of graphene, which prevented it from transforming into different carbonaceous structures or rolling into nanotubes [1].

Graphene, a remarkable material, may be functionalized with diverse useful functional groups to its surface. Hydrophilic assemblies enable the synthesis of nanocomposites with other materials. Graphene's exceptional properties like high surface area,

H. S. Biswas (✉)
Department of Chemistry, Surendranath College, Kolkata, West Bengal 700009, India
e-mail: harishankarb7@gmail.com

M. El Ghzaoui et al. (eds.), *Next Generation Wireless Communication*, Signals and Communication Technology, https://doi.org/10.1007/978-3-031-56144-3_28

biocompatibility, electrical conductivity, and chemical tolerance, make it an excellent matrix for nanocomposites [7]. The pioneering work of R. S. Ruoff's research group at Northwestern University in 2006 marked the beginning of graphene-based nanocomposites. Graphene sheets containing inorganic nanomaterials significantly enhance the electrical conductance of the resulting nanocomposite, leading to the formation of new types of materials.

Various synthetic methods, including in situ formation, ex situ hybridization, chemical reduction, electrochemical deposition, solvothermal/hydrothermal process, and sonochemical methods, can be employed to synthesize these nanocomposite [8]. The incorporation of different functional constituents allows for harnessing the essential various properties of graphene and its derivatives. Over time, numerous carbon and graphene-based nanocomposites have been successfully synthesized using different inorganic nanomaterials like gold, silver, tin, copper, cobalt, various metal oxides like Fe_3O_4, ZnO, SnO_2, TiO_4, MnO_2, NiO organic crystal, metal organic frameworks (MOFs), polymers, biomaterials, and carbon nanotubes (CNTs).

These nanocomposites find diverse applications in energy storage, solar cells, supercapacitors, pseudocapacitors, lithium-ion batteries, memory devices, sensing platforms, Raman enhancement, molecular imaging, catalysis, and drug delivery. The current advances in graphene-based nanocomposite materials can be summarized in several review articles, highlighting their significance and potential in various fields [9].

This breakthrough in graphene research covered the way for the advances of graphene-based nanomaterials, where graphene is integrated into various materials to enhance their properties. Graphene-based nanocomposites have since become a substantial area of research and development, allowing scientists and engineers to explore the vast potential of this unique material in practical applications. Extraordinary and diverse properties exhibited by these materials, spanning from the visible to the terahertz regions, have covered the way for innovative applications in various optical devices. Throughout this systematic review, we investigated the fundamental principles, synthesis methods, and unique optical characteristics of graphene-based materials [10].

28.1.1 Background of the Study

The discovery and development of new materials have historically catalyzed significant advancements across various scientific and technical fields. These materials often become the foundation for revolutionary technologies, impacting society by enhancing existing applications and even enabling entirely novel solutions to previously unsolvable problems. Among these revolutionary materials is graphene, it is a single layer of carbon sheet organized as a hexagonal honeycomb lattice structure, constituting a two-dimensional wonder material with extraordinary electrical, thermal, mechanical, and optical properties [11]. Graphene's emergence has marked a transformative chapter in the realm of optics, showcasing its potential to revolutionize the way we harness and manipulate light. Graphene's unique combination of mechanical strength, electronic conductivity, and optical properties has encouraged intensive research across the optical spectrum. From visible light to infrared and even terahertz regions, graphene has become a focal point of investigation for scientists and engineers seeking innovative solutions to optical challenges [12]. When graphene is ingeniously manipulated, altered, or integrated with other materials at the nanoscale, it gives rise to a diverse family of materials collectively known as graphene-based nanomaterials. These materials inherit graphene's exceptional properties while exhibiting additional characteristics, making them highly versatile and applicable across a wide range of optical technologies.

In this systematic review, our exploration investigates the latest strides and achievements in harnessing graphene-based nanomaterials for a multitude of optical spectrum applications. By encompassing the visible, infrared, and terahertz regions, this review encapsulates an inclusive understanding of the field. We navigate through the fundamental principles underpinning graphene-based materials, exploring their synthesis methods and elucidating their unique optical properties. A specific focus is placed on their applications in photodetectors and photovoltaics, shedding light on their potential to revolutionize these crucial technologies. By illuminating the promising future that graphene-based nanomaterials hold for advancing optical technology, this review aims to not only present the current state of the field but also to inspire further research and innovation. However, amidst the promises lie challenges. The transition from laboratory-scale experiments to scalable, real-world applications requires overcoming hurdles such as stability, scalability, and integration. As researchers push the boundaries of what is possible, addressing these challenges becomes topmost. By critically examining the obstacles and envisioning potential solutions, this review not only highlights the current achievements but also covers the way for the future of graphene-based nanomaterials in the field of optics.

28.1.2 Structure of Graphene and Graphene-Based Nanomaterial

Graphene is a 2D structure single-layer honeycomb structure hexagonal lattice, resembling a honeycomb pattern [13]. Each carbon atom within the lattice forms strong covalent bonds with three neighboring carbon atoms, resulting in a highly stable and exceptionally strong material. Graphene's unique properties stem from its hexagonal honeycomb fashion and the presence of loosely bound electrons that create an "electron sea." A graphene nanocomposite refers to a material where graphene is combined with other substances to enhance or modify its properties [14]. The specific structure of a graphene nanocomposite can vary widely, as it depends on the additional materials incorporated and the intended application. The graphene component provides exceptional strength, electrical conductivity, and thermal conductivity

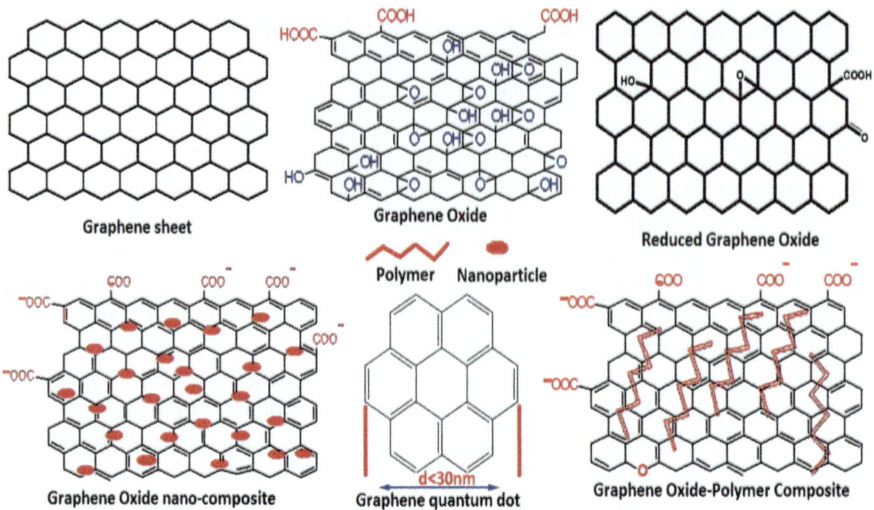

Scheme 28.1 Schematic drawing of the graphene sheet, graphene oxide, graphene oxide nanocomposite, reduced graphene, graphene quantum dot, and graphene oxide polymer composite

to the composite material, making it valuable in various industries and research fields. Scheme 28.1 described the various structures of graphene and graphene-based materials.

28.2 Graphene Synthesis Technique

Various methods can be employed to synthesize graphene layer on different substrates, with the preparation method playing a crucial role in determining the resulting purity, size, transport properties, and crystallinity of the graphene product. In the initial stages, diverse techniques were employed for producing thin graphitic films. Thin layers of graphite were created on a transition metal surface through the precipitation method in the 1970s [15, 16]. Subsequently, in 1975, multi-layer graphite was synthesized on a single-crystal platinum surface using the chemical decomposition method. Despite these advancements, the term "graphene" did not gain immediate recognition in scientific circles, possibly due to various factors [17]. From the 1980s onward, there has been explosive growth in surface characterization techniques, particularly at the micro- and nanoscale. Presently, highly advanced techniques are available for the preparation of nanostructured graphene [18]. Some useful methods include electron beam lithography, chemical synthesis, electrochemical preparation, reduction of graphene oxide, the catalytic transformation of fullerene, microwave-assisted hydrothermal, (MAH), ultrasonic exfoliation, hydrothermal methods, and soft-template method, etc.

28.2.1 Exfoliation Process

In 2004, a groundbreaking development occurred in the field of graphene research for the first time mechanically exfoliated from graphite. Graphene synthesized through this exfoliation method exhibits exceptionally high electron mobility and possesses the lowest number of defects among all graphene synthesis techniques. This method is not only cost-effective but also plays a fundamental role in sparking unpredictable interest and enthusiasm in the scientific community regarding graphene. The graphene flakes obtained through mechanical exfoliation have proven to be invaluable in understanding its fundamental structural characteristics. The process of preparing graphene through mechanical exfoliation involves several techniques such as shearing, the use of adhesive tapes, the wedge-based technique, graphite oxidation–reduction, electrochemical synthesis, solvent-aided methods, and application of immiscible liquids and molten salts. These techniques collectively contribute to the successful exfoliation of graphene, enabling researchers to explore its exclusive properties with potential applications in different fields [19].

28.2.2 Graphene by Adhesive Tape Method

Andre Geim and his collaborators pioneered in introduction of the adhesive tape method for the isolation of graphene layers from graphite sheets [20]. The process typically involves multiple exfoliation steps to obtain a monolayer or ultra-pure graphene sheet, with each step resulting in a reduction of layers until only a single layer remains. Following the exfoliation process, these graphene flakes are frequently deposited onto a silicon plate or another suitable surface. Crystallites, if their size surpasses 1 mm, can be observed with the naked eye [21].

28.2.3 Graphene-Oxide Reduction

Production of monolayer graphene sheet by reduction of graphene oxide was documented by P. Boehm in 1962 [22, 23]. The rapid heating and exploitation of graphite oxide resulted in the creation of a dispersed carbon powder containing a lower percentage of graphene or rGO. Single-layer sheets of graphite oxide, when reduced, can yield graphene sheets, a process often achieved through reduction using hydrazine in argon/hydrogen. Over time, this method has been refined to synthesize rGO with a coherent carbon framework, allowing the entire removal of functional groups. Notably, the charge carrier mobility of this rGO has been measured at over 1000 cm (393.70 inches)/Vs [24]. The characterization of rGO has been carried out using various spectroscopic techniques [25, 26].

28.2.4 Shearing

In 2004, an innovative method was developed, enabling the creation of graphene-infused, unoxidized liquids and defects free from graphite powder. This process involved high-shear mixers generating local shear rates exceeding 10×10^4. Remarkably, this technique wasn't limited to graphene; it extended to diverse two-dimensional materials like BN, MoS_2, and various layered crystals. This innovation marked a significant advancement, opening avenues for manipulating a range of two-dimensional substances, and revolutionizing materials science and technological applications [27–30].

28.2.5 Graphene by Solvent-Aided Process

Graphite powder was dispersed in a suitable aqueous medium through sonication, forming graphene sheets. Graphene is separated using a centrifuge from graphite [31]. Initially, the prepared graphene concentration was approximately 0.01 mg/ml in N-methyl pyrrolidone (NMP), later improved to 2.1 mg/ml [32]. By using an appropriate ionic solution as a dispersing agent in an aqueous medium, a concentration of 5.33 mg/ml was achieved [33]. However, the method's yield remained low due to graphene sheets restacking owing to van der Waals forces. This restocking hindrance was overcome using surfactant to the solvent before sonication. The surfactant, adsorbed to the graphene, prevented restacking, resulting in a higher graphene concentration. Nonetheless, the careful removal of the surfactant is necessary in this technique.

28.2.6 Graphene by Using Immiscible Liquids

Macro-scale graphene sheets can be produced by a technique involving the use of two immiscible liquids, such as water and heptane, which is employed for graphite sonication. During this process, graphene sheets are adsorbed at the high-energy interface between heptane and water, effectively preventing restacking. Subsequently, the solvents can be evaporated, leaving behind graphene films. These films have been found to be approximately 95% transparent and conductive, as indicated in reference [34].

28.2.7 Electrochemical Syntheses

Graphene synthesis was achieved by the reduction of graphite particles in the molten salts, and various carbon-based nanostructures, as mentioned in reference [35]. In the presence of molten lithium chloride, H^+ ions discharge on the cathodically polarized graphite, leading to the intercalation and peeling of graphene films. This process results in the formation of single-crystalline graphene nanosheets with a lateral size of up to 100 μm, exhibiting a high grade of crystallinity with remarkable thermal stability, as highlighted in reference [36].

28.2.8 Graphene by Hydrothermal Self-Assemble Process

Graphene sheets can be produced using sugars such as fructose, and glucose through a substrate-free hydrothermal self-assembly method known as the "Tang-Lau Method." [37] This eco-friendly process enables precise control of thickness, ranging from single layers to multilayers. It stands out for its simplicity and environmental friendliness, making it a preferable choice over conventional methods, as stated in reference [38].

28.2.9 Epitaxy

Epitaxy involves depositing a crystalline overlayer onto a crystalline substrate, where graphene layers attach to surfaces through van der Waals interactions. This preserves graphene's 2-D structure and electronic band properties [39, 40]. An excellent illustration is epitaxial graphene on silicon carbide, where silicon carbide reduces graphite to graphene at high temperatures and low pressures [41]. A variety of epitaxial methods are utilized in diverse applications. These techniques encompass the use of metal substrates, sodium ethoxide pyrolysis, roll to roll processes, cold wall epitaxy, chemical vapor deposition (CVD), nanotube slicing, intercalation, carbon dioxide reduction, spin coating, supersonic spray, and laser-based methods. Each of these approaches contributes to the development and refinement of epitaxial growth processes, enabling a broad range of applications across different industries [42–59].

28.2.10 Microwave-Assisted Reduction

In 2012, a groundbreaking one-step synthesis of graphene was achieved through a microwave-assisted method [60]. This approach yielded graphene with minimal oxygen content, eliminating the need for additional treatment and omitting the use of the KMnO4 in the reaction mixture. The creation of the defect was controllable by adjusting the microwave duration, akin to an improved version of Hummer's method. Notably, this modified process utilized microwave heating instead of conventional methods.

28.2.11 Chemical Vapor Deposition

A crucial technique is the chemical vapor deposition for producing top-quality solid materials by exposing a wafer to gaseous compounds. This method is using in the semiconductor industries to create graphene thin films and various nanomaterials. It is a proven process for depositing graphene on an industrial scale, producing thin films with substantial surface areas. The process involves decomposing gaseous compounds on a solid surface, although it may also generate volatile by-products that need to be removed [61, 62].

28.3 Graphite Powder to Graphene Oxide Journey

The transformation from humble graphite powder to cutting-edge graphene oxide represents a remarkable journey at the forefront of modern materials science. Graphite, a naturally occurring form of carbon, has been used for centuries in various applications, from writing instruments to lubricants. However, the discovery of graphene, a single layer of carbon atoms arranged in a hexagonal lattice, ushered in a new era of scientific exploration. Graphene's exceptional mechanical, electrical, and thermal properties captured the imagination of researchers worldwide. To harness its potential, scientists developed innovative methods to isolate and manipulate graphene oxide, a derivative of graphene that exhibits extraordinary properties of its own. This journey involves intricate processes and intricate techniques, revealing the ingenuity of scientists and the boundless possibilities of materials engineering.

The expedition begins with graphite powder, a crystalline form of carbon consisting of layers of graphene stacked together. Historically, graphite was primarily valued for its slippery nature, making it an ideal lubricant and, most famously, the "lead" in pencils. However, the advent of groundbreaking research in the late twentieth century led to the isolation of individual graphene layers from graphite. This pioneering workmade by Andre Geim and Konstantin Novoseloy won the Nobel Prize in Physics in 2010 and opened the door to a myriad of applications. Scientists soon realized that the isolation of graphene was only the first step in a transformative journey.

Graphene oxide emerged as a pivotal milestone in this expedition. Graphene oxide is a chemically modified form of graphene, wherein oxygen-containing functional groups are introduced onto the graphene surface [63]. These functional groups, such as epoxides, hydroxyls, and carboxyls, render graphene oxide highly hydrophilic, allowing it to disperse well in water—a property not inherent to pristine graphene [64]. The production of graphene oxide involves a series of chemical treatments, including oxidation and exfoliation, to break down the layered structure of graphite into individual graphene oxide sheets.

Oxidation of graphite typically begins with the treatment of graphite powder with strong oxidizing agents, like H_2SO_4, and $KMnO_4$. This process introduces oxygen-containing functional groups onto the graphene layers, disrupting the π-electron system and rendering the material hydrophilic. The resulting product, graphite oxide, is a precursor to graphene oxide. Further exfoliation is achieved by subjecting graphite oxide to sonication or mechanical shear forces, breaking down the stacked layers into individual graphene oxide sheets.

Graphene oxide's unique properties have sparked immense interest in diverse fields. Its hydrophilicity allows for facile processing and incorporation into various composites, coatings, and films. Moreover, graphene oxide exhibits exceptional mechanical strength, electrical conductivity, and thermal stability, making it an ideal candidate for advanced materials and devices. In the realm of electronics, graphene oxide has been explored for flexible and transparent conductive films, energy storage devices, and sensors. Its biocompatibility and functionalizability in biomedical applications have paved the way for drug delivery systems, biosensors, and tissue engineering scaffolds [8].

Beyond its intrinsic properties, graphene oxide's chemical versatility is a driving force behind its widespread applications. Researchers have developed numerous techniques to further modify graphene oxide, tailoring its properties for specific applications. Functionalization with various molecules, polymers, or nanoparticles enhances its performance in targeted fields, such as catalysis, environmental remediation, and optoelectronics.

Graphite powder to graphene oxide journey epitomizes the interdisciplinary nature of modern science. Chemists, physicists, engineers, and materials scientists collaborate to refine synthesis techniques, understand fundamental properties, and explore novel applications. This expedition also underscores the importance of innovation in materials engineering, where the manipulation of atomic and molecular structures leads to groundbreaking advancements. As the journey continues, scientists delve deeper into the realm of two-dimensional materials, exploring graphene oxide, its derivatives, and other layered materials like transition metal dichalcogenides [10].

28.4 Overview of Graphene Nanocomposite

Graphene nanocomposites, a promising branch of nanotechnology, involve the controlled synthesis, operation, characterizations, and applications of materials where at least one dimension is less than 1000 nm. In the twenty-first century, graphene nanocomposites stand as an ideal of potential within materials science, influencing physics, chemistry, biology, and medicinal chemistry. The given materials encompass a diverse range of characteristics that can be finely adjusted, opening the door to more affordable biomedical products, high-performance materials, cutting-edge electronic devices, and cosmetic fields. The utilization of graphene nanotechnology in the commercial realm holds the potential for significant technological progress, elevating the overall quality of life and yielding social and environmental advantages on a global scale. This groundbreaking field is poised to transform numerous sectors, such as organic synthesis, inorganic materials, energy storage, semiconductors, and biotechnology, fostering innovation and progress in each domain. At the core of graphene nanocomposites lies the concept of integrating nanoscale graphene particles into standard matrix materials. The term "nano" originates from the Greek word meaning "dwarf" or very small, reflecting the minute scale at which these materials operate. This nanoscale dimensionality places graphene nanocomposites between the realms of quantum phenomena and atomic structures, exhibiting behavior distinct from conventional bulk materials. The laws of atomic physics strongly influence the properties of nanomaterials, leading to unique chemical and physical characteristics that find novel applications.

Two primary categories nanostructures are engineered and environmental. Engineered nanoparticles, as the name suggests, are synthetically constructed and form the basis of graphene nanocomposites [65]. These nanoparticles possess remarkable properties that can revolutionize various industries. This overview delves into the exploration of diverse types of engineered nanoparticles within the realm of graphene nanocomposites. The insights gained from these studies illuminate the significant role these nanomaterials are poised to play in future societal advancements. As researchers continue to push the boundaries of graphene nanocomposites, the possible applications and benefits of this technology are boundless, promising a future shaped by innovation and scientific ingenuity.

28.5 In Optical Field

Graphene, a hexagonal lattice of a single layer of carbon atoms has stored immense interest in the scientific community due to its extraordinary optical properties. These properties have covered the way for the development of various graphene-based nanomaterials with unique applications in optoelectronics, photonics, and other advanced technologies. Graphene is known for its exceptional transparency, allowing more than 97.7% of visible light to pass through. This property, coupled with its high electrical conductivity, makes it an ideal candidate for transparent conductive films used in applications such as touchscreens and solar cells. Moreover, graphene exhibits strong absorption of light across a broad range of wavelengths, including ultraviolet, visible, and infrared regions [69, 70]. This absorption property is harnessed in photodetectors and sensors. Graphene supports surface plasmon resonances, collective oscillations of electrons at the interface of graphene and a dielectric medium [71, 72]. These resonances can be tuned across a wide spectral range by varying the carrier concentration in graphene, enabling applications in surface-enhanced Raman spectroscopy (SERS) and biosensing. Plasmonic graphene nanostructures enhance the local electromagnetic field, leading to highly sensitive detection of molecules and biomolecules [73]. Graphene exhibits strong nonlinear optical responses due to its unique electronic band structure. Nonlinear optical effects, such as harmonic generation and Kerr effect, have been observed in graphene-based materials. These effects are crucial for developing ultrafast photonic devices, including mode-locked lasers and optical modulators, enabling high-speed data communication and signal processing. Graphene's electrical conductivity can be dynamically modulated by applying an external electric field, allowing active control of its optical properties [74, 75]. This tunability is exploited in devices like electro-optic modulators, where the refractive index of graphene is modified in real-time, enabling high-speed optical communication systems and signal processing applications. Graphene itself does not exhibit significant intrinsic photoluminescence due to its zero bandgap. However, graphene quantum dots (GQDs) and other graphene-based nanomaterials, such as graphene oxide quantum dots (GOQDs), display strong photoluminescence [76–78]. These materials find applications in bioimaging, light-emitting diodes (LEDs), and displays. Additionally, heterostructures combining graphene with other 2D materials, such as transition metal dichalcogenides (TMDs), result in intriguing light-emitting properties with potential applications in optoelectronic devices. Graphene-based materials can support guided waves for their high carrier movement and low loss. Graphene-based waveguides and photodetectors have been developed, enabling the integration of graphene into photonic circuits. These devices are compact, energy-efficient, and compatible with existing silicon-based technology, promising advancements in on-chip optical communication and signal-processing systems [79, 80].

28.6 Various Applications

28.6.1 Graphene as Transparency Conductive Films

Exceptional electric conduction and high transparency of graphene indeed position it as an ideal candidate for transparent conductive films. These films have been used in various fields in different industries, for graphene's unique properties. The advantages of using graphene-based films over traditional indium tin oxide (ITO) in these applications are significant [81, 82]. Graphene-based films are inherently flexible, making them suitable for curved surfaces and flexible electronic devices. This flexibility is particularly valuable in applications like flexible displays, rollable screens, and wearable technology. Graphene is known for its exceptional mechanical strength and durability. It can withstand bending and stretching without compromising its electrical conductivity [83]. Scheme 28.2 describes various applications of Graphene.

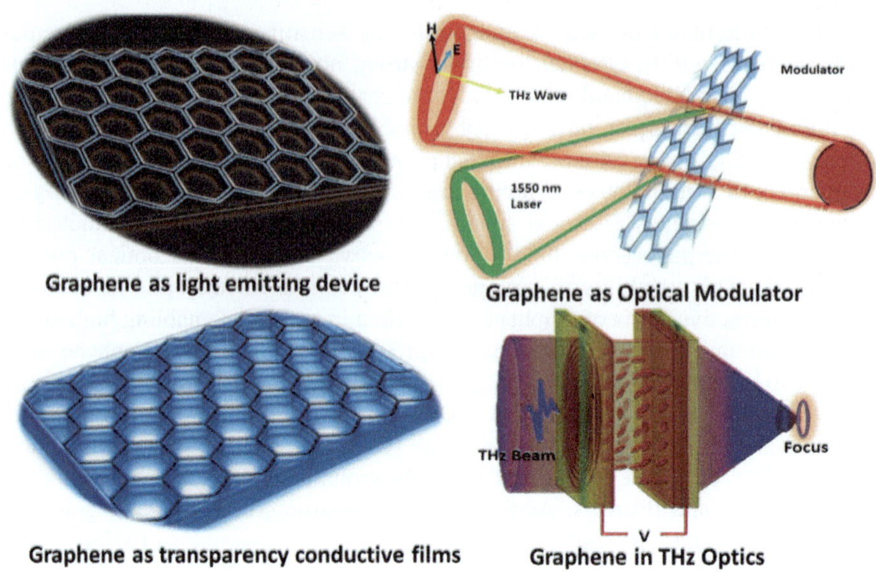

Scheme 28.2 Schematic drawing of some applications

28.6.2 Light-Emitting Devices

Graphene-based materials have indeed demonstrated remarkable potential in revolutionizing the field of light-emitting devices, particularly to develop of organic light-emitting diodes and quantum dot LEDs [84, 85]. This emergence of graphene in such applications has not only promised but delivered substantial benefits in terms of efficiency, longevity, and overall performance, thereby establishing themselves as strong contenders in the display and lighting industries. One of the most notable advantages of incorporating graphene into light-emitting devices is the significant enhancement in efficiency. Graphene's exceptional electron transport properties and high electron mobility make it an ideal candidate for improving charge injection and transport within these devices. This results in higher electron–hole recombination rates, leading to brighter and more energy-efficient displays and lighting systems. OLEDs, in particular, have seen improvements in power efficiency due to graphene's assistance in electron and hole transport.

Graphene's exceptional mechanical strength and chemical stability contribute to the longevity of light-emitting devices. It acts as a protective layer, shielding the sensitive organic materials within OLEDs from environmental factors such as moisture and oxygen, which can degrade their performance over time. This extended lifespan not only benefits consumers but also reduces waste and maintenance costs, making graphene-based devices more economically attractive. Graphene can be integrated into flexible and transparent substrates, allowing for the development of bendable and transparent OLEDs and quantum dot LEDs [85, 87]. This versatility opens up new avenues for applications, including wearable technology, foldable displays, and innovative lighting designs that were previously challenging to achieve with traditional materials. Another critical aspect of graphene's contribution to light-emitting devices is its exceptional thermal conductivity. It efficiently dissipates heat generated during operation, preventing overheating and ensuring stable performance over extended periods. This is particularly important for high-brightness applications where heat management is crucial. As research progresses, the production of graphene-based materials has become more cost-effective. This means that the integration of graphene into light-emitting devices is becoming increasingly economically viable, further bolstering their competitiveness in the market. Graphene-based materials often reduce the need for rare and toxic materials that are traditionally used in LED and OLED manufacturing. This makes these devices more environmentally friendly, aligning with the growing demand for sustainable technologies. In the case of quantum dot LEDs, graphene can be used to tailor the emission spectra, enabling the creation of displays and lighting systems with specific color rendering and efficiency characteristics to meet the demands of various applications.

28.6.3 Graphene Optical Modulator

Graphene-based nanocomposites have garnered significant attention in the field of optical modulators due to their unique properties and potential applications. Optical modulators are devices that control the intensity, phase, or polarization of light signals, making them crucial components in various optical communication systems, imaging devices, and optical sensing technologies [88]. Here are some key applications and advantages of using graphene-based nanocomposites as optical modulators. Graphene-based nanocomposites exhibit exceptionally fast carrier mobility and response times, making them suitable for high-speed optical modulation. They can modulate light signals in the gigahertz (GHz) range, which is essential for data transmission in telecommunications and data centers. Graphene's unique electronic band structure allows for broadband optical modulation across a wide range of wavelengths, from the ultraviolet to the infrared. This versatility is advantageous for applications in spectroscopy and optical sensing. Graphene-based optical modulators can be integrated into compact and miniaturized devices, enabling the development of smaller and more efficient optical communication systems and photonic integrated circuits (PICs). Graphene-based modulators can operate with low power consumption, making them energy-efficient. This is crucial for extending the battery life of portable devices and reducing energy costs in data centers. By applying an external voltage or current, the optical properties of graphene can be dynamically tuned, allowing for on-the-fly adjustment of modulation depth and frequency. This tunability is valuable in adaptive optical systems. Graphene-based optical modulators can be easily integrated with other photonic components, such as waveguides, detectors, and amplifiers, to create multifunctional photonic devices [89, 90].

Graphene's nonlinear optical properties enable the generation of new optical frequencies through processes like four-wave mixing. This can be exploited for applications in frequency comb generation and nonlinear optics. Graphene-based nanocomposites can be engineered to be highly sensitive to external stimuli, such as temperature, strain, or gas molecules. This sensitivity makes them suitable for optical sensors and environmental monitoring [91, 92]. Graphene-based optical modulators have potential applications in biomedical imaging, where rapid and precise modulation of light can improve imaging resolution and contrast. These modulators can be used in defense and security applications, such as lidar systems, secure communication, and remote sensing. Despite their numerous advantages, there are also challenges associated with the practical implementation of graphene-based optical modulators, including fabrication techniques, material stability, and cost considerations. Researchers continue to work on addressing these challenges to unlock the full potential of graphene-based nanocomposites in optical modulation technologies.

28.6.4 Graphene in IR Application

The infrared (IR) spectrum, which spans from approximately 700 nm to 1 mm in wavelength, is a crucial portion of the electromagnetic spectrum with wide-ranging applications in fields such as communication, sensing, and imaging. In recent years, graphene-based nanomaterials have made remarkable strides in harnessing and manipulating IR radiation, opening up exciting possibilities for enhanced performance in various applications.

Graphene's exceptional optical properties, including its high conductivity and tunable plasmonic behavior, have made it a promising candidate for enhancing data transmission in the IR range. By integrating graphene-based devices into optical communication systems, it is possible to achieve faster data rates and improved signal quality. Graphene-based IR sensors are highly sensitive to various gases and can be used for environmental monitoring, industrial safety, and medical diagnostics. They enable real-time detection of trace amounts of gases, which is critical in applications like air quality monitoring and detecting harmful emissions. Graphene's biocompatibility and surface-enhanced infrared absorption properties have paved the way for advanced biosensors [93]. These sensors can detect molecular and cellular changes, making them valuable tools in health care for disease diagnosis and monitoring. Thermal Imaging: Graphene-based materials can be integrated into thermal imaging devices to enhance their sensitivity and resolution. This is especially useful in applications like surveillance, industrial inspections, and firefighting, where detecting temperature variations is crucial. In the field of medical imaging, graphene-based IR detectors can provide higher spatial resolution and sensitivity, improving the accuracy of techniques like infrared spectroscopy and thermography.

Graphene-based materials can be used in thermophotovoltaic devices to convert IR radiation from heat sources into electricity. This technology has the potential to improve the efficiency of energy harvesting from waste heat in various industries [94, 95]. Graphene's unique electrical and optical properties enable it to act as an excellent substrate for studying the IR absorption properties of other materials. Researchers can use graphene-based setups to characterize and analyze the IR spectra of molecules and compounds in a controlled environment. Graphene-based metamaterials have shown promise in the development of cloaking devices that can render objects invisible to IR sensors. This has applications in military stealth technology and security.

28.6.5 Graphene in IR Photodetector

Graphene's remarkable properties, including its high carrier mobility and broad spectral response, position it as an exceptional material for infrared (IR) photodetectors. These detectors have the unique capability to operate efficiently at room temperature, which provides significant advantages in terms of cost-effectiveness and seamless integration into existing systems. This groundbreaking technology has found widespread applications in different, such as night vision, remote sensing, and environmental monitoring. In night vision applications, graphene-based IR photodetectors offer enhanced sensitivity to the infrared spectrum, enabling clearer and more detailed imagery even in low-light conditions. This improved performance can greatly benefit military and security operations, as well as wildlife observation and surveillance. Remote-sensing applications benefit from graphene's ability to detect a broad range of IR wavelengths. This enables the monitoring of natural phenomena like temperature variations, gas emissions, and vegetation health. In agriculture, for instance, graphene-based IR detectors can provide crucial information for optimizing crop management and resource allocation. Environmental monitoring also benefits from the versatility of graphene-based IR photodetectors [96, 97]. These detectors can help in tracking pollutants, monitoring climate changes, and assessing air and water quality. Their capacity to operate at room temperature simplifies the deployment of monitoring networks, making them more cost-effective and accessible for a wide range of applications.

28.6.6 Graphene Plasmonics: IR Advancements

Graphene's remarkable ability to support surface plasmon resonances in the infrared (IR) region has spurred the emergence of a fascinating field known as graphene plasmonics. In this field, researchers harness the unique properties of graphene to manipulate light at the nanoscale, opening up exciting possibilities in various technological domains. One of the key advantages of graphene plasmonics lies in its capacity to confine and amplify electromagnetic fields on the nanometer scale. This allows for the precise control and manipulation of light at dimensions far smaller than the wavelength of light itself. Such subwavelength optics have profound implications for various applications. One of the most promising applications is in the realm of sensing. Graphene plasmonic sensors can detect and analyze molecules and particles at extremely low concentrations, making them invaluable in medical diagnostics and environmental monitoring. These sensors take advantage of the sensitivity of graphene plasmons to changes in the local environment, enabling the detection of even single molecules. Additionally, graphene plasmonics can significantly enhance the efficiency of photodetectors and solar cells [75]. By integrating graphene into these devices, researchers can improve light absorption, collection, and conversion, ultimately leading to more efficient energy harvesting and detection technologies. Furthermore, the exceptional properties of graphene plasmons can be broken for the development of ultra-compact photonic circuits, enabling the manipulation of light for data transmission and processing at unprecedented scales [98–100]. This holds the potential to revolutionize information technology and telecommunications.

28.6.7 Graphene IR Metamaterial Advancement

Graphene-based metamaterials have revolutionized the manipulation of infrared (IR) radiation. Through precise engineering of their geometry and composition, scientists wield unprecedented control over the propagation and dispersion of IR waves. These metamaterials are indispensable in a range of groundbreaking applications, from rendering objects invisible through cloaking devices to achieving superlenses that surpass the limitations of conventional optics. Moreover, they enable precise beam steering, offering immense potential in telecommunications and sensing technologies. The versatility and remarkable properties of graphene-based metamaterials continue to push the boundaries of what's possible in IR manipulation, opening up new horizons for advanced technologies [101, 102].

28.6.8 Graphene in THz Optics

The terahertz (THz) spectrum, covering wavelengths from 1 mm to 100 μm, holds immense promise as a frontier in optics. Emerging as a critical player in this spectral range are graphene-based nanomaterials, offering novel opportunities in security, imaging, and spectroscopy. These remarkable materials exhibit exceptional electronic and thermal properties, making them ideal for THz device development [103]. Their potential applications include high-resolution imaging for concealed threat detection, precise spectroscopic analysis of biomolecules, and breakthroughs in communications technology. As research advances, graphene's role in the THz realm continues to expand, promising groundbreaking innovations and transformative solutions across various fields.

Graphene's THz Applications

Graphene's exceptional nonlinear optical properties, stemming from its unique electronic structure, make it an ideal candidate for generating terahertz (THz) radiation using techniques like difference frequency generation and photon rectification. These THz sources are invaluable in various fields, including security scanning and medical imaging. Their ability to emit high-frequency, low-energy THz waves enables non-invasive and highly precise imaging, promising breakthroughs in healthcare diagnostics and enhancing security measures. As research in graphene continues to advance, it holds great promise for revolutionizing THz technology and its wide-ranging applications [104].

Graphene THz Detect

Graphene-based THz detectors harness the exceptional carrier mobility of graphene to achieve rapid and highly sensitive detection of terahertz (THz) radiation [105]. This breakthrough technology holds immense promise across various domains, including wireless communication, where it can enable ultra-fast data transfer rates. Additionally, it finds utility in spectroscopy, enabling precise molecular analysis, and in security screening, offering enhanced threat detection capabilities. These versatile detectors mark a significant stride toward unlocking the full potential of THz technology for numerous real-world applications.

28.7 Conclusion Outlook and Remarks

In conclusion, the exploration of graphene-based nanomaterials in the realm of optics has undeniably opened new horizons in scientific research and technological advancements. The diverse and extraordinary properties exhibited by these materials, spanning from the visible to the terahertz regions, have paved the way for innovative applications in various optical devices. Throughout this systematic review, we delved into the fundamental principles, synthesis methods, and unique optical characteristics of graphene-based materials. Additionally, we extensively examined their applications in photodetectors and photovoltaics, shedding light on the immense potential they offer in revolutionizing optical technology. The integration of graphene-based nanomaterials into photodetectors has shown tremendous promise in enhancing sensitivity and response speed, thereby enabling more efficient and precise detection of light across different wavelengths. In the realm of photovoltaics, these materials have demonstrated remarkable light-absorption capabilities, contributing to the development of highly efficient solar cells. Moreover, the exceptional optical belongings of graphene-based nanostructure materials have also been applied in other areas, such as modulators, lenses, and waveguides, offering unprecedented opportunities for manipulating light in advanced optical systems. However, despite the remarkable progress made, challenges persist in the practical implementation of graphene-based nanomaterials. Issues related to large-scale production, stability, and integration into existing technologies need to be addressed to fully harness their potential. Additionally, further research is essential to explore novel synthesis methods and functionalization techniques that can tailor the properties of graphene-based materials for specific optical applications. Looking ahead, the future of graphene-based nanomaterials in the optical spectrum appears incredibly promising. Continued research efforts, coupled with interdisciplinary collaborations, will likely lead to breakthroughs that address current limitations and propel these materials into mainstream applications. As we overcome challenges and refine our understanding of these materials, we can anticipate witnessing their widespread adoption in various sectors, ranging from telecommunications and imaging technologies to renewable energy solutions. In essence, this systematic review underscores the transformative impact of graphene-based nanomaterials on optical technology and emphasizes the need for sustained research and development initiatives. By capitalizing on the unique properties of these materials and addressing existing challenges, we can usher in a new era of optical innovations, ultimately benefiting society by enabling more efficient, reliable, and advanced optical devices for diverse applications.

Acknowledgements Hari Shanker Biswas (author) thanks to University Grant Commission (UGC) for sanctioning minor project funding number F.PSW-140/15-16. The author also thank, Dr. Indranil Kal, Principal, Surendranath College, Dr. Suchandra Chatterjee, Dr. Lalita Das, Dr. Harisadhan Ghosh, Dr. Apurba Biswas, Dr. Noor Salam, Department of Chemistry, Surendranath College for various suggestion and fruitful discussions.

Conflict of Interest There is no conflict of interest.

References

1. Novoselov, K.S., Geim, A.K., Morozov, S.V., Jiang, D., Zhang, Y., Dubonos, S.V., et al.: Electric field effect in atomically thin carbon films. Science **306**(5696), 666–669 (2004)
2. Geim, A.K., Novoselov, K.S.: The rise of graphene. Nat. Mater. **6**(3), 183–191 (2007)
3. Das, S.S.: Graphene: A Platform for Sensing Applications. Wiley (2014). ISBN: 978-3-527-33517-6
4. Abergel, D.S.L., Russell, A.J., Fal'ko, V.I.: Visibility of graphene flakes on a dielectric substrate. Appl. Phys. Lett. **96**(23), 233110 (2010)
5. Catania, F., Marras, E., Giorcelli, M., Jagdale, P., Lavagna, L., Tagliaferro, A., Bartoli, M.: Appl. Sci. **11**, 614 (2021)
6. Novoselov, K.S., Geim, A.K., Morozov, S.V., Jiang, D., Katsnelson, M.I., Grigorieva, I.V., et al.: Two-dimensional gas of massless Dirac fermions in graphene. Nature **438**(7065), 197–200 (2005)
7. Smith, A.T., LaChance, A.M., Zeng, S., Liu, B., Sun, L.: Synthesis, properties, and applications of graphene oxide/reduced graphene oxide and their nanocomposites. Nano Mater. Sci. **1**(1), 31–47 (2019)
8. Li, J., Zeng, H., Zeng, Z., Zeng, Y., Xie, T.: Promising graphene-based nanomaterials and their biomedical applications and potential risks: a comprehensive review. ACS Biomater. Sci. Eng. **7**(12), 5363–5396 (2021)

9. Tiwari, S.K., Thakur, A.K., Thakur, A.K., De Adhikari, A., Zhu, Y., Wang, N.: Current research of graphene-based nanocomposites and their application for supercapacitors. Nanomaterials **10**(10), 2046 (2020)
10. Partha, K., Jayan, J.S., Madhavikutty, A.S., Sreeram, P.R., Saritha, A., Ito, T., Tiwary, C.S.: Prospective applications of two-dimensional materials beyond laboratory frontiers: iSci. Rev. **26**(5), 106671 (2023)
11. Zhang, S., Li, Z., Xing, F.: Review of polarization optical devices based on graphene materials. Int. J. Mol. Sci. **21**(5), 1608 (2020)
12. Ilyakov, I., Ponomaryov, A., Reig, D.S., Murphy, C., Mehew, J.D., de Oliveira, T.V.A.G., Prajapati, G.L., Arshad, A., Deinert, J.-C., Craciun, M.F., Russo, S., Kovalev, S., Tielrooij, K.-J.: Ultrafast tunable terahertz-to-visible light conversion through thermal radiation from graphene metamaterials. Nano Lett. **23**(9), 3872–3878 (2023)
13. Gura, L., Brinker, M., Marschalik, P., Kala, F., Junkes, B., Junkes, H., Heyde, M., Freund, H.-J.: The real honeycomb structure—from the macroscopic down to the atomic scale. J. Appl. Phys. **133**, 215305 (2023)
14. Singh, S., Hasan, M.R., Sharma, P., Narang, J.: Graphene nanomaterials: the wondering material from synthesis to applications. Sensors Int. **3**, 100190 (2022)
15. Yin, Z., Wu, S., Zhou, X., Huang, X., Zhang, Q., Boey, F., Zhang, H.: Electrochemical deposition of ZnO nanorods on transparent reduced graphene oxide electrodes for hybrid solar cells. Small **6**, 307–312 (2010)
16. Wu, S., Yin, Z., He, Q., Huang, X., Zhou, X., Zhang, H.: Electrochemical deposition of semiconductor oxides on reduced graphene oxide-based flexible, transparent, and conductive electrodes. J. Phys. Chem. C **114**, 11816–11821 (2010)
17. Mandal, P., Molla, R.A., Chattopadhyay, A.P., Poddar, S., Biswas, H.S.: GO-APTES-Cu (II) schiff base complex as efficient heterogeneous catalyst for aerobic decarboxylation reaction of phenylacetic acids. Inorg. Chem. Commun. **144**, 109825 (2022)
18. Muszynski, R., Seger, B., Kamat, P.V.: Decorating graphene sheets with gold nanoparticles. J. Phys. Chem. C **112**, 5263–5266 (2008)
19. Chem, T.J., Moosa, A.A., Abed, M.: Graphene preparation and graphite exfoliation. Turk. J. Chem. **45**(3), 493–519 (2021)
20. Shi, W., Zhu, J., Sim, D.H., Tay, Y.Y., Lu, Z.Y., Zhang, X.J., Zhang, H., Hng, H.H., Yan, Q.: Achieving high specific charge capacitances in Fe_3O_4/reduced graphene oxide nanocomposites. J. Mater. Chem. **21**, 3422–3427 (2011)
21. Zhu, J., Zhu, T., Zhou, X., Zhang, Y., Lou, X.W., Chen, X., Chen, H., Zhang, H., Hng, H.H., Ma, J., Yan, Q.: Facile synthesis of metal oxide/reduced graphene oxide hybrids with high lithium storage capacity and stable cyclability. Nanoscale **3**, 1084–1089 (2011)
22. Han, T.H., Lee, W.J., Lee, D.H., Kim, J.E., Choi, E.Y., Kim, S.O.: Peptide/graphene hybrid assembly into core/shell nanowires. Adv. Mater. **22**, 2060–2064 (2010)
23. Huang, X., Li, S., Huang, Y., Wu, S., Zhou, X., Li, S., Gan, C.L., Boey, F., Mirkin, C.A., Zhang, H.: Synthesis of hexagonal close-packed gold nanostructures. Nat. Commun. **2**, 292–297 (2011)
24. Jahan, M., Bao, Q., Yang, J.-X., Loh, K.P.: Structure-directing role of graphene in the synthesis of metal−organic framework nanowire. J. Am. Chem. Soc. **132**, 14487–14495 (2010)
25. Petit, C., Bandosz, T.J.: Enhanced adsorption of ammonia on metal-organic framework/graphite oxide composites: analysis of surface interactions. Adv. Funct. Mater. **20**, 111–118 (2010)
26. Petit, C., Burress, J., Bandosz, T.J.: The synthesis and characterization of copper-based metal–organic framework/graphite oxide composites. Carbon **49**, 563–572 (2011)
27. Dong, X., Li, B., Wei, A., Cao, X., Chan-Park, M.B., Zhang, H., Li, L.-J., Huang, W., Chen, P.: One-step growth of graphene-carbon nanotube hybrid materials by chemical vapor deposition. Carbon **49**, 2944–2949 (2011)
28. Fan, Z., Yan, J., Zhi, L., Zhang, Q., Wei, T., Feng, J., Zhang, M., Qian, W., Wei, F.: A three-dimensional carbon nanotube/graphene sandwich and its application as electrode in supercapacitors. Adv. Mater. **22**, 3723–3728 (2010)

29. Wang, Y., Li, Z., Hu, D., Lin, C.-T., Li, J., Lin, Y.: Aptamer/graphene oxide nanocomplex for in situ molecular probing in living cells. J. Am. Chem. Soc. **132**, 9274–9276 (2010)
30. Lu, C.H., Yang, H.H., Zhu, C.L., Chen, X., Chen, G.N.: A graphene platform for sensing biomolecules. Angew. Chem. Int. Ed. **121**, 4879–4881 (2009)
31. Zhang, L.L., Zhou, R., Zhao, X.S.: Graphene-based materials as supercapacitor electrodes. J. Mater. Chem. **20**, 5983–5989 (2010)
32. Vivekchand, S., Rout, C., Subrahmanyam, K., Govindaraj, A., Rao, C.: Graphene-based electrochemical supercapacitors. J. Chem. Sci. **120**, 9–14 (2008)
33. Wang, X., Zhi, L., Mullen, K.: Transparent, conductive graphene electrodes for dye-sensitized solar cells. Nano Lett. **8**, 323–329 (2007)
34. Guo, C.X., Yang, H.B., Sheng, Z.M., Lu, Z.S., Song, Q.L., Li, C.M.: Layered graphene/quantum dots for photovoltaic devices. Angew. Chem. Int. Ed. **49**, 3014–3021 (2010)
35. Chang, H., Lv, X., Zhang, H., Li, J.: Quantum dots sensitized graphene: in situ growth and application in photoelectrochemical cells. Electrochem. Commun. **12**, 483–488 (2010)
36. Wang, P., Jiang, T., Zhu, C., Zhai, Y., Wang, D., Dong, S.: One-step, solvothermal synthesis of graphene-CdS and graphene-ZnS quantum dot nanocomposites and their interesting photovoltaic properties. Nano Res. **3**, 794–799 (2010)
37. Liu, J., Shuping, C., Yanan, L., Bijing, Z.: Progress in preparation, haracterization, surface functional modification of graphene oxide: a review. J. Saudi Chem. Soc. **26**(6), 101560 (2022)
38. Mandal, P., Bala, S., Poddar, S., Sarkar, S., Biswas, H.S.: Fabrication of graphene-Fe_3O_4-polypyrrole based ternary material as an electrode for pseudocapacitor application. Mater. Today: Proc. **65**(2), 1001–1010 (2022)
39. Navarro, C.G., Weitz, R.T., Bittner, A.M., Scolari, M., Mews, A., Burghard, M., Kern, K.: Electronic transport properties of individual chemically reduced graphene oxide sheets. Nano Lett. **7**, 3499–3503 (2007)
40. Zhang, L., Liang, J., Huang, Y., Ma, Y., Wang, Y., Chen, Y.: Size-controlled synthesis of graphene oxide sheets on a large scale using chemical exfoliation. Carbon Carbon **47**, 3365–3368 (2009)
41. Xie, L., Ling, X., Fang, Y., Zhang, J., Liu, Z.: Graphene as a substrate to suppress fluorescence in resonance raman spectroscop. J. Am. Chem. Soc. **131**, 9890–9891 (2009)
42. Wang, Z., Zhang, J., Chen, P., Zhou, X., Yang, Y., Wu, S., Niu, L., Han, Y., Wang, L., Boey, F., Zhang, Q., Liedberg, B., Zhang, H.: Label-free, electrochemical detection of methicillin-resistant staphylococcus aureus DNA with reduced graphene oxide-modified electrodes. Biosens. Bioelectron. **26**, 3881–3887 (2011)
43. Schedin, F., Geim, A.K., Morozov, S.V., Hill, E.W., Blake, P., Katsnelson, M.I., Novoselov, K.S.: Detection of individual gas molecules adsorbed on graphene. Nat. Mater. **6**, 652–657 (2007)
44. Cai, W., Chen, X.: Nanoplatforms for targeted molecular imaging in living subjects. Small **3**, 1840–1854 (2007)
45. Peng, C., Hu, W., Zhou, Y., Fan, C., Huang, Q.: Intracellular imaging with a graphene-based fluorescent probe. Small **6**, 1686–1692 (2010)
46. Scheuermann, G.M., Rumi, L., Steurer, P., Bannwarth, W., Mülhaupt, R.: Palladium nanoparticles on graphite oxide and its functionalized graphene derivatives as highly active catalysts for the Suzuki−Miyaura coupling reaction. J. Am. Chem. Soc. **131**, 8262–8270 (2009)
47. Singh, R.N., Awasthi, R.: Graphene support for enhanced electrocatalytic activity of Pd for alcohol oxidation. Catal. Sci. Technol. **1**, 778–783 (2011)
48. Zhang, N., Qiu, H., Liu, Y., Wang, W., Li, Y., Wang, X., Gao, J.: Fabrication of gold nanoparticle/graphene oxide nanocomposites and their excellent catalytic performance. J. Mater. Chem. **21**, 11080–11083 (2011)
49. He, H., Gao, C.: Graphene nanosheets decorated with Pd, Pt, Au, and Ag nanoparticles: synthesis, characterization, and catalysis applications. Sci. China Chem. **54**, 397–404 (2011)
50. Li, Y., Fan, X., Qi, J., Ji, J., Wang, S., Zhang, G., Zhang, F.: Gold nanoparticles–graphene hybrids as active catalysts for Suzuki reaction. Mater. Res. Bull. **45**, 1413–1418 (2010)

51. Mastalir, A., Kiraly, Z., Patzko, A., Dekany, I., Argentiere, P.L.: Synthesis and catalytic application of Pd nanoparticles in graphite oxide. Carbon **46**, 1631–1637 (2008)
52. Dreyer, D.R., Bielawski, W.: Carbocatalysis: Heterogeneous carbons finding utility in synthetic chemistry. Chem. Sci. **2**, 1233–1240 (2011)
53. Long, Y., Zhang, C., Wang, X., Gao, J., Wang, W., Liu, Y.: Oxidation of SO2 to SO3 catalyzed by graphene oxide foams. J. Mater. Chem. **21**, 13934–13941 (2011)
54. Pyun, J.: Graphene oxide as catalyst: application of carbon materials beyond nanotechnology. Angew. Chem. Int. Ed. **50**, 46–54 (2011)
55. Jia, H.P., Dreyer, D.R., Bielawski, C.W.: C-H oxidation using graphite oxide. Tetrahedron **67**, 4431–4434 (2011)
56. Akhavan, O., Ghaderi, E., Esfandiar, A.: Wrapping bacteria by graphene nanosheets for isolation from environment, reactivation by sonication, and inactivation by near-infrared irradiation. J. Phys. Chem. B **115**, 6279–6288 (2011)
57. Eizenberg, M., Blakely, J.M.: Carbon monolayer phase condensation on Ni(111). Surf. Sci. **82**, 228–236 (1979)
58. Eizenberg, M., Blakely, J.M.: Carbon interaction with nickel surfaces: monolayer formation and structural stability. J. Chem. Phys. **71**, 3467–3477 (1979)
59. Lang, B.: A LEED study of the deposition of carbon on platinum crystal surfaces. Surf. Sci. **53**, 317–329 (1975)
60. Novoselov, K.S., Geim, A.K., Morozov, S.V., Jiang, D., Zhang, Y., Dubonos, S.V., Grigorieva, I.V., Firsov, A.A.: Electric field effect in atomically thin carbon films. Science **306**, 666–669 (2004)
61. Abbas, Q., Shinde, P.A., Abdelkareem, M.A., Alami, A.H., Mirzaeian, M., Yadav, A., Olabi, A.G.: Graphene synthesis techniques and environmental applications. Materials **15**(21), 7804 (2022)
62. Saeed, M., Alshammari, Y., Majeed, S.A., Al-Nasrallah, E.: Chemical vapour deposition of graphene—synthesis, characterisation, and applications: a review. Molecules **25**(17), 3856 (2020)
63. Biswas, H.S., Bala, S., Kundu, A.K., Saha, I., Poddar, S., Sarkar, S., Mandal, P.: Tuned synthesis and designed characterization of graphene oxide thin film. Inorg Chem Commun **139**, 109356 (2022)
64. Qiu, C., Jiang, L., Gao, Y., Sheng, L.: Effects of oxygen-containing functional groups on carbon materials in supercapacitors: a review. Mater. Des. **230**, 111952 (2023)
65. Baig, N., Kammakakam, I., Falath, W.: Nanomaterials: a review of synthesis methods, properties, recent progress, and challenges. Mater. Adv. **2**, 1821–1871 (2021)
66. Worku, A.K., Ayele, D.W.: Recent advances of graphene-based materials for emerging technologies. Results Chem. **5**, 100971 (2023)
67. Abdel-Salam, A.I., Awad, M.M., Soliman, T.S., Khalid, A.: The effect of graphene on structure and optical properties of CdSe nanoparticles for optoelectronic application. J. Alloys Comp. **898**, 162946 (2022)
68. Tiwari, S.K., Sahoo, S., Wang, N., Huczko, A.: Graphene research and their outputs: Status and prospect. J. Sci. Adv. Mater. Dev. **5**(1), 10–29 (2020)
69. Cheng, X., Wang, Y.: Enhanced ultraviolet absorption in graphene by aluminum and magnesium hole-arrays. Sci. Rep. **11**, 8516 (2021)
70. Yan-Li, X., Li, H.-X., Zhou, C.-B., Xiao, X.-S., Bai, Z.-C., Zhang, Z.-P., Qin, S.-J.: The ultraviolet absorption of graphene in the Tamm state. Optik **219**, 165015 (2020)
71. Cui, L., Wang, J., Sun, M.: Graphene plasmon for optoelectronics. Rev. Phys. **6**, 100054 (2021)
72. Zhang, Z., Lee, Y., Haque, M.F., et al.: Plasmonic sensors based on graphene and graphene hybrid materials. Nano Converg. **9**, 28 (2022)
73. Nurrohman, D.T., Chiu, N.-F.: A review of graphene-based surface Plasmon resonance and surface-enhanced Raman scattering biosensors: current status and future prospects. Nanomaterials **11**(1), 216 (2021)

74. Zhang, C., Wang, Y., Yao, Y., Tian, L., Geng, Z., Yang, Y., Jiang, J., He, X.: Active control of electromagnetically induced transparency based on terahertz hybrid metal-graphene metamaterials for slow light applications. Optik **200**, 163398 (2020)
75. Ullah, Z., Witjaksono, G., Nawi, I., Tansu, N., Khattak, M.I., Junaid, M.: A review on the development of tunable graphene nanoantennas for terahertz optoelectronic and Plasmonic applications. Sensors **20**(5), 1401 (2020)
76. Kurniawan, D., Weng, R.-J., Chen, Y.-Y., Rahardja, M.R., Nanaricka, Z.C., Chiang, W.-H.: Recent advances in the graphene quantum dot-based biological and environmental sensors. Sensors Actuators Rep. **4**, 100130 (2022)
77. Kalluri, A., Dharmadhikari, B., Debnath, D., Patra, P., Kumar, C.V.: Advances in structural modifications and properties of graphene quantum dots for biomedical applications. ACS Omega **8**(24), 21358–21376 (2023)
78. Zhao, C., Song, X., Liu, Y., et al.: Synthesis of graphene quantum dots and their applications in drug delivery. J. Nanobiotechnol. **18**, 142 (2020)
79. Ren, T., Loh, K.P.: On-chip integrated photonic circuits based on two-dimensional materials and hexagonal boron nitride as the optical confinement layer. J. Appl. Phys. **125**, 230901 (2019)
80. Nguyen, B.H., Nguyen, V.H.: Advances in graphene-based optoelectronics, plasmonics and photonics. Adv. Nat. Sci Nanosci. Nanotechnol. **7**, 013002 (2016)
81. Christodoulou, C., Wolter, B., Ioakeimidis, A., Chouliaras, G., Wiesner, S., Lauermann, I., Centeno, A., Zurutuza, A., Fostiropoulos, K.: Surface analysis and surface doping of graphene on indium-tin-oxide. Thin Solid Films **682**, 57–62 (2019)
82. Nguyen, B.H., Nguyen, V.H.: Promising applications of graphene and graphene-based nanostructures. Adv. Nat. Sci Nanosci. Nanotechnol. **7**, 023002 (2016)
83. Das, T., Sharma, B.K., Katiyar, A.K., Ahn, J.-H.: Graphene-based flexible and wearable electronics. J. Semicond. **39**(1), 011007 (2018)
84. Junaid, M., Md Khir, M.H., Witjaksono, G., Ullah, Z., Tansu, N., Saheed, M.S.M., Kumar, P., Hing Wah, L., Magsi, S.A., Siddiqui, M.A.: A review on graphene-based light emitting functional devices. Molecules **25**, 4217 (2020)
85. Gaurav, A., Jain, A., Tripathi, S.K.: Review on fluorescent carbon/graphene quantum dots: promising material for energy storage and next-generation light-emitting diodes. Materials **15**(22) (2022)
86. Aydin, E., Altinkaya, C., Smirnov, Y., Yaqin, M.A., Zanoni, K.P.S., Paliwal, A., Firdaus, Y., Allen, T.G., Anthopoulos, T.D., Bolink, H.J., Morales-Masis, M., De Wolf, S.: Sputtered transparent electrodes for optoelectronic devices: induced damage and mitigation strategies. Matter **4**(11), 3549–3584 (2021)
87. Sang, M., Shin, J., Kim, K., Yu, K.J.: Electronic and thermal properties of graphene and recent advances in graphene based electronics applications. Nanomaterials **9**, 374 (2019)
88. Esfandiari, M., Lalbakhsh, A., Shehni, P.N., Jarchi, S., Ghaffari-Miab, M., Mahtaj, H.N., Reisenfeld, S., Alibakhshikenari, M., Koziel, S., Szczepanski, S.: Recent and emerging applications of graphene-based metamaterials in electromagnetics. Mater. Des. **221**, 110920 (2022)
89. Yang, H., Wang, Y., Tiu, Z.C., Tan, S.J., Yuan, L., Zhang, H.: All-optical modulation technology based on 2d layered materials. Micromachines **13**(1), 92 (2022)
90. Kumbhakar, P., Gowda, C.C., Tiwary, C.S.: Advance optical properties and emerging applications of 2D materials. Front Mater Sec. Thin Solid Films **8** (2021)
91. Demongodin, P., El Dirani, H., Kerdilès, S., Lhuillier, J., Wood, T., Sciancalepore, C., Monat, C.: Pulsed four-wave mixing at telecom wavelengths in Si_3N_4 waveguides locally covered by graphene. Nanomaterials **13**, 451 (2023)
92. Ferrand, A., Siaj, M., Claverie, J.P.: Graphene, the Swiss Army knife of nanomaterials science. ACS Appl. Nano Mater. **3**(8), 7305–7313 (2020)
93. Lee, C.W., Eom, T.H., Cho, S.H., Jang, H.W.: Chemical sensors based on graphene and 2D graphene analogs. Adv. Sensor Res. **2**, 2200057 (2023)

94. Zhang, X., Ang, Y.S., Du, J.-Y., Chen, J., Ang, L.K.: Graphene-based thermionic-thermoradiative solar cells: concept, efficiency limit, and optimum design. J. Clean. Prod. **242**, 118444 (2020)
95. Sheng, Y.: Enhancement of a graphene-based near-field thermophotovoltaic system by optimization algorithms and dynamic regulations. Photonics **10**, 137 (2023)
96. Yang, J., Yang, Q., Zhang, Y., Wei, X., Shi, H.: Graphene nanowalls in photodetectors. RSC Adv. **13**, 22838–22862 (2023)
97. Chi, H., Murali, K., Li, T., Thomas, S.: Recent advances in graphene based photoresponsive materials. Prog. Nat. Sci. Mater. Int. **29**(6), 603–611 (2019)
98. Lin, S., Lu, Y., Xu, J., Feng, S., Li, J.: High performance graphene/semiconductor van der Waals heterostructure optoelectronic devices. Nano Energy **40**, 122–148 (2017)
99. Yan, N., Qiu, Y., He, X., Tang, X., Hao, Q., Chen, M.: Plasmonic enhanced nanocrystal infrared photodetectors. Materials **16**, 3216 (2023)
100. Bao, Y., Fang, Z.: Plasmon-enhanced photodetection in nanostructures. Nanotechnol. Rev. **4**(4), 325–336 (2015)
101. Cynthia, S., Ahmed, R., Islam, S., Ali, K., Hossain, M.: Graphene based hyperbolic metamaterial for tunable mid-infrared biosensing. RSC Adv. **11**(14), 7938–7945 (2021)
102. Ye, M., Zha, J., Tan, C., Crozier, K.B.: Graphene-based mid-infrared photodetectors using metamaterials and related concepts. Appl. Phys. Rev. **8**, 031303 (2021)
103. Li, Y., Tantiwanichapan, K., Swan, A.K., Paiella, R.: Graphene plasmonic devices for terahertz optoelectronics. Nanophotonics **9**(7), 1901–1920 (2020)
104. Bi, H., Yang, M., You, R.: Advances in terahertz metasurface graphene for biosensing and application. Discover Nano **18**, 63 (2023)
105. Jin, M., Wang, Y., Chai, M., Chen, C., Zhao, Z., He, T.: Terahertz detectors based on carbon nanomaterials advanced functional materials. Wiley (2021)

Chapter 29
Conducting Polymer Coated Nanomaterial: An Advanced Material for Optoelectronic Devices

Kajal Gupta, Paresh Chandra Jana, and Ajit Kumar Meikap

29.1 Introduction

For a smart society it is recommended that they have some smart materials for use in different optoelectronic devices. These can show advanced properties. Scientists are dedicatedly searching some advanced and most economic smart materials to develop the optoelectronic devices. Smart materials having nanomaterials are best choice and particularly nanomaterials of Gold, Silver, etc. are very important choice for making of modern optoelectronic devices. On the other hand, conducting polymers with polyaromatic backbones are important materials. Composite of nanomaterials of silver, gold, etc. with conducting polymers like polyaniline, polypyrrole, and polyacetylene are very interesting for optoelectronic properties. Thus interest obviously lying on these materials for their remarkable optical and electrical properties and give some light on their proper use. In this chapter polyaniline–gold and polyaniline–silver nanocomposite are considered. Remarkable increase in fluorescence, electrical conductivity and optical absorbance makes these materials hot choices in optoelectronic devices. Thus variation of composition and condition of preparation of composite and notice of the variation of optical and electrical properties keeping in mind their proper use should be our objective.

K. Gupta (✉)
Department of Chemistry, Nistarini College, Purulia 723101, India
e-mail: kajaldgpnit@gmail.com

P. C. Jana
Department of Physics and Technophysics, VU, Midnapore 721102, India

K. Gupta · A. K. Meikap
Department of Physics, NIT, Durgapur 713209, India

M. El Ghzaoui et al. (eds.), *Next Generation Wireless Communication*, Signals and Communication Technology, https://doi.org/10.1007/978-3-031-56144-3_29

29.2 Advanced Materials

Some organic polymers with aromatic backbone like polyaniline (PANI), polythiophene (PT), etc. (structures are given in Fig. 29.1) are capable of conducting electricity [1] though most of the polymers are insulating in nature.

Different unique properties like chemical, optoelectronic, and magnetic properties. CPs can be effectively applied in electronic devices in making batteries, electrodes, EMI shielding material, sensors, coatings, LEDs, etc. [2–7]. CPs can be synthesized easily in the laboratory with good yield and their precursors are very cheap. CPs have good levels of thermal stability and are comparatively less toxic [8–11]. An extended π-conjugation is found in the backbone of polymer chain and electrons of these π-bond are responsible for the electrical conduction of the polymers. Formation of the π-bond in CPs may be described as follows.

Carbon atoms of CPs are sp^2 hybridized and three sp^2 hybridized orbitals of a carbon atom remain engaged in forming three strong σ bonds with adjacent carbon atoms. These σ bonds form the polymer backbone. Unhybridized p-orbital of a carbon atom forms a π-bond with the unhybridized p-orbital of another adjacent carbon atom. Large band gap and low concentration of charge carriers at normal temperature make CPs weak conductor (In pure state conductivity $= 10^{-13}$ S cm^{-1}). It was suggested that conductivity could be increased from insulating to metallic range by increasing carrier concentration, using different oxidizing or reducing agents. Range of conductivity of polyaniline, polyacetylene, polypyrrole, and polythiophene with proper doping is 10–10^5 S cm^{-1} [12]. Iodine doped trans-polyacetylene has conductivity 1.5×10^5 S cm^{-1} at room temperature [13]. There are some serious limitations of these CPs. Most of them are insoluble, infusible, reactive, and brittle. CPs may be used as commercial materials if their processibility is improved, conductivity is enhanced, and structural defects are reduced. Mostly used CPs are polyaniline and polypyrrole and are described below.

29.2.1 Polyaniline (PANI)

In 1834 scientist Runge first synthesized polyaniline as a black powder. Oxidation of aniline with suitable oxidizing agent in presence of acid results in the formation of

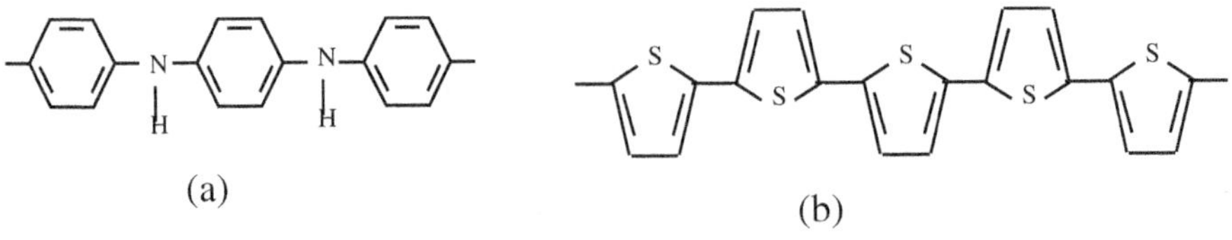

N N H H (a) S S S S S (b)

Fig. 29.1 Structure of **a** PANI and **b** PT [1]

polyaniline [14]. Structure analysis was done by Fritzsche [15]. Molecular formula of polyaniline may be of following type (given in Fig. 29.2).

$(1 - y)$ represents oxidation state of the polyaniline having values in the range $0 \leq (1 - y) \leq 1$. When $y = 1$, 0.5, and 0.0 then the oxidation states will be Leucoemeraldine, Emeraldine, and Perni-graniline base, respectively. Structures of different oxidation states of polyaniline are given in Fig. 29.3.

Leucoemeraldine base is pale yellow, emeraldine salt is bright green, and pernigraniline base is dark blue or violet in color [16, 17]. It can be synthesized in a simple process from aniline precursor. It remains stable upto 300–350 °C and so its thermal stability is quite satisfactory. It is much more conductive in nature than other CPs. Good optoelectronic, magnetic, and tunable chemical properties are obtained

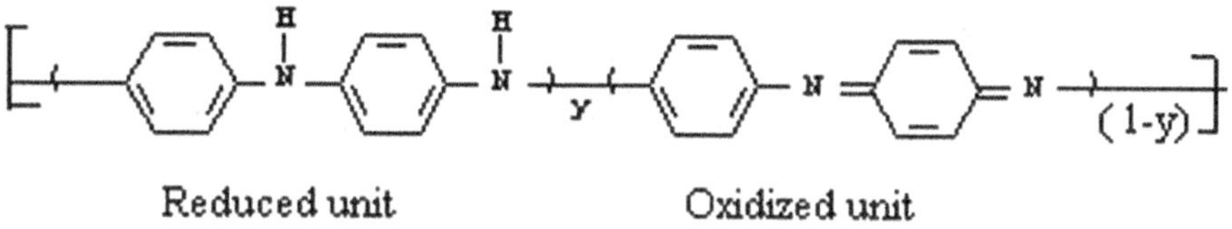

Reduced unit Oxidized unit

Fig. 29.2 Molecular formula of polyaniline

Leucoemeraldine (pale yellow)

Emeraldine base (blue)

$-H^+$ $+H^+$

Emeraldine salt (green)

Pernigraniline (violet)

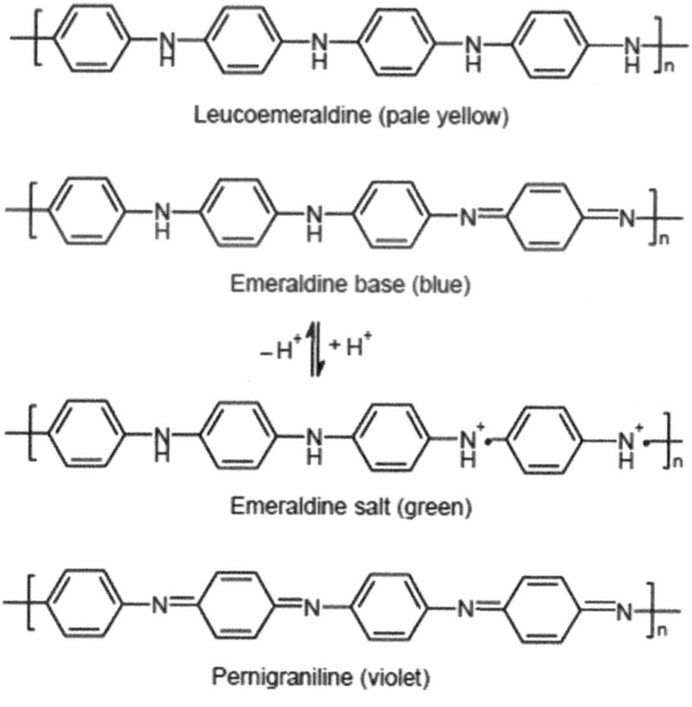

Fig. 29.3 Four different forms of PANI [1]

in polyaniline [18, 19]. Doping with suitable dopant makes PANI applicable for particular use. Of the four forms of PANI, the emeraldine base is allowed for doping with dopants [20]. Proton (H^+) doping to the non-conducting emeraldine base of polyaniline makes it a green conducting emeraldine salt. Interconversion of different oxidation state is given in Fig. 29.4.

Emeraldine base has poor conductivity (~10^{-10} S cm^{-1}) and its conductivity can be increased to ~ 5 S cm^{-1} by protonating with 1 M aqueous acid [21–23]. This protonation occurs in nitrogen atom of every imine groups of emeraldine base and becomes stable through resonance. The $-C_6H_4$ ring of aniline and its N atom becomes chemically equivalent and they remain as the poly-semiquinone radical cation. There were a lot of experimental and theoretical support like EPR, magnetic measurements, and theoretical modeling toward the existence of the protonated structure of polyaniline [24–27]. Heteroatoms like nitrogen in both polypyrrole and polyaniline as well as sulfur in polythiophene present in CPs are responsible for electrical charge transport and also some interesting optoelectronic properties [28, 29]. There were some properties which depend on the factors like (1) reaction temperature, (2) oxidizing agents, (3) aniline/ oxidant mole ratio (k), (4) time of polymerization, (5) effect of protonic

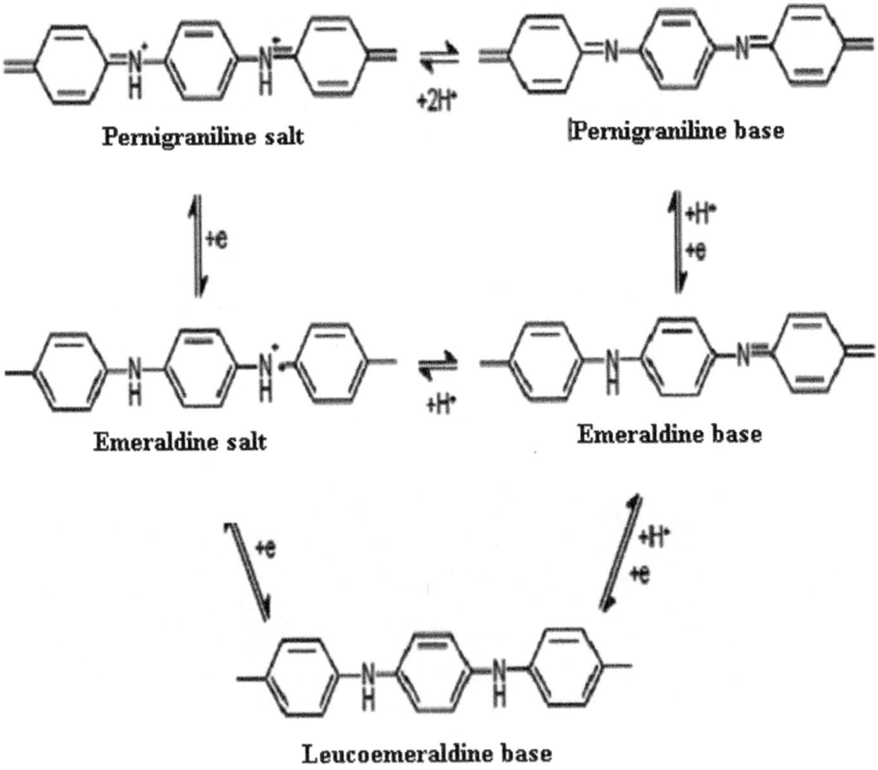

Pernigraniline salt

Pernigraniline base

Emeraldine salt

Emeraldine base

Leucoemeraldine base

Fig. 29.4 Interconversion of the different oxidation states of polyaniline [1]

acid, and (6) concentration of the protonated acid. Polyaniline-camphor sulfonic acid composite has higher solubility and processibility, and it has improved electrical transport properties [30–33] with conductivity in the range 100–400 S cm^{-1}. Conductivity of para-toluene sulfonic (PTSA) and sulfosalicylic acid (SSA) doped polyaniline 300 K are 7.8 and 2.0 S cm^{-1}, respectively [34].

29.2.2 Polypyrrole

Another widely investigated CP is Polypyrrole (PPy). Its mechanical and electrochemical stability are quite satisfactory. PPy can be prepared by some very easy methods like chemical polymerization, electrochemical, and enzyme catalyzed polymerization [35, 36]. It has relatively good conductivity. It has aromatic, quinoid, polaronic, and bipolaronic structure and structures can be established by FTIR, XPS (X-ray photoelectron spectroscopy), NMR spectroscopy, etc. [37–39]. It can be applied successfully in different printed circuit, heating films, chip carriers, film keyboards, electromagnetic shielding, etc. [40].

Effective variation of conductivity and optical properties of polypyrrole can be done with suitable doping. [41–43]. Electrical conductivity of PPy can be enhanced to many folds even at room temperature. Room temperature conductivity of doped PPy (polymerized electrochemically at – 20 °C to – 30 °C) can be increased from 200 S cm^{-1} to more than 1000 S cm^{-1} by stretching [44, 45]. Conductivity of phosphorous hexafluoride doped polypyrrole at – 40 C and at 300 K of varying thickness of the film lie in the range 100–400 S cm^{-1} and 20–50 S cm^{-1}, receptively [46].

29.2.3 Inorganic Organic Hybrid Nanomaterial

Interest in the development of hybrid type of material based on some inorganic and organic substance is growing rapidly for the wide range of application. Nano-sized materials of inorganic and organic in nature are very much demanding for their interesting optoelectronic properties [47]. In designing various optoelectronic devices, normally the particle size dependent properties are accepted [48]. Surface Plasmon resonance (SPR) is the characteristics of metal nanomaterials like silver, gold etc. Peaks and line widths of SPR varies strongly with size and shape of the nanomaterials [49–51], metals [52], and medium [53]. Optical property depends on dielectric medium. Polyaniline, polypyrrole (organic counterpart), etc. are good choices as dielectric medium [54].

29.3 Synthesis

Several methods of preparation of nanomaterials coated with conducting polymers are there and are physical, chemical, and biological methods. Blending of nanomaterials are done with CPs. Nanomaterials can also be prepared by using microbes or phytochemicals obtained from plants and they are then coated with some CPs. In chemical route, nanomaterials are prepared and are in-situ coated with some CPs. Silver–polyaniline nanocomposite is prepared in two step [55]. Silver nanoparticles are synthesized from $AgNO_3$ solution by reducing with sodium citrate. Formation of silver nanoparticles are indicated by pale yellow coloration of the solution. Trisodium citrate is used here for reducing the nitrate salt of silver and stabilizing silver nanoparticles. Coating of silver nanoparticles is done by polyaniline. For this, precursor aniline is acidified with required acid and is mixed with silver nanoparticles. Precooled oxidant like ammonium peroxodisulphate solution is poured drop by drop to the mixture at 0 °C. The solution gradually becomes green in color. By centrifuging the mixture at 10,000 rpm solid mass is obtained as a residue. Monomer, oligomers and excess of oxidant are removed by washing with solvent until the filtrate turns colorless.

Hybrid material polyaniline nanorod–gold can be synthesized as follows. At first, acidified (with HCl) aniline (of milimolar strength) solution is taken in a conical flask at 0 °C and precooled ammonium peroxodisulphate solution is added. Gradually a green colored solution is formed indicating the formation of polyaniline. One hour centrifugation at about ten thousand rotation per minute is done to obtain solid precipitate and to remove the impurities, and it is then rinsed with water and acetone and ethanol. In second step, this precipitate is mixed with 3% NH_4OH with continuous stirring. Emeraldine base (EB) is thus obtained. Dry EB and chloroauric acid are mixed in 25:1 (*w/w*) and poured in water under sonication. Desired nanocomposite is obtained on centrifugation [56].

29.4 Characterization

Transmission electron microscopy helps to know the size, morphology of samples. Powdered XRD is done from two theta 10°–90° for the identification of phase of the composite. Thermal stability of the composite is compared with the polymer itself within 25–700 °C. A pellet of composite with KBr is taken in a FTIR-NEXUS, NICOLET to record if there is any structural change in the range of 500–4000 cm^{-1}. Double beam spectrophotometer is used to record optical absorption using appropriate solvent. Photoluminescence is taken by some FL spectrophotometer. Standard four probe or two probe methods are employed to measure direct or alternating current electrical conductivity within 5–400 K.

29.4.1 Morphology

Several morphologies like spherical, tubular, rod like, of the hybrid samples are obtained from their TEM image. Well resolved, circular shaped grains are obtained for polyaniline–silver hybrid material (given in Fig. 29.5). Nanoparticles of silver of diameter approximately 50 nm are embedded within polyaniline. Presence of a strong chemical force acting among the silver nanoparticle and the nitrogen atom of the polyaniline is responsible for the adherence of silver nanoparticles [55].

Polyaniline–gold hybrid material is obtained with rod like structures (given in Fig. 29.5) with diameter 60 nm and several micrometers in length. From the energy dispersive spectroscopy (EDX) presence of Au can be confirmed [56].

Fig. 29.5 Microgram of **a** polyaniline–silver nanocomposite, **b** polyaniline nanorod–gold composite

29.4.2 XRD Pattern

XRD pattern indicates the presence of inorganic nanomaterials. Presence of Au and Ag can be identified from the peaks of XRD pattern. Conducting polymer polyaniline has a broad peak at $2\theta = 25.03°$ which arises due to the regular repetition of aniline [55, 56]. Characteristic peaks are obtained at $2\theta = 25°, 38°, 45°, 65^0, 73°, 77°$ in X-Ray diffractogram of polyaniline–silver nanocomposite. These peaks arises due to reflections from (111), (200), (220), (311) planes of an fcc lattice (JCPDS no. 011164) for silver nanoparticles there. For calculating the particle or grain size standard Scherrer's equation can be used and the equation is given in elsewhere [55]. Particle size of this composite is 45–55 nm [55].

Peaks at $2\theta = 38.1, 44.36, 64.4, 77.4$ represent reflections from (111), (200), (220), (311) planes of a fcc lattice (JCPDS no. 011172) and they are for the Au nanoparticles in the composite [56].

29.4.3 Structural Analysis Using FTIR

Structural characterization conducting polymer like polyaniline and its hybrid with inorganic nanomaterials can be under stood from the FTIR spectrums. From the FTIR spectrum (given in Fig. 29.6) bands for polyaniline are obtained at 1590, 1490, 1310, 1130, 820 cm^{-1}. Bands at 1590, 1490 cm^{-1} indicate C=C and C=N stretching modes of vibration, 1310 cm^{-1} is for C–N stretching for the different units of polyaniline. Quinoid unit of polyaniline can be assigned by 1130 cm^{-1}. 820 cm^{-1} is indicating C–C and C–H stretching and 680 cm^{-1} is for C–H vibration. There occur a shifting of bands to 1560, 1485, 1290, 1120, and 800 cm^{-1}, respectively, in composite indicating Ag PANI interaction. Higher conjugation length of the polyaniline–silver composite is the cause of increasing intensity of the FTIR spectrum [55].

29.5 Mechanism of Formation

Mechanism of formation of these nanocomposite may be explained as follows. Protonation of amine group of aniline occurs with protonic acid like nitric acid and forms the aniline cation. This cation attracts negatively charged citrate capped nanoparticles by strong electrostatic force of attraction before the polymerization and thereby formation of the complex of aniline cation and silver occurs. This complex gets polymerize on addition of APS as an oxidant. Silver nanoparticles may catalyze the process of polymerization [55].

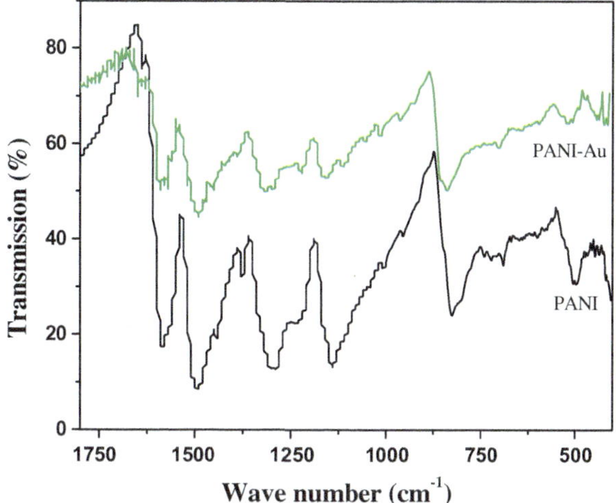

Fig. 29.6 Spectrum shows different peaks in IR of polyaniline nanorod–gold nanocomposite

29.6 Optical Property

29.6.1 Optical Absorption Spectrum

Two absorption bands are obtained around 290 and 600 nm for π–π^* and n–π^* transition, respectively [57, 58]. In polyaniline–silver hybrid nanocomposite, a red shift occurs for the π–π^* transition to 315 nm and whereas a blue shift occurs for the n–π^* absorption band to 540–600 nm with low intensity. Excited electrons of silver and gold exhibit a SPR band and for silver it is obtained at 380 nm in the nanocomposite. Silver to amine/imine units interaction makes a change in the electronic band. This can be understood from the FTIR spectrum as well [55]. Band gap calculated for polyaniline and composite are 2.8, 2.1 eV, respectively. This decrease in band gap may be due to the presence of Ag nanoparticles [55].

Whereas in hybrid type of nanocomposite of polyaniline nanorod–gold a red shift is obtained to 425 nm for the π–π^* transition and a red shift to 710–725 nm for the n–π^* transition (given in Fig. 29.7). A Plasmon resonance band is obtained at around 520 nm for gold [59]. Band gap of polyaniline nanorod–gold hybrid varies from 1.91 to 12.0 eV. This decrease may be for the gold nanoparticles [56].

29.6.2 Photoluminescence (PL)

PL is obtained in the hybrid composite of silver and gold. When the composite of silver is excited at wave length 330 nm, emission is found at 440 nm and increase in PL is for π–π^* transition [60]. Extended π-conjugation, ordered benzenoid and quinoid units of the nanocomposite are the cause of forming the singlet exciton and a light is

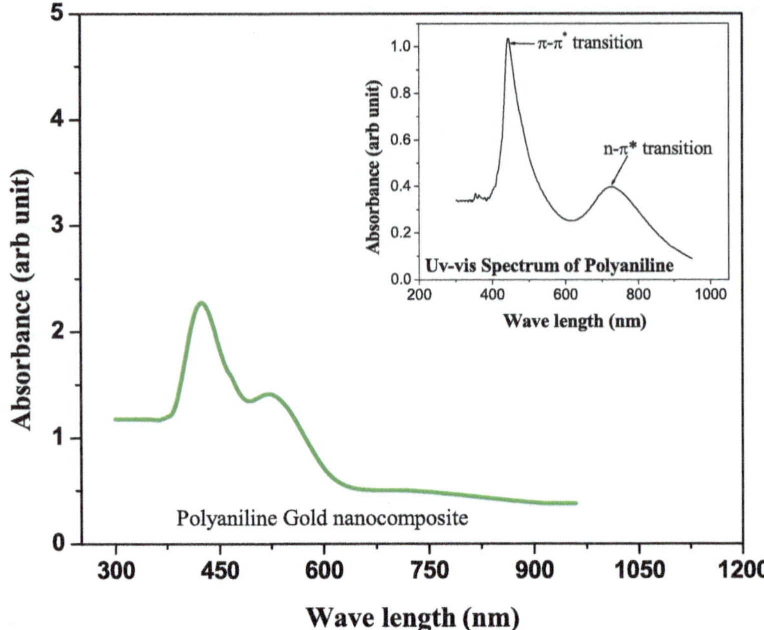

Fig. 29.7 Absorbance versus wavelength graph of polyaniline (in the inset) and polyaniline–gold nanocomposite

emitted when singlet exciton decays to its ground state [61]. Delocalization length of singlet exciton of composite and the π-conjugation length are almost comparable to each other [61]. This singlet exciton is responsible for the luminescence in this case as the triplet exciton remain confined. Spin flip is forbidden in conjugated polymers, thus singlet exciton has the sole responsibility for PL emission [62]. Higher PL emission from the nanocomposite is obtained for its higher π conjugation.

29.7 Electrical Property

Room temperature direct current conductivity of polyaniline is 3.0×10^{-2} S cm^{-1}. Ten times increase in conductivity is obtained for the polyaniline–silver nanocomposite. A probable cause of this may be tunneling of electrons though the nanoparticles of the Ag. Appreciable increase in conductivity is obtained when more amount of silver nanoparticle is added with polyaniline. More the amount of silver nanoparticles more will be concentration of free electrons and thus probability of tunneling will be more resulting in the increased conductivity. Since the conductivity or electronic tunneling in this case depends on the activation energy so calculation of activation energy is necessary and for this Arrhenius equation is required. Using Arrhenius equation activation energy is calculated and it varies from 62 to 64 meV.

A decrease in resistivity decreases from 5.2×10^3 Ω m to 1.6×10^{-2} Ω m at 25 °C [56]. Thus conductivity in case of the composite increases and this may be due to the gold nanoparticles present in the polymer matrix. Gold nanoparticles form metallic islands in the polyaniline matrix. Enlargement of metallic island occurs when more and more Au is injected and as a result of it localization length increases. Increase in localization increases conductivity. So it may be a cause for increasing the conductivity of the gold–polyaniline nanocomposite. Mott's VRH model can be used to explain the fact.

29.8 Conclusion

Conducting polymer like polyaniline coated nanomaterials like silver and gold nanoparticles may be considered as an advanced material. These materials are also most economic. Improved optical absorption and photoluminescence of these materials can be applied in the optical devices for sensing and other applications. Improved optical and electrical property of these materials may also be suitable for application in modern optoelectronic devices. Variation of CPs and variation of metals will be a challenge for the future application in optoelectronic devices.

References

1. Skotheim, T.A., Elsenbaumer, R.L., Reynolds, J.R. (eds.): Hand book of conducting polymers, p. 945. Marcel Dekker Inc., New York (1998)
2. Aphesteguy, J.C., Bercoff, P.G., Jacobo, S.E.: Phys. B **398**, 200 (2007)
3. Naarman, H.: Science and application of conducting polymers. Adam Hilger, Bristol (1991)
4. Joo, J., Epstein, A.: Appl. Phys. Lett. **65**, 2278 (1994)
5. Osaheni, J.A., Jenekhe, S.A., Vanherzeele, H., Meth, J.S., Sun, Y., MacDiarmid, A.G.: J. Phys. Chem. **96**, 2830 (1992)
6. Asim, N., Radiman, S., Bin Yarmo, M.A.: Mater. Lett. **62**, 1044 (2008)
7. Nalwa H.S. (ed.): Hand book of organic conductive molecules and polymers, vol. 2, Chapter 2. Wiley, New York (1997)
8. Gupta, C.M., Umare, S.S.: Macromolecules **25**, 138 (1992)
9. Mizzoguchi, K.: Synth. Met. **119**, 35 (2001)
10. Li, X., Gao, Y., Zhang, X., Gong, J., Sun, Y., Zheng, X., Qu, L.: Mater. Lett. **62**, 2237 (2008)
11. Liang, L., Liu, J.L., Windisch, C.F., Exarhos, G.J., Lin, Y.H.: Angew. Chem. Int. Ed. Engl. **41**, 3665 (2002)
12. Chiang, C.K., Fincher, C.R., Park, Y.W., Heeger, A.J., Shirakawa, H., Louis, E.J., Gau, S.C., MacDiarmid, A.G.: Phys. Rev. Lett. **39**, 1098 (1977)
13. Basescu, N., Liu, Z.-X., Moses, D., Heeger, A.J., Naarmann, H., Theophilou, N.: Nature **327**, 403 (1987)
14. Letheby, H.: J. Chem. Soc. **15**, 161 (1862)
15. Fritzsche, J.: J. Prakt. Chem. **20**, 453 (1940); **28**, 198 (1943)
16. Stejskal, J., Kratochvil, P., Jenkins, A.D.: Collect. Czech. Chem. Commun. **60**, 1747 (1995)
17. Lindfors, T., Ivaska, A.: J. Electroanal. Chem. **535**, 65 (2002)
18. Pillalamarri, S.K., Blum, F.D., Tokuhiro, A.T., Bertino, M.F.: Chem. Mater. **17**, 5941 (2005)

19. Haba, Y., Segal, E., Narkis, M., Titelman, G.I., Siegmann, A.: Synth. Met. **110**, 189 (2000)
20. Kulkarnai, V.M., Kasi Viswanath, A., Marimuthu, R., Seth, T.: J. Polym. Sci.: Part A **42**, 2043 (2004)
21. Epstein, A.J., Ginder, G.M., Zuo, F., Bigelow, R.W., Woo, S.H., Tanner, D.B., Richter, A.F., Huang, W.S., MacDiarmid, A.G.: Synth. Met. **18**, 303 (1987)
22. Ginder, G.M., Richter, A.F., MacDiarmid, A.G., Epstain, A.J.: Solid State Commun. **63**, 97 (1987)
23. .Ginder, J.M., Epstein, A.J., Bigelow, R.W., Richter, A.F., Macdiarmid, A.G.: Bull. Am. Phys. Soc. **31**, 582 (1986)
24. Krinichnyi, V.I., Tokarev, S.V., Roth, H.-K., Schrödner, M., Wessling, B.: Synth. Met. **156**, 1368 (2006)
25. Choi, H.Y., Mele, E.J.: Phys. Rev. Lett. **59**, 2188 (1987)
26. MacDiarmid, A.G., Chiang, J.C., Richter, A.F., Epstein, A.J.: Synth. Met. **18**, 285 (1987)
27. Epstein, A.J., Ginder, J.M., Zuo, F., Woo, H.S., Tanner, D.B., Richter, A.F., Angelopoulous, M., Huang, W.S., MacDiarmid, A.G.: Synth. Met. **21**, 63 (1987)
28. Linschitz, H., Rennert, J., Korn, T.M.: J. Am. Chem. Soc. **76**, 5839 (1954)
29. Rice, M.J., Mele, E.J.: Phys. Rev. Lett. **49**, 1455 (1982)
30. Cao, Y., Smith, P., Heeger, A.J.: Synth. Met. **48**, 91 (1993)
31. Cao, Y., Heeger, A.J.: Synth. Met. **52**, 193 (1993)
32. Cao, Y., Treacy, G.M., Smith, P., Heeger, A.J.: Appl. Phys. Lett. **60**, 2711 (1992)
33. Cao, Y., Smith, P.: Polymer **34**, 3139 (1993)
34. Raghunathan, A., Rangarajan, G., Trivedy, D.C.: Synth. Met. **81**, 39 (1996)
35. DallOlio, A., Dascola, Y., Varacco, V., Bocchi, V.: C. R. Acad. Sci. Ser. C **267**, 433 (1968)
36. Gardini, C.P.: Adv. Heterocyclic Chem. **15**, 67 (1973)
37. Novak, P.: Electrochim. Acta **37**, 1227 (1992)
38. Street, G.B., Clark, T.C., Geiss, R.H., Lee, V.Y., Nazzal, A., Pflunger, P., Scott, J.C.: J. Phys. (Paris) C **3**, 599 (1983)
39. Lei, J., Martin, C.R.: Synth. Met. **48**, 331 (1992)
40. Naarmann, H.: J. Polym. Sci. Polym. Symp. **75**, 53 (1993)
41. Aguilar-Hernandez, J., Potje-Kamloth, K.: Phys. Chem. Chem. Phys. **1**, 1735 (1999)
42. Phillips, G., Suresh, R., Waidman, J., Kumar, J., I-Chen, J., Tripathy, S., Huang, J.C.: J. Appl. Phys. **69**, 899 (1991)
43. Ion, R.M., Scarlat, F., Scarlet, F., Niculescu, V.I.R.: J. Optoelectron. Adv. Mater. **5**, 109 (2003)
44. Kang, E.T., Neoh, K.G., Tan, T.C., Ong, Y.K.: Polymer **27**, 1958 (1986)
45. Kang, E.T., Neoh, K.G., Tan, T.C., Ong, Y.K.: J. Macromal. Sci. Chem. A **24**, 631 (1987)
46. Yoon, C.O., Reghu, M., Moses, D., Heeger, A.J.: Phys. Rev. B **49**, 10851 (1994)
47. Alivisatos, A.P.: Science **271**, 933 (1996)
48. Ellingson, R.J., Beard, M.C., Johnson, J.C., Yu, P., Micic, O.I., Nozik, A.J., Shabaev, A., Efros, A.L.: Nano Lett. **5**, 865 (2005)
49. Berciaud, S., Cognet, L., Tamarat, P., Lounis, B.: Nano Lett. **5**, 515 (2005)
50. Kelly, K.L., Coronado, E., Zhao, L.L., Schatz, G.C.: J. Phys. Chem. B **107**, 668 (2003)
51. Jin, R.C., Cao, Y.W., Mirkin, C.A., Kelly, K.L., Schatz, G.C., Zheng, J.G.: Science **294**, 1901 (2001)
52. Mulvaney, P.: Langmuir **12**, 788 (1996)
53. Kossyrev, P.A., Yin, A., Cloutier, S.G., Cardimona, D.A., Huang, D., Alsing, P.M., Xu, J.M.: Nano Lett. **5**, 1978 (2005)
54. Jing, S., Xing, S., Zhao, C.: Mater. Lett. **62**, 41 (2008)
55. Gupta, K., Jana, P.C., Meikap, A.K.: Synth. Met. **160**, 1566–1573 (2010)
56. Gupta, K., Jana, P.C., Meikap, A.K.: Solid State Sci. **14**, 324–329 (2012)
57. Erdem, E., Karakisla, M., Sacak, M.: Euro. Polym. J. **40**, 785 (2004)
58. Shihai, D., Hui, M., Zhang, W.: J. Appl. Polym. Sci. **109**, 2842 (2008)
59. Bhattacharya, S., Srivastava, A.: Proceed. Ind. Acad. Sci. **115**, 613 (2003)
60. Shimano, J.Y., MacDiarmid, A.G.: Synth. Met. **123**, 251 (2001)
61. Wohlgenannt, M., Vardeny, Z.V.: J. Phys: Condens. Mater. **15**, R83 (2003)
62. Kohler, A., Wilson, J.S., Friend, R.H.: Adv. Mater. **14**, 701 (2002)

Chapter 30
Machine-Learning-Enhanced Polarization Splitter in Silicon-Integrated Dual-Core Photonic Crystal Fiber

Lavanya Anbazhagan, R. Jansi, and Sudhanya P.

30.1 Introduction

The advent of Photonic Crystal Fibers (PCF) has revolutionized photonics, offering unprecedented versatility in tailoring optical properties to a myriad of applications, including essential components like polarization filters, couplers, and wavelength coupler-splitters, integral to integrated optical systems. Among these, dual core (DC) PCF couplers stand out, providing design flexibility and shorter coupling lengths compared to conventional fiber couplers, holding immense promise for efficient wavelength coupler/splitters crucial in modern optical communication systems. Infusing DC-PCFs with materials like liquids, metals, and nematic liquid crystals (NLC) enhances coupling characteristics, yielding significantly shortened coupling lengths and unlocking new horizons in optical device miniaturization and performance improvement.

In the realm of DC-PCF polarization splitters, literature emphasizes optimizing geometrical parameters for superior polarization separation efficiency. Extensive numerical simulations and experimental validations guide this optimization, offering valuable predictive insights and real-world performance assessment. Researchers focus on refining PCF structures to enhance polarization splitter functionality, enriching the field of optical components and communication systems. This iterative

A. Lavanya · R. Jansi · P. Sudhanya (✉)
Department of Electronics and Communication Engineering, Faculty of Engineering and Technology, SRM Institute of Science and Technology, Kattankulathur 603203, India
e-mail: sudhanya.p@gmail.com

A. Lavanya
e-mail: lavanyaanbazhagan04@gmail.com

R. Jansi
e-mail: jansir@srmist.edu.in

M. El Ghzaoui et al. (eds.), *Next Generation Wireless Communication*, Signals and Communication Technology, https://doi.org/10.1007/978-3-031-56144-3_30

approach facilitates continual advancements and practical applications of polarization splitter [1]. The D-shaped PCF is engineered with a planar surface on one side, featuring a coating of metal to induce surface plasmons. This innovative device serves dual purposes, functioning as either a filter or a sensor based on the specific application requirements. The surface plasmon phenomenon enriches its capabilities, broadening its potential utility in the realm of optical technology [2]. The D-shaped PCF comprises two closely spaced cores exhibiting different effective indices, facilitating coupling between them through the evanescent field. This polarization-dependent coupling results in a distinct phase shift between polarization states, enabling effective separation of the two polarization states. The operational mechanism of the polarization beam splitter (PBS) relies on this differential phase shift, underscoring its significance in polarization-based optical applications [3]. A significant advantage of this approach lies in its seamless integration with established semiconductor processing techniques commonly employed for gallium arsenide (GaAs) materials fabrication. This compatibility facilitates the harmonious integration of polarization splitter devices with other GaAs-based optical components, including lasers and detectors. This result in compact and efficient optical platform, underlining the potential for streamlined and optimized optical systems using GaAs technology [4].

An ultra-broadband polarization splitter, utilizing a three-core PCF with a modulation core, is designed to efficiently divide an incident light beam into its two orthogonal polarization components, encompassing TE and TM modes, across a wide spectrum of wavelengths. The inclusion of a modulation core in the three-core configuration enhances the device's versatility. Additionally, the three-core PCF offers a substantial mode area, promoting reduced nonlinearities and high power handling capacities, further enhancing its performance and applicability [5]. A polarization splitter built upon a dual-core holey fiber with liquid crystal-filled is designed to partition an incident light beam into two output beams characterized by orthogonal polarization states. This device leverages a specialized type of optical fiber known as holey fiber, featuring air holes running along its length. The strategic arrangement of these air holes creates a distinct refractive index profile, enabling efficient guidance and manipulation of light. Holey fibers are extensively applied in optical sensing and measurement contexts, particularly in scenarios demanding precise control over polarization for optimal performance and accuracy [6]. In summary, the ultra-short PCF-based polarization splitter stands as a compact and efficient device with versatile applications in optical communications. Particularly, it finds utility in optical communication applications like wavelength division multiplexing (WDM) systems and fiber optic gyroscopes. The combination of its small form factor and high efficiency in polarization splitting renders it an appealing choice for seamless integration with various photonic components on a chip, paving the way for enhanced optical system integration and performance [7].

The dual-porous fiber-based broadband terahertz polarization splitter is a device engineered to effectively divide the polarization of terahertz (THz) waves into two orthogonal orientations with minimal loss and across a broad bandwidth. This innovative device capitalizes on the birefringent characteristic of dual-porous fiber, featuring two distinct pore types arranged in a periodic pattern. This arrangement creates a disparity in refractive index between the two polarizations, enabling efficient polarization splitting in the terahertz spectrum [8]. The fiber's structure comprises a central porous core surrounded by a regular arrangement of air holes in the cladding region. To begin the design process, the initial task involves identifying the ideal parameters for the porous core, including its diameter, porosity, and refractive index. Subsequently, the focus shifts to determining the optimal parameters for the air holes in the cladding region. Once the optimal structural parameters for both the core and cladding are established, the subsequent step involves a comprehensive evaluation of the fiber's performance as a polarization beam-splitter [9]. A DC-PCF based terahertz (THz) polarization beam splitter (PBS) stands as a device engineered to effectively divide an incoming THz beam into two orthogonal polarizations with exceptional efficiency across a broad spectrum of frequencies. Notably, this PBS achieves ultra-high extinction ratios, making it a highly desirable choice for applications demanding precise and high-quality polarization splitting of THz waves. Its superior efficiency and broad operating bandwidth position it as an attractive solution for high-speed THz communication systems and various other applications where reliable and efficient polarization splitting of THz waves is imperative [10]. The polarization splitter described employs a dual-core soft glass PCF enhanced with a micron-scale gold wire, effectively splitting an incident light beam into two orthogonal polarizations with notable efficiency. The innovative design includes a modification of the dual-core soft glass PCF structure by embedding a micron-scale gold wire between the two cores. In the simulation process, the optical properties of this structure are meticulously modeled using advanced numerical techniques such as FEM or FDTD methods. The simulation outcomes offer valuable insights into the electromagnetic fields and optical properties of the fiber, aiding in identifying potential avenues for enhancing the device's performance [11]. In recent research [12], a silicon nanowire-integrated DC-pentagonal PCF has been proposed as an efficient polarization-independent wavelength splitter. Employing advanced analysis methods like the FEM and coupled-mode theory (CMT), the study optimizes the PCF's structural parameters to achieve a coupling length ratio of 2. The resulting design exhibits notable reductions in cross-talk (CT) and expanded bandwidths for both X- and Y-polarization modes at critical wavelengths, showcasing promise for integration into advanced optical systems.

The integration of machine learning (ML) in the realm of PCF design has brought about a revolutionary paradigm shift. ML algorithms, including neural networks and genetic algorithms, play a pivotal role in optimizing and customizing the PCF geometry and structure to attain specific optical properties. This advancement empowers automated exploration of extensive design spaces, facilitating precise parameter selection such as hole size, arrangement, and materials to achieve desired characteristics like dispersion, nonlinearities, and birefringence. The synergy of ML and PCF design expedites innovation, providing tailored fibers for a myriad of applications encompassing telecommunications, sensing, and high-power laser systems [13]. A pioneering utilization of feed-forward multilayer perceptron artificial neural networks, revolutionizing the prediction of 12 crucial optical parameters for silica-based PCF is explored [14]. The optimized approach demonstrates exceptional accuracy, exceeding conventional numerical simulations by over 99.9% in speed.

In this research, our primary focus is directed towards the meticulous design of a compact PCF based polarization-splitter. We delve into the intricacies of DC-PCF, utilizing the principles of CMT tailored for this specific configuration. Our overarching objective is the optimization of structural parameters within this fiber design to significantly enhance coupling efficiency at the critical wavelength of 1.55 μm. This optimization endeavor holds immense promise for revolutionizing optical device compactness and operational efficiency, thus driving notable advancements in the field. To achieve our objective, we adopt a comprehensive and interdisciplinary approach that seamlessly integrates the finite element method (FEM) and advanced machine-learning algorithms. This collaborative approach empowers us to automate and streamline the fine-tuning process for the DC-PCF's structural parameters. The ultimate aim is to achieve a specific coupling length ratio, critical for optimal performance. By leveraging computational simulations in harmony with machine-learning-driven optimization, we aspire to unlock the full potential of DC-PCF technology. This, in turn, will catalyze the creation of highly compact and efficient optical devices, fostering a new era in photonics and optical communication.

30.2 Material and Design

Figure 30.1 illustrates a proposed design for a PCF-based polarization splitter. The design consists of a hexagonal arrangement of air holes, with a specific modification involving the removal of two adjacent air holes in the first cladding layer to achieve a DC configuration. After meticulous analysis, the optimal structural parameters were determined: an air hole diameter (d) of 1.2 μm, a pitch between air holes (Λ) of 2.15 μm, and an air filling fraction (d/Λ) of 0.56. To enhance the birefringence of the structure, a suggestion is to replace one of the air holes in the core with silicon material, with a diameter (d_1) of 0.18 μm.

30.3 Numerical Analysis

To ascertain the effective refractive index and field distribution within the envisaged PCF, a solution to Maxwell's wave equation is pursued. This computational approach allows for a comprehensive understanding of the fiber's optical properties and the behavior of electromagnetic fields within the structure [15].

$$\nabla \times (\nabla \times E) - k_0^2 \epsilon_r E = 0 \tag{30.1}$$

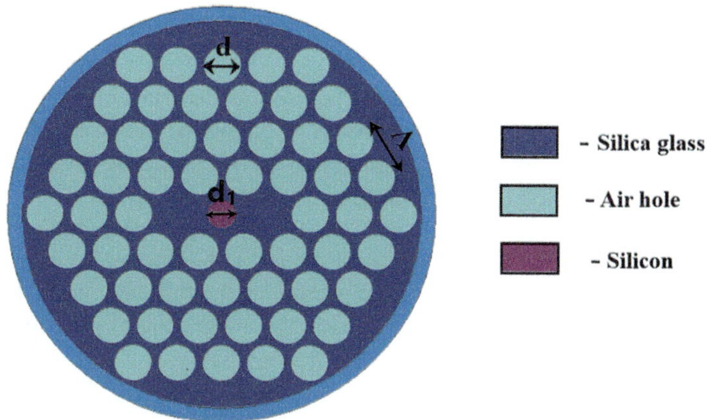

Fig. 30.1 Schematic representation of the proposed polarization splitter-based PCF

The modal field within a DC-PCF can be expressed as a combination of odd and even modes, in accordance with CMT. The coupling length (CL) in this context represents the minimum length at which optimal power transfer occurs between the two cores. Mathematically, the expression for the coupling length (CL) is given by [16]:

$$\mathrm{CL}(\lambda)_i = \frac{\pi}{\beta_i^{\mathrm{even}} - \beta_i^{\mathrm{odd}}} = \frac{\lambda}{2\left(n_i^{\mathrm{even}} - n_i^{\mathrm{odd}}\right)} \tag{30.2}$$

In this equation, the variable "*i*" represents the polarization mode (*X* or *Y*), while λ refers to the wavelength of the incident light. The effective refractive indices of the even mode and odd mode for the *X* or *Y* polarization mode are denoted by n^{even} and $n^{\mathrm{odd},}$ respectively.

To quantify the relationship between the coupling length (CL) of the *X*-polarized mode and that of the *Y*-polarized mode at a wavelength of 1.55 μm, we introduce the term CLR (coupling length ratio). The formula to compute CLR is expressed as follows:

$$\mathrm{CLR}_\lambda = \frac{\mathrm{CL}_x}{\mathrm{CL}_y} \tag{30.3}$$

Cross-talk is defined as the amplitude of the undesired wavelength that remains present at the output of the core alongside the desired wavelength. It can be quantitatively expressed as follows:

$$\mathrm{CT} = 10\log_{10}\left(\frac{P_{\mathrm{out}}^x}{P_{\mathrm{out}}^y}\right) \tag{30.4}$$

30.4 Results and Discussions

In the analysis of our proposed polarization splitter, we leveraged the powerful computational tool, COMSOL Multiphysics 5.2a, which is based on the FEM. Employing the CMT, we explored the intricate behavior of the DC-PCF under study. Our investigation unveiled the fascinating phenomenon of superposition within the PCF, yielding four distinct super modes: *X*-even, *Y*-even, *X*-odd, and *Y*-odd. These modes manifest due to the interplay of core modes with either identical or opposite phases. Figure 30.2 in our analysis visually captures the intricate modal field distribution of these super modes, shedding light on the underlying physics and behavior of our proposed polarization splitter.

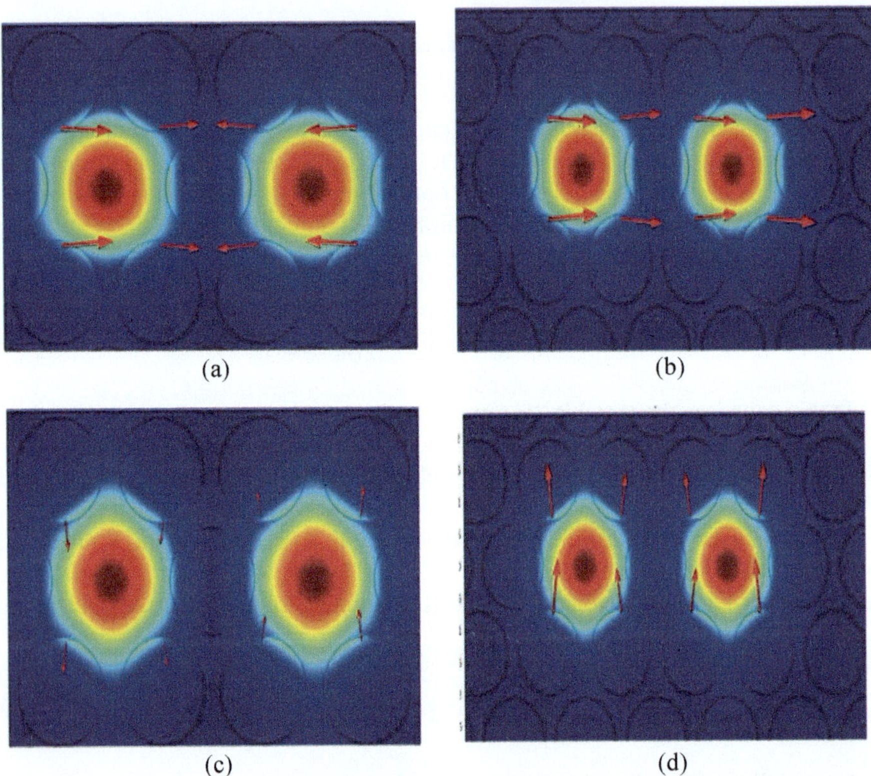

Fig. 30.2 Visual representation of the modal field distribution for the super modes **a** *X*-odd, **b** *X*-even, **b** *Y*-odd, and **d** *Y*-even

30.4.1 Optimization of PCF Design Parameters

When illuminating the dual-core fiber with input light, power coupling from one core to another transpires owing to differing phase shifts encountered during propagation. Efficient separation of input comprising two wavelengths at the output of the cores is achieved with a short coupling length when the CLR remains at 2. To optimize the performance, a meticulous analysis is conducted to determine the ideal values of structural parameters: pitch (Λ), air hole diameter (d), and silicon nanowire diameter (d_1) in the proposed PCF, ensuring the CLR attains the desired value of 2. This optimization process is pivotal in achieving superior performance and functionality of the PCF-based polarization splitter.

Influence of Pitch (Λ)

In order to assess the impact of the pitch value on the CLR at a wavelength of 1.55 μm for both X- and Y-polarization, a systematic study was conducted. The pitch value was varied within a range of 1.51–1.63 μm, maintaining a consistent diameter for the air hole at 1.2 μm. The resulting CLR values were then presented in Fig. 30.3.

It was observed that an increase in the pitch value led to a proportional increase in the separation between the two cores. Consequently, this elongated the propagation length, resulting in an augmentation of both the CL and CLR. This experimental analysis sheds light on the pivotal role the pitch value plays in tailoring the polarization behavior of the PCF, which is vital for optimizing its performance in various applications, particularly at the crucial wavelength of 1.55 μm.

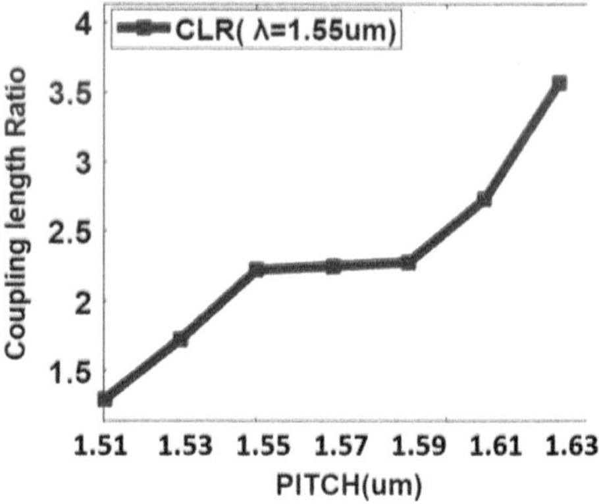

Fig. 30.3 Influence of the pitch (Λ)

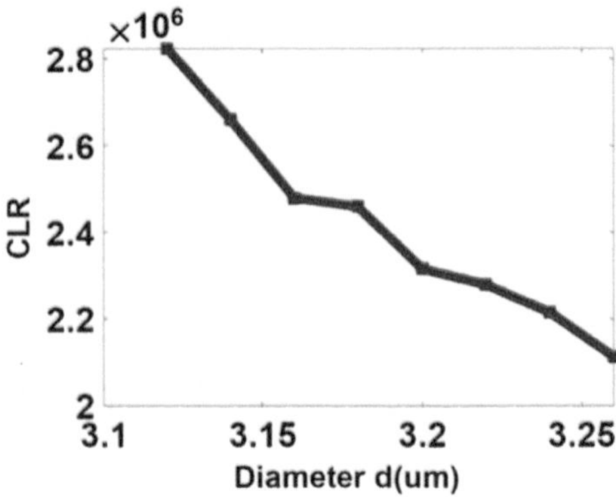

Fig. 30.4 Influence of the cladding air hole diameter on CLR

Influence of Cladding Air-Hole Diameter (*D*)

To determine the optimal cladding air hole diameter (d) for achieving a coupling ratio (CLR) of 2 while maintaining a consistent pitch of 2.15 μm, a systematic study was conducted. The value of d was varied within the range of 3.15–3.25 μm. The resulting CLR values were computed at a wavelength of 1.55 μm and showcased in Fig. 30.4.

It was observed that with an increase in d, the guiding mode became more confined to the core, reducing the likelihood of leakage to the cladding region. This confinement enhancement led to improved coupling efficiency and a subsequent decrease in CL. The plot in Fig. 30.4 clearly illustrates the inverse relationship between CLR and d, further emphasizing the critical role of d in optimizing the performance of the PCF-based polarization splitter.

Influence of Silicon Diameter d_1

In the proposed PCF, a silicon nanowire is strategically embedded at the center to create separation between the two cores. The impact of the diameter of the silicon nanowire (d_1) on the CLR for both X- and Y-polarization is thoroughly examined. Keeping the pitch (Λ) constant at 2.15 μm and maintaining the cladding air hole diameter (d) at 1.2 μm, CLR values are calculated for varying d_1 diameters, as illustrated in Fig. 30.5.

With an increase in the value of d_1, the separation between the two cores widens, resulting in an increase in both CL and CLR, as depicted in Fig. 30.5. For the specific case of achieving a CLR of 2, the optimal d_1 is determined to be 0.18 and 0.175 μm for X- and Y-polarization, respectively. Therefore, the diameter of d_1 is selected as 0.18 μm, effectively yielding CLR values of 2.004 and 2.112 for X- and Y-polarization, respectively, further validating the design and optimization of the proposed PCF-based polarization splitter.

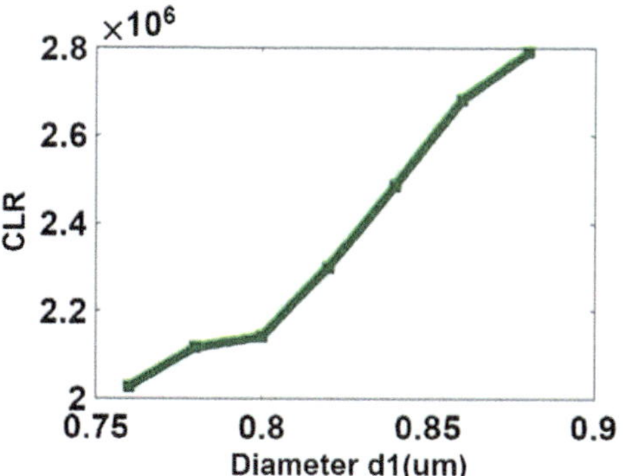

Fig. 30.5 Influence of diameter of silicon in the core on the CLR

30.4.2 Features of the Proposed PCF

The most effective structural parameters, derived from our analysis for the proposed PCF-based polarization splitter, are a pitch (Λ) of 2.15 μm, a cladding air hole diameter (d) of 1.2 μm, and a silicon nanowire diameter (d_1) of 0.18 μm. In this optimized dual-core PCF design, four super modes (x-even, x-odd, y-even, and y-odd) are excited, necessitating consideration of the effective refractive index for a comprehensive wavelength-dependent analysis. The effective refractive index, illustrated in Fig. 30.6a for the DC configuration embedded with air holes and in Fig. 30.6b for Si core, portrays its dependency on the wavelength.

The DC pentagonal PCF structure is recognized for its inherent asymmetry, leading to mode degeneracy. As the wavelength is elevated, the effective refractive index difference between even and odd modes becomes more pronounced for both X- and Y-polarization modes. This disparity is clearly evident in Fig. 30.6a and b, underscoring its heightened significance particularly in the Si core configuration.

The computation of coupling length (CL) at different wavelengths is portrayed in Fig. 30.7, considering both the air hole core and silicon core while varying the wavelength.

The transfer of energy between the two cores mandates a more extensive CL for lower wavelengths due to the tightly confined modes within the core. Conversely, an increase in wavelength leads to reduced mode confinement, consequently causing a decrease in CL. Notably, given the parallel alignment of the cores along the x-axis, the X-polarized mode experiences a more robust coupling, resulting in a shorter CL compared to the Y-polarized mode. It is essential to highlight that CL is inversely related to the difference between the refractive indices of the odd and even modes. The PCF embedded with silicon exhibits a more substantial disparity between odd and even modes, leading to a shorter CL in comparison with the air hole core.

The CL values acquired at the specified wavelength of 1.55 μm for the proposed silicon core PCF are presented in Table 30.1.

These results affirm the applicability of the proposed PCF for efficient polarization splitting operations.

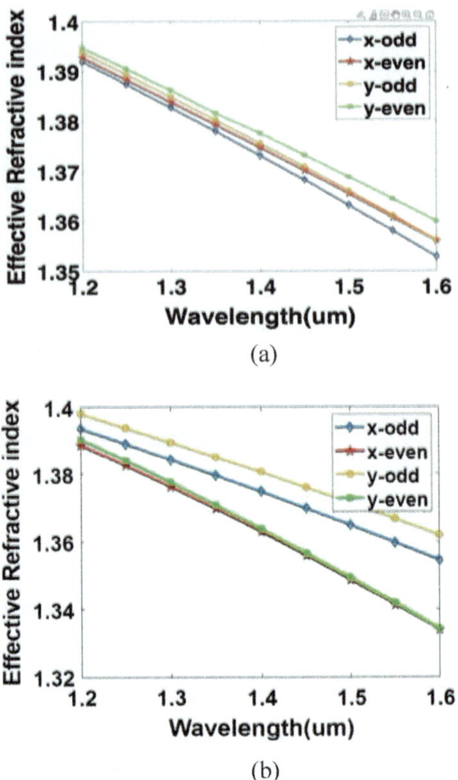

Fig. 30.6 Effective refractive index (ERI) characteristics of the proposed PCF with **a** an air-hole core and **b** a silicon (Si) core

30.4.3 Cross-Talk Performance of the Proposed PCF

The Cross-talk (CT) serves as a pivotal parameter in quantifying the extent of undesired wavelengths present alongside the intended wavelength. Within the context of the envisaged polarization splitter based on PCF, this parameter assumes crucial importance.

In the specific scenario described, upon introducing light containing a particular wavelength into core A of the proposed PCF-based polarization splitter, the maximum power output is achieved from both core A and core B at precisely 1.55 μm, covering both polarization modes. However, it remains imperative to consider the potential cross talk (CT) that may manifest from spectral components around the undesired wavelength due to fabrication tolerances. These tolerances, inherent to the fabrication process, can introduce variations, necessitating a thorough evaluation and management of the CT.

It is essential to account for and comprehensively assess the CT to gain insights into the influence of unwanted wavelengths on the performance of the polarization splitter. Understanding the CT enables necessary adjustments and fine-tuning, ensuring optimal efficiency and functionality (Fig. 30.8).

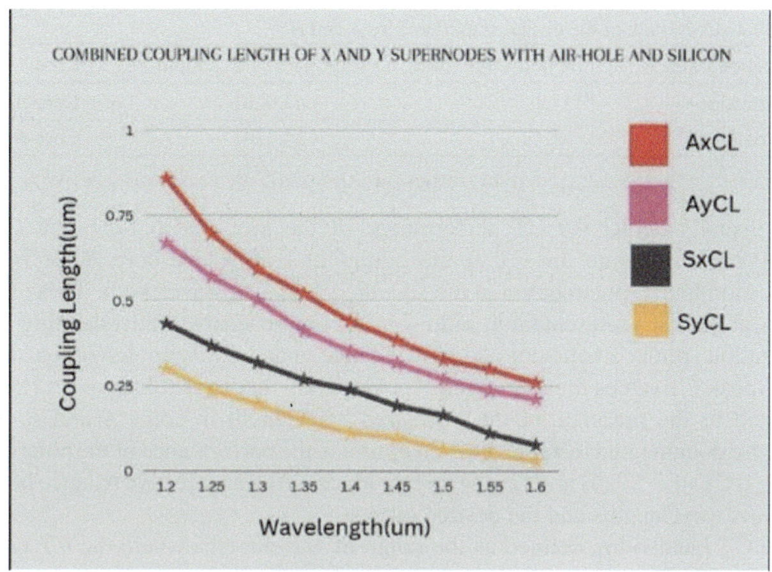

Fig. 30.7 CL characteristics as a function of wavelength

Table 30.1 CL and CLR obtained for the proposed PCF at 1.55 μm

Polarization mode	CL at 1.55 um	CLR = CL_x/CL_y
X-polarization	112.1	2.1
Y-polarization	51.12	

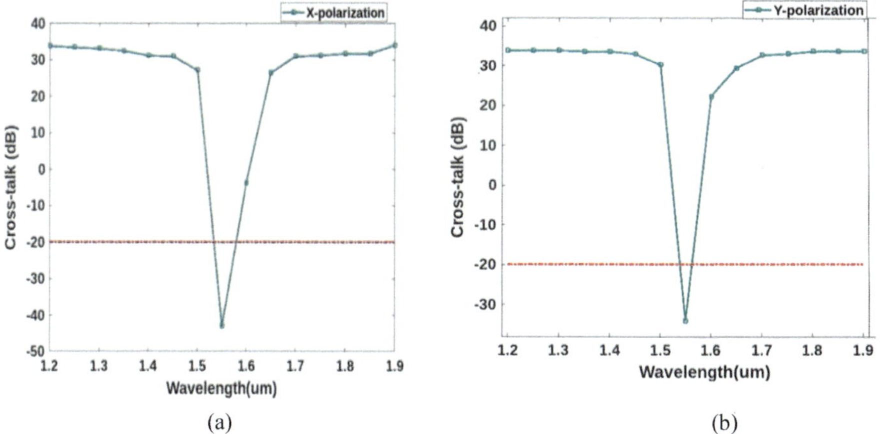

Fig. 30.8 CT efficiency of the polarization splitter at **a** core A (X-polarization) and **b** core B (Y-polarization)

Table 30.2 Cross-talk of the designed PCF at core *A* and *B*

Polarization mode	Device length (um)	CT (dB) at 1.55um	Core
X-polarization	112.1	– 43.0261	Core-*A*
Y-polarization	51.12	– 34.1291	Core-*B*

To visually depict the CT performance of the proposed polarization splitter, Fig. 30.8a, b illustrate the CT at the output of core *A* and core *B* for both *X*-polarization and *Y*-polarization at the specified 1.55 μm wavelength. These figures offer a graphical representation, aiding in a comprehensive understanding of the polarization splitter's behavior and performance under distinct polarization states.

The peak CT values for both *X* and *Y* polarization modes are observed at 1.55 μm, attributed to the presence of the undesired wavelength in cores *A* and *B*. These values are documented in Table 30.2. To optimize the performance of the polarization splitter, a CT of – 20 dB is set as the threshold, ensuring an effective balance between unwanted wavelengths and the desired output.

The CT bandwidth, defined as the range of wavelengths where the CT remains below – 20 dB, is critical for assessing the effectiveness of the wavelength splitter. This bandwidth signifies the operational range within which the polarization splitter demonstrates desirable performance by minimizing the impact of undesired wavelengths.

Table 30.2 encapsulates the CT values and their corresponding CT bandwidth, providing a comprehensive view of the polarization splitter's performance across different wavelengths.

The analysis unequivocally indicates that the proposed PCF necessitates varying lengths to effectively operate as a polarization splitting device for the *X* and *Y* polarization modes.

30.5 Implementing Machine Learning

Utilizing machine-learning algorithms for the optimization of PCF offers a more efficient alternative to conventional techniques. This approach involves collecting and structuring data on PCF properties and parameters, engineering relevant features, selecting appropriate machine-learning models (e.g., regression, classification, neural networks), training and validating the model, and evaluating its performance [17]. By employing the trained model to predict PCF properties based on input parameters, and integrating it with an optimization algorithm, the PCF design can be effectively optimized to meet desired performance metrics. This iterative process allows for the automation and acceleration of the PCF optimization, ultimately achieving optimal designs for various applications.

The foundation of a robust machine-learning model lies in a high-quality dataset that embodies attributes such as data completeness, consistency, relevance, and accuracy. These qualities are fundamental for effective model training and predictive accuracy. Additionally, addressing class imbalances in classification tasks is vital to ensure fair and reliable model outcomes.

In the realm of predictive modeling, the Multilayer Perceptron Regression Algorithm (MLP Regressor) proves to be highly effective. This algorithm stands out for its ability to capture intricate non-linear relationships within the data, making it particularly useful in scenarios where complex patterns exist. Notably, omitting an activation function in the output layer enhances the MLP Regressor's performance. The backpropagation mechanism, a core element of the MLP Regressor, optimizes model weights during training. This optimization process significantly contributes to enhancing the predictive performance of the model. When combined with a well-curated dataset, leveraging the strengths of the MLP Regressor holds immense promise for constructing highly accurate predictive models across diverse domains. This amalgamation of a powerful algorithm and quality data can revolutionize predictions and insights, ultimately advancing progress and understanding in various fields.

In Fig. 30.9, we visualize the process of optimizing structural parameters using the MLP Regressor. The inputs for this optimization are the air hole radius, pitch, and silicon core radius, while the output is represented by the coupling length ratio. The MLP Regressor, a powerful machine-learning algorithm, is harnessed to understand and model the intricate relationships that exist between the input parameters and the desired coupling length ratio. By employing this algorithm, we can fine-tune the structural dimensions of the PCF, aiming to enhance its coupling characteristics effectively.

To facilitate analysis and modeling, we compiled a dataset encompassing three crucial input parameters and one output parameter. The dataset incorporates various desired output node quantities, meticulously organized and stored in an Excel sheet for further analysis and modeling endeavors. This structured dataset serves as a

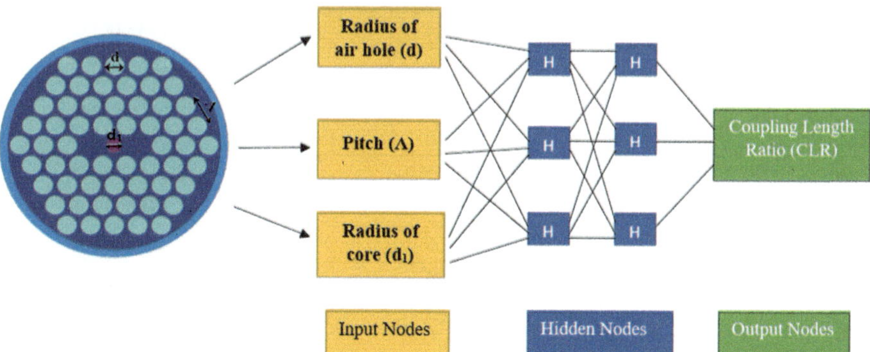

Fig. 30.9 Optimization of structural parameters using machine-learning algorithm

fundamental resource for training and optimizing machine-learning models tailored to the specific attributes and behavior of hexagonal PCFs.

All essential input and output variables have been successfully compiled and stored in CSV/XLSX format within the dataset as tabulated in Table 30.3. These values have undergone a scaling process, rendering them suitable for utilization as inputs to our machine-learning model. During model training and evaluation, these scaled values will fuel the learning process, enabling the model to discern patterns and make accurate predictions. Upon obtaining model predictions, an inverse transform will be performed to restore the values to their original scale, providing us with real and interpretable results. This crucial step ensures that our model's output is easily understandable and applicable within the context of the original problem.

Table 30.3 Input dataset to MLP regressor

Pitch (Λ)	Core radius (r_1)	Cladding air hole (r)	CLR
1.55	0.4	0.5	2.129032258
1.55	0.41	0.51	2.142857143
1.55	0.42	0.53	2.192307692
1.55	0.43	0.52	2.173913043
1.55	0.44	0.54	2.166666667
1.55	0.45	0.58	2.157894737
1.55	0.46	0.57	2.2
1.55	0.47	0.56	2.166666667
1.54	0.4	1.46	2.125
1.53	0.4	1.47	2.125
1.52	0.4	1.49	2.125
1.51	0.4	1.48	2.125
1.5	0.4	1.53	2.161290323
1.49	0.4	1.51	2.15625
1.48	0.4	1.52	2.1875
1.47	0.4	1.54	2.193548387
1.53	0.4	1.55	2.166666667
1.55	0.42	1.55	2.2
1.57	0.44	1.55	2.136363636
1.59	0.46	1.55	2.210526316
1.52	0.45	1.55	2.136363636
1.54	0.43	1.55	2.260869565
1.56	0.41	1.55	2.185185185
1.51	0.39	1.55	2.151515152

30.5.1 Program Code

The program given below showcases a comprehensive approach to predictive modeling using the Multilayer Perceptron Regressor (MLP Regressor) in the domain of PCF [18].

```
import numpy as np import pandas as pd
from sklearn.model_selection import train_test_split from
sklearn.neural_network import MLPRegressor
from sklearn.preprocessing import MinMaxScaler
scale = MinMaxScaler(feature_range=(0, 1))
num_inputs = 3
num_outputs = 1
datafile = pd.read_excel('/content/drive/MyDrive/googlecollab/
Copy of 2VARIABLE.ipynb')
scale.fit(datafile)
datafile_scale = scaler.transform(df)
x_data = datafile_scale [:,range(0, num_inputs)]
y_data=datafile_scale[:,range(num_inputs,num_inputs+num_
outputs)]
X_traindata,X_testdata,y_traindata,y_testdata=train_test_
split(x_data,y_data,test_size = 0.1)
epochs = 1000
mlp = MLPRegressor(shuffle=True, random_state=1,
        max_iter=epochs)
mlp.fit(X_traindata, y_traindata)
def prediction(data):
predicted_output = mlp.predict(data)
final =np.concatenate((data, predicted_output.reshape(-1,1)),
axis=1)
return scaler.inverse_transform(final)
print(prediction(X_testdata))
def predict_on_user_input(user_input):
user_input = np.append(user_input, 0).reshape(1,-1) user_input =
scaler.transform(user_input)
return prediction(user_input[:,0:num_inputs])
output = predict_on_user_input ([1.92,1.6,1.20])
print('output: ', output)
```

The provided code intricately orchestrates the essential stages of data preprocessing and machine-learning model implementation. It commences by importing fundamental Python libraries, such as pandas for efficient data manipulation and scikit-learn for robust machine-learning functionalities. A pivotal aspect of this code is the instantiation of a MinMaxScaler object, a fundamental preprocessing tool that standardizes the dataset, scaling it within the [0, 1] range—a critical requisite for ensuring optimal model performance. Moreover, the code mandates the specification of the number of input and output features, paramount for effective data partitioning and model training. Subsequently, the dataset, assumed to hold indispensable data for the modeling process, is extracted from an Excel file. Employing the initialized MinMaxScaler, the dataset undergoes a crucial scaling process, harmonizing

```
[[1.55      0.47      0.56      2.16876185]
 [1.52      0.45      1.55      2.17797055]
 [1.59      0.46      1.55      2.17762379]]
output:  [[1.92      1.6       1.2       2.50194135]]
```

Fig. 30.10 Output of the machine-learning model

the data and amplifying the model's accuracy. The code concludes with the definition of a multilayer perceptron (MLP) regressor model, its training on the designated training set, and subsequent predictions on both the test set and user-input data—forming a comprehensive and adaptable framework for predictive modeling endeavors (Fig. 30.10).

The output is calculated based on the user-provided inputs, specifically with a pitch of 1.92, a core material radius (r_1) of 1.6, and a cladding airhole radius (r) of 1.2. Upon processing these inputs through the model, the predicted Coupling Length Ratio (CLR) is determined to be approximately 2.50194135. This predictive outcome showcases the model's ability to estimate the coupling length ratio, a critical parameter in the context of the given application or system. It underscores the importance of accurate modeling and how varying input parameters can influence the resultant predictions, aiding in better understanding and decision-making within the domain of PCF.

30.6 Conclusion

This research presents a groundbreaking silicon embedded DC hexagonal PCF-based polarization splitter that effectively separates the desired 1.55 μm wavelength for both Xpol and Ypol modes. The optimization of key structural parameters achieves a CLR of 2, crucial for efficient polarization splitting. The designed PCF, only 112 μm in length, achieves impressive Crosstalk (CT) efficiencies of approximately − 43.0261 dB for *X*-polarized and − 34.1291 dB for *Y*-polarized modes at 1.55 μm. This outperforms existing literature and positions the polarization splitter as a compact, high-efficiency solution for optical communication systems and sensors.

Furthermore, the study underscores the vast potential of PCF in advancing photonics technologies. Opportunities for future research span diverse areas including novel PCF designs, specialty PCFs, advanced integration techniques, and applications in quantum photonics. The limitations of the current machine-learning model are acknowledged, prompting future exploration of alternative algorithms to enhance predictive accuracy. This research serves as a solid foundation, paving the way for further innovation and application of PCFs in cutting-edge photonics technologies.

References

1. Rajeswari, D., Raja, A.S., Selvendran, S.: Design and analysis of polarization splitter based on dual-core photonic crystal fiber. Optik **144** (2017)
2. Selvendran, S., Divya, J., Sivananthan Raja, A., Sivasubramanian, A., Itapu, S.: A reconfigurable surface-Plasmon-based filter/sensor using D-shaped photonic crystal fiber. Micromachines **13**(6), 917 (2022)
3. Bai, Y., Hao, R., Yuan, B., Cao, B.: An ultrashort length and high extinction ratio polarization beam splitter based on dual-core PCF. J. Opt. **50**, 257–263 (2021)
4. Ding, Y., Liu, C., Yang, L., Lv, J., Fu, G., Li, X., Liu, Q., Wang, F., Sun, T., Chu, P.K.: Efficient photonic crystal fiber polarization splitters composed of gallium arsenide and nematic liquid crystals. Mod. Phys. Lett. B **35**(04), 2150077 (2021)
5. Zhao, T., Lou, S., Wang, X., Zhou, M., Lian, Z.: Ultrabroadband polarization splitter based on three-core photonic crystal fiber with a modulation core. Appl. Opt. **55**(23), 6428–6434 (2016)
6. Wang, E.L., Jiang, H.M., Xie, K., Chen, C., Hu, Z.J.: Polarization splitter based on dual core liquid crystal-filled holey fiber. J. Appl. Phys. **120**(11), 114501 (2016)
7. Chiang, J.S., Sun, N.H., Lin, S.C., Liu, W.F.: Analysis of an ultrashort PCF-based polarization splitter. J. Lightwave Technol. **28**(5), 707–713 (2009)
8. Li, S., Zhang, H., Bai, J., Liu, W., Jiang, Z., Chang, S.: Dual-porous fiber-based low loss broadband terahertz polarization splitter. IEEE Photonics Technol. Lett. **26**(14), 1399–1402 (2014)
9. Reyes-Vera, E., Usuga-Restrepo, J., Jimenez-Durango, C., Montoya-Cardona, J., Gomez-Cardona, N.: Design of low-loss and highly birefringent porous-core photonic crystal fiber and its application to terahertz polarization beam splitter. IEEE Photonics J. **10**(4), 1–13 (2018)
10. Zhang, J.X.: Dual-core PCF-based THz polarization beam splitter with broad bandwidth and ultra-high extinction ratio. Optik **251**, 168425 (2022)
11. Liu, Q., Li, S., Wang, X., Shi, M.: Theoretical simulation of a polarization splitter based on dual-core soft glass PCF with micron-scale gold wire. Chin. Phys. B, **25**(12), 124210 (2016). https://doi.org/10.1088/1674-1056/25/12/124210
12. Anbazhagan, L., Ganesan, G.: Polarization-independent wavelength splitter based on silicon nanowire embedded dual-core pentagonal lattice photonic crystal fiber for optical communication systems. Opt. Eng. **60**(4), 047103 (2021). https://doi.org/10.1117/1.OE.60.4.047103
13. Chen, S., Wang, X., Ren, S., Yang, J., Zhang, Y., Wang, G.: Collaborative photonic crystal fiber property optimization: a new paradigm for reverse design. IEEE Photonics Technol. Lett. **35**(19), 1035–1038 (2023). https://doi.org/10.1109/LPT.2023.3296674
14. Jabin, M.A., Fok, M.P.: Prediction of 12 photonic crystal fiber optical properties using MLP in deep learning. IEEE Photonics Technol. Lett. **34**(7), 391–394 (2022). https://doi.org/10.1109/LPT.2022.3157266
15. Knight, J.C., Birks, T.A., Atkin, D.M., Russell, P.S.J.: Pure silica single-mode fibre with hexagonal photonic crystal cladding. In: Optical Fiber Communication Conference, Paper PD3, Feb. 1996, p. PD3 23 (1996)
16. Agrawal, G.P.: Nonlinear Fiber Optics. Elsevier Science (2007). [Online]. Available: https://books.google.co.in/books?id=b5S0JqHMoxAC
17. Jansi, R., Amutha, R.: Hierarchical evolutionary classification framework for human action recognition using sparse dictionary optimization. Swarm Evol. Comput. **1**(63), 100873 (2021)
18. Chugh, S., Gulistan, A., Ghosh, S., Rahman, B.M.A.: Machine learning approach for computing optical properties of a photonic crystal fiber. Opt. Express **27**, 36414–36425 (2019)

Chapter 31
Progress in Birefringent Material-Based Achromatic Phase Modulators

Nilanjan Mukhopadhyay, Anirban Patra, and Arijit Saha

31.1 Introduction

In polarization optical device designing, the critical factor is birefringence, also known as double refraction, generally found in an anisotropic medium. In an anisotropic medium, unpolarized light splits into two rays, each traveling with different velocities and having orthogonal polarization states. The working of birefringent-based devices like retarders, optical filters, compensators, etc., is based on introducing a specific phase shift between the o-ray and e-ray that leads to a change in the polarization state of the output light wave.

A stack of quarter-wave plates exhibiting an achromatic nature is in high demand since the wavelength dependency of a single quarter-wave plate is increased. Earlier, several achromatic combinations were proposed containing two or more plates covering various wavelength regions, focused on particular applications. Such an achromatic combination based on the same material, presented by Pancharatnam, covers the 400–700 nm wavelength range [1]. Several other systems for different spectrum regions have been proposed by Hariharan and many others [2, 3]. A combination of more than two plates of the same material has been submitted by Hariharan and Malacara, which behaves as an achromatic half-wave retarder [4]. A super achromatic zero-order wave plate having more than two wave plates has been studied [5]. In the 400–700 nm range, Harris, Ammann, and Chang have designed an achromatic phase modulator combination [6]. Two plates of different materials have been

N. Mukhopadhyay (✉)
Department of ECE, Global Institute of Management and Technology, Krishnagar, Nadia, India
e-mail: nilu.opt@gmail.com

A. Patra
Department of ECE, JIS College of Engineering, Kalyani, Nadia, West Bengal, India
e-mail: anirban.patra@jiscollege.ac.in

A. Saha
Dum Dum Motijheel Rabindra Mahavidyalaya, Kolkata, West Bengal, India

M. El Ghzaoui et al. (eds.), *Next Generation Wireless Communication*, Signals and Communication Technology, https://doi.org/10.1007/978-3-031-56144-3_31

designed over a limited wavelength range that show achromatic behavior [7]. Using a stack of three crystalline quartz plates, an achromatic system has been designed recently in the visible range [8].

A lot of studies have been done on the performance of birefringent networks. The procedure for synthesizing the birefringent network consists of identical cascaded birefringent crystals between an input and an output polarizer, which has been studied by Harris et al. [9–15]. The synthesis of the network considers a lossless network of equal length and also a double-pass network. In a linear birefringent network, circular birefringent material has been studied [16]. Pochi et al. described the equivalent circuit of a general birefringent network, which is nothing but a liquid crystal cell (LC) comparable to a waveplate followed by a polarization rotator [17]. The method for the measurement of minor phase differences using optical compensators has been discussed by Jerrard [18]. Optical retarders play a vital role in an imaging system as they take care of the phase shift and, in turn, the polarization state of the incoming light. The system's performance depends very much on wavelength, as referred to in Eq. (31.1), considering the relation between phase shift and wavelength. This phase change will affect the polarization state. Therefore, it is required to have a constant phase shift over the operating wavelength range. This phenomenon is known as achromaticity. Now, to design this achromatic retarder system, there is a definite need for the birefringent network, which consists of quarter and half waveplates. The realization of an achromatic phase modulator by the juxtaposition of two crystal plates of exact nature was first proposed by Destriau et al. [19]. A stack of three sapphire birefringent plates to form an achromatic half-wave plate in the millimeter wavelength band has been reported [20]. Clarke has studied the effects of multiple-beam interference within a series of waveplates [21].

Phase modulators, e.g., phase retarders, have lots of application areas. Recent developments suggest that for defense applications, SWIR (Short wave infrared range) is very useful for achieving high-resolution infrared imaging [22, 23]. In astronomical polarimetry, using broadband achromatic quarter-wave phase retarders is very useful [24]. Demands for the designing of highly efficient optical systems in the visible to infrared spectrum region are for the remote sensing applications [25] and also to collect hyper-spectral images [26]. A demanding remote sensing technique based on circular polarization spectroscopy [27] is an attractive field for researchers. For cancer tissue detection, NIR (Near-infrared range) high-resolution spectroscopy for polarization imaging with the help of a quarter-wave plate is very useful [28]. NASA uses an achromatic quarter-wave retarder for the Multiangle Spectro-Polarimetric Imager (MSPI) [29]. For the determination of the polarization imaging of the body fluids, achromatic phase modulators can be used.

In the present chapter, recent developments in phase retarder design have been illustrated. The next section describes the mathematical relationships required for retarder designing. In Sect. 31.3, the creation of phase retarders based on wave plates of the same birefringent materials has been discussed, which is followed by a design based on different birefringent materials in Sect. 31.4, followed by conclusion in Sect. 31.5.

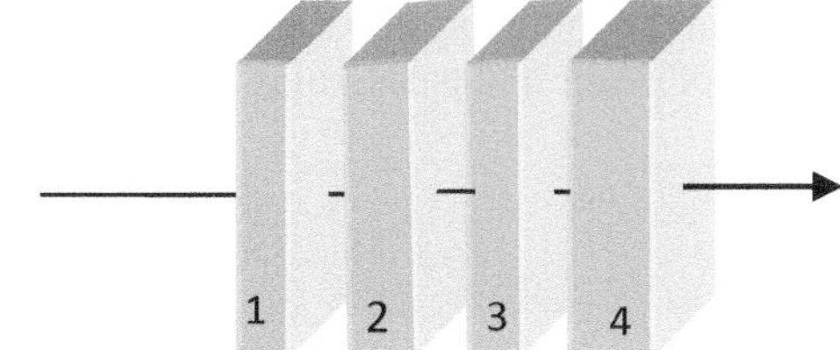

Fig. 31.1 Block diagram of different birefringent material-based four-plate achromatic quarter-wave systems with thicknesses in sequential order

31.2 Mathematical Aspects for Retarder Designing

Introduction of a retardation plate in the path of light, a phase difference is produced due to the path difference introduced between orthogonal components of the ray. In this section, we will discuss the necessary theoretical background and the proposed methodology for designing a super achromatic retarder.

31.2.1 Conventional Approach

Path differences introduced in this process will be taken into account for the retardation calculation. Consider an optical system of M number of wave plates with thickness as $d_1, d_2, \ldots, d_M$ respectively with N number of design wavelength, i.e., $\lambda_1, \lambda_2, \ldots, \lambda_N$. For a quarter-wave plate combination, plate thickness is obtained from the equation shown below [30]:

$$\begin{vmatrix} \triangle\, n_{11} & \triangle\, n_{12} & .. & \triangle\, n_{1M} \\ \triangle\, n_{21} & \triangle\, n_{22} & .. & \triangle\, n_{2M} \\ .. & .. & .. & .. \\ \triangle\, n_{N1} & \triangle\, n_{N2} & .. & \triangle\, n_{NM} \end{vmatrix} \begin{vmatrix} d_1 \\ d_2 \\ .. \\ d_M \end{vmatrix} = \pm\frac{1}{4} \begin{vmatrix} \lambda_1 \\ \lambda_2 \\ .. \\ \lambda_N \end{vmatrix} \tag{31.1}$$

where $\triangle\, n_{NM}$ is the value of the birefringence of the material of the Mth plate at the design wavelength λ_N. Using Cramer's rule, plate thickness is computed from Eq. (31.1) [31]. The achromatic combination acts as a $\pi/2$ phase shift plate (Fig. 31.1).

31.2.2 Calculation of Plate Thickness

With the proper values of the birefringence at the designing wavelengths, the plate thickness can be obtained from Eq. (31.1). After deciding the behavior of phase modulator, the next step is to consider a few discrete wavelengths over the wavelength range with known values of crystal birefringence taken from [32].

From the following equation, the overall retardation value can be calculated with M no. of plates at the designing wavelength (i.e., λ_i) [30].

$$\varphi = \frac{2\pi}{\lambda} \sum_{j=1}^{M} \triangle\, n_{i,j} d_j \tag{31.2}$$

where the birefringence of the plates is $\triangle\, n_{i,j}$ and wavelength of interest is λ_i.

31.3 Achromatic Phase Retarder Having Plates of the Same Birefringent Materials

In the present section, a few of the recently proposed achromatic phase retarders having combinations of different numbers of waveplates have been discussed. Various design aspects and the overall performance of the systems have been illustrated.

31.3.1 Superachromatic Phase Modulator with the Same Materials in the Visible Range

In the present study, a system having three waveplates of crystalline Quartz, of which two waveplates are quarter-wave retarder (QWR), and the middle wave plate is half-wave retarder (HWR), and the designing wavelength is 575 nm, has been designed [33]. The wavelength region of interest range is over visible range. The thickness of the QWR and HWR are 15.73 and 31.47 μm, respectively. The material has good transmittance. At the orientation angle of 67.5° for the central plate and 90° for the last wave plate, excellent overall achromatic characteristics are obtained. The schematic diagram is illustrated in Fig. 31.2.

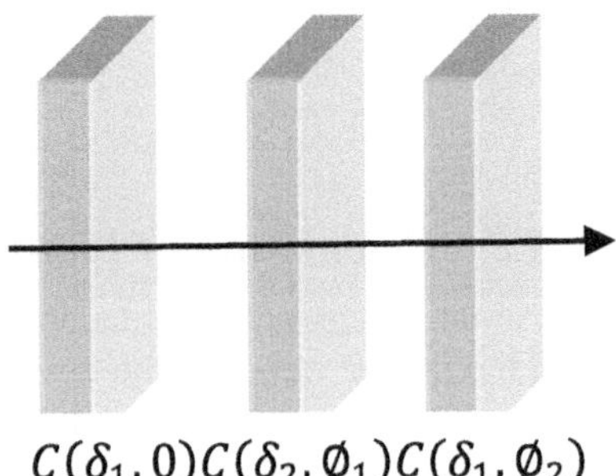

Fig. 31.2 Crystalline quartz-based achromatic phase modulator [33]

Theory

The total phase Δ and orientation θ of the system can be calculated using Jones calculus, as shown in Eq. (31.3), by multiplying those matrices in a particular order [33].

$$C(\Delta, \theta) = C(\delta_1, \phi_2)C(\delta_2, \phi_1)C(\delta_1, 0) \tag{31.3}$$

After the mathematical formulation, the overall retardation can be expressed as in Eq. (31.4), and also the overall azimuth ω is shown in Eq. (31.5) [33].

$$\tan^2 \frac{\Delta}{2} = \left(\frac{|\mathrm{Im}A|^2 + |\mathrm{Im}B|^2}{|\mathrm{Re}A|^2 + |\mathrm{Re}B|^2} \right) \tag{31.4}$$

and

$$\tan 2\omega = \left(\frac{B - B^*}{A - A^*} \right) \tag{31.5}$$

In Fig. 31.3, the retardation variation of the system with wavelength has been illustrated and compared with that of the earlier proposed setup. The present result indicates that, the variation of phase and equivalent fast axis are found to be $\pm$ 2.26°and $\pm$ 1.05° respectively. At the same time, it is substantially improved to $\pm$ 0.13° for the proposed system, as shown by curve 1 in Fig. 31.3. In Fig. 31.4, the azimuth of the setup has been compared with that of the Pancharatnam setup. Figure 31.5, represents the change in phase over the wavelength range, and the retardation varies from maximum + 0.0025% to minimum -0.388%. If we add a tolerance of $\pm$ 11% to the thickness value, only a 0.05% maximum variation in retardation is obtained, as shown in Fig. 31.6. Also, for birefringence, $\pm$ 10% tolerance value will vary overall retardance value to 0.056%, as observed in Fig. 31.7.

In summary, a superachromatic phase modulator using three waveplates of crystalline Quartz for 500–700 nm wavelength range in the visible spectrum has been designed, and the deviation of retardance is within $\pm$ 0.13° covering the intended wavelength range. All the calculations have been done based on Jones calculus. Also, the change of azimuth of the overall system is very small over the range. The dependencies on thickness and birefringence have also been studied by considering some tolerance ranges. However, the present system deals with a drawback that has been discussed in the following proposed strategies.

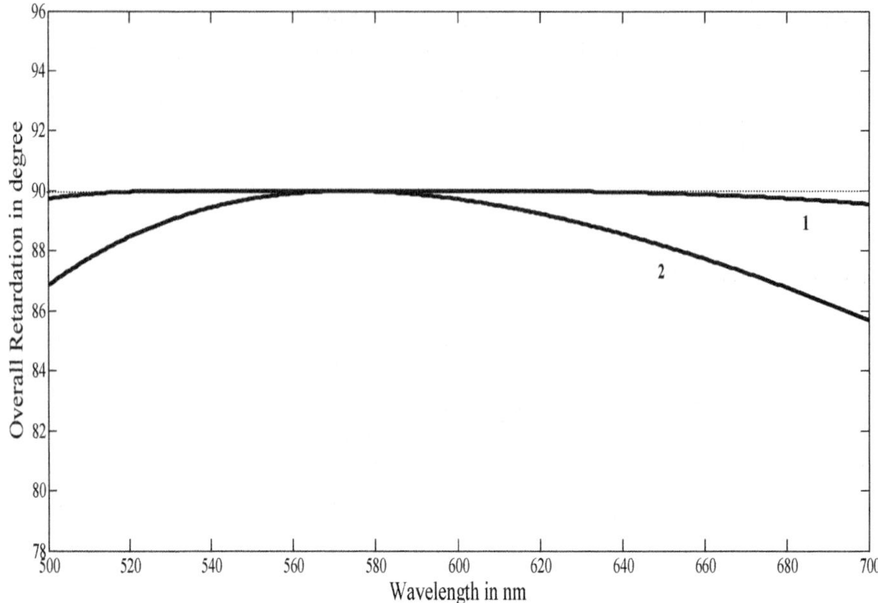

Fig. 31.3 Total phase change of the setup (curve1) versus wavelength when $\varnothing_1 = 67.5^\circ$ and $\varnothing_2 = 90^\circ$. The total phase change of the Pancharatnam setup (curve 2) is also shown for comparison [33]

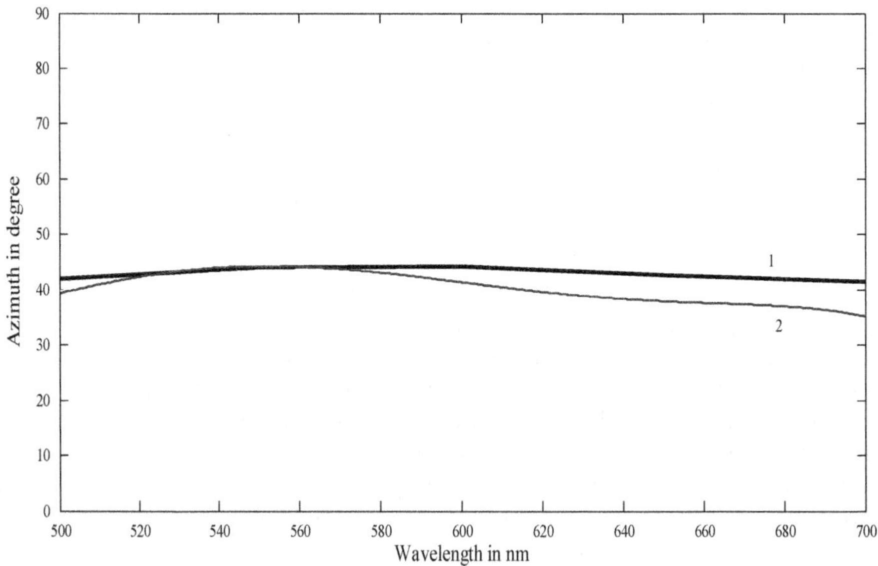

Fig. 31.4 Change of azimuth the setup (curve 2) and that of the Pancharatnam setup (curve 1) [33]

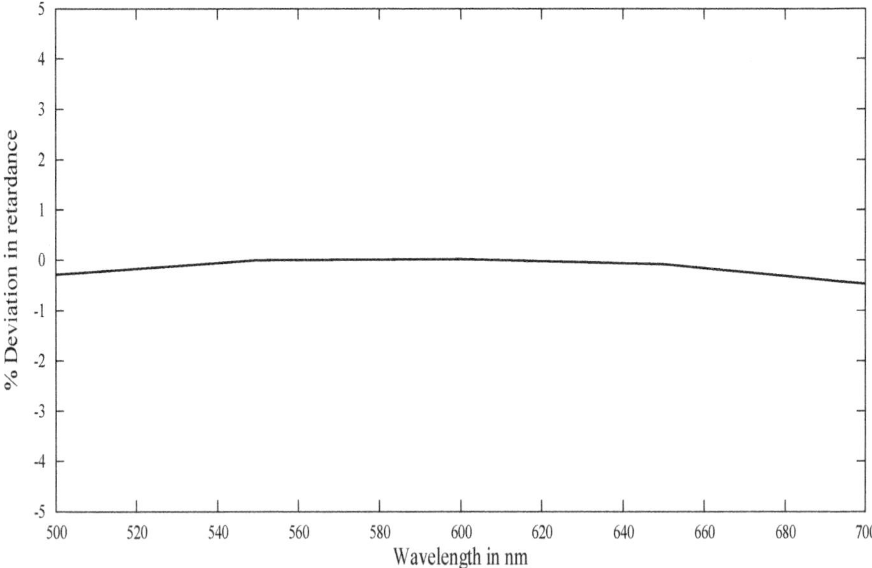

Fig. 31.5 Change of phase with change of wavelength when the angle of rotation of central plate is at $\varnothing_1 = 67.5°$ and rotation angle of last plate (i.e., QWR) is at $\varnothing_2 = 90°$ [33]

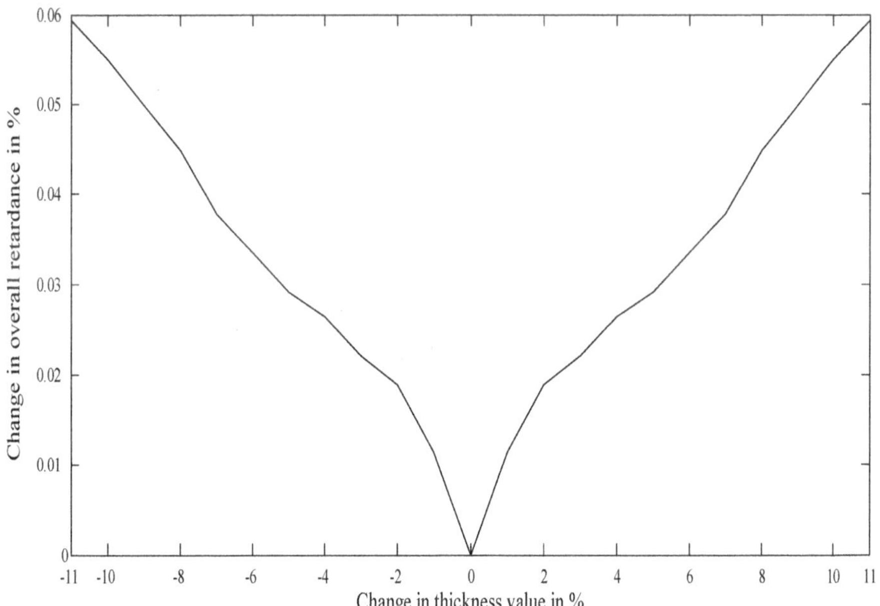

Fig. 31.6 Percentage variation of overall retardance with thickness value

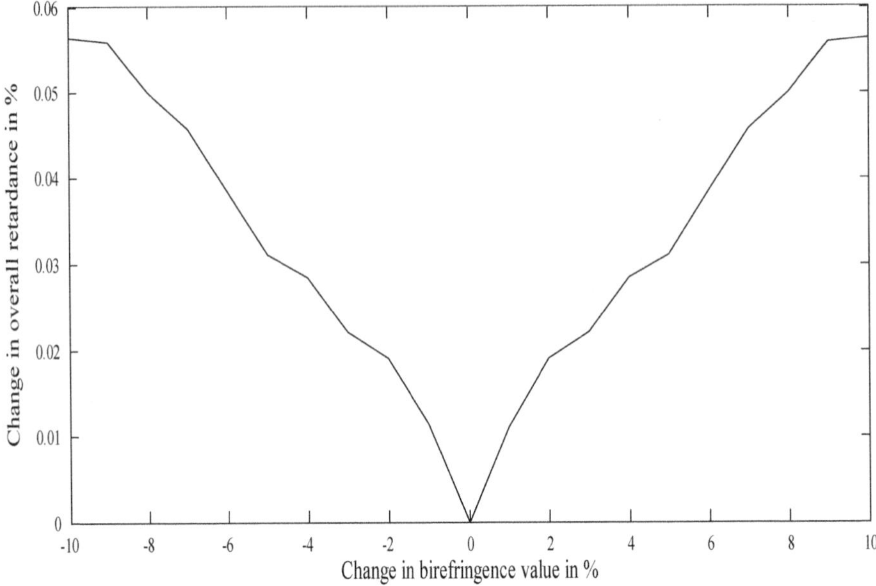

Fig. 31.7 Percentage variation of overall retardance with birefringence

31.4 Achromatic Phase Modulator with Different Birefringent Materials

31.4.1 Superachromatic Phase Modulator Retarder Using More Than Two Birefringent Plates Covering Visible to SWIR Range

In the present study, a quarter-wave phase modulator behaving having KDP, crystalline Quartz, ADP and ZnS has been designed [34]. An approach to create a super achromatic retarder using novel birefringent materials has been performed. The principal axes of all four waveplates are parallel to each other in this case. These retarder materials have high transmittance. The operating range of the designed system is over visible to short wave infrared wavelength region of 400 nm–2100 nm. The design is illustrated in Fig. 31.8. The thicknesses of the four plates are 130.6 μm, 162.6 μm, 133.7 μm, and 238.1 μm, respectively.

Theory

Considering the plate thicknesses to be d_X, d_Y, $d_{Z,}$ and $d_{K,}$ respectively, so that the combination has $\pi/2$ phase shift whose thicknesses can be obtained from Eq. (31.5) considering the birefringence value of each material with four design wavelengths, i.e., 0.45 μm, 0.8 μm, 1.4 μm, and 2.1 μm respectively.

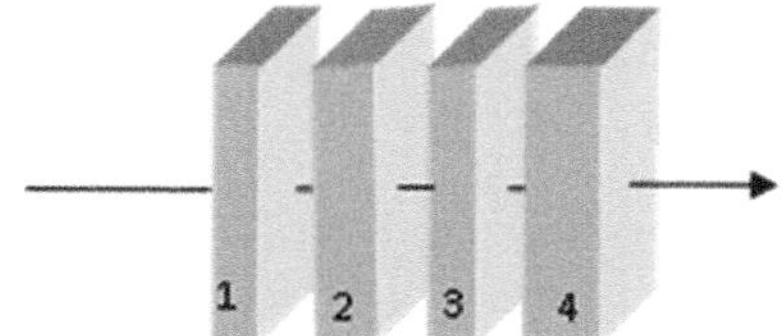

Fig. 31.8 Block diagram of KDP (**1**), crystalline Quartz (**2**), ADP (**3**), and ZnS (**4**)-based achromatic phase retarder [34]

$$\begin{vmatrix} \Delta\, n_{X1} & \Delta\, n_{Y1} & \Delta\, n_{Z1} & \Delta\, n_{K1} \\ \Delta\, n_{X2} & \Delta\, n_{Y2} & \Delta\, n_{Z2} & \Delta\, n_{K2} \\ \Delta\, n_{X3} & \Delta\, n_{Y3} & \Delta\, n_{Z3} & \Delta\, n_{K3} \\ \Delta\, n_{X4} & \Delta\, n_{Y4} & \Delta\, n_{Z4} & \Delta\, n_{K4} \end{vmatrix} \begin{vmatrix} d_X \\ d_Y \\ d_Z \\ d_K \end{vmatrix} = \pm\frac{1}{4} \begin{vmatrix} \lambda_X \\ \lambda_Y \\ \lambda_Z \\ \lambda_K \end{vmatrix} \tag{31.6}$$

where birefringence values of all the materials are denoted by Δn_{X1}, Δn_{X2}, Δn_{X3}, Δn_{X4} (Plate1), Δn_{Y1}, Δn_{Y2}, Δn_{Y3}, Δn_{Y4} (Plate2), $\Delta n_{Z1}\Delta n_{Z2}$, Δn_{Z3}, Δn_{Z4} (Plate3), and Δn_{K1}, Δn_{K2}, Δn_{K3}, Δn_{K4} (Plate4). The overall phase of the cascaded system is given by,

$$\varphi = (2\pi/\lambda) \times [(\Delta\, n_1 \times d_X) + (\Delta\, n_2 \times d_Y) + (\Delta\, n_3 \times d_Z) + (\Delta\, n_4 \times d_K)] \tag{31.7}$$

In this case, a combination of four birefringent plates operates over a broad spectrum of 400–2100 nm.

The variation of retardation in visible, NIR and SWIR are plotted in Figs. 31.9, 31.10, and 31.11, respectively. Over this range of wavelengths, $\pm$ 0.34° retardation variation in visible, $\pm$ 2.05° in NIR and $\pm$ 2.57° in SWIR range. Figure 31.12 represents the change in retardation value which is within $\pm$ 2.25°. Figure 31.13 represents the change in percentage retardation. The variation of overall retardation with the departure of the thickness of each plate from the designed value has been illustrated in Fig. 31.14. It can be observed that for phase deviation within $\pm$ 1° tolerance range of the thickness values for each plate are 130.5–131 μm (KDP), 162.2–162.8 μm (Quartz), 133.6–133.9 μm (ADP), and 237.7–238.6 μm (ZnS) respectively. If we consider extreme tolerance values of thickness on both sides, then the effect on overall retardation has been shown in Fig. 31.15. The operating temperature range where the maximum deviation is within $\pm$ 2.25° is – 30 to 150 °C, as shown in Fig. 31.16.

31.4.2 Superachromatic Phase Modulator Using Different Materials in the SWIR Range

In the present study, a superachromatic phase modulator has been designed to cover the wavelength range 1000–2300 nm with plates of Calcite, crystalline Quartz, KDP, and Rutile [35]. These birefringent materials have excellent values of transmittance,

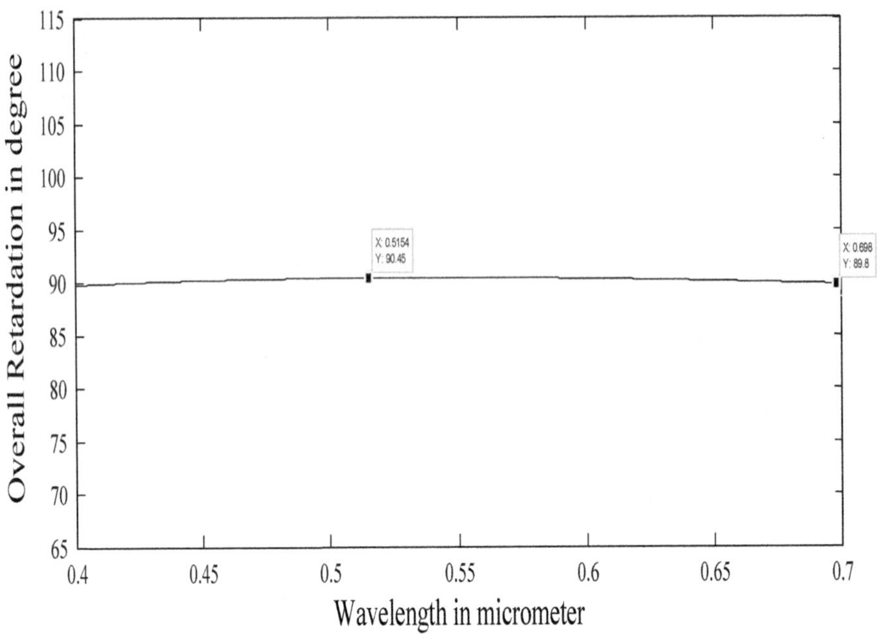

Fig. 31.9 Obtained retardation variation in the visible range [34]

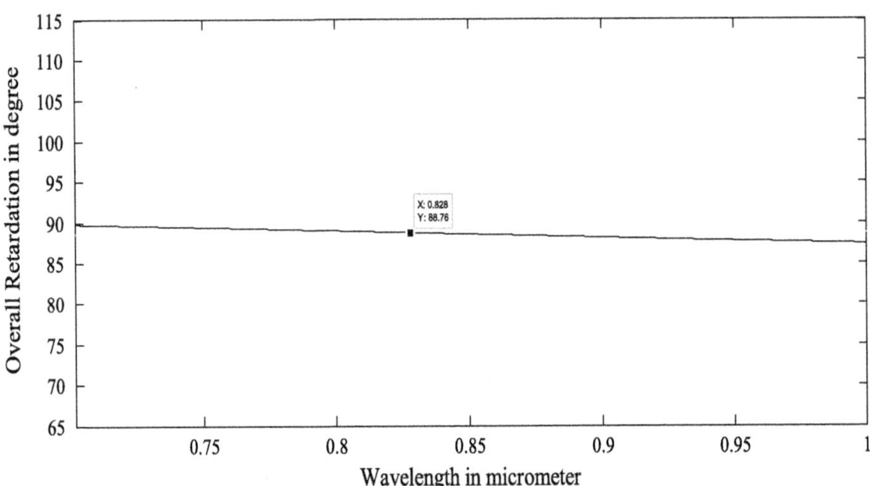

Fig. 31.10 Obtained retardation variation in the near-infrared range [34]

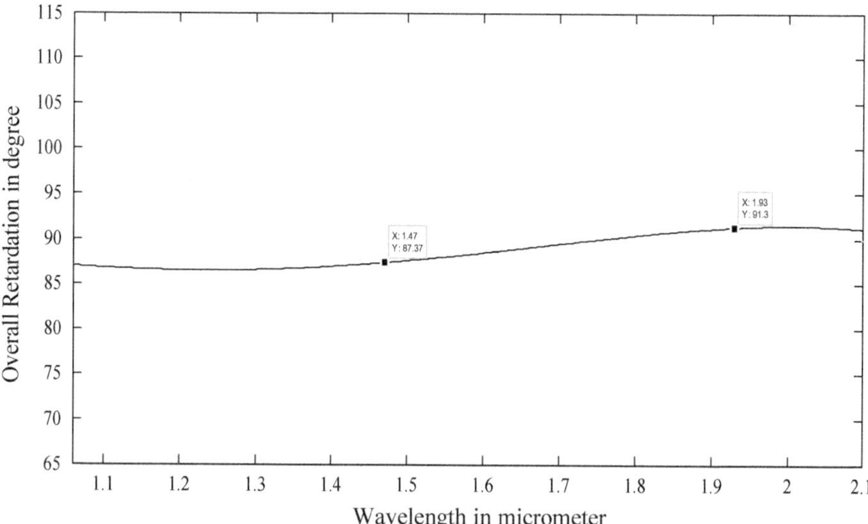

Fig. 31.11 Variation of retardation in short-wave infrared range [34]

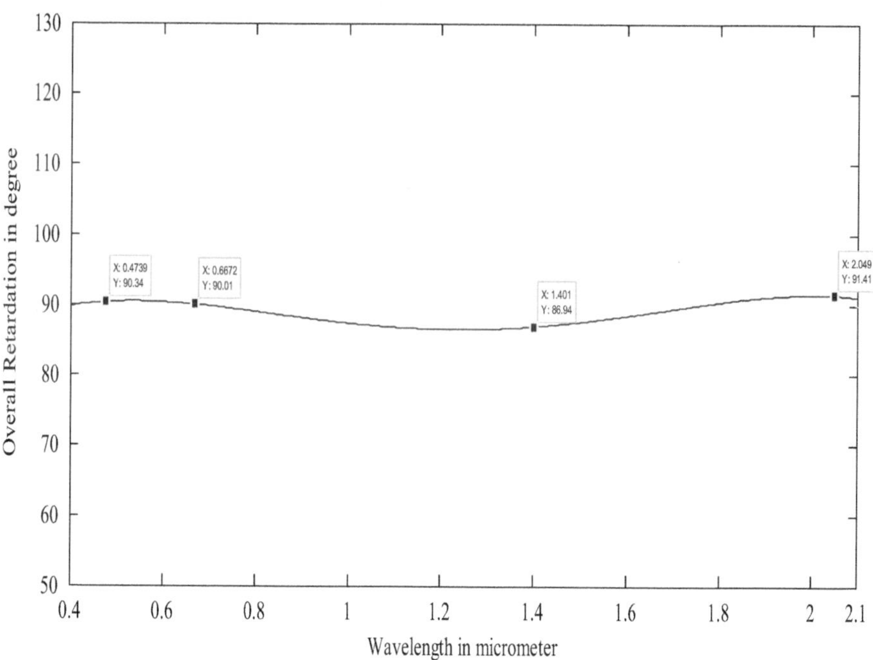

Fig. 31.12 Overall retardation over the desired wavelength range [34]

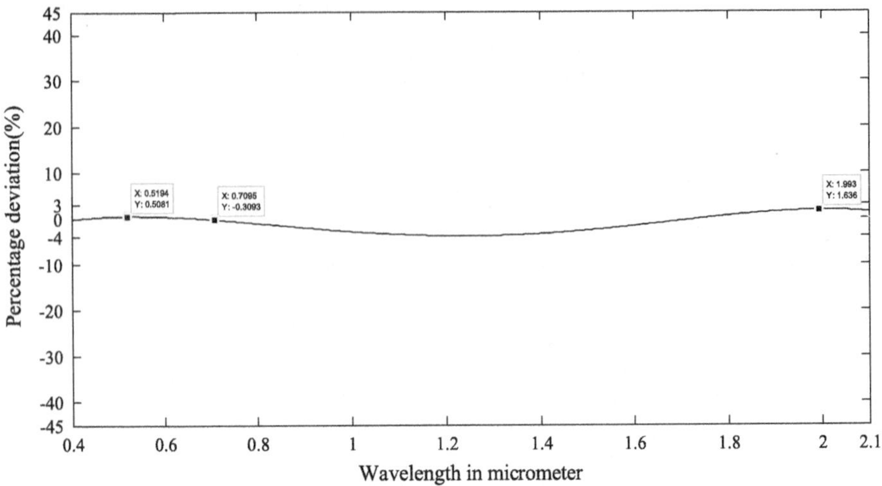

Fig. 31.13 Change of phase retardation over intended spectrum [34]

hygroscopicity, brittleness, etc. The construction of the retarder is illustrated in Fig. 31.17.

Theory

Considering the thicknesses of the birefringent plates as d_X, d_Y, d_Z and d_K so that it behaves as a $\pi/2$ phase shift plate. The thicknesses of the plates have been calculated from Eq. (31.5) considering the birefringence value of each material with four design wavelengths, i.e., 1.0 μm, 1.5 μm, 2.0 μm, and 2.3 μm respectively. The overall phase of the cascaded system is given by Eq. (31.6).

Over 1000–2300 nm wavelength range an achromatic a $\pi/2$ phase shift retarder with different birefringent material-based plates is designed to exhibit achromatism over a that range. Compared to the earlier planned phase modulator used in the AirMSPI system on NASA ER-2 reported with a deviation of retardation value $90° \pm 10°$ our proposed four-plate system results in retardance variation within $90° \pm 1.25°$ with improved temperature range of operation suggesting a suitable replacement over the previous one.

The proposed system exhibits a retardation deviation of $90° \pm 1.25°$ throughout the spectrum. That indicates the improvement of achromatism. Figure 31.18 represents the effects of thickness variation of each plate on overall phase retardation. It can be observed that for phase deviation within $\pm 1°$ tolerance range of the thickness values for each plate are 158.8–159.1 μm (Calcite), 136.4–136.8 μm (Quartz), 164.6–165.2 μm (KDP), and 114.7–115.2 μm (Rutile) respectively. If we consider extreme tolerance values of thickness, then the effect on overall retardation has been shown in Fig 31.19. The operating temperature range where the maximum deviation is within $\pm 1.25°$ is – 30 to 145 °C, as in Fig 31.20.

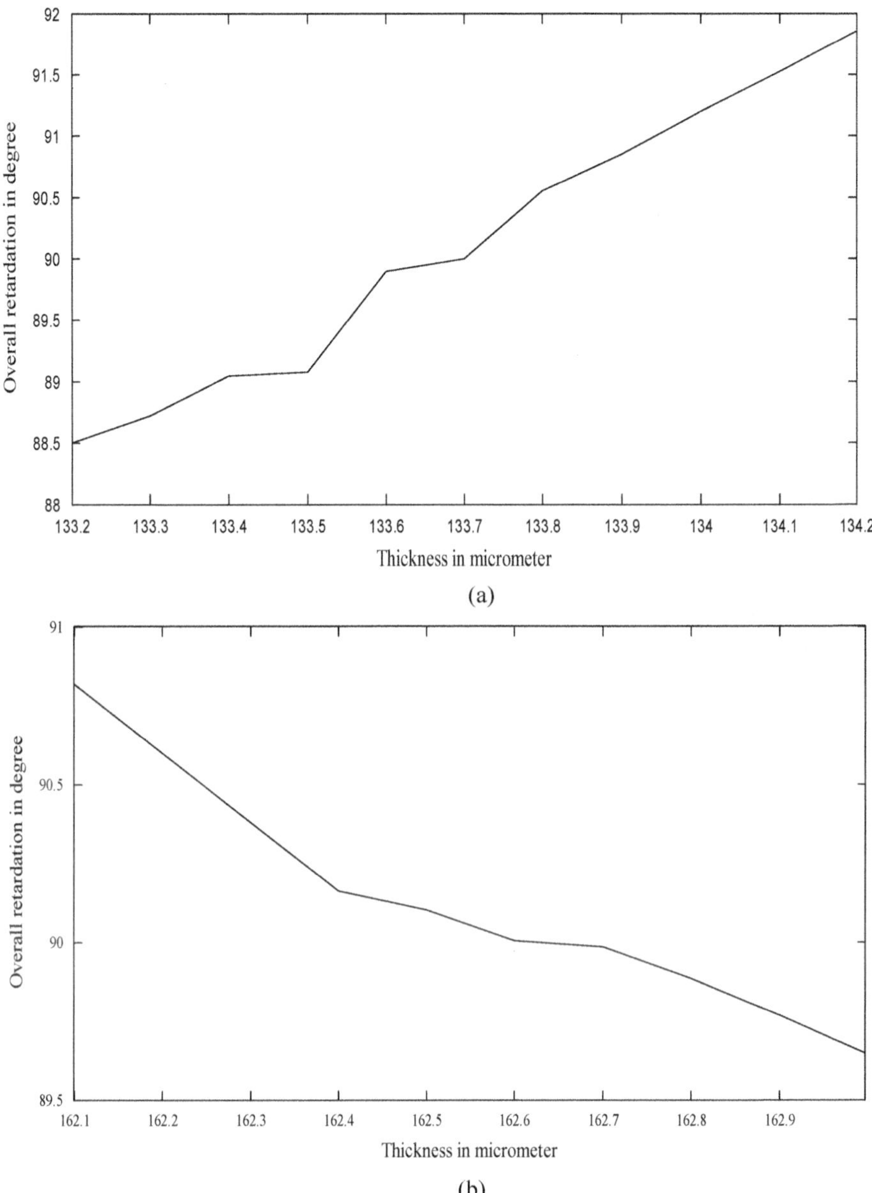

Fig. 31.14 Variation of retardation of plates of **a** KDP, **b** quartz, **c** ADP, and **d** ZnS

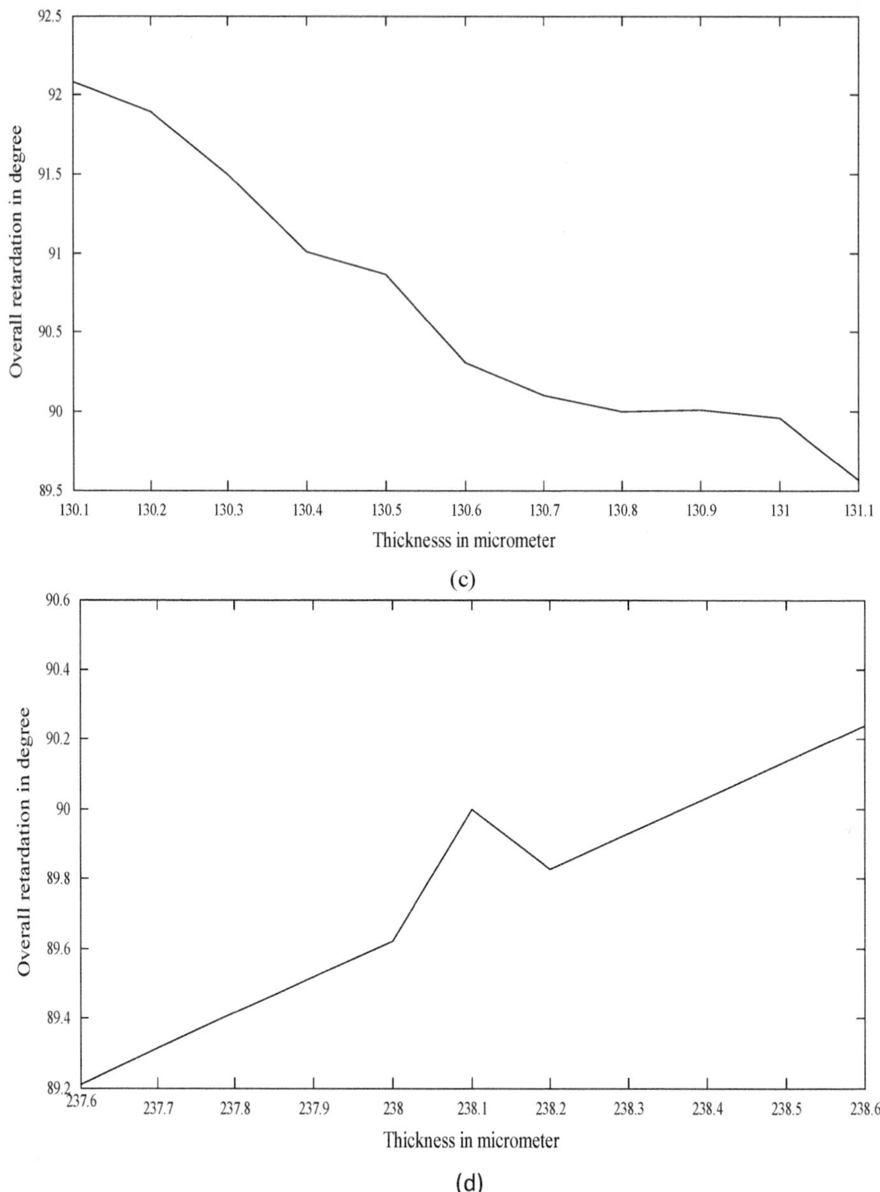

Fig. 31.14 (continued)

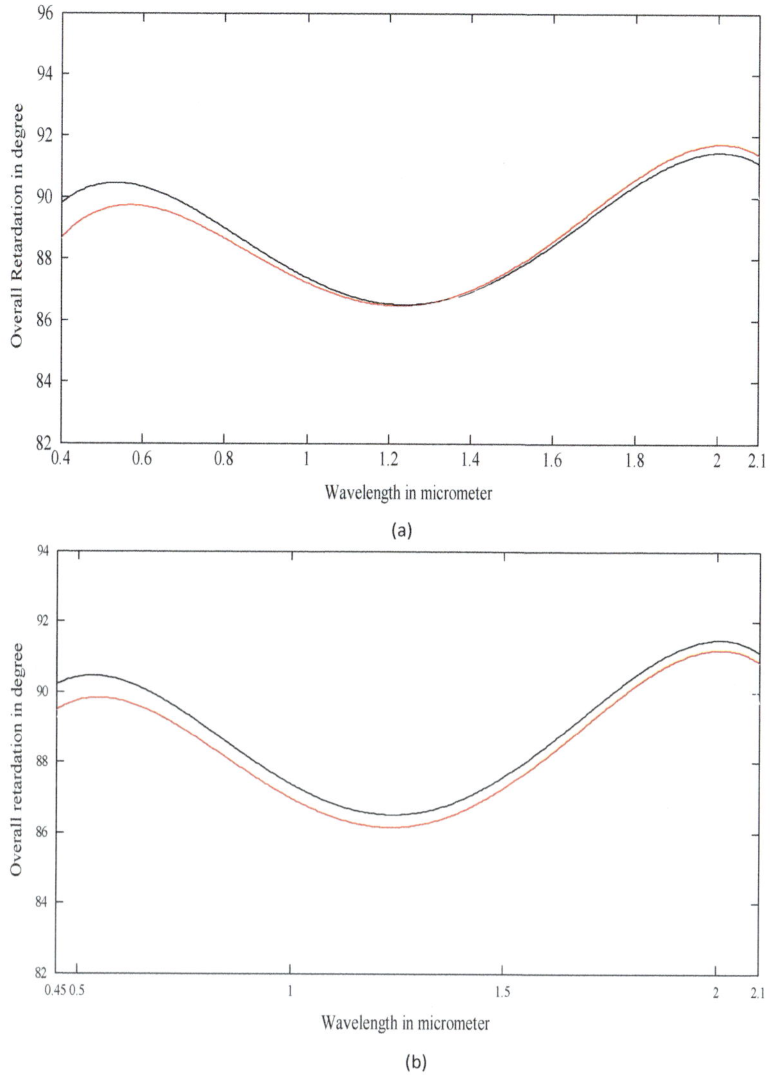

Fig. 31.15 Variation of overall retardation with the thickness of plates considering extreme tolerance values **a** on the positive side **b** on the negative side where the red curve represents the new deviation

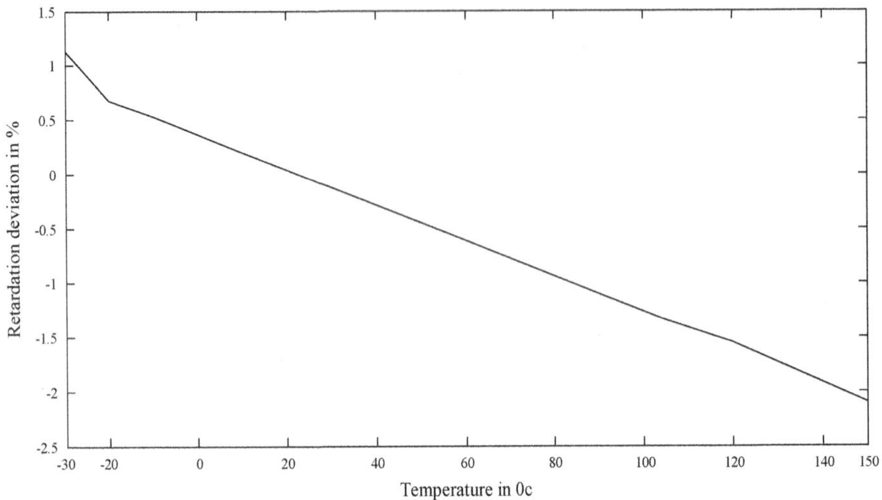

Fig. 31.16 Percentage change in overall retardance (%) versus temperature (°C) at design wavelength 800 nm

Fig. 31.17 Block diagram of Calcite (**1**), crystalline Quartz (**2**), KDP (**3**), and Rutile (**4**)-based phase retarder system

31.4.3 Superachromatic Quarter-Wave Retarder Design with Optimum Thickness of Waveplates Using Flower Pollination Algorithm

Superachromatic Phase Retarder in NIR Range Using Optimization Technique

This phase retarder system is designed to operate over the near-infrared range that consists of four waveplates of different birefringent materials viz. Sapphire, CdSe, Calcite, and Rutile have different thicknesses [36]. A recently proposed optimization technique is utilized to obtain the minimum thickness. The schematic diagram is illustrated in Fig. 31.21. The transmission axes of the waveplates are aligned at the same line.

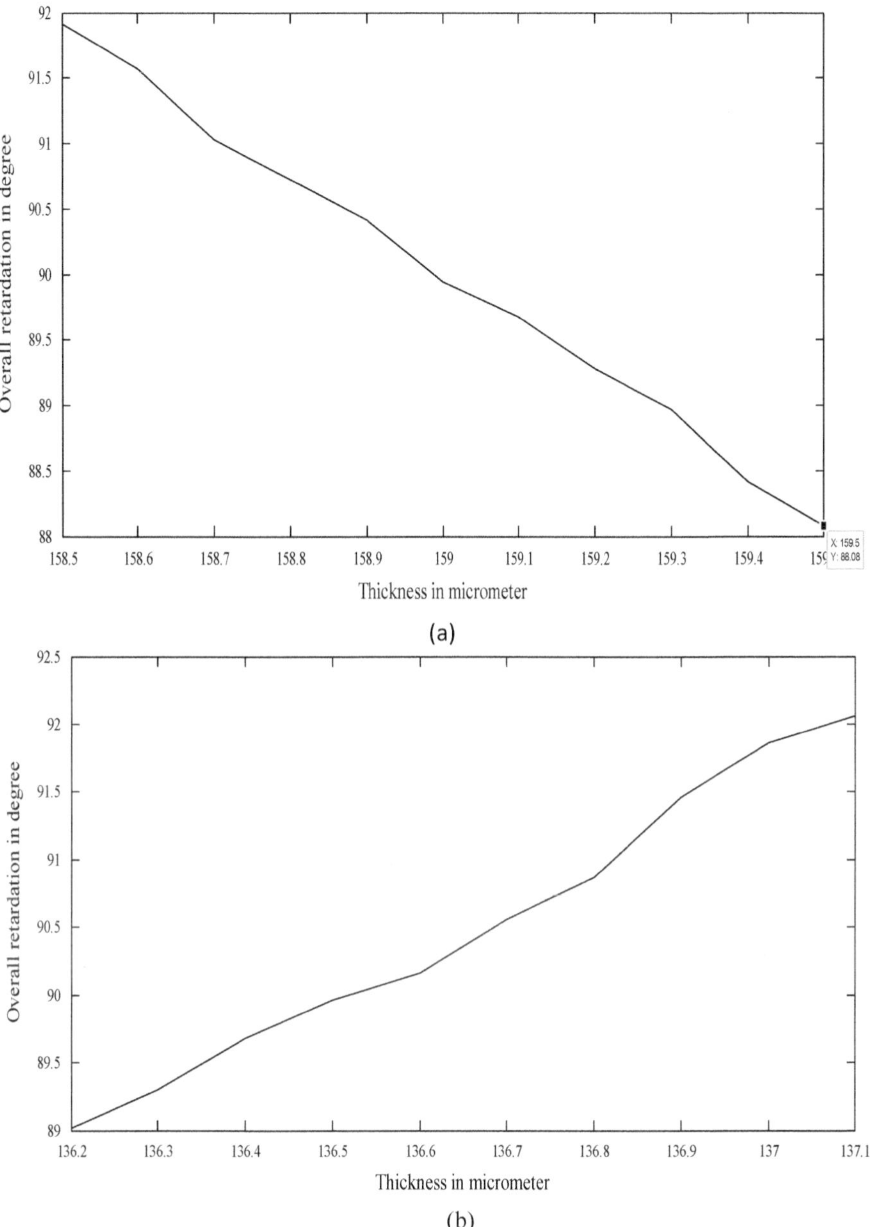

Fig. 31.18 Variation of obtained retardation with the change in plate thickness **a** calcite, **b** quartz, **c** KDP, and **d** Rutile

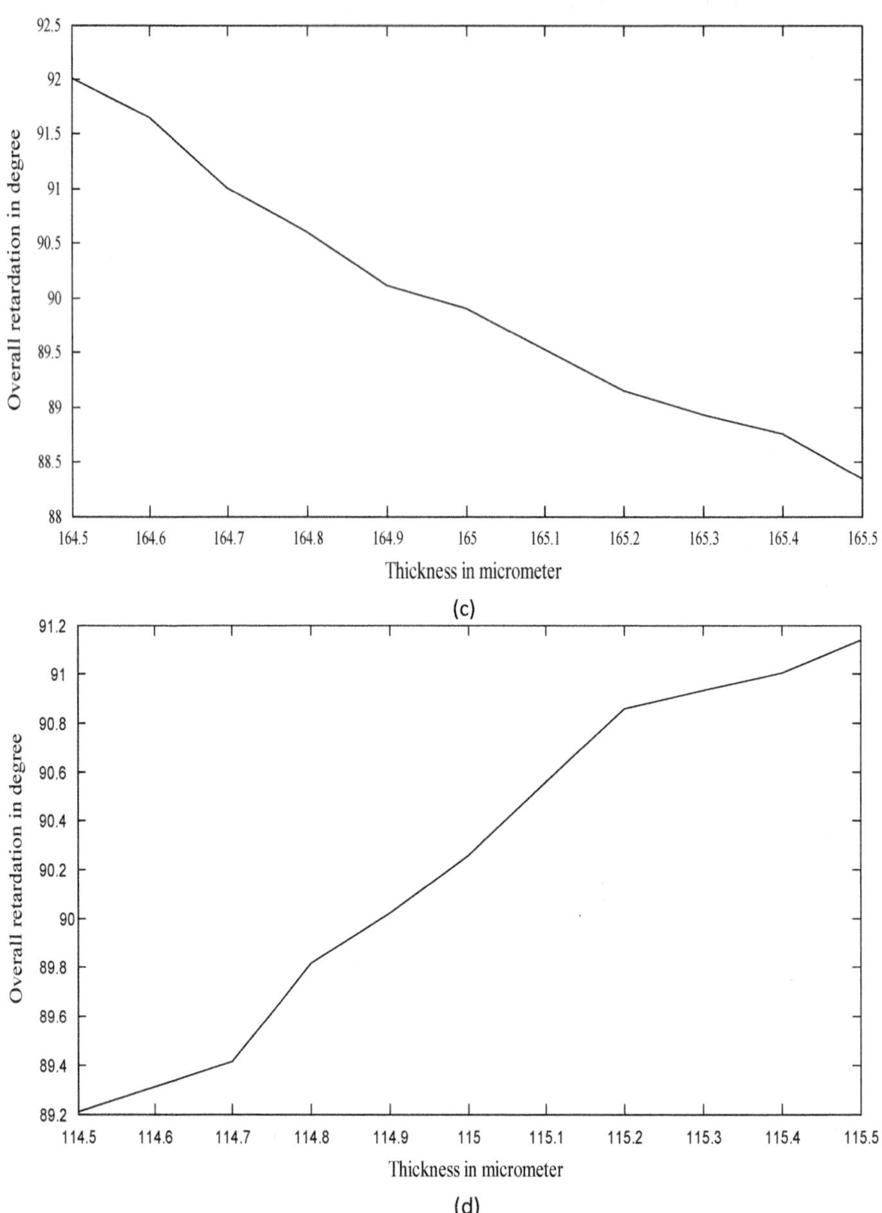

Fig. 31.18 (continued)

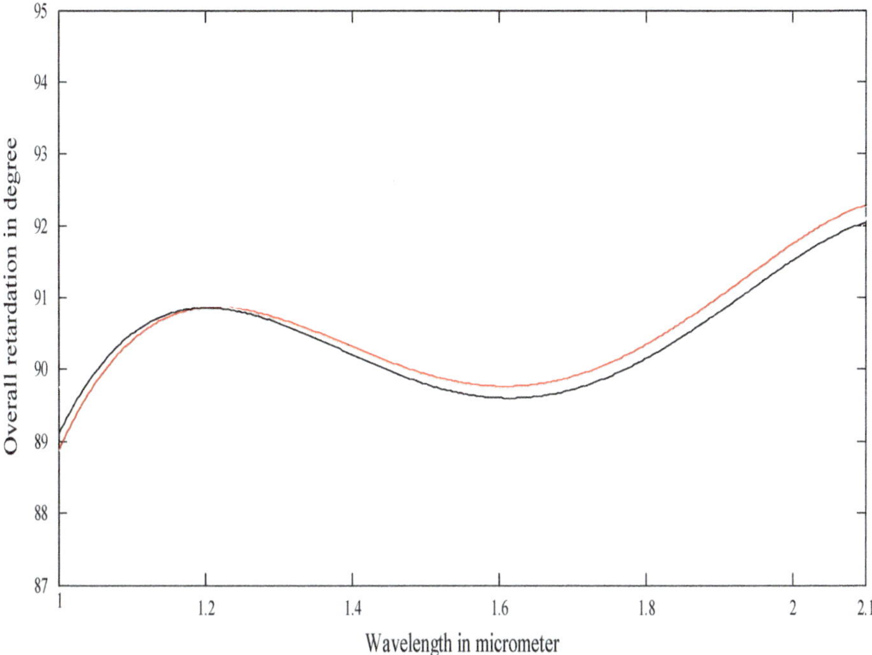

Fig. 31.19 Variation of overall retardation with thickness of plates considering extreme tolerance values where the red curve represents the new deviation

Theory

For optimizing purposes, an objective function f is defined, which represents the change in obtained retardation from the intended value is written as follows [36]:

$$f = \sum_{i=1}^{N} \left(\varphi_{\mathrm{Exp},i} - \varphi_{\mathrm{Cal},i}\right)^2 \tag{31.8}$$

where $\varphi_{\mathrm{Exp},i}$ is the intended retardation and $\varphi_{\mathrm{Cal},i}$ is obtained retardation at a particular wavelength λ_i. For the designing purpose wavelengths are chosen to be 0.8 μm, 1.2 μm, 1.6 μm, and 2.0 μm respectively. Computed optimal thickness values are 120.8 μm, 180.0 μm, 17.29 μm, and 2.0 μm, respectively.

In this optimization methodology, the number of iterations is 5000. The overall retardation is calculated and a deviation of general retardation within ± 0.9°. It is reported that deviation of achromatism is within + 1.16% and – 0.995%. The thickness of each plate varies from 120.3 to 121.0 μm (sapphire), 179.9–180.4 μm (cdse), 17.0–17.5 μm (Calcite) and 1.9–2.1 μm (Rutile) then the overall phase retardation varies within ± 1° as observed from Fig. 31.22. Variation of general retardation with the thickness of plates considering extreme tolerance values is referred to in

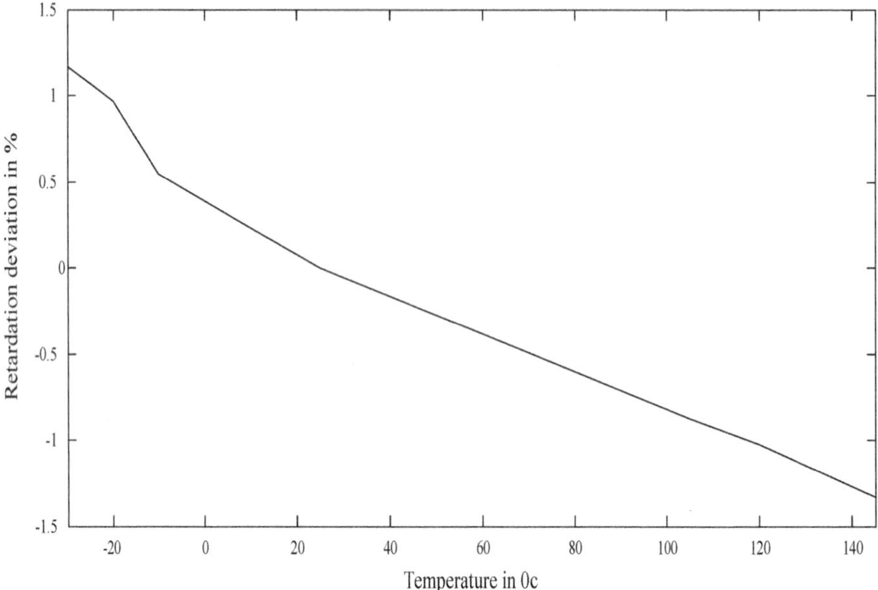

Fig. 31.20 Percentage change in overall retardance (%) versus temperature (°C) at design wavelength 1000 nm

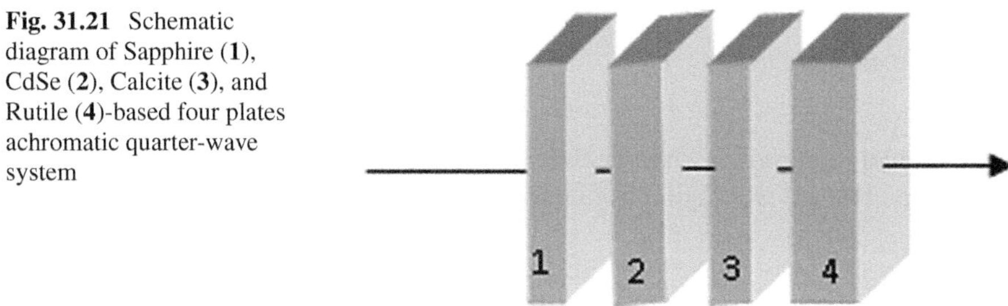

Fig. 31.21 Schematic diagram of Sapphire (**1**), CdSe (**2**), Calcite (**3**), and Rutile (**4**)-based four plates achromatic quarter-wave system

Fig. 31.23. The temperature range within which the system shows super achromatic behavior is from – 30 to 135 °C, as observed in Fig. 31.24.

Thus, a nature-inspired algorithm has been employed in designing achromatic phase retarder: the construction and the characteristics of three waveplates. Sapphire, CdSe, and Calcite using FPA have been discussed. The deviation of overall retardation is within + 1.04°, and – 0.9° has been observed. The computational results suggest a significant improvement in achromatic behavior.

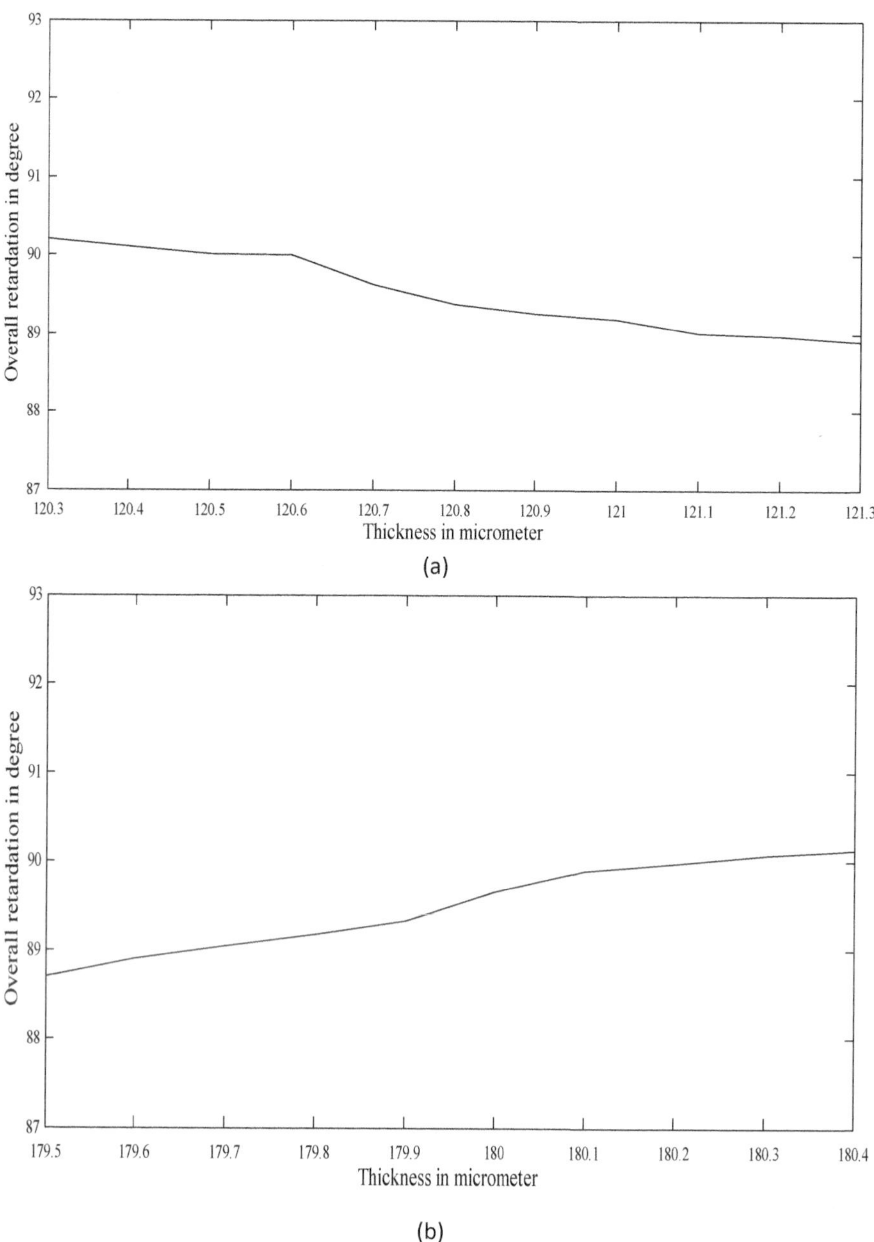

Fig. 31.22 Variation of overall retardation with the thickness of plates **a** sapphire, **b** case, **c** calcite, and **d** Rutile

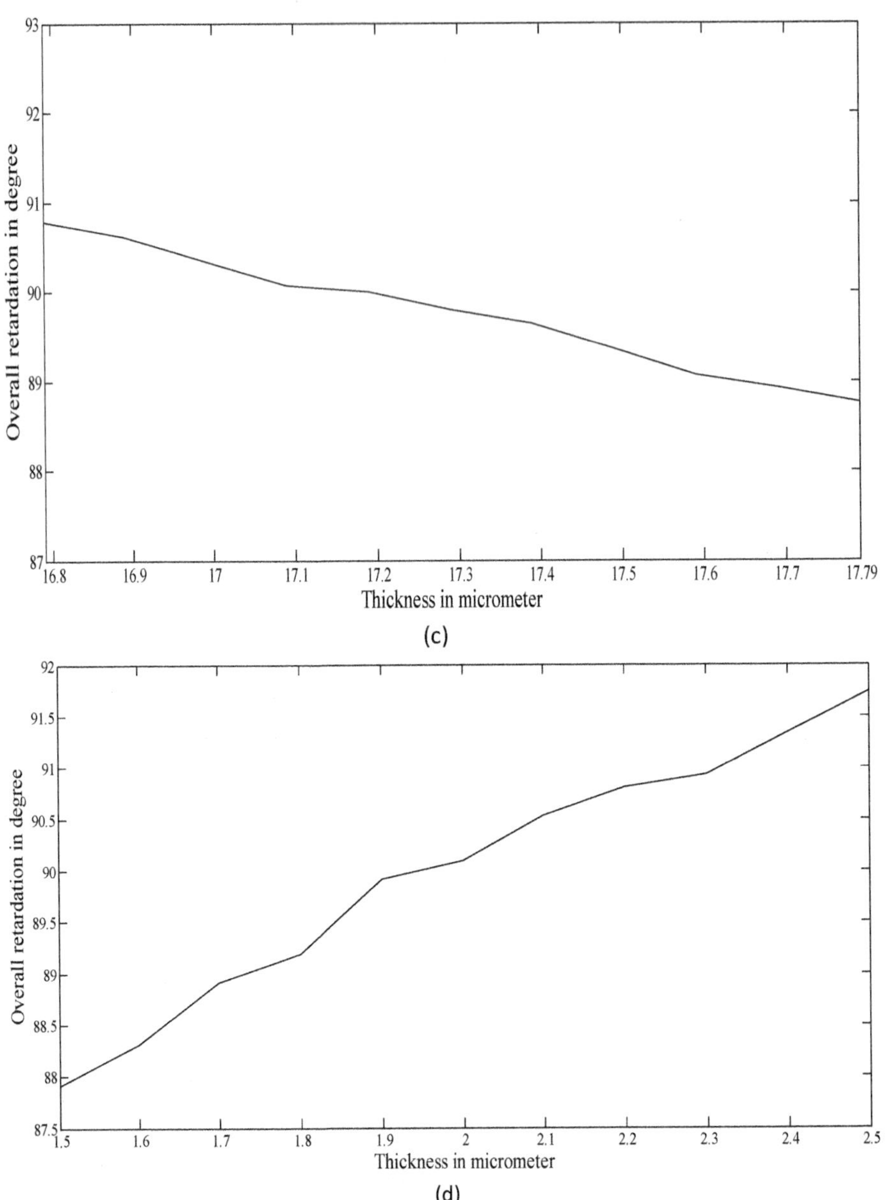

Fig. 31.22 (continued)

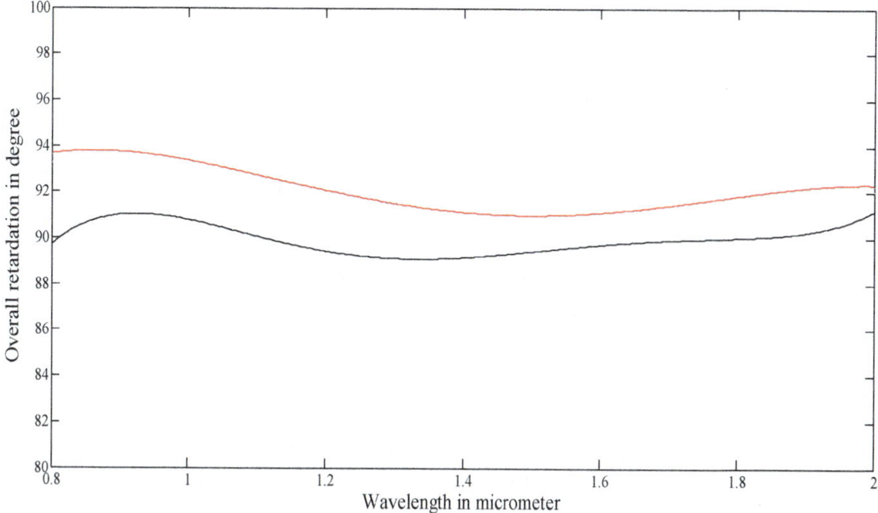

Fig. 31.23 Variation of overall retardation with thickness of plates considering extreme tolerance values where the red curve represents the new deviation

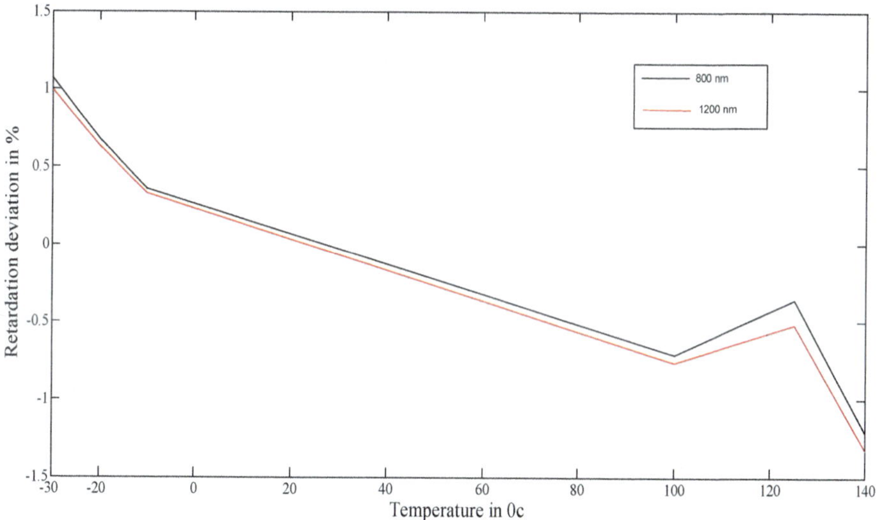

Fig. 31.24 Percentage change in overall retardance (%) versus temperature (°C) for 800 nm (black curve) and 1200 nm (red curve) wavelengths

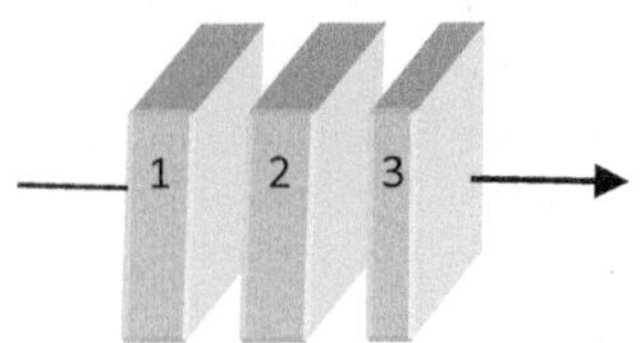

Fig. 31.25 Block diagram of Sapphire (**1**), CdSe (**2**) and Calcite (**3**)-based three plates achromatic quarter-wave system [30]

31.4.4 Proposed System with Different Birefringent Materials Behaving as Superachromatic Retarder in 800–2000 nm Wavelength Range

In the present study, another approach to designing superachromatic quarter-wave retarders with optimum thickness of waveplates using an evolutionary optimization technique, i.e., Flower pollination algorithm, has been employed [30]. The application of the optimizing tool is for precise measurement of thickness because if it deviates from the calculated value, then the overall retardance will start shifting from the target value. The proposed system has three different materials, viz. Sapphire; CdSe; Calcite operating over 800–2000 nm wavelength range. The schematic diagram is illustrated in Fig. 31.25.

Theory

In a system having M-plate optical retarders, the overall retardation at the designing wavelength (i.e., λ_i) is computed using the following equation,

$$\varphi_i = \frac{2\pi}{\lambda} \sum_{j=1}^{M} \Delta\, n_{i,j} d_j \tag{31.9}$$

where for a design wavelength λ_i, $\Delta\, n_{i,j}$ represents the birefringence value of the jth plates and λ is the wavelength range under consideration. The change in retardation value from the target retardation at design wavelength is defined as objective function f that has to be optimized as referred to in Eq. (31.7). The three design wavelengths are $\lambda_1 = 0.8\,\mu\text{m}$, $\lambda_2 = 1.4\,\mu\text{m}$ and $\lambda_3 = 2.0\,\mu\text{m}$ respectively. Computed optimal thickness values are 99.19 μm, 181.47 μm, and 15.32 μm, respectively.

After calculating the overall retardation, a plot against the wavelength, considering the thickness values up to two decimal places, has been illustrated in Fig. 31.26 that exhibits phase deviation within ± 0.96°. In Fig. 31.27, the effect of changing the iteration number on the retardation deviation has been established, where it is observed that variation reduces with the increasing number of iterations. The impact of changing the thickness of the plates on retardation has also been studied by introducing a tolerance of around 0.1% to the thickness values that result in retardation variation within ± 1%, as shown in Fig. 31.28. The proposed system has an overall transmittance of around 55% at the three design wavelengths. A study on the speed of convergence of the FPA technique is illustrated in Fig. 31.29. Within the temperature range from – 20 to 120 °C, the superachromatic behavior of the system can be obtained as referred to in Fig. 31.30.

Thus, a new optimization technique inspired by the pollination of flowers, i.e., the flower pollination algorithm, has been used in designing an achromatic phase retarder. Construction and the characteristics of a three waveplates system using FPA that consists of different materials, viz. Sapphire, CdSe, and Calcite have been

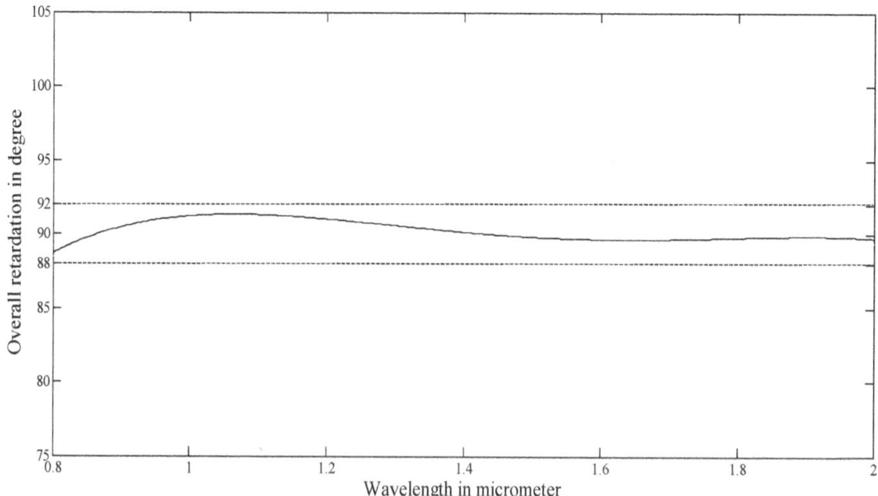

Fig. 31.26 Overall retardation with wavelength with optimized thickness values [30]

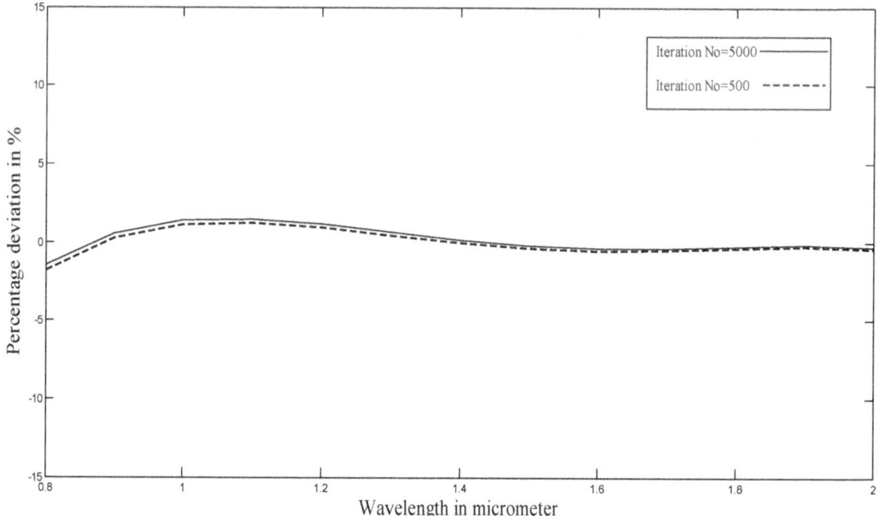

Fig. 31.27 Change in phase deviation with wavelength [30]

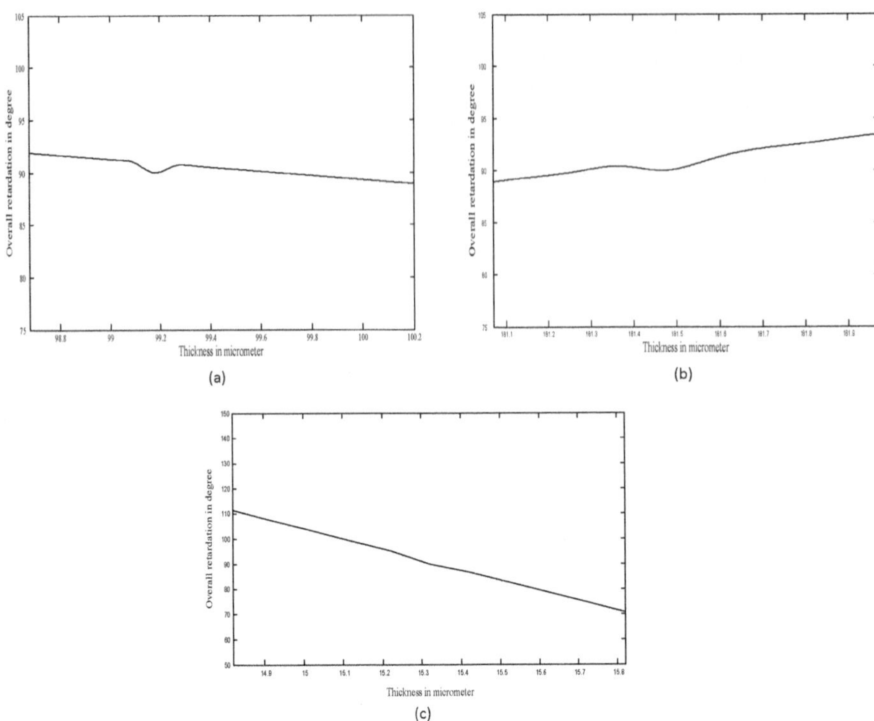

Fig. 31.28 Variation of overall retardation with the thickness of plates **a** Sapphire, **b** CdSe, and **c** Calcite

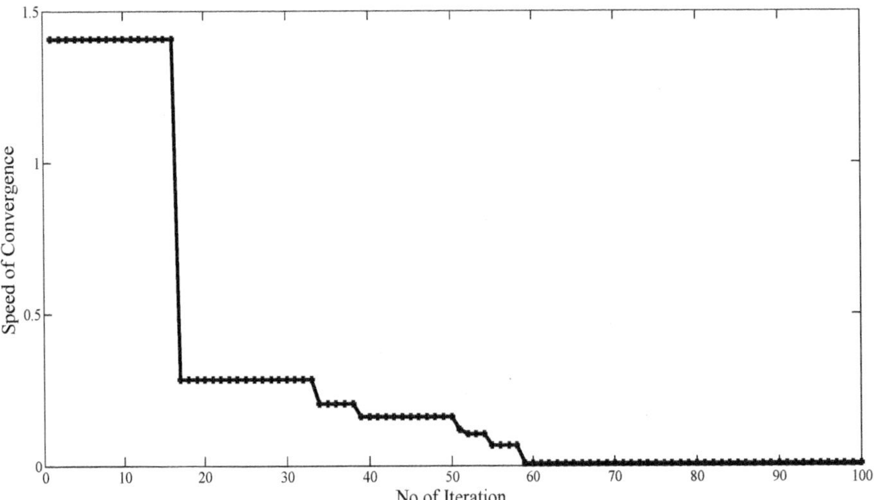

Fig. 31.29 Variation of convergence speed with the number of iterations in 800-2000 nm wavelength range [30]

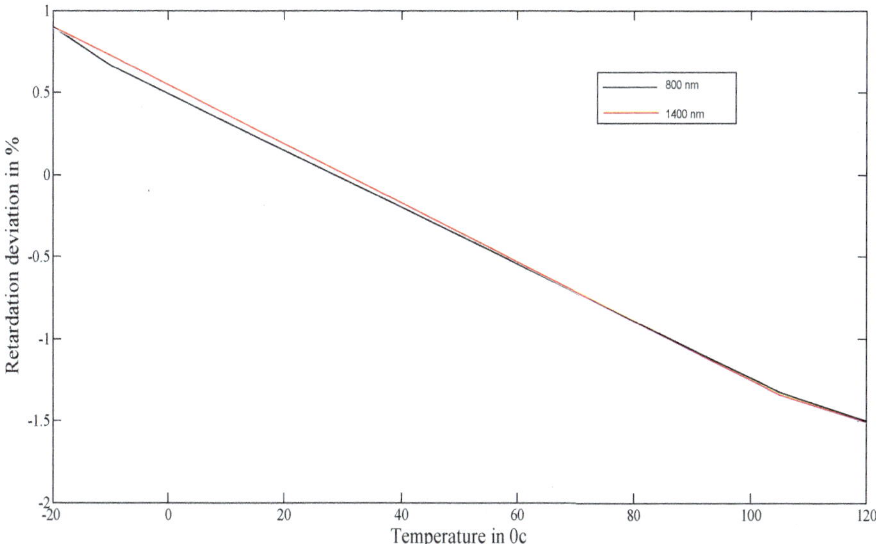

Fig. 31.30 Percentage change in overall retardance (%) versus temperature (°C) for 800 nm (black curve) and 1400 nm (red curve) wavelengths

discussed in detail, and the simulation results illustrate that the deviation is within $\pm$ 0.96° in NIR. Preliminary computations reveal that optimal thickness values will lead to better achromatic behavior.

31.5 Conclusion

Following the preliminary work done so far in the areas of polarization optics, which can be enriched by the ingenious use of birefringent networks, in the present dissertation, some work has been done in designing superachromatic retarders that have the potential to exhibit good achromatic behavior over an extensive wavelength range including the whole visible region and near-infrared region. Construction and characteristics of a quarter-wave retarder having three waveplates of crystalline Quartz, of which the first and last waveplates are quarter-wave retarders, and the central wave plate is half-wave retarder orientated at a suitable angle to achieve superachromatic behavior in visible range have been proposed. Zero-order super achromatic quarter-wave phase retarders based on combinations of waveplates from different birefringent materials have been submitted. Also, a new optimization technique, i.e., flower pollination algorithm, is used to obtain the minimum thickness to improve the achromaticity of the proposed quarter-wave phase retarders. We suggest using this technique in finding the optimum performance of devices we have undertaken for our study. In the present dissertation, we have also proposed a birefringent network that will act as a wavelength filter having a narrow transmission band in the visible range. With the exploration of a terahertz or millimeter wavelength range, the future scope of the investigation includes the development of new birefringent networks or optical devices in this range for specific applications.

References

1. Pancharatnam, S.: Achromatic combinations of birefringent plates. Part II. An achromatic quarter wave plate. In: Proceedings of the Indian Academy of Sciences, vol. 41, no. 4, Sec. A, p. 137 (1955)
2. Hariharan, P: Achromatic retarder using Quartz and mica. Meas. Sci. Technol. **6**, 1078 (1995)
3. Passily, N., et al.: Achromatic phase retardation by subwavelength gratings in total internal reflection. J. Opt. A Pure Appl. Opt. **10**, 015001(6pp) (2008)
4. Hariharan, P., Malacara, D.: A simple achromatic half-wave retarder. J. Modern Opt. **41**(1), 15–18 (1994)
5. Samoylov, A.V., Samoylov, V.S.: Achromatic and super-achromaticzero-order waveplates. In: Proceedings of LFNM 2003, IEEE, pp. 119–121 (2003)
6. McIntyre, C.M., Harris, S.E.: Achromatic wave plates for the visiblespectrum. JOSA **58**(12), 1575–1580 (1968)
7. Hariharan, P.: Broadband super-achromatic retarders. Meas. Sci. Technol. **9**, 1678–1681 (1998)
8. Saha, A., Bhattacharya, K., Chakraborty, A.K.: Reconfigurableachromatic half-wave and quarter-wave retarder in near infraredusing crystalline quartz plates. Opt. Eng. **50**(3), 034004–1–034004–4 (2011)
9. Harris, S.E., Ammann, E.O., Chang, I.C.: Optical network synthesis using birefringent crystals. I. Synthesis of lossless networks of equal-length crystals. J. Opt. Soc. Am. **54**, 1267–1278 (1964)
10. Ammann, E.O., Chang, I.C.: Optical network synthesis using birefringent crystals. II. Synthesis of networks containing one crystal, optical compensator, and polarizer per stage. J. Opt. Soc. Am. **55**, 835–841 (1965)
11. Ammann, E.O.: Optical network synthesis using birefringent crystals. HI. Some general properties of lossless birefringent networks. J. Opt. Soc. Am. **56**, 943–951 (1966)
12. Ammann, E.O.: Optical network synthesis using birefringent crystals. IV. Synthesis of lossless double-pass networks. J. Opt. Soc. Am. **56**, 952–955 (1966)
13. Ammann, E.O., Yarborough, J.M.: Optical network synthesis using birefringent crystals. V. Synthesis of lossless networks containing equal-length crystals and compensators. J. Opt. Soc. Am. **56**, 1746–1752 (1966)
14. Ammann, E.O., Yarborough, J.M.: Optical network synthesis using birefringent crystals. VI. Additional techniques for the synthesis of lossless double-pass networks. J. Opt. Soc. Am. **57**, 349–350 (1967)
15. Ammann, E.O.: Synthesis of optical birefringent networks. Prog. Opt. **9**, 123–177 (1971)
16. Chakraborty, B.: The use of circular birefringent material as an inter stage element of a linear birefringent network. J. Appl. Phys. **64**, 2847–2853 (1988)
17. Pochi, Y., Claire, G.: Equivalent circuit of a general birefringent network. In: Optics of Liquid Crystal Displays, 2nd edn, , pp. 186–192, 201–207, 254–282. Wiley (2009)
18. Jerrard, H.G.: Optical compensators for measurement of elliptical polarization. J. Opt. Soc. Am. **38**, 35–57 (1948)
19. Destriau, G., Prouteau, J.: Realization of an almost achromatic quarter-wave by juxtaposition of two crystalline plates of the same nature. J. Phys. Radium **10**, 53–55 (1949)
20. Hanany, S., Hubmayr, J., Johnson, B.R., Matsumura, T., Oxley, P., Thibodeau, M.: Millimeter-wave achromatic half-wave plate. Appl. Opt. **44**, 4666–4670 (2005)

21. Clarice, D.: Interference effects in Pancharatnam wave plates. J. Opt. A Pure Appl. Opt. **6**, 1047–1051 (2004)
22. Gardner, M.C., et al.: Electro-Optical and Infrared Systems: Technology and Applications. In: Proceedings of SPIE 9987 vol. 13, p. 99870C (2016)
23. Vizgaitis, J.N., et al.: Proceedings of SPIE 9822, p. 982201 (2016)
24. Hough, J.H.: Astronomical Polarimetry: Current Status and Future Directions ASP Conference Series, vol. 343 (2005)
25. Sabins, F.F.: Third Edition, 11-Jun-2007—Technology & Engineering—494 p. Waveland Press
26. Bauer, T., Thome, H., Eisenhammer, T.: Proceedings of SPIE 9241, Sensors, Systems, and Next-Generation Satellites, vol. 28, p. 92411K (2014)
27. Sparksa, W.B., Hough, J., et al.: Proceedings of the National Academy of Sciences of the United States of America, vol. 106, no. 19, pp. 7816–7821
28. Huang, Z., McWilliams, A., Lui, H., et al.: Int. J. Cancer **107**, 1047–1052 (2003)
29. Diner, D.J., et al.: UV-SWIR Achromatic Quarter Wave Retarder for the Multiangle Spectropolarimetric Imager (MSPI); College of Optical Sciences, University of Arizona, Tucson, AZ 85721 USA
30. Mukhopadhyay, N., Mandal, S., Bhattacharya, K., Saha, A.: Design of a superachromatic quarter-wave retarder for near-infrared region using flower pollination algorithm. Opt. Eng. **58**(9), 095101 (2019). https://doi.org/10.1117/1.OE.58.9.095101
31. Saha, A., Bhattacharya, K., Chakraborty, A.K.: Pramana. J. Phys. **77**(4) (2011)
32. Bennet, J.M.: Polarizers. In: Handbook of optics edited by M Bass, vol. 2, Chap. 3, Table 8, pp. 3.46–3.49. McGraw-Hill, New York (1995)
33. Mukhopadhyay, N., Saha, A. and Bhattacharya, K.: A study on superachromatism of quarter-wave retarder for visible range. J. Opt. Technol. **87**(11), 638–641 (2020). ISSN: 1070–9762. https://doi.org/10.1364/JOT.87.000638
34. Mukhopadhyay, N., Saha, A., Bhattacharya, K.: Super-achromatic quarter-wave phase retarder forvisible, near infrared and short wave infrared region applications. Opt. Spectrosc. **128**(8), 1199–1204. ISSN: 0030-400X
35. Mukhopadhyay, N., Saha, A., Bhattacharya, K.: Multi-crystal achromatic quarter wave retarder for the air-multiangle spectropolarimetric imager (AirMSPI) in SWIR region. In: IEEE Conference Proceeding of 2nd International Conference on Electronics, Materials Engineering & Nano-Technology (IEMENTech2018), pp. 1–3,2018, INSPEC Accession Number:18128933, Electronic ISBN:978-1-5386-5550-4. https://doi.org/10.1109/IEMENTECH.2018.8465355
36. Mukhopadhyay, N., Saha, A., Bhattacharya, K.: Design of a broadband achromatic quarter-wave phase retarder for near infrared spectrum using Flower Pollination Algorithm. Int. J. Innov. Knowl. Concepts (IJIKC). **7**(1), 239–242. ISSN: 2454-2415 (2019)

Chapter 32
A Highly Stable Perfect Metamaterial Absorber Based on the Plasmonic Effect of Metamaterial Nano-cells for Optical Spectrum Applications

Mohammed Berka, Tanvir Islam, K. Vasu Babu, Sudipta Das, and Zoubir Mahdjoub

32.1 Introduction

With the strong competition for the manufacture and modernization of devices of different types between the countries around the world, energy consumption has become a matter of concern to most of these countries. That's why; the search for alternative energies in recent years has become necessary to confront these problems and obstacles. Renewable energies based on photovoltaic [1, 2] or wind turbines [3, 4] have demonstrated in a remarkable way their effectiveness for the continuation and development of the industry with lower costs. The contribution of solar electricity in the world's total electricity generation is currently small percentage of total world energy production (0.2%) (Corresponding to 72 GWp capacity of solar plants as

M. Berka
Department of Electrotechnic, University Mustapha Stambouli of Mascara, Mascara, Algeria

T. Islam
Department of Electrical and Computer Engineering, University of Houston, Houston, TX 77204, USA

K. Vasu Babu
Department of Electronics and Communication Engineering, BVRIT Hyderabad College of Engineering for Women, Bachupally, Telangana, India

S. Das (✉)
Department of Electronics and Communication Engineering, IMPS College of Engineering and Technology, West Bengal, Malda 732103, India
e-mail: sudipta.das1985@gmail.com

M. Berka · Z. Mahdjoub
Department of Electronic/EPO Laboratory, University Djillali Liabes of Sidi Bel Abbès, Sidi Bel Abbès, Algeria

M. El Ghzaoui et al. (eds.), *Next Generation Wireless Communication*, Signals and Communication Technology, https://doi.org/10.1007/978-3-031-56144-3_32

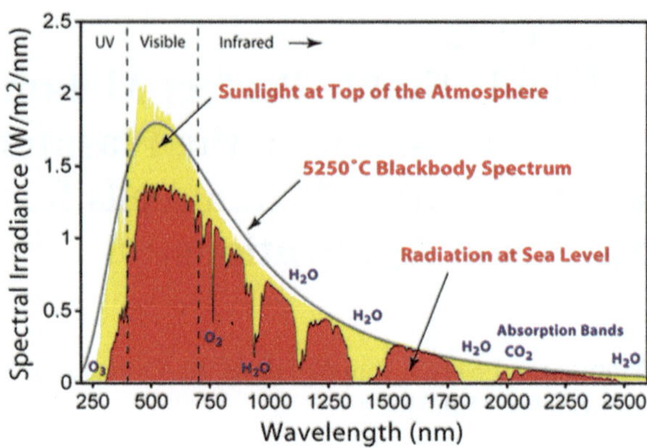

Fig. 32.1 Solar spectrum Google images [6]

compare to 3600 GWp capacity of generated power plants). During the last several years, the average annual growth rates of renewable energy capacity have been 70% [5]. The wavelength of the Sun lies mainly between 300 and 3000 nm, which imply that the frequency is between 100 and 1000 THz by referring Fig. 32.1 [6]. The working of usual solar cells operates when the light photons reach the solar panel and are absorbed by a semiconductor or some other kind of material.

The absorption of light in the form of incident EM waves at nanometric wavelengths is relatively sensitive to the nature of these waves themselves. The materials constituting such an absorber and the way in which the design is constructed represent the determining factors for having the necessary adaptation [7] of the absorbent structure to the environment of evolution. In the last few years, scholars have been able to improve the performance of absorbers by proposing new structures based on metamaterials. This new class of materials is known for its extraordinary physical characteristics which do not exist in nature. Perhaps the most important of these features is the possibility of having mediums or environments of simultaneously negative permittivity and permeability [8, 9], therefore negative refraction indices [10]. This makes metamaterials too much in demand, whether for microwave systems like filters [11–14], antennas [15–18], C-, X-, and Ku- band absorbers [19–22] or terahertz structures like optical filters [23, 24], photodetectors [25], optical sensors [26–28], etc.

In the THz regime, Metamaterial Absorbers (MMAs) of all shapes and sizes have found their featured niche by meeting most requirements of designers. To achieve the desired absorption rates, the designers played in a particular way on the sensitive chord of the patch shape because of its great importance in creating the surface plasmonic polariton (SPP) [29–31]. MIM structures have shown their effectiveness in generating this kind of plasmons, when the incident EM waves (with their too small wavelengths) fall on the surface of the patch and refract on the dielectrics [32, 33]. Replacing noble metals with other less expensive ones (like titanium (Ti), nickel (Ni), cobalt (Co), and tungsten (W)) made it possible to maintain the same level of performance with very acceptable absorption rates. Add to this the possibility of easily implanting these metals on the surfaces of dielectrics such as silicon-dioxide, gallium-arsenide, and titanium-dioxide [34–36].

Because of the importance of this field in enriching modern research and innovations, the authors of this chapter decided to present a new contribution that includes the latest findings of their research in the hope that it will have a positive impact on progress in this field. Here, a new metamaterial absorber of simple design is proposed in order to be able to exploit it in applications covering the visible to NIR spectrum. The proposed absorber has a compact size with a period of 350 nm and a thickness of 100 nm (which are optimized according to the design method). The designed PMMA is represented by a MIM structure containing an intermediate dielectric layer of TiO_2 sandwiched by two metallic layers of tungsten. The chosen patch shape contributed remarkably to creating surface plasmon resonances, which directly provided the near-perfect absorption rates with an average absorption of the order of 92.32% over the entire spectrum. The study and analysis of suggested PMMA have been based on the nano-cell constituting the overall absorbent. All its characteristics were also discussed.

The main contributions of this chapter are as follows:

1. Design of nano-scaled cell metamaterial perfect absorber for a diversity of applications in the visible to near-infrared (NIR) spectrum.
2. MIM configuration with simplicity of the design to create plasmonic resonance characteristic.
3. Use of available metals and dielectrics for the design of the absorber, which gives it credibility, particularly the tendency to realize it.
4. Miniature dimensions which offer low weight and low cost.
5. Insensitivity of the PMMA to polarization with stability to oblique incidence up to 60° over a wide band for TE and TM modes.
6. High absorption rates over a wide range while maintaining good performance.

32.2 Related Works

Since the first proposal to study and design them by the physicist Veselago in 1967 [37], metamaterials have opened up great fields to scholars and designers for the development of devices and systems for different areas. The first metamaterial structure in (3-D) is made by Pendry [38]; this structure shows a magnetic resonance for negative permeability and low dimensions. In the early 2000s, metamaterials were used for the design of circuits evolving in microwave regime. After that, the trend toward light and infrared waves requires the use of MIM configuration for multilayer structures. Several works and research have been presented to enrich this important field. The work conducted in [39] showed the importance and indispensability of metamaterial absorbers for the improvement of solar systems in which, a MMA based on hexagonal ring resonator has been suggested. This absorber had a compact size of $66 \times 66\ nm^2$; it offered a maximum absorption of 99% with an angular stability up to 70° for an operating band of 380–2500 nm. In [40], a triple-band absorber has been designed for THz applications. The proposed MMA had three absorption peaks of 98.76, 99.58, and 92.39% for the resonances of 15.68, 37.48 and 39.55 THz, respectively. It also had remarkable stability against an oblique incidence of up to 90°. An absorber based on the MIM configuration and using tungsten as metal and silicon-dioxide as dielectric has been proposed in [41]. The MMA designed in this reference has been studied on the 441–998 THz spectrum. The maximum absorption of 99.9% was obtained at the resonance of 700 THz. This absorber also showed stability over a wide range of incidence angles. In order to meet the sensing needs, an MMA was reported in [42]. This absorber provided absorption above 90% in the range of 2.65–6.76 THz with a bandwidth of 4.11 THz. The insensitivity to incidence of the proposed absorbent was justified up to 40°. The authors of the article published in [43] suggested a metamaterial absorber based on a plus-shaped resonator. The proposed structure consists of three layers of Tungsten, Rexolite (Cross linked polystyrene, PS), and Nickel; one is held on top of the other, respectively. This MMA provided a maximum absorption of 99% over the 545–628 THz range. In [44], a new MMA has been proposed for infrared applications. This absorber is formed by seven layers of which two of them were metallic (gold and tungsten) and four others were dielectric (HFO_2 and AL_2O_3) which are all implemented on the upper face of the silicon substrate. The designed structure which had nanometric dimensions showed almost perfect absorption over a wide range. A novel petal-shaped tungsten resonator was used in [45] for the design of a metasurface implemented on the top face of the silicon-dioxide. The optical characteristics of this metasurface were exploited to obtain an MMA capable of absorbing with quality in the visible spectrum. This MMA had almost perfect absorption over the wavelength range from 456.62 to 677.73 nm. It had remarkable stability for an incidence up to 60°. Other work concerning optical absorbers based on metamaterials targeting the same frequency spectra have been reported in [46–50] for various applications.

32.3 Theoretical Aspects of Metamaterial Absorbers

The light falls in the form of EM incident waves obliquely on the different structures, whether metal or dielectric. Consequently, propagation modes can be generated such as TE and TM mode. Furthermore, if the dimensions of the structure are small enough compared to the wavelength, surface plasmonic resonances (SPR) are created in the patch-substrate interface [51]. This phenomenon can certainly contribute to the absorption of light with high quality. The mechanism of light absorption by metamaterial absorbers according to the MIM configuration is represented by Fig. 32.2.

To understand the absorption mechanism, the analysis of EM waves becomes indispensable. First of all, the EM wave is always associated with an electromagnetic field. This last is a physical behavior that is produced in a space due to time-varying

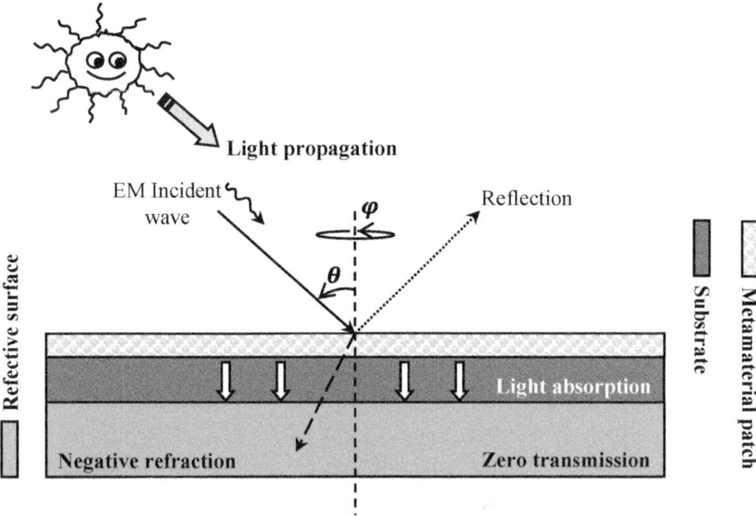

Fig. 32.2 Light absorption according to the MIM configuration

electric charges and represents the interaction between electric and magnetic fields. Unlike static charges that can only produce static electric fields in space, time-varying electric charges are one of sources for the rise of magnetic fields, which in turn produce time-varying electric fields. This is summarized in the four time-varying Maxwell's equations given in differential form [52, 53].

$$\begin{cases} \nabla \times \boldsymbol{E} = -\mu \frac{\partial \boldsymbol{H}}{\partial t} \\ \nabla \times \boldsymbol{H} = \boldsymbol{J}(\boldsymbol{t}) + \varepsilon \frac{\partial \boldsymbol{E}}{\partial t} \\ \nabla . \boldsymbol{B} = 0 \\ \nabla . \boldsymbol{E} = \frac{\rho(t)}{\varepsilon} \end{cases} \tag{32.1}$$

where ∇ is the differential operator defined in the Cartesiȃn system by its three coordinates $\nabla = \left[\frac{\partial}{\partial x}, \frac{\partial}{\partial y}, \frac{\partial}{\partial z} \right]$ .and ρ is time-varying volume charge density, ε and μ are the electric permittivity and magnetic permeability, respectively, $\boldsymbol{J}$ is the time-varying electric current density in a medium, $\boldsymbol{D}$ and $\boldsymbol{B}$ are time-varying electric and magnetic flux densities, respectively, and $\boldsymbol{E}$ and $\boldsymbol{H}$ are time-varying electric and magnetic field intensities, respectively. The solutions of the first two equations of system (1) can define the incident wave and the reflected wave of the electromagnetic field according to Eq. (32.2).

$$\begin{cases} E(\boldsymbol{r}, t) = \boldsymbol{E}_1 e^{j(\omega t - \boldsymbol{k}.\boldsymbol{r})} + \boldsymbol{E}_2 e^{j(\omega t + \boldsymbol{k}.\boldsymbol{r})} \\ H(r, t) = \boldsymbol{H}_1 e^{j(\omega t - \boldsymbol{k}.\boldsymbol{r})} + \boldsymbol{H}_2 e^{j(\omega t + \boldsymbol{k}.\boldsymbol{r})} \end{cases} \tag{32.2}$$

where $\boldsymbol{k}$ and $\boldsymbol{r}$ represent the wavenumber and the position vector, respectively. When the electromagnetic field propagates in the structure studied, the notions of permittivity and permeability (hence the refractive index) become essential to define the relationships existing between the fields and the electric ($\boldsymbol{D}$) and magnetic ($\boldsymbol{B}$) density fluxes. Equation (32.3) can represent the relationship thus discussed.

$$\begin{cases} D(t) = \varepsilon(t)E(t) \\ B(t) = \mu(t)H(t) \end{cases} \tag{32.3}$$

Returning to the wave number k (which also determines the propagation direction of the EM wave), its module is defined by.

$$k = \beta - j\alpha \tag{32.4}$$

In Eq. (32.4), β refers to phase constant (rad/m) and α refers to losses factor (Np/m). In metallic layers (where there is considerable conductivity σ), the condition ($\sigma/\omega\varepsilon \gg 1$) is justified. Then Eq. (32.4) becomes,

$$k = \omega\sqrt{\mu\varepsilon}\sqrt{-j\sigma/\omega\varepsilon} = \sqrt{\sigma\mu\omega/2}(1-j) \tag{32.5}$$

So,

$$\alpha = \beta = \sqrt{\sigma\mu\pi f} \tag{32.6}$$

In the THz regime, the metamaterial absorber is considered as a junction of two ports. At each port, the incident and reflected EM powers can define (depending on the scattering parameters S_{ij}) the transmission and reflection coefficients depending on wavelength or frequency, respectively, by the following expression [54].

$$\begin{cases} R(\lambda) = |S_{11}(\lambda)|^2 \\ T(\lambda) = |S_{21}(\lambda)|^2 \end{cases} \tag{32.7}$$

where

$$\begin{cases} S_{11}(\lambda) = \sqrt{\dfrac{P(R)}{P_{\text{incident}}}} \\ S_{21}(\lambda) = \sqrt{\dfrac{P(T)}{P_{\text{incident}}}} \end{cases} \tag{32.8}$$

In Eq. (32.8), $P(R)$ and $P(T)$ represent the electromagnetic power of the reflected and the transmitted wave, respectively, while P_{incident} represents the EM power of the incident wave. If all these parameters are obtained, the absorption of MMA is calculated according to the following expression [55].

$$A(\lambda) = 1 - R(\lambda) - T(\lambda) = 1 - |S_{11}(\lambda)|^2 - |S_{21}(\lambda)|^2 \tag{32.9}$$

32.4 PMMA Design

32.4.1 Proposed Configuration

The performances of the proposed absorber are extracted from the nano-cell representing our PMMA. This nano-cell is represented according to the MIM configuration comprising three stacked layers including the intermediate layer which represents the dielectric substrate in titanium-dioxide (TiO_2). This layer separates the two other metallic layers of tungsten (W) at the top and bottom sides. TiO_2 has a high dielectric constant measured at room temperature. It is of the order of 63.7 with a dielectric temperature coefficient of – 579 ppm/°C [56]. For his part, the Tungsten is a grayish-white lustrous metal, which is a solid at room temperature; it has excellent corrosion resistance and is attacked only slightly by most mineral acids due to its relatively high conductivity of $1.82 \times 10^{+7}$ (S/m) [57]. The PMMA represented by an (4×4) array and the nano-cell with its geometric parameters are represented in Fig. 32.3. The key parameters considered when designing the nano-cell are the period ($P =$ 350 nm) and the three patch parameters which are the nano-gap, the spacing between the two rings of the patch and the diameters of the different slit circles. The three patch parameters can be chosen according to the following equality system.

$$\begin{cases} g = g_1 = w \\ e = 2S \\ D_2 = D_3 = 3D_1 \end{cases} \tag{32.10}$$

The dimensions of the nano-cell patch which based on Eq. (32.10) are summarized in Table 32.1.

32.4.2 Simulation Arrangement

The PMMA is represented by a (4×4) array of nano-cells. The overall structure is periodic with period P which is the same as the length and width of each nano-cell. The spectral frequency responses of the reflection and transmission are calculated and represented by a numerical simulator. For our study, we used the High Frequency Structure Simulator (HFSS) which is suitable for periodic optical structures of dimensions on the nano-scale. Since the nano-cell represents the overall absorber is periodic, it was necessary to use the boundary conditions in the simulator such as the Floquet port, MASTER, and SLAVE on the different faces of the nano-cell as shown in Fig. 32.4.

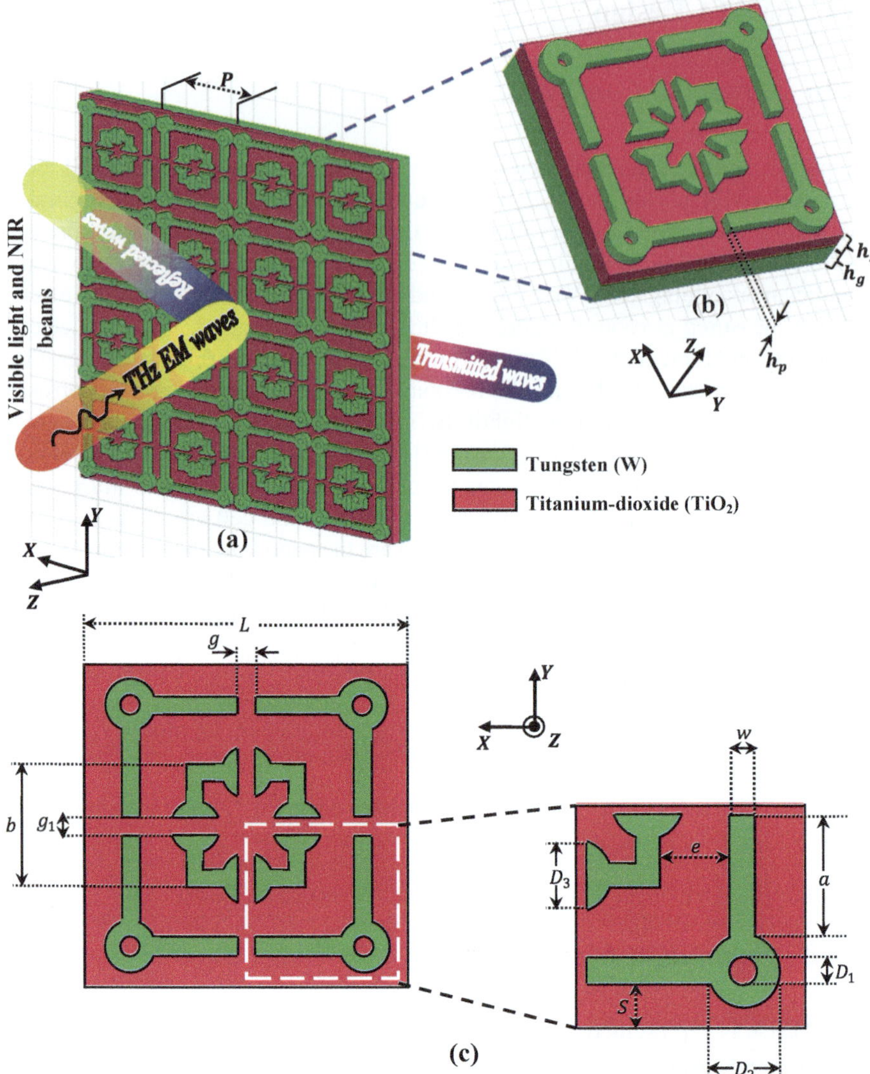

Fig. 32.3 Representation of the proposed PMMA **a** (4 × 4) periodic array **b** nano-cell in perspective **c** top view of the patch with its parameters

Table32.1 Parametric values of the proposed nano-cell

Parameter	a	b	e	w	D_1	S	L	h_s	h_g	h_p
Values (nm)	95	130	60	20	20	30	350	30	50	20

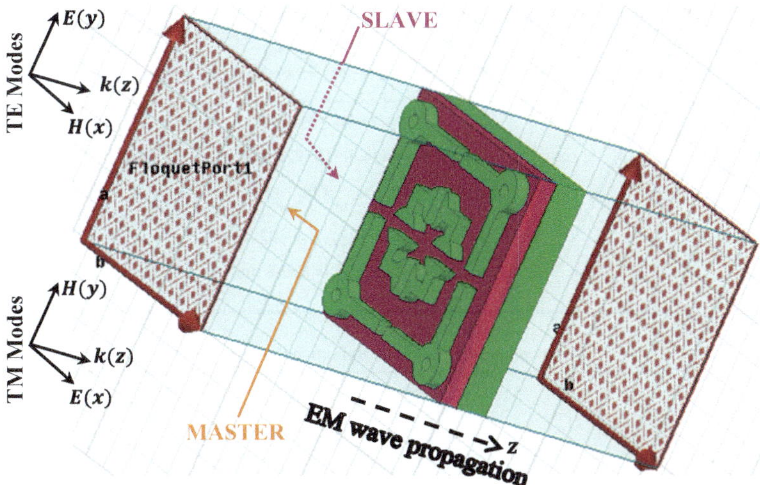

Fig. 32.4 Boundary conditions in HFSS

32.5 Results and Discussion

32.5.1 Frequency Spectra of the Proposed PMMA

From Eq. (32.9), reflection and transmission can characterize the absorption of the nano-cell. For our study on the visible spectrum toward near-infrared (NIR), we represented the different spectra over the wavelength range 400–1400 nm for a normal incidence ($\theta = 0°$). The spectral features are shown in Fig. 32.5a, b.

In Fig. 32.5a, we observe zero transmission on the visible and NIR spectrum. This characteristic is due to the tungsten (W) layer implemented on the bottom side of the TiO_2 substrate. On the other hand, the reflection (linear values) and the absorption are symmetrical to each other. Returning to the absorption, we notice that it is above 80% over a broad spectrum ranging from 478.4 to 1050.4 nm for the first bandwidth and from 1270.6 to 1400 nm for the second one. Absorption peaks are also obtained at wavelengths of 602.4 and 670.8 nm; these peaks represent the maximum absorption of the order of 99.95 and 99.37% and are due to surface plasmon resonances; propagating (PSPR) for the first peak and localized (LSPR) for the second. It should also be noted that there is a third peak located in the NIR, it was of the order of 91.55% and it has been obtained at the wavelength of 1383.2 nm. In Fig. 32.5b, the phase of the reflection coefficient is varied between $-\pi$ and $+\pi$. We note that this phase is minimum in the visible spectrum bandwidth and maximum in the NIR. The average absorption of our PMMA is of the order of 92.32% over the entire spectrum, and this characteristic is calculated from the following expression of Eq. .(32.11) [58].

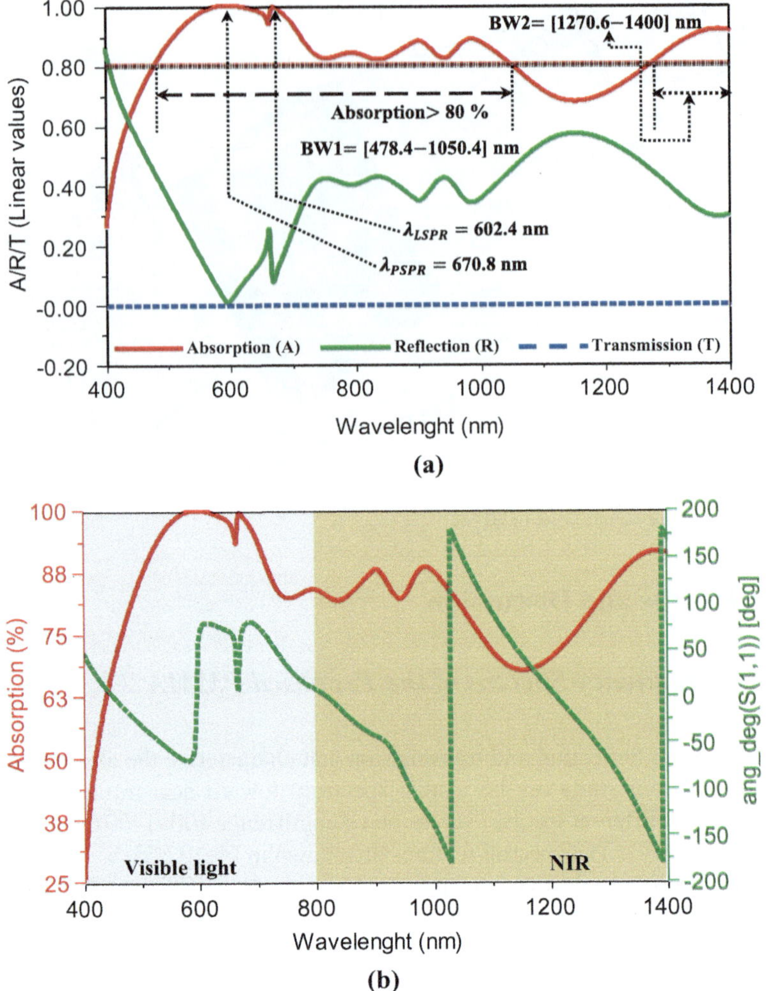

Fig. 32.5 Absorption spectra **a** absorption, reflection and transmission **b** absorption with reflection phase

$$A(\lambda)_{\text{avg}} = \frac{1}{\lambda_u - \lambda_l} \int_{\lambda_l}^{\lambda_u} A(\lambda)\mathrm{d}\lambda \tag{32.11}$$

Here, λ_l and λ_u represent the lower and higher wavelength, respectively.

32.5.2 PMMA Adaptation

The adaptation of the absorber to the free space represents one of the most desired characteristics during the design. When the absorber is matched, the reflection toward space will be zero, which makes it possible to have perfect absorption rates [59], possibly justified by Eq. (32.9). The matching of our PMMA is discussed based on its impedance, either the electrical impedance (expressed in Ω), or the normalized impedance defined by the following relationship.

$$Z_n(\lambda) = \frac{Z(\lambda)}{Z_0} = \sqrt{\frac{\mu_{\text{eff}}(\lambda)}{\varepsilon_{\text{eff}}(\lambda)}} \tag{32.12}$$

In Eq. (32.12), Z_0 represents the free space impedance ($Z_0 = 120\,\pi \approx 377\,\Omega$). If we introduce the transmission and reflection of the absorber, the normalized impedance is written in the following form [60].

$$Z_n = \sqrt{\frac{(1+S_{11})^2 - S_{21}^2}{(1-S_{11})^2 - S_{21}^2}} = \frac{1+S_{11}}{1-S_{11}}(S_{21}=0) \tag{32.13}$$

The absorption characteristic can be represented by Fig. 32.6.

The normalized impedance of our PMMA has a real part and an imaginary one, and these two quantities are represented in Fig. 32.7.

As can be observed in Fig. 32.7, the two real and imaginary parts of Z_n have reduced values between − 2 and 4. More precisely, at plasmonic resonances associated with wavelengths λ_{PSPR} and λ_{LSPR}, the real part of the normalized impedance has a unit value ($\text{Re}\{Z_n\} = 1$) and the imaginary part is zero ($\text{Im}\{Z_n\} = 0$). This result can justify the adaptation of our PMMA on the one hand and the obtaining of almost perfect absorption rates on the other hand.

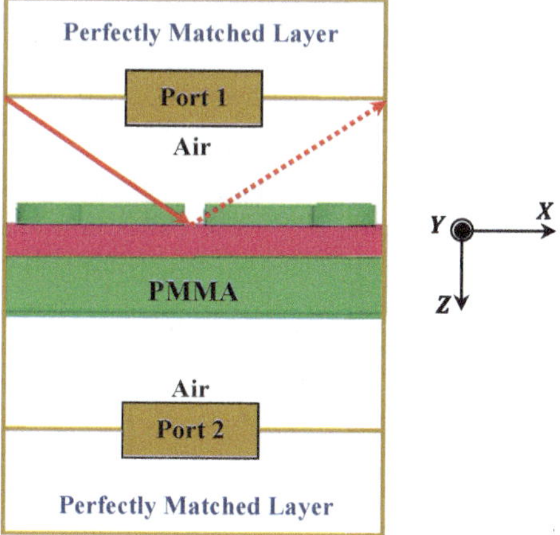

Fig. 32.6 Matching mechanism of the proposed PMMA

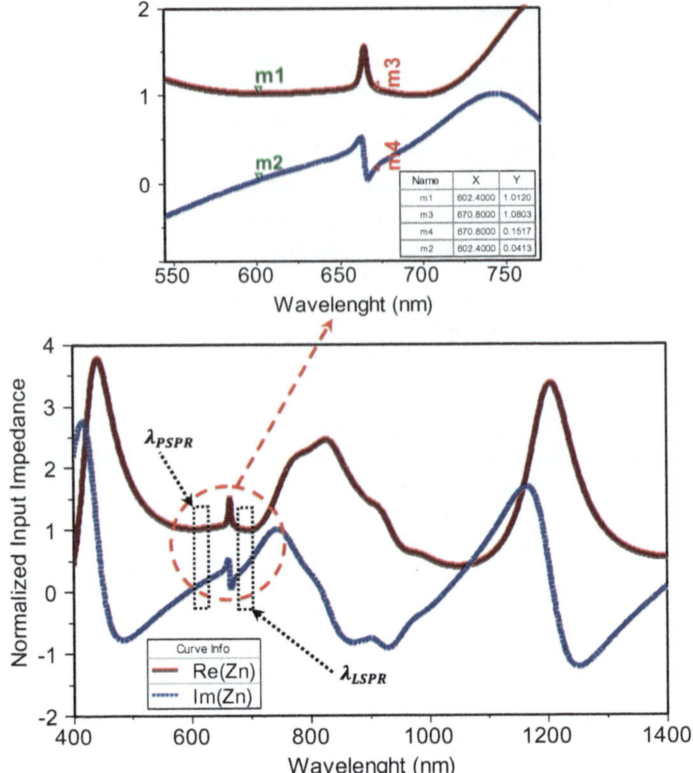

Fig. 32.7 Variation of the real and imaginary parts of the normalized impedance according to the wavelengths

32.5.3 Extraction of Effective Parameters

The effective parameters of such an absorber which are the permittivity, the permeability, and the refractive index are often brought into play during the electromagnetic characterization of this absorber [61]. The study of the variation of these parameters as a function of frequency and therefore as a function of wavelengths amounts to characterizing its real and imaginary parts according to the reflection and transmission coefficients [62]. The extraction of these parameters in the optical spectrum as well as the microwave spectrum is done using the Nicolson-Ross-Weir (NRW) method [63, 64] in subsequent iterations.

The addition (V_1) and subtraction (V_2) of the scattering parameters are given by,

$$\begin{cases} V_1 = |S_{11}| + |S_{21}| \\ V_2 = |S_{21}| - |S_{11}| \end{cases} \tag{32.14}$$

The transmission coefficient is defined as a function of V_1 and V_2 by,

$$X = \frac{1 - V_1 V_2}{V_1 - V_2} \tag{32.15}$$

For the reflection coefficient too,

$$\Gamma = X \pm \sqrt{X^2 - 1} \tag{32.16}$$

The transmission coefficient becomes,

$$X = \frac{V_1 - \Gamma}{1 - V_1 \Gamma} \tag{32.17}$$

Eventually,

$$\text{Permittivity, } \varepsilon_{\text{eff}}(\lambda) = \frac{\lambda}{j\pi h_s} \times \frac{1 - (S_{21} + S_{11})}{1 + (S_{21} + S_{11})} \tag{32.18}$$

$$\text{Permeability, } \mu_{\text{eff}}(\lambda) = \frac{\lambda}{j\pi h_s} \times \frac{1 - (S_{21} - S_{11})}{1 + (S_{21} - S_{11})} \tag{32.19}$$

Refractive index,

$$n_{\text{eff}}(\lambda) = \sqrt{\varepsilon_{\text{eff}}(\lambda)\mu_{\text{eff}}(\lambda)} \tag{32.20}$$

There in, $c = 3 \times 10^{+8}$(m/s) is the velocity of light and h_s is thickness of the substrate.

The variation of the effective parameters of our PMMA as a function of wavelengths is shown in Fig. 32.8 on the spectrum 400–1400 nm for both TE and TM modes. First of all, the spectral characteristic of each parameter is the same for the two modes studied TE and TM. In Fig. 32.8a, we notice that the real part of the permittivity of the absorber changes their sign in both spectra; visible and NIR. At propagating and localized plasmonic resonances, is negative ($\text{Re}\{\varepsilon_{\text{eff}}\} < 0$) which justifies the plasmonic effect. In Fig. 32.8b, the imaginary part of the permeability is positive on the visible and NIR spectrum, but the imaginary part is negative, especially around plasmon resonances. Moving to Fig. 32.8c, both real and imaginary parts of the refractive index are negative around plasmonic resonances. This characteristic is the most essential for our PMMA; it can justify the left-handed behavior (LHM) of our absorber.

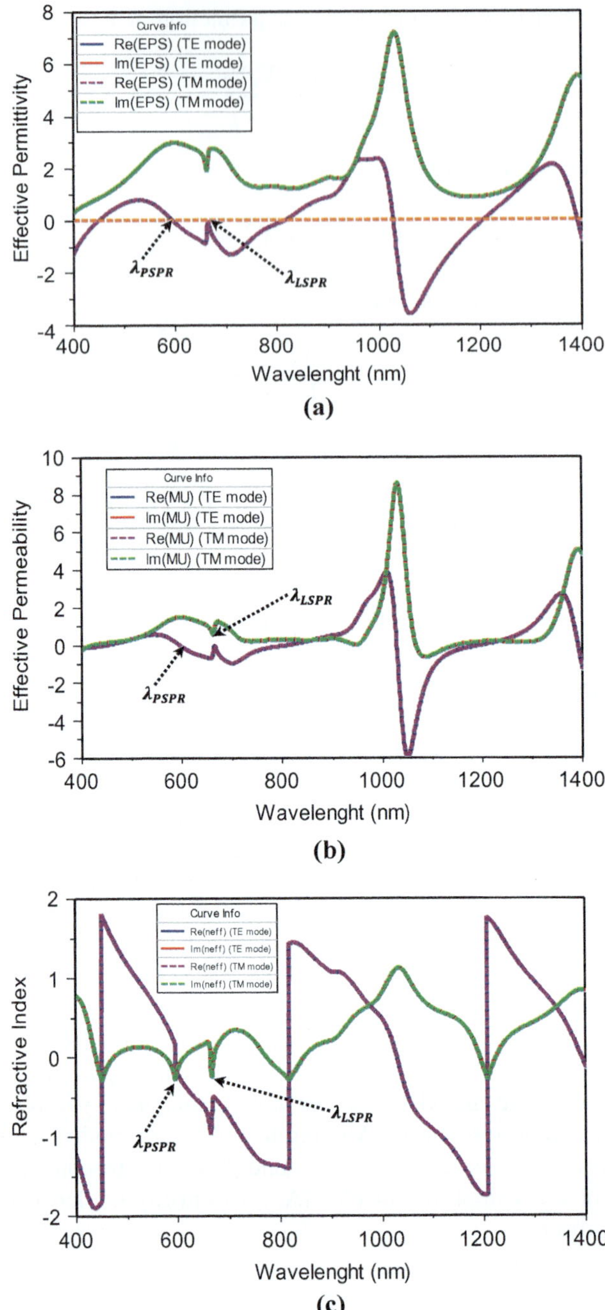

Fig. 32.8 Effective parameters for TE and TM mode **a** permittivity, **b** permeability, and **c** refractive index

32.5.4 Polarization Analysis

In this section, we present a detailed analysis of the polarization of our absorber. Polarization of PMMA concerns azimuthally polarization as a function of angle (φ) and elevation polarization as a function of angle of oblique incidence (θ). To better characterize PMMA, the polarization analysis concerns both TE and TM modes. The variation of the normalized impedance of our PMMA can characterize the variation of the reflection coefficients of TE and TM polarized waves according to the following relationship [65].

$$\Gamma_{\mathrm{TE}} = \Gamma = \frac{Z_{\mathrm{n}} \cos\theta_{\mathrm{i}} - \cos\theta_{\mathrm{t}}}{Z_{\mathrm{n}} \cos\theta_{\mathrm{i}} + \cos\theta_{\mathrm{t}}} \tag{32.21}$$

$$\Gamma_{\mathrm{TM}} = \Gamma_{//} = \frac{Z_{\mathrm{n}} \cos\theta_{\mathrm{t}} - \cos\theta_{\mathrm{i}}}{Z_{\mathrm{n}} \cos\theta_{\mathrm{t}} + \cos\theta_{\mathrm{i}}} \tag{32.22}$$

where θ_{i} and θ_{t}, stand for incident and transmitting angle, they are related by Snell's laws follows,

$$\frac{Z_0}{Z} = \frac{\sin\theta_{\mathrm{t}}}{\sin\theta_{\mathrm{i}}} \tag{32.23}$$

From both Fig. 32.8a, b, we can notice that the absorption performance is not affected by the polarization angle (φ) for the two studied modes. If by comparing the two absorption characteristics for the two modes, we observe a slight difference at the level of the two regions A and B representing the lower and higher frequencies, respectively (Fig. 32.9).

In Fig. 32.10a, b, the spectral characteristic of the elevation polarization is shown for the TE and TM mode, respectively. For the TE mode shown in Fig. 32.10a, we notice that the absorption rates are considerable for all oblique angles of incidence located between 0° and 60°. Beyond the angle ($\theta = 60°$), the absorption level was reduced below 80%. This characteristic can justify the angular stability of our PMMA up to the angle of 60° (for TE mode). For the TE mode shown in Fig. 32.10b, we have almost the same absorption levels, except for the angle ($\theta = 60°$) where we observe that the absorption rates are below 80% over the majority of the NIR spectrum. From this characteristic, we can say that our PMMA is less stable in front of θ for the TM mode compared to the TE mode. It should be noted that the absorption at the incidence of 90° is almost zero for both modes, and this is well justified if returning to the theoretical calculations (from Eqs. (32.21) and (32.22)).

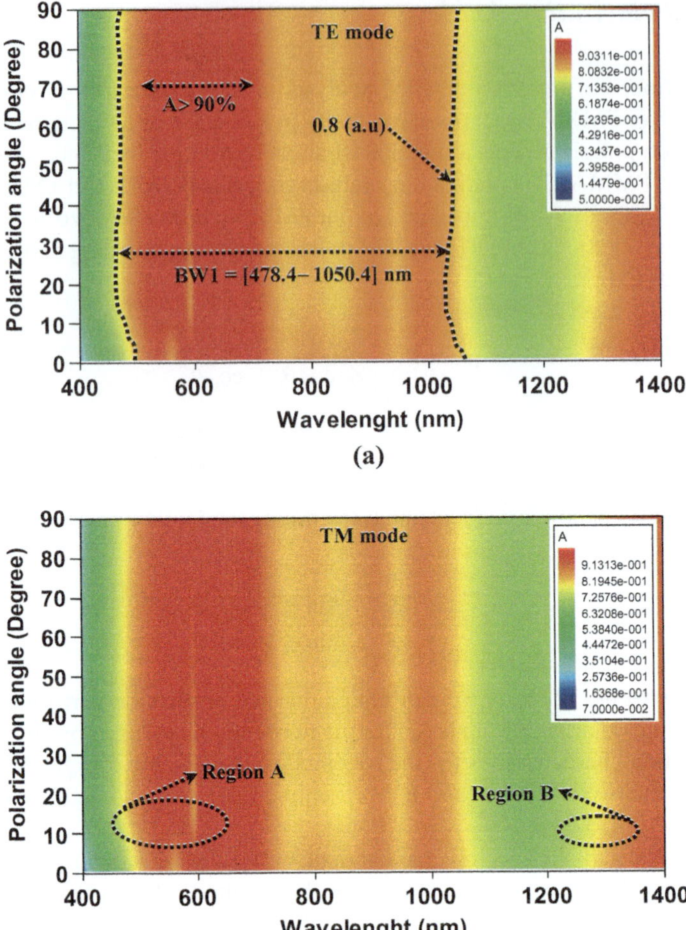

Fig. 32.9 Characteristic of azimuthally polarization **a** TE mode **b** TM mode

32.5.5 Surface Current Distribution on PMMA

Generally, the surface current expressed in (A/m^2) also represents an essential feature for understanding the electromagnetic behavior of such an absorber. In the nanoscale, the currents entwined in the surface of the patch (in the top side) can explain the plasmonic effect of the structure in the metal–dielectric interface. The representation of surface currents (for both modes) at plasmonic frequencies is therefore necessary to recapitulate all of this.

In Fig. 32.11, the surface current distribution is shown at the two plasmon resonances for the TE and TM mode. For the TE mode distribution shown in Fig. 32.11a,

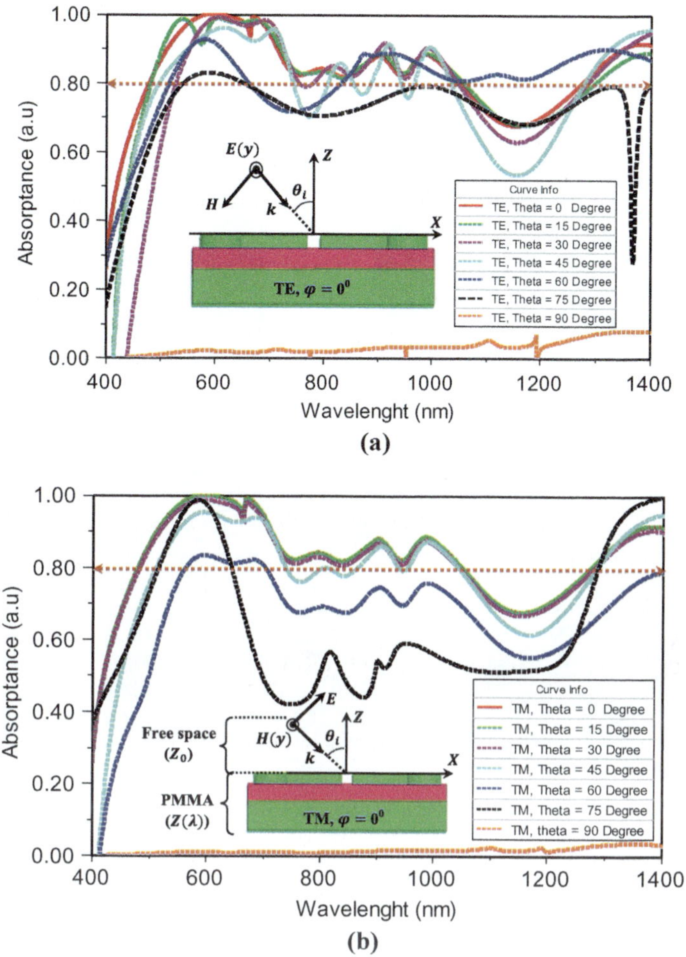

Fig. 32.10 Characteristic of elevation polarization **a** TE mode **b** TM mode

we note that the current concentrations are relatively low at the wavelength of 602.4 nm corresponds to the resonance (PSPR). The high concentration for this case has been observed at the level of the inner ring, especially in the nano-gaps of the inner circles. This is justified by the plasmonic effect of the first resonance which is due to the shape of the ring itself. At the second resonance of 670.8 nm corresponds to the (LSPR), the current propagates to the outer ring with considerable concentrations. For the TM mode characteristic shown in Fig. 32.11b, the current propagates in different directions on the patch surface (for both resonances). This feature can explain the creation of magnetic dipoles contributing to obtaining high absorption rates around plasmonic resonances.

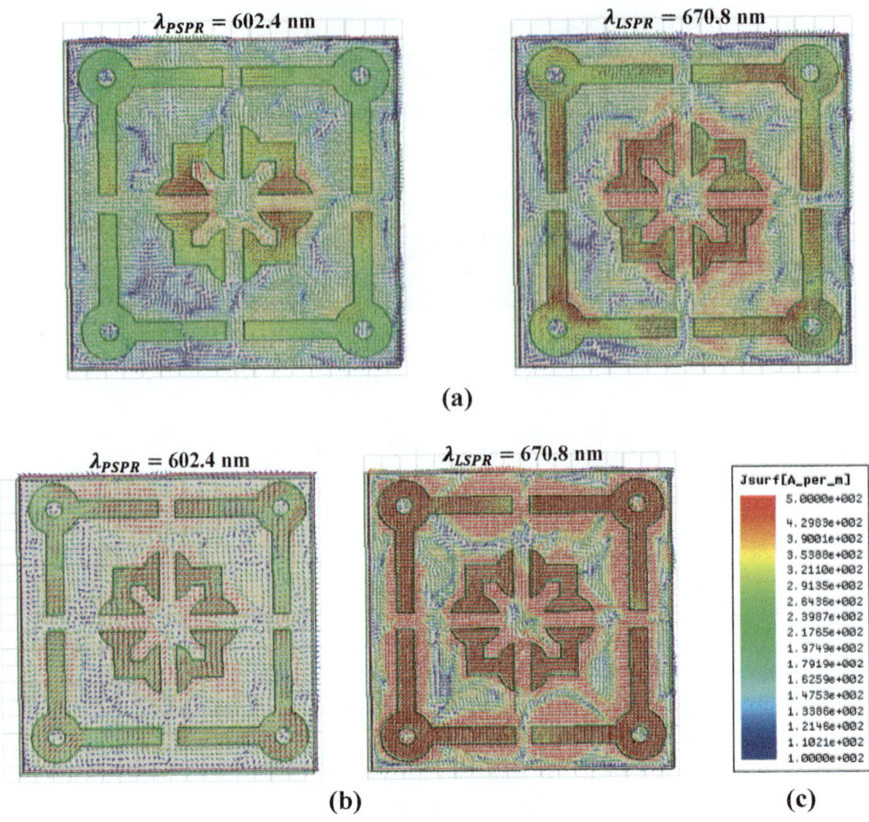

Fig. 32.11 Distribution of the surface current on the top side of the patch at plasmonic resonances for **a** TE mode **b** TM mode **c** A/m scale

32.5.6 Influence on Absorption Performance

The performance of our PMMA can be influenced by several geometric parameters that come into play during the design. In this section, we focused on the nano-split gap and the size of the nano-cell denoted by the period of the absorber.

As can be seen in Fig. 32.12, the absorption rates of our PMMA are different from one value of g (and g_1) to another (across the entire simulation spectrum). We notice that when the nano-gap increases, the absorption rates have been reduced. In the case where the patch does not have nano-gaps, we notice that the absorption is clearly affected (below 80%). This comes down to the effect of electromagnetic coupling in the two rings; inner and outer. The design according to this result was made for nano-split gaps of the order of 20 nm.

According to the curves in Fig. 32.13 representing the different absorptions of the different sizes of the nano-cell, we can say that the resonance of the absorber has been shifted toward the blue when the size (the period) decreases. Given the high

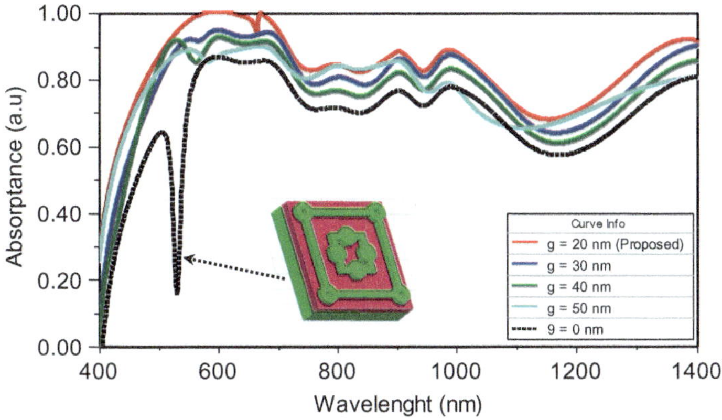

Fig. 32.12 Different absorption spectral characteristics according to various nano-gaps values

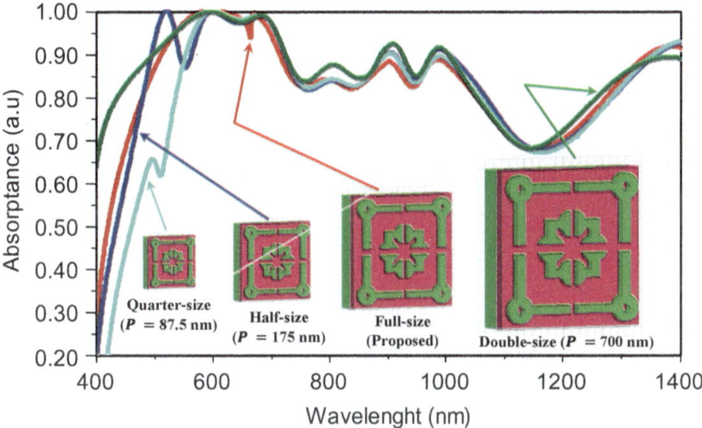

Fig. 32.13 Different absorption spectral characteristics according to various nano-cell sizes values

absorption rates for the period of 350 nm, our design was based on this size of the nano-cell.

32.5.7 Equivalent Circuit Model

We discussed the adaptation of our PMMA (in Sect. 5.2) based on its normalized impedance Z_n. The reflectance R of the absorber is expressed as a function of S_{11} as follows:

$$R = |S_{11}|^2 \tag{32.24}$$

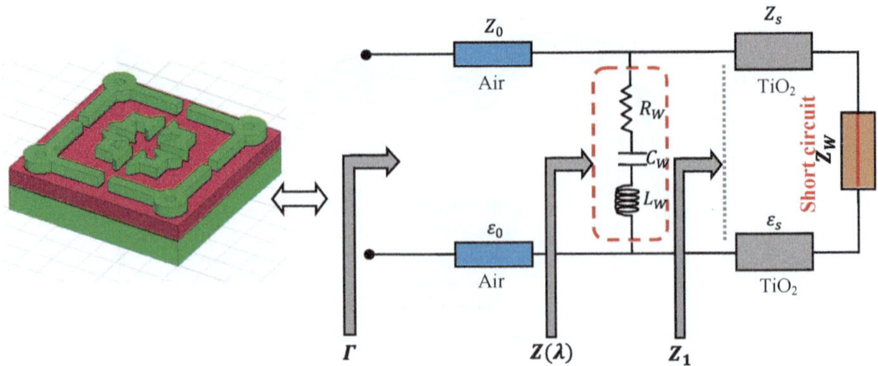

Fig. 32.14 The nano-cell with its equivalent circuit model

On the other side,

$$S_{11} = \frac{Z(\lambda) - Z_0}{Z(\lambda) + Z_0} \tag{32.25}$$

So,

$$R = \left(\frac{Z(\lambda) - Z_0}{Z(\lambda) + Z_0}\right)^2 \tag{32.26}$$

By introducing the notions of PMMA impedance and free space impedance, transmission line theory (in the THz regime) can explain the physical aspect of Eq. (32.26). The proposed PMMA can be represented by a RLC resonator in series branch. The reflective tungsten layer implemented on the bottom side of the substrate in turn can be represented by a short circuit [66] (reflection effect). Figure 32.14 shows the equivalent circuit model of our PMMA.

The impedance of the titanium-dioxide substrate and its input impedance are given by Liu et al. [67].

$$\begin{cases} Z_1 = jZ_d \mathrm{tg}(k_s h_s) \\ Z_s = \sqrt{\frac{\mu_0}{\varepsilon_0 \varepsilon_s}} \\ k_s = 2\pi f \sqrt{\mu_0 \varepsilon_0 \varepsilon_s} \end{cases} \tag{32.27}$$

The impedance of the patch layer (in tungsten W) is given by.

$$Z_W = R_W + jX_W = R_W + j\left(2\pi f L_W - \frac{1}{2\pi f C_W}\right) \tag{32.28}$$

When $X_W = 0$, so $Z_W = R_W$.

The impedance of the proposed PMMA becomes,

$$Z(\lambda) = \frac{Z_1 Z_W}{Z_1 + Z_W} \quad (32.29)$$

From the equivalent circuit model,

$$\Gamma = \frac{Z(\lambda) - Z_0}{Z(\lambda) + Z_0} = S_{11} \quad (32.30)$$

Eventually, the maximum absorption is given by.

$$A(\lambda) = 1 - R(\lambda) = 1 - |S_{11}(\lambda)|^2 = 1 - \Gamma^2 = 1 - \left(\frac{Z(\lambda) - Z_0}{Z(\lambda) + Z_0}\right)^2 \quad (32.31)$$

So,

$$A_{\mathrm{Max}}(\lambda) = \frac{4}{\frac{Z(\lambda)}{Z_0} + \frac{Z_0}{Z(\lambda)} + 2} \quad (32.32)$$

32.6 Comparison with State of the Art

In order to validate the results obtained in this chapter, we propose in this section a comparison of the performances of our PMMA with those existing in the state of the art for the same frequency spectrum and the same applications. This performance comparison is listed in Table 32.3.

Based on Table 32.3, our PMMA is well placed in the state of the art of solar absorbers based on metamaterials. From the point of view of dimensions, our design is the most compact compared to those reported in [68–70, 73]. If discussing the stability of the absorbers, our proposed PMMA is stable compared to the absorbers proposed in [70–72]. The other absorbent structures proposed in [68, 69, 73] have practically the same stability of our absorber. If discussing the bandwidths, it is clear that the proposed PMMA has two wide bandwidths compared to all the works listed in Table 32.3, except the design proposed in reference [71]. So from this comparison, the absorber proposed in this chapter which is first of all an almost perfect solar absorber is strongly recommended for optical filtering, sensing applications, and especially for harvesting solar energy.

Table 32.3 Comparative performance analysis of the proposed PMMA with the state of the art

Ref. no	Used materials	Dimensions (nm)	Maximum absorption (%)	Angular stability ($A > 70\%$)	Bandwidth (nm)
[68]	Ni, SiO_2	300 × 300 × 270	Above 97	≤ 60°	400–800
[69]	W, SiO_2	950 × 950 × 240.13	99	≤ 60°	380–750
[70]	Al, GaAs	610 × 610 × 137	94.16	≤ 50°	591.54–04.40
[71]	Ti–SiO_2	250 × 250 × 250	N/A	≤ 45°	250–1600
[72]	W, SiO_2	200 × 200 × 65	Above 80	≤ 40°	400–750
[73]	Au, SiO_2	1000 × 1000 × 644	99.90	≤ 60°	520–540
Proposed design	W, TiO_2	350 × 350 × 100	99.95	≤ 60°	478.4–1050.4 and 1270.4–1400

32.7 Fabrication Possibilities

The metamaterial absorbers fabrication with nanometric dimensions represents the final and most sensitive step to justify the performances obtained based on the design. In the visible and NIR spectrum where the resonances are in THz, several techniques have been tried to achieve the desired precisions. Recently, the thin film growth technique based on metal–organic chemical vapor deposition (MOCVD) [74] and chemical vapor deposition (CVD) technique [75] have shown their efficiencies during manufacturing especially for precision. These two techniques are also very popular for minimizing lattice mismatch according to the proposed periods [76]. Another technique that was widely used for nanoscale structures is the molecular beam epitaxy (MBE) [77]. This technique has also shown its effectiveness for a significant number of works [78–80]. Next comes the procedure of etching on the surfaces of the used dielectric. This step is embodied by lithography technique [81]. The main objective of this step is to obtain the final shape of the proposed patch with the desired dimensions based on the preparation of the TiO_2 substrate, the deposition of the tungsten, the etching, and the resist removal and cleaning.

32.8 Conclusions

In this chapter, we have presented and designed a high-performance perfect metamaterial absorber for applications in visible and near-infrared spectrum. To achieve the desired goals, we have presented a comprehensive study on solar absorbers. The theoretical aspect of this kind of optical structures is also presented. For the design of our PMMA, we proposed a simple and clear method which based on the shape and dimensions of the absorber itself. The simplicity of our design method can be summarized or represented by the simplicity of the shape of the proposed nanoscale patch. The proposed PMMA is made up of three layers corresponding to the MIM configuration; tungsten-titanium dioxide-tungsten stacked one above the other for an overall thickness of 100 nm. The patch is formed by two square-shaped rings each; the inner one has four circles in the middle of its arms with uniform gaps. The outer ring has four circles of the same size with circular slots in the centers. The study of our absorber concerns the visible and NIR spectrum, the simulation was carried out on the wavelength spectrum ranging from 400 to 1400 nm. The performance of the solar absorber has shown its efficiency and reliability through simulation outcomes. Considerable absorption above 80% over wide bandwidth in both visible and NIR regions with an average absorption of 95.32% and a maximum absorption of around 99.95% obtained at the plasmonic resonance with a wavelength of 602.4 nm. Angular stability was one of the most emphasized performances of our proposed PMMA for TE and TM modes. The absorption rates obtained for oblique incidences up to 60° were considerable, the majority of them being above 80% as well. The metamaterial absorber proposed and investigated here is suitable and worthy to be used for various optical window applications, especially for solar energy harvesting. The future recommendations of this work carried out in this chapter are the reduction of the dimensions of the basic cells without as many remarkable changes in the shape of the patches which could disrupt the electromagnetic performances.

Acknowledgements This work was supported by the General Directorate of Scientific Research and Technological Development (DGRSDT), the body affiliated to the Algerian Ministry of Higher Education and Scientific Research. Also, the authors extend their appreciation to Prof. Dr. Zoubir Mahdjoub president of PRFU Project No A25N01UN220120200001 and Dr. Sudipta Das, for their supervision and observations for the improvement of this work.

References

1. Azzeddine, H.A., Chaouch, D.E., Berka, M., Hebali, M., Larbaoui, A., Tioursi, M.: Fuel cell grid connected system with active power generation and reactive power compensation features. Przeglad Elektrotechniczny **11**(48), 124–127 (2020). https://doi.org/10.15199/48.2020.11.25
2. Zouggaret, A., Azzeddine, H.A., Chaouch, D., Berka, M., Hebali, M., Bellil, M., Djilani Kobibi, Y.I.: An efficient fuel cell maximum power point tracker based on an adaptive neural fuzzy inference system. Przeglad Elektrotechniczny **99**(2), 135–139 (2023). https://doi.org/10.15199/48.2023.02.23
3. Zhang, B., Jin, Y., Cheng, S., Zheng, Y., Chamorro, L.P.: On the dynamics of a model wind turbine under passive tower oscillations. Appl. Energy **311**(1), 118608 (2022). https://doi.org/10.1016/j.apenergy.2022.118608
4. Gloe, A., Jauch, C., Räther, T.: Grid support with wind turbines: the case of the 2019 blackout in Flensburg. Energies **14**(6), 1697 (2021). https://doi.org/10.3390/en14061697
5. Renewable Energy Policy Network for the 21st century, Renewables 2012, Global Status report (2012)

6. https://socratic.org/questions/radiation-from-the-sun-often-gets-trapped-in-the-earth-s-atmosphere
7. Berka, M., Yacine, R.A., Bendaoudi, A., Mahdjoub, Z.: Triangular split ring resonators for X-band applications and operations. J. Nano Electron. Phys. **14**(01), 1007 (2022). https://doi.org/10.21272/jnep.14(1).01007
8. Estep, N., Askarpour, A., Shvets, S.: Transmission-line model and propagation in a negative-index, parallel-plate metamaterial to boost electron-beam interaction. Antenna Propag. IEEE Trans. **62**, 3212–3221 (2014)
9. Liu, N., Guo, H., Fu, L., Kaiser, S.: Three dimensional photonic metamaterials at optical frequencies. Nat. Mater **7**, 31–37 (2008)
10. Zhu, Y., Li, Z., Hao, Z., Di Marco, C., Maturavongsadit, P., Hao, Y., et al.: Optical conductivity-based ultrasensitive mid-infrared biosensing on a hybrid metasurface. Light Sci Appl. **7**, 67 (2018). https://doi.org/10.1038/s41377-018-0066-1
11. Berka, M., Bendaoudi, A., Benkhallouk, K., et al.: Designing of tri-band bandpass microwave filter based on (E–Z) inter-coupled tapered metamaterial resonators for C- and X-band applications and operations. Appl. Phys. A **128**, 1112 (2022). https://doi.org/10.1007/s00339-022-06242-0
12. Rouabhi, A.Y., Berka, M., Benadaoudi, A., et al.: Investigation of dual-band bandpass filter inspired by a pair of square coupled interlinked asymmetric tapered metamaterial resonator for X-band microwave applications. Bull. Mater. Sci. **45**, 118 (2022). https://doi.org/10.1007/s12034-022-02693-6
13. Berka, M., Azzeddine, H.A., Bendaoudi, A., et al.: Dual-band bandpass filter based on electromagnetic coupling of twin square metamaterial resonators (SRRs) and complementary resonator (CSRR) for wireless communications. J. Electron. Mater. **50**, 4887–4895 (2021). https://doi.org/10.1007/s11664-021-09024-1
14. Berka, M., Mahdjoub, Z., Hebali, M.: New design of dual-band bandpass microwave filter based on electromagnetic effect of metamaterial resonators. J. Electr. Eng. **69**(4), 311–316 (2019). https://doi.org/10.2478/jee-2018-0044
15. Mohammed, B., Mourad, H., Amina, B., Özkaya, U., Zoubir, M.: Design of miniaturized planar antenna based on complementary metamaterial resonators dedicated to wireless communication handheld devices. In: 2nd International Conference on Innovative Academic Studies: Konya, Turkey (2023)
16. Bendaoudi, A., Berka, M., Debab, M., Mahdjoub, Z.: Effects of ground plane on a square graphene ribbon patch antenna designed on a high permittivity substrate with PBG structures. Frequenz (2023). https://doi.org/10.1515/freq-2022-0149
17. Berka, M., Bendaoudi, A., Benkhallouk, K., Mahdjoub, Z., Rouabhi, A.Y.: Enhanced 4-port MIMO antenna based on electromagnetic mutual coupling reduction using CSRRs metamaterials for sub-6 GHz 5G applications. J. Nano Electron. Phys. **14**(6), 06021 (2022). https://doi.org/10.21272/jnep.14(6).06021
18. Benkhallouk, K., Bendaoudi, A., Berka, M., Mahdjoub, Z.: Enhanced radiation characteristics of regular dodecagon split ring resonator (D SRR)-based microstrip patch antenna employing dielectric superstrate for THz applications. J. Eng. Appl. Sci. **69**(1) (2022). https://doi.org/10.1186/s44147-022-00119-y
19. Berka, M., Fellah, B., Islam, T., et al.: A triple band square-interlinked-rhombus shaped metamaterial absorber for microwave C and X-band utilizations. Opt. Quant. Electron. **55**, 1223 (2023). https://doi.org/10.1007/s11082-023-05511-3
20. Jahan, M.I., Iqbal Faruque, M.R., Hossain, Md.B., Khandaker, M.U., et al.: Quad-band metamaterial perfect absorber with high shielding effectiveness using double X-shaped ring resonator. Materials **16**, 4405 (2023). https://doi.org/10.3390/ma16124405
21. Berka, M., Özkaya, U., Islam, T., et al.: A miniaturized folded square split ring resonator cell based dual band polarization insensitive metamaterial absorber for C- and Ku-band applications. Opt. Quant. Electron. **55**, 699 (2023). https://doi.org/10.1007/s11082-023-04954-y

22. Bennaoum, M., Berka, M., Bendaoudi, A., et al.: Investigation of a near-perfect quad-band polarization-insensitive metamaterial absorber based on dual-T circular shaped resonator array designed on a silicon substrate for C-. X- Ku-Bands Appl. Silicon **15**, 699–712 (2023). https://doi.org/10.1007/s12633-022-02038-2
23. Chao, X., Wang, Y., Huang, F., Chen, Z.: A new design of ultra-thin wide stop-band metamaterial filter for terahertz waves. Opt. Quantum Electron. **55**, 800 (2023). https://doi.org/10.1007/s11082-023-05064-5
24. Elizabeth Caroline, B., Sagadevan, K., Danasegaran, S.K., Kumar, S.: Characterization of a pentagonal CSRR bandpass filter for terahertz applications. J. Electron. Mater. **51**, 5405–5416 (2022). https://doi.org/10.1007/s11664-022-09779-1
25. Zhang, L., Lu, W., Qian, R., Wang, H., Xu, H., Zhu, L., An, Z.: Highly responsive silicon-based hot-electron photodetector with self-aligned metamaterial interdigital electrodes. Appl. Phys. Lett. **122**(3) (2023). https://doi.org/10.1063/5.0133705
26. Shen, S., Liu, X., Shen, Y., Qu, J., Pickwell-MacPherson, E., et al.: Recent advances in the development of materials for terahertz metamaterial sensing. Adv. Opt. Mater. **10**(1) (2021). https://doi.org/10.1002/adom.202101008
27. Fusco, Z., Taheri, M., Bo, R., Tran-Phu, T., Chen, H., et al.: Non-periodic epsilon-near-zero metamaterials at visible wavelengths for efficient non-resonant optical sensing. Nano Lett. **20**(5), 3970–3977 (2020). https://doi.org/10.1021/acs.nanolett.0c01095
28. Bhati, R., Malik, A.K.: Ultra-efficient terahertz metamaterial sensor. Results Opt. **8**, 100236 (2022). https://doi.org/10.1016/j.rio.2022.100236
29. Quynh Anh, N.P., Hieu, N.N.: Surface plasmon-polaritons in anisotropic hyperbolic metamaterials. J. Nonlinear Opt. Phys. Mater. **32**(02) (2023). https://doi.org/10.1142/S0218863523500121
30. Gric, T., Rafailov, E.: Propagation of surface plasmon polaritons at the interface of metal-free metamaterial anisotropic semiconductor inclusions. Optik **254**, 168678 (2022). https://doi.org/10.1016/j.ijleo.2022.168678
31. Trofimov, A., Gric, T., Hess, O.: Three-layered nanostructured metamaterials for surface plasmon polariton guiding. J. Math. Chem. **57**, 190–201 (2019). https://doi.org/10.1007/s10910-018-0943-0
32. Zhang, G., Liu, J., Dong, D., Chen, C., Liu, Y., Fu, Y.: Controlling the directional excitation of surface plasmon polaritons using tunable non-Hermitian metasurfaces. Opt. Lett. **48**(16), 4296–4299 (2023). https://doi.org/10.1364/OL.499760
33. Liu, R., Zha, Z., Shafi, M., Li, C., Yang, W., Xu, S., Liu, M., Jiang, S.: Bulk plasmon polariton in hyperbolic metamaterials excited by multilayer nanoparticles for surface-enhanced Raman scattering (SERS) sensing. Nanophotonics **10**(11), 2949–2958 (2021). https://doi.org/10.1515/nanoph-2021-0301
34. Li, Y., Chen, Q., Wu, B., Shi, L., Tang, P., Du, G., Liu, G.: Broadband perfect metamaterial absorber based on the gallium arsenide grating complex structure. Results Phys. **15**, 102760 (2019). https://doi.org/10.1016/j.rinp.2019.102760
35. Ding, F., Dai, J., Chen, Y., et al.: Broadband near-infrared metamaterial absorbers utilizing highly lossy metals. Sci. Rep. **6**, 39445 (2016). https://doi.org/10.1038/srep39445
36. Areed, N.F.F., EL-Wasif, Z., Obayya, S.S.A.: Nearly perfect metamaterial plasmonic absorbers for solar energy applications. Opt. Quant. Electron. **50**, 197 (2018). https://doi.org/10.1007/s11082-018-1464-6
37. Veselago, V.G.: The electrodynamics of substances with simultaneously negative values of ε and μ. Soviet Phys. Uspekhi **10**, 509–514 (1968)
38. Marquez, R., Medina, F., Raffi, R., Phys. Rev. B. 65 (2002)
39. Hakim, M.L., Hanif, A., Alam, T., Islam, M.T., Arshad, H., Soliman, M.S., Albadran, S.M., Islam, M.S.: Ultrawideband polarization-independent nanoarchitectonics: a perfect metamaterial absorber for visible and infrared optical window applications. Nanomaterials **12**, 2849 (2022). https://doi.org/10.3390/nano12162849
40. Abdulkarim, Y.I., Özkan Alkurt, F., Awl, H.N., Altıntaş, O., Muhammadsharif, F.F., Appasani, B., Bakır, M., Karaaslan, M., Taouzari, M., Dong, J.: A symmetrical terahertz triple-band metamaterial absorber using a four-capacitance loaded complementary circular split ring resonator

and an ultra-thin ZnSe substrate. Symmetry **14**, 1477 (2022). https://doi.org/10.3390/sym14071477

41. Abishek, E., Azhagar, E., Esakkimuthu, M., Arumugam, K.: Design and evaluation of ultra-broadband metamaterial absorber for energy harvesting applications. Appl. Comput. Electromagnet. Soc. J. (ACES) **38**(6), 416–423 (2023). https://doi.org/10.13052/2023.ACES.J.380606
42. Ri, K.-J., Ri, C.-H., Ri, S.-Y.: Ultra-broadband terahertz metamaterial absorber using a simple design method. Opt. Commun. **515**(15), 128191 (2022). https://doi.org/10.1016/j.optcom.2022.128191
43. Ünal, E., Bağmancı, M., Karaaslan, M., Akgol, O., Sabah, C.: Strong absorption of solar energy by using wide band metamaterial absorber designed with plus-shaped resonators. Int. J. Mod. Phys. B **32**, 1850275 (2018). https://doi.org/10.1142/S0217979218502752
44. Rasoga, O., Dragoman, D., Dinescu, A., et al.: Tuning the infrared resonance of thermal emission from metasurfaces working in near-infrared. Sci. Rep. **13**, 7499 (2023). https://doi.org/10.1038/s41598-023-34741-4
45. Kumar, R., Singh, B.K., Pandey, P.C.: Broadband metamaterial absorber in the visible region using a petal-shaped resonator for solar cell applications. Physica E Low-Dimension. Syst. Nanostruct. **142**, 115327 (2022). https://doi.org/10.1016/j.physe.2022.115327
46. Wu, Z., Tian, J., Yang, R.: A graphene based dual-band metamaterial absorber for TE polarized THz wave. Micro Nanostruct. **168**, 207331 (2022). https://doi.org/10.1016/j.micrna.2022.207331
47. Rajakani, K.: Design of a seven-band perfect metamaterial absorber for THz sensing applications. Pramana J. Phys. **96**, 205 (2022). https://doi.org/10.1007/s12043-022-02443-7
48. Anwar, S., Khan, Q., Ali, G., et al.: Triple-band terahertz metamaterial absorber with enhanced sensing capabilities. Eur. Phys. J. D **77**, 69 (2023). https://doi.org/10.1140/epjd/s10053-023-00658-w
49. Shabanpour, J., Beyraghi, S., Oraizi, H.: Reconfigurable honeycomb metamaterial absorber having incident angular stability. Sci. Rep. **10**, 14920 (2020). https://doi.org/10.1038/s41598-020-72105-4
50. Rengasamy, S., Natarajan, R., Kuttathati Srinivasan, V.: A miniaturized penta-band metamaterial THz absorber for sensing and imaging applications. Plasmonics **18**, 1933–1940 (2023). https://doi.org/10.1007/s11468-023-01914-w
51. Cheng, Y., Chen, F., Luo, H.: Plasmonic chiral metasurface absorber based on bilayer fourfold twisted semicircle nanostructure at optical frequency. Nanoscale Res. Lett. **16**, 12 (2021). https://doi.org/10.1186/s11671-021-03474-6
52. Wartak, M.S., Tsakmakidis, K.L., Hess, O.: Introduction to metamaterials. Phys. Can. **67**(1), 30–34 (2011)
53. Moniruzzaman, M., Islam, M.T., Islam, M.R., Misran, N., Samsuzzaman, M.: Coupled ring split ring resonator (CR-SRR) based epsilon negative metamaterial for multiband wireless communications with high effective medium ratio. Results Phys. **18**, 103248 (2020). https://doi.org/10.1016/j.rinp.2020.103248
54. Wang, Z., Wei, Y., Liu, Z., Duan, G., Yang, D., Cheng, P.: Perfect solar absorber with extremely low infrared emissivity. Photonics **9**, 574 (2022). https://doi.org/10.3390/photonics9080574
55. Khuyen, X., et al.: Ultra-subwavelength thickness for dual/ tripleband metamaterial absorber at very low frequency. Sci. Rep. **8**, 11632 (2018). https://doi.org/10.1038/s41598-018-29896-4
56. Wypych, A., Bobowska, I., Tracz, M., Opasinska, A., Kadlubowski, S., Krzywania-Kaliszewska, A., Grobelny, J., Wojciechowski, P.: Dielectric properties and characterization of titanium dioxide obtained by different chemistry methods. J. Nanomater. **2014** | Article ID 124814 | (2014). https://doi.org/10.1155/2014/124814
57. Bilal, R.M.H., et al.: Elliptical metallic rings-shaped fractal metamaterial absorber in the visible regime. Sci. Rep. **10**, 14035 (2020). https://doi.org/10.1038/s41598-020-71032-8
58. Sayed, S.I., Mahmoud, K.R., Mubarak, R.I.: Design and optimization of broadband metamaterial absorber based on manganese for visible applications. Sci. Rep. **13**, 11937 (2023). https://doi.org/10.1038/s41598-023-38263-x

59. Bilal, R.M.H., Saeed, M.A., Naveed, M.A., Zubair, M., Mehmood, M.Q., Massoud, Y.: Nickel-based high-bandwidth nanostructured metamaterial absorber for visible and infrared spectrum. Nanomaterials **12**, 3356 (2022). https://doi.org/10.3390/nano12193356
60. Lai, R., Chen, H., Zhou, Z., Yi, Z., Tang, B., Chen, J., Yi, Y., Tang, C., Zhang, J., Sun, T.: Design of a penta-band graphene-based terahertz metamaterial absorber with fine sensing performance. Micromachines **14**, 1802 (2023). https://doi.org/10.3390/mi14091802
61. Rahmanshahi, M., Noori Kourani, S., Golmohammadi, S., et al.: A tunable perfect THz metamaterial absorber with three absorption peaks based on nonstructured graphene. Plasmonics **16**, 1665–1676 (2021). https://doi.org/10.1007/s11468-021-01432-7
62. Chahkoutahi, A., Emami, F.: Design of novel sensitive terahertz metamaterial absorbers based on graphene-plasmonic nanostructures. Opt. Quant. Electron. **53**, 491 (2021). https://doi.org/10.1007/s11082-021-03142-0
63. Nicolson, A.M., Ross, G.F.: Measurement of the intrinsic properties of materials by time domain techniques. IEEE Trans. Instrum. Meas. **19**, 377–382 (1970). https://doi.org/10.1109/TIM.1970.4313932
64. Weir, W.B.: Automatic measurement of complex dielectric constant and permeability at microwave frequencies. Proc. IEEE **62**, 33–36 (1974). https://doi.org/10.1109/PROC.1974.9382
65. Mehrabi, S., Bilal, R.M.H., Naveed, M.A., Ali, M.M.: Ultra-broadband nanostructured metamaterial absorber based on stacked square layers of TiN/TiO_2. Opt. Mater. Express **12**, 2199–2211 (2022). https://doi.org/10.1364/OME.459766
66. Barzegar-Parizi, S., Khavasi, A.: Designing dual-band absorbers by graphene/metallic metasurfaces. IEEE J. Quantum Electron. **55**, 1–8 (2019). https://doi.org/10.1109/JQE.2019.2901992
67. Liu, W., Liu, C., Wang, J.X., Lv, J.W., Lv, Y., Yang, L., An, N., Yi, Z., Liu, Q., Hu, C.J., et al.: Surface plasmon resonance sensor composed of microstructured optical fibers for monitoring of external and internal environments in biological and environmental sensing. Results Phys. **47**, 106365 (2023). https://doi.org/10.1016/j.rinp.2023.106365
68. Kumar, R., Singh, B.K., Pandey, P.C.: Cone-shaped resonator-based highly efficient broadband metamaterial absorber. Opt. Quant. Electron. **55**, 579 (2023). https://doi.org/10.1007/s11082-023-04865-y
69. Hakim, M.L., et al.: Wide-oblique-incident-angle stable polarization insensitive ultra-wideband metamaterial perfect absorber for visible optical wavelength applications. Materials **15**(6), 2201 (2022). https://doi.org/10.3390/ma15062201
70. Kumar, R., Singh, B.K., Tiwari, R.K. and Pandey, P.C.: Perfect selective metamaterial absorber with thin-film of GaAs layer in the visible region for solar cell applications. Opt. Quant. Electron **54**, 416 (2022). https://doi.org/10.1007/s11082-022-03798-2
71. Li, F., et al.: Polarization-dependent wideband metamaterial absorber for ultraviolet to near-infrared spectral range applications. Opt. Express **30**(15), 25974–25984 (2022). https://doi.org/10.1364/OE.458572
72. Bilal, R.M.H., et al.: Ultrathin broadband metasurface-based absorber comprised of tungsten nanowires. Results Phys. **19**, 103471 (2020). https://doi.org/10.1016/j.rinp.2020.103471
73. Zhang, Y., Yi, Z., Wang, X., Chu, P., Yao, W., Zhou, Z., Cheng, S., et al.: Dual band visible metamaterial absorbers based on four identical ring patches. Phys. E Low Dimens. Syst. Nanostruct. **127**, 114526 (2021). https://doi.org/10.1016/j.physe.2020.114526
74. Jeong, H., Kim, D.Y., Kim, J., et al.: Wafer-scale and selective-area growth of high-quality hexagonal boron nitride on Ni(111) by metal-organic chemical vapor deposition. Sci. Rep. **9**, 5736 (2019). https://doi.org/10.1038/s41598-019-42236-4
75. Serp, P., Kalck, P., Feurer, R.: Chemical vapor deposition methods for the controlled preparation of supported catalytic materials. Chem. Rev. **102**(9), 3085–3128 (2002). https://doi.org/10.1021/cr9903508
76. Park, S.H., Lee, H.S., Ahn, H.S., et al.: Crystal phase control of ε-Ga_2O_3 fabricated using by metal-organic chemical vapor deposition. J. Korean Phys. Soc. **74**, 502–507 (2019). https://doi.org/10.3938/jkps.74.502

77. Goray, L.I., Pirogov, E.V., Sobolev, M.S., et al.: Matched X-ray reflectometry and diffractometry of super-multiperiod heterostructures grown by molecular beam epitaxy. Semiconductors **53**, 1910–1913 (2019). https://doi.org/10.1134/S1063782619140082
78. Elakkiya, A., Nath, S.S., Vanitha, M.: Design, fabrication and characterization of terahertz five-band metamaterial absorber. Pramana J. Phys. **97**, 112 (2023). https://doi.org/10.1007/s12043-023-02583-4
79. Alex-Amor, A., Palomares-Caballero, Á., Molero, C.: 3-D Metamaterials: trends on applied designs, computational methods and fabrication techniques. Electronics **11**(3), 410 (2022). https://doi.org/10.3390/electronics11030410
80. Squires, A.D., Gao, X., Du, J., et al.: Electrically tuneable terahertz metasurface enabled by a graphene/gold bilayer structure. Commun. Mater. **3**, 56 (2022). https://doi.org/10.1038/s43246-022-00279-7
81. Huang, L., Xu, K., Yuan, D., et al.: Sub-wavelength patterned pulse laser lithography for efficient fabrication of large-area metasurfaces. Nat. Commun. **13**, 5823 (2022). https://doi.org/10.1038/s41467-022-33644-8

Part IV
Recent Developments in Optical Wireless Communication Technology

Chapter 33
Video Encryption Using Diffraction Grating and QR Code in Optical Frequency Domain

Anirban Patra, Aniruddha Ghosh, Mainuck Das, Anirban Ghosal, Nilanjan Mukhopadhyay, and Arijit Saha

33.1 Introduction

Industry 5.0 basically deals with cloud manufacturing and big data processing which is greatly dependent on the Internet. A major problem of both processes is securing the information from intruders. Hence encryption of images is necessary to protect the information. To encrypt the information, researchers discussed a lot of techniques for the encryption of images in both spatial as well as in frequency domains. In very recent applications, the optical process especially phase and amplitude grating has been applied to encrypt and multiplex images. The images are multiplexed by varying spatial frequencies and modulation angles in the optical system with the help of a random phase matrix [1–4]. The researchers also used different coding systems (Cyclic Coding, Hamming Code, and Linear Block Code) for the protection of remote sensing and medical information [5–7]. Apart from many techniques, QR coding is also widely used to encrypt images and text for the last few years due to many advantages like optimization of storage space, etc. Biomedical signals (ECG) are encrypted using the QR code [8]. Here, ECG signal samples are converted into the binary pair and then transformed into ciphertext which is further transformed into QR a code. To reduce error, single-shot ptychography encoding along with a QR code is used to encrypt images [9]. To hide secret information, researchers used

A. Patra (✉) · A. Ghosh · M. Das · A. Ghosal
Department of ECE, JIS College of Engineering, Kalyani, West Bengal, India
e-mail: anitublu@gmail.com

N. Mukhopadhyay
Department. of ECE, Global Institute of Management and Technology, Krishnanagar, West Bengal, India

A. Saha
Dumdum Motijheel Rabindra Mahavidyalaya, Kolkata, West Bengal, India

M. El Ghzaoui et al. (eds.), *Next Generation Wireless Communication*, Signals and Communication Technology, https://doi.org/10.1007/978-3-031-56144-3_33

QR codes in the steganography method [9]. Here, QR codes are easily decoded · but the secret information can only be decrypted using a public key. Like in many systems, in the optical domain the use of the QR code is mainly for the reduction of speckle noise [10], to enhance further security [11], improved signal to noise ratio and optimize space, etc. [12]. However, these systems are unable to reduce the noise effect beyond a certain value. New techniques are proposed by researchers, which are based on coding and a barcode scheme, Bose–Choudhury–Hocquenghem (BCH) coding to achieve almost noise-free information [13, 14]. In an optical system, polarization is used in the Fresnel domain with the help of a QR code to secure the information [15]. This system provides a satisfactory result in the case of hologram encryption. To verify e-documents as well as to compress the information lossless, this code is used as a scheme [16]. This system overcomes the size limitation which is the drawback of a few QR-related methods. To reduce cross-talk from grayscale images, one image is converted into a QR code [17]. This system is effective in reducing the silhouette problem from filtered images. In other techniques, binary images are encrypted using run-length coding [18], gyrator transforms [19], and computational ghost imaging method [20] with the QR code. In the ghost imaging process, the information is converted into the QR code followed by encoding of the QR code with the help of ghost imaging. Real values of color images are coded with a QR code in the optical domain in addition to the subtraction process [21, 22]. In most optical encryption-based systems, speckle noise is one of the main problems that reduces the performance of the system. Researchers used different techniques to overcome this problem using QR codes [23]. The use of Shearlet Transform along with spiral phase transform provides excellent results in noise attacks too [24]. With Hadamard transformation, QR is used in the watermarking process also [25]. Steganography, which is related to watermarking, is also performed by QR codes [26]. To hide information, the Fresnel transform is applied with the Least Significant Bits subbands. In a Discrete Wavelet transform-based steganographic process with QR, researchers show excellent recovery (high Peak Signal to Noise Ratio) of the images [27]. QR codes are also helpful in compressing three-dimensional images where a double random phase encoding process is used to do so [28]. In a modified optical process, the spectrum is transformed with double random phase encoding to improve security [29]. QR is used in the watermarking of medical images also [30].

Though various processes of the image are widely used in many areas, and the application of image processing is limited in the pharmaceutical industry. Recently few researchers proposed methods in the pharmaceutical industry to enhance performance and monitor quality [31–36]. Some of them used different thresholding processes in addition to non-linear filtering, Walsh Transform, etc. for characterization [37–40]. Modern image processing systems use neural networks and deep learning processes to handle a large amount of data along with other advantages [41–43].

In this paper, the selected pharmaceutical video is initially clipped into sequential frames. The frames are converted into the corresponding QR codes. The QR codes are placed in a canvas which is to be encrypted. The canvas is then multiplied with the help of a phase matrix before transmission. At the receiving end, the canvas is decrypted to recover the QR codes followed by decoding to get back the video frames. The extracted frames are added sequentially to get back the original video. The correlation coefficient method is used to check the quality of the extracted frames. In our research work, we have used a pharmaceutical video to explain the effectiveness of our proposed method.

33.2 Methodology

The selected pharmaceutical video is marked by $v(x, y)$.

Video $v(x, y)$ is divided into the frames denoted by $f_1(x, y) to f_n(x, y)$[n = last video frame number].

Let, the phase matrix is denoted by P.

The canvas is marked by C where frames $f_1(x, y) to f_n(x, y)$ are arranged sequentially.

Canvas C is multiplied by the phase matrix P to encrypt.

Encrypted object $E = P * C$.

The summary of the total process is discussed with the help of a flow chart (Fig. 33.1).

In this research work, we have used a 30 fps (frame per second) pharmaceutical video. Here, we have shown only 16 frames to discuss our proposed algorithms.

The entire work is done using the MATLAB 2015 and Python platform. Python is used to generate the QR codes while MATLAB is used for the rest work.

To measure quality, we used the correlation coefficient method. The correlation coefficient method calculates neighboring pixel values. If the value is close to 1, it indicates a better quality. As this system is used in the optical frequency domain, hence encryption is better than many existing techniques.

Flow Chart:

See Fig. 33.1.

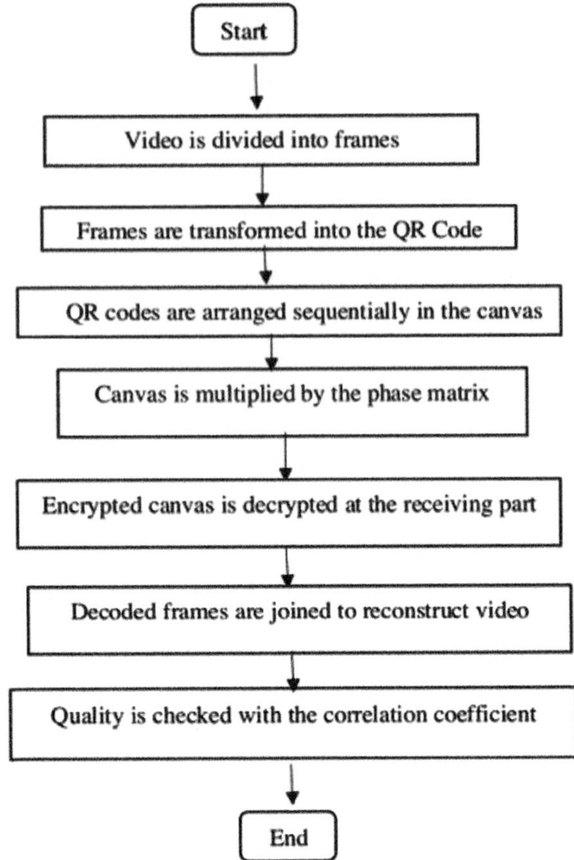

Fig. 33.1 Summary of the process

33.3 Result

The size of the selected video frames is 800 × 1200. 16 frames are used to discuss our algorithms. The video frames are displayed in Fig. 33.2.

Each frame is converted into the QR code which is shown in Fig. 33.3. The size of each QR code is 200 × 200. Though we have shown only 16 frames in the figure but it is applicable to more number of frames too.

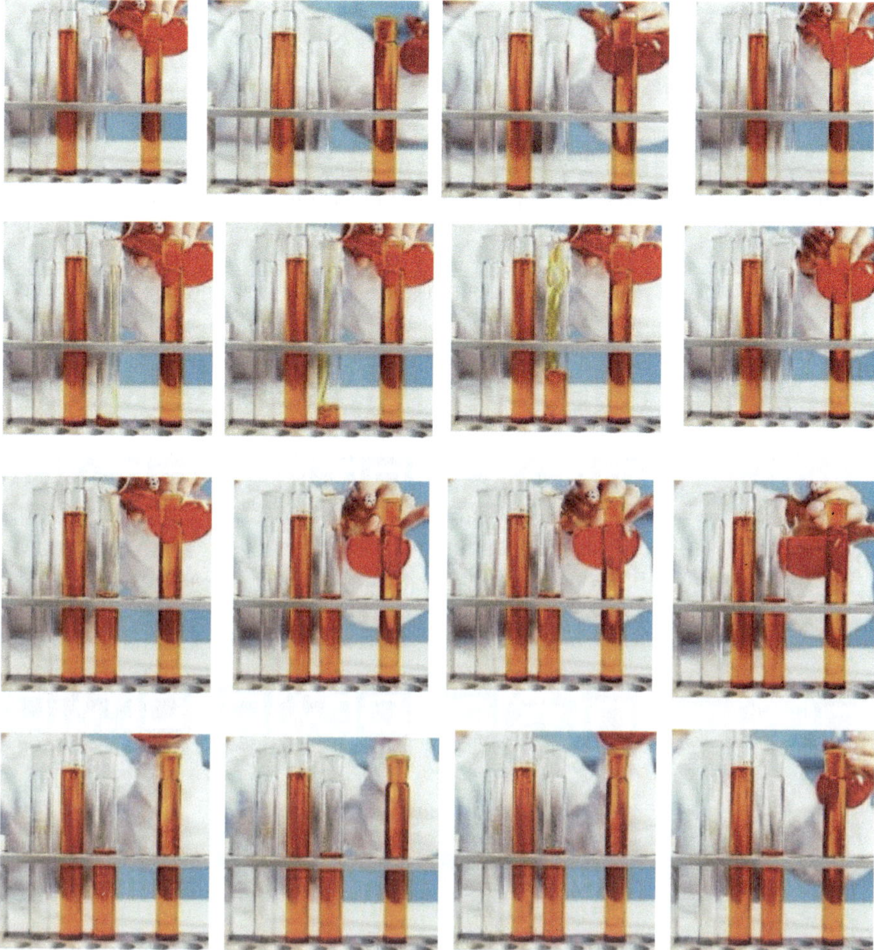

Fig. 33.2 Video frames

In Fig. 33.3, sequential representation of the QR codes is displayed. As we are dealing with only 16 frames, hence 16 QR codes are represented in the diagram.

The converted QR codes are placed in the canvas (shown in Fig. 33.4a) and the encrypted canvas is displayed in Fig. 33.4b. The size of the encrypted canvas is 850 × 850.

Fig. 33.3 Converted QR Codes of Video Frames

Figure 33.4a displays the sequential arrangement of the QR code in the canvas. We arranged them in this way so that it would be easy to encrypt them using the random phase matrix.

The extracted frames are displayed in Fig. 33.5. The size of the extracted video frame is the same as the original frame (800×1200).

Figure 33.5 represent the 16 extracted frames. During this process, we used PSNR to measure the quality of the video frames which is shown in Tables 33.1 and 33.2.

(a)

(b)

Fig. 33.4 **a** Canvas. **b** Encrypted canvas

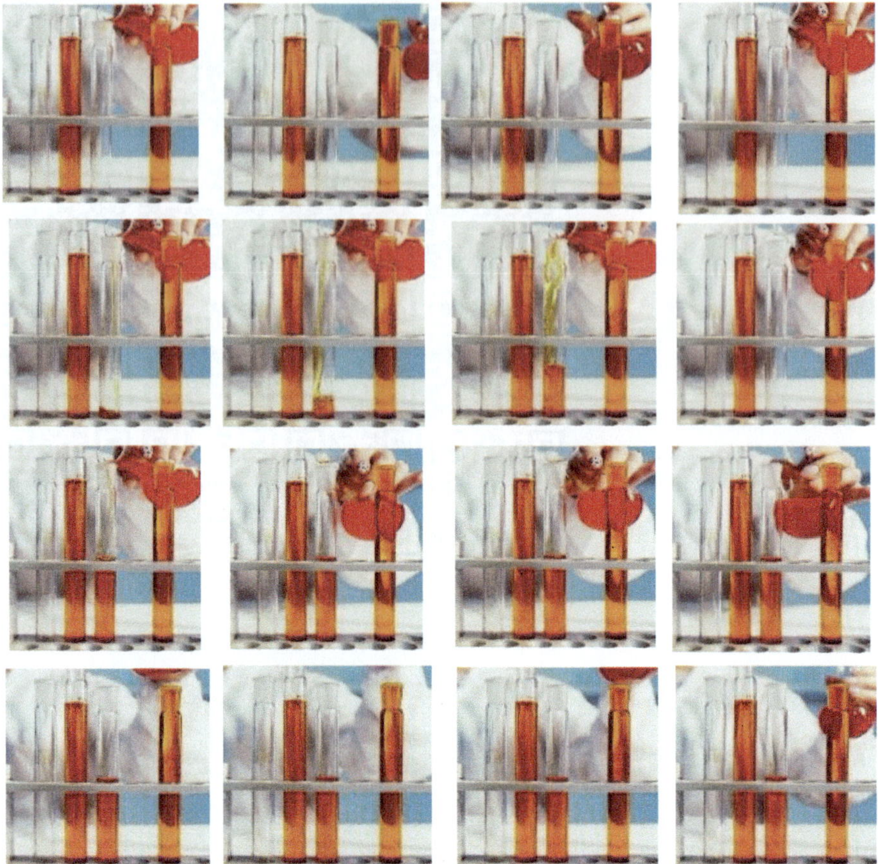

Fig. 33.5 Recovered video frames

Table 33.1 Correlation coefficient calculation

Frame No.	Correlation coefficient
1	0.976
2	0.988
3	0.964
4	0.992
5	0.969
6	0.991
7	0.982
8	0.958
9	0.945
10	0.951
11	0.972
12	0.981
13	0.962
14	0.986
15	0.951
16	0.972

Table 33.2 Comparison table

Ref. No	Drawback	Our method
[2]	A limited number of images	Used for a large number of images
[3]	Grayscale images	Color images
[10]	Time-consuming	Takes less than a second to process
[10]	Noisy	Almost free from noise

33.4 Conclusion

The modern pharmaceutical industry uses the cloud manufacturing process to deal with Industry 5.0. To do this, they have to process a large amount of data through the Internet which is a threat to security. Hence there is a need to encrypt the data. This paper shows one of the advanced applications of the QR code for the encryption of pharmaceutical video. We encrypted the video frames using the QR code and phase matrix. The correlation coefficient proves the excellent quality of the output images. It is not time-consuming also (takes only 0.726 s to process a 1-s video). Many researchers worked with grayscale images whereas we have worked with color images which are suitable for many practical applications.

References

1. Patra, A., Saha, A., Bhattacharya, K.: Compression and multiplexing of medical images using optical image processing. In: Computational Intelligence and Its Applications in Healthcare, pp. 63–71 (2020)
2. Patra, A., Saha, A.: Kallol Bhattacharya; Multiplexing and encryption of images using phase grating and random phase mask. Opt. Eng. **59**(3), 033105 (2020)
3. Patra, A., Saha, A., Bhattacharya, K.: High-resolution image multiplexing using amplitude grating for remote sensing applications. Opt. Eng. **60**(7), 073104–1–11 (2021)
4. Patra, A., Saha, A., Bhattacharya, K.: Efficient storage and encryption of 32-slice CT scan images using phase grating. Arab. J. Sci. Eng. **47**(6) (2022)
5. Bose, B., Dey, D., Sengupta, A., Mulchandani, N., Patra, A.: A novel medical image encryption using cyclic coding in Covid-19 pandemic situation. J. Phys. Conf. Ser. **1797**, 012035 (2021)
6. Patra, A., Das, M., Ghosal, A., Ghosh, A., Kushary, I., Roy, S., Chakraborty, D.: Remote sensing image encryption and error detection using hamming code. J. Phys. Conf. Ser. **2286**, 012018 (2022)
7. Patra, A., Saha, A., Bhattacharya, K.: Compression of high-resolution space video using phase grating. J. Indian Soc. Remote Sens. **59** (2023)
8. Mathivanan, P., Ganesh, A.B., Venkatesan, R.: QR code–based ECG signal encryption/ decryption algorithm. CRYPTOLOGIA (2019)
9. Zhu, Y., Xu, W., Shi, Y.: High-capacity encryption system based on single-shot-ptychography encoding and QR code. Opt. Commun. **435**, 426-432 (2019)
10. Alajmi, M., Elashry, I., El-Sayed, H.S., Allah, O.S.F.: Steganography of encrypted messages inside valid QR codes. IEEE Access **8**, 27861–27873 (2020)
11. Barrera, J.F., Mira, A., Torroba, R.: Optical encryption and QR codes: secure and noise-free information retrieval. Opt. Express **21**, 5373–5378 (2013)
12. Barrera, J.F., Mira-Agudelo, A., Torroba, R.: Experimental QR code optical encryption: noise-free data recovering. Opt. Lett. **39**, 3074–3077 (2014)
13. Cheremkhin, P.A., Krasnov, V.V., Rodin, V.G., Starikov, R.S.: QR code optical encryption using spatially incoherent illumination. Laser Phys. Lett. **14**, 026202 (2017)
14. Jiao, S., Jin, Z., Zhou, C., Zou, W., Li, X.: Is QR code an optimal data container in optical encryption systems from an error-correction coding perspective? J. Opt. Soc. Am. A Opt. Image Sci. Vis. **35**(1), A23–A29 (2018)
15. Lin, C., Shen, X., Li, B.: Four-dimensional key design in amplitude, phase, polarization and distance for optical encryption based on polarization digital holography and QR code. Opt. Express **22**, 20727–20739 (2014)
16. Ali, A.M., Farhan, A.K.: Enhancement of QR code capacity by encrypted lossless compression technology for verification of secure E-document. IEEE Access **8**, 27448–27458 (2020)
17. Qin, Y., Wang, H., Wang, Z., Gong, Q., Wang, D.: Encryption of QR code and grayscale image in interference-based scheme with high-quality retrieval and silhouette problem removal. Opt. Lasers Eng. **84**, 62–73 (2016)
18. Qin, Y., Wang, Z., Wang, H., Gong, Q.: Binary image encryption in a joint transform correlator scheme by aid of run-length encoding and QR code. Opt. Laser Technol. **103**, 93–98 (2018)
19. Wei, Y., Yan, A., Dong, J., Hu, Z., Zhang, J.: Optical image encryption using QR code and multilevel fingerprints in gyrator transform domains. Opt. Commun. **403**, 62–67 (2017)
20. Zhao, S., Wang, L., Liang, W., Cheng, W., Gong, L.: High-performance optical encryption based on computational ghost imaging with QR code and compressive sensing technique. Opt. Commun. **353**, 90–95 (2015)
21. Deng, X.: Optical image encryption based on real-valued coding and subtracting with the help of QR code. Opt. Commun. **349**, 48–53 (2015)
22. Deng, X., Zhu, X..: A simple and practical color image encryption with the help of QR code. Optica Applicata **45**(4), 513–21 (2015)
23. Jiao, S., Zou, W., Li, X.: QR code-based noise-free optical encryption and decryption of a grayscale image. Opt. Commun. **387**, 235–40 (2017)

24. Kumar, R., Bhaduri, B., Hennelly, B.: QR code-based non-linear image encryption using shearlet transform and spiral phase transform. J. Mod. Opt. **65**(3), 321–30 (2018)
25. Ramya, V., Gopinath, G.: Review on quick response codes in the field of information security. In: IEEE—International Conference on Advances in Engineering and Technology (ICAET2014) (2014)
26. Maheswari, S.U., Hemanth, D.J.: Frequency domain QR code-based image steganography using Fresnelet transforms. Int. J. Electron. Commun. (AEÜ) 1434–8411 (2015)
27. Islam, Md.W., al Zahir, S.: A novel QR code guided image stenographic technique. In: IEEE International Conference on Consumer Electronics (ICCE) (2013)
28. Markman, A., Wang, J., Javidi, B.: Three-dimensional integral imaging displays using a quick-response encoded elemental image array. Optica **1**, 332–335 (2014)
29. Wang, Z., Zhang, S., Liu, H., Qin, Y.: Single-intensity-recording optical encryption technique based on phase retrieval algorithm and QR code. Opt. Commun. **332**, 36–41 (2014)
30. Seenivasagam, V., Velumani, R.: A QR code based zero-watermarking scheme for authentication of medical images in teleradiology cloud. Comput. Math. Methods Med. **2013**, 516465 (2013)
31. Karloff, A.C., Scott, N.E., Muscedere, R.: A flexible design for a cost effective, high throughput inspection system for pharmaceutical capsules. In: IEEE International Conference Industrial Technology (ICIT'2008), April 21–24 (2008)
32. Kaur, H., Garg, E.N.: Inspection of defective pharmaceutical capsules using harris algorithm. Int. J. Adv. Electron. Eng., 165–166 (2012)
33. Zhengtao, Z., Xiongyi, Y., Liuqian, H., De, W.: Fast capsule image segmentation based on linear region growing. Comput. Sci. Autom. Eng. (CSAE) (2011)
34. Derganc, J., Likar, B., Bernard, R., Tomaževič, D., Pernuš, F.: Real-time automated visual inspection of color tablets in pharmaceutical blisters. Sci. Dir. Real-time Imag. **9**, 113–124 (2003)
35. Bansal, A., Bajpai, R., Saini, J.P.: Simulation of image enhancement techniques using Matlab. In: First Asia International Conference on Modeling and Simulation (AMS' 07), pp. 296–301 (2007)
36. Blanco, D., et al.: Image-based characterization of powder flow to predict the success of pharmaceutical minitablet manufacturing. Int. J. Pharm. (2020)
37. Sezgin, M., Sankur, B.: Survey over image thresholding techniques and quantitative performance evaluation. J. Electron. Imaging **13**(1), 146–168 (2004)
38. Koker, R., Sari, Y.: Neural Network based automatic threshold selection for an industrial vision system. In: Proceedings of International Conference on Signal Processing, pp. 523–525 (2003)
39. Rajkumar, S., Gupta, S.: Analysis of non-linear filtering techniques based on quantitative metrics using different images. Int. J. Comput. Appl. (0975–8887) **52**(6) (2012)
40. Kekre, H.B., Mishra, D.: Four walsh transform sectors feature vectors for image retrieval from image databases. Int. J. Comput. Sci. Inf. Technol. (IJCSIT) **1**(2), 33–37 (2010)
41. Anwar, S.M., et al.: Medical image analysis using convolutional neural networks: a review. J. Med. Syst. **42**(11) (2018)
42. Badhe, A., Chang, S.: Fast image classification by boosting fuzzy classifier. Neural Netw. Mach. Learn. **1**(1) (2016)
43. Bernal, J., et al.: Deep convolutional neural networks for brain image analysis on magnetic resonance imaging: a review. Artif. Intell. Med. **95** (2019)

Chapter 34
All-Optical Binary Decrementer Using Terahertz Optical Asymmetric Demultiplexer Switches

Dilip Kumar Gayen

34.1 Introduction

Photonic systems for all-optical processing of information [1–4] are expanding rapidly and have enormous promise for applications in a wide range of sectors. These applications include optical sensing, image processing, and optical computing, as well as ultra-high-speed optical telecommunications and metrology. The tremendous bandwidth provided by the all-optical domain, combined with the intrinsic parallelism of optical technologies, enables unparalleled processing rates that would be unreachable by other means. Improvements in modulation forms, as well as advances in continuous communications using optical fibers, have increased fascination with the development of optical signal processing methods, particularly those involving photonic phase manipulation. When compared to typical electrical systems that depend exclusively on signal intensity, the use of optical phase for signal processing offers significant advantages. Optics comes up as a prospective choice for processing information, data, and images. Because of its inherent parallelism, optics has been harnessed as the information-carrying signal in the processing of all-optical information as well as data. These technologies have prompted the study of all-optical systems in various systems for fundamental logic, arithmetic, and image processing activities. This chapter digs into the multidimensional realm of all-optical signal processing, revealing optics' potential to transform how we manage data and information. Improvements in modulation designs, as well as advances in unified communication using optical fibers have had an impact as catalysts for increased interest in optical signal processing methods, particularly those focusing on the optical phase. The use of the optical phase for signal processing offers various benefits over traditional electrical processing approaches that depend largely on signal strength. In

D. K. Gayen (✉)
Department of Computer Science and Engineering, College of Engineering and Management, Kolaghat, KTPP Township, Purba Medinipur 721171, West Bengal, India
e-mail: dilipgayen@cemk.ac.in

M. El Ghzaoui et al. (eds.), *Next Generation Wireless Communication*, Signals and Communication Technology, https://doi.org/10.1007/978-3-031-56144-3_34

the realms of information, data, and image processing, optics appears as a possible possibility. Optics has found its place as a medium for conveying information in the context of all-optical data and information processing, relying on its inherent advantages in parallelism. Many strategies for investigating fundamental logic and memory operations have evolved, with all-optical systems being used to perform algebraic, arithmetic, and image processing. A slew of all-optical digital circuits designed to execute various arithmetic operations have been proposed by researchers. In the last few decades, the invention of several ultra-fast throughout-optical circuits has aroused the curiosity of academics all around the world, owing to the potential they represent for the future of optical computing and information processing [5–7]. The TOAD paired with the Sagnac gate with Semiconductor Optical Amplifier (SOA) assistance stands out among these suggested techniques. This combination efficiently combines quick switching times and low noise figures while providing integration and practicality, establishing it as a viable challenger in the area of comparable technologies for optical time division multiplexing (OTDM) [8–10]. A variety of appealing qualities distinguish the TOAD. These features include rapid switching times, an excellent repetition rate, minimal energy usage, low latency, a high tolerance to noise and jitter, small physical dimensions, outstanding thermal stability, and powerful nonlinear characteristics. These characteristics, taken together, enable TOAD to be effectively used with the environment of very high velocity optical communications transmission and reception [11–13]. Furthermore, the intrinsic integration potential of TOAD holds the promise of consistent and scalable production. This capacity lays the road for its dependable large-scale manufacturing, making it not only financially viable but also placing it as a competitive solution when compared to other buffering alternatives [14–16]. The TOAD-based switch is a popular choice in the area of all-optical switches because of its adaptability in performing numerous functions inside the all-optical domain. In this chapter, we introduce our novel contribution, an all-optical decrementer circuit that leverages the capabilities of optical switch based on TOAD. The utilization of each of the transmitted and reflected ports on the TOAD device to achieve the necessary decrementation is a unique feature of our design. The operational principles and performance of the circuit are thoroughly explored, and this is supported by a solid theoretical foundation. In addition, we undertake comprehensive numerical simulations to evaluate the efficacy and practicality of our suggested strategy. This chapter delves into the capabilities of TOAD-based switches in providing enhanced functionality inside the all-optical domain, focusing on the specific context of an all-optical decrementer circuit.

34.2 Interferometric Optical Switch Utilizing TOAD Technology

Two 3 dB couplers make up an interferometric switch in most cases, which are critical to the device's functionality. The first coupler divides the incoming optical signal into two identically lengthened arms. The input beam can traverse these arms independently since they provide unique channels for it. The phase discrepancy among the light rays is produced by the insertion of an index change, which generates a change in phase among the two sides of the interferometer. In order to reconnect the optical beams and divide the signal once more, a second coupler is positioned strategically. Interference patterns are produced by this phase difference and can be either destructive or constructive. Thus, the electricity at alternative output ports is either maximum or minimal. The device effectively operates as an optical switch by suitably choosing the output beam depending on these interference patterns. A nonlinear interferometer-based optical switch allows for the control and modification of one optical signal by another through internal nonlinear interactions in a substance. Initially, the coming switching signal is divided across the interferometer's arms.

The meticulous rebalancing of this interferometer ensures that, in without the presence of a controlling signal, the input signal will escape from a designated destination port. Most importantly, the presence of a strong control pulse modifies the medium's refractive index. In this setup, optical fiber serves as the medium, while an asymmetrically positioned SOA is placed strategically within the fiber loop. Notably, the SOA's contribution in this case is not amplification, but rather the introduction of a significant phase shift in the signal. This phase shift is critical for achieving the necessary optical signal control and switching. TOAD technology takes advantage of the strong but relatively low photonic irregularities present in materials made from semiconductors, which enable the wavelength-based separation of control and pulses of signal. TOAD accomplishes this with an astonishingly low switching energy requirement of less than one pico-Joule (pJ) [17]. As shown in Fig. 34.1, the TOAD equipment comprises of a loop reflector plus an intra-loop two by two coupler with a fifty percent to fifty percent split ratio.

Within the loop, a purposefully offset SOA adjacent to the loop's midway coexists with a control pulse (CP) of an alternate wavelength than the entering signal. The optical circulator stands out among optical components. This piece of equipment allows light to travel sequentially through all of its ports while preventing light from returning to the source port. An optical circulator's performance is determined by its unique design and material composition, with typical insertion losses of approximately − 1 decibel (dB), isolation levels greater than forty dB and losses influenced by polarization smaller than 0.1. The optical system is made up of a fiber loop that is joined at its base by a fifty: fifty optical coupler. When a frequency λ_1 incoming signal (IS) comes, it separates into two equal halves. One component propagates counterclockwise (CCW), while the other goes clockwise (CW) around the loop. These two segments go in opposing directions around the loop and finally meet at the same coupler. Whether destructive or constructive interaction occurs at this

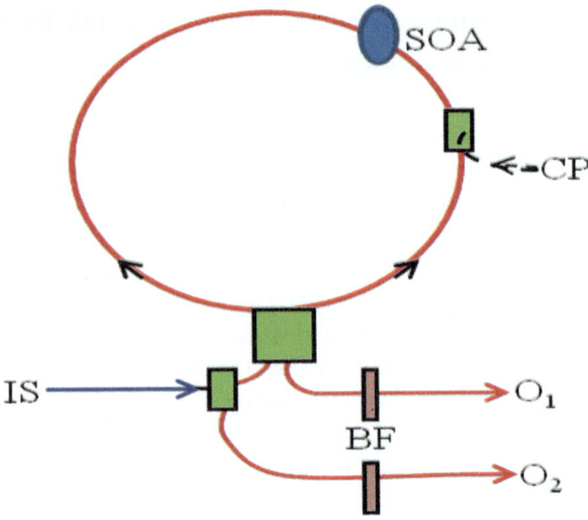

Fig. 34.1 Optical switch (TOAD), IS: Incoming signal, CP: Control pulse, SOA: Semiconductor optical amplifier, BF: Band-pass filter, O_1 and O_2: Output

coupler depends on the phase mismatch between the CCW and CW segments. As shown in Fig. 34.1, this phase contrast is able to change through the introduction of a reference illumination at wavelength λ_2 into the loop through a coupler positioned inside the fiber ring. A Gaussian CP with a full width half-length duration is administered almost concurrently with the data signals. The SOA is deliberately placed within the fiber loop at x, enabling the CW and CCW segments to interact at two unique places within the SOA. As a result of this interaction, the CW and CCW segments have a relative phase difference ($\Delta\theta$). Depending on the phase difference, either constructive or destructive interference occurs, directing the signal to the lower or upper port. The way that power is represented at the O_1 and O_2 terminals is able to be found in references [17–21]. A simplified illustration of a TOAD-based switch is shown in Fig. 34.2.

These formulas are able to use to indicate the powers at the optical switch's O_1 and O_2 ports:

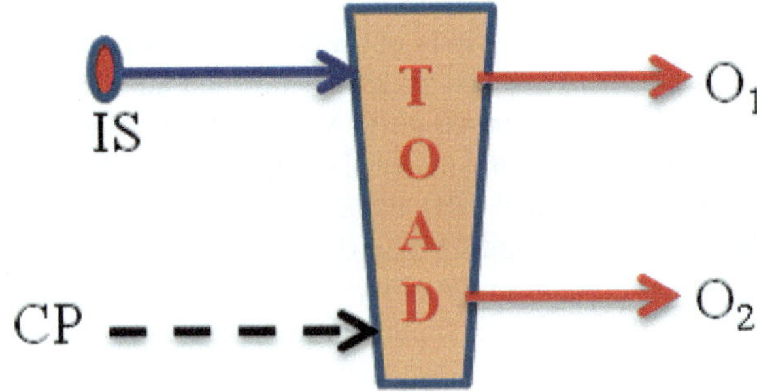

Fig. 34.2 Block diagram of TOAD

$$P_{O1} = \frac{P_{\text{IS}}}{2}\{A_{\text{CW}}(t) + A_{\text{CCW}}(t) \\ -2 * \text{sqrt}[A_{\text{CW}}(t) * A_{\text{CCW}}(t)] * \cos(\Delta\theta)\} \quad (34.1)$$

$$P_{O2} = \frac{P_{\text{IS}}}{2}\{A_{\text{CW}}(t) + A_{\text{CCW}}(t) \\ +2 * \text{sqrt}[A_{\text{CW}}(t) * A_{\text{CCW}}(t)] * \cos(\Delta\theta)\} \quad (34.2)$$

The gains for clockwise and counterclockwise pulses are reflected in $A_{\text{CW}}, A_{\text{CCW}}$, and $\Delta\theta$. in the aforementioned equations, in addition to the variation in phase among components. When there is no CP, the IS passes through the optical loop along with counter-propagates around the loop, crossing the SOA in various times. It travels through unsaturated tiny amplifier gain (A_{ss}) on this route. Both counterclockwise and clockwise pulses attain the same gain—that is, the A_{CW} gain—at the input coupler equals gain of A_{CCW}. As a result, $\Delta\theta$ equals to 0 and the power at the upper port (O_1) is marked as P_{O1} equals to 0, whilst the power at the lower port (O_2) is designated as P_{O2} equals to $P_{\text{IS}}(t) * A_{\text{SS}}$. This means that the information is correctly bounced back toward its origin if there isn't a CP. When a CP is present, the IS travels through the optical loop and counter-propagates around the loop, traversing the SOA at various points. They encounter divergent gain saturation profiles along the way, resulting in separate gains for counterclockwise and clockwise pulses, resulting in gain of A_{CW} $\neq$ gain of A_{CCW}. As a result, when the phase difference ($\Delta\theta$) is nearly equal to π (pi), they meet at the coupler for input. In this design, the electrical supply at the lower port (O_2) approaches zero, whereas the power at the upper port (O_1) may be calculated using Eq. 34.1. At the receiving end of a TOAD-based switch, the CP can be filtered out and the IS can be sent using a band-pass or wavelength filter. A wavelength or band-pass filter is an important part of optical systems and devices. Its objective is to selectively allow some wavelengths to pass through while blocking or attenuating others. This filtering method is critical in situations requiring precise wavelength control or isolation. In optical communication systems, wavelength filters are often used to divide channels in dense wavelength-division multiplexing configurations. Multiple optical signals of various wavelengths are sent simultaneously via a single optical cable in these systems. Wavelength filters keep each signal within its prescribed wavelength range, eliminating interference and allowing for more effective data transfer. Band-pass filters, on the other hand, allow only a certain band or range of wavelengths to pass through while blocking all others. They are commonly employed in spectroscopy, optical sensors, and other applications where identifying and isolating a certain spectral region is critical for precise measurements or detection. In essence, wavelength and band-pass filters are critical in managing the optical spectrum and allowing various optical technologies to operate effectively and precisely. Their capacity to regulate and isolate certain wavelengths is critical in a variety of optical systems and applications.

34.3 All-Optical Decrementer

All-optical decrementers are devices or circuits that perform the operation of reducing a binary number by one in the optical domain. They enable fast and energy-efficient operations and are crucial parts of optical information and communication systems. All-optical decrementers take an optical signal as input and manipulate it to produce a decremented output signal, without the need for electrical conversion or processing. This is achieved by leveraging the unique properties of optical signals, such as their speed and parallelism. The advancement of next-generation laser computing and systems for signal processing depends on all-optical decrementers. Numerous applications, such as frequency dividers, arithmetic units, and counters, can be implemented with them. The decrementer is a basic digital component that may reduce a value by one within a single clock cycle. It is widely used in digital systems and has a wide range of applications, including program counters and frequency dividers. Table 34.1 displays the proposed decrementer circuit's truth table, along with its operating features.

The expressions for each output can be derived from the truth table as follows:

$$y_0 = x_1\overline{x_0} + x_2\overline{x_1}\,\overline{x_0} \tag{34.3}$$

$$y_1 = x_1 x_0 + x_2\overline{x_1}\,\overline{x_0} \tag{34.4}$$

$$y_2 = x_2 x_1 + x_2\overline{x_1} x_0 \tag{34.5}$$

These expressions represent the relationship between the inputs ($x_2x_1x_0$) and the corresponding outputs ($y_2y_1y_0$) of the decrementer circuit. We designed a binary decrementer circuit using five photonic switches derived from TOADs, designated in Fig. 34.3 as W_1–W_5. We used Eqs. 34.3–34.5 as our basis. When there is no CP, the IS is routed via the switch's lower channel; when there is a CP, the IS is routed via the switch's upper channel. This setup allows us to control the decrementing process optically, offering a flexible and efficient solution for binary decrementation.

Table 34.1 Truth table of proposed decrementer

Inputs ($x_2x_1x_0$)	Outputs ($y_2y_1y_0$)
001	000
010	001
011	010
100	011
101	100
110	101
111	110

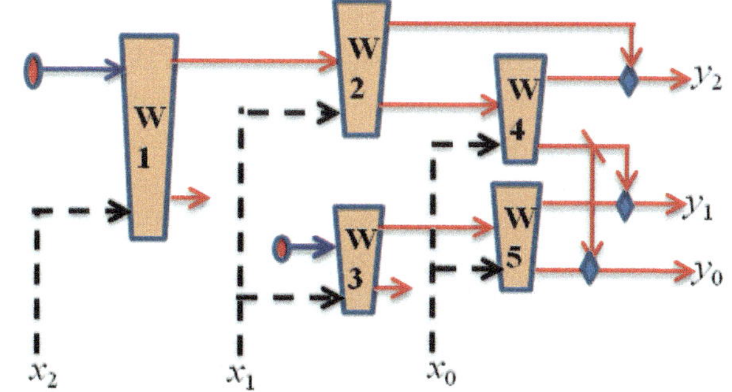

Fig. 34.3 All-optical binary decrementer circuit, : Incoming signal (IS), W_1–W_5: Switches based on TOAD, $x_2x_1x_0$: Binary inputs, : Beam combiner, : Beam splitter, and $y_2y_1y_0$: Binary inputs

Let us now imagine the existence of a continuous source of pulsed light, possibly a laser generator. The luminescent emitted by this constant pulsed source serves as our incoming signal. Additionally, our control signals also take the form of light signals. When a specific switch remains inactive due to the absence of the incoming signal, it does not produce any output, regardless of whether a control pulse is present or absent. In our circuit design, we've incorporated one beam splitter and three beam combiners (BC) to facilitate the necessary routing and manipulation of these optical signals. In the initial stage of our circuit design, specifically for switch W_1, we ensure the presence of an incoming signal, while applying control pulse x_2. In this configuration, only the upper output from switch W_1 is directed to W_2 as the incoming signal, and the lower port remains unused. According to the TOAD-based functioning principle of the switch, the switch's upper port W_1 exclusively receives light when x_2 equals 1. Moving on to the second stage of our circuit design, which encompasses switches, W_2 and W_3, we consistently maintain an incoming signal for switch W_3. Control pulse x_1 is applied to both switches W_2 and W_3. Within this stage, the lower output from W_2 serves as the incoming signal for W_4, while the upper output of W_2 contributes to the final output. In the case of switch W_3, only the upper output is directed to W_5 as the incoming signal, with no utilization of the lower port. In the ultimate stage of our circuit design, we introduce two switches, namely W_4 and W_5. Control pulse x_0 is applied to both of these switches. In this configuration, the lower outputs of W_5 and W_4 are merged to produce the output y_0 using a beam combiner. Simultaneously, the upper output of switch W_5 and the lower output of switch W_4 are combined through a beam combiner, resulting in output y_1. Finally, the upper output of switches W_2 and W_4 is coupled with an additional BC to produce the result y_2. The ultimate output of the decrementer circuit is represented as $y_2y_1y_0$. This design incorporates three control pulses, denoted as x_2, x_1, and x_0, each capable of assuming binary values 1 and 0. A control pulse is regarded as in the "one-state" whenever a beam of light is visible and in the "zero-state" whenever there isn't a beam of light. Various operational scenarios and cases are elaborated in further detail.

Case 1: Assume $x_2 = 0$, $x_1 = 0$, and $x_0 = 1$

In this scenario, the incoming light from IS first encounters switch W_1. Here, with $x_2 = 0$, the control pulse x_2 is absent, following the switching principle that light passes through the lower port, and no light emerges from the upper port. Since the upper port of switch W_1 is connected to W_2, switch W_2 remains inactive. Proceeding to the second level, where $x_1 = 0$, the control pulse x_1 is also absent. As switch W_2 is inactive, it does not produce any output. For switch W_3, with $x_1 = 0$, there is no light at the higher port while the light is coming from the bottom port. In the final stage, when $x_0 = 0$, signifying the absence of control pulse x_0, both switches W_4 and W_5 remain inactive, resulting in no light output from this stage. Consequently, no light reaches outputs y_2, y_1, and y_0. Therefore, the final output is 000 ($y_2y_1y_0$).

Case 2: When $x_2 = 0$, $x_1 = 1$, and $x_0 = 0$

In this case, the light originating from IS is initially directed to switch W_1. With $x_2 = 0$, the control pulse x_2 is not present, adhering to the bottom port allows the light to exit according to the switching concept, with no light emerging from the upper port. Since the upper port of W_1 is connected to W_2, switch W_2 remains inactive. Moving to the second stage, where $x_1 = 1$, indicating the presence of control pulse x_1, switch W_2, however, remains inactive, yielding no output. For switch W_3, given the existence of control pulse $x_1 = 1$, the upper port is where the light appears, subsequently reaching W_5. In the final stage, with $x_0 = 0$, implying the absence of control pulse x_0, switch W_4 remains inactive. Nevertheless, in the case of switch W_5, the IS exists, even though control pulse x_0 is not present. Consequently, light only emerges from the lower port of W_5, resulting in light being present solely in y_0 (i.e., $y_0 = 1$), with no other outputs, y_2 and y_1, receiving any illumination. Thus, the resultant product is 001 ($y_2y_1y_0$).

Case 3: When $x_2 = 0$, $x_1 = 1$, and $x_0 = 1$

In this case, the light originating from IS is initially directed to switch W_1. With $x_2 = 0$, control pulse x_2 is absent, in accordance to the switching theory in which the bottom port is where light exits, and no light emerges from the upper port. Since the upper port of W_1 is connected to W_2, switch W_2 remains inactive. Progressing to the second stage, with control pulse $x_1 = 1$, switch W_2, though inactive, yields no output. For switch W_3, given the existence of control pulse $x_1 = 1$, the top port is where the light appears, subsequently reaching W_5. In the final stage, with control pulse $x_0 = 1$, switch W_4 remains inactive. However, in the case of switch W_5, the IS exists, and control pulse x_0 is further there. Consequently, light emerges only from the upper port of W_5, ultimately reaching y_1 (i.e., $y_1 = 1$), with no other outputs, y_2 and y_0, receiving any illumination. Thus, the resultant product is 010 ($y_2y_1y_0$).

Case 4: When $x_2 = 1$, $x_1 = 0$, and $x_0 = 0$

In this case, the light from IS is initially directed to switch W_1. With $x_2 = 1$, control pulse x_2 is present. The switching concept states that light comes from the higher port and is directed to W_2, activating switch W_2. In the second stage, with control pulse $x_1 = 0$, switch W_2 is active, and as control pulse x_1 is absent, light emerges from the lower port, falling onto switch W_4. For switch W_3, with control pulse $x_1 = 0$, light emerges from the lower port, and no light emerges from the upper port, rendering switch W_5 inactive. In the final stage, with control pulse $x_0 = 0$, switch W_4 is inactive, but for switch W_5, there is an inbound signal, while control pulse x_0 is not present. Consequently, light only emerges from the lower port of W_5. Thus, the light only reaches y_1 and y_0, resulting in $y_1 = 1$ and $y_0 = 1$, with no light received at y_2, i.e., $y_2 = 0$. Therefore, the final output is 011 ($y_2 y_1 y_0$).

Case 5: When $x_2 = 1$, $x_1 = 0$, and $x_0 = 1$

In this scenario, the light from IS first reaches switch W_1. With $x_2 = 1$, control pulse x_2 is present, causing the light to emerge through the upper port and feed into W_2, thus activating switch W_2. In the second stage, with control pulse $x_1 = 0$, switch W_2 remains active, and as control pulse x_1 is absent. The lower port emits light, which lands on switch W_4. On switch W_4, light emerges from the lower port. For switch W_3, with control pulse $x_1 = 0$, light emerges from the lower port, and nothing illuminates the topmost port, rendering switch W_5 inactive. In the final stage, with control pulse $x_0 = 1$, switch W_4 is inactive, but for switch W_5, the incoming signal is present, while control pulse $\times$ 0 is present. Consequently, light solely emerges from the upper port of W_4. This means that the light reaches only y_2 (i.e., $y_2 = 1$), with no other outputs, y_1 and y_0, receiving any light (i.e., $y_1 = 0$ and $y_0 = 0$). Hence, the final output is 100 ($y_2 y_1 y_0$).

Case 6: When $x_2 = 1$, $x_1 = 1$, and $x_0 = 0$

In this scenario, the light from IS first encounters switch W_1. With $x_2 = 1$, control pulse x_2 is present, causing the light to emerge through the upper port and feed into W_2, thus activating switch W_2. In the second stage, with control pulse $x_1 = 1$, both switches W_2 and W_3 are active, and as control pulse x_1 is present, light emerges from the upper port of switches W_2 and W_3. This means that the light reaches y_2 (i.e., $y_2 = 1$). From the upper port of switch W_3, the light falls onto switch W_5, with no light received by switch W_4, rendering it inactive. In the final stage, with control pulse $x_0 = 0$, switch W_4 remains inactive, but for switch W_5, CP (x_0) is not present, but the IS is present. Consequently, illumination only emerges from the lower port of W_5, reaching y_0 (i.e., $y_0 = 1$). In this case, only output y_1 does not receives any illumination (i.e., $y_1 = 0$). Thus, the resultant product is 101 ($y_2 y_1 y_0$).

Case 7: When $x_2 = 1$, $x_1 = 1$, and $x_0 = 1$

In this case, the light from IS initially reaches switch W_1. With $x_2 = 1$, control pulse x_2 is present, causing the light to emerge through the upper port and feed into W_2, thus activating switch W_2. In the second stage, with control pulse $x_1 = 1$, both switches W_2 and W_3 are active, and as control pulse x_1 is present, light emerges from the upper port of switches W_2 and W_3. This means that the light reaches y_2 (i.e., $y_2 = 1$). From the upper port of switch W_3, the light falls onto switch W_5, with no light received by switch W_4, rendering it inactive. In the final stage, with control pulse $x_0 = 1$, switch W_4 remains inactive, but for switch W_5, the incoming signal is present, while control pulse x_0 is present. Consequently, light only emerges from the upper port of W_5, reaching y_1 (i.e., $y_1 = 1$). In this case, only output y_0 does not receives any illumination (i.e., $y_0 = 0$). Thus, the resultant product is 110 ($y_2 y_1 y_0$).

Indeed, it is clear from the preceding explanation that the optical circuit created using optical switches provides a viable option for building a binary decrementer circuit capable of processing any three-bit binary number. The capacity of the circuit to alter light signals using control pulses x_2, x_1, and x_0 demonstrates its adaptability and promise in binary decremental operations. This optical technique corresponds to the ever-increasing demand for high-speed data processing and optical computing in a variety of disciplines, making it a viable option for future applications in digital systems and beyond.

34.4 Circuit Performance Analysis and Discussion

To closely match real-world settings, the simulation used parameter values taken from experimental literature [22–25]. These settings were carefully chosen to adequately represent the circuit and provide relevant results. Specifically, the parameters chosen for the simulation were: 20 dB (decibel) a gain in tiny signals (A_{ss}) and a control energy of 70 fJ (femto-Joule), with other parameters having various values. These parameter settings were critical for accurate and representative simulations, which provided information about the circuit's performance under realistic conditions. In the simulation, two sets of waveforms were analyzed. First, the input waveforms, represented as power (mW) on the y_{axis} and time (ps) on the x_{axis}, were examined. These waveforms depicted the power variations over time at the input of the circuit.

Second, the output waveforms were studied. These waveforms displayed power (W) on the y_{axis} and time (ps) on the x_{axis}, illustrating how the power levels changed over time at the circuit's output. Figure 34.4 depicts both of these waveforms. Analyzing both the input and output waveforms allowed for a comprehensive understanding of the circuit's behavior and its impact on the power of the optical signals throughout the simulation.

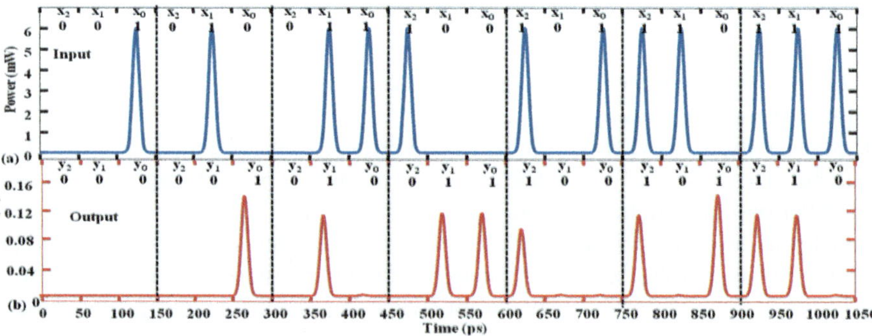

Fig. 34.4 Simulated waveforms, **a** input waveforms [Power (mW) waveforms over time] and **b** output waveforms [Power (W) waveforms over time]

Each input is represented by a three-bit binary value, denoted as $x_2x_1x_0$, while each output corresponds to a three-bit binary value, indicated as $y_2y_1y_0$. The total duration for both the input and output waveforms is 150 ps, with each data value lasting for 50 ps. The first input waveform, as shown in Fig. 34.4a, is 001 ($x_2x_1x_0$), and it spans a duration of 150 ps. The corresponding output waveform, depicted in Fig. 34.4b, is 000 ($y_2y_1y_0$) and shares the same 150 ps duration. Subsequently, the next input waveform in Fig. 34.4a is 010 ($x_2x_1x_0$), lasting for 150 ps, commencing at 150 ps, and concluding at 300 ps. Correspondingly, the output waveform in Fig. 34.4b is 001 ($y_2y_1y_0$), also spanning 150 ps, with the same time frame. This pattern continues for successive input waveforms in Fig. 34.4a, each lasting 150 ps with varying start and end times. Figure 34.4b is the associated resultant waveforms, exhibit the same 150 ps duration, but with different time intervals. The consistency between the input and output waveforms, as observed in Fig. 34.4, confirms that the proposed optical circuit accurately produces the desired outputs for a range of input scenarios. The Quality-factor (*Q–F*), a measure computed as $Q - F = (\text{Avg}_{P1} - \text{Avg}_{P0})/(\text{Std}_{Q0} + \text{Std}_{Q1})$, is used to evaluate the performance of the decrementer circuit. In this formula, Avg_{P1} and Avg_{P0} represent, respectively, the average of the device's outcomes at logic high level and logic low as well. While Std_{Q0} and Std_{Q1} signify the standard variation of the device's terminals at logic minimal and logic elevated, respectively. A desirable *Q–F* value, preferably greater than 6, is required to provide a decreased bit error rate. Investigations have been conducted to evaluate the impact of various parameters on *Q–F*, such as control pulse energy and tiny signal gain. Figure 34.5 illustrates how the CP's energy grows, the *Q–F* curve climbs steadily until it reaches its peak just before the midpoint of the investigated range. Beyond this point, as control pulse energy grows, the slope reverses and the *Q–F* begins a slow decrease, eventually becoming intolerable after exceeding a particular energy threshold.

Notably, for control energies within a tolerable range and significantly lower than the SOA saturation energy, the circuit's performance remains satisfactory. This

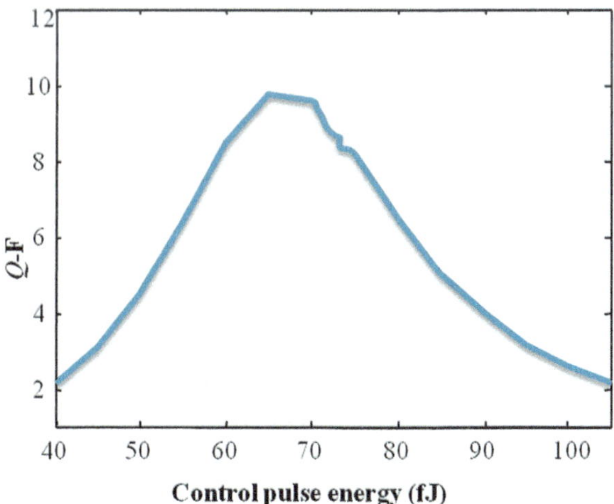

Fig. 34.5 Analyzing *Q–F* dependency on CP energy

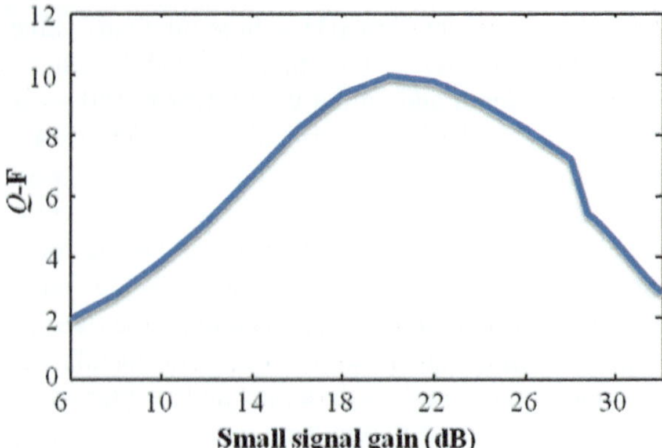

Fig. 34.6 Dependence of *Q–F* on SOA small signal gain

discovery is consistent with published criteria for the required degree of SOA saturation for all-optical logic applications and substantially verifies the suggested simulation model's dependability. According to Fig. 34.5, the greatest $Q - F$ is reached at around 70 fJ of control pulse energy. The allowable range of tiny signal gain within which the Q-F remains above 8 is seen in Fig. 34.6.

This adaptability indicates that the parameter may be easily selected by employing a source of current and introducing the required quantity of carriers according to the distinct use needs. Figure 34.6 shows that the maximum *Q–F* is obtained at around 20 dB of modest signal gain. A *Q–F* of 8.16 is obtained at this stage, demonstrating the outstanding quality of the circuit's output. This finding emphasizes the circuit's ability to provide high-quality outcomes and demonstrates the possibility for its usefulness in multiple contexts. The circuit's versatility and ability to attain excellent output quality, as demonstrated by the wide range of tiny signal gain values that preserve the ideal *Q–F*, as well as the peak performance seen at 20 dB, make it a potential candidate for a variety of practical circumstances.

The eye-diagram is a visual depiction of the circuit's output across the repetition period of the input, capturing fluctuations as inputs shift from one state to another. It's worth noting that the eye-diagram illustrated in Fig. 34.7 is not the same as the typical eye-diagrams widely utilized in point-to-point communication systems. Traditional eye-diagrams incorporate useful components such as noise sources and transmission effects inside optical fibers into the analysis, making them more thorough.

In this scenario, the relative eye opening (*E–O*) is crucial in deciding the circuit's performance. Formula $E - \mathrm{O} = (\mathrm{Min}_{P1} - \mathrm{Max}_{P0})/\mathrm{Min}_{P1}$ is used to compute *E–O*, where Max_{P0} and Min_{P1} indicate the highest and minimum power levels of the circuit outputs at low (logic 0) and high (logic 1) states, respectively. In this instance, the *E–O* result is very impressive, with a computed value of 98.43%. This high *E–O* % suggests a clean transmission with a low bit error rate, indicating the circuit's outstanding performance and output quality. While this eye-diagram does not follow the standard pattern, it nonetheless gives useful information about the circuit's efficacy and applicability for practical applications in optical data processing.

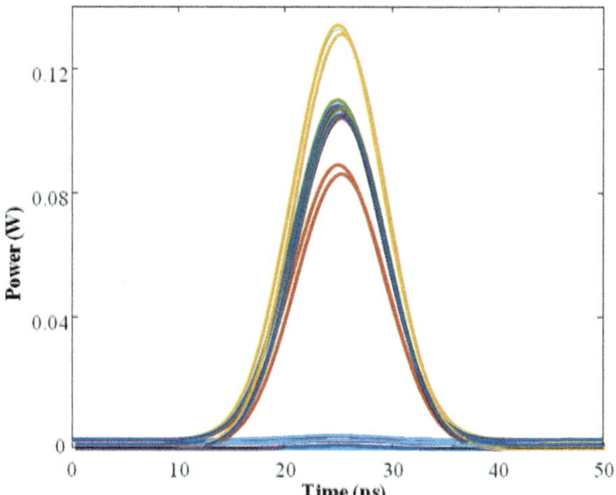

Fig. 34.7 Simulated output waveforms with *E–O*

Our simulation findings demonstrate the efficacy of our suggested model in executing all-optical decrementer operations. We investigated several key variables that significantly affect the circuit's performance in this evaluation, including the energy and width of the CP, loop asymmetry, the gain of signal of the SOA, and other variables. The energy of the CP is very important in defining the level of saturation of SOA gain necessary for optimal performance. We selected to keep the control pulse energy within the range of 60–80 femto-Joules (fJ) to strike a compromise between energy economy and operational efficacy. Setting the small signal gain (A_{ss}) to 20 dB was purposeful, since larger values have a significant influence on SOA dynamics. The SOA gets increasingly saturated as tiny signal gain increases, resulting in less energy-dependent gain. 70 fJ per control pulse appeared as the optimal energy range for obtaining the best contrast in this scenario. Within the SOA, carrier lifespan is a significant component impacting operational performance. For fast connectivity optics logic processes, the gain time to recovery of the SOA is crucial. As a result, advances in traditional Large-scale SOAs are necessary, maybe through gain recovery reduction approaches. We chose a carrier lifetime of 50 picoseconds (ps) to guarantee because the duration is shorter than the bit, allowing complete gain recovery between successive pulses. The eccentricity of the loop is another critical element, and we chose 30 ps on purpose. With this decision, approaching the top limit of the carrier lifespan leads in better performance. The counterclockwise pulse is adequately delayed in comparison to its clockwise counterpart in this setup, allowing for the generation of a considerable gain and phase difference separately. These precise parameter adjustments lead to our all-optical decrementer circuit's flawless functioning and excellent performance.

34.5 Conclusion

In conclusion, the all-optical binary decrementer circuit, which uses a mixture of TOAD-based optical switches, offers a flexible solution for binary number decrementing. The circuit conducts binary subtraction by manipulating the switch states in a controlled sequence, producing the required output dependent on the states of control pulses x_2, $x_{1,}$ and x_0. We investigated seven alternative scenarios, each reflecting a different combination of control pulses, and methodically examined the circuit's reaction in each of these circumstances. The results from each scenario demonstrate the circuit's capacity to conduct binary decrementation successfully. The study presents novel ways based on TOAD technology for developing integrated photonic systems for decrementer operations. It should be noted that the discussions offered here are based on a skewed model. A number of design considerations need to be made in to achieve experimentally apply this strategy and produce realistic results. These are essential for success and include suppression proportion, cross-talk, asymmetric modulation of phase, duration interruption, and transmission coordination. The numerical simulation results offered support the suggested technique by providing light on the change of Quality-factor (Q–F) in relation to control pulse energy and gain recovery time. This innovative technique promises increased processing speed and accuracy. It is a fundamental building component for increasingly complicated all-optical circuits with improved functions, opening the door for broader applications in the domain of integrated optical systems. The all-optical decrementer circuit is a valuable tool in digital systems, serving a variety of purposes, most notably in computer architecture and digital signal processing. The described instances collectively offer insight on the circuit's performance across various control pulse settings, highlighting its adaptability and versatility in binary decrementation applications. This study serves as a springboard for further development and enhancement of all-optical circuits within the context of digital systems and optical computing applications.

References

1. Wu, Y.D., Huang, M.L., Chen, M.H., Tasy, R.Z.: All-optical switch based on the local nonlinear Mach-Zehnder interferometer. Opt. Express **15**(16), 9883–9892 (2007)
2. Das, K., Das, P.P., Mukhopadhyay, S.: A new approach of binary addition and subtraction by non-linear material based switching technique. J. Phys. **64**(2), 239–247 (2005)
3. Phongsanam, P., Mitatha, S., Teeka, C., Yupapin, P.P.: All-optical half-adder / half-subtractor using dark bright soliton conversion control. Microw. Opt. Technol. Lett. **53**(7), 1541–1544 (2011)
4. Lin, D., Xia, K., Li, J., Li, R., Ueda, K., Li, G., Li, X.: Efficient, high -power, and radially polarized fiber laser. Opt. Lett. **35**(13), 2290–2292 (2010)
5. Kozicki, B., Maruta, A., Kitayama, K.: Experimental demonstration of optical performance monitoring for RZ-DPSK signals using delay tap sampling method. Opt. Express **16**(6), 3566–3576 (2008)

6. Savage, S., Robinson, B., Hamilton, S., Ippen, E.: Wavelength maintaining polarization insensitive all-optical 3R regenerator. Opt. Express **14**(5), 1748–1754 (2006)
7. Ramaswami, R., Sivarajan, K.N.: Optical Networks: A Practical Perspective. Morgan Kaufmann, San Mateo (2001)
8. Jonker, R.J.W., Schimmel, R.C., Waardt, H., de: Experimental assessment of the gating functionality of a TOAD operating at 1310 nm in high speed OTDM systems. In: Proceedings of Symposium IEEE/LEOS Benelux Chapter, pp. 171–174 (2002)
9. Gnauck, A.H., Raybon, G., Bernasconi, P.G., Leuthold, J., Doerr, C.R. Stulz, L.W.: 1 Tbit/s (6×170.6 Gbit/s) transmission over 2000 km NZDF using OTDM and RZ-DPSK format. IEEE Photonics Technol. Lett. **15**(11), 1618–1620 (2003)
10. Hamilton, S.A., Robinson, B.S., Murphy, T.E., Savage, S.J., Ippen, E.P.: 100 Gbit/s optical time division multiplexed networks. J. Lightwave Technol. **20**(12), 2086–2100 (2002)
11. Zoiros, K.E., Papadopoulos, G., Houbavlis, T., Kanellos, G.T.: Theoretical analysis and performance investigation of ultra-fast all-optical Boolean XOR gate with semiconductor optical amplifier-assisted Sagnac interferometer. Opt. Commun. **258**(2), 114–134 (2006)
12. Liu, Y., Tangdiongga, E., Li, Z., Zhang, S., de Waardt, H., Khoe, G.D., Dorren, H.J.S.: Error free all-optical wavelength conversion at 160 Gbit/s using single semiconductor optical amplifier and an optical band pass filter. J. Lightwave Technol. **24**(1), 230–236 (2006)
13. Glesk, I., Runser, R.J., Prucnal, P.R.: New generation of devices for all-optical communications. Acta Phys. Slovaca **51**(2), 151–162 (2001)
14. Minh, H.L., Ghassemlooy, Z., Ng, W.P.: All-optical flip-flop based on a symmetric Mach-Zehnder switch with a feedback loop and multiple forward set/reset signals. Opt. Eng. **46**(4), 40501–40503 (2007)
15. Huybrechts, K., Morthier, G., Baets, R.: Fast all-optical flip-flop based on a single distributed feedback laser diode. Opt. Express **16**(15), 11405–11410 (2008)
16. Zhang, S., Lenstra, D., Liu, Y., Ju, H., Li, Z., Khoe, G.D., Dorren, H.J.S.: Multi-state optical flip-flop memory based on ring lasers coupled through the same gain medium. Opt. Commun. **270**(1), 85–95 (2007)
17. Sokoloff, J.P., Prucnal, P.R., Glesk, I., Kane, M.: A terahertz optical asymmetric demultiplexer (TOAD). IEEE Photonics Technol. Lett. **5**(7), 787–790 (1993)
18. Gayen, D.K.: Optical arithmetic operation using optical demultiplexer. Circ. Syst. **7**(11), 3485–3493 (2016)
19. Bhattachryya, A., Gayen, D. K.: All-optical n-bit binary to two's complement converter with the help of semiconductor optical amplifier-assisted Sagnac switch. J. Mech. Continua Math. Sci. **17**(1), 117–125 (2022)
20. Gayen, D.K.: Optical parallel half adder using semiconductor optical amplifier-assisted Sagnac gates. J. Mech. Continua Math. Sci. **17**(4), 1–7 (2022)
21. Gayen, D.K.: All -Optical 3:8 Decoder with the help of terahertz optical asymmetric demultiplexer. Opt. Photonics J. **6**(7), 184–192 (2016)
22. Hall, K.L., Rauschenbach, K.A.: 100 Gbit/s bitwise logic. Opt. Lett. **23**(16), 1271–1273 (1998)
23. Stubkjaer, K.E.: Semiconductor optical amplifier based all-optical gates for high speed optical processing. IEEE J. Sel. Top. Quantum Electron. **6**(6), 1428–1435 (2000)
24. Chattopadhyay, T., Gayen, D.K.: All-optical two's complement number conversion scheme without binary addition. IET Optoelectron. **11**(1), 1–7 (2017)
25. Siarkos, T., Zoiros, K.E., Nastou, D.: On the feasibility of full pattern-operated all-optical XOR gate with single semiconductor optical amplifier-based ultrafast nonlinear interferometer. Opt. Commun. **282**(14), 2729–2740 (2009)

Chapter 35
Software Defined Optical Wireless Network with AI

T. Perarasi, K. Shoukath Ali, and M. Leeban Moses

35.1 Introduction

Enterprise networks can now handle security and policy at the flow level thanks to SDN's centralised management approach, as demonstrated by initiatives like Ethane. The way we design and manage networks has been completely changed by this architectural shift, which provides a different viewpoint for addressing a range of networking issues and adjusting to new technologies like 5G and the Internet of Things [1].

Moreover, the integration of artificial intelligence (AI) into SDN is an exciting avenue for innovation. AI technologies can enhance network monitoring, anomaly detection, predictive analytics, and automated decision-making, leading to intelligent and autonomous network operations [2]. The synergy of AI and SDN has the potential to revolutionize network management, paving the way for advanced network applications and services.

However, overview and survey articles about SDN-based optical networks (SDONs) are in relatively short supply. Of particular note is the comprehensive survey carried out by Zhang et al. on flexible optical networking in core backbone networks using Orthogonal Frequency Division Multiplexing (OFDM) [3]. An introduction to SDN and network virtualization is given by Gupta et al., with an emphasis on how these ideas can be applied to optical networking [4]. Through the analysis of white papers from top networking companies their review primarily looks at industry efforts.

When it comes to geographically dispersed data centres (DCs), the data centre network usually consists of several domains and technologies, such as Ethernet access

T. Perarasi (✉) · M. Leeban Moses
Department of ECE, Bannari Amman Institute of Technology, Erode, Tamil Nadu 638401, India
e-mail: pera.sweety@gmail.com

K. Shoukath Ali
Presidency University, Itgalpura, Rajanukunte, Yelahanka, Bengaluru, Karnataka 560064, India

M. El Ghzaoui et al. (eds.), *Next Generation Wireless Communication*, Signals and Communication Technology, https://doi.org/10.1007/978-3-031-56144-3_35

networks and optical transport networks. An SDN controller capable of supporting extended protocols is installed in each network domain. With a hierarchical architecture, the control plane is organised so that local domain resources are managed by individual domain controllers and cross-domain orchestration is handled by a central controller, also known as the "father". A great deal of research has been done on the basic concepts of SDN. Network management in SDN has been investigated, and the field of SDN security has been covered in depth. Additionally, the investigation of SDN-based satellite networking is covered.

In summary, SDN's core principles involve centralizing network control, segregate control and data planes, and enabling programmability among networks. This approach simplifies network management and offers scalability, making it suitable for modern network challenges. Additionally, the integration of AI further enhances SDN's capabilities, promising intelligent and automated network operations.

35.2 Software Defined Network (SDN)

SDN represents an innovative method for managing networks, providing streamlined solutions for intricate functions such as traffic engineering and network optimization. By combining network control, it makes it easier to manage network resources dynamically and clearly distinguishes among control and data planes. A mere flexible and responsive network architecture is the outcome of this.

SDN adoption is contingent upon its ability to address problems that are outside the purview of conventional networking protocols and architectures. SDN has already been deployed in data centres by well-known businesses like Microsoft and Google, proving its usefulness in actual situations.

According to the Open Networking Foundation (ONF), three-layer SDN architecture: the application layer, control layer, and infrastructure layer. Applications for network control and monitoring reside in the application layer, where they communicate with the underlying infrastructure via abstractions. The control layer communicates with the application and infrastructure layers and is in charge of network control functions. The South Bound Interface (SBI) between the control and infrastructure layers is frequently implemented using the OpenFlow protocol. Each layer in this layered model functions independently, enabling the existence of multiple solutions within a single layer. Programmable devices, also known as network elements, forwarding elements, or network devices, make up infrastructure layer. SBI (Data-Controller Plane Interface) and the North Bound Interface (NBI) (Application-Controller Plane Interface) are the main interfaces that link the SDN layers. These interfaces help the layers communicate and abstract from one another. To link the SDN architecture with external network architectures, two more interfaces are defined. These are called the East Bound Interface (EBI) and West Bound Interface (WBI), or collectively, Intermediate Controller Plane Interfaces [5].

Based on the architecture of computer systems, the SDN abstraction layering is organized into three levels: the application layer is layered first, control layer is

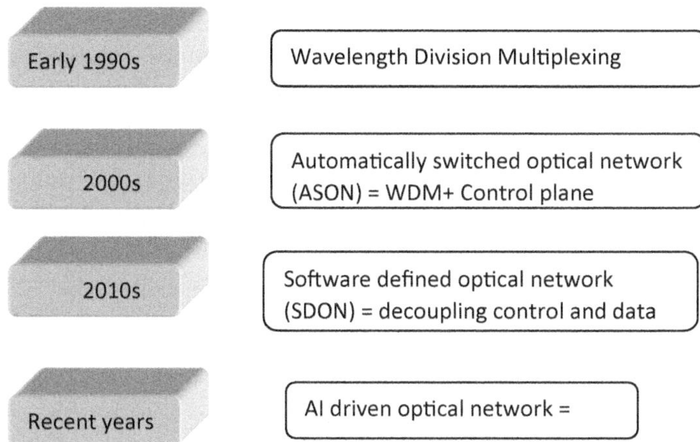

Fig. 35.1 Growth of optical networks

layered in the middle, and the infrastructure layer at the bottom. The term "South Bound Interface" (SBI) refers to interface between infrastructure and control layer, whereas "North Bound Interface" (NBI) refers to interface among application and control layer. These interfaces are being developed, especially the SBI, using a number of emerging standards like the OpenFlow protocol.

In SDN, application layer comparable to software using computers in order to carry out tasks. In contrast, the operating system (OS) of a computer serves as the model for the control layer. It controls the computer's resources, offers an abstraction layer to make device interfaces easier. Device drivers provides uniform interface for diverse device types by abstracting the specifics of interacting with different devices and integrating them into the operating system of a computer. Comparably, in the SDN model, the control layer's functionality and the unified SBI work together to act as a "device driver" for connecting to infrastructure layer devices like packet switches.

The large transmission capacities of optical networks make them indispensable to modern information technology and its growth is depicted in Fig. 35.1. Controlling optical networks is complex, though, because of the unique features of optical transmission. SDONs seek to take advantage of the flexibility offered by SDN control.

35.2.1 Infrastructure Layer

The environmental forwarding data traffic (data plane), which can be implemented in virtual or physical hardware, is included in the infrastructure layer of SDN. The South Bound Interface (SBI) allows data plane's network elements to reveal their capabilities to control plane. Decision-making powers for network operations like switching and routing are ingrained in the network infrastructure in conventional networking since the control mechanisms are integrated into the infrastructure itself. Furthermore, in traditional network elements, these forwarding actions are usually established automatically using self-evaluated and specific algorithms. Because of this, traditional network element configurations are typically difficult to change without interfering with services, which reduces the network's flexibility.

A logical node called the controller is centrally controlled in SDN, as opposed to the infrastructure, which is where autonomous control functions like neighbor discovery and forwarding algorithms are moved. The network elements function as "dumb switches" in this decoupled architecture, simply carrying out the controller's instructions. The complexity of network elements is decreased and their reconfigurability is enhanced by separating the control functions from the infrastructure.

Control Layer

South Bound Interfaces (SBIs) are used by the control layer at SDN to programme and configure network elements like switches. Logic in the form of the SDN controller decides how to configure the network infrastructure according to the needs of application layer apps. In order to effectively oversee the network, SDN controllers are able to collect data from the SDN infrastructure, including flow statistics, topology details, neighbor relationships, and link status. We often refer to the software as the Network Operating System (NOS), which is responsible for implementing the SDN controller. An Open Network Operating System (ONOS) that uses distributed control architecture in order to run over wide area networks (WANs) lies as a suitable support for SDN-based NOS.

Application Layer

Network apps and services use control plane to implement its functions among physical or virtual infrastructure make up the application layer in SDN. Fault restoration, provisioning, and network topology discovery are a few types of network applications. The SDN controller helps the SDN applications implement their functionalities by giving them an abstracted view of the network. Applications at the application layer may also include specialized force that need to be processed in big data centers or advanced network management features like network data analytics.

Orchestration Layer

The importance of the orchestration layer has grown with the complexity of SDN systems, despite it not being historically regarded as primary SDN architectural layers. A single SDN controller is nothing bit hierarchical structure of several SDN controllers, or a group arranged "flat" are all coordinated by the SDN orchestrator. SDN orchestrator is a post decision maker whereas SDN controller is a logically centralized control entity that immediately controls the network elements by signaling predefined control actions or rules.

Using input from users, the SDN orchestrator can request manual recommendations or use SDN applications to help make automated decisions. The network elements receive these orchestration decisions—which involve configurations and actions—through the South Bound Interfaces (SBIs) and SDN controllers. Compared to the SDN controllers, the SDN orchestrator offers more comprehensive coordination and management functionalities. It is conceptualised as a separate abstracted layer situated above the application and control layers.

For example, within cloud networks, cloud SDN orchestrator creates and destroy Virtual Machines (VMs) across various network domains and layers according to workload requirements. Conversely, SDN controllers concentrate on executing particular network operations, like path computation and routing.

35.3 Software Defined Optical Network (SDON)

Next, we have Software Defined Optical Networks (SDON), which provide centralised scheduling and management of diverse optical layer resources according to user or operator specifications. Programmable software and dynamic customization are used by SDONs to provide flexible service provisioning, efficient resource usage, and quick response times to requests. They enable programmable tuning of optical network elements by supporting programmable transmission devices, control strategy, and service processing [6].

Within an SDON in Fig. 35.2, the OpenFlow protocol extended for optical transmission devices is used by the optical controller to analyse the underlying optical transmission equipment. It gathers network topology information, stores it in an optical database, and extracts status information from optical devices (ON-TED). Tasks like resource allocation and lightpath creation and deletion are managed by the Optical Transport Network (OTN) manager. The multi-domain controller receives notification from the optical network controller, which encapsulates the network status and topology data. The multi-domain controller tracks network health and integrates network data gathered by the optical controller via the Southbound Control Virtual Network Interface (CVNI) interface [7]. It parses application requests using the Northbound Restful API and is composed of several function modules and a resource integration module.

The application monitor, service selection engine (SSE), application resource manager, request resolver, virtual network manager, policy generator, policy analysing engine, Path Computation Element (PCE), and wrapper are among the function modules in the multi-domain controller. Within the SDON architecture, each module plays a distinct role in managing and coordinating requests and resources.

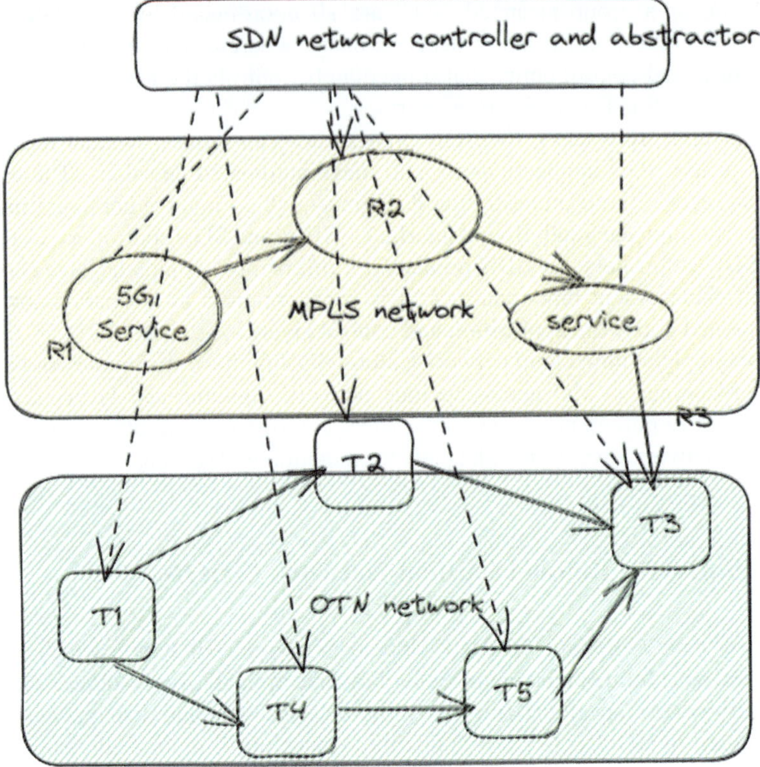

Fig. 35.2 SDON architecture

Overall, higher-level coordination, management, and programmability in SDN systems are made possible by the SDN orchestrator and SDON, allowing for effective resource utilization and adaptable service provisioning across various network domains and layers [8, 9].

35.4 SDN Interfaces

35.4.1 Northbound Interfaces (NBIs)/Application-Controller Plane Interfaces (A-CPI)

Northbound Interfaces are logical interfaces that link the SDN controller to application-layer software entities. The RESTful API, which adheres to the Representational State Transfer (REST) principles, is one popular kind of NBI. REST is an architectural style for software that encourages scalability, flexibility, and interoperability. A RESTful API in the context of SDN enables multiple SDN applications to interact with a single SDN controller and coexist. It possesses the following qualities:

- **Client–Server**: SDN apps function as clients, and the controller serves as a server to facilitate communication between the controller and other applications.
- **Stateless**: The client, or application, controls and preserves the state, while the server, or controller, executes commands from the client.
- **Caching**: By storing data temporarily, the client can decrease the number of times it communicates with the server, increasing scalability and performance.
- **Uniform/Interface Contract**: Every service that makes use of the REST API adheres to the same technical interface, which includes using JSON or XML as their primary data format.
- **Layered System**: This type of architecture allows for modularity and flexibility because the interface only handles the immediate node.

35.4.2 Southbound Interfaces (SBIs)

Alternatively referred to as Data-Controller Plane Interfaces (D-CPI), Southbound Interfaces are logical interfaces that link network components functioning at the infrastructure layer (data plane) to the SDN controller. Many SBI protocols have been put forth, but they are usually incompatible with one another. Several instances of SBIs are as follows:

- **OpenFlow Protocol**: This protocol allows OpenFlow switching elements (SDN switches) and controllers to communicate with one another. Through the specification of matching criteria and corresponding actions to be performed by the switches, it enables the controller to define and manage packet flows.
- **Path Computation Element Protocol**: (PCEP makes it easier for Path Computation Element (PCE) in the controller and the Path Computation Client (PCC) in network elements to communicate.
- **Network Configuration (NETCONF) Protocol**: This protocol offers methods for setting up, altering, and removing network device configurations. It employs XML to encrypt protocol messages and configuration data, and remote procedure calls are used to carry out actions on devices
- **Border Gateway Protocol**: Link State Distribution (BGP-LS) Protocol: BGP-LS is an addition to BGP that enables the central controller to collect data from network components about topology and bandwidth usage.
- **Network virtualization**: This technique creates one or more virtual networks that run on top of the underlying physical network infrastructure by abstracting it. It makes it possible to create autonomous virtual networks in a flexible way with set resource allocations. A specific type of network virtualization known as "Network Function Virtualization" (NFV) involves the implementation of network functions as cloud-based or data center-based software entities.

35.4.3 *Overview of Optical Networking*

Optical Switching Paradigms: Transmission channels carrying optical signals or signals in the optical domain are used by optical networks to function. Circuit, packet, and burst switching are among the granularities at which optical switching can be executed.

- **Circuit Switching**: Circuit switching in space, wavelength, band, or time can be achieved through optical means. Wavelengths within the optical spectrum are arranged on either a rigid or pliable wavelength grid. Flexible grid standards provide greater freedom in selecting the spectral width and center frequency compared to fixed grid standards, which dictate specific center frequencies. The flexible grid is used by Elastic Optical Networks (aeons) to better utilize the optical spectrum; however, when light paths are built and taken down, they can also cause spectral fragmentation, which lowers the efficiency of spectrum utilizations.
- **In optical networks**, packet switching entails utilizing header fields in optics to switch individual packets. Several components, including optical synchronization, demultiplexing, and multiplexing, optical switch fabric, optical buffering, and optical packet forwarding table computation and lookup, are needed for an all-optical packet switch. Nevertheless, some of these functions are frequently carried out in the electrical domain in real-world applications. For instance, packet forwarding table computation and lookup are frequently carried out electrically.
- In **optical packet switching**, a packet might require optical buffering when there is competition for a destination port. Nevertheless, fiber delay line-based optical buffering may not always be feasible. Fiber optic cables with predetermined lengths set up to add a delay to the optical signal are called fiber delay lines. Deflection routing, in which the packet is redirected to an output might or might not lead to intended destiny, is an alternative to optical buffering. Alternatively, the packet can be dropped.

Alternatively, optical networks that use burst switching can overcome the difficulties caused by spikes in traffic by doing away with the need for buffering, forwarding table lookup, and computation of optical packet forwarding tables. Rapidly creating transient circuit to transfer more packets combined into burst is made possible by burst switching. Sending a control packet across the network determines the burst's light path, after which the burst is sent on the temporary circuit without the need for packet lookup or buffering along the way. By preventing network resources from being wasted during idle times, this method maximizes the use of optical circuits.

Peripheral networks like data centers, residential local area networks, and cellular wireless networks are connected by optical access networks. They are essential in establishing the internet connection between data centers, mobile devices, and end devices. Metropolitan area networks link backbone networks to optical access networks inside a metropolitan area. Utilizing optical networking technologies, MANs typically have a ring or star topology.

Across national or international distances, backbone networks—also referred to as wide area networks—interconnect individual MANs. In order to enable dependable and effective data transfer between various regions, backbone networks usually use high-speed optical transmission links and a mesh architecture.

35.5 Artificial Intelligence (AI)

Artificial Intelligence (AI) is a fast-expanding domain that encompasses many subfields, such as machine learning (ML), optimisation, planning, decision making, knowledge representation, and meta-heuristic algorithms. Alan Turing's Turing test defines intelligence as requiring a computer to respond to questions in a way that makes it impossible to tell it apart from a human.

A summer workshop at Dartmouth College led by Claude Shannon and Martin Minsky launched the field of artificial intelligence research in the mid-1950s. McCulloch and Pitts' 1943 model served as the basis for the first artificial neural network, which is a common feature in artificial intelligence. Some of the subfields of artificial intelligence (AI) that have emerged as a result of the continuous evolution of AI. To advance AI, researchers have consistently improved already-existing techniques and put forth fresh, hybrid intelligent strategies [10].

A crucial element of artificial intelligence is machine learning, in which a clever machine gains knowledge from experience and applies it to enhance its functionality. The learning methods can be divided into four main groups:

1. **Supervised learning**: A training dataset with input–output pairs serves as the source of pre-defined knowledge for these techniques. A function that converts inputs into suitable outputs is learned by the system.
2. **Artificial neural networks (NNs)**: NNs can adapt to data without explicitly stating a model; they are modelled after biological learning systems. They can be trained with supervised algorithms and can approximate any function. NNs may experience over-fitting, though, if the model contains an excessive number of parameters.
3. **Support vector machines (SVMs)**: SVMs leverage the margin between classes to identify linear separators. They have a strong resistance to over-fitting and can represent complex functions. Cases that are not linearly separable are handled by kernel methods.
4. **Decision trees (DTs)**: DTs offer comprehensible models and are useful in classification issues. To create forecasts, they run a series of tests on the input properties. In the event that the data lacks pattern, DTs may experience over-fitting.

Ensemble methods enhance learning algorithm performance by combining predictions from many approaches. Two well-liked ensemble techniques that combine the outputs of separate classifiers to produce composite classifiers are bagging and boosting.

Deep learning is one of the potent machine learning techniques that makes use of multi-level representations for tasks like detection and classification. It makes it possible for machines to autonomously identify representations from unprocessed data, which makes complex functions learnable [11]. In tasks including object detection, audio recognition, and natural language understanding, deep learning models demonstrated remarkable performance. RNNs are appropriate for sequential or time-series problems, whereas DNNs are deep feed-forward multilayer networks. When processing input in the form of arrays, like photographs, CNNs work especially well [12].

All things considered, machine learning is essential to artificial intelligence (AI), and diverse methods and algorithms are used to address diverse issues and enhance system functionality.

35.5.1 Unsupervised Learning

When pre-defined knowledge or labelled data are unavailable, unsupervised learning techniques are applied. Finding patterns and structures in the supplied data is the main goal of these techniques. Clustering is one type of unsupervised learning in which related data points are grouped together according to a predetermined distance metric, like the cosine, Jaccard, or Euclidean distance.

A well-known method in unsupervised learning is K-means clustering. It needs to know ahead of time how many clusters ('k') are wanted. Until convergence is achieved, the procedure iteratively updates the centroids using the allocated data points. However, the initial cluster selection might affect the quality of the results; choosing the wrong value for 'k' can produce subpar results. Soft K-means, another name for fuzzy C-means clustering, enables data points to be part of several clusters with different levels of membership.

Based on artificial neural networks, self-organizing maps (SOMs) are another well-liked unsupervised learning method. High-dimensional input data is mapped by SOMs to a lower-dimensional representation known as a SOM map. Building and rearranging the map using input data is part of the training process. Finding the winning neuron or node in the map allows one to classify a fresh input vector.

Systems having unobservable (hidden) states are represented statistically using Hidden Markov Models (HMMs). According to HMMs, a Markov process exists in which the likelihood of entering a new state is solely dependent on the one that before it. Sets of states, an output alphabet, transition probabilities, observation probabilities, and beginning state probabilities make up HMMs. HMM parameters can be taught either under supervision or without it. One popular unsupervised training approach for HMMs is the Baum-Welch algorithm.

A hidden layer and an input layer make up the two layers of restricted Boltzmann machines (RBMs), which are stochastic artificial neural networks. No connections exist between neurons in the same layer in RBMs due to their constrained connectivity. A key element of deep belief networks (DBNs) is the RBM, which is frequently utilised for feature extraction.

Generative models are used to extract high-order correlations from the input data in unsupervised deep-learning techniques. A top layer is usually needed for discriminative tasks in generative models, and an unsupervised pre-training step is usually needed to extract data structures. Two popular deep generative models in the context of deep learning are stacked auto-encoders (SAEs) and DBNs. An unsupervised, greedy layer-wise technique is used for pre-training of SAEs, and labelled examples are used for supervised training during the fine-tuning step. By layer-by-layer training, stacked RBMs are used in DBNs to derive hierarchical representations of the training set [13–15].

35.5.2 Reinforcement Learning (RL)

A learning strategy known as reinforcement learning (RL) teaches a system by using the reinforcements it receives from its surroundings. Based on how well it performs, the system is rewarded or punished, and it updates its knowledge with this data. RL is predicated on the Markovian property, which asserts influenced by current state.

In regulated Markovian environments, agents can behave optimally thanks to a type of model-free reinforcement learning known as Q-learning. Finding the best course of action that maximises Q—the total expected reward—for carrying out activities in particular situations is the aim. The incremental dynamic programming technique known as Q-learning converges under specific circumstances.

High-dimensional state and action spaces can be used for decision-making when RL and deep learning approaches are combined in deep RL. Deep RL approximates the ideal policy by using deep neural networks (DNNs) [16]. It makes use of representation learning, which includes learning from raw data using convolutional neural networks (CNNs). One notable example of effective deep reinforcement learning (deep RL) is AlphaGo.

Learning from both labelled and unlabelled data is referred to as semi-supervised learning. Because it employs a combination of more unlabelled data and some labelled data, it falls somewhere between supervised and unsupervised learning. Because it gains from the limited quantity of labelled data, semi-supervised learning can perform better than unsupervised learning. When acquiring labelled data is expensive or impracticable, it is very helpful.

Meta-heuristic algorithms are general-purpose, nature-inspired optimisation techniques. They are used to resolve challenging optimisation issues that deterministic approaches are unable to resolve effectively. The ant colony optimisation method, genetic algorithms, simulated annealing, and whale optimisation algorithm are a few instances of meta-heuristic algorithms. Numerous combinatorial, nonlinear, and multi-modal optimisation issues may be resolved with these strategies [17–19]. To improve performance, hybrid strategies that draw on the advantages of many meta-heuristics have been created. Parameter adjustment is necessary for optimal performance, as meta-heuristics cannot provide optimal solutions.

35.5.3 AI Algorithms

ACO, or Ant Colony Optimisation, meta-heuristic algorithm that draws inspiration from ants' foraging habits. Problems involving combinatorial optimisation are resolved using it. Ants in ACO begin by haphazardly scouring the issue space. They leave a chemical pheromone trail behind them that directs other ants to more advantageous locations. Pheromone concentration rises in the directions leading to optimal solutions. Premature convergence is avoided in part via pheromone evaporation.

Darwinian theory-based optimisation algorithms fall under the category of evolutionary algorithms, or EAs. Genetic Programming, Evolutionary Programming, Genetic Algorithms (GA), and Evolution Strategies are some of these. A population of chromosomes evolves in GA, where solutions are represented as chromosomes. In order to provide novel solutions, crossover and mutation processes are used. When putting GAs into practise, selection tactics and the selection of crossover and mutation operators are crucial.

Inspired by the swarming behaviour of birds, Particle Swarm Optimisation (PSO) is an additional population-based meta-heuristic method. Each particle in the swarm has a location and a velocity, and they are produced randomly in the search space. While PSO has been effective in locating the best areas of the search space, it is devoid of a method for convergent on ideal regions.

The collective behaviour of honeybee colonies serves as the inspiration for swarm intelligence-based optimisation algorithms known as Bee Colony Optimisation (BCO) methods. Within this class, one popular algorithm is Artificial Bee Colony (ABC). Worker bees use a waggle dance to communicate where food is available. ABC uses a neighbourhood source creation process in addition to having global search capabilities.

The Whale Optimisation Algorithm (WOA) mimics humpback whale social behaviour in the context of hunting. Bubble-net feeding is a technique used by humpback whales to catch their food. On comparison, WOA has demonstrated superior performance and mathematical modelling of this behaviour.

Firefly Optimisation (FFO)-population-based meta-heuristic, inspired by the social and communicative behaviours of many fireflies. More vibrant fireflies draw in other fireflies, and their allure diminishes with distance. Global firefly communication is essential to FFO.

Bat Algorithm (BA), technique that draws inspiration from microbats' echolocation habits. In order to locate prey, bats use ultrasonic sound and modify the intensity and rate of emission. BA integrates ideas from local search and PSO.

Initially, a population-based technique called Teaching-Learning-Based Optimisation (TLBO) was presented for restricted mechanical design optimisation issues. The population is made up of a group of students, and the teacher who provides

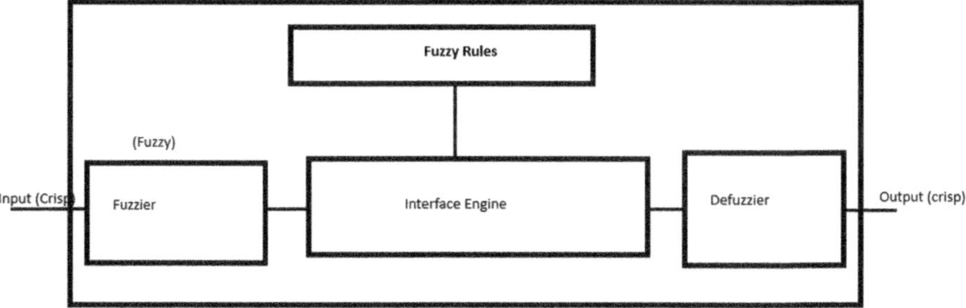

Fig. 35.3 Structure of a typical fuzzy inference system

the best answer is chosen. By using teacher and student methods, TLBO iteratively improves the answers [20, 21].

These meta-heuristic algorithms offer alternative methods for addressing optimisation issues and have been used to identify near-optimal or ideal solutions across a range of fields.

A population-based algorithm called GWO (Grey Wolf Optimisation) mimics the social structure and hunting techniques of real grey wolves. Alphas, betas, deltas, and omegas are the first four levels of leaders into which the grey wolves are arranged in this algorithm. Decisions for the pack are made by the alphas, who are the highest-ranking leaders. The GWO algorithm simulates the hunting strategy of grey wolves, which involves circling and attacking their target.

Based on fuzzy set theory, fuzzy inference systems (FIS) as in Fig. 35.3 are used to translate input variables to suitable output values. Fuzzy logic differs from standard binary logic in that it permits membership or degrees of truth between 0. Fuzzy rules with linguistic variables and values are used in fuzzy systems. Using membership functions, these rules are applied to the input variables, and the inference engine assesses results to determine the output. The fuzzy result is defuzzed using techniques like the weighted average or centre of gravity to get a crisp number.

Mamdani's method and Sugeno's method are the two inference approaches that are frequently applied in fuzzy systems. Sugeno's approach depicts the consequent as a single point, whereas Mamdani's approach employs fuzzy membership functions. Since Sugeno's approach eliminates the requirement to locate a shape's centroid, it is thought to be more computationally efficient.

Fuzzy systems offer knowledge representation and explanation capabilities similar to those of humans. They are unable to pick up new skills or adjust to other surroundings, though. When creating these models, tuning fuzzy sets and rules is frequently necessary. Hybrid approaches, such neuro-fuzzy systems, combine fuzzy systems' representation and explanation skills with neural networks' learning capabilities.

35.6 AI in SDN

Software-Defined Networking (SDN) has seen a broad use of AI and ML techniques, such as machine learning (ML), fuzzy systems (FSs), and meta-heuristics, to a variety of networking issues. The scientific community as well as industry have begun to pay close attention to and use these methods. Within the SDN paradigm, we will now address the specific application domains for each technique.

SVMs in SDN: SVMs have been a popular choice for SDN intrusion detection systems. SVMs are used by these systems to identify several kinds of threats, including as floods and DDoS attacks. In these situations, SVMs have demonstrated better accuracy and lower false positive rates than other classifiers. To improve the effectiveness of intrusion detection systems, feature selection strategies like the ID3 DT method have been coupled with SVMs.

DTs and Ensemble Methods in SDN: In SDN, DTs have been applied to tasks including traffic categorization, malware detection, and decision-making. SDN has also made use of ensemble approaches, which blend many models to increase prediction accuracy. Both accuracy and performance have showed promise with these techniques.

There are two main components to their AI module:

a. **Traffic Classification: ** The first section classifies network traffic using the Support Vector Machine (SVM) algorithm, which enables it to identify the kind of traffic.
b. **Estimator: ** In order to guarantee Quality of Service (QoS) and Quality of Experience (QoE), the SDN controller receives guidance from the estimator in the second part.

According to their experiments, there was an average 70% decrease in jitter and a drop in losses from 9.07% to roughly 1.16%. Additionally, with a 77% accuracy rate, the SVM algorithm demonstrated effectiveness in identifying critical traffic.

35.6.1 Supervised Deep Learning in SDN

Deep learning techniques have shown more promise than traditional machine learning (ML) techniques in Software-Defined Networking (SDN). An intrusion detection system was proposed by Tang et al. using a basic Deep Neural Network (DNN). With an accuracy of 75.75%, the deep learning method outperformed other supervised machine learning (ML) algorithms, including naive Bayes, SVM, and Decision Trees. In a subsequent study, the same authors extended their earlier findings by substituting a GRU-RNN for the basic deep learning model. This new model beat SVM, NB Tree, and their previous DNN-based model with an accuracy of 89%.

"Deep Flow," a system for SDNs provide measurement strategies for essential path, was developed by Lazaris and Prasanna. When switch resources restrict the placement of flow counters, it employs an LSTM-RNN to forecast the magnitude of remaining flows. According to analysis for replicating Google's B4 network architecture was 3.9%, whereas the MAPE for estimating actual flow volumes on CAIDA traces was 12%.

35.6.2 WOA in SDN

Particle Swarm Optimization (PSO) in SDN

The MAP-SDN framework was presented by Farshin and Sharifian and is intended for use in SDN-based cloud data centres that support a variety of service classes. It is used for assignment and provisioning. The WOA and PSOGSA algorithms' performances were compared in the study. The exploratory feature of the gravitational search algorithm is combined with the exploitation capabilities of PSO in PSOGSA. The trials' outcomes showed that WOA produced better results with less calculation time and showed higher accuracy.

Firefly Algorithm (FFA) in SDN

Sahoo et al. tackled the controller placement problem in SDN-based Wide Area Networks (WAN) by using both PSO and the Firefly algorithm (FFA). Three parameters were taken into consideration: multi-path connection between switches and controllers, inter-controller latency, and controller-to-switch latency. An "average delay rise" metric was also included in the study to quantify the additional delay caused by primary path failure. The conclusions of the experiment showed that FFA generated accurate and efficient results.

Teaching–Learning-Based Optimization (TLBO) in SDN

SDN-based Quality of Service (QoS) for remote telesurgery applications was made possible by Mohammadi and Javidan through the use of TLBO. Using network resource status as a basis, the suggested model calculated the best route between the surgeon and the operating room. TLBO outperformed a shortest-path-based strategy in terms of packet loss ratio, peak signal-to-noise ratio (PSNR), end-to-end latency, and structural similarity index (SSIM).

35.6.3 Enabling Technologies in the Device Aspect

Motivated by their strong capacity for self-learning, machine learning methods allow for adaptive EDFA operations with respect to dynamic input power and non-flat gain spectrum. The use of the multi-layer perceptron (MLP) is one of the methods that have been suggested for adaptive EDFA functioning. One method uses MLP to adaptively set EDFA parameters with the goal of maximising the NF-GF trade-off for every EDFA in the group. The neighbour amplifiers are kept in balance by an iterative updating mechanism, which produces power performance that is around 3 dBm with a 0.1 dB variation.

A different approach also uses a machine learning technique known as Kernelized Bayesian Regression (KBR) to estimate power excursions in cascade EDFAs throughout the optical path. Using radial basis functions (RBF), the trained model predicts slot choices by adding or removing particular wavelengths in order to minimise power variance among neighbouring lines. In addition, a variation of this study uses a ridge regression model to automatically modify EDFA power when various defragmentation strategies are utilised. A dynamic power adjustment approach for diverse circumstances is provided by the trained model, which also detects power excursions in cascade EDFAs during different defragmentation operations. With this approach, power excursion problems arising from spectral configurations that change dynamically may be efficiently addressed, with minimal power excursion often being less than 1%.

Dynamic optical networks provide several issues, including non-flat power gain impact and random noise introduction. Switched light paths and variable input power have a significant impact on EDFA characteristics. The MLP-based EDFA control scheme offers a simple method for modifying the parameters of individual EDFAs in optical links; however, it is predicated on the oversimplified supposition that global optimal configuration for optical link does not always correspond to the optimal configuration parameters for optimal signals. More investigation and development could be necessary to guarantee the anticipated optical link power performance and enhance stability and interpretability. In order to overcome limitations associated with predicted output power, this technique's implementation might also benefit from taking amplifier energy consumption into account and placing power attenuators during optical communications.

35.6.4 AI Based Optical Network Traffic Prediction

Network traffic in today's optical networks is growing at a very rapid rate. A greater requirement for flexible resource allocation is being created by the increased range of services provided, which is also causing extremely dynamic traffic patterns. Traffic prediction becomes vitally necessary in order to handle this dynamic traffic effectively while optimising resource allocation. Integrating traffic prediction into optical networks is done primarily for two reasons.

First of all, traffic prediction is essential for enabling more accurate resource allocation, which lowers operating expenses. In order to guarantee network performance and dependability, traditional resource allocation frequently depends on peak situations. But this method often leads to significant waste and excessive resource redundancy. In the future, transport networks' resource allocation should be adaptable enough to change dynamically in response to particular service requirements, allowing for a more precisely calibrated distribution of available resources. To improve network stability, improve user experiences, and lower operational expenditures (OPEX), precise traffic forecast and rapid resource modifications are crucial [22, 23].

However, traffic prediction plays a critical role in identifying and predicting abnormalities in traffic, which guarantees steady network performance. It is possible to proactively identify and anticipate any irregularities in network traffic by using traffic prediction. This helps to guarantee the safe and stable functioning of the network by allowing network operation and maintenance staff to get timely notifications and make educated choices.

From an industry data perspective, utilizing AI algorithms for traffic prediction is entirely feasible. Firstly, user behaviour and service bandwidth fluctuations within optical networks often exhibit certain regularities and periodic patterns. These characteristics create opportunities to predict future data based on historical data. Furthermore, during network operation and maintenance, a significant amount of historical traffic data is gathered in the field of transmission networks. With this enormous dataset at their disposal, internet service providers (ISPs) may create precise traffic forecast models. In addition, a number of studies have shown that the majority of machine learning algorithms, including random forests and neural networks, are highly effective at learning nonlinear connections and have significant resistance to data noise [24]. More precise traffic forecasts may be made by integrating this large information with the right AI algorithms.

AI-based traffic prediction typically uses supervised learning, which involves four steps:

a. Gather historical traffic data and define the input time window.
b. Train a supervised learning model using historical data as input and future traffic volume as output.
c. Evaluate the trained model, verify its performance, and identify the hyperparameters and input dimensions.
d. Deploy the trained model to predict future traffic data without relying on supervisory data.

In an optical data centre network, LSTM is used to forecast how much time is still needed for traffic aggregation. For jobs requiring sequential data as input, LSTM is specially designed. Each hidden unit takes into consideration the output of the preceding hidden unit in real time in addition to receiving input data. The amount of time and held time information (HTI) used determines the average remaining life (MRL) of the traffic [25, 26]. Within data centres, resource reallocation or allocation depends on knowing the average remaining lifespan. However, HTI information for traffic is frequently unavailable because of the diversity and heterogeneity of data centre applications.

35.7 AI-Powered Resource Allocation in Optical Networks

Optical transport networks have special network properties that make multi-dimensional physical resource allocation difficult. These resources might include modulation formats, wavelengths, spectrums, fibre linkages, and fibre cores [27].

A large number of resources are frequently wasted in the traditional method of exchanging network dependability for redundancy. AI algorithms must be used for accurate resource allocation in light of the exponential rise in traffic and the increasing scarcity of network resources. Artificial intelligence (AI)-based optimum resource allocation may take into consideration a variety of user needs (like latency) and operator costs (like bandwidth use and load balancing), resulting in exact allocations that save expenses and improve resource utilisation efficiency. AI-based resource allocation reduces the requirement for data plane operation and mathematical optimisation theories, in contrast to traditional research that frequently rely on rule-based policies based on domain expertise. This method improves scalability while also lessening the effort for networking specialists and operators [28]. The additional benefit of AI-based resource allocation is that it speeds up online processing, which lowers computation time spent on resource allocation. When it comes to using AI techniques for resource allocation, there are basically two approaches.

35.7.1 Supervised Learning for Resource Allocation Strategy

Using this method, resource allocation techniques are coordinated with supervised learning. In order to do this, supervised learning is used on a large number of examples that are obtained via solving conventional optimisation models. Next, the resource allocation strategy of the conventional model is analysed using the supervised learning technique to identify the underlying patterns. Supervised learning training requires a lot of computational power, although it may be done offline, especially in data centres with plenty of processing power. The trained model computes result quickly in an online environment [29].

35.7.2 *Optical Network Routing Assignment*

The assignment of optical connections to optical channels in optical networks is essential during higher transmission. Conventional algorithms such as Shortest Path First (SPF) frequently lead to highly increased blocking rates and poor use of network resources. Large-scale networks may not be able to use heuristic routing assignment due to its high processing complexity, notwithstanding its effectiveness [30–32].

35.7.3 *Supervised Learning-Based Routing*

Route assignment is handled by classification or regression techniques. Precomputed route assignments and traffic statistics are included in the training dataset. In software-defined optical networks, Troia et al. presented a classification-based routing decision process for allocating the best routes to meet the demands of the flow of traffic at that moment. REST APIs are used to gather traffic requests for machine learning-driven routing calculations. Traffic matrices are then categorised in order to determine the best route assignments [33]. The categorization outcomes match the optimal routing solutions that have already been calculated. Optimal routing sets are produced using the machine learning method logistic regression, which is taught using a variety of traffic data. With this method, network reconfiguration is possible in real time and results are obtained quickly—usually in 80 ms [34].

35.7.4 *Intra-Domain Learning with LSTM*

Long Short-Term Memory (LSTM) is used to infer routing patterns from previous route datasets in multi-domain network environments, where controllers may not have access to all intra-domain network information because of security restrictions. Without requiring intra-domain data, the LSTM-based model is able to construct traffic paths that are practical. The route sets that make up the supervised output of the LSTM are precomputed using backward recursive PCE-based computation (BRPC). When computing routes between domains, this technique eliminates the necessity for intra-domain information [35].

35.7.5 *Reinforcement Learning for Decision-Making*

Decision-making activities can be used to frame routing difficulties. In these cases, learning models interact with the network to determine the best rules depending on certain network states and are rewarded for their efforts. Optimal paths inside optical networks may be found by applying reinforcement learning, which is frequently used for decision-making tasks. Optical Burst Switching (OBS) networks present a concept based on reinforcement learning. This method uses reinforcement learning agents on every node, which are each outfitted with a Q-table that keeps pairs of neighbours and destinations together with their associated Q-values. In a reinforcement learning architecture, this Q-table is crucial for choosing the traffic's next hop. This RL-based technique significantly outperforms conventional shortest path-based techniques [36, 37].

Routing and Spectrum Assignment (RSA) is the phrase used to describe the same activity in Elastic Optical Networks (AEON). The assignment of transmission wavelengths and routing channels to distinct traffic flows is the focus of RWA. In contrast, the goal of RSA is to set up optical channels for traffic transmission that have several spectrum time slots. Allocating spectrum guarantees that the time slots in the spectrum are assigned to the correct optical pathways upon request. Moreover, modulation types may be used for resource distribution in AEON depending on variables like perceived signal quality and transmission distance. RMSA is a common term used to describe this combined resource allocation. Notably, the necessary data rate is accommodated by EON's flexible spectrum distribution. Network slicing and virtual network embedding both benefit from the use of RSA and RWA. It's crucial to remember that RSA and RWA problems are both NP-complete.

35.7.6 Supervised Learning and Integer Linear Programming (ILP) for RWA

The ideal RWA scheme is the result of treating RWA as a classification assignment. This categorization issue is resolved by using logistic regression techniques and Deep Neural Networks (DNNs) [38]. Additionally, the precomputation of ideal RWA schemes is done by Integer Linear Programming (ILP). The input characteristics consist of traffic matrices, connection capacity, wavelength utilisation, network topology, and other network status information. The ideal RWA setup is the expected result. The ideal RWA solutions and the network state are utilised for training the DNN and logistic regression models.

Machine Learning-Based RWA with Physical Constraints

In a method for machine learning-based RWA that considers physical disabilities is presented. In order to keep traffic from being rejected, this technique makes sure that the Quality of Transmission (QoT) and wavelength continuity constraints are satisfied. In particular, a traffic flow's wavelength needs to be constant along its course, and the connected lightpath's QoT needs to fulfil performance standards in order to be used [39]. This technique calculates a lightpath's QoT while accounting for different impairments and determines if it meets the necessary level to qualify the lightpath for usage in the future.

Deep Reinforcement Learning for Online RMSA

Optimal online RMSA schemes inside aeons are learned while processing each light path request for optical RMSA. Taking into account the distinct features of the RMSA issue, two training techniques for Deep RMSA are suggested in order to train deep Q-learning. The better blocking rate performance of Deep RMSA over heuristic methods is illustrated by numerical findings and diagrammatically in Fig. 35.4.

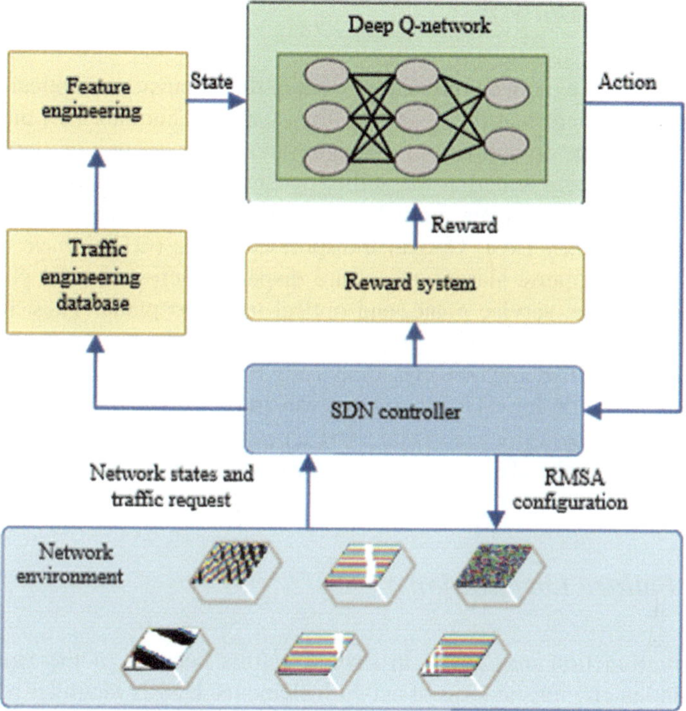

Fig. 35.4 Deep RNN Architecture

35.8 AI Based Optical Network Failure Management

35.8.1 Network Scale and Alarm Analysis

The tasks of identifying failures and evaluating a high volume of alarms become critical due to scaling of devices. Optimising AI methods can help comprehend the complex and nonlinear relationships among localization data, failure classifications, and failure features. Consequently, failure recovery is made easier with the help of this failure information [40].

35.8.2 Alarm Analysis

Alarms have become more common as a result of the expansion of optical networks; in a given week, there may be over a million alarms. There are two main factors that contribute to the complexity of managing this sheer volume of alarms. A true network fault, to begin with, has the ability to set off multiple alarms, which could lead to redundancy. Moreover, a notable incidence of false alarms unconnected to particular defects may exist. Optical transport networks (OTNs) have a complex alarm landscape because alarm sources are dispersed across several planes, such as the control plane, service plane, and optical transport plane. Thus, it becomes imperative to use effective tools to preprocess this large amount of alarm data. Alarm attributes are quantified and then classified using the K-means algorithm as part of a preprocessing method for OTN alarms that was implemented [41]. The importance of the alarm attributes is then assessed using a neural network (NN) algorithm. This method compresses the alarm data by giving each alarm an importance score.

35.8.3 Failure Localization

A single malfunction may result in several alarms because of the vast network topology and interconnectedness of network elements. DNEN identifies real failures by extracting deep, hidden failure characteristics. It starts with a set of neural networks and keeps creating new ones by crossover and mutation until the model satisfies the required level of accuracy. With the least amount of computational complexity, the DNEN-based approach provides the best localization accuracy, according to experimental results using actual optical network data [42].

35.8.4 Hybrid Failure Localization Method

A hybrid method for localising multiple failures in radio and optical wireless networks using a Hopfield Neural Network (HNN) was presented by Yang et al. This technique identifies the precise failure nodes and links by analysing alarms, network topology, and service data. The bipartite graph is established by modelling the relationship between detected failures and alarms. HNN's fast processing speed is leveraged to solve combinatorial optimisation problems by expediting the computation of the bipartite graph [43–45]. Computation time is reduced in the training process of the HNN-based approach thanks to parallel computing [46, 47].

35.8.5 Data Visualization for Failure Localization

K-means clustering aids in the improvement of human operational capabilities through the use of data visualisation in failure localization. K-means algorithm cluster paths based on features such as the maximum Bit Error Rate (BER) and BER trends. Operators can identify the source of performance degradation in each optical link by using different colours to distinguish different centroids within these clusters [48–50].

35.9 Conclusion

A massive exploration on the software defined networks incorporated with optical network terminology is done. This chapter will definitely provide a deep insight on the Software defined radio optical network. Various AI algorithm is reviewed and tried to link with the upcoming SRON techniques. The SDN architecture is clearly explained and the growth in technology is highlighted.

Over time, AI will increasingly enable networks to continually learn, self-optimize, and even predict and rectify service degradations before they occur. AI-SDOWN can parse through thousands of network devices to verify that all devices have the latest software image and look for potential vulnerabilities in device configuration.

References

1. Yang, H., Wang, B.H., Yao, Q.Y., et al.: Efficient hybrid multi-faults location based on hopfield neural network in 5G coexisting radio and optical wireless networks. IEEE Trans. Cogn. Commun. Netw. **5**, 1218–1228 (2019)
2. Li, B.J., Lu, W., Zhu, Z.Q.: Deep-NFVOrch: leveraging deep reinforcement learning to achieve adaptive vNF service chaining in DCI-EONs. J. Opt. Commun. Netw. **12**, A18 (2020)
3. Wang, D.S., Lou, L.Q., Zhang, M., et al.: Dealing with alarms in optical networks using an intelligent system. IEEE Access **7**, 97760–97770 (2019)
4. Gupta, A., Jha, R.K.: A survey of 5G network: architecture and emerging technologies. IEEE Access **3**, 1206–1232 (2015)
5. Thyagaturu, A.S., Mercian, A., McGarry, M.P., et al.: Software defined optical networks (SDONs): a comprehensive survey. IEEE Commun. Surv. Tut. **18**, 2738–2786 (2016)
6. Dvalos, J.E., Barn, B.: A survey on algorithmic aspects of virtual optical network embedding for cloud networks. IEEE Access **6**, 20893–20906 (2018)
7. Gu, R.T., Zhang, S.Z., Ji, Y.F., et al.: Network slicing and efficient ONU migration for reliable communications in converged vehicular and fixed access network. Vehicular Commun. **11**, 57–67 (2018)
8. Jin, W.Q., Gu, R.T., Tan, Y.X., et al.: Proactive grooming with delay optimization in sliceable elastic optical network. IEEE Access **7**, 105030–105040 (2019)
9. Chen, X.L., Proietti, R., Ben Yoo, S.J.: Building autonomic elastic optical networks with deep reinforcement learning. IEEE Commun. Mag. **57**, 20–26 (2019)
10. Chen, X.L., Li, B.J., Proietti, R., et al.: DeepRMSA: a deep reinforcement learning framework for routing, modulation and spectrum assignment in elastic optical networks. J. Lightw. Technol. **37**, 4155–4163 (2019)
11. Schmidhuber, J.: Deep learning in neural networks: an overview. Neural Netw. **61**, 85–117 (2015)
12. LeCun, Y., Bengio, Y., Hinton, G.: Deep learning. Nature **521**, 436–444 (2015)
13. Martin, I., Troia, S., Hernandez, J.A., et al.: Machine learning-based routing and wavelength assignment in software defined optical networks. IEEE Trans. Netw. Serv. Manage. **16**, 871–883 (2019)
14. Wang, D.S., Wang, M.Y., Zhang, M., et al.: Cost-effective and data size-adaptive OPM at intermediated node using convolutional neural network-based image processor. Opt. Express **27**, 9403–9941 (2019)
15. Choi, H.Y., Takushima, Y., Chung, Y.C.: Optical performance monitoring technique using asynchronous amplitude and phase histograms. Opt. Express **17**, 23953–23958 (2009)
16. Ji, Y.F., Zhang, J.W., Xiao, Y.M., et al.: 5G flexible optical transport networks with large-capacity, low-latency and high-efficiency. China Commun. **16**, 19–32 (2019)
17. Tao, Z.N., Dou, L., Yan, W.Z., et al.: Multiplier-free intra channel nonlinearity compensating algorithm operating at symbol rate. J. Lightw. Technol. **29**, 2570–2576 (2011)
18. Rottondi, C., Barletta, L., Giusti, A., et al.: Machine-learning method for quality of transmission prediction of unestablished light paths. J. Opt. Commun. Netw. **10**, 286–297 (2018)
19. Morais, R.M., Pedro, J.: Machine learning models for estimating quality of transmission in DWDM networks. J. Opt. Commun. Netw. **10**, 84–99 (2018)
20. Mata, J., de Miguel, I., Duran, R.J., et al.: Artificial intelligence (AI) methods in optical networks: a comprehensive survey. Opt. Switch. Netw. **28**, 43–57 (2018)
21. Ip, E., Kahn, J.M.: Compensation of dispersion and nonlinear impairments using digital backpropagation. J. Lightw. Technol. **26**, 3416–3425 (2008)
22. Li, Z., Jian, Z., Cheng, L., et al.: Signed chromatic dispersion monitoring of 100Gbit/s CS-RZ DQPSK signal by evaluating the asymmetry ratio of delay tap sampling. Opt. Express **18**, 3149–3157 (2010)
23. Luis, R.S., Teixeira, A., Monteiro, P.: Optical signal-to-noise ratio estimation using reference asynchronous histograms. J. Lightw. Technol. **27**, 731–743 (2009)

24. Kozicki, B., Takuya, O., Hidehiko, T.: Optical performance monitoring of phase-modulated signals using asynchronous amplitude histogram analysis. J. Lightw. Technol. **26**, 1353–1361 (2008)
25. Dong, Z.H., Khan, F.N., Sui, Q., et al.: Optical performance monitoring: a review of current and future technologies. J. Lightw. Technol. **34**, 2 (2016)
26. Shariati, B., Ruiz, M., Comellas, J., et al.: Learning from the optical spectrum: failure detection and identification. J. Lightw. Technol. **37**, 433–440 (2019)
27. Khan, F.N., Yu, Y., Tan, M.C., et al.: Experimental demonstration of joint OSNR monitoring and modulation format identification using asynchronous single channel sampling. Opt. Express **23**, 30337–30346 (2015)
28. Wang, D.S., Zhang, M., Li, Z., et al.: Modulation format recognition and OSNR estimation using CNN-based deep learning. IEEE Photon. Technol. Lett. **29**, 19 (2017)
29. Wang, D.S., Zhang, M., Li, J., et al.: Intelligent constellation diagram analyzer using convolutional neural network-based deep learning. Opt. Express **25**, 17150–17166 (2017)
30. Ji, Y.F., et al.: Sci. China Inf. Sci. **63**, 160301, 22 (2020)
31. Li, J., Zhang, M., Wang, D.S., et al.: Joint atmospheric turbulence detection and adaptive demodulation technique using the CNN for the OAM-FSO communication. Opt. Express **26**, 10494–10508 (2018)
32. Shen, S.R.T., Meng, K., Lau, P.T.A., et al.: Optical performance monitoring using artificial neural network trained with asynchronous amplitude histograms. IEEE Photon. Technol. Lett. **22**, 1665–1667 (2010)
33. Panayiotou, T., Chatzis, S.P., Ellinas, G.: Performance analysis of a data-driven quality-of-transmission decision approach on a dynamic multicast-capable metro optical network. J. Opt. Commun. Netw. **9**, 98–108 (2017)
34. Proietti, R., Chen, X.L., Zhang, K.Q., et al.: Experimental demonstration of machine-learning-aided QoT estimation in multi-domain elastic optical networks with alien wavelengths. J. Opt. Commun. Netw. **11**, 1–10 (2019)
35. Wang, Z.L., Zhang, M., Wang, D.S., et al.: Failure prediction using machine learning and time series in optical network. Opt. Express **25**, 18553–18565 (2017)
36. Wen, B., Shenai, R., Sivalingam, K.: Routing, wavelength and time-slot-assignment algorithms for wavelength-routed optical WDM/TDM networks. J. Lightw. Technol. **23**, 2598–2609 (2005)
37. Furdek, M., Natalino, C., Schiano, M., et al.: Experiment-based detection of service disruption attacks in optical networks using data analytics and unsupervised learning. In: Proceedings of SPIE (2019)
38. Sartzetakis, I., Christodoulopoulos, K.K., Varvarigos, E.M.: Accurate quality of transmission estimation with machine learning. J. Opt. Commun. Netw. **11**, 140–150 (2019)
39. Leung, H.C., Leung, C.S., Wong, E.W.M., et al.: Extreme learning machine for estimating blocking probability of buffer less OBS/OPS networks. J. Opt. Commun. Netw. **9**, 682–692 (2017)
40. Li, J., Wang, D., Zhang, M.: Low-complexity adaptive chromatic dispersion estimation scheme using machine learning for coherent long-reach passive optical networks. IEEE Photon J. **11**, 1–11 (2019)
41. Singh, S.K., Jukan, A.: Machine-learning-based prediction for resource (re)allocation in optical data centre networks. J. Opt. Commun. Netw. **10**, D12 (2018)
42. Caballero, F.J.V., Ives, D.J., Laperle, C., et al.: Machine learning based linear and nonlinear noise estimation. J. Opt. Commun. Netw. **10**, 42–51 (2018)
43. Zhuge, Q.B., Zeng, X.B., Lun, H.Z., et al.: Application of machine learning in fiber nonlinearity modelling and monitoring for elastic optical networks. J. Lightw. Technol. **37**, 3055–3063 (2019)
44. Rafique, D., Velasco, L.: Machine learning for network automation: overview, architecture, and applications. J. Opt. Commun. Netw. **10**, 126 (2018)
45. Wang, D.S., Zhang, M., Fu, M., et al.: Nonlinearity mitigation using a machine learning detector based on k-nearest neighbours. IEEE Photon. Technol. Lett. **28**, 19 (2016)

46. Cote, D.: Using machine learning in communication networks. J. Opt. Commun. Netw. **10**, 100–109 (2018)
47. Rafique, D., Szyrkowiec, T., Grieser, H., et al.: Cognitive assurance architecture for optical network fault management. J. Lightw. Technol. **36**, 1443–1450 (2018)
48. Cavazzoni, C., Barosco, V., D'Alessandro, A., et al.: The IP/MPLS over ASON/GMPLS test bed of the IST project LION. J. Lightw. Technol. **21**, 2791–2803 (2003)
49. Christodoulopoulos, K., Manousakis, K., Varvarigos, E.: Offline routing and wavelength assignment in transparent WDM networks. IEEE/ACM Trans. Netw. **18**, 1557–1570 (2010)
50. Berthold, J.E., Ong, L.Y.: Next-generation optical network architecture and multidomain issues. Proc. IEEE **100**, 1130–1139 (2012)

Chapter 36
Enhancing Optical MIMO System Performance Through Optical Multiplexing and Amplification

Nourddine Elhajrat, Fouad Essahlaoui, Omar Eloutassi, and Mohammed Halimi

36.1 Introduction

Currently, the telecommunications sector grapples with significant challenges, notably security concerns and an escalating need for high-speed services and technologies like Internet connectivity, as well as audio and video transmission, resulting in data rates reaching several gigahertz per second [1–4]. Furthermore, there's been a pressing need to enhance transmission network capacities to support internet-based voice and video protocols for a large user base. Effectively addressing these demands now requires deploying communication networks spanning several kilometers while maintaining high-speed interfaces [5, 6]. Nevertheless, conventional wired infrastructures, including copper coaxial cables and twisted pair cables, display significant signal attenuation, typically around 20 dB per 100 m.

In the field of telecommunication technology, optical communication networks relying on optical fibers have become crucial. These fibers offer multiple advantages compared to coaxial cables and conventional wireless links [7, 8], including enhanced transmission security, minimal signal loss, vast bandwidth capacity, reduced susceptibility to electromagnetic interference, and cost-effectiveness. Among these advantages, the benefits of heightened transmission security, low signal loss, extensive

N. Elhajrat (✉) · M. Halimi
Optoelectronics and Energetic Techniques Applied (OETA) Team, FST, University Moulay Ismail, B.P. 509, Boutalamine, Errachidia, Morocco
e-mail: n.elhajrat@edu.umi.ac.ma

F. Essahlaoui
LRETTI: Laboratory of Renewable Energies, Treatment and Transmission of Information, Faculty of Science and Technology, Moulay Ismail University, Errachidia, Morocco

O. Eloutassi
Materiaux et Energie Renouvelable Team, FS, University Moulay Ismail, Meknes, Morocco

M. El Ghzaoui et al. (eds.), *Next Generation Wireless Communication*, Signals and Communication Technology, https://doi.org/10.1007/978-3-031-56144-3_36

bandwidth, reduced electromagnetic interference, and economical deployment are particularly notable. Furthermore, multimode fibers (MMF) present unique advantages over single-mode fibers (SMF) due to their larger core diameter, enabling the transport of light through multiple propagation modes and simplifying the integration of optoelectronic transmitters and receivers [9]. Graded-index multimode fibers characterized by a 62.5 μm core diameter and a 2% refractive index normalization between the fiber core and the cladding have gained widespread usage in local area networks due to their high bandwidth capabilities [10].

From the viewpoint of information theory, multimode fiber MMF, in particular graded-index fiber GIMMF, offers very high transmission capacity by exiting several modes inside the fiber cores. For maximize this bandwidth, employing optical multiplexing techniques becomes crucial. Among these methods, spatial multiplexing within the wavelength-division multiplexing (WDM) framework stands out [10–12]. This approach not only enhances network capacity but also optimizes bandwidth usage. Practically, WDM enables high-speed data transmission from individual users over a single optical fiber, each utilizing specific wavelengths [13]. However, it presents challenges such as accurate long-distance data transmission, generating suitable optical carriers, and increased complexity in transmission systems employing multimode fibers MMF.

Indeed, Optical Time Division Multiplexing technique (OTDM) has been a significant development in optical communication, particularly in the 1990s and beyond [14]. The development and application of OTDM have played a pivotal role in expanding the capabilities of optical communication, allowing for the efficient and high-speed transmission of data over optical fiber networks. This technology has been a cornerstone in meeting the ever-increasing demands for bandwidth and data capacity in modern telecommunications [15–17].

Spatial multiplexing is a communications technique that involves the use of several independent transmission channels on the same physical link. This helps increase transmission capacity and improve network performance by allowing multiple data streams to be transmitted simultaneously without significant interference between them. Spatial multiplexing is commonly used in wireless communication technologies, optical networks and other communication fields to optimize resource utilization and increase network efficiency. In this context, the MGDM technique is a recent spatial multiplexing technique. Indeed, spatial multiplexing is a communication technique that involves the use of several independent transmission channels on the same physical link, exploiting spatial aspects such as the direction or physical position of the signals. This increases transmission capacity and improves network performance by enabling the simultaneous transmission of several data streams without significant interference between them. Spatial multiplexing is commonly used in wireless communication technologies, optical networks and other areas of communication to optimize resource utilization and increase network efficiency [18].

Moreover, several optical amplifiers have been developed, including erbium-doped fiber amplifiers EDFA, solid-state optical amplifiers SOA and Raman amplifiers [19, 20]. Notably, since its introduction in 1987, EDFA has been widely used to amplify optical signals in fiber-optic telecommunications networks operating in the 1550 nm wavelength range, offering high gain with low noise. This preference is due to its notable features, such as high gain, saturation power, and minimal noise.

The present chapter introduces an inventive optical transmission system architecture employing a 3 × 3 Multiple-Input Multiple-Output setup, utilizing the spatial multiplexing technique known as MGDM. This design achieves a noteworthy data transmission rate of 3.6 Gb/s across 3 subchannels, spanning a distance of 2000 m using multimode fiber (MMF). Additionally, an enhanced version of this architecture integrates two erbium-doped fiber amplifiers to alleviate interference between different transmission modes within the MMF, resulting in significant transmission performance improvements. The proposed system offers high data rates suitable for extensive MMF-based networks, effectively covering distances of up to 2000 m while minimizing chromatic and modal dispersion.

36.2 Optical Communication System

The MGDM (Mode Group Diversity Multiplexing) transmission system is carefully designed to optimize data transmission in optical communication networks. It centers on deploying N lasers, known as optical transmitters, at the input. At the output, there are M photodiodes functioning as receivers. Crucially, in this setup, the number of receivers, denoted as M, must be equal to or greater than the number of transmitters, denoted as N ($M \geq N$), as illustrated in Fig. 36.1.

At the input side of the MGDM system, each optical transmitter plays a crucial role by activating a specific group of GM modes within the multimode fiber (MMF). These modes are responsible for carrying the data signals, and what sets them apart is the introduction of data with distinct spatial offsets. Essentially, this means that each transmitter is transmitting data with a unique signature, a spatial fingerprint if you will, within the MMF.

In a Multimode Graded-Index Fiber (MMF) system context, the incident field at the MMF's input face is characterized as a Gaussian field, which is defined by four essential parameters. These parameters play a crucial role in determining how the

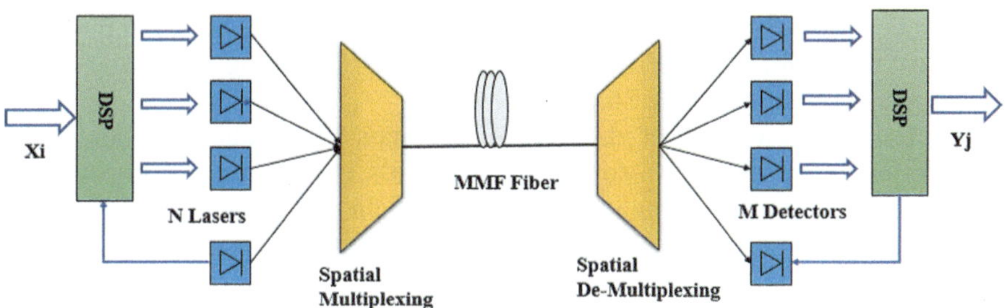

Fig. 36.1 Representation of the elements and process in an optical communication transmission chain

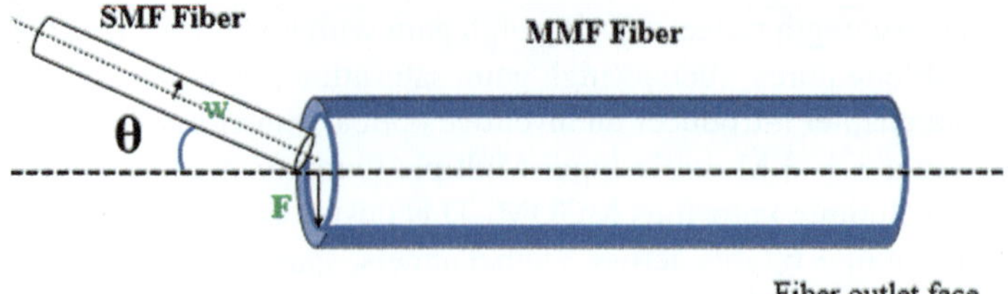

Fig. 36.2 Synoptic diagram of power input parameters

Table 36.1 Input and output configurations in a (3 × 3)-MGDM system

Symbol	Parameter	Value	Unit
θ	Angular offset	4	(°)
F	Radial offset	0–13–26	μm
w	Spot size	4	μm
λ	Wavelength	1550	nm

modes within the MMF are excited, and they can be enumerated as: Wavelength λ Radial offset F, Angular offset θ, and Spot size w [2, 18].

Figure 36.2 provides a synoptic diagram illustrating the critical power injection parameters at the input facet of the optical fiber (Spot size w, Radial offset F, Angular offset θ). This diagram is specifically designed for the implementation of MGDM (Mode Group Diversity Multiplexing) spatial multiplexing, a technique that enhances optical communication.

The parameters carefully chosen for spatial detection and injection phases were employed to uphold the transmission quality ratio in each subsystem. These specific parameters are organized and detailed in Table 36.1.

36.3 Design of Optical Communication Systems

To enhance MMF communication performance and guarantee high optical transmission quality over long distances, a (3 × 3)-MIMO architecture has been proposed, using on the spatial multiplexing technology and exploiting input and output parameters described in Table 36.1.

For transmit a variety of data types, encompassing audio, text and video other forms of information, a sophisticated process is demonstrated in Fig. 36.2. This process involves manipulating the emitted light wave from a LASER diode by generating pulsed electrical signals through an external Mach–Zehnder MZN modulator.

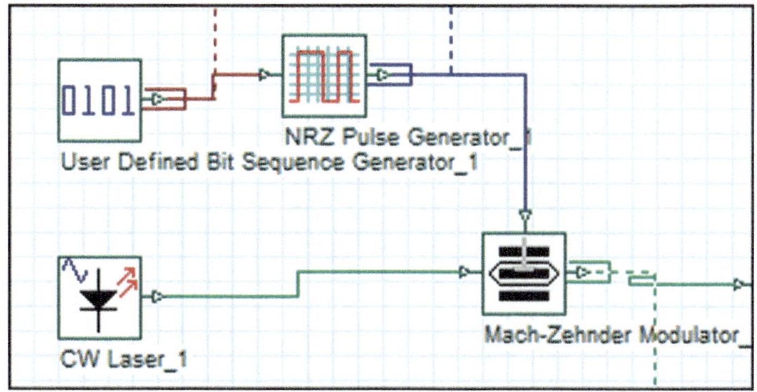

Fig. 36.3 Data transmission using pulsed electrical signals and Mach–Zehnder modulator

- **LASER Diode Emission:** The journey begins with the LASER diode, depicted in the figure. This diode serves as the source of the optical signal, emitting a continuous wave of light.
- **Pulsed Electrical Signals:** To convey digital data, pulsed electrical signals are generated externally. These signals represent the binary information that encode the digital data. The timing and duration of these pulses determine the data being transmitted.
- **Mach–Zehnder MZN Modulator:** A crucial component of this process is the Mach–Zehnder MZN modulator, prominently displayed in the diagram. This external modulator takes the pulsed electrical signals as input and uses them to modulate the intensity of the LASER diode's emitted light wave.
- **Light Modulation:** As the pulsed electrical signals are applied to the modulator, they exert control over the light wave. Depending on the signal's state, the intensity of the light wave varies accordingly. This modulation of light intensity encodes the digital data onto the optical signal (Fig. 36.3).

This depiction highlights a transmission part where entire bit sequences, functioning at a data rate of 1.2Gbits/s, were produced and subsequently transmitted via 3 LASER diodes situated at the input facet of an MMF fiber.

The sequences were generated and transmitted using the OptiSystem simulation software, with the signal power range at the MMF input observed, ranging from a maximum amplitude of 1.4957 dBm to a minimum of − 31.504 dBm, as shown in Fig. 36.4.

In reception phase of the transmission chain, a complex process takes place to extract, filter, and analyze the signals transmitted from the output facet of the multimode fiber MMF, as shown in Fig. 36.5.

This process is illustrated in Fig. 36.4 and encompasses the following key steps:

- **PIN-Type Photodetectors:** At the termination point of the multimode fiber (MMF), three PIN-type photodetectors are strategically positioned to capture the

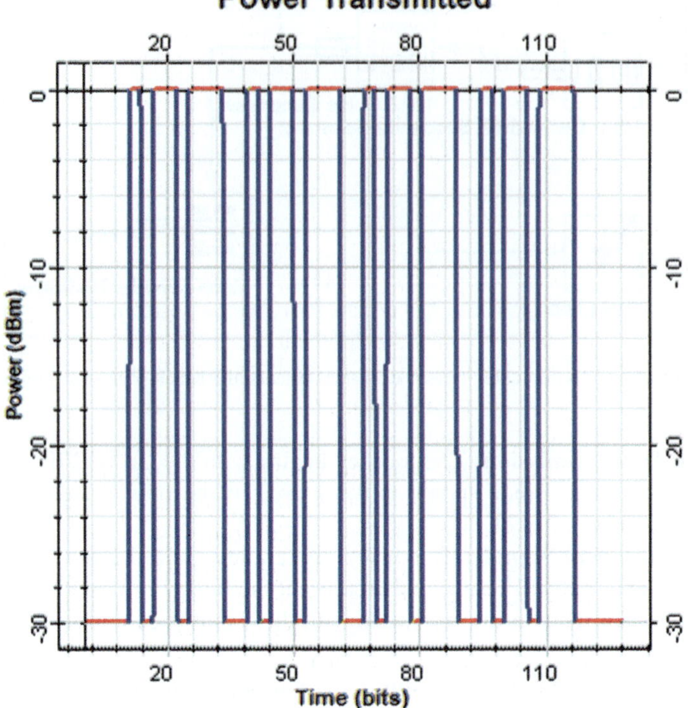

Fig. 36.4 Signal power transmitted in dB

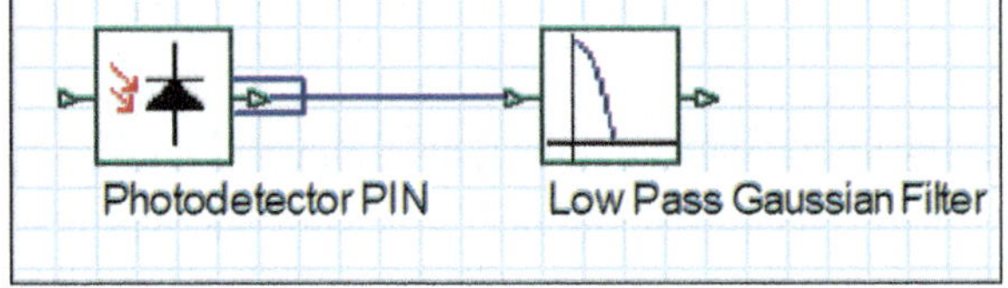

Fig. 36.5 Reception and Signal Processing

incoming optical signals. Each photodetector is dedicated to detecting one of the three transmitted signals.

- **Bandpass Filters:** The detected optical signals are next routed through bandpass filters. These filters serve the essential function of isolating specific frequency components within the detected signals. By doing so, they effectively separate and enhance the signal of interest for further processing.
- **Baseband Signal Generation:** Following the bandpass filtration, the processed signals are now commonly referred to as baseband signals. To prepare these baseband signals for analysis and decoding, they undergo an additional filtering stage. Specifically, each subchannel's baseband signal is directed through a dedicated

low-pass filter. The specific cutoff frequency of these low-pass filters is adaptive, dependent upon the input data rate. This adaptability ensures that the filtered signals remain within the desired frequency range.

- **Signal Analysis:** Once the baseband signals have been filtered and conditioned, they are ready for analysis. These signals are subjected to comprehensive examination and evaluation to determine their quality and integrity. The performance assessment aims to assess the overall functionality of the reception system.

36.4 Performance Analysis of a (3 × 3) MIMO System

36.4.1 Long-Haul Optical Communication Chain

This section conducts an extensive performance analysis of a three-input, three-output optical transmission chain, employing identical input and output spatial parameters utilized in the Mode Group Diversity Multiplexing MGDM technique.

In addition, this chain, based on a GI-MMF (Gradient-Index Multimode Fiber) with a length of $L = 2000$ m, has been thoroughly evaluated.

Figure 36.6 illustrates the proposed configuration, providing a visual view of the key components and signal paths within the system. The main objective of this performance study is to obtain valuable information on the suitability of this MGDM transmission chain in different optical communication scenarios.

Figure 36.7 demonstrates the fluctuation in signal strength observed at the output of the 2000-m transmission chain. In fact, amplitude serves as a measure of signal intensity or strength. In this context, the minimum amplitude represents its lowest power intensity, while the maximum amplitude indicates the highest power intensity

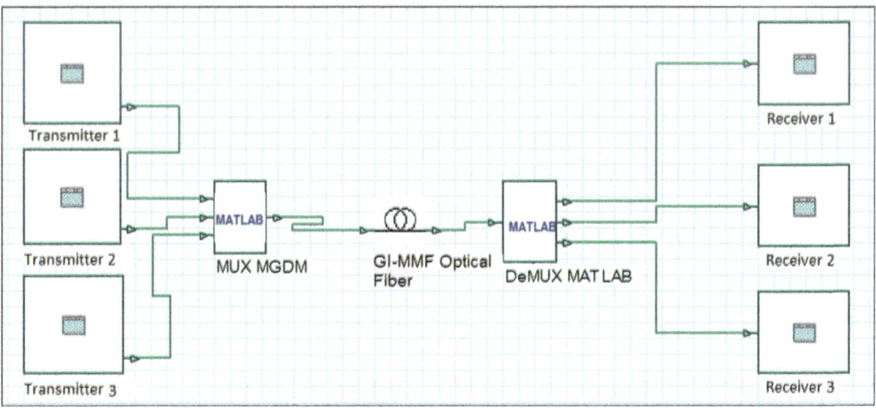

Fig. 36.6 Conception of a (3 × 3) MGDM system in an optical transmission chain

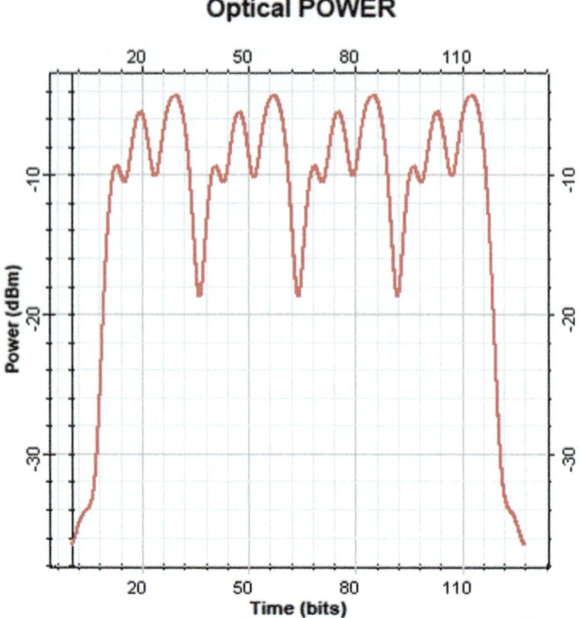

Fig. 36.7 Output signal power for a 2000 m distance chain

of the output signal. These power levels are commonly quantified in dBm (decibels per milliwatt), reflecting signal power relative to a reference value of 1 milliwatt.

The measurements presented in Fig. 36.7 reveal significant insights into the optical power characteristics of the system. The maximum optical power, approximately − 2.7291 dBm (0.00038658 watts), signifies the peak power level achieved within the system, while the minimum optical power, roughly − 38.157 dBm (1.8176e−5 watts), represents the lowest detected power. This range of optical power levels underscores the dynamic nature of signal strength within the system, with the maximum denoting strong signal intensity and the minimum indicating weak signal levels. In conclusion, these measurements indicate that the performance of the 2000-m optical communication chain is compromised. The observed increase in modal dispersions and the significant signal attenuations along the communication chain contribute to this poor performance. Modal dispersions can lead to signal distortion and degradation over longer distances, while signal attenuation results in a reduction of signal strength, both of which impact the overall quality and reliability of optical communication.

36.4.2 Three Optical Communication Subchains

This section presents a solution aimed at overcoming the challenges associated with modal and chromatic dispersion in the communication chain and effect of interference

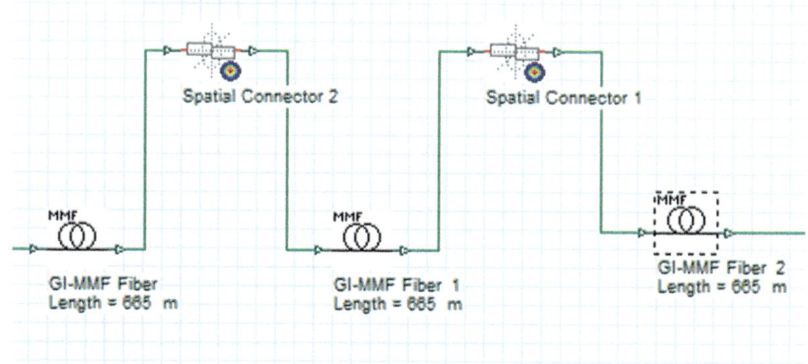

Fig. 36.8 Proposed modification to the (3 × 3)-MGDM system

present between the different modes along the MMF fiber. Our proposed approach entails dividing the communication chain into three distinct sub-chains, each with manageable and acceptable distances. This segmentation strategy empowers us to exercise better control over dispersion effects, leading to more effective mitigation. The ultimate goal is to enhance the overall performance and reliability of the optical communication system. The visual representation of this proposed architecture is depicted in Fig. 36.8, providing a clear illustration of our innovative approach.

To divide the communication chain into three sub-chains, each with a length of 665 m, we employed two space connectors strategically placed to regenerate the optical signal injection parameters. This approach was implemented to ensure that each subchain remains within manageable and acceptable distances.

Figure 36.9 illustrates the variation in signal strength at the output of the modified transmission chain, spanning a total distance of 2000 m (split into three segments of 665 m each). Amplitude, within this context, serves as a measure to gauge the intensity or potency of the signal. The maximum amplitude represents the peak intensity of the output signal power, while the minimum amplitude indicates its lowest intensity.

Measurements presented in Fig. 36.9 reveal significant insights into the optical power characteristics of the system. The maximum optical power, approximately − 3.3858 dBm (0.00033157 watts), signifies the peak power level achieved within the system, while the minimum optical power, roughly − 39.032 dBm (− 1.5599e−5 watts), represents the lowest detected power. This range of optical power levels underscores the dynamic nature of signal strength within the system, with the maximum denoting strong signal intensity and the minimum indicating weak signal levels.

In conclusion, the measurements reveal a notable improvement in both signal quality and the performance of the modified optical communication chain, which spans a total length of 2000 m. This improvement is attributed to the mitigation of modal dispersions and the reduction in significant signal attenuations that were observed in the initial configuration. While the decomposition of the main chain has

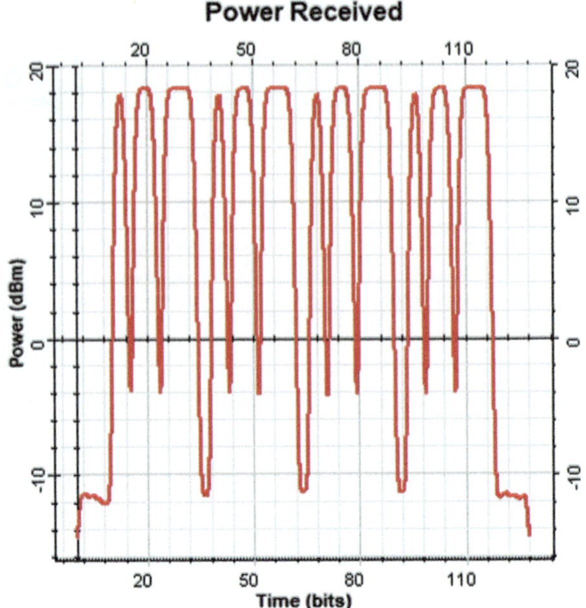

Fig. 36.9 Output signal power for (3 × 665 m) distance chain

yielded positive results, it's important to note that the system still exhibits weaknesses related to optical signal attenuations and the potential for interferences during transmission.

36.4.3 Assessment of Optical Amplification in the Proposed System

For enhancement of transmission quality within the proposed system requires the mitigation of interference between various modes during optical fiber transmission. To address this, the EDFA optical amplifiers have been employed, as illustrated in Fig. 36.10. These EDFAs are widely utilized in long distance and high capacity optical communication systems [20], as well as in some laser systems, medical equipment, and scientific instrumentation. EDFAs are favored due to their advantages, such as low noise and the capacity to amplify a broad spectrum of wavelengths.

In this part, a comprehensive comparative analysis is outlined, highlighting the distinctions between a (3 × 3) MIMO system and its amplified counterpart across a consistent transmission chain length of (3 × 665 m).

Indeed, Fig. 36.11 illustrates the variation in signal strength at the output of the amplified transmission chain covering a total length of 2000 m (divided into three segments of 665 m each). In this scenario, the measurement of amplitude plays a critical role in assessing signal intensity or strength. Specifically, the maximum

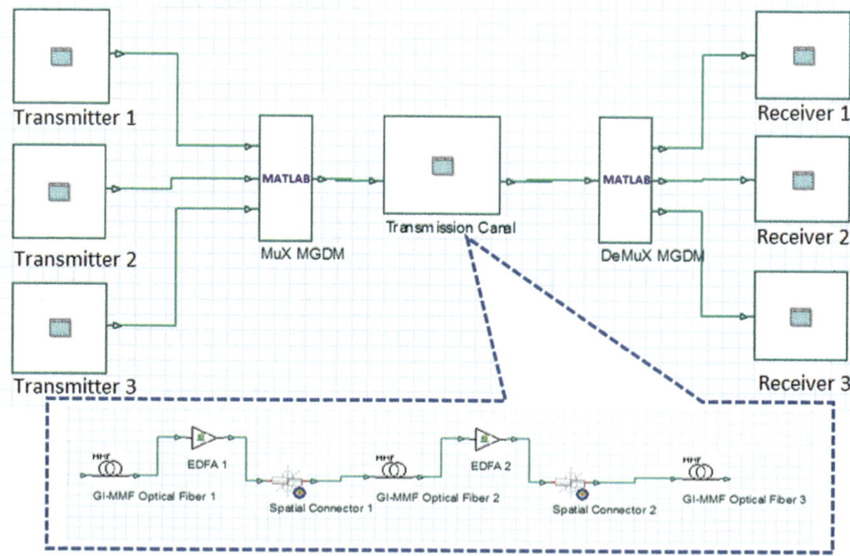

Fig. 36.10 Modified (3 × 3) MGDM system transmission chain incorporating optical amplifier

amplitude serves as an indicator of the peak intensity within the output signal power, while the minimum amplitude signifies the lowest observed intensity within the output signal power.

Measurements shown in Fig. 36.11 reveal important information about the optical power characteristics of the system. Maximum optical power, at around 18.261 dBm (0.048133 watts), represents the maximum power level achieved by the system, while minimum optical power, at around − 18.012 dBm (-0.0022678 watts). In conclusion, the results obtained from the proposed system, which incorporates two EDFA (Erbium-Doped Fiber Amplifier) optical amplifiers, demonstrate very good signal quality and excellent transmission performance for the optimized configuration with a total length of 2000 m. The inclusion of EDFA amplifiers has significantly enhanced the signal strength and overall system performance, ensuring robust and reliable optical communication over an extended distance.

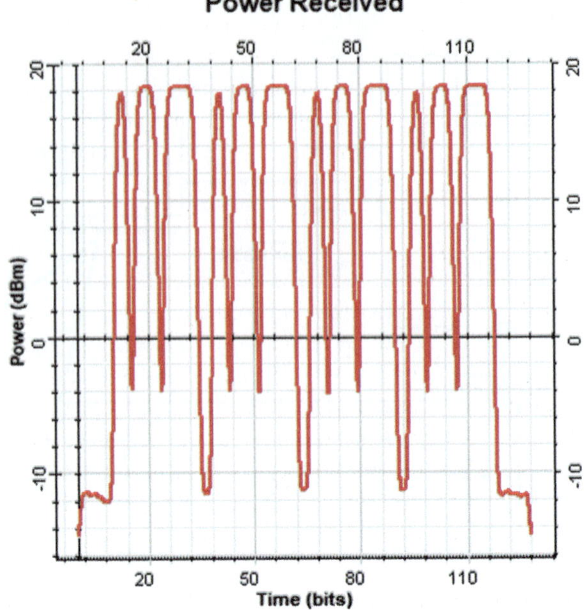

Fig. 36.11 Output signal power for (3 × 665 m) distance chain with two optical amplifiers

36.5 Conclusion

The conclusion serves as a pivotal point in underlining the remarkable benefits and the remarkable effectiveness of the proposed MGDM-MIMO optical transmission architecture. This innovative solution has successfully surmounted the intricate challenges associated with transmitting data over multi-mode optical fiber, marking a significant advancement in the field of optical communication. This distinctive architectural configuration, operating as a 3 × 3 system, is a testament to meticulous engineering. It divides the transmission into subchannels, each stretching across 665 m, collectively achieving an impressive total transmission rate of approximately 3.6 Gbit/s. Notably, each subchannel boasts a rate of 1.2 Gbit/s, showcasing the architecture's capacity to deliver high-speed data transmission. Furthermore, the strategic incorporation of erbium-doped fiber amplifiers (EDFA) adds a pivotal dimension to the system's capabilities. The genesis of this architecture is rooted in the critical need for high-performance optical communication, particularly over extended distances. In essence, it provides a viable solution for scenarios where the challenges of multi-mode optical fiber transmission once posed insurmountable barriers. In practical terms, this optimized architecture promises to redefine the landscape of optical communication. Whether deployed in long-distance network scenarios, intricate optical link configurations, or addressing the demands of high-speed optical communication across industries, it consistently delivers heightened performance, steadfast signal quality, and heightened reliability. The road ahead holds the promise of novel advancements, underpinned by the transformative capabilities of this architectural triumph in optical communication.

References

1. Elhajrat, N., Essahlaoui, F., Eloutassi, O., Halimi, M.: Towards an optical MIMO system performance improvement using MGDM multiplexing and EDFA optical amplification. Results Opt. **13**, 100526 (2023). https://doi.org/10.1016/j.rio.2023.100526
2. Elhajrat, N., Essahlaoui, F., Eloutassi, O., Abbassi, A.: Adaptation of the MIMO Architecture Over MMF Fiber Using MGDM Multiplexing (2020)
3. Li, K., Chen, X., Mishra, S.K., Hurley, J.E., Stone, J.S., Li, M.-J.: Modal delay and modal bandwidth measurements of bi-modal optical fibers through a frequency domain method. Opt. Fiber Technol. **55**, 102145 (2020). https://doi.org/10.1016/j.yofte.2020.102145
4. El Ghzaoui, M., Das, S., Lenka, T.R., Biswas, A.: Terahertz wireless communication components and system technologies. In: Terahertz Wireless Communication Components and System Technologies, pp. 1–310 (2022)
5. Krishnamoorthy, A.V., Thacker, H.D., Torudbakken, O., Muller, S., Srinivasan, A., Decker, P.J., Opheim, H., Cunningham, J.E., Shubin, I., Zheng, X., Dignum, M., Raj, K., Rongved, E., Penumatcha, R.: From chip to cloud: optical interconnects in engineered systems. J. Lightw. Technol. **35**, 3103–3115 (2017). https://doi.org/10.1109/JLT.2016.2642822
6. Chen, X., Li, K., Zakharian, A.R., Hurley, J.E., Stone, J.S., Coleman, D., Liu, J., Wu, Q., Li, M.-J.: A fiber modal adapter for upgrading 850 nm multimode fiber links to 1310 nm single-mode transmission. Opt. Fiber Technol. **56**, 102210 (2020). https://doi.org/10.1016/j.yofte.2020.102210
7. Yabre, G.: Influence of core diameter on the 3-dB bandwidth of graded-index optical fibers. J. Lightw. Technol. **18**, 668–676 (2000). https://doi.org/10.1109/50.842081
8. Sonne, A.: Designing and controlling a single-mode dual concentric core fiber with nearly zero ultra-flattened chromatic dispersion for linear or nonlinear application. Optik **231**, 166375 (2021). https://doi.org/10.1016/j.ijleo.2021.166375
9. Awad, M., Dayoub, I., Hamouda, W., Rouvaen, J.-M.: Adaptation of the mode group diversity multiplexing technique for radio signal transmission over multimode fiber. J. Opt. Commun. Netw. **3**, 1 (2011). https://doi.org/10.1364/JOCN.3.000001
10. Sim, D.H., Takushima, Y., Chung, Y.C.: High-speed multimode fiber transmission by using mode-field matched center-launching technique. J. Lightw. Technol. **27**, 1018–1026 (2009). https://doi.org/10.1109/JLT.2008.2005040
11. Panicker, R.A., Wilde, J.P., Kahn, J.M., Welch, D.F., Lyubomirsky, I.: 10$\, \times\,$10 Gb/s DWDM transmission through 2.2-km multimode fiber using adaptive optics. IEEE Photon. Technol. Lett. 19, 1154–1156 (2007) . 10.1109/LPT.2007.901437
12. Chung, H.S., Chang, S.H., Kim, K.: 6 x 86 Gb/s WDM transmission over 2 km multimode fiber using center launching technique and multi-level modulation. Opt. Express **17**, 8098 (2009). https://doi.org/10.1364/OE.17.008098
13. Bhattacharjee, R., Dey, P., Saha, A.: An improved hybrid OTDM-WDM transmission system for effective nonlinearity mitigation utilizing Ti:PPLN waveguide based OPC module. Optik **219**, 165241 (2020). https://doi.org/10.1016/j.ijleo.2020.165241
14. Lerman, L., Zehou, O., Ortonne, N., Sbidian, E., Ferkal, S., Hodel, J., Calvo-Bota, E., Blanc-Durand, P., Luciani, A., Wolkenstein, P., Itti, E.: Intérêt de la TEP/TDM au 18F-FDG dans la neurofibromatose de type 1, expérience du centre national de référence Henri-Mondor sur 10 ans. Méd. Nucléaire. **43**, 370–380 (2019). https://doi.org/10.1016/j.mednuc.2019.05.001
15. Zhu, K., Zhou, B., Wu, H., Shang, C., Lu, L., Adeel, M., Yan, Y., Zhao, Z., Tam, H.-Y., Lu, C.: Multipath distributed acoustic sensing system based on phase-sensitive optical time-domain

reflectometry with frequency division multiplexing technique. Opt. Lasers Eng. **142**, 106593 (2021). https://doi.org/10.1016/j.optlaseng.2021.106593
16. Miao, X., Bi, M., Li, L., Fu, Y., Hu, W.: Comparative investigation on 10G-class and 25G-class receivers for O-band DML-based 25/50 Gbps TDM-PON. Opt. Fiber Technol. **54**, 102105 (2020). https://doi.org/10.1016/j.yofte.2019.102105
17. Feng, N., Sun, X.: Nyquist four-level pulse amplitude modulation scheme (PAM-4) based on hierarchical modulation in IM/DD-TDM PON with hybrid equalization. Opt. Commun. **457**, (2020)
18. Elhajrat, N., Elouatssi, O., El Abbassi, A., Essahlaoui, F., Aftatah, M.: Study of optical MIMO transmission systems using the MGDM multiplexing technique. MAS **14**, 34 (2019). 10.5539/mas.v14n1p34
19. Kaur, A., Bhamrah, M.S., Atieh, A.: SOA/EDFA/RAMAN optical amplifier for DWDM systems at the edge of L & U wavelength bands. Opt. Fiber Technol. **52,** 101971 (2019). https://doi.org/10.1016/j.yofte.2019.101971
20. Akcesme, O., Yucel, M., Burunkaya, M.: The design and implementation of a software based gain control for EDFAs used in long-haul optical networks. Optik **239**, 166850 (2021). 10.1016/j.ijleo.2021.166850

Chapter 37
Fostering Advanced Optical Wireless Communication: Approaches for Addressing 5G/6G, IoT, Industry 4.0, and WLANs

R. Praveenkumar, S. D. Vijayakumar, G. Vijayakumari, and V. Kumar

37.1 Introduction

In the ever-evolving landscape of communication technology, several key concepts have emerged, each playing a crucial role in shaping the way we connect, communicate, and exchange data. This introduction provides a brief overview of these concepts as represented in Fig. 37.1, including Free-Space Optical Communication (FSO), Visible Light Communication (VLC), 5G /6G networks, the Internet of Things (IoT), Industry 4.0, and Wireless Local Area Networks (WLANs).

37.1.1 Free-Space Optical Communication (FSO)

Free-Space Optical Communication, or FSO, represents a cutting-edge approach to data transmission. Unlike traditional communication methods that rely on radio waves or wired connections, FSO leverages the power of light to transmit data through the atmosphere. Utilizing laser beams, FSO systems establish high-speed optical links between two points without the need for physical cables. This technology is particularly valuable in scenarios where traditional communication infrastructure is impractical or costly, such as urban areas with RF spectrum congestion or remote locations where laying cables is challenging.

R. Praveenkumar (✉) · S. D. Vijayakumar · G. Vijayakumari · V. Kumar
Department of ECE, Builders Engineering College, Kangeyam Taluk, Nathakadaiyur, Tirupur 638108, India
e-mail: rpraveenster@gmail.com

M. El Ghzaoui et al. (eds.), *Next Generation Wireless Communication*, Signals and Communication Technology, https://doi.org/10.1007/978-3-031-56144-3_37

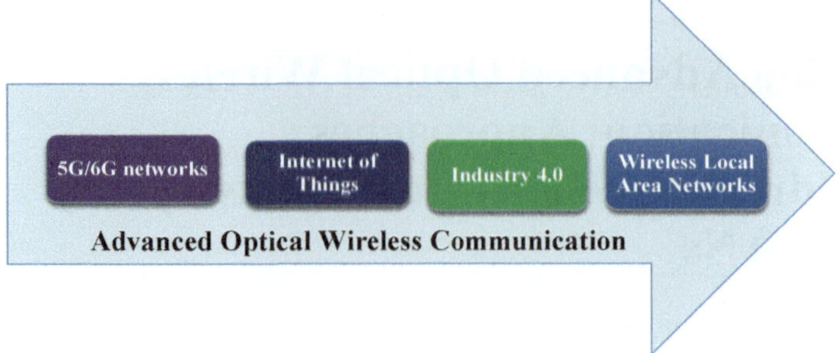

Fig. 37.1 Overview of advanced optical wireless communication

37.1.2 Visible Light Communication (VLC)

Visible Light Communication, or VLC, is a revolutionary concept that harnesses visible light to transmit data. In VLC, LED light sources are used not only to provide illumination but also as data transmitters. By modulating the intensity of light at speeds imperceptible to the human eye, VLC enables wireless communication within confined spaces like homes, offices, and industrial facilities. VLC has the potential to transform indoor connectivity, offering high-speed, energy-efficient data transmission while simultaneously providing illumination, making it a cornerstone technology for smart buildings and the Internet of Things (IoT).

37.1.3 5G /6G Networks

The evolution of wireless communication networks has reached unprecedented heights with the advent of 5G and the anticipation of 6G. These next-generation networks promise remarkable enhancements in data rates, latency, and connectivity. 5G networks are designed to accommodate the growing demand for high-bandwidth applications, from augmented reality (AR) and virtual reality (VR) to the proliferation of IoT devices. As for 6G, it's envisioned to push the boundaries even further, ushering in the era of holographic communication, advanced AI integration, and global-scale connectivity at speeds and reliability levels we can hardly imagine today.

37.1.4 Internet of Things (IoT)

The Internet of Things, or IoT, is a transformative paradigm that's reshaping industries and everyday life. IoT refers to the network of interconnected devices, sensors, and objects capable of collecting and exchanging data over the internet. These devices range from smart thermostats and wearable fitness trackers to industrial sensors and autonomous vehicles. IoT is poised to revolutionize various sectors, including healthcare, agriculture, transportation, and smart cities, by enabling real-time data monitoring, automation, and decision-making.

37.1.5 Industry 4.0

Industry 4.0, often dubbed the Fourth Industrial Revolution, represents the integration of digital technologies into manufacturing and industrial processes. It encompasses concepts like automation, artificial intelligence (AI), big data, and the Industrial Internet of Things (IIoT). Industry 4.0 aims to create smart factories and production processes that are highly efficient, flexible, and adaptive. By leveraging advanced communication technologies, including FSO and VLC, Industry 4.0 enables real-time monitoring, predictive maintenance, and seamless communication within manufacturing environments, leading to increased productivity and reduced operational costs.

37.1.6 Wireless Local Area Networks (WLANs)

Wireless Local Area Networks, or WLANs, are ubiquitous in our modern world. These networks use wireless communication technologies, typically based on the IEEE 802.11 standard, to connect devices within a limited geographical area. WLANs are commonly found in homes, offices, airports, and public spaces, providing wireless internet access to laptops, smartphones, and IoT devices. With the growing demand for faster and more reliable connectivity, WLANs continue to evolve, incorporating technologies like FSO and VLC to enhance data rates, security, and indoor navigation.

In the pages that follow, we will delve deeper into these concepts, exploring their capabilities, challenges, and the transformative impact they have on our interconnected world. Each of these technologies represents a piece of the puzzle in our quest for faster, more efficient, and more secure communication. Moreover, the subsequent portions of the content explore enhanced communication solutions:

- In Sect. 37.2, the challenges in 5G /6G, IoT, Industry 4.0, and WLANs are outlined.
- Sections 37.3 and 37.4 delve into Free-Space Optical Communication (FSO) and Visible Light Communication (VLC).
- Section 37.5 establishes the correlation between the communication process and 5G /6G networks, the Internet of Things (IoT), Industry 4.0, and Wireless Local Area Networks (WLANs).

- Sections 37.6– 37.9 address Optical Wireless Solutions through various approaches.
- Section 37.10 focuses on advanced optical technologies.
- Finally in Sect. 37.11, the document concludes with a discussion on Conclusion and Future Work.

37.2 Challenges in 5G /6G, IoT, Industry 4.0, and WLANs

The advent of 5G and the anticipated emergence of 6G necessitate the resolution of a fresh set of critical challenges, not only for network infrastructure but also to accommodate the ever-evolving needs of society. This evolving landscape calls for innovative architectural designs and technical solutions. In this document, we outline a series of prospective challenges and suggest potential technical elements to tackle these challenges in the context of forthcoming 6G technology [1]. The proliferation of IoT applications brings about significant security, ethical, privacy, and legal complexities that exert a substantial influence on our daily lives. This includes examining the concept of trust and the potential hurdles associated with smart contracts, alongside a myriad of use cases. This perspective offers valuable insights into how IoT's threats and vulnerabilities shape our experiences and interactions [2]. Industry 4.0 initiatives hold the potential to reshape entire business systems by revolutionizing the processes of product design, manufacturing, distribution, and disposal. However, Industry 4.0 remains a relatively new concept in developing nations, particularly in a country like India. There is a pressing need for a precise definition to ensure a comprehensive understanding and practical application of Industry 4.0 principles in the business context [3]. Recognizing the primary challenges associated with Industry 4.0 initiatives is crucial. This recognition serves as the foundation for a thorough analysis of these challenges, enabling a prioritization that facilitates the effective implementation of Industry 4.0 concepts within the context of supply chain sustainability in emerging economies, with a specific focus on the Indian manufacturing industry. Within the spectrum of Wireless Local Area Networks (WLANs), a unique and intricate research domain is emerging-WLANs established and managed by a community of users. These networks, often termed as Community-Based WLANs or simply wireless communities, take shape organically from the ground up. Consequently, they introduce a host of theoretical and practical complexities that pose significant obstacles to their growth and widespread adoption. These challenges primarily revolve around three domains: technology, business/economic, and social/individual aspects [4].

Optical wireless communication has the potential to address several challenges in 5G /6G, IoT, Industry 4.0, and WLANs. Here's how it can help rectify the above challenges.

37.2.1 High Data Rates and Spectrum Efficiency

Challenge: The demand for higher data rates in 5G /6G networks and IoT applications requires efficient use of available spectrum.

Rectification: Optical wireless communication, with its wide bandwidth capabilities, can provide high data rates without spectrum congestion. Multiple optical channels can be used simultaneously, maximizing spectral efficiency.

37.2.2 Interference and Coexistence

Challenge: Coexistence and interference management are complex in crowded RF environments.

Rectification: Optical communication uses focused beams, reducing interference. Optical beams can be precisely directed, minimizing cross-talk and ensuring secure data transmission.

37.2.3 Security and Privacy

Challenge: Ensuring data security and privacy in IoT and Industry 4.0 is critical.

Rectification: Optical communication offers inherent security benefits due to its narrow beams, making eavesdropping more challenging. Additionally, optical encryption technologies can be implemented to enhance security further.

37.2.4 Scalability

Challenge: Scalability is crucial in IoT, where a vast number of devices need to connect seamlessly.

Rectification: Optical wireless communication can support a large number of devices simultaneously, making it suitable for scalable IoT deployments. Visible Light Communication (VLC) in particular, with its low-power requirements, is ideal for connecting numerous IoT devices.

37.2.5 Line of Sight and Obstacles

Challenge: Maintaining line of sight and dealing with obstacles in RF communication can be problematic.

Rectification: Optical communication requires clear line of sight, but it can also be directed more precisely, allowing for obstacle avoidance. In Industry 4.0 applications, machinery and equipment can be designed to accommodate optical communication needs.

37.2.6 Energy Efficiency

Challenge: IoT devices often rely on battery power, necessitating energy-efficient communication.

Rectification: Optical communication can be energy-efficient, particularly when using low-power optical sources like LEDs. This is well-suited for IoT devices where power conservation is essential.

37.2.7 Integration and Standardization

Challenge: Integrating new technologies into existing infrastructure and adhering to standards can be complex.

Rectification: Optical communication can be integrated into existing networks, complementing RF communication where necessary. Standardization efforts are ongoing to ensure interoperability.

37.2.8 Cost and Scalability

Challenge: Cost-effective solutions that can scale are vital for widespread adoption.

Rectification: Optical components have become more cost-effective over time, and economies of scale can be achieved as adoption increases. Scalability is a natural feature of optical communication, making it suitable for growing networks.

37.2.9 Environmental Factors

Challenge: Environmental conditions like fog or rain can affect communication.

Rectification: Optical communication systems can be designed with robustness to environmental factors. Adaptive optical systems can adjust parameters to account for changing conditions.

Regulatory Considerations

Challenge: Navigating regulatory frameworks can be challenging for new communication technologies.

Rectification: Engaging with regulatory bodies and demonstrating the benefits of optical communication, including its security and efficiency, can help shape favorable regulations.

In summary, optical wireless communication, whether through Free-Space Optical Communication (FSO) or Visible Light Communication (VLC), offers solutions to many challenges faced by 5G /6G, IoT, Industry 4.0, and WLANs. Its advantages in terms of data rates, security, scalability, and energy efficiency make it a promising technology to complement and enhance these communication domains. Addressing these challenges in 5G /6G networks, IoT, Industry 4.0, and WLANs requires ongoing research, innovation, collaboration among stakeholders, and the development of new technologies and standards. As these technologies continue to evolve, overcoming these challenges will be crucial to unlocking their full potential and delivering the promised benefits to society and industries.

37.3 Free-Space Optical Communication (FSO)

Free-space optical communication is widely regarded as the forefront wireless communication technology of the future, and numerous ongoing endeavors are dedicated to augmenting its capabilities. The primary focus revolves around enhancing the performance of FSO communication, and a pivotal avenue for advancement lies in refining photo detectors to efficiently capture light within the optical communication wavelength spectrum. This is especially critical within FSO systems, where the modulated beam traverses through open space devoid of optical fiber cables. A notable challenge in FSO communication is managing beam divergence, which can be effectively mitigated by transitioning to an operating wavelength of 1.55 μm [5]. This approach signifies an inventive method of data transmission that harnesses the speed and precision of light. Unlike traditional communication methods reliant on electrical signals or radio waves, FSO utilizes laser beams as the medium for transmitting information. Often referred to as "optical wireless communication,"

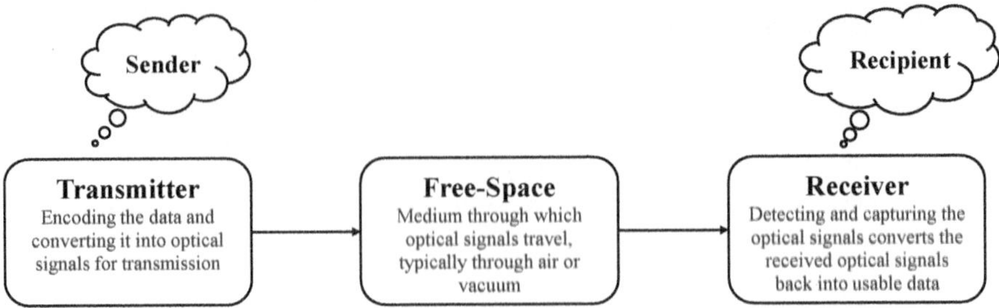

Fig. 37.2 Operational block diagram of free-space optical communication (FSO)

this cutting-edge technology employs light as the conveyance mechanism for data. At the core of FSO technology are highly focused laser beams, which function as data carriers. Information is encoded by modulating the intensity of these laser beams. The transmission process involves emitting laser beams from a transmitter, typically equipped with a telescope or optical system, to a receiver situated at the opposite end of the communication link. The receiver collects and deciphers the optical signals to recover the transmitted data which is clearly given in Fig. 37.2.

37.3.1 Key Features and Advantages of FSO

High Data Rates

FSO offers remarkable data transmission rates, often reaching gigabits per second (Gbps) and beyond. This high-speed capability makes it a compelling choice for applications requiring rapid data transfer, such as backhaul for wireless networks and data centers.

Low Latency

FSO communication is known for its low latency, meaning data travels with minimal delay. This feature is particularly crucial in scenarios like financial trading, autonomous vehicles, and real-time remote control systems.

Security and Interference Resistance

The use of narrow laser beams in FSO provides inherent security benefits. It is challenging for eavesdroppers to intercept data, and FSO systems are less susceptible to interference compared to RF-based communication.

Immunity to Electromagnetic Interference

FSO is immune to electromagnetic interference, which can disrupt RF-based communication. This immunity is particularly valuable in urban areas and other environments where RF spectrum congestion is a concern.

Ease of Deployment

FSO systems are relatively straightforward to deploy, especially in scenarios where laying physical cables is impractical or cost-prohibitive. The technology is particularly useful for establishing rapid communication links during disaster recovery or in remote areas with limited infrastructure.

37.3.2 Challenges and Considerations

While FSO offers numerous advantages, it also faces some challenges:

Atmospheric Conditions

FSO performance can be affected by atmospheric conditions such as fog, rain, and snow. These weather-related phenomena can scatter or absorb light, leading to signal degradation.

Line of Sight

FSO requires an unobstructed line of sight between the transmitter and receiver. Obstacles like buildings or hills can disrupt communication.

Distance Limitations

FSO is most effective over relatively short to medium distances, typically up to several kilometers. Beyond these ranges, signal attenuation becomes a limiting factor.

Alignment and Tracking

Precise alignment and tracking mechanisms are necessary to maintain the laser beams' connection over longer distances. Any misalignment can lead to signal loss.

Regulatory Considerations

FSO systems may be subject to regulatory restrictions and require licenses due to the use of laser technology.

37.3.3 Applications of FSO

Applications of FSO technology span various domains, encompassing.

Wireless Network Backhaul

FSO technology can function as a high-capacity backhaul solution for 4G, 5G, and upcoming wireless networks, effectively mitigating RF spectrum congestion and delivering rapid connectivity.

Data Center Connectivity

FSO offers the means to link data centers via low-latency, high-bandwidth connections, facilitating swift data exchange and communication.

Disaster Recovery

In disaster-prone regions, FSO systems are deployed to swiftly re-establish communication links when conventional infrastructure is compromised or damaged.

Secure Communication

FSO's utilization of narrow beams renders it suitable for secure communication, particularly in sectors such as government, military, and finance, where data privacy and security are paramount.

Campus Networking

FSO is employed to achieve high-speed connectivity within academic institutions, corporate settings, and research facilities, enhancing communication capabilities.

As FSO technology continues to advance and address its challenges, it holds promise as a key player in the future of high-speed, secure, and efficient optical wireless communication. Its unique ability to transmit data using beams of light opens up possibilities for a wide range of applications in our increasingly connected world.

37.4 Visible Light Communication (VLC)

Visible Light Communication (VLC), also known as Illuminating Connectivity through Light, stands as an emerging technology that harnesses visible light for data transmission and communication purposes. The functional blocks of VLC represented in Fig. 37.3 which performs in contrast to conventional wireless communication methods reliant on radio waves or microwaves, VLC utilizes the visible light spectrum as a conduit for transmitting data. This innovative approach facilitates data exchange through light-emitting diodes (LEDs), presenting a promising solution for a multitude of applications. VLC functions by modulating the intensity of light emitted by LEDs at a rapid rate, often beyond the range of human visual perception. This modulation process encodes data, enabling its transmission through light pulses. Specialized receivers, equipped with photodetectors, capture these light signals and subsequently decode them into usable data. VLC is versatile and applicable in both indoor and outdoor communication scenarios. As an emerging technology, VLC holds substantial potential to revolutionize wireless communication. It can complement traditional radio frequency communication methods while introducing novel applications for mobile wireless devices. Among the most compelling applications are high-speed downlink communication in residential and office environments and precise indoor positioning within retail stores. However, the widespread commercial adoption of VLC devices relies on concurrent advancements in robust and efficient engineering solutions and the methodical implementation of step-by-step commercialization strategies [6].

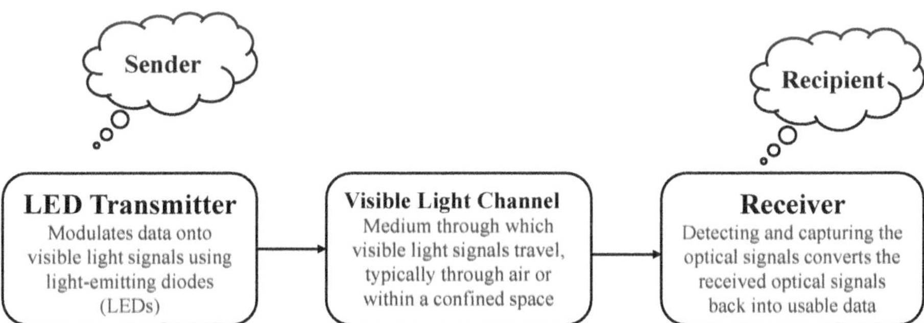

Fig. 37.3 Operational block diagram of visible light communication (VLC)

37.4.1 Key Features and Advantages of VLC

High Data Rates

VLC can achieve impressive data transmission rates, comparable to or exceeding those of traditional Wi-Fi technologies. This capability is essential for applications demanding fast and reliable connectivity.

Energy Efficiency

VLC is energy-efficient, as it utilizes existing lighting infrastructure, primarily LEDs. This feature is particularly advantageous for battery-powered devices and IoT applications.

Security

VLC offers inherent security benefits because visible light signals are confined to specific areas. Unauthorized interception of data is challenging, enhancing the overall security of communication.

Interference-Free

Unlike radio-frequency communication, VLC operates in an uncluttered spectrum, reducing the risk of interference from neighboring devices or networks.

Indoor Localization

VLC can provide precise indoor localization and positioning, making it valuable for applications like indoor navigation and asset tracking.

No RF Interference

VLC can be employed in environments sensitive to radio frequency interference, such as hospitals and aircraft, without causing disruption.

37.4.2 Challenges and Considerations

While VLC presents numerous advantages, it also faces some challenges.

Line of Sight

VLC relies on a direct line of sight between the transmitter (LED) and receiver (photodetector). Obstacles in the optical path can disrupt communication.

Short Range

VLC is most effective over relatively short distances due to the limited range of visible light. Extending coverage for outdoor applications may require additional infrastructure.

Lighting Conditions

Variations in ambient lighting conditions, such as daylight or changes in LED brightness, can affect VLC performance.

Compatibility

Widespread adoption of VLC may require compatible devices and infrastructure, which can take time to implement.

37.4.3 Applications of VLC

VLC technology finds applications in various domains, including.

Indoor Wireless Communication

VLC is used to provide high-speed wireless connectivity in indoor environments like homes, offices, and retail spaces.

IoT Connectivity

VLC serves as an energy-efficient communication solution for IoT devices, allowing them to exchange data in smart homes, smart cities, and industrial settings.

Retail and Marketing

Retailers use VLC for location-based services, customer engagement, and indoor navigation through smartphone apps.

Transportation

VLC can enhance communication in transportation systems, including vehicle-to-vehicle (V2V) and vehicle-to-infrastructure (V2I) communication.

Healthcare

VLC ensures interference-free communication in healthcare facilities, supporting medical device connectivity and patient monitoring.

Underwater Communication

VLC's properties also make it suitable for underwater communication, where radio waves cannot propagate effectively.

As VLC technology continues to advance and address its challenges, it holds promise as a versatile and efficient means of communication in an increasingly connected world. Its ability to provide high-speed, secure, and energy-efficient connectivity through visible light makes it a valuable addition to the realm of wireless communication.

37.5 FSO and VLC in 5G /6G Networks, the Internet of Things (IoT), Industry 4.0, and Wireless Local Area Networks (WLANs)

Free-Space Optical Communication (FSO) and Visible Light Communication (VLC) are two innovative technologies that can significantly impact various domains, including 5G /6G networks, the Internet of Things (IoT), Industry 4.0, and Wireless Local Area Networks (WLANs). Here's how FSO and VLC can be applied in these contexts as reflected in Fig. 37.4.

37.5.1 FSO and VLC in 5G /6G Networks

FSO can serve as a high-capacity backhaul solution for 5G and 6G networks, especially in urban areas where RF spectrum congestion is a pressing issue. VLC can complement FSO by providing indoor and short-range connectivity. FSO's low latency is crucial for 6G applications that demand real-time communication, such as holographic communication and remote control of autonomous systems.

37.5.2 FSO and VLC in IoT

VLC is well-suited for short-range, low-power IoT communication within smart homes, industrial facilities, and healthcare settings. It provides energy-efficient

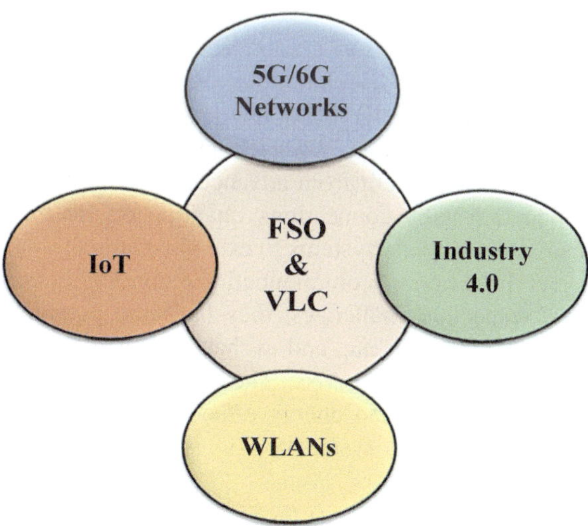

Fig. 37.4 FSO and VLC functional domains

data transmission while also serving as an illumination source. VLC's capability to offer precise indoor localization and tracking of IoT devices can enhance asset management and location-based services.

37.5.3 FSO and VLC in Industry 4.0

The Seamless Communication in Manufacturing is employed where the FSO can enable secure and high-speed communication within manufacturing facilities and warehouses, reducing the risk of RF interference that can disrupt operations. VLC can complement this by offering short-range connectivity between machines and sensors. FSO's low latency is valuable for real-time monitoring of production processes, allowing for immediate adjustments and optimizations.

37.5.4 FSO and VLC in WLANs

FSO can augment existing WLANs in high-density environments like airports, stadiums, and urban centers. Temporary or permanent optical links can enhance data rates and secure communication. VLC can be used to provide indoor navigation and location-based services within WLAN-covered areas, improving user experiences in shopping malls, museums, and airports.

37.5.5 Hybrid Solutions

Combining FSO and VLC can create hybrid communication solutions that leverage the strengths of both technologies. For instance, FSO can handle long-range outdoor communication in 5G /6G networks, while VLC can handle indoor and short-range connections, ensuring seamless connectivity across diverse environments. Both FSO and VLC can benefit from advanced technologies like beam forming and tracking to maintain stable connections, quantum key distribution for enhanced security, and distributed antenna systems to extend coverage.

As these optical communication technologies continue to evolve and overcome their respective challenges, they have the potential to revolutionize the way we connect, communicate, and exchange data in the realms of 5G /6G networks, IoT, Industry 4.0, and WLANs. Their unique features, including high data rates, low latency, security, and energy efficiency, position them as key enablers for future communication ecosystems.

37.6 Optical Wireless Solutions in 5G /6G Networks

Optical wireless solutions play a crucial role in enhancing the performance and capacity of 5G and future 6G networks. These technologies leverage optical signals, such as infrared (IR) or visible light, for data transmission and can provide several advantages in terms of data rates, security, and energy efficiency. Researchers are currently directing their efforts towards establishing standardized specifications for 5G /6G networks [7]. Here's an overview of optical wireless solutions in 5G /6G networks.

37.6.1 Li-Fi (Light Fidelity)

Li-Fi is a technology that uses visible light communication (VLC) or infrared signals to transmit data. It operates by modulating LED or laser light sources at high speeds to transmit data. Advantages include extremely high data rates (tens of Gbps), immunity to radio frequency interference, and improved security (difficult to intercept signals outside the line of sight). Li-Fi can be used for indoor wireless communication, providing high-speed connectivity in areas where Wi-Fi may be congested or less secure.

37.6.2 Optical Wireless Backhaul

Optical wireless links can serve as backhaul connections in 5G and 6G networks, connecting base stations to the core network or data centers. These links can offer high capacity, low latency, and quick deployment in scenarios where laying optical fiber is challenging or costly. They are especially useful in dense urban environments or remote areas.

37.6.3 Quantum Key Distribution (QKD)

Quantum communication technologies, including QKD, can enhance the security of 5G /6G networks. QKD uses the principles of quantum mechanics to enable secure key exchange, making it extremely difficult for eavesdroppers to intercept communication. Integrating QKD with optical wireless solutions can provide a high level of security for network communications.

37.6.4 Terahertz (THz) Communications

THz waves, part of the electromagnetic spectrum between microwaves and infrared, have the potential for extremely high data rates. Research is ongoing to harness THz technology for wireless communication in 6G networks, which may provide multi-terabit-per-second data rates.

37.6.5 Network Synchronization

Optical wireless links can help synchronize 5G /6G networks, which require precise timing for various applications, including autonomous vehicles and industrial automation. Optical time synchronization using accurate optical clocks can ensure network reliability and low-latency communication.

37.6.6 Energy Efficiency

Optical wireless solutions are generally more energy-efficient than traditional radio-based communication, making them suitable for green and sustainable network designs.

To sum up, optical wireless solutions present numerous advantages for 5G and 6G networks, encompassing high data speeds, heightened security, and energy conservation. These innovations hold promise for meeting the increasing requirements for connectivity and data transfer in the forthcoming era of wireless networks. Nonetheless, they also bring their distinct challenges, such as the necessity for a direct line of sight and vulnerability to atmospheric conditions, necessitating meticulous consideration in the planning and implementation of network infrastructure.

37.7 Optical Wireless Solutions for IoT Deployments

Optical wireless solutions can be highly beneficial for Internet of Things (IoT) deployments, offering unique advantages in terms of data communication, energy efficiency, and security. The ongoing advancement of the Internet of Things (IoT) necessitates the exploration of creative solutions and technologies to effectively materialize the IoT vision. A burgeoning connectivity technology that holds promise for IoT deployment is Optical Wireless Communication (OWC). OWC involves the utilization of light beams emitted by light sources to encode information [8]. Here's an overview of how optical wireless solutions can be applied in IoT scenarios.

37.7.1 High Data Rates

Optical wireless communication, such as Li-Fi (Light Fidelity), can provide significantly higher data rates compared to traditional radio frequency (RF) technologies like Wi-Fi or Bluetooth. This makes it suitable for IoT applications that require rapid data transfer, such as real-time sensor data or high-definition video streaming in smart cities or industrial settings.

37.7.2 Low Latency

Optical wireless communication typically offers low latency, which is crucial for IoT applications that require real-time responsiveness, such as autonomous vehicles, industrial automation, and healthcare monitoring.

37.7.3 Energy Efficiency

Optical wireless solutions can be highly energy-efficient, especially in scenarios where devices need to operate on battery power for extended periods. Transmitting data using light sources consumes less energy compared to traditional RF technologies, which can be advantageous for IoT sensors and devices with limited power resources.

37.7.4 Security

Optical wireless communication is inherently more secure than RF-based solutions because it is confined to a physical line of sight. This makes it difficult for eavesdroppers to intercept data, enhancing the overall security of IoT deployments. Additionally, encryption techniques can be applied to further secure the communication.

37.7.5 Interference Immunity

Optical wireless communication is less susceptible to interference from other wireless devices or electromagnetic sources, making it suitable for IoT deployments in crowded RF environments.

37.7.6 Indoor and Controlled Environments

Optical wireless solutions like Li-Fi are well-suited for indoor IoT deployments where lighting fixtures can double as communication nodes. This can simplify the deployment of IoT devices and reduce infrastructure costs.

37.7.7 Healthcare and Sensitive Environments

In healthcare and other sensitive environments, optical wireless solutions can be used to transmit data without electromagnetic interference concerns. This is important for medical IoT devices and equipment.

37.7.8 Smart Agriculture

In agriculture IoT applications, optical wireless solutions can be used for precision farming, where sensors and drones communicate data regarding soil conditions, crop health, and environmental factors.

37.7.9 Environmental Monitoring

Optical wireless communication can be employed in environmental monitoring systems, allowing IoT devices to transmit data in remote or challenging terrains where RF signals may have limited reach.

37.7.10 Retail and Location-Based Services

Optical wireless solutions can enable precise location-based services and customer engagement in retail environments, improving the shopping experience.

While optical wireless solutions offer many advantages for IoT deployments, it's important to note that they also come with limitations, such as the requirement for line-of-sight communication and potential signal blockage by physical obstacles. Therefore, the choice of communication technology for IoT deployments should be based on the specific requirements and constraints of the application. In some cases, a hybrid approach that combines optical wireless with other wireless technologies may provide the best solution for IoT connectivity.

37.8 Industry 4.0: Revolutionizing Manufacturing with Optical Wireless Solutions

Industry 4.0, also known as the Fourth Industrial Revolution, is transforming the manufacturing sector by incorporating advanced technologies to improve efficiency, productivity, and flexibility. Optical wireless solutions play a significant role in this revolution by offering unique advantages in the context of modern manufacturing processes.

37.8.1 High-Speed Data Transfer

Optical wireless technologies like Li-Fi can transmit data at extremely high speeds, enabling real-time communication between machines, sensors, and control systems. This high-speed data transfer is essential for efficient monitoring, control, and coordination of manufacturing processes.

37.8.2 Low Latency

Optical wireless communication typically provides low-latency connectivity, which is critical for real-time automation and robotics applications in manufacturing. This minimizes delays in data transmission and decision-making, improving overall system responsiveness.

37.8.3 Network Reliability

Optical wireless solutions are less susceptible to interference from electromagnetic sources, making them more reliable in industrial environments where electronic noise can disrupt traditional wireless communication methods.

37.8.4 Energy Efficiency

Optical wireless communication is energy-efficient, which is essential for battery-powered devices and sensors used in smart manufacturing. Reduced power consumption can extend the operational life of these devices and reduce maintenance requirements.

37.8.5 Security

Optical wireless communication offers enhanced security as it operates within a physical line of sight. This reduces the risk of data interception or tampering, critical for protecting sensitive manufacturing data and intellectual property.

37.8.6 Flexibility and Scalability

Optical wireless solutions are flexible and scalable, making it easier to adapt manufacturing processes to changing requirements. They can be deployed quickly and reconfigured as needed, reducing downtime during upgrades or expansions.

37.8.7 Indoor Localization

Optical wireless technologies can be used for precise indoor localization of assets, equipment, and personnel in manufacturing facilities. This capability improves asset tracking, inventory management, and worker safety.

37.8.8 Maintenance and Predictive Analytics

Optical wireless solutions enable the collection of real-time data from sensors on machinery and equipment. This data can be analyzed using machine learning algorithms to predict maintenance needs, reducing downtime and increasing overall equipment effectiveness (OEE).

37.8.9 Quality Control

High-speed optical wireless communication can support vision systems and cameras for quality control and inspection processes. It allows for rapid image and data transfer, facilitating quick decision-making based on inspection results.

37.8.10 Human–Machine Interaction

Optical wireless solutions can enhance human–machine interaction through augmented reality (AR) and virtual reality (VR) applications. This enables remote monitoring, training, and troubleshooting of manufacturing processes.

37.8.11 Supply Chain Integration

Optical wireless technologies can improve supply chain visibility by enabling real-time tracking of raw materials, work-in-progress, and finished goods. This transparency helps optimize inventory management and reduce lead times.

Finally, optical wireless solutions are revolutionizing manufacturing in the Industry 4.0 era by providing high-speed, low-latency, and secure communication for smart factories. They support automation, data analytics, predictive maintenance, and other key elements of modern manufacturing, ultimately leading to increased efficiency, reduced costs, and improved product quality. Manufacturers adopting these technologies are better positioned to compete in a rapidly evolving global market.

37.9 Optical-Based Solutions for WLANs

Optical-based solutions in Wireless Local Area Networks (WLANs) are emerging technologies that use optical signals, such as visible light or infrared (IR), for wireless data communication within a local area. The Optical Wireless Communication underscores the benefits and diverse array of potential uses of this emerging technology [9]. These solutions offer several advantages and applications in WLANs.

37.9.1 Indoor Navigation and Positioning

Optical-based solutions can be used for precise indoor positioning and navigation systems in WLANs. Visible light positioning (VLP) and VLC can help users or devices determine their exact location within an indoor environment. This technology is useful in applications like indoor wayfinding, asset tracking, and location-based services.

37.9.2 Smart Lighting Integration

Optical-based WLAN solutions can be integrated with smart lighting systems. LED light fixtures can serve dual purposes, providing both illumination and data communication, which is particularly useful in indoor environments. This integration reduces the need for separate data communication infrastructure.

37.9.3 Low Interference and Secure Communication

Optical signals used in WLANs are less susceptible to interference from other electronic devices, leading to more reliable and stable connections. The line-of-sight nature of optical communication adds an extra layer of security, making it challenging for unauthorized users to intercept data.

37.9.4 Healthcare and Sensitive Environments

Optical-based WLANs are suitable for use in healthcare facilities and other sensitive environments where RF interference could disrupt critical equipment or pose security risks. For example, optical solutions can be used for data communication in operating rooms or patient monitoring systems.

37.9.5 Energy Efficiency

Optical-based solutions can be more energy-efficient than traditional RF-based Wi-Fi, which is important for battery-powered devices and sensors in WLANs. Reduced power consumption can extend the operational life of devices and reduce maintenance requirements.

37.9.6 High-Density Environments

In areas with high device density, such as conference rooms, auditoriums, or stadiums, optical-based solutions can offer better performance and less interference compared to traditional Wi-Fi.

37.9.7 Data Security in Critical Applications

Critical applications, such as defense or financial services, can benefit from optical-based WLANs due to their enhanced security features, making it difficult for cyber-attacks to compromise data transmission.

37.9.8 Underwater Communications

Optical-based WLANs can also be employed in underwater environments, where radio frequency signals are ineffective due to high absorption in water. Submersible optical modems are used for underwater data communication in applications like oceanography and underwater robotics The ocean depths hold significant potential for upcoming industrial development and various applications. The effectiveness of robotic autonomous systems in this context will rely heavily on the establishment of a dependable communication channel, facilitating collaborative operations between operators and equipment, whether they are working together or located on the ocean's surface [10]. While optical-based solutions in WLANs offer numerous advantages, they also have limitations, such as limited range and the need for line-of-sight communication. Therefore, the choice of technology should be based on specific use cases and requirements. In some scenarios, a hybrid approach that combines optical-based solutions with traditional Wi-Fi may provide the most comprehensive WLAN coverage and capabilities.

37.10 Advanced Optical Communication Technologies

Advanced optical communication technologies are at the forefront of the telecommunications industry, driving innovation and enabling faster, more secure, and more reliable data transmission. Here's an overview of some of these advanced technologies.

37.10.1 Advanced Modulation Schemes

Advanced modulation schemes, such as Quadrature Amplitude Modulation (QAM) and Orthogonal Frequency Division Multiplexing (OFDM), increase the data-carrying capacity of optical fiber networks. These schemes enable higher data rates by encoding more bits per symbol and mitigating signal impairments.

37.10.2 Beamforming and Tracking

Beamforming technology focuses optical signals to specific locations or receivers. Beam tracking adjusts the direction of the optical beam to maintain a stable connection, even when the receiver or transmitter is in motion. These technologies are vital for optical wireless communication, such as free-space optical links or Li-Fi, to maintain strong connections.

37.10.3 Hybrid Solutions

Hybrid optical communication solutions combine optical and radio frequency (RF) technologies to optimize network performance. They offer a flexible approach, enabling communication in various scenarios, from indoor environments to outdoor backhaul connections.

37.10.4 Quantum Key Distribution (QKD)

QKD leverages the principles of quantum mechanics to provide ultra-secure key exchange for encrypted communication. It offers unconditional security by detecting any eavesdropping attempts. QKD is crucial for safeguarding sensitive data in sectors like finance, government, and healthcare.

37.10.5 Distributed Antenna Systems (DAS)

DAS extend wireless coverage and capacity in indoor or large outdoor areas by distributing wireless signals through a network of antennas. Optical fiber connections are commonly used to link remote antennas, reducing signal loss and latency.

37.10.6 Li-Fi (Light Fidelity)

Li-Fi uses visible light or infrared for wireless data communication. It offers high-speed, secure, and energy-efficient connectivity, particularly in settings like smart cities, offices, and manufacturing facilities.

37.10.7 Miniaturization

Miniaturization involves shrinking optical components and devices, making them more compact and suitable for various applications. Miniaturized optical transceivers are used in data centers and edge computing to increase data transmission speed and capacity.

37.10.8 Terrestrial and Satellite Integration

Integrating optical communication with satellite networks enhances global connectivity. Optical inter-satellite links (OISLs) enable high-capacity data transfer between satellites in orbit, improving data relay and Earth observation missions.

These advanced optical communication technologies collectively contribute to the development of faster, more secure, and more efficient communication networks. They are essential for meeting the growing demands of data-intensive applications, such as 5G and 6G networks, cloud computing, IoT, and beyond. As technology continues to evolve, further advancements in these areas are expected to shape the future of global communications.

37.11 Conclusion and Future Work

In the rapidly evolving landscape of data-driven applications and interconnected devices, the need for innovative communication solutions is more evident than ever. In this exploration, we have delved into the capabilities of optical wireless solutions, which include free-space optical communication (FSO) and visible light communication (VLC). These technologies have been examined for their potential to meet the varied challenges and demands presented by 5G /6G networks, the Internet of Things (IoT), Industry 4.0, and Wireless Local Area Networks (WLANs). In the realm of 5G /6G networks, optical wireless solutions have gained prominence as an attractive choice for strengthening backhaul connections and enhancing network density, especially in urban areas dealing with congestion in the radio frequency (RF) spectrum. Their ability to offer minimal latency and exceptional reliability is well-suited to meet the rigorous requirements of forthcoming 6G applications, heralding an era of high-capacity and ultra-low-latency connectivity. In the realm of IoT implementations, optical wireless communication, particularly Visible Light Communication (VLC), offers an appealing option for transmitting data over short distances with minimal power consumption in smart homes, industrial environments, and healthcare establishments. Utilizing visible light not only enhances communication efficiency but also supports indoor localization and tracking of IoT devices, offering the potential for enhanced efficiency and convenience. In the realm of Industry 4.0, optical wireless solutions hold the promise of delivering uninterrupted and interference-resistant communication within manufacturing plants and warehouses. The utilization of optical wireless technologies stands to greatly benefit real-time monitoring, asset tracking, and the interaction between robotic systems. In the context of Wireless Local Area Networks (WLANs), optical wireless communication can act as a valuable supplement to current networks, particularly in densely populated areas. Temporary or permanent optical connections have the capability to improve data speeds, indoor navigation, location-based services, and secure communication, effectively addressing the requirements for data security and dependability. The research and development related to Standardization and Interoperability, Enhanced Security, Scalability, Integration with 5G and Beyond, Energy Efficiency, Standardization for Manufacturing, User Experience, Terrestrial and Satellite Integration are pursued, which will evolve optical wireless communication to offer more innovative solutions to meet the growing demands of our increasingly connected world.

References

1. Wikström, G., et al.: Challenges and Technologies for 6G, 2nd 6G Wireless Summit (6G SUMMIT), pp. 1–5. Levi, Finland (2020)
2. Karale, A.: The challenges of IoT addressing security, ethics, privacy, and laws. Internet of Things **15**, 100420 (2021). ISSN 2542-6605
3. Luthra, S., Mangla, S.K.: Evaluating challenges to Industry 4.0 initiatives for supply chain sustainability in emerging economies. Process Safety Environ. Protect. **117**, 168–179 (2018). ISSN 0957-5820
4. Bina, M., Giaglis, G.M.: Emerging issues in researching community-based WLANS. J. Comput. Inf. Syst. **46**(1), 9–16 (2005)
5. Abdi, G., Golmohammadi, S.: Absorption enhancement in photodetectors based on T-shaped coupled nano-antenna for free-space optical communications applications. In: 4th West Asian Symposium on Optical and Millimeter-wave Wireless Communications (WASOWC), pp. 01–05. Iran, Islamic Republic of Tabriz (2022)
6. Jovicic, A., Li, J., Richardson, T.: Visible light communication: opportunities, challenges and the path to market. IEEE Commun. Mag. **51**(12), 26–32 (2013)
7. Chowdhury, M.Z., Shahjalal, M., Hasan, M.K., Jang, Y.M.: The role of optical wireless communication technologies in 5G /6G and IoT solutions: prospects, directions, and challenges. Appl. Sci. **9**(20), 4367 (2019)
8. Hamza, A., Tripp, T.: Optical Wireless Communication for the Internet of Things: Advances, Challenges, and Opportunities. Tech Rxiv. Preprint (2020)
9. Uysal, M., Nouri, H.: Optical wireless communications—an emerging technology. In: 16th International Conference on Transparent Optical Networks (ICTON), Graz, Austria, pp. 1–7 (2014)
10. Poncela, J., Aguayo, M.C., Otero, P.: Wireless underwater communications. Wireless Pers. Commun. **64**, 547–560 (2012)

Printed by Printforce, the Netherlands